BIOTECHNOLOGY IN AGRICULTURE, INDUSTRY AND MEDICINE

FOCUS ON CHITOSAN RESEARCH

BIOTECHNOLOGY IN AGRICULTURE, INDUSTRY AND MEDICINE

Additional books in this series can be found on Nova's website under the Series tab.

Additional E-books in this series can be found on Nova's website under the E-books tab.

BIOTECHNOLOGY IN AGRICULTURE, INDUSTRY AND MEDICINE

FOCUS ON CHITOSAN RESEARCH

ARTHUR N. FERGUSON
AND
AMY G. O'NEILL
EDITORS

Nova Science Publishers, Inc.
New York

For permission to use material from this book please contact us:
Telephone 631-231-7269; Fax 631-231-8175
Web Site: http://www.novapublishers.com

Additional color graphics may be available in the e-book version of this book.

LIBRARY OF CONGRESS CATALOGING-IN-PUBLICATION DATA

Focus on chitosan research / editors, Arthur N. Ferguson and Amy G. O'Neill.
p. ; cm.
Includes bibliographical references and index.
ISBN 978-1-61324-454-8 (hardcover)
1. Chitosan--Biotechnology. 2. Biomedical materials. I. Ferguson, Arthur N. II. O'Neill, Amy G.
[DNLM: 1. Chitosan--chemistry. 2. Biocompatible Materials--chemistry. QU 83]
TP248.65.C55F63 2011
610.28--dc22
2011010934

Published by Nova Science Publishers, Inc. † New York

CONTENTS

Preface **vii**

Chapter 1 Porous Structure and Adsorption Behaviours of Chitosan **1**
O. K. Krasilnikova, O. V. Solovtsova, E. V. Khozina, T. Y. Grankina, N. V. Serebryakova and S. M. Shinkarev

Chapter 2 Chitosan: A Promising Biomaterial for Tissue Engineering and Hemorrhage Control **49**
Nan Cheng, Xudong Cao and Henry T. Peng

Chapter 3 The Applications of Chitosan with Different Modifications in Tissue Engineering **83**
Gan Wang, Qiang Ao, Yandao Gong and Xiufang Zhang

Chapter 4 Enhanced Biochemical Efficacy of Oligochitosans and Partially Depolymerized Chitosans **115**
Riccardo A. A. Muzzarell

Chapter 5 Thermal Relaxation Properties and Glass Transition Phenomena in Chitosan Films **141**
J. Betzabe González-Campos, Evgen Prokhorov, G. Luna-Bárcenas, L. Chacón-García and Rosa E. N. del Río-Torres

Chapter 6 Biological Activities of Chitin, Chitosan and Respective Oligomers **171**
F.reni K. Tavaria, João Fernandes , Alice Santos-Silva, Sara Baptista da Silva, Bruno Sarmento and Manuela Pintado

Chapter 7 The Potential of Chitosan in Drug Delivery Systems **199**
Sara Baptista da Silva, João Fernandes, Freni Tavira, Manuela Pintado, and Bruno Sarmento

Chapter 8 Solid State Synthesis and Modification of Chitosan **223**
Tatiana A. Akopova, Alexander N. Zelenetskii and Alexander N. Ozerin

Chapter 9 Chitosan Derived Smart Materials **255**
Ashutosh Tiwari, Dohiko Terada, Chiaki Yoshikawa and Hisatoshi Kobayashi

Chapter 10 Chitosan Copolymers and IPNs for Controlled Drug Release **289**
P. P. Kundu, Vinay Sharma and Kamlesh Kumari

Chapter 11 Biomedical Applications of Hydrogels Based in the Chemical Modification of Chitosan **317**
E. A. Elizalde-Peña, G. Luna-Bárcenas, E. Prokhorov, J. E. Gough, C. Velasquillo-Martínez, C. E. Schmidt and I. Sanchez

Chapter 12 A Spectroscopic Study of the Interaction of Metal Ions with Biomaterials **349**
Yassin Jeilani' *Beatriz H. Cardelino and Natarajan Ravi*

Chapter 13 Chitosan: Properties and Its Pharmaceutical and Biomedical Aspects **377**
Ali Demir Sezer

Chapter 14 Combined Use of Biopolymer Chitosan and Enzyme Tyrosinase for Removal of Bisphenol A and its Derivatives from Aqueous Solutions **399**
Kazunori Yamada, Mizuho Suzuki, Ayumi Kashiwada and Kiyomi Matsuda

Chapter 15 Chitosan Usage in Synthesis of Hydrogel-Biomaterials with Controlled Properties for Biomedical/Biotechnological Applications **427**
Eugenia Dumitra Teodor', *Simona Carmen Lițescu, Cristian Petcu and Gabriel Lucian Radu*

Index **455**

PREFACE

Chitosan is a partially deacetylated derivative of chitin, a natural polysaccharide extracted from crustaceans, insects and certain fungi. Owing to its unique properties such as biodegradability, biocompatability, biological activity and capacity of forming polyelectrolyte complex with anionic polyelectrolytes, chitosan has been widely applied in the food and cosmetics industry, as well as the biomedical field in relation to tissue engineering, and the pharmaceutical industry relating to drug delivery. This book gathers current research from around the globe in the study of chitosan.

Chapter 1 - Interrelation between the structure and adsorption properties towards to water and nitrogen vapors and Ni, Cu and Ag cations of the chitosan samples modified by freeze-drying after precipitation with NaOH or Na_2CO_3 or drying in air, as well as cross-linking is studied. The investigations by IR spectroscopy, X-ray diffraction and pulsed NMR show the influence of the preliminary treatments on the supramolecular structure. The more ordered structure is observed in the freeze-dried samples of chitosan compared with the air-dried sample. The parameters of porous structure of the chitosan samples calculated from the isotherms of water vapors adsorption and the nitrogen vapors low temperature adsorption are compared. It is revealed that the isotherms of water vapors adsorption on the chitosan samples are more sensitive to the structural modification of polymers. The efficiency of different models based on the Flory-Huggins, Dubinin-Astakhov, and Dubinin-Serpinsky equations for fitting the experimental isotherms of water vapor adsorption is evaluated. The Dubinin-Serpinsky equitation 2 based on a model of two-stage localized adsorption of water molecules describes the experimental water vapor adsorption on the chitosan samples in the full range of relative pressures. In the range of small relative pressures, water molecules interact with chitosan polar groups which serve as the primary adsorption centers (PAC). Then, water molecules adsorbed on PAC become the secondary adsorption centers for further adsorption. As a result, at relatively high pressures the water clusters are formed. In the range of low water fillings of the chitosan adsorbents, the amount of primary adsorption centers can be evaluated by the comparative method.

The freeze-dried chitosan samples display the increased adsorption capacity towards to water and nitrogen vapors, as well as to metals cations. The cross-linking of the freeze-dried chitosan samples by glutaraldehyde reduces the adsorption capacity through the blockage of amino-groups. Cross-linking the chitosan films leads to the growth of adsorption capacities towards to water vapors. Thus, the analysis of the isotherms of water vapors adsorption specifies that structural changes of the chitosan samples after freeze drying and cross-linking resulting in a growth of the amount of accessible adsorption centers.

The adsorption capacities and the rates of adsorption of metal cations as Ni, Cu and Ag onto the modified chitosan samples are examined. The Langmuir, Freundlich and Dubinin-Radushkevich adsorption models are used to describe the isotherms of metal cations adsorption. The kinetics experimental data properly correlate with the pseudo-first-order kinetic model. The modification of the chitosan structure by freeze-drying noticeably increases the adsorption capacities towards to metal cations. The adsorption capacity of the freeze-dried chitosan sample towards to nickel cations is as high as 4.5 mmol/g that is threefold higher than that of the air-dried chitosan sample.

The IR-VIS spectroscopic experiments performed for the copper cations adsorption on the chitosan samples at high adsorption values allowed to advance a probable model of copper-chitosan complex that is a "bridge" structure based on a binuclear copper complex which bounds two chitosan macromolecules, so that every Cu (II) cation of a pair is associated with one amino group in axial position which belongs to different chitosan chains.

Chapter 2 - Chitosan, a natural polysaccharide, and its derivatives are most important biomaterials because of their distinct physicochemical and biological properties such as biocompatibility, biodegradability, non-toxicity, antimicrobial and hemostatic properties. In addition, the existence of reactive functional groups in the chitosan molecule makes it also possible to tailor the chitosan material for many specific tissue engineering and hemorrhage control applications. In the first part of this chapter, we will focus on the properties of chitosan and various modifications to improve its properties to enhance drug delivery and tissue regeneration. The modification of chitosan will be introduced from two aspects: chemical and physical modifications, i.e. grafting with other small molecules or biomolecules and blending chitosan with other different biomaterials, respectively. The applications of chitosan, especially its derivatives in drug delivery and tissue regeneration targeting different organs, such as cartilage, bone, skin, blood vessel, nerve system and liver will be discussed. Current improvements of these different artificial substitutes (based on chitosan) in different organs by the authors and other researchers will be reviewed. In the second part of the chapter, we will review the development of chitosan-based biomaterials for hemorrhage control in trauma and surgical injuries. Both chemical and physical modifications of chitosan will be discussed. Finally, challenges and future directions to use chitosan as a biomaterial for tissue engineering and hemorrhage control will be elaborated.

Chapter 3 - Chitosan, a cationic polymer, is the N-deacetylated derivative of chitin, which is abundant in nature. Chitosan has drawn much attention because of its good biocompatibility, biodegradability, low toxicity, and ability to be fabricated into various forms in tissue engineering. Because of the interesting biological properties, the utilizations of chitosan have been sufficiently developed. To make a breakthrough for the exploiture, modification will be a key point. Here we described several modifications including gamma irradiation, blending with PEG, PHB, gelatin, butyl bromide, collagen, polycation materials, nano-hydroxyapatite, coupled with GRGDS peptide, and carboxymethylation. Then the effect of the different modifications on chitosan cell affinity and biocompatibility has been discussed. Tissue engineering has emerged as promising field to treat the loss or malfunction of a tissue. A scaffold in tissue engineering plays an important role in accommodating and stimulating new tissue growth. Chitosan is one the widely used natural polymers in tissue engineering. Here the authors also discussed the applications of chitosan with different forms in tissue engineering including cartilage tissue, bone tissue, nerve tissue, vascular tissue, liver tissue, and skin tissue. All the descriptions have shown that the application of chitosan has a good prospect.

Chapter 4 - It is being realized that the performances of chitosans depend in most cases on their essential characteristic properties, such as the degree of acetylation and the average molecular weight. Crystalline polymorphic form, degree of crystallinity, polydispersity of molecular weight, zeta potential and other properties are also important. The selection of a certain chitosan for a particular application has consequences in terms of degree of efficacy, reproducibility of the results and suitability for scaling up in various areas such as pre-clinical tests, adoption for pharmaceutical formulations, environment protection. The requirements for reproducible quality of tailor-made chitosans become even more stringent when they are chemically or enzymatically modified.

Macrophages, fibroblasts and other human cells respond to the presence of chitosans in significantly different ways depending on their DA and MW, in terms of phenotypes, secretory repertoire, efficacy of their receptors, and final capacity to demonstrate superior biochemical significance compared to other polysaccharides. This is particularly true in wound healing, bone defect repair, control over inflammation, exertion of immune-stimulation, efficacy in drug delivery, anti-cholesterolemic capacity.

Chitosans are also antioxidants and antistress agents, capable to scavenge free radicals in a mammalian organism: this action too is controlled by DA and MW, chelating activity, and capacity to interact with biochemicals like nitric oxide, superoxide dismutase and glutathione peroxidase. Chitosans of proper chain size lend themselves to be targeted to a tumor, where they suppress the proliferation of tumor cells or at least reduce their viability; moreover they exert anti-metastatic action.

Chitosans are also active on the fungal and bacterial cells, and the selection of the proper size of the chitosans permits to optimize the results against pathogens, in consideration of the multiplicity of mechanisms of action of oligochitosans and partially hydrolyzed chitosans, using a combination of approaches, including in vitro assays, killing kinetics, cellular leakage measurement, membrane potential estimation, and electron microscopy, in addition to transcriptional response analysis.

This review includes concise but detailed well-assessed protocols for the preparation of oligochitosans and partially hydrolyzed chitosans, mainly by using sodium nitrite, sonication, hydrogen peroxide, electrolysis, and hydrolytic unspecific enzymes.

Chapter 5 - In this chapter thermal relaxation properties of chitosan neutralized and non-neutralized films will be reported as a function of water content using dynamical mechanical analysis and dielectric spectroscopy in the temperature range from 20 to 250°C. Three relaxation processes have been observed in different temperature and frequencies ranges. For the first time, the low frequency α-relaxation associated with the glass-rubber transition, which relate to a plasticizing effect of water, has been detected by this technique in both chitosan forms in the temperature range 20-70°C and for moisture contents between 0.5 to 10 wt %. On this basis, the glass transition temperature was estimated in the range 86-102°C which shifts to higher temperature with decreasing moisture content and the glass transition vanishes in a dry material. A second low frequency relaxation was observed from 80°C to the onset of thermal degradation (240°C) and identified as the sigma-relaxation often associated with the hopping motion of ions in the disordered structure of the biomaterial. This relaxation exhibits a normal Arrhenius-type temperature dependence with activation energy of 86-88 kJ/mol and it is independent of water content. The non-neutralized chitosan possess higher ion mobility than the neutralized one as determined by the frequency location of the σ-relaxation. A high frequency (10^4 -10^8 Hz) secondary β-relaxation in neutralized and non-neutralized chitosan, related to side group motions by means of the glucosidic linkage is observed in the temperature range 20-120°C and moisture contents less than 3 wt % with Arrhenius activation energy of 46.0-48.5 kJ/mol. In the films with higher moisture contents this relaxation is not well resolved due to a superposition of two relaxation processes: beta and beta wet relaxations that merge into one common β-relaxation process.

Chapter 6 - Chitin and chitosan have been receiving great attention as a novel functional material for their excellent biological properties such as biodegradation, immunological, antioxidant and antibacterial activities. However, its use has been scarcely developed due to its high molecular weight, which causes poor water solubility and high viscosity of the solutions. Compared with ordinary chitosan, chitosan oligomers have much improved water-solubility and some special biological functions due to the high absorption rate at the intestinal level, thus permitting its entrance in blood circulation and distribution all over the body. The potential application of chitin and chitosan, and respective oligomers, is multidimensional and it has been found in food and nutrition, biotechnology, material science, drug delivery, agriculture and environmental protection, and recently gene therapy too. In the realm of this chapter, we will focus on some biological activities attributed to those molecules, such as, antimicrobial and antifungal activities, wound healing, antioxidant, anti-inflammatory, anti-carcinogenic and anti-coagulant.

Chapter 7 - Chitosan has been exploited in promising drug delivery systems to release drugs in the target place, at the appropriate time and dosage, improving their bioavailability and, consequently, the therapeutic efficiency. On top of that, also allows the protection, control and release of biotech drugs such as peptides, genes, vaccines, antigens as well as synthetic drugs, which are highly limited to cross biological barriers and reach the target site, without collateral damage. Besides being easily manipulated, chitosan has important features, which make it unique. Its hydrosolubility and positive charge enable negative interactions with charged polymers, macromolecules, and polyanions when in contact with aqueous environment. Its exceptional biological properties such as mucoadhesiveness, permeation-enhancing effect across biological surfaces, biocompatibility and non-toxicity/non-antigenic make it useful for transmucosal drug delivery, improving absorption of the paracellular route. However, a better and future application of chitosan as therapeutic agent clearly depends on the design of appropriate carriers. As revised in this chapter, chitosan has been used in preparing films, beads, intragastric floating tablets, microparticles and nanoparticles for applications in the pharmaceutical field, surgeries and bone restructuring. Controlled drug delivery technology represents one of the borders of science, which involves a multidisciplinary scientific approach, contributing to human health care. Chitosan, with all its chemical and biological multipotential emerged as a promising solution for different conditions and medical applications, overcoming common problems of other drug carriers. As also described in this chapter, several clinical trials involving chitosan-based delivery systems are on the roll to explore in clinical the advantages of this exceptional material.

Chapter 8 - This chapter deals with a novel method of chemical modification of polysaccharides, in particular solid-state synthesis of chitosan and its composites with biocompatible synthetic polymers. The technique is based on a variety of chemical and physical transformations at conditions of plastic flow, which are realized in polymeric solids at pressure and shear deformation. The extrusion in the solid state is one of the most environmentally friendly methods of polysaccharide modification, since it does not require the presence of any catalysts, initiators, as well as organic solvents. It is desirable for biomedical applications, as it allows avoiding toxic remains over final composite materials. At the same time it provides numerous possibilities to circumvent many processing obstacles typical for preparation of natural polymers-based hybrid materials and to achieve better results compared to those of reactions in a melt or in a solution. Compared with traditional methods for design and manufacture of hybrid polysaccharide-based materials, solid-state reactive blending is a relatively simple, cost effective, and convenient way to improve the composite properties. In particular, grafting of polyester moieties onto chitosan chains was found to occur under selected deformation conditions. These polymeric materials demonstrate an amphiphilic behavior with dispersion propensity in organic solvents and could be promising for various biomedical applications. Microfibers containing chitosan up to 40% can be obtained by electrospinning of these dispersions. A uniform structure of the obtained polylactide/chitosan materials is confirmed by improved mechanical properties of films as compared with a model molten system with the same composition. Also it was shown recently that chitosan-g-poly(vinyl alcohol) copolymers can be produced by simultaneous alkaline solid-state deacetylation of chitin and poly(vinyl acetate) in an extruder. The main feature of these co-polymeric systems is solubility in neutral water at low and moderate temperatures. The obtained graft-copolymers possess excellent ability to stabilize nano-scale particles of both organic-inorganic origins. The dependence of chemical composition, morphologies and macroscopic properties on the ratios of co-extruded components as well as on the processing conditions is discussed.

Chapter 9 - Chitosan (CHIT), a non-toxic, biocompatible, and biodegradable natural polymer finds tremendous commercial application in its native and modified form. In the recent years, considerable efforts have been devoted to fabricate CHIT based smart materials with adopting attractive approaches, for examples, the ability to alter texture in a controlled mode via external stimulus including stress, electric or magnetic fields, temperature, moisture and pH. In the chapter, we have discussed the recent studies of the preparation of CHIT based smart materials from their various perspectives with put their special attention in the field of medical diagnostic and treatments. The modifications are illustrated basically on the stimuli-responsive CHIT as molecular device materials, biomimetic materials, hybrid-type composite materials, functionalised polymers, supermolecular systems, information- and energy-transfer materials, environmentally friendly materials, *etc.* at synergetic levels along their striking applications in both strategic and civil sectors.

Chapter 10 - Chitosan is a linear aminopolysaccharide obtained from the alkaline deacetylation of chitin. Chitosan is both biocompatible and biodegradable, making it an attractive material for use in drug delivery, gene delivery, cell culture, and tissue engineering applications. The achievement of predictable and reproducible release of an active agent into a specific environment over an extended period of time has significant advantage. A number of biodegradable polymers are potentially useful for this purpose, including synthetic as well as natural substances. Because chitosan is a swellable polymer with properties that can be tailored by varying the degree of deacetylation and molecular weight, it has been extensively investigated for use as a retardant polymer in matrix tablet formulations. The purpose of this review is to take a close look at the applications of chitosan IPNs and copolymers in controlled drug release. Based upon the present research and available products, some new and futuristic approaches in this area are thoroughly discussed.

Chapter 11 - The interest for finding better materials with the objective to use them as implants, has led to search in the mixture of natural polymers for a source to satisfy this necessity. This chapter presents the information about the synthesis, characterization, and some applications of hydrogels based in the chemical reaction between the hybrid, natural-synthetic, Chitosan-g-Glycidyl Methacrylate (CTS-g-GMA), of cationic nature, with water-soluble anionic polymers, such as Xanthan gum (X) and Hyaluronic acid (HA). The formed polyelectrolyte complexes, due to electrostatic attraction between the polymers, have improved properties when compared to hybrid CTS-g-GMA, which provides a wide range of applications in the biomedical field. All materials have been characterized by different analytical techniques such as infrared spectroscopy (FTIR), X-ray diffraction (XRD), thermal analysis (DSC and TGA), and the results were compared to the precursor materials (chitosan, X, HA, and CTS-g-GMA). Due to the HA nature, the film obtained from this reaction has been assessed for use as a patch for wound healing; whereas the properties showed by the hydrogels obtained from the reaction with X, make them very promising for applications in the treatment of recovery of spinal cord injuries. Cell culture was performed in all materials; different cell types were seeded and the viability has been quantified by the DNA (proliferation) assay, over several time intervals. The analysis showed satisfactory results of the (CTS-g-GMA)-X when compared to pure chitosan. Peroxide and interleukin-1β (IL-1β) assays have been performed to analyze the inflammatory response caused by biomaterials. Results show a moderate inflammatory response of our hydrogels when compared to raw chitosan. The implant of the hydrogels [(CTS-g-GMA)-X] in Wistar rats was performed after injuring the spinal cord by a laminectomy. The somatosensory evoked potentials (SEP) obtained by electric stimulation onto peripheral nerves were registered in the corresponding central nervous system areas, showing a successful recovery after 30 days of the implant. The results are promising and strongly support the future use of these hydrogels as scaffolds for tissue engineering and recovering.

Chapter 12 - The fundamental electrical and magnetic interactions between iron ions and biomaterials were investigated using an experimental approach, ^{57}Fe Mössbauer spectroscopy, and ab initio computational methods. A conventional spin-Hamiltonian approach adopted for the data analysis of the Mössbauer data showed that the metal ion in the Fe-chitosan complex is in the high-spin ferric state and that it has an internal magnetic field of approximately 440 kG, at the nucleus. The magnitude of the internal field arises from the predominant Fermi-contact interaction of the high-spin ferric species with N/O ligands. Based on the analysis of the experimental data, a scheme for the Fe-chitosan complex has been proposed. Similar analyses of spectral data were performed for other biomaterials such as glucose, cellobiose, and glucosamine. Studies pertaining to metal ion interaction with monomers such as glucosamine, glucose, and chondritin sulfate indicate that the oxidation state of the metal ion does depend on the pH of the reaction medium. Stability of any oxidation state is attributed to the presence or absence of a glycosidic linkage between sugar units.

To further probe the geometry, energy, and details of bonding of these Fe complexes, density functional theory (DFT) computations were performed, using Becke's three-parameter functionals with Lee Yang-Parr's correlation correction (B3LYP), together with the largest suitable basis sets available in the Gaussian quantum mechanical program. Four hexa-coordinated iron compounds were studied: (a) Complex 1 - Fe(II) with β-D-glucopyranose (glucose); (b) Complex 2 - Fe(II) with amino-2-deoxy-β-D-glucose chitosamine (glucosamine); (c) Complex 3 - Fe(II) with protonated glucosamine; and (d) Complex 4 - Fe(III) with protonated glucosamine. In all four cases, the iron atom was coordinated to oxygen atoms of two water molecules placed on an axial position. In addition, the iron atom was coordinated to two hydroxyl oxygen atoms of two glucose molecules in Complex 1, to two hydroxyl oxygen atoms of two protonated glucosamine molecules in Complexes 3 and 4, and with one hydroxyl oxygen and one amine nitrogen from two glucosamine molecules in Complex 2. In all four complexes, the two monosaccharides were rotated with respect to each other by 180°, both around the axis perpendicular to the molecular plane and around the molecular plane. Complex 1 was studied with low and high-spin electron configurations, whereas Complexes 2 and 3 were studied only with high-spin configurations. Predictions of Mössbauer chemical isomer shifts (δ_{Fe}) and electric quadrupole interaction (ΔE_Q) for these four complexes were obtained from standard curves based on experimental values.

Chapter 13 - Chitosan, a natural based nontoxic cationic polysaccharide polymer obtained by alkaline deacetylation of chitin, presents excellent properties such as biodegradability, antibacterial and wound-healing activity and immunological properties. These properties make chitosan a good candidate for the development of conventional and novel drug delivery systems. Chitosan has been used as a polymer for a controlled delivery system, gene delivery, scaffold, haemostatic action in wound healing, cell culture and cosmetic applications. Chitosan has become the focus of major interest in recent years because it has applications in not only the drug industry but also the agriculture, textile and paper industries. On the other hand, use of this biopolymer as a pharmaceutical excipient by different dose and for a number of applications is not new, but it still appears to be present in marketed drugs. The development of new delivery systems for controlled release of drugs is one of the most interesting areas of research in the pharmaceutical sciences. Nano and microparticles can be used for controlled release of different biological substances and drugs such as plasmids, hormones, peptide and proteins, antibiotics and vaccines. In this field, chitosan particular systems and chitosan matrixes, prepared by using different methods, can be used for encapsulating the drugs. Moreover, there has been interest in the chemical modification and PEGylated chitosan in order to improve its solubility and applications. Representatives of these novel chitosan modified polymers are carboxymethyl chitosan, thiolated chitosan, succinate and phthalate of chitosan salts and trimethyl-chitosan. The main chemical modifications of chitosan that have been proposed in the literature are reviewed in this chapter. Furthermore, recent studies suggested that chitosan and its derivatives are promising candidates as a supporting material for tissue engineering applications owing to their porous structure, gel forming properties, ease of chemical modification, high affinity to in vivo macromolecules. In general, this review provides an overview of using chitosan in pharmaceutical and biomedical applications.

Chapter 14 - In this chapter, the availability of chitosan was systematically investigated for removal of bisphenol A (BPA, 2,2-bis(hydroxyphenyl)propane) through the tyrosinase-catalyzed quinone oxidation. First, the process parameters, such as the hydrogen peroxide (H2O2)-to-BPA ratio, pH value, temperature, and tyrosinase dose, were discussed in detail for the enzymatic quinone oxidation of BPA. Tyrosinase-catalyzed quinone oxidation of BPA was found to be effectively enhanced by adding H2O2, and the optimum conditions for BPA at 0.3 mM were determined to be pH 7.0 and 40°C in the presence of H2O2 at 0.3 mM ([H2O2]/[BPA]=1.0). Removal of BPA from aqueous solutions was accomplished by adsorption of enzymatically generated quinone derivatives on chitosan beads. The use of chitosan in the form of porous beads was found to be more effective than that in the form of powders and solutions because heterogeneous removal of BPA with chitosan beads was much faster than homogeneous removal of BPA with chitosan solutions, and quinone conversion for the heterogeneous system with chitosan beads was a little higher than that for the heterogeneous system with chitosan powders. The removal efficiency was enhanced by increasing the amount of chitosan beads dispersed in the BPA solutions and BPA was completely removed by quinone adsorption in the presence of chitosan beads more than 0.10 cm^3/cm^3. In addition, a variety of bisphenol derivatives were completely or effectively removed by the procedure constructed in this study, although the enzyme dose and/or the amount of chitosan beads were further increased as necessary for some of the bisphenol derivatives used.

Chapter 15 - Over the past 40 years a greater attention has been focused on development of controlled and sustained drug delivery systems. The goals in designing these systems are to reduce the frequency of dosing or to increase effectiveness of the drug by localization at the site of the action, decreasing the dose required or providing uniform drug delivery. Polymers have been the keys to the great majority to design biomaterials for drug delivery systems. Chitosan, a natural biopolymer, has been extensively used for its potential in the development of biomaterials due to the polymeric cationic character and gel and film forming properties.

Hydrogels preformed by chemical cross-linking or physical interactions form three-dimensional, hydrophilic, polymeric networks capable of imbibing large amounts of water or biological fluid. The association of magnetic nanoparticles, which could be controlled by an exterior magnetic field and have dimensions to facilitate their penetration in cells/tissues, with hydrogel type biopolymeric shells confer them compatibility and the capacity to retain and deliver bioactive substances.

In this chapter we present a synthesis of our works, from the last 5 years, in this domain of chitosan usage for synthesis of biomaterials suitable in obtaining of drug delivery systems. In our works we used mainly chitosan with slight addition of hyaluronic acid for obtaining biomaterials with controlled properties. Chitosan and hyaluronic acid are two natural biopolymers with similar structure and special biological characteristics, easily to process in porous scaffolds, films/pellicles and beads. They are biocompatible, biodegradable, promote cell migration and cell adhesion and electrostatic interact, having a great potential in the development of drug delivery systems and in tissue engineering.

In a first stage, we obtained biopolymeric pellicles by casting method using mixtures of chitosan and hyaluronic acid solutions. Chitosan/hyaluronic acid pellicles crosslinked with sodium citrate could be used for enzymes immobilization on the surface or by entrapping in the polymer network. The pellicles with immobilized enzymes are stable during several months and could be used multiple times without a notable decrease of the immobilized enzymatic activity. The ultrastructure of simple chitosan pellicles, chitosan/hyaluronic acid pellicles and enzymes immobilized on chitosan/hyaluronic acid pellicles were examined by confocal scanning laser microscopy.

More recently, in other works, we developed a new system based on magnetic nanoparticles covered in a layer-by-layer technique with a biocompatible hydrogel from chitosan and hyaluronic acid capable of vectoring support for biologic active agents (L-asparaginase, protease inhibitor, e.g.). Characterization of size and morphology of the obtained hydrogel-magnetic nanoparticles with entrapped L-asparaginase/protease inhibitor was made using dynamic light scattering method, transmission electron microscopy and confocal microscopy.

The structure of magnetic nanoparticles coated with hydrogel was characterized by Fourier transformed infrared spectroscopy. The biocompatibility of nanoparticles was evaluated and also the interactions with microorganisms.

The characteristics of developed pellicles and nanocomposite materials made them suitable to be used for medical or biotechnological purposes.

In: Focus on Chitosan Research
Editors: Arthur N. Ferguson and Amy G. O'Neill
ISBN 978-1-61324-454-8

Chapter 1

POROUS STRUCTURE AND ADSORPTION BEHAVIOURS OF CHITOSAN

O. K. Krasilnikova, O. V. Solovtsova, E. V. Khozina, T. Y. Grankina, N. V. Serebryakova and S. M. Shinkarev[1]*
A.N. Frumkin Institute of Physical Chemistry and Electrochemistry RAS, Moscow, Russia

ABSTRACT

Interrelation between the structure and adsorption properties towards to water and nitrogen vapors and Ni, Cu and Ag cations of the chitosan samples modified by freeze-drying after precipitation with NaOH or Na_2CO_3 or drying in air, as well as cross-linking is studied. The investigations by IR spectroscopy, X-ray diffraction and pulsed NMR show the influence of the preliminary treatments on the supramolecular structure. The more ordered structure is observed in the freeze-dried samples of chitosan compared with the air-dried sample. The parameters of porous structure of the chitosan samples calculated from the isotherms of water vapors adsorption and the nitrogen vapors low temperature adsorption are compared. It is revealed that the isotherms of water vapors adsorption on the chitosan samples are more sensitive to the structural modification of polymers. The efficiency of different models based on the Flory-Huggins, Dubinin-Astakhov, and Dubinin-Serpinsky equations for fitting the experimental isotherms of water vapor adsorption is evaluated. The Dubinin-Serpinsky equitation 2 based on a model of two-stage localized adsorption of water molecules describes the experimental water vapor adsorption on the chitosan samples in the full range of relative pressures. In the range of small relative pressures, water molecules interact with chitosan polar groups which serve as the primary adsorption centers (PAC). Then, water molecules adsorbed on PAC become the secondary adsorption centers for further adsorption. As a result, at relatively high pressures the water clusters are formed. In the range of low water fillings

[1] 119991 Moscow, Leninsky Prospect., 31, e-mail: albert-voloshchuk@rambler.ru.
* All-Russian scientific-research and technological Institute of biological industry, Moscow Region, Russia.

of the chitosan adsorbents, the amount of primary adsorption centers can be evaluated by the comparative method.

The freeze-dried chitosan samples display the increased adsorption capacity towards to water and nitrogen vapors, as well as to metals cations. The cross-linking of the freeze-dried chitosan samples by glutaraldehyde reduces the adsorption capacity through the blockage of amino-groups. Cross-linking the chitosan films leads to the growth of adsorption capacities towards to water vapors. Thus, the analysis of the isotherms of water vapors adsorption specifies that structural changes of the chitosan samples after freeze drying and cross-linking resulting in a growth of the amount of accessible adsorption centers.

The adsorption capacities and the rates of adsorption of metal cations as Ni, Cu and Ag onto the modified chitosan samples are examined. The Langmuir, Freundlich and Dubinin-Radushkevich adsorption models are used to describe the isotherms of metal cations adsorption. The kinetics experimental data properly correlate with the pseudo-first-order kinetic model. The modification of the chitosan structure by freeze-drying noticeably increases the adsorption capacities towards to metal cations. The adsorption capacity of the freeze-dried chitosan sample towards to nickel cations is as high as 4.5 mmol/g that is threefold higher than that of the air-dried chitosan sample.

The IR-VIS spectroscopic experiments performed for the copper cations adsorption on the chitosan samples at high adsorption values allowed to advance a probable model of copper-chitosan complex that is a "bridge" structure based on a binuclear copper complex which bounds two chitosan macromolecules, so that every Cu (II) cation of a pair is associated with one amino group in axial position which belongs to different chitosan chains.

I. Introduction

Chitosan [1] is a widely spread natural polymer represent in shells of insects and Crustaceans; from a chemical point of view, it is partly deacetylated poly(glucosamine) (Figure 1). This nontoxic and biodegradable copolymer of β-(1-4) linked – 2 – acetamido-2-deoxy-d-glucopyranose and 2-amino-2-deoxy-D-glucopyranosee) is produced by deacetylation of chitin and is a renewable natural polymer.

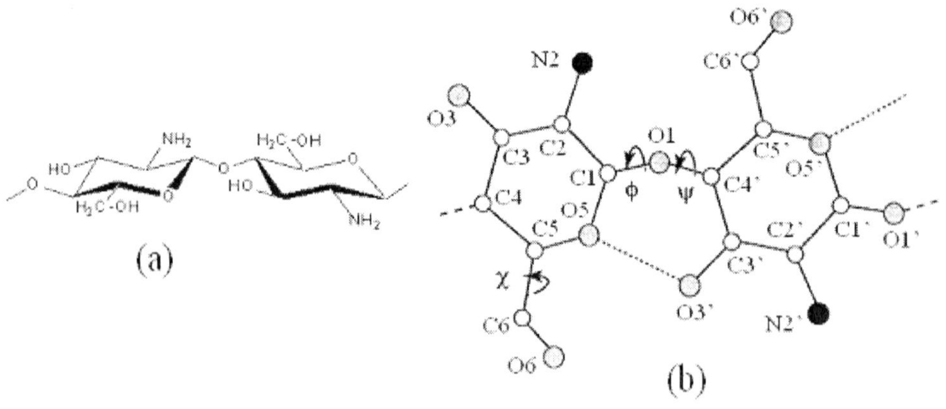

Figure 1. Chemical form (a) and structure (b) of chitosan together with atomic numbering. Dashed lines denote O3---O5 hydrogen bonds. Two dihedral angles (φ, ψ) that define the main chain conformation and one dihedral angle χ that defines the O6 orientation are also shown [2].

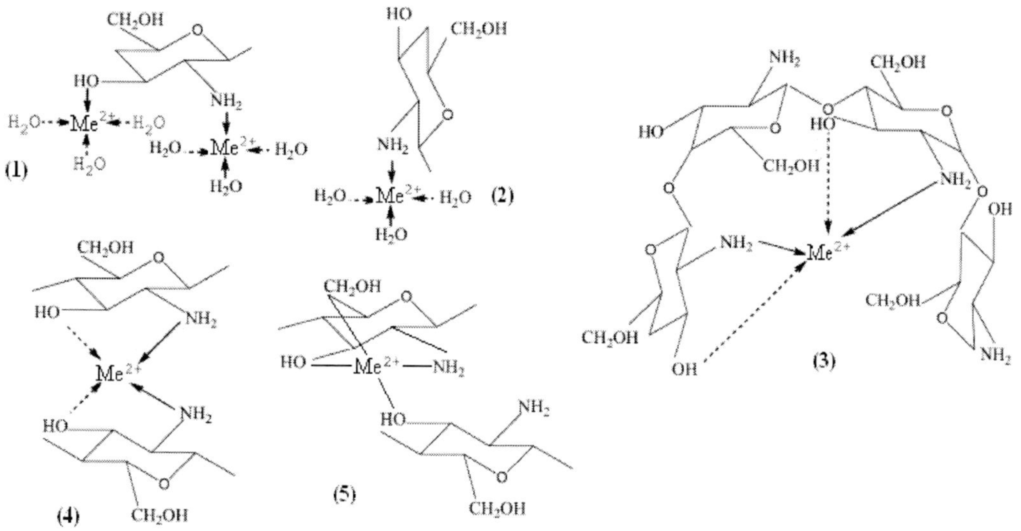

Figure 2. The reasonable structures of Chitosan-Me complexes: structures 1, 2 are possible in terms of the "pendant" model; structures 3, 4, 5 are suggested by the "bridge" model.

At present, the potentialities of chitosan and its derivatives for adsorption and concentration of metal ions and for preparation of pharmaceutical products containing metal nanoclusters, catalysts, and metal nanoparticles are being extensively discussed [3]. Indeed, chitosan is an excellent natural adsorbent for metal cations due to presences of the amino (-NH_2) and hydroxyl (-OH) groups. For instance, immobilization of copper cations on a water swollen chitosan matrix gives gelatinous hydrophilic catalysts that are highly active and selective in liquid-phase oxidation and combine the advantages of both the homogeneous and heterogeneous catalysts [1].

Moreover, with the purpose of increasing the adsorption capability, chitosan may be exposed to chemical and physical modifications. Apart from increasing of the adsorption activity, physical modifications allow the expansion of the porous network due to gel transformation, which eventually changing the crystallinity of the adsorbent. As for chemical modification, it also enhances the adsorption properties while preventing the dissolution chitosan in high acidity solutions and improving the mechanical properties of chitosan. Chemical modification by cross-linking may include grafting the new functional groups and acetylation of chitosan, thereby changing the crystalline structure of chitosan and enhancing its resistance against the acids, alkali and other chemicals. The most commonly used cross-linked agents are glutaraldegide (GLA), epichlorohydrine (ECH) and ethylene glycol diglycidyl ether (EGDE) [4].

There are many methods for removing heavy metals from water [5]. Among them the most widely used are chemical and electric precipitations, membrane separation, and adsorption. However, the economically advantageous chemical precipitation with -OH and S_2–ions appears to be inefficient for dilute solutions. The ion exchange and reverse osmosis methods are fairly efficient; but they are expensive. While cheap wastes of agricultural industry, clays, and natural polymer adsorbents, e.g., cellulose and chitosan, may be applied in adsorption methods of water purification. Because of the presence of amino groups, the ability of chitosan to form complexes with transition metals allows it to be used for selective removal of metal cations and radioactive nuclides from solutions for industrial use even at extremely low concentrations, when other methods are inefficient. Chitosan adsorbs the transition metal cations, but it does not adsorb the cations of alkali and alkaline-earth metals at all. So, chitosan can be used as a selective adsorbent in many processes. Although the adsorption of heavy metal cations by chitosan and its derivatives plays a leading part in many nanotechnological processes, the complexation mechanism, especially the detailed structures of the resulting complexes between chitosan and heavy metal cations, has been left undetermined due to lack of experimental data. Indeed, the interpretation of the experiments of the adsorption of heavy metal cations by the chelating polymer chitosan implies that not only the measuring the amount of adsorbed metal and plotting corresponding adsorption isotherm, but also a supramolecular organization of the metal cation – chitosan complex (i.e., the adsorption mechanism) has to be found out.

At present, two models of the metal–chitosan complex are mainly discussed. Figure 2 shows the reasonable structures of adsorption complexes of metal with chitosan expected in view of these two models. In terms of a so-called “bridge” model (Figure 2, scheme 3-5), the metal (e.g., copper) cation may be coordinated by four amino groups (the maximum value) of one or more chitosan chains [6, 7]. It is pointed out that chitosan forms a unique complex with copper. The structure of this complex is close to $CuNH_2(OH)_2$. According to coordination sphere of copper, the fourth site can be occupied by either a water molecule or OH group in C-3 position and copper bonds to three oxygen atoms and one nitrogen atom, forming square-planar or tetrahedral geometry. The other, “pendant” model (Figure 2, scheme 1-2) was proposed from an analysis of x-ray diffraction data for metal–chitosan (in particular, copper–chitosan) complexes [8, 9]. According to the latter “pendant” model (Figure 2 structure 1, 2), metal ions are considered to be suspended on the amino groups of one chitosan chain. The calculations of the energy of interactions between copper cations and chitosan olygomers indicate the most stable complexes based on “bridge model” (Figure 2, structure 3-5) [10].

However, both the models are equally dubious for lack of experimental evidence and complicated interpretation of experimental data. Apparently, these difficulties are partly due to the influence of the conditions of the synthesis and preparation of a structurally flexible polysaccharide adsorbent, which greatly swells in a solvent.

Many attempts have been made to establish the correlation between the structure of swelling polymers and their adsorption ability. The main problem is in continuous structural changes of swelling polymers that accompany the adsorption process. In particular, the interpretation of data of water adsorption on swelling polymers must take into account that displacements of polymer chains due to the interactions with water molecules lead to opening formerly disclosed or new adsorption centers. The traditional approach of studying the parameters of porous structure of adsorbent based on low temperature nitrogen vapors (or other non-polar molecules as argon, krypton and etc.) adsorption data show that the specific surface of swelling polymers is small and the porosity is insignificant, so it does not give the knowledge about the adsorption properties of swelling polymers towards to water vapors or metal cations from water solutions. At the same time it is known that chitosan is able to absorb the considerable quantity of water vapors due to hydrogen bonding with polar groups of polymers [11], and metal cations as it was mentioned above.

The water vapor adsorption on chitosan is caused by the changes in chemical potentials both of water and polymer. The changes in chemical potential of swelling adsorbent are defined by the changes of entropy of chitosan macromolecules due to adsorption of water. In other words, the adsorbed water molecules shift polymer chains, thereby forming a new supramolecular structure of chitosan and, as a consequence, determining the adsorption properties. Thus, the adsorption of water on swelling chitosan adsorbents is sensitive to the structure of polymers and is important for characterization their adsorption properties towards to metal cations from water solutions, too. Different models based on the Flory-Huggins theory of polymer solution [13], as well the Dubinin-Astakhov [12] and the Dubinin-Serpinsky [12] adsorption equations are tested for describing the data of equilibrium adsorption of water vapors by different chitosan samples.

The results of the investigation of adsorption of transition metal ions: silver, nickel and cooper, by the differently modified chitosan samples are considered. Thereto, the parameters of porous structure of chitosan samples are estimated from the analysis of water vapor adsorption isotherms and compared with the data obtained from the isotherms of low temperature nitrogen vapors adsorption. The effect of drying conditions and swelling on supramolecular structure of chitosan is studied by FTIR and UV-VIS spectroscopy, X-ray diffraction and pulsed NMR methods. Finally, the isotherms of adsorption of metal ions on the chitosan samples obtained in different dehydration conditions and of different cross-linking extents are interpreted in the terms of several models based on Langmuir [12], Freundlich [12], and Dubinin-Radushkevich [12] equations. The kinetics of metal cations adsorption on the chitosan samples are examined in terms of well known models [14, 15]).

II. Experimental

2.1.Materials

The salts of heavy metals (nitrate nickel, sulfate copper, nitrate silver) and inorganic chemicals (glutaraldehyde, ethylene glycol diglycidyl ether, acetic acid and all) were supplied by *Acros*. All the reagents used are analytical-reagent grade. Deionized by membrane filter water is used to prepare all the solutions under study.

The descriptions and designations of the chitosan samples with different degrees of deacetylation (DD) and molecular mass (MM) are reported in Table 1. The samples were produced from the shells of Kamchatka crabs at the pilot plant of All-Russia Research and Technological Institute for Biological Industry. The DD and MM values were determined by the back potentiometric titration and viscosimetry, respectively [16].

2.2. Preparation of Chitosan Samples

In order to establish the effect of drying procedures employed for sample preparation, the chitosan with a DD of 87.1% and an MM of 287 kDa dried in air at 353 K (sample 1) is used. Other experimental samples were obtained through the preparation of a 3% solution of this chitosan sample in a 1% acetic acid followed by its thorough filtration. Part of this solution was applied to reprecipitate chitosan with a 3% NaOH solution added under stirring until pH of the mixture became equal to 7.4. A residue chitosan thus obtained was washed with deionized water, frozen to 323 K and freeze-dried in a device TG-50 setup (sample 2). Another part of the solution was used to precipitate chitosan with a 3% Na_2CO_3 solution added to reach pH equal to 7.4. Following the precipitation, the gel was subsequently freeze-dried (sample 5).

The cross-linked samples 3 and 4 were obtained from the freeze-dried chitosan sample 2 precipitated by alkali through the reaction with *glutaraldehyde* (GA) in different conditions (Figure3, scheme a). Sample 3 was prepared by cross-linking with GA in solution. For this purpose the freeze-dried sample 2 (5 g) was soaked in 75 ml of a 0.01% GA solution for 24 hours at room temperature. Then the cross-linked product was carefully washed by distilled water and dried. Sample 4 was prepared through cross-linking in vapors of GA by soaking the 5-g sample 2 in vapors of a 25% GA solution for 24 hours. After that procedure the chitosan sample 4 was dried on air to a constant weight.

For the sake of comparison, the water vapor adsorption on the native chitosan films (sample 6), and various chitosan films cross-linked by GA, ethylene glycol diglycidyl ether (EGDE), samples 7 and 8 (see Figure3, scheme b), respectively, are examined, as well. The details of the cross-linking procedure are described by Krasil'nikova [17]. Chitosan flakes were dissolved into aqueous acetic acid solution to prepare 1% of aqueous acetic acid solution. The solution in 0.05 M GA was mixed with the chitosan solution to obtain a molar ratio chitosan/GA equal to 0.7. The mixture was stirred at room temperature for 24 hours. The resulting solution was filtered and left standing for 24 hours at room temperature. Then the cross-linked chitosan films were prepared by dry phase inversion.

Chitosan-EGDE films were prepared according to the following procedure. A solution 5 % EGDE was added to the 1% chitosan solution in order to achieve a 3.0 molar ratio of EGDE/chitosan. The resulting solution was stirred and heated up to 323-333 K for 3 hours for cross-linking. The cross-linked chitosan films (sample 8) were prepared through dry phase inversion over 24 hours.

Table 1. Characterization of the chitosan samples

No	MM, KDa	DD, %	Reprecipitation agent	Drying method	Additional treatment	Degree of cristallinity, %	Amount of PAC after evacuation, mmol/g	
							293 K	373 K
1	287.3	87.1	NaOH	Air drying	-	38.1	0.54	–
2	287.3	87.1	NaOH	Freeze-drying	-	76.9	1.25	1.80
3	287.3	87.1	NaOH	Freeze-drying	Cross-linking in GA solution	-	0.83	1.16
4	287.3	87.1	NaOH	Freeze-drying	Cross-linking in GA vapors	-	0.33	0.41
5	287.3	87.1	Na_2CO_3	Freeze-drying		65.5	0.21	0.83
6	400.0	95.0	NaOH	Air drying	Film	36.0	0.23	0.36
7	400.0	95.0	NaOH	Air drying	Film cross-linked by GA	45.0	0.07	–
8	400.0	95.0	NaOH	Air drying	Film cross-linked by EGDE	-	0.23	0.48

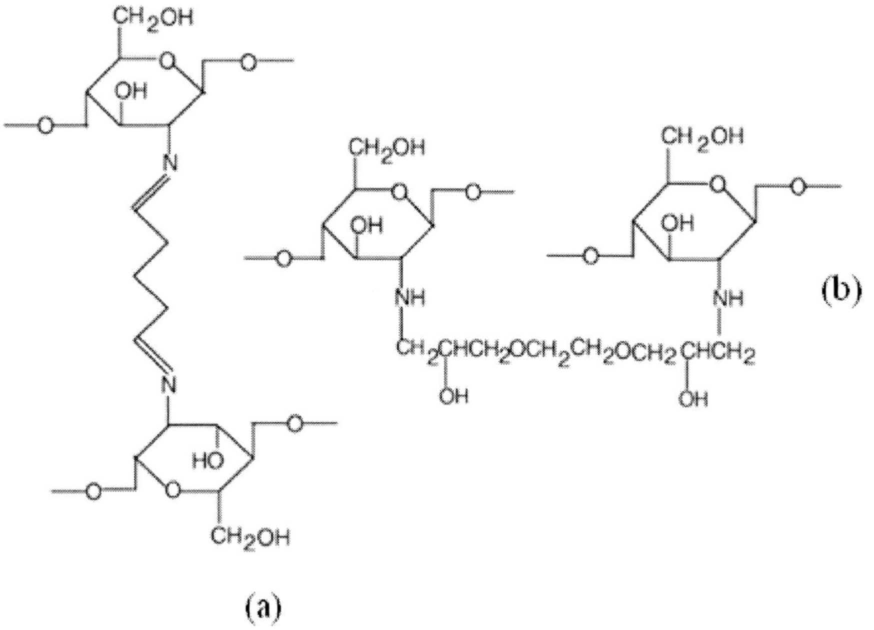

Figure 3. Schematic representation of the chitosan and cross-linked chitosan beads: (a) - chitosan-glutaraldehyde, (b) - chitosan-ethylene glycol diglycidyl ether [4].

2.3. Nitrogen and Water Vapors Adsorption Experiments

The investigation of low temperature nitrogen adsorption was performed using the Accelerated Surface Area and Porosimetry system, ASAP 2020, (Micromeritics, USA). The total specific area was determined by the BET method [12].

The water vapor adsorption isotherms on the chitosan samples (Table 1) are measured at 293 K by means of a vacuum gravimetric setup equipped with a quartz spring McBain balance (sensitivity of 10 μg at a sample weight of 100 mg), with a precision of adsorption value measurements up to $3 \cdot 10^{-4}$ g/g. The equilibrium water pressure is determined by means of U-manometer with cathetometer KM-8 with an accuracy of 0.1 millimeters of mercury. Before the first adsorption experiment, the samples are outgassed up to a constant weight at 20°C in order to remove pre-adsorbed water. After the first adsorption experiment the samples were outgassed under vacuum at 373 K to remove the adsorbed water. Then water vapor adsorption experiment was repeated.

To establish the mechanism of water vapors adsorption on the chitosan samples, the comparative method based on comparison of the adsorption data obtained for the samples under study with non-porous graphitized carbon black ("Vulkan-7H", graphitized at 2800 °C) as the reference adsorbent with known water adsorption mechanism and characteristics is used [18]. It is convenient to take the adsorption on the surface of graphitized carbon black expressed in the a/a_m unites as abscissa, and the adsorption on the chitosan sample measured at the same relative pressures is used as the ordinate. The comparative plots are straight lines from the coordinate origin. Their ordinates at a/a_m=1 correspond to the amount of primary adsorption centers (PAC) per 1 g of chitosan [18].

2.4. Measuring of Metal Cations Adsorption on the Chitosan Samples

The adsorption isotherms of metal cations on the chitosan samples 1, 2, 5 (see Table 1) are measured by the bath method.

The weighed chitosan samples are put directly in contact with the aqueous solutions with different concentrations of nickel nitrate, copper sulfate or silver nitrate and kept 10 days under room temperature. Such duration was found by kinetic measurements to be sufficient for establishing the adsorption equilibrium for nickel, cooper and silver ions. Then, the solutions are filtered and analyzed.

The equilibrium concentrations of copper and silver cation solutions in the filtrates are measured with an accuracy of 5% by potentiometric titration. This technique involves the measurement of the potential difference (emf) between the indicator and reference electrodes. For cooper, the titration is carried out in a solution of $CuSO_4 \times 5H_2O$ with an EKOTEST-2000 ionometer using Ekom-Cu cooper-selective electrode and EVL-1M 3.1 silver-chloride reference electrode filled with saturated solution of KCl. For silver, the titration is carried out in a solution of $AgNO_3$ with an EKOTEST-2000 ionometer using Ekom-Ag silver-selective electrode and EVL-1M 3.1 silver-chloride reference electrode filled with saturated solution of KCl.

The equilibrium nickel cation concentrations are measured with an accuracy of 5% in the filtrates by compleximetric titration with ethylenediaminetetraacetic acid solution containing murexide as an indicator [19].

2.5. Structural Investigations

The structure of both the polymer samples and the adsorption polymer-metal complexes are studied by FTIR diffuse reflection and UV-VIS spectroscopy methods. Spectra are measured using a Perkin-Elmer 2000 FTIR instrument in the range of 15000–30 cm^{-1} and a Lambda -35 spectrophotometer in the range of 200- 1100 nm.

The X-ray diffraction patterns of powdered chitosan flakes and films are recorded with a DRON-3M diffractometer in the transmission mode in the angle range of $2\theta=5°–50°$ using nickel-filtered $CuK\alpha_{\alpha}$ radiation. The degrees of crystallinity of the samples determined from X-ray diffraction patterns are estimated according the method [20]. The crystallinity of the chitosan samples is calculated as the relation of a total area of the X-ray diffraction patterns to the area corresponding to the crystalline phase.

The scan-electron microscopy (SEM) is used to study the morphological peculiarities of the chitosan samples after different preliminary treatment. Figure 4 shows the images of the samples dried in different conditions made by means of scanning electron microscope *JSM-U3 (LEOL, Japan)*. The images show that the air-dried chitosan, sample 1, differed by lamellate structure (Figure 4, *a*) from the spongy freeze-dried samples 2 and 5 (Figure 4, *b*, *c*). In addition, the occlusion of crystalline are observed for the freeze-dried chitosan precipitated in NaOH, while the sample 5 precipitated in Na_2CO_3 has a bulged surface.

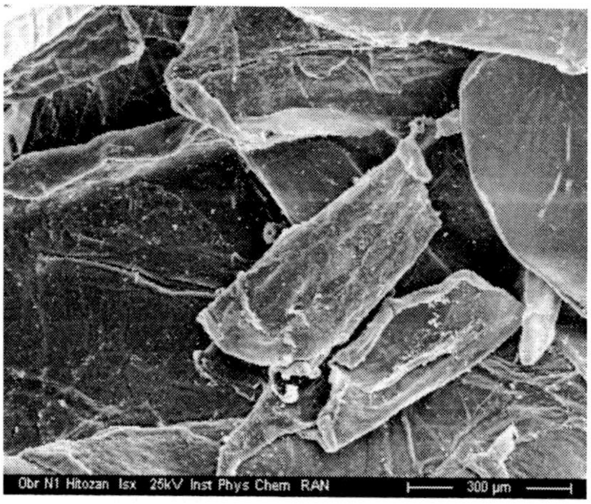

a.

b.

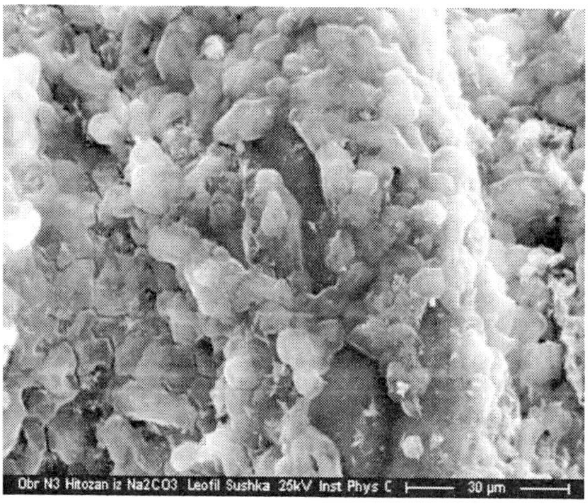

c.

Figure 4. SEM images of the chitosan samples 1 (a); 2 (*b*) and 5 (*c*).

The nuclear transverse relaxation parameters are measured by means of a 19 MHz ^{1}H-nmr relaxometry instrument designed in Department of Molecular Physics, Kazan state University at 303 K. The combined analysis of both free induction decay (FID) signals followed the sequence of RF pulses: $90_x°$-τ-90°-τ-"solid echo" [21] and spin-echo decays from the CPMG experiments [22, 23] allow getting a complete set of spin-spin relaxation times. The accuracy of detecting the amplitudes of free induction decays or spin echoes is not more than 5 %. Self-diffusion coefficients of water molecules adsorbed in the chitosan samples are measured by NMR with pulsed magnetic field gradient (PMFG) at 303 K by means of stimulated echo sequence [24] at ^{1}H frequency of 50 MHz. The maximum value of PMFG was 60 T/m.

III. Results and Discussion

3.1. The Effect of Drying and Cross-Linking Chitosan Samples on their Structural Features

3.1.1. Crystallographic Study of Chitosan Samples

The X-ray diffraction patterns of samples 1, 2, and 5 are reported in Figure 5. The initial air-dried chitosan (sample 1) demonstrates the characteristic peaks at 2θ = 10.9 and 19.8° attributed to the crystalline and amorphous phases of polymer, respectively. It is obvious that the dehydration conditions have a strong effect on the structure of the chitosan samples. The crystallinity values of freeze-dried samples (76.9 % and 65.5% for samples 2 and 5, respectively) determined according [20] exceed that for the initial air-dried sample 1 (38.1 %). Moreover, the X-ray diffraction patterns for the alkali- and soda-precipitated samples are unlike each other that can be attributed to the different conformations of polymer chains in the conditions with different pH [25]. The observed increase of crystallinity of the freeze-dried chitosan samples can be related to the fact that, the freezing of the equilibrium swollen polymer hydrogel, which was precipitated due to varying pH, results in fixation of the ordered structure of hydrogel, and its compression and deformation during dehydration are markedly diminished. Recently the same effect has been found by *Jaworska* [20], while the inverse effect was observed by *Jaworska* for non-precipitated chitosan [20]. *Valentin et al* reported that the supercritical drying of chitosan hydrogel also preserves the structure of hydrogels in the dried sample [26]. The changes in the arrangement of chitosan chains due to hydration – dehydration were determined recently by *Okuyama et al* by the X-ray fiber diffraction method and the linked-atom least-squares method [2]. It has been found that the air-dried chitosan has a double helical conformation. Two chains are packed in the antiparallel directions into a $P2_12_12_1$ orthorhombic cell with a = 0.807 nm, b = 0.844 nm, and c = 1.034 nm. The neighboring parallel chains are linked by the O(6)–N(2) hydrogen bonds to form a sheet structure. Two antiparallel chains make a unit cell containing no crystallization water. The hydrated chitosan crystallizes into an orthorhombic unit cell with the parameters a = 0.895 nm, b = 1.697 nm, and c = 1.034 nm. The double helix forms O(3)–O(5) hydrogen bonds with the *gt* orientation of O(6). The unit cell of hydrated chitosan comprises four chains and eight water molecules. The N(2)–O(6) hydrogen bonds between the adjacent chains aligned with the axis b make the layer parallel to the plane bc. Each plane with respect to the neighboring ones has a symmetry of 2_1 along the axis *b* and the layers are stacked along the axis *a*. The adjacent planes are connected by water molecules (as if they are small pillars) bridging the neighboring layers. On dehydration, two water molecules are removed, the chains shift along the axis *a*, and the hydrogen bonds between these chains break down. This is accompanied by compression along the axis *b* and expansion along the axis *a*, which makes a new layered structure in which one chain is shifted by *c*/4 along the axis *c*.

In contrast to the air drying, the freeze drying of chitosan sample results in a cryogel that partly retains the rather loose crystal structure of hydrogel which may facilitate the complex formation with cations of transition metals. These chitosan cryogels have a substantially greater number of active adsorption sites accessible for adsorption of water molecules compared with the air-dried sample.

The films that were cross-linked by GA and DGA are also characterized by the increased crystallinity values compared with the initial chitosan film (sample 6). According X-ray diffraction pattern reported in Figure 6 the two crystal structures coexist in the initial chitosan film –phases I and II, as it was observed recently [27, 28]. These structures are characterized by an identical repeating period along the macromolecular axis (or fiber axis), *c*, equal to 1.03 HM, and differ from each other by the orthonormal parameters of a rectangular unit cell of chitosan *a* and *b*. Large majority of chitosan chains are involved into the crystal phase II with the structural organization close to chitin [27, 28]. It has been found that the cross-linking chitosan by GA resulted in the increase of crystallinity from 36 to 45 % with expanding the size of crystallites from 65 up to 75 Å that was calculated from the most intensive reflex (100). In other words, the cross-linking of chitosan film leads to the ordering of structure.

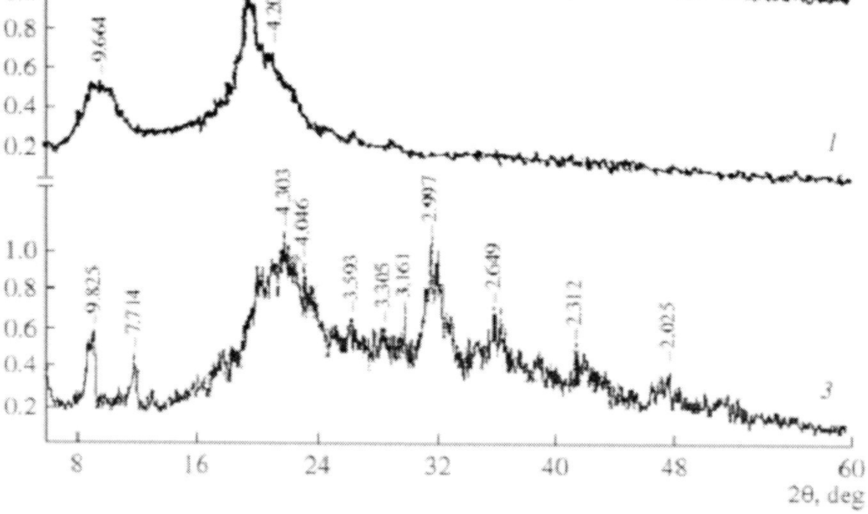

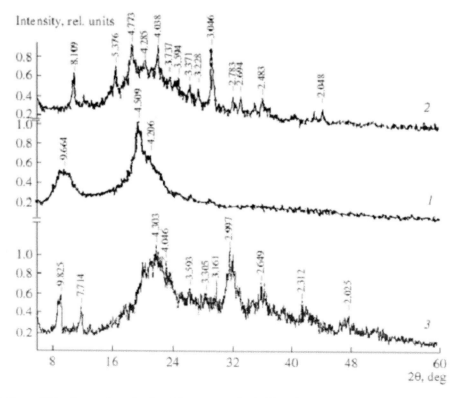

Figure 5. X-ray diffraction pattern for the chitosan samples 1 (*1*), 2 (*2*), 5(*3*).

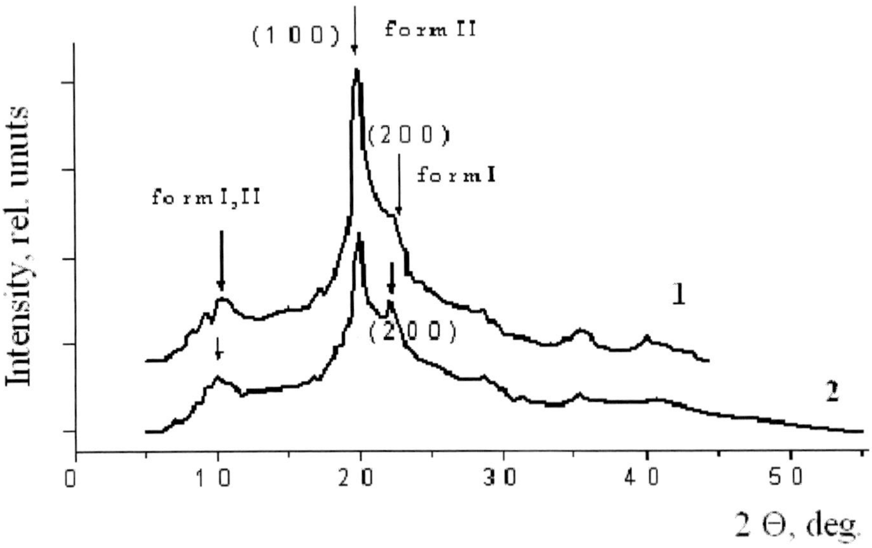

Figure 6. X-ray diffraction patterns of: 1- the initial chitosan film (sample 6) and 2 – the chitosan films cross-linked by GA (sample 7).

Although the crystalline domains of chitosan is often considered not accessible to water molecules [e.g. Despond, 29] by our opinion, the increase of crystallinity of chitosan adsorbents does not indicate directly to the enhancement or abatement of their adsorption properties towards to water and metal cations as well. It should be noted that the swollen hydrogel of chitosan is transparent. The ordered or disordered arrangements of polymer chains that ensure the maximal associability of adsorption centers is a key factor for adsorption of both water and metal cations.

3.1.2. FTIR Spectroscopy of the Chitosan Samples of Different Modifications

It can be seen from Figure 7 that the general spectral pattern changes dramatically when passing from the sample 1 (curve *1*) to the sample 2 (curve *2*). The most pronounced change is a shift of the band attributed to the stretching vibrations of hydrogen-bonded OH-, NH_2-groups and water molecules from 3499 to 3403 cm $^{-1}$. This shift suggests a structural change in the network of hydrogen bonds and increase in the energies of hydrogen bonds in the sample that can be related to the freeze-drying. In addition, the pronounced peaks at 2935, 2885, and 2854 cm $^{-1}$ attributed to C–H stretching vibrations in the polymer molecule appear in the spectrum for the freeze-dried sample instead of the diffuse absorption band at 2899 cm-1 (curve *1*). Analogously, the peaks at 1434, 1382, 1342, and 1327 cm^{-1} in the range related with C–H bending vibrations become more pronounced. Apparently, the discrete absorption at the frequencies of C–H stretching vibrations results from the ordering of chitosan structure. In the range of C–O stretching vibrations, the corresponding absorption bands shift from 1167 to 1156 cm^{-1} (ν_{as} (C–O–C) in the pyranose ring) and from 1129 to 1113 cm^{-1} (ν_{as}(C–O–C) between pyranose ring). These shifts suggest the conformational changes in polymer structure resulted from the freeze-drying. In addition, the relative intensity of the absorption at 1030–1060 cm^{-1} increases substantially.

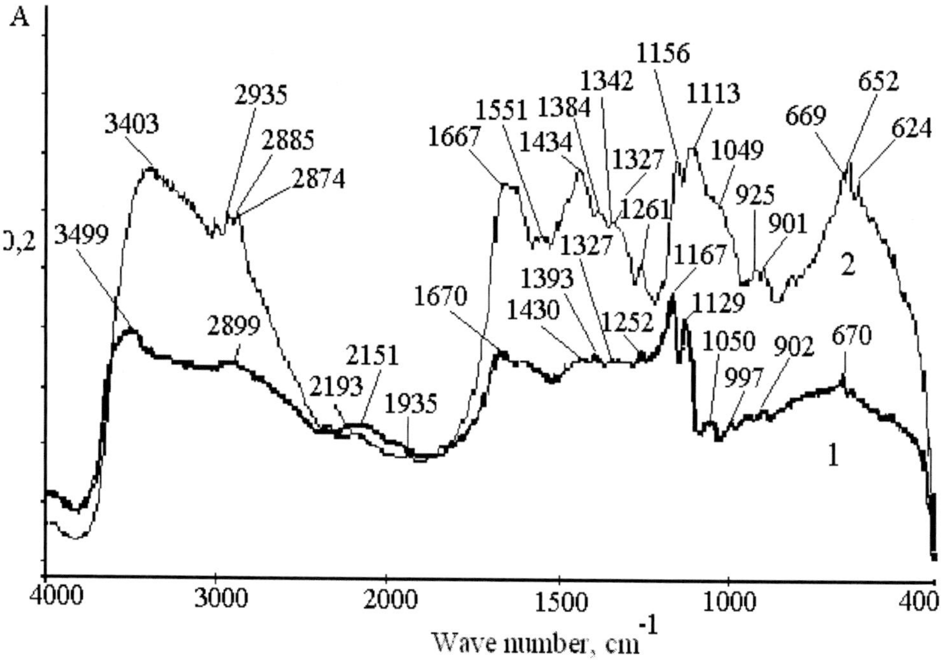

Figure 7. The IR diffuse reflection spectra of chitosan samples 1 (curve 1) and 2 (curve 2).

The bands in this range are assigned to the bending vibrations of primary alcohol groups –CH_2–OH and the asymmetric stretching vibrations of the –C–O–C– bridge connecting the pyranose rings. The peak at 900 cm^{-1} is assigned to the symmetric vibrations of the bridge; however, the peak at 900 cm^{-1} can be also assigned to the >CH–OH group. The freeze-dried sample shows more intensive absorption in this range (at least two bands at 925 and 901 cm^{-1} can be observed, see Figure 7) compared with the air-dried chitosan. The observed changes are due to the fact that OH groups of the polymer are not involved in hydrogen bonding or because chitosan molecules are bent more strongly (the angle of the bridge between the pyranose rings is smaller). The bands at 1650–1670 cm^{-1}, which are attributable to the stretching vibrations of N–H bonds in the primary NH_2-groups in chitosan samples, also become more intense after the freeze-drying. Thus, the spectral changes include the more intense absorption bands for all groups of polymer sample, the lower-frequency bands assigned both to the stretching vibrations of the pyranose ring and the hydrogen bonds, and the distinct bands characteristic of the main chitosan groups in the crystal state (instead of the wide featureless absorption). Hence, one can conclude from the spectroscopic data that freeze-drying results in removing a larger amount of the adsorbed water compared with drying in air. Thus, the resulting supramolecular structure of chitosan is determined likely by intramolecular hydrogen bonds. Obviously, the observed changes in the IR spectra point to the ordered supramolecular structure of the polymer chains of the freeze-dried chitosan sample. This result agrees with the X-ray powder diffraction data obtained for the chitosan samples under study and presented above.

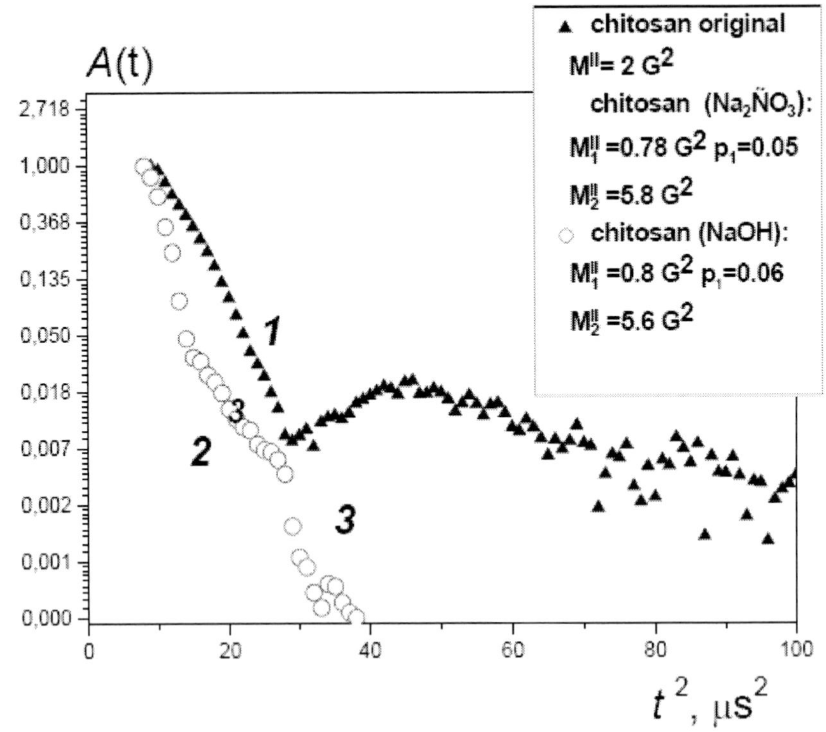

Figure 8. FID signals observed in the chitosan samples, *1* - sample 1, *2* - sample 2 and *3* - sample 5.

3.1.3. Pulsed NMR Measurements in the Chitosan Samples Dried in Different Conditions

The FID signals measured for the chitosan samples with and without water (the last are shown in Figure 8) at 303 K are fitted by a sum:

$$A(t)/A(0) = \sum_{i=1}^{2} p_{Si}F_{Si}(t) + \sum_{i=1}^{3} p_{Li}\exp(-t/T_{2i}), \tag{1}$$

where $A(t)$ – FID amplitude at the moment t; F_S – the relaxation function of immobile protons of adsorbents, T_2 – the spin-spin relaxation times of mobile protons influenced by molecular mobility and interactions, $p_{S,L}$ – the relative fractions of the components in the total signal, respectively.

The oscillating function F_S:

$$F_{SA}(t) = \exp\left(\frac{t^2a^2}{2}\right)\frac{\sin(bt)}{bt}, \tag{2}$$

with fitting parameters of $b^2/3= 1.3\times10^{10}\ s^{-2}$ and $a=0.6\times10^{10}\ s^{-2}$ is found to describe FID signal in the air-dried sample 1. The oscillating FID signals characterize the nuclear relaxation in rigid crystalline and glassy systems [30, 31] with non-random molecular motion. One of the main parameter that gives the information about the mobility in these rigid systems is a second moment of NMR line, M^{II}, defined for oscillating FID as $M^{II} = a^2 + \frac{1}{3}b^2$. From the physical point of view, the second moment can be seen as the square of the average local magnetic field, H_{loc}, induced by dipole-dipole interaction at the position of resonant nuclei,

$$M^{II}=H_{loc}^{2}, \text{ where } H_{loc}=\sum b_{ij}\times 0.5(1\text{-}3\cos^2\theta_{ij})(3\, I_{iz}\, I_{jz} - \mathbf{I}_i\mathbf{I}_j) \text{ and } b_{ij}=(\mu_0/4\pi)(\gamma_H^{2}\hbar/r_{ij}^{\ 3}) \tag{3}$$

where θ_{ij} is the angle between the vector $\mathbf{r}_{ij}$ joining a spin i with a spin j, γ_H – magnetogyric ratio of the proton [30, 31].

Thus, the second moment is a measure of dipolar interactions between protons (either C-H, N-H or O-H of chitosan molecules [32]) thus forming dipoles and is sensitive to the "packing" of the protons in the sample and provides the information about non-random molecular motion, as well.

It is found that the second moment for the air-dried chitosan is equal to 2 G^2. The effect of the drying procedure is obvious when the FID signals observed in the air-dried and freeze-dried samples are compared (Figure 8). Instead the oscillating function in the air-dried chitosan (curve 1), the FID signals in both freeze-dried samples (curves 2 and 3 on Figure 8) are described by a sum of two Gaussian relaxation functions F_{SG}:

$$F_S(t) = p_1 \exp\left(\frac{t^2 M_1^{II}}{2}\right) + p_2 \exp\left(\frac{t^2 M_2^{II}}{2}\right), \tag{4}$$

with M_1^{II}=0.78 G^2 and 0.8 G^2, and M_2^{II} = 5.8 G^2, 5.6 G^2 for the chitosan samples precipitated with soda and alkali, correspondingly.

These distinctions indicate an existence of rather solid structure with homogeneous distribution of hydrogen atoms (protons) in the air-dried chitosan sample 1. Unlike the sample 1 there are two phases with various mobility and arrangement of chitosan chains in the freeze-dried samples 2 and 5. The maximum values of M_2^{II} equal to 5.8 ÷ 5.6 G^2 compared to that in the sample 1 indicate the more rigid phase existing in both the freeze-dried samples. The relative fractions of protons in the rigid phase are about 95%. These facts are consistent with the above X-ray diffraction data that the freeze-drying procedure causes the enhancement of the ordering of the chitosan supramolecular structure. The second moments with the less values M_1^{II} equal to 0.78 ÷ 0.8 G^2 indicates the phase with the other "packing" of more mobile protons where the interproton distances are larger than in more rigid part; as it follows from the Eq.6. It should be noted that the experimental values of $M_{1,2}^{II}$ are less than a theoretical value of 13.9 G^2 calculated for rigid chitosan. The lower experimental values of M^{II} reflect the local motions of chitosan macromolecules that average the dipole-dipole interactions. For example, the rotation of one amino group induces the total decrease of second moment over 4.76÷5.79 G^2 [33] and reorientation of CH_2OH groups reduces their contribution by about 2 G^2 [34].

When water is adsorbed on the chitosan samples, the contribution of mobile protons in FID increases from p_m=0.02 (initial samples) up to p_m=0.51 (water mass content - 62 wt. %), at the same time the disordering of the system takes place with accompanying acceleration of molecular motion. The contribution of water or chitosan protons can be calculated from the known weight percentages of water and chitosan. At the temperature of measurements the experimental fraction of mobile protons, p_m, are less than calculated value of the relative contribution of water into FID signal (see Table 2).

This occurs because the separation of the NMR signal by the fitting procedure is not based on discrimination between polymer and water protons. This separation is temperature and water content dependent because the molecular mobility is dependent on these two factors. The less fractions of mobile protons compared to the calculated values is due to some part of water molecules that are tightly coordinated to chitosan. The number of water molecules tightly bound with chitosan macromolecule, n_{bw}/n_{ch}, can be calculated from these discrepancy and is in the range from 1 to about 5 per monomer unite for water content from 12 to 60 wt.%, respectively (see Table 2).

Table 2. The calculated and experimental relative contributions of mobile protons into FID signal

$mass_w/mass_{ch}$	p_{water} calculated	Sample 1		Sample 5		Sample 2	
		p_m experiment	n_{bw}/n_{ch}	p_m experiment	n_{bw}/n_{ch}	p_m experiment	n_{bw}/n_{ch}
1.6	0.81	0.51	4,7	0.51	4,7	0.54	4.2
0.8	0.68	0.33	3.3	0.30	3.1	0.36	3
0.6	0.62	0.27	2.7	0.23	3.1	-	-
0.4	0.51	0.20	1.9	0.25	1.6	0.29	1.3
0.3	0.44	-	-	0.23	1.1	0.13	1.7
0.22	0.37	-	-	-	-	0.11	1.2
0.2	0.35	0.06	1.3	0.06	1.3	0.08	1.2
0.16	0.3	0.04	1.1	-	-	-	-

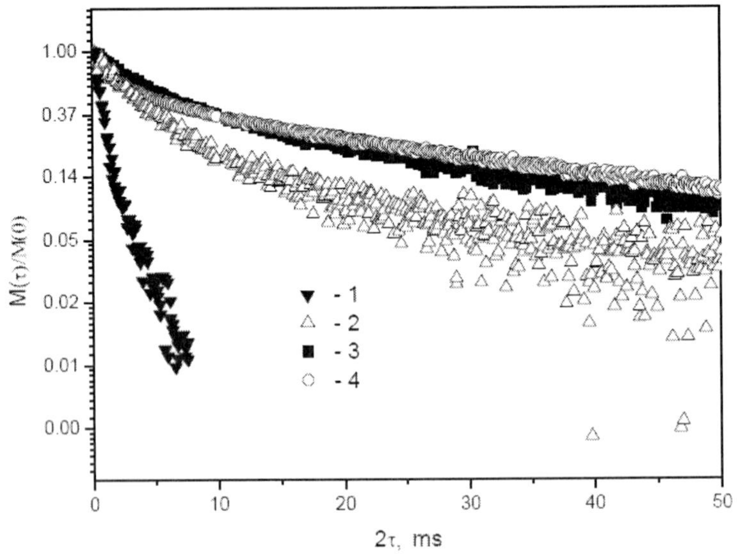

Figure 9. Magnetization decays in CPMG experiments for the sample 5 with different water mass content: 1 - w=17%, 2 - w=28%; 3 - w=44%; 4- w=60%.

The increase of adsorbed water leads to increasing contribution of mobile protons into FID signal, as it can be seen from Figure 9. Moreover, the distribution of the values of T_2 characterizing mobile protons of water is observed in the adsorption systems with water content more than 20 wt. %. The larger values of T_2 of water adsorbed on the freeze-dried samples compared with that in the air-dried chitosan correspond to larger extend of freedom and mobility of adsorbed water in swollen adsorbents.

If one considers the chitosan samples as a porous media, according the Brownstein-Tarr theory [35] in condition of fast diffusion, a ratio $T_2 \sim x$ (x – a pore size) is valid, so the distribution of spin-spin relaxation times of adsorbed water $p(T_2)$ reflects the distribution of water in the pores of different sizes or indirectly the distribution of pore sizes $p(x)$ in chitosan. The comparison of $p(T_2)$ observed in adsorption chitosan systems indicated that the freeze-dried resulted in a wider pore distribution in chitosan. The proposal that the minimal value of T_{2min} corresponds to the relaxation of water molecule in the minimal pore with a size equal to water molecule ~ 0.3 nm [36], the largest pore size can be calculated using a proportion

$$\frac{T_{2\max}}{x_{\max}} = \frac{T_{2\min}}{x_{\min}}. \qquad (5)$$

For the content of water from 23 up 62 %, the largest pore sizes observed in chitosan samples are in the range from 5.6 to 7.1 nm (for soda-precipitated sample); from 3.8 nm to 9 nm (for alkali precipitated sample). The increase of the largest pore size with water content in chitosan samples is probably due to swelling of polymer. In contrast to the freeze-dried chitosan filled by water, the value of $T_{2\max}$ observed for water in the sample 1 is rather small (Figure 10). The application of the proportion (7) gives that the largest pore size in the air-dried chitosan is not more than 0.8 nm (in the sample 1 water content is 60 wt%).

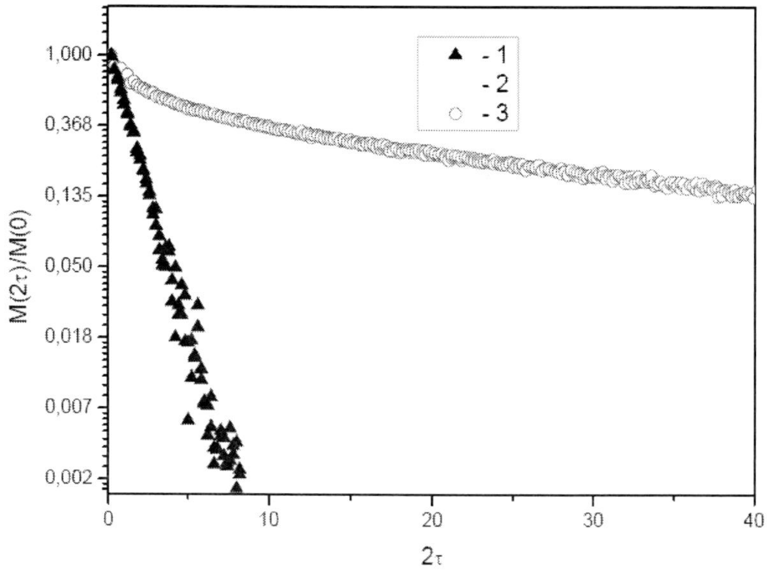

Figure 10. Magnetization decays in CPMG experiments for the chitosan samples: sample 1 (curve 1); sample 2 (2) and sample 5 (3) with equal content of adsorbed water (62%).

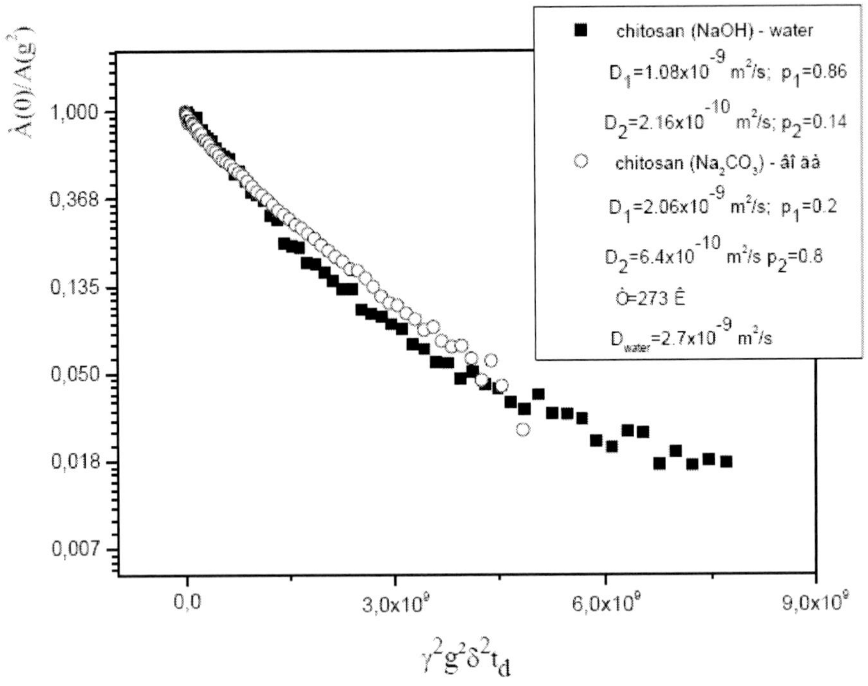

Figure 11. Spin-echo decays observed in the alkali- (sample 2 - ■) and the soda-precipitaded (sample 5 - o) chitosan samples with 60% water, the diffusion time t_d=5 ms.

Indeed, the loose macromolecular packing of the chitosan cryogel is adequate to the mesoporous structure with a larger volume accessible for water compared with the air-dried sample. The increase of water adsorption leads to increasing relative fraction of water in the largest mesopores from 2 to 42 %.

The distribution of the sizes of pores in the freeze-dried chitosan samples where water molecules are adsorbed determines the existence of a set of self-diffusion coefficients of adsorbed water. Figure 11 shows the diffusion decays of stimulated spin-echo obtained in the PMFG experiment for two freeze-dried chitosan samples with ~ 62% water.

The time conditions of PMFG experiments allow to measure only water molecules as polymer macromolecules with the part of water that is tightly bound with them are invisible due to the short relaxation times. Also the time parameters of PMFG experiments exceed the times of displacements of pore size dimensions in several orders: $t_d \sim 10^{-3}$s >> t = $(\langle x_{max}^2(t)\rangle/6D_{min})^{1/2} \sim 10^{-9}$ s, where x_{max} – the largest pore size defined from T_2 measurements.

Generally it is known the mean square displacement of molecules in porous medium is less than in the bulk state so that for a self-diffusion coefficient of adsorbed molecules is valid:

$$D_{ads} = {D_0}/{\xi} \qquad (6)$$

where ξ - is a tortuosity of diffusion path in porous system dependent on geometry of pores, in particularly on the porosity. Thus the different values of self-diffusion coefficient correspond to adsorbed water molecules in the chitosan phases with different porosities. The shape of diffusion decays does not depend on the diffusion time in the interval of 5-50 ms, so the phases with different porosities are isolated and the exchange processes averaging the diffusion heterogeneity occur in the periods of time less than 5 ms. So, one can estimate the maximum limit of dimensions of each phases $L_{1,2}$ with the help of Einstein ratio $<r^2(t)>=6Dt_d$ between the mean square displacements of water molecules and the minimum diffusion time. In our case the minimum experimental diffusion time was 5 ms, so for the freeze-dried chitosan samples with 60% water with two SDC values of water one can obtain the maximum limits of extension of the phases with relatively high and low porosities, consequently: $L_1=<r_1^2(t)>^{1/2}=(6D_{max})\ t_d= 5.7\times10^{-6}$ m; $L_2=<r_2\ ^2(t)>^{1/2}=(6D_{min})\ t_d=2.5\times10^{-6}$ m – in the alkali precipitated sample and $L_1= 7.8\times10^{-6}$ m, $L_2= 4.4\times10^{-6}$ m in the soda precipitated chitosan sample.

The fraction of fastest water molecules with D_{max} in the alkali precipitated sample is equal to 0.8 that is larger than in the soda precipitated chitosan, so the fraction of the high porosity phases of in the former chitosan is larger too.

As for the air-dried chitosan sample, the short values of relaxation times do not allow to perform PMFG self-diffusion measurements even for the sample with 60% water.

Thus the performed pulsed nmr experiments on measuring of spin-spin relaxation parameters and self-diffusion point to the differences both in the arrangement of macromolecules (ordered or disordered supramolecular structure with different packing of chains) in the chitosan samples dried in different conditions and mobility of water molecules adsorbed on the chitosan samples, as well. The results obtained are in accordance with the results of X-ray diffraction and IR-spectroscopy investigations.

3.2. Estimating of Porous Structure of the Chitosan Samples

3.2.1. Adsorption of Nitrogen Vapors at 77 K

The studies of adsorption properties of hydrophilic swelling natural polymers, e.g. the partially-crystallized chitosan, have recently gained increasing attention [13]. At the same time the estimation of porous parameters of such adsorbents using the data of physical adsorption of non-polar molecules, as nitrogen, argon, krypton and some hydrocarbons, gives minor values of specific surface and negligible porosity. *Breuer, M. and Gocho, H.* [37, 38] found the specific surface of chitosan equal to 0.3 m^2/g.

The adsorption isotherms of nitrogen vapors at 77 K on the chitosan samples dried at different conditions are shown in Figure 12. The values of specific surfaces of the chitosan samples calculated by applying the BET theory are small. Nevertheless, the BET surface determined for the air-dried chitosan sample is less than 0.012 m^2/g and differs from the corresponding values for the lyophilized chitosan samples. The specific surface of the soda-precipitated chitosan without evacuation is equal to 0.13 m^2/g, after evacuation at heating up to 373 K this value increases up to 4.87 m^2/g. The specific value determined for the lyophilized alkali-precipitated chitosan sample after evacuation at heating up to 373 K is obtained to be equal to 21.27 m^2/g. Thus, the freeze drying of the chitosan samples results in the increase of their specific surfaces. It should be noted that *Di Renzo et al* used super critical drying to prepare the chitosan samples with the specific surface area, calculated from the low temperature adsorption of nitrogen vapors, equal to 175 m^2/g [39].

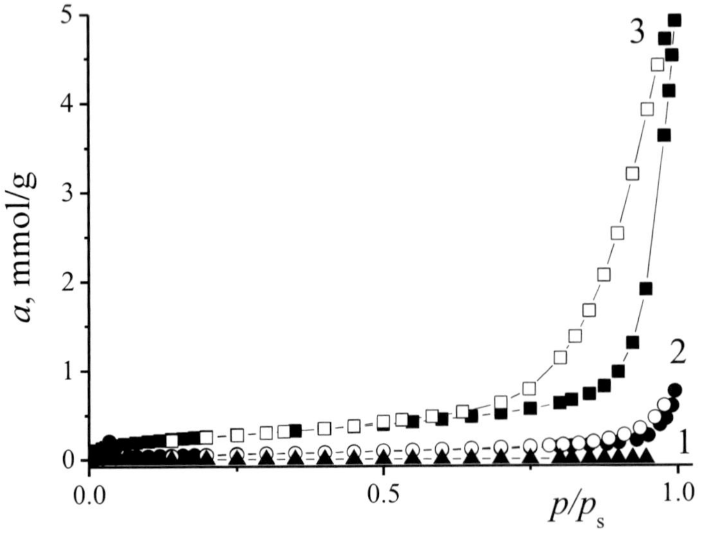

Figure 12. Experimental isotherms of nitrogen vapors adsorption at 77 K on the chitosan samples dehydrated at 373 K under vacuum: *1* – sample 1; *2* – sample 5; *3* – sample 2 (see Table 1). Filled symbols show the adsorption curve, unfilled symbols – the desorption data.

It means that all functional groups or adsorption centers (up to 5.8 mmol/g) of the chitosan samples which were dried in critical conditions become completely accessible to small nitrogen molecules.

3.2.2. Water Vapors Adsorption on the Chitosan Samples

Although the adsorption capacities of the chitosan samples towards to non-polar nitrogen molecules are not high, the amounts of adsorbed polar molecules, e.g. of water, on the chitosan sample are large due to increasing of free volume of chitosan under swelling. Thus, the adsorption ability of polymer adsorbents as well as their free adsorption volume can be determined from the water vapors adsorption data not from the nitrogen vapors adsorption isotherm. Let us consider the experimental data on water vapor adsorption by the chitosan samples 1, 2, and 5, which have equal DD and MM values, but are precipitated and dried under different conditions (Figure 13). The water vapors isotherms have sigmoid shape and can be attributed to the Type II according Brunauer classification. After the first adsorption experiment was performed, the freeze-dried chitosan samples were heated under vacuum at 373 K in order to remove adsorbed water. Then, the repeatedly measured isotherms of water vapor adsorption virtually coincide for the samples precipitated with soda and alkali (the samples 5, 2, curves *1, 2*), and the water adsorption increased, exceeding the values for the air-dried chitosan sample (curve 5). It obvious that, at low relative pressures $p/p_s < 0.4$, the adsorption isotherms of the lyophilized polymer samples are quite close to the isotherm of the initial air-dried sample, when the adsorption of water molecules is going on PAC. It is known [18] that all water vapors adsorption S-shaped isotherms correspond to the mechanism of adsorption on PAC. The amount of PAC can be derived from the comparative plots presented on Figure 14.

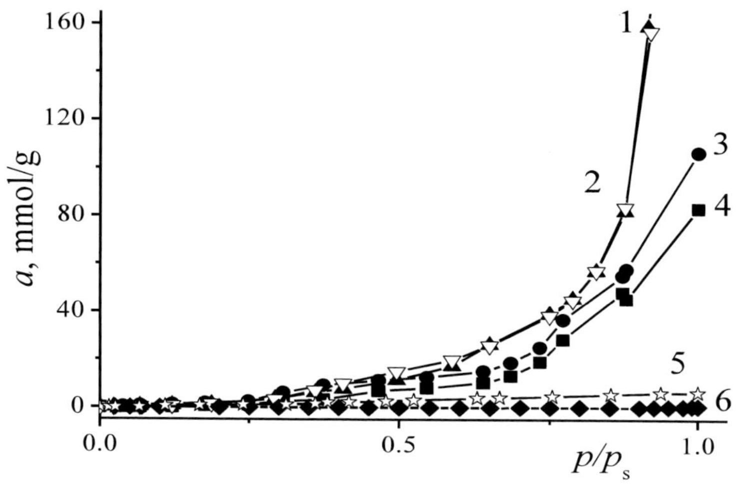

Figure 13. The water vapor adsorption isotherms for the chitosan samples modified by freeze drying at different conditions: curve *1* – sample 5, evacuated at heating up to 373 K after water adsorption; *2* – sample 2, evacuated at heating up to 373 K after water adsorption; *3* – sample 2; *4* – sample 5; *5* – sample 1; *6* – the sample of graphitized carbon black Vulkan-7H (2800) used for comparison.

The water adsorption values on the chitosan samples under study are plotted as a function of the adsorption value for water on the surface of a standard nonporous carbon adsorbent, i.e., Vulkan-7H graphitized at 2800 K carbon black, at the same values of relative pressures. According to the fundamental mechanism of the water adsorption on carbon adsorbents the initial water adsorption is likely to be due to the oxygen containing groups which may act as the primary adsorption centers.

For the nonporous and mesoporous samples with the surfaces of similar chemical nature the comparative plots are the straight lines starting from the coordinates origin. The ordinates at a/a_m=1 correspond to the amount of PAC per gram of adsorbent [18]. The inflection upward in the comparative plots of water adsorption indicates on the changes in adsorption of water molecules on PAC and merging of clusters of water molecules.

The PACs amounts were calculated from the comparative plots on Figure 14 for the freeze-dried samples employing the cluster model [18] (Table 1). Note that the cluster model fits for the samples under study when the degrees of pore filling are low, and volume swelling of polymers is insignificant. In this case the comparative plots are linear. According to the data calculated by the comparative method, the PACs amounts for water are larger for the freeze-dried chitosan samples (Table 1). The largest amount of adsorption centers for water molecules are inherent in the biopolymer samples, which were lyophilized and dehydrated under vacuum at 373 K, irrespective of the substance used for their precipitation. At the same time, there is a certain discrepancy between the not heated and evacuated chitosan samples that differ by the precipitation methods. The PACs amount is larger for the alkali-precipitated sample compared with the data for the soda-precipitated sample. For the samples heated under vacuum, the PACs amount is almost twofold larger than that for the unheated samples. Therefore, it can be assumed that in the case of the unheated samples containing residual water, the adsorption proceeds due to these preliminarily adsorbed water molecules that act as the secondary adsorption sites. The PACs amount is very small in the air-dried sample, a fact that is related to the low accessibility of the primary adsorption sites in it.

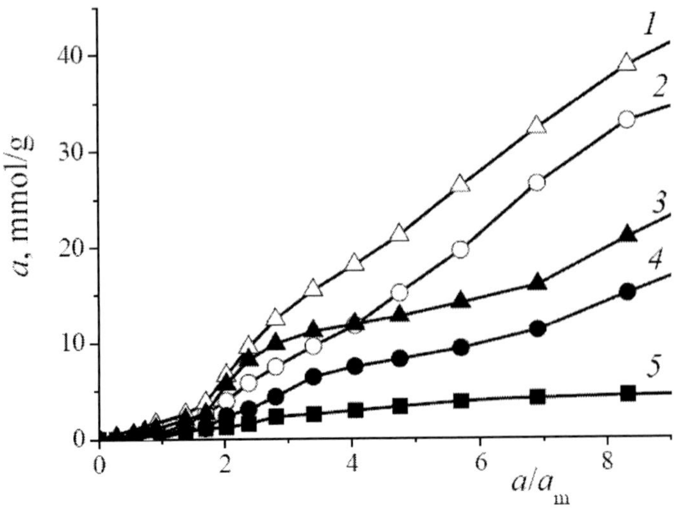

Figure 14. Comparative plots of water vapor adsorption isotherms in the chitosan samples: 5 dehydrated at 373 K (*1*), sample 2 dehydrated at 373 K(*2*), sample 2 (*3*), sample 5 (*4*), and sample 1 (*5*).

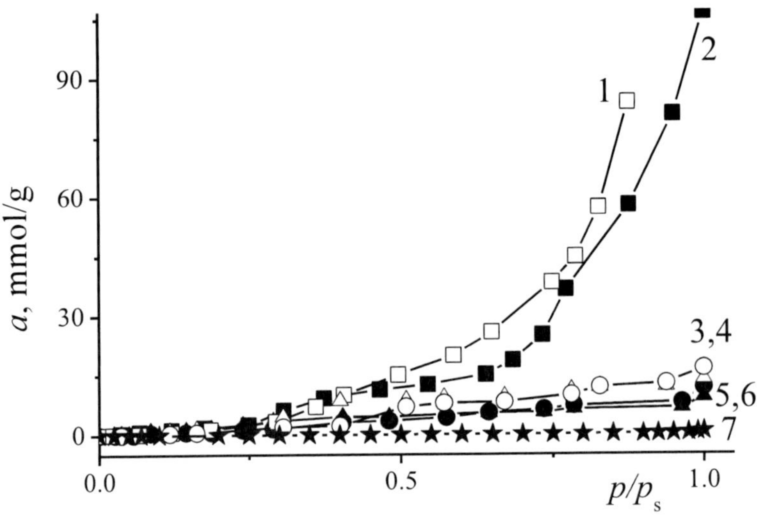

Figure 15. The water vapor adsorption isotherms for the chitosan samples that were cross-linked after freeze-drying: curve *1* – sample 2 dehydrated at 373 K; *2* – sample 2; *3* – sample 3 dehydrated at at 373 K; *4* – sample 4 dehydrated at at 373 K; *5* – sample 4; *6* - sample 3; *7* – graphitized black carbon Vulkan 7H (2800) used for comparison.

In contrast to the range of low pressures, when the relative pressures exceed 0.6 the isotherms for the various chitosan samples become drastically different from each other (Figure 13). The maximum adsorption values for the freeze-dried samples are as large as 100 mmol/g and more, thus being tenfold larger than that for the air-dried sample. So large amounts of water molecules adsorbed on the freeze-dried chitosan samples are attributed to the changes of polymer supramolecular structure resulted from the freeze-drying.

The cross-linking of the freeze-dried samples by GA (Figure 15, curves 5 and 6) results in reducing of adsorption capacity of chitosan, but even in this case this value is significant compared with the air-dried sample 1.

It follows from the adsorption data shown on Figure 16, the cross-linking of the films by GA and DGA results in the increase of their adsorption capacity toward to water in contrast to the cross-linked freeze-dried samples. The cross-linking of films by EGDE enhances the adsorption capacity in the interval of small relative pressures from $p/p_S = 0.2$, while cross-linking by GA has no influence on the adsorption isotherm of the chitosan films up to $p/p_S =$ 0.6.

It is obvious that both freeze-drying and evacuation at 373 K result in the essential increase of the amount of PAC from 0.54 up to 1.25 and 1.80 mmol/g (373 K). Cross-linking of the chitosan sample 2 in liquid and vapor phases reduces the amount of PAC to 0.83 and 0.33 mmol/g, respectively. While the amount of PAC in the air-died chitosan sample is 0.54 mmol/g. Thus the preliminary freeze-drying of the chitosan sample helps to preserve the adsorption activity. The larger amount of PAC in the chitosan sample cross-linked in GLA solution points to the advantage of this method of cross-linking. Similar dependence is observed for the chitosan films (see data for the samples 6-8). Only cross-linking of the chitosan film in EGDE reduces the amount of PAC, but the evacuation at 373 K increases this value. Moreover, the adsorption of water on the chitosan film cross-linked by GLA increases in the range of high relative pressures of water vapors compared with the initial chitosan film. Thus, the cross-linking of both freeze-dried chitosan samples and chitosan films have distinguishing characteristics that define the resulting properties of chitosan. As amino-groups of chitosan serve as PAC, so the most perspective are the methods of cross-linking by that the minimal amount of amino-groups are blocked.

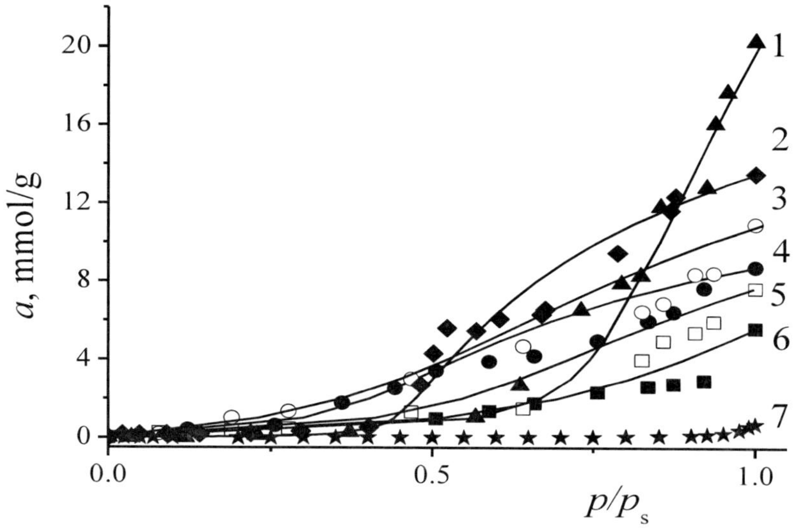

Figure 16. The water vapor adsorption isotherms for the chitosan cross-linked films: curve *1* – sample 7; *2* – sample 9; *3* – sample 8 dehydrated at 373 K; *4* – sample 8; *5* – sample 6 dehydrated at 373 K; *6*– sample 6; *7* – the graphitized soot Vulkan 7H(2800) used for comparison. Solid curves are calculated by the Dubinin-Serpinsky-2 equation [40].

3.2.3. Models of Water Vapor Adsorption on Swelling Polymers. The approximation of Water Adsorption Isotherms in the Chitosan Samples

The water vapor adsorption on swelling polymer adsorbents, such as chitosan, differs from the adsorption of non-polar substances because of the ability to change the adsorption features due to swelling. In general, some models are proposed for interpretation of water vapors adsorption isotherms for swelling polymers. The models used for description of water adsorption on polymers are based either on the thermodynamic theories of polymer solutions (as Flory-Huggins theory) or complicated modifications of Langmuir or BET equations. In all cases the adsorption isotherms are divided in two ranges of relative pressures. In the range of low relative pressures adsorption of water molecules proceeds on PAC. Polymers remain in glass state up to the point of inflection of S-shaped isotherm and water molecules occur in a free volume between polymer chains like a molecule in microporous volume. In this case the interactions polymer – adsorbed molecules prevail. The first stage of the adsorption process can be described either by Langmuir equation of localized adsorption on PAC [41] as it was done for interpretation of water adsorption on polyvinylpyrrolidone, polyvinylcaprolactam, polyvinyl alcohol up to the transition from glass to high elasticity [42]. The amount of PACs in polymer adsorbents can be determined from Langmuir equation or from the comparative plots [18] as it was done above in *3.2.2* section.

The essential growth of adsorption values is observed for the second stage of adsorption at higher relative pressures and is accompanied by swelling of polymer. Unlike active carbons with rigid structure, the water adsorption on amorphous-crystalline polymers, such as chitosan, leads to the changes in their structure and physics-chemical properties. Different approaches are applied for interpretation of the experimental data in this stage of water adsorption on polymer. The Flory-Huggins equation derived from the thermodynamic theory of diluted athermic polymer solutions is used extensively for description of water adsorption on the second stage while Langmuir equation is applied for analysis of the first stage [12, 42].

The approach based on the equations derived from BET theory that assuming (i) the condensation of an infinite number of layers from the vapor phase onto the adsorbent surfaces and (ii) condensation – evaporation properties of the molecules in the second and higher layers equal to those of the liquid state. As classically observed for hydrophilic glassy polymers [ref. 5, 6, 15 in 29] water adsorption at low pressures is well described by this model. Nevertheless, it fails to represent water sorption at high pressures. The modified BET model considering a limited number of adsorbed layers was applied, and the resulting three-parameter equation is effective for describing the water sorption over the relative pressure range 0–0.8. While the Guggenheim-Anderson-Boer (GAB) equation [43] may be mentioned among the most efficient equations used for the predictions of the water adsorption on polymers. This model assumes that the heat of adsorption to be less than the corresponding heat of condensation. The parameter of water content corresponding to saturation of all primary adsorption sites by one water molecule (formerly called the monolayer in BET theory) is used as well as the parameter that correcting properties of the multilayer molecules with respect to the bulk liquid. The main advantage of this model is in using only three parameters with unambiguous physical meaning. Nevertheless, the fitting performance of the GAB equation to adsorption data depends on the regression method used. It has been found that for the chitosan samples the non-linear regression method applies better for fitting the water adsorption data compared with the polynomial regression which was more adequate to the chitin [44].

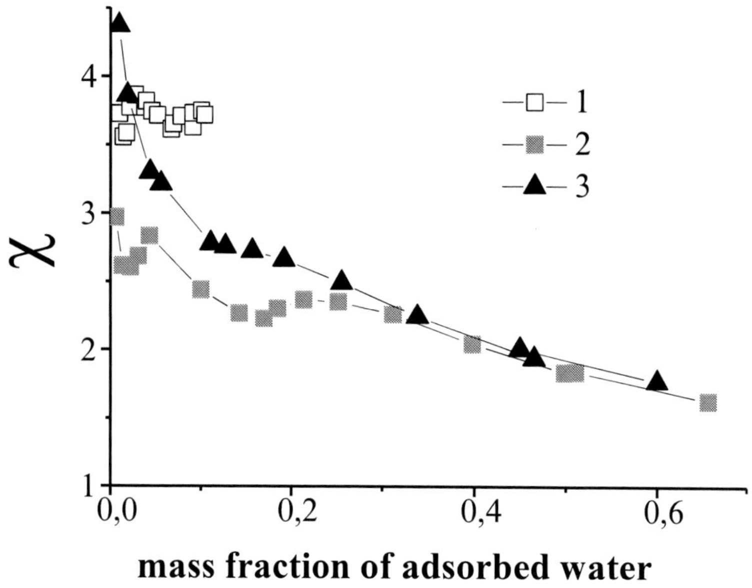

Figure 17. The dependence of the Flory-Huggins interaction parameter χ calculated from the adsorption of water on the chitosan samples on the relative mass fraction of adsorbed water on: sample 1 – *1*; sample 2 – *2*; sample 5 –*3*.

In any case the model that pretends to best approximation of data of water adsorption on polymer must take into account the changes of adsorption properties with increasing pressure, notably the rate of adsorption and the amount of PAC. Indeed, in the range of small relative pressures every models mentioned agree that water molecules form hydrogen bonds with PAC of polymer [45]. Polar groups such as amino- and hydroxyl groups of the chitosan samples can serve as PAC for water molecules. Then each water molecule adsorbed on PAC acts as a secondary adsorption center. With increasing water vapor pressure, the clusters that are formed on primary adsorbed water molecules begin to grow with subsequent merging [45].

The loose structure of chitosan samples in glass state is similar to microporous structure, as the interspaces between macromolecules are comparable with the sizes of water molecule. Thus the adsorption equations by Dubinin-Astakhov and Dubinin-Serpinsky for approximating the water isotherms in microporous adsorbents must be applied in the attempts to interpret the experimental data for the chitosan samples.

Let's consider the experimental data for water adsorption on the chitosan samples under study and compare their interpretations based on some of the above mentioned approaches.

a) The approach based on the Flory-Huggins lattice model of polymer solutions [13] gives the equation for describing the water adsorption on polymers:

$$\ln\frac{p_1}{p_1^0} = \ln(1-\varphi_2) + \varphi_2 + \chi_1\varphi_2^2 , \qquad (7)$$

where $\frac{p_1}{p_1^0}$ is the relative pressure of solvent (adsorbing molecules) vapor; φ_2 – volumetric fraction of polymer, χ_1 - the Flory-Huggins interaction parameter of the solvent (adsorbed molecules).

Figure 17 shows the Flory-Huggins interaction parameters calculated from the water adsorption isotherms on the chitosan samples as a function of mass fraction of adsorbed water. It is obvious that in the interaction parameter χ determined in the chitosan samples exceeds 1, so the thermodynamic affinity of water to polymer in all cases is bad, especially to air-dried chitosan. Freeze drying of the chitosan improves the affinity of water to the polymer that exhibits in the less values of χ in the freeze-dried samples compared with that for sample 1. The reducing value of χ observed with increase of water content in the samples 2 and 5 can be attributed to enhancing interactions between polymer and water molecules. The complicated character of the function $\chi(w_{water})$ observed in the chitosan samples is likely connected with the structure transition of polymer from glass state under water adsorption.

It should be noted that the most important factor which limits the range of confidence is the Flory-Huggins interaction parameter χ. It is usually assumed to be constant for a given polymer-solvent pair. However, this assumption was generally proven incorrect by different experimental and theoretical works [45] that showed a dependence of χ on the volumetric fraction of polymer.

Thus, in order to construct the calculated adsorption isotherms of water in the chitosan samples, the averaged value of the Flory-Huggins interaction parameter χ is used. Figures 18 show the experimental water adsorption isotherms on the chitosan samples 1, 2 and 5 compared with the water isotherms calculated from Flory-Huggins equation (7) with a constant value of χ.

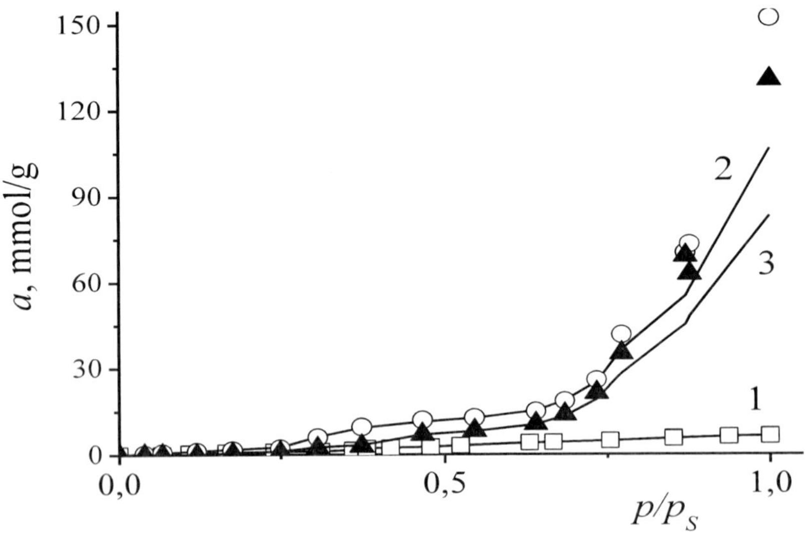

Figure 18. The comparison of the Flory-Huggins model (symbols) with the experimental (solid curves) water adsorption isotherms on the chitosan samples 1 (curve 1, □), 2 (curve 2, o) and 5 (curve 3, ▲).

The analysis of Figure 18 shows that experimental adsorption data for the freeze-dried chitosan samples 2 and 5 can be fitted by the Flory-Huggins equation in the ranges of relative pressures up to p/p_s=0.75 (Figure 18) and in the whole range of p/p_s for the air-dried sample.

b) The Dubinin-Serpinsky equation 2 [40].

As the water adsorption isotherms on the chitosan samples are S-shaped like the same for carbon adsorbents, the Dubinin-Serpinsky equation 2 (DS2) can be used for interpretation the experimental data. The main advantages of the DS2 equation are in applying the parameters that have a clear physical meaning, and accounting the changes in the amount of adsorption centers in the course of water adsorption:

$$h = \frac{a}{c(a_0 + a)(1 - ka)}, \qquad (8)$$

where a_0 – the amount of PACs, $h = p/p_s$, c - constant of interactions characterizing the relation of rates constants of water adsorption and desorption on PAC, k – parameter of variation of the amount of active adsorption centers during filling the volume accessible for adsorption. It should be noted that the adsorption value of water adsorbed a_0 on PAC differs from the value determined from the comparative plots (Figure 13). The reason is in the difference of the range the water adsorption isotherms that are covered by estimating these parameters with the DS2 equation (the whole interval of relative pressures) and the comparative method (the low-pressure range).

Plotting the water vapors adsorption isotherms on the chitosan samples in linear form [46] allowed to estimate the parameters of equation DS2. For this purpose equation (8) was expressed as:

$$\frac{\frac{a}{h} - a_s}{a - a_s} = (A_2 - A_3 a_s) - A_3 a, \qquad (9)$$

where $A_3 = tg\,\alpha$, $A_2 = z_0 + A_3 a_S$, $A_1 = a_S - A_2 a_S + A_3 a_S^2$,

with $z_0 = \frac{\frac{a}{h} - a_s}{a - a_s}$ for $a \to 0$, $a_S = a$ at $p/p_S = 1$

Then, from the determined values of A_1, A_2, A_3, one can calculate parameters c, k and a_0 using the expression:

$$c = \frac{1}{2}\left[A_2 + \sqrt{A_2^2 + 4A_1 A_3}\right], \qquad (10)$$

where $a_0 = A_1/c$ and $k = a_3/c$.

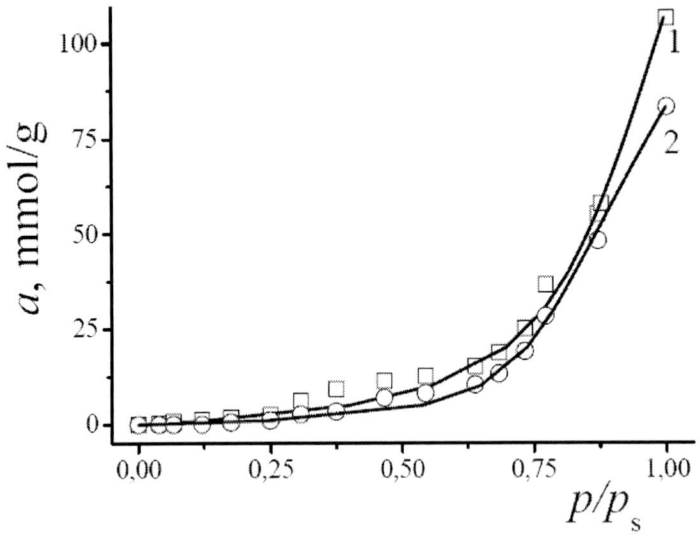

Figure 19. The experimental (□ – sample 2, o - sample 5) and calculated from the DS2 equation (solid lines 1 - sample 2 and 2 – sample 5) water vapor adsorption isotherms on the chitosan samples 2 and 5.

Table 3. The parameters of the DS2 equation used for approximation of water vapors adsorption isotherms of the chitosan samples

№	sample	a_0, mmol/g	c	k, g/mmol	a_S, mmol/g
1	1	0.93	2.00	0.087	6.45
2	2	6.17	1.13	0.0015	106.60
3	2(373 K)	3.70	1.49	0.0043	332.09
4	3	0.95	3.50	0.080	9.25
5	3(373 K)	1.23	4.08	0.056	13.756
6	4	1.31	1.72	0.041	11.74
7	4(373 K)	0.73	2.30	0.035	16.59
8	5	2.22	1.30	0.0030	83.23
9	5(373 K)	5.68	1.45	0.0010	332.09
10	6	0.88	1.077	0.035	5.60
11	6(373 K)	0.71	1.50	0.051	7.65
12	7	0.39	1.29	0.012	20.30
13	8	0.66	2.15	0.065	8.73
14	8(373 K)	1.90	1.56	0.042	10.95

The parameters a_0, c, and k were calculated for all the experimental water adsorption isotherms and these values are presented in Table 3. The comparison of the experimental and calculated water vapor adsorption isotherms for the chitosan samples 2 and 5 (see Figure 19) confirms the suitability the DS2 equation for fitting and interpretation of the data of water adsorption on polymer samples. As it obvious from Figure 19 that the experimental water adsorption isotherms are fitted by equation DS2 over the entire range of relative pressures $p/p_S = 0 - 1.0$.

c) *The Dubinin-Astakhov equation (DA)* [47] for approximation of water adsorption isotherm on the chitosan samples

The Dubinin-Astakhov equation with a variable index *n*:

$$\ln\frac{a}{a_{0d}} = -\left(\frac{RT\cdot\ln p_S/p}{E_{0d}}\right)^n \tag{11}$$

is often used for fitting the concave adsorption isotherms, e.g. water adsorption on chitosan. Here a_{0d} – a limit value of adsorption, mmol/g, E_{0d} – the effective adsorption energy.

Figure 20 shows the results of experimental water adsorption data for the chitosan sample 2 recalculated with various index *n* in coordinates of the DA equation. It is occurred that the minimal deviation of the experimental data from a linearity is obtained when *n*=0.3. The regression coefficient R=0.032 for *n*=0.3, while for *n*=0.5: R=0.035, and *n*=0.7: R=0.040).

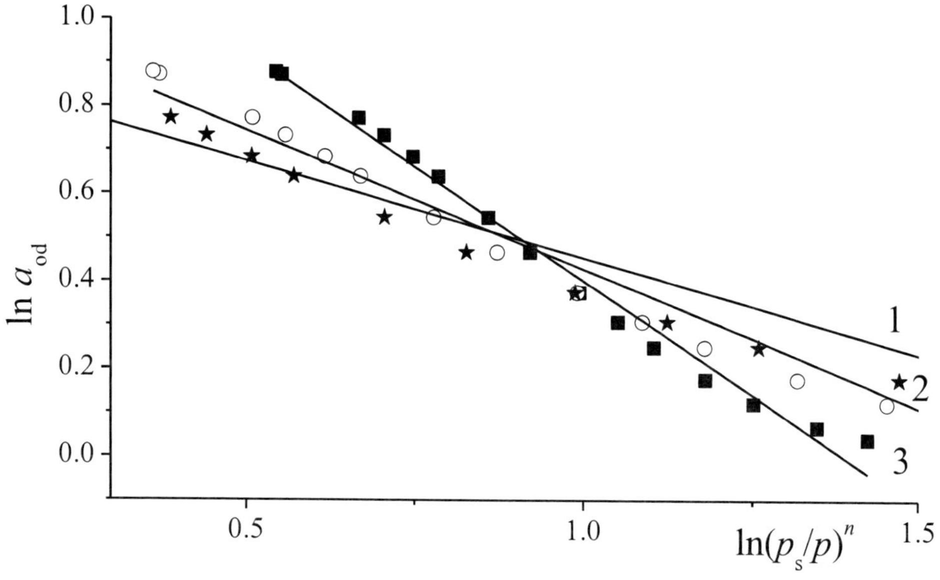

Figure 20. The water adsorption isotherms on the chitosan sample 2 in coordinates of the DA equation for different values of *n*: 0.7 - 1; 0.5 – 2; 0.3– 3.

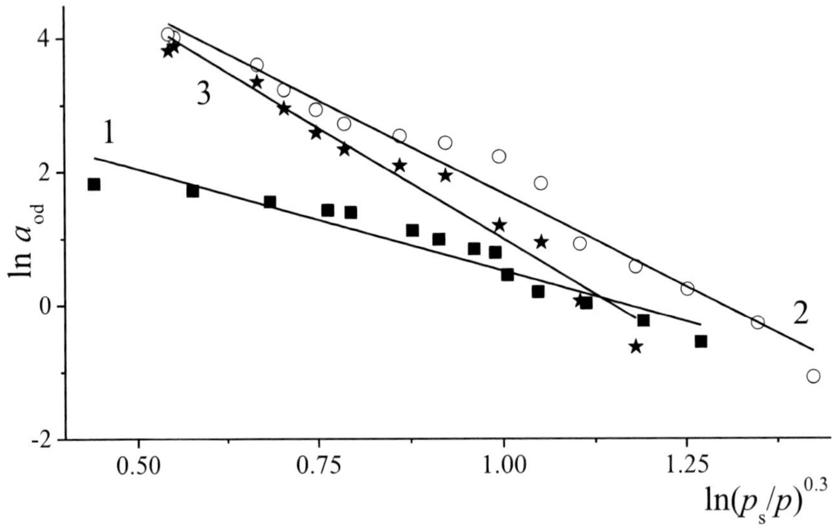

Figure 21. The water adsorption isotherms on the chitosan samples 1 (*1*), 2 (*2*) and 5 (*3*) in coordinates of the DA equation for *n*=0.3 .

Table 4. The parameters of the DA equation for the water adsorption on the chitosan samples

Sample	a_{0d}, mmol/g	E_{0d}, kJ/mol
1	6.23	1.4
2	48.0	1.0
5	57.4	1.5

The values of the adsorption energy and limit value of adsorption of water on the chitosan samples which differ from each other by the drying conditions were calculated from the DA equation for *n*=0.3 and are shown in Table 5

The values of parameter a_{0d} from Table 4 show the increase of adsorption capacities of the chitosan samples caused by the freeze-drying compared with the air-dried sample. The closeness of the adsorption energies of different chitosan samples points to same nature of adsorption centers in these chitosan samples.

3.4. Adsorption of Nickel, Copper and Silver Cations

3.4.1. The Equilibrium Studies

The isotherms of nickel cations adsorption by the chitosan samples dried in different conditions are shown on Figure 22. It is obvious that the alkali-precipitated sample 2 possesses the adsorption capacity of 4.5 mmol/g towards to nickel cations, although the adsorption capacity of the air-dried chitosan (sample 1) does not exceed 1.5 mmol/g.

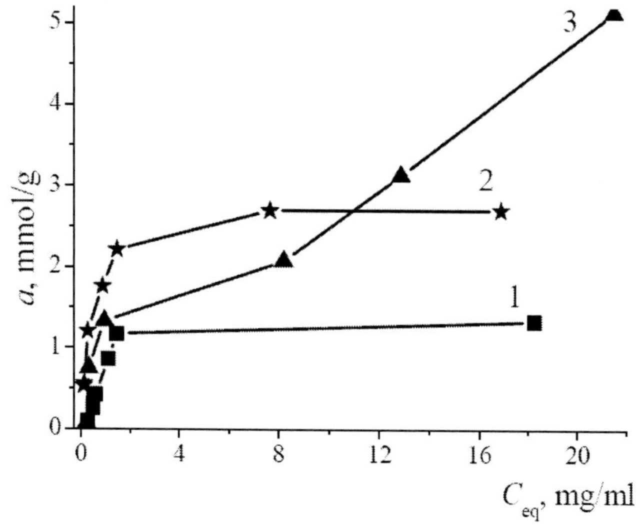

Figure 22. Isotherms of nickel cations adsorption on the chitosan samples 1 (*1*); 5 (*2*); 2 (*3*).

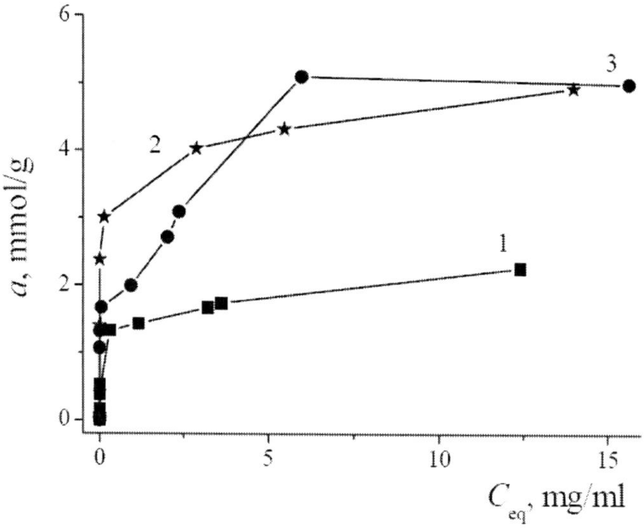

Figure 23. Isotherms of cooper adsorption by the chitosan samples 1 (*1*); 2 (*2*); 5 (*3*).

The isotherms of copper cations adsorption on the samples 1, 2 and 5 are shown in Figure 23. The least steep isotherm relates to the sample 1; its adsorption capacity does not exceed 1.8 mmol/g. The steeper isotherm relates to the sample 2; its adsorption capacity reaches up to 5.1 mmol/g. The steepest isotherm relates to the sample 3; its limiting sorption capacity is equal to 4.0 mmol/g, which is probably due to the influence of the carboxy anion on the precipitation of the polymer from an acidic medium with soda. A comparison of the copper sorption capacities with the literature data shows that the freeze-dried chitosan samples are the most active adsorbents towards to cooper cations. *Ng* [48] and *Juang* [49] found the limiting sorption capacities of chitosan towards to copper cations equal to 2.75 and 2.5 mmol/g, respectively.

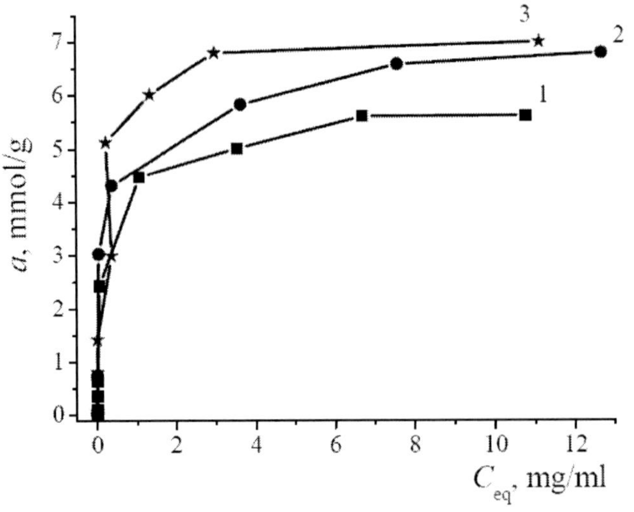

Figure 24. Isotherms of silver adsorption by the chitosan samples 1 (*1*), 2 (*2*), 5 (*3*).

Table 5. The parameters of the Freundlich equation (k_f and *n*) used for approximation of copper adsorption on the chitosan samples

Sample	*n*	k_f,
1	0,35	7,07
2	0,25	13,09
5	0,17	8,87

The degree of binding the active amino groups of the chitosan samples by copper cations reaches almost 95% for the sample 2, while that for the sample 1 it is only equal to 34% (table 6). Apparently, such increase in the sorption activity is due to ordering its supramolecular structure resulted from the freeze-drying. The tabulated data show that the freeze-drying gives rise to the degree of the crystallinity of the chitosan sample. The water adsorption capacities determined from the water adsorption isotherms also unambiguously indicate the large structural differences between the freeze-dried and the air-dried chitosan samples.

The Langmuir, Freundlich [12] and Dubinin–Radushkevich [12] equations are used in order to describe the adsorption isotherms and determine the parameters characterizing the sorption process. The Langmuir and Freundlich equations occur to be suitable only in a narrow concentration range, as it can be seen from Table 5 with the calculated parameters of the Freundlich equation:

$$\lg a = \lg k_f + 1/n \lg C_{eq} , \quad (12)$$

where k_f is a constant value, $1/n$ is an exponent
and Figure 25 which reports the adsorption isotherms of copper on the chitosan samples in coordinates of the Freundlich equation (Figure 25, a) and Langmuir equations (Figure 25, b).

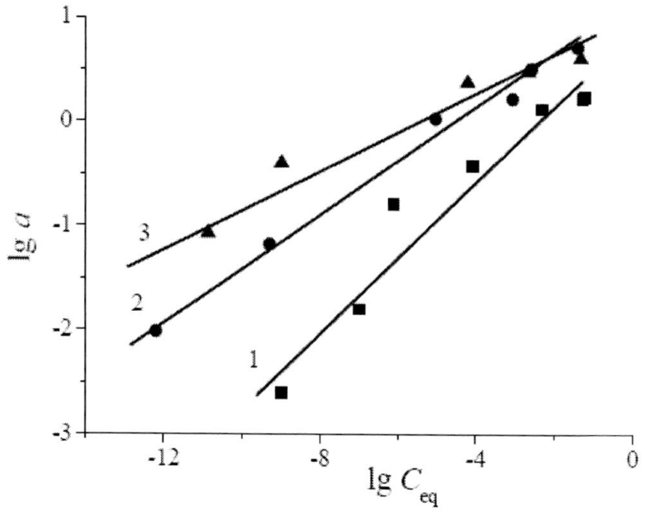

a.

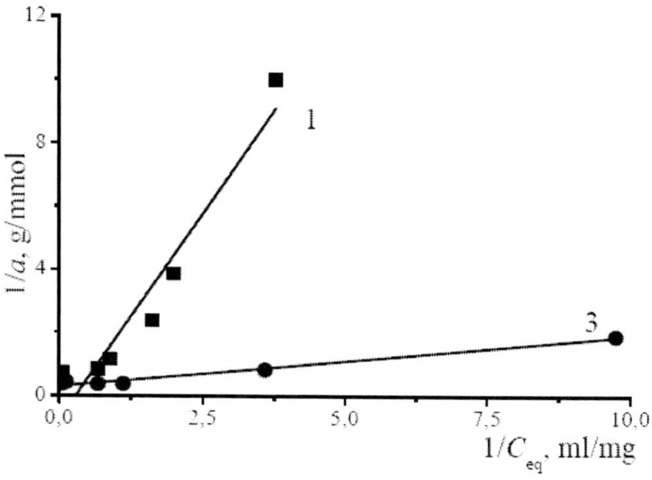

b.

Figure 25. The isotherms adsorption of copper cations (a) in the coordinates of the Freundlich equation; nickel cations (b) in coordinates of the Langmuir equation for the chitosan samples 1 (*1*), 2(*2*) and 5(*3*).

The adsorption isotherms were described more exactly by the Dubinin–Radushkevich equation in the form proposed for the adsorption from the liquid phase [50, 51]. Unlike the classic Dubinin–Radushkevich equation, the modified Dubinin–Radushkevich equation contains the effective energy of adsorption E_s (instead of the characteristic adsorption energy E_0), and corresponds to a particular adsorptive and, in contrast to E_0, is not the characteristic of the adsorbent.

$$\ln(a/a_0) = -\left[\frac{RT\ln(C_S/C_{eq}}{E_s}\right]^2 \tag{13}$$

here, a is the equilibrium adsorption (mmol/g) at the temperature T, and the equilibrium concentration C_{eq} (mg/ml); a_0 is the limiting adsorption (mmol/g); R is the universal gas constant (J/(mol K)); C_S is the concentration of a saturated solution (mg/ml); and E_S is the effective energy of adsorption of copper or nickel cations from the solution in water, kJ/mol.

The adsorption isotherms of nickel and cooper cations in the coordinates of the Dubinin–Radushkevich equation for the different samples of chitosan are shown in Figures 26-28. All the isotherms are linear over a wide concentration range. The parameters of the Dubinin–Radushkevich equation determined from these isotherms are given in Table 4.

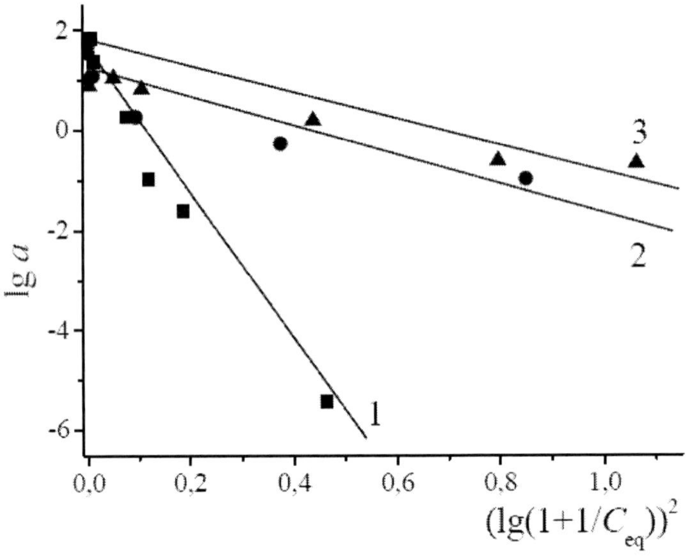

Figure 26. The isotherms of nickel cation adsorption on the chitosan samples 1 (*1*); 2 (*2*) and 5(*3*) plotted in the Dubinin–Radushkevich equation coordinates.

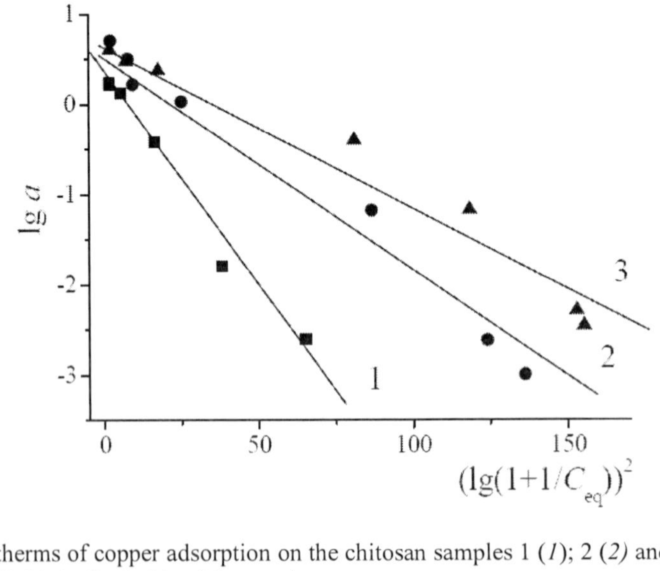

Figure 27. The isotherms of copper adsorption on the chitosan samples 1 (*1*); 2 (*2*) and 5 (*3*) in the coordinates of the Dubinin–Radushkevich equation.

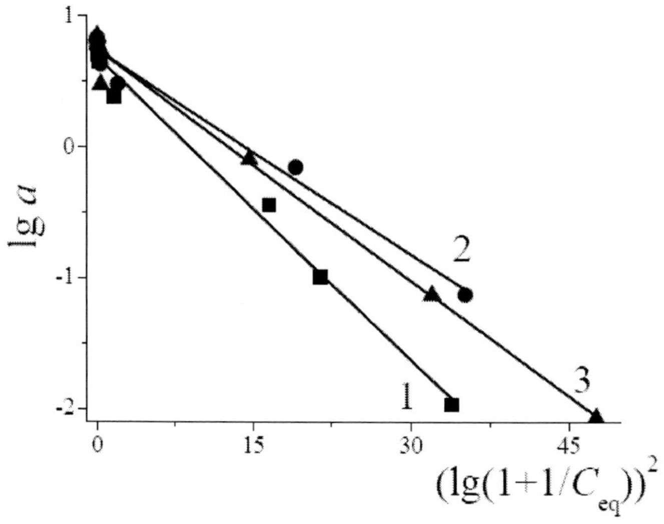

Figure 28. The isotherms of silver adsorption on the chitosan samples 1 (*1*); 2 (*2)* and 5 (*3*) in the coordinates of the Dubinin–Radushkevich equation.

Table 6. The parameters of the Dubinin–Radushkevich equation and degree of binding the active amino groups of the chitosan samples by metals cations

Sample	Adsorption capacity, mmol/g			E_S, кJ/мol, according Dubinin-Radushkevich equation			Degree of binding cation/NH_2		
	Ni	Cu	Ag	Ni	Cu	Ag	Ni/NH_2	Cu/NH_2	Ag/NH_2
1	1.5	1.8	5.6	3.5	23.4	11.2	0.25	0.49	1.047
2	4.5	5.1	6.8	3.5	26.9	14.7	0.96	0.93	1.26
5	2.5	4.0	10.3	6.6	33.1	17.8	0.50	0.92	1.30

It can be seen in Table 6 that the energies of the adsorption sites for nickel and copper cations of samples precipitated differently but dried in the same way (freeze-drying) differ substantially. At the same time, the characteristic energies of adsorption of samples 2 and 1 (air-dried sample), which have the different supramolecular structures also differ largely. The parameters of the Dubinin–Radushkevich equation (the adsorption energy and a_0) are highest for the soda-precipitated chitosan sample and less high for the alkali-precipitated sample (Table 6). These parameters for air-dried sample are almost two times lower.

The high energy of both nickel and copper ion adsorption sites of the freeze-dried chitosan samples in comparison with the air-dried sample may be related to the fact that, during adsorption, metal cations can interact with several amino groups of different glucopyranose rings belonging to either one or several adjacent polymer chains [52, 53]. The mutual location of chitosan macromolecules differs depending on the drying conditions. This, in turn, must lead to the formation of nickel or copper cations adsorption sites with different energies.

As it follows from the data of Table 4 the values of adsorption capacity of the freeze-dried chitosan sponges towards to the metal cations under study are very high compared with the data of a number of investigations [49, 54-56]. Thus, *Juang et al* [54, 55] prepared the chitosan samples with the adsorption capacities towards to copper cations not larger than 2,75 mmol/g (compare with 5.1.and 4.0 mmol/g for the samples 2 and 5). We have not found so many investigations devoted to the adsorption of nickel and silver cations on the chitosan samples. *Juang and Shao* have studied the chitosan-based samples with the adsorption capacity towards to nickel cations of 0,75 mmol/g [49], while *Yoshizuka et al* obtained the adsorption capacities of the chitosan samples towards to silver equal to 0,4 mmol/g [56].

3.4.2. The Kinetics Studies

The kinetics of the adsorption process is studied by carrying out a set of experiments at constant temperature (298 K) and monitoring the amount of adsorbed cations with time. The well known models [14, 15] are tested for description of the adsorption kinetics:

a) Pseudo-first-order rate kinetics equation by Lagergren [14, 15]:

$$\log(a_{eq}-a_t)=\log a_0-k_1\times t \qquad (14)$$

where a_{eq} and a_t are the amounts of adsorbed species per unit mass at equilibrium and at any point of time t, respectively, and k_1 is the constant of pseudo-first-order adsorption rate. The values of k_1 can be calculated by the slope of a linear plot of $\log(a_{eq} - a_t)$ versus t (Figure 29) and are listed in Table 7.

The relative amounts of adsorbed metal cations (the relation between the concentration of metal ion at the moment of time t and the equilibrium concentration versus $t^{0.5}$) for the kinetic curves for copper ions at different initial concentration are shown on the Figure 30. The uptake of ions reaches a significant value within the first 40-60 minutes and continues to grow up slowly during 240 min until the equilibrium values are attained.

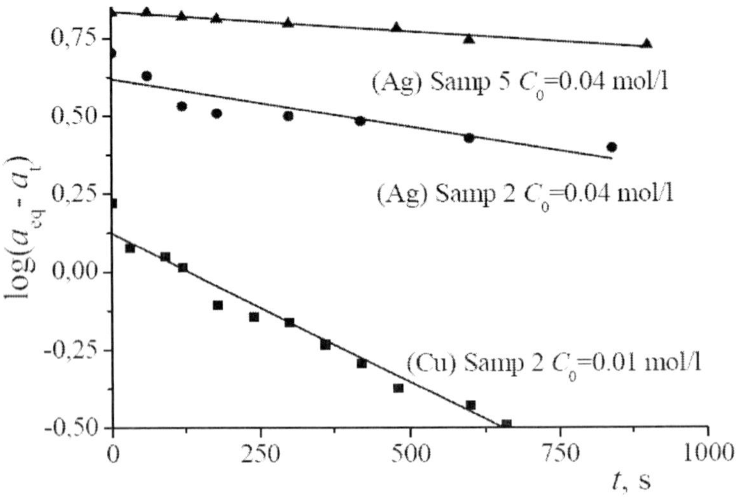

Figure 29. The pseudo-first-order model for Ag and Cu cations adsorbed onto the chitosan samples at different concentration.

Table 7. The parameters of the pseudo-first-order rate equation (14) with the mean-square deviation R^2

Cation	Sample	C_0, mol/l	a_{eq}, mmol/g	R^2	$k_1 \times 10^4$, s^{-1}
Cu	1	0.01	0.705	0.98	6.0
		0.02	4.007	0.88	4.6
	2	0.01	1.66	0.88	4.6
	5	0.01	2.4	0.93	4.0
		0.02	5.7	0.69	8.0
Ag	1	0.1	4.86	0.88	0.6
		0.08	5.61	0.99	0.7
		0.03	4.47	0.95	0.2
	2	0.1	6.8	0.55	3.3
		0.0806	6.6	0.94	5.5
		0.0431	5.04	0.88	7.1
	5	0.1	9.7	0.98	2.9
		0.0436	6,0	0,90	6.8
		0.0691	12,7	0,94	3.9

After equilibrium is reached, no measurable change in pH is observed. The initial uptake rates for copper cations adsorption for the freeze-dried chitosan samples are more than for the air-dried chitosan sample. Both the high specific surface BET of the freeze-dried sample and the large amount of the adsorption sites available for adsorption are responsible for the high rate of metal cations adsorption.

When the adsorption sites are filled up, the adsorption becomes slower and the kinetics becomes more dependent on the rate at which ions transfer from the bulk phase to the actual adsorption sites.

The Lagergren pseudo-first-order plots are fitted successfully by linear function with the regression coefficient varying from 0.55 to 0.98 (Figure 29). The pseudo-first-order rate constant varies between 0.2×10 and 6.8×10 c (Table 7). The differences still existing might be due to the actual process are not fitted by the simple first-order or second order kinetics models.

Indeed, the second-order kinetics model is applied when the value of ln a_{eq} is not equal to the intercept of the first-order plot based on Eq 13. The second-order kinetics is given by the equation [57]:

$$\text{(b) } t/a_t = 1/(k_2 \times a_{eq}^2) + (1/a_{eq})t \,, \qquad (15)$$

where the product $(k_2 \times a_{eq}^2)$ is attributed to the initial adsorption rate at $t \to 0$. The plot of t/a_t versus t gives a straight line which allows the calculation of the values of a_{eq} and k_2.

The attempts to apply the second-order constants to the experimental data occur to be ineffective because the dependence of t/a_t on t is non-linear.

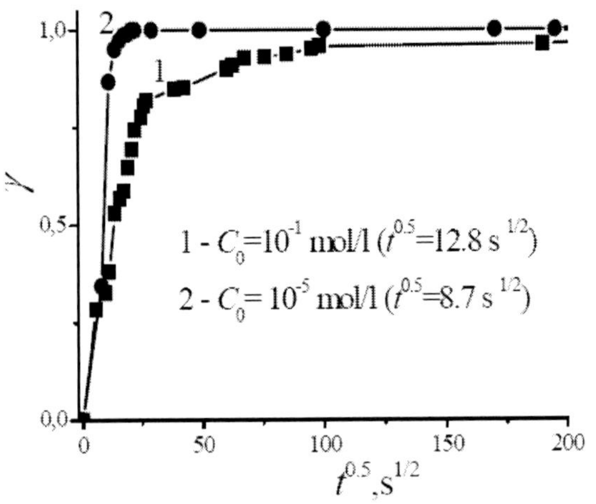

a.

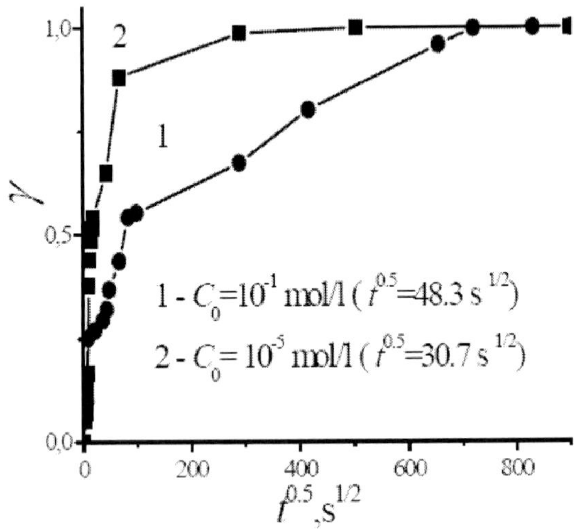

b.

Figure 30. Copper sorption kinetics on chitosan samples: (a) – sample 2; (b) – sample 1, at different initial concentration of copper cations in solutions.

The intraparticle diffusion plots ($a/a_{eq}=\gamma$ versus $t^{0.5}$) are shown on Figure 30. The plots indicate that the experimental data do not obey completely to the model of intraparticle diffusion. So the intraparticle diffusion is likely not a dominant controlling factor of the kinetic of the adsorption process.

The above results clearly show that the kinetics of the metal ions adsorption on the chitosan samples cannot be interpretated by a simple mechanism. It is likely, that the adsorption of Cu or Ag ions on the chitosan samples go by several different completion mechanisms. The same conclusion is followed from the IR spectroscopy data mentioned later. The pseudo-first-order model gives better results for approximation the kinetic curves of adsorption of metal cations on the chitosan samples. At the same time the essential influence of the diffusion processes, in particular intraparticle, on the metal cation adsorption kinetics should be taken into account. Most probably the nature and energy of the forming metal cation-polymer complex have an effect on the rates of metal cation adsorption.

Table 8. The kinetic parameters of copper cation adsorption on the chitosan samples

Sample	C_0, mol/l	$t^{0.5}$, $s^{1/2}$
1	10^{-1}	48.3
1	10^{-5}	30.7
2	10^{-1}	12.8
2	10^{-5}	8.7
5	10^{-1}	12.3
5	10^{-5}	3.7

The analysis of kinetics data of copper cations adsorption on the chitosan sample points to the onset of equilibrium is faster for the smaller initial concentration of copper cations in the solutions. It should be noted that the values of $t^{0.5}$ are smaller for the metal cations adsorption on the freeze-dried samples 2 and 5 in three times compared with the air-dried sample 1, it means the larger rate of metal cation adsorption process in the samples 2 and 5 (Table 8).

3.4.3. IR and UV-VIS Spectroscopy Investigations of Cu(II) Cations Adsorption on the Chitosan Samples

The foregoing data assert that the conditions of drying determine the adsorption behaviors of the chitosan samples towards to metal cations. Thus, the active polymer-metal complexes are considered for the chitosan samples dried in different conditions as well as for different ratios of amino groups to metal cations, i.e., for different content of adsorbed metal species. For the purpose of determination of the possible structure of the metal cation – chitosan complex the data of IR and UV-VIS spectroscopy experiments performed for the chitosan samples with adsorbed copper cations have to be considered.

Let's consider the spectra of the chitosan samples with the adsorbed Cu (II) species. Figure 31 reports the IR spectra of a diffuse reflection for the air-dried chitosan sample 1 before (curve 1) and after adsorption of copper ions from the solutions of various concentrations (curves 2 and 3). It is obvious that the spectrum for samples after the contact with low concentration solution of Cu (II) species (curve 2) differs insignificantly from that obtained for the sample 1 (curve 1, Figure 31). However, the increase of concentration of Cu (II) species in solution causes the considerable changes in the spectrum (curve 3, Figure 31). Instead of a wide band with a maximum at 3499 cm^{-1} attributed to the stretching vibrations of bonds O–H and N–H linked by hydrogen bonds, three pronounced narrow peaks at 3589, 3565 and 3410 cm^{-1} are observed (curve 3, Figure 31). The bands attributed to the deformation vibrations of O–H and N–H bonds shifts from 1670 cm^{-1} to 1654 cm^{-1}. Moreover, a new band is clearly exhibited in the range of 1559 -1552 cm^{-1} relating to the deformation vibrations of NH_2 groups which form a coordination bond with copper ions. It is known that in various amine compounds of copper the frequencies of the stretching vibrations of N–Cu bonds occur in the range of 420–500 cm^{-1} [58]. Such a band observed at 450 cm^{-1} in the spectrum 3 is likely refers to the stretching vibrations of bonds N–Cu of the chitosan complexes with Cu (II) species.

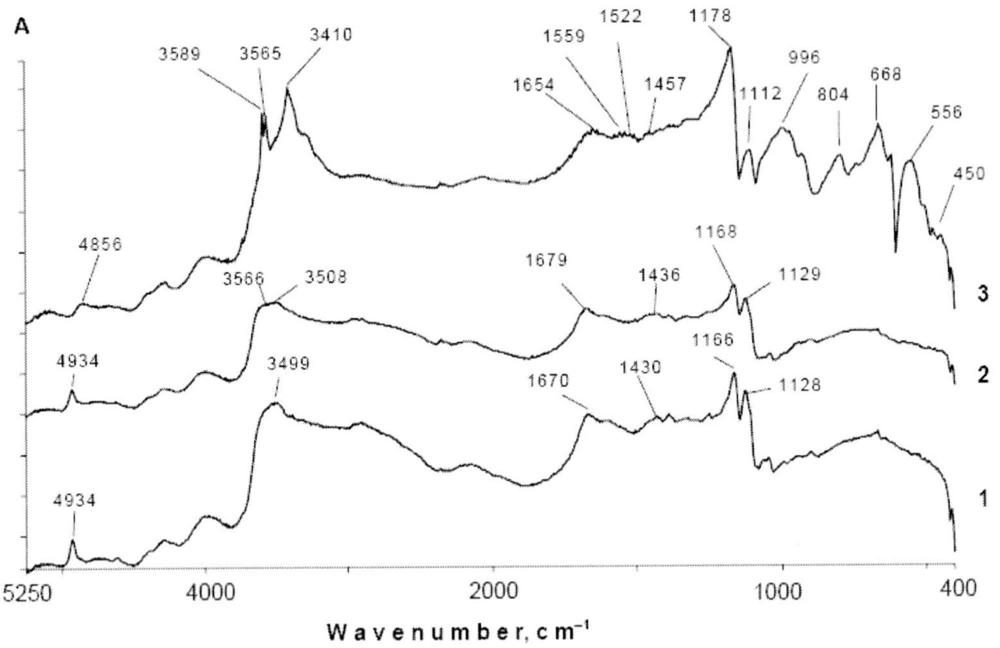

Figure 31. IR diffuse reflectance spectra of the air-dried chitosan before (curve 1) and after the contact with Cu (II) solution of concentration 10^{-4} mol/l (spectrum 2) and 10^{-2} mol/l (spectrum 3).

The changes of the combination band at 4934 cm^{-1} attributed to amino-groups (Figure 31, curve 1) indicate to the participation of amino groups of the chitosan samples in bonding of copper cations. This band represents a combination of stretching and deformation vibrations of the bonds of primary N-H group. At low copper concentration the intensity of this band decreases (curve 2, Figure 31) as a result of interaction of chitosan and copper cations. This band decreases completely from the spectrum after the contact of the chitosan sample with the high-concentration of copper (Figure 31, curve 3). Instead of it the unsharp band appears at 4856 cm^{-1}. The shift over 78 cm^{-1} to lower values is caused by reducing the frequencies of deformation and stretching vibrations of NH_2 group resulted from the formation of the complex with copper cations. The complete vanishing of a starting combination band at 4934 cm^{-1} in the spectrum 3 (Figure31) testifies that all amino groups of the chitosan samples participate in bonding with copper cations. The other components of the forming complexes are likely the OH^- ions and water molecules which are responsible for the bands of absorption of stretching vibrations O–H at 3589, 3565 and 3410 cm^{-1}, respectively.

Further the adsorption of copper cations on the freeze-dried chitosan sample is under consideration. Figure 32 reported the infrared spectra of the freeze-dried chitosan sample 2 before (spectrum 1) and after the contact with the copper solutions of with the concentrations of the Cu(II) cations (spectra 2 and 3). Spectrum 1a relating to the air-dried sample 1 is shown for comparison. It is obvious that the spectrum 1 attributed to the freeze-dried chitosan (sample 2) substantially changes after the contact with the low concentration solutions of Cu(II) (see spectrum 2) which occur to be similar to the spectrum 1a of the air-dried chitosan sample 1. One can conclude that the adsorption of Cu (II) cations from the low-concentration solutions of Cu(II) leads to the relaxation of freeze-dried chitosan sample to that of the air-dried chitosan. It concerns essentially the hydrogen bonding system. The value of the adsorption of Cu(II) species on the sample 2 is small (see Table 6) and displays in occurrence of a weak absorption band at 1560 cm^{-1}. The contact of the freeze-dried sample 2 with the copper solutions of higher concentration of Cu(II) leads to the appearance of much more intensive band in the range of 1530-86 cm^{-1} that is attributed to the vibrations of amino groups bounded with copper ions (see spectrum 3). In addition, a band related to a grouping of pyranose ring —C—O—C— of the freeze-dried chitosan sample shifts from the position at 1156 cm^{-1} (curve 1) to 1176 cm^{-1} (curve 3). The band with a maximum at 3403 cm^{-1}, which is attributed to the stretching vibrations of OH and NH groups bounded with hydrogen bondings, shifts from 3403 to 3392 cm^{-1} with the shoulders at 3320 and 3280 cm^{-1}. The bands of the deformation oscillations of both OH and NH bonds shift from 1670 cm^{-1} to 1649 cm^{-1}. The positions of the bands and relations of their intensities in the range of 1250-1470 cm^{-1} which are related to deformation vibrations of bonds CH mixed with the deformation vibrations of OH groups of chitosan are also changed. The band at 442 cm^{-1} refers to the bond of N–H in the chitosan-copper cation complexes (see spectrum 3) [58-61]. The combination band of amino group (ν (NH) + δ(NH)) is displaced from 4935 cm^{-1} to 4868 cm^{-1} similar to the air-dried chitosan sample after the contact with the copper solution. The displacements of the combination band to lower value is caused by lowering both the deformation and stretch vibrations frequencies of N–H bond under the complex formation by means of the chitosan amino groups and copper cations. The complete disappearance of the starting band at 4935 cm^{-1} testifies that all amino groups have reacted to copper cations. Thus under the copper cations adsorption on the chitosan samples, the coordination compounds of copper and chitosan amino groups are formed.

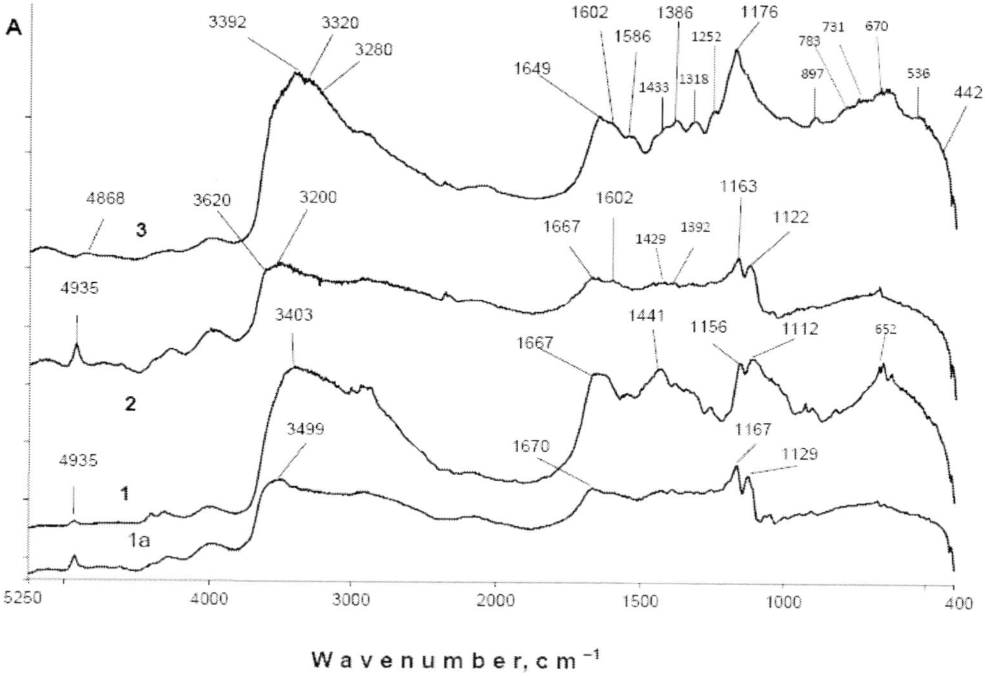

Figure 32. IR diffuse reflectance spectra of the freeze-dried chitosan sample 2 before (spectrum 1) and after the contact with the solution of copper (II) with concentration equal to 2×10^{-4} (spectrum 2) and 10^{-2} mol/l (spectrum 3). The spectrum 1a refers to the air-dried sample 1 for comparison.

However, the maximal amounts of copper adsorption on the air-dried and freeze-dried chitosan samples differ in two times from each other. The origin of this discrepancy is probably in the distinctions of structure of the copper complexes that depends on the pretreatment of chitosan samples. It should be noted that IR spectroscopy experiments are not enough for the exact determination of the difference between the complexes formed by amino groups of chitosan and copper cations in such complicated systems. In order to solve this problem the UV-VIS spectroscopy data are to be taken into account. Figure 33 shows the spectra of Cu (II) species adsorbed on the air-dried chitosan sample (spectrum 1) and on the freeze-dried sample (spectrum 2). In the spectrum 1 for the air-dried sample a broad structured absorption band is observed in the range of 16000-13000 cm^{-1} with the maximum at 14230 cm^{-1}. The corresponding absorption band in the spectra of the copper-chitosan complexes is relevant to the copper complexes of planar-square type, in particular observed for ligands which contain amino groups, and refer to the d-d transitions. The similar absorption band was referred by *Chiessi et al* [62] to the complexes of a planar-square type with the local D_{4h} symmetry in which a single chitosan amino group and three hydroxyl groups act as ligands. Insignificant asymmetry of the band observed in the range of low energies and structureless absorption in the UV range indicate the presence of some amount of the complexes with distorted planar-square symmetry and/or the presence of small amount of the impurity complexes of another structure.

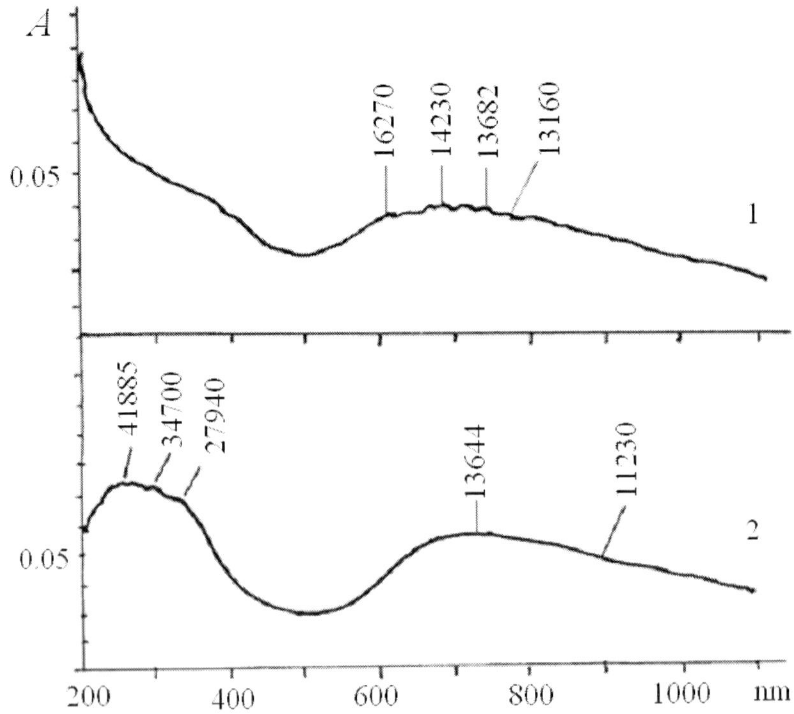

Figure 33. UV-VIS-NIR Spectra of the air-dried (curve 10 and freeze-dried (curve 2) chitosan samples after the contact with copper solutions of concentration $C_0 = 10^{-2}$ mol/liter. The numbers above the curves designate the maximum values of absorption bands expressed in inverse centimeters.

The spectrum freeze-dried sample 2 (Figure 33, curve 2) contains a wide asymmetric band with a maximum at 13640 cm^{-1} and an additional maximum at 11230 cm^{-1}. Moreover, in UV range the intensive absorption with the peaks at 27940, 34700 and 41885 cm^{-1} is observed. Such a spectrum, and especially the presence of the band at 27940 cm^{-1}, is peculiar for the binuclear copper complexes [59].

Thus one can conclude that the conditions of pretreatment of the chitosan samples influence both on the amount of adsorbed metal species and structure of the forming complexes. The data obtained allow proposing the following idealized model structures of the copper cations – chitosan complexes. Thus in the case of the highest adsorption of copper cations on the air-dried chitosan sample 1 the ratio of the amount of the adsorbed copper species to the amount of amino groups of the chitosan sample is equal to 0.49 (see Table 6), so that approximately two amino groups per one copper atom. As it was mentioned above, the copper cations form the planar-square complex (or close to it) with the amino groups of chitosan. So, one can conclude that in the case of the highest adsorption of copper cations on the air-dried chitosan sample the planar-square complexes as $Cu(NH_2)_2(OH)_2$ are formed with the axial group NH_2 of each complex belong to different chitosan macromolecules. As it follows from corresponding calculations such a configuration is the most stable for the copper adsorption on chitosan [10].

As for the freeze-dried chitosan sample, it should be noted that the adsorption of copper on chitosan from the high-concentrated solution gives the samples with the ratio of the amount of adsorbed copper cations to amino groups of chitosan equal to 0.93, or approximately one amino group per one copper atom. It has been calculated by [10] that the other stable configuration can be the "pendant" complex (see Figure 34 (a)), according which every amino group is associated with one Cu (II) ion. But our experiments show that the most probable is the “bridge” structure where a binuclear copper complex bounds two chitosan macromolecules, so that every Cu (II) cation of a pair is associated with one amino group in axial position which belongs to different chitosan chains (Figure.34 (b). In the limiting case, when all amino groups are involved in the complex with copper cations [$-NH_2\leftrightarrow Cu-Cu\leftrightarrow NH_2-$], the ratio $Cu/NH_2=1$ is also valid.

Thus, the copper cation adsorption on the chitosan samples from high-concentrated copper solution can be considered as a process of cross-linking linear polymer into 3D structure by the bridges composed of mono- or binuclear coordination compounds of Cu(II) cations. Probably, in the case when the binuclear copper complex serves as a cross-linking agent, the resulted chitosan sample has probably the pores with the larger sizes compared with the case of mononuclear cooper complex as the cross-linking agent.

It should be noted that the similar results were found in the course of the spectroscopy experiments of Ni and Ag cations adsorption on the chitosan samples. The metal cation adsorption data leads to transformation of polymer – metal cation structure as the amount of introduced species increases. In conclusion the structure of polymer-metal cation complex is determined by the following factors as: metal cation concentration in the initial solution, nature of metal cation and supramolecular structure of polymer adsorbent.

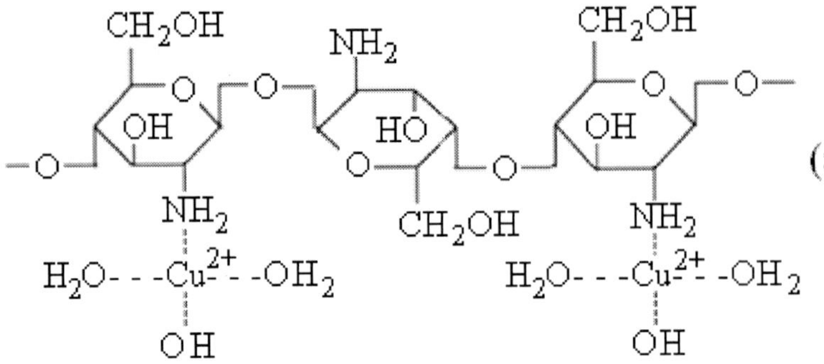
(a)

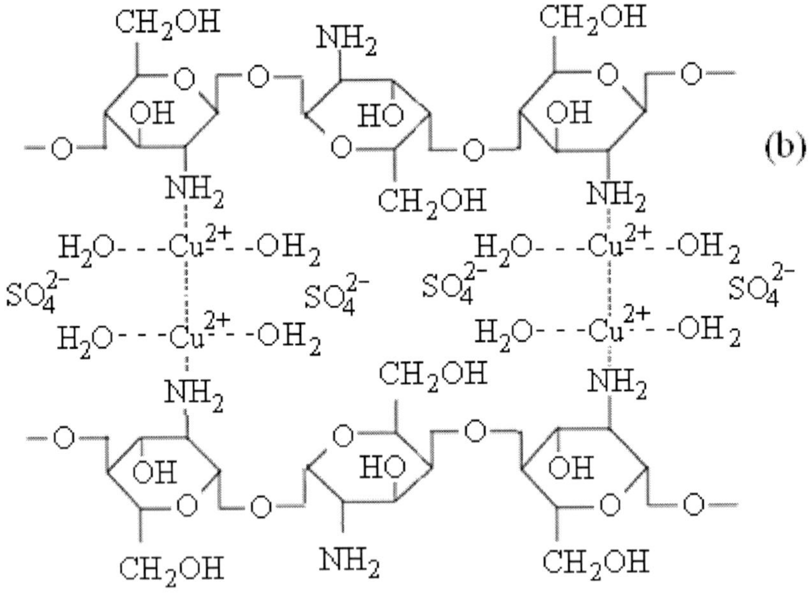
(b)

Figure 34. The models of copper cations – chitosan complexes: the planar-square complex (a) with the ratio $Cu/NH_2=1/2$ in limited case (pedant model) and the bridge binuclear complex (b) with the ratio $Cu/NH_2=1/1$ in limited case.

IV. Conclusions

1. Investigations of the chitosan samples by IR spectroscopy, x-ray diffraction and pulsed nmr methods indicated the influence of drying conditions on supramolecular structure of chitosan. It has been shown that freeze-drying of chitosan resulted in the changes of polymer chains locations leading to ordering of macromolecules, reconstruction of system of hydrogen bonds and the changes of porosities.
2. These changes in supramolecular structure cause the substantial enhancement of adsorption ability of the freeze-dried chitosan samples in respect to water vapors. Water vapor adsorption data for the chitosan samples have been fitted by applying the models based on different equations as: Flory-Huggins, Dubinin-Astakhov and Dubinin-Serpinsky (DS2). The DS2 equation based on the two-step model that

assumes the existence of both primary and secondary adsorption centers approximate the water adsorption isotherms in the chitosan samples in the range of relative pressures from 0.1 to 1.0.

The amounts of primary adsorption centers for water in the chitosan samples have been estimated by means of the comparative method.

3. Different models based on the Freundlich and Langmuir equations have been testified for approximating the adsorption of metal cations (Ni, Cu, Ag) on the chitosan samples. The adsorption of nickel, copper and silver cations by chitosan enhanced after the preliminary freeze-drying as it follows from the parameters of the Dubinin–Radushkevich equation used for the adsorption isotherm approximation. The freeze-drying and different pH resulted in different location of polymer chains and, as a consequence, to different locations of amino groups of glucopyranose rings that likely serve as the adsorption sites of metal cations with different energies.
4. The IR-VIS spectroscopic experiments performed for the copper cations adsorption on the chitosan samples at high adsorption values allowed suggesting a model of the copper-chitosan complex. It has been shown that the most probable model of the copper cation – chitosan complex is the "bridge" structure based on a binuclear copper complex bounds two chitosan macromolecules, so that every Cu (II) cation of a pair is associated with one amino group in axial position which belongs to different chitosan chains.

ACKNOWLEGMENTS

The authors are grateful to A.D. Aliev, S.D. Artamonova, T.P. Puranova, G.A. Vichoreva for assisting with experiments.

REFERENCES

[1] Kucherov, A.V.; Kramareva, N.V.; Finashina, E.D.; et al. *J. Mol. Catal A Chem.* 2003, 198, 377-389.

[2] Okuyama, K; Noguchi, K; Miyazawa, T; Yui, T; Ogawa, K. *Macromol* 1997, 30, 5849-5855.

[3] Muzzarelli, R.A.A., Chitin, Oxford: Pergamon; 1977.

[4] Wan Ngah, W.S.; Endud, C.S.; Mayanar, R. *React. and Functional Polym.* 2002, 50, 181-190.

[5] Smirnov, A.D. Sorbtsionnaya ochistka vody (Sorption Water Treatment), Leningrad: Khimiya, 1982.

[6] Yaku, F.; Muraki, E.; Tsuchiya, K.; et al. *Cellul. Chem. Technol.* 1977, 11, 421-430.

[7] Schlick, S. *Macromol.* 1986, 19, 192-195.

[8] Ogawa, K.; Oka, K.; Miyanishi, T.; Hirano, S. in Chitosan and Related Enzymes, Zikakis, J.P.; Ed.; Orlando: Academic, 1984, 327-345.

[9] Domard, A. *Int. J. Biol. Macromol.* 1987, 9, 98-104.

[10] Terreux, R.; Domard, M.; Viton, C.; Domard, A. *Biomacromol.* 2006, 7, 31-37.

[11] Despond, S.; Espuche, E.; Domard, A. *J. Polym. Sci.* Part B Polym Physics 2001, 39, 3114–3127.
[12] Gregg S. Sing K.S.W. Adsorption, Surface area and porosity: New York: Academic, 1982.
[13] Flory, P. Principles of Polymer Chemistry; Cornell University Press, Ithaca, N.Y., 1953
[14] Lagergren, S.K. Sven Vetenskapsakad Handl 1898, 24, 1-39.
[15] Ho, Y.S. *Scientometrics* 2004, 59, 171-177.
[16] Gamzazade, A.E.; Shlimak, B.M.; Sklyar, A.M.; Stykova, E.V.; Pavlova, S.-S.A.; Rogozin, S.V. *Acta Polym.* 1985, 36, 421-424.
[17] Krasil'nikova, O.K., Artamonova, S.D., Vikhoreva, G.A., et al., Abstracts of Papers, VI Mezhdunar. konf. "Novye dostizheniya v isledovanii khitina i khitozana". (VI Int. Conf. "Progress in Investigation of Chitin and Chitosan"), Shchelkovo, 2001, 295.
[18] Vartapetyan, R.Sh; Voloshchuk, A.M. *Russian Chem. Rev.* 1995, 64, 985-1001.
[19] Schwarzenbach, G. and Flaschka, H. Complexometric Titrations. Stuttgart: Ferdinand Enke; 1965.
[20] Jaworska, M.; Sakurai, K.; Gaudon, P.; Guibal, E. *Polym. Int*.2003, 52, 198-205.
[21] Levitt, M.H.; Freeman R. *J. Magn. Res.* 1981, 43, 65-72.
[22] Carr, H.Y.; Purcell E.M. *Phy. Rev.* 1954, 94, 630-638.
[23] Meiboom, S.; Gill, D.; *Rev. Sci. Instr.* 1958, 29, 6881-6885.
[24] Stejescal, E.O.; Tanner, J.E. *J. Chem. Phys.* 1965, 42, 288-292.
[25] Gerente, C.; Lee, V.K.C.; Cloirec, P.Le.; McKay, G. *Environ. Sci. Technol.* 2005, 37, 41-127.
[26] Valentin, R.; Molviger, K.; Viton, C.; Domard, A.; Di Renzo. F. *Biomacromol.* 2005, 6, 2785-2792.
[27] Cilipotkina, M.V.; Tager, A.A.; Petrov, B.S.; Pustobaeva G. Visokomol Soed 1962, 4, 1844-1850.
[28] Kotelnikova, T.A. Sorption and chromatographic process 2008, 8, 50-59.
[29] Desmond, S.; Espuche, E.; Domard, A. *J. Polym. Sci.* Part B Polym Physics 2001, 39, 3114–3127.
[30] Abragam, A. The Principles of Nuclear Magnetism, U. P., Oxford 1961.
[31] Derbyshire, W. and all. *J. Magn. Reson.* 2004, 168, 278-283.
[32] Aerberhardt, K.; Bui, Q.D.; Normand, V. *Biomacromolec.* 2007, 8, 1038-1042.
[33] Aoarwal, S.C.; Gupta R.C.; Current Science, Letters to the Editor, 1972, 41, 631-632.
[34] Froneman, S.; Reynhardt, E.C. *J. Phys.* D: Appl Phys 1991, 24, 387-391.
[35] Brownstein, K.R.; Tarr, C.E. *J. Magn. Reson* 1977, 26, 17-24.
[36] Ainscough, A.N.; Dollimore, D. Langmuir 1987, 3, 708-713.
[37] Breuer, M.; Buras, E.M.; Fooksoon, A. In Water in Polymers; Rowland, S.R.; Ed.; American Chem. Soc., Washington, D.C. 1980, pp. 304-314.
[38] Gocho, H.; Shimizu, H.; Tanioka, A.; Chou, T.J.; Nakajima, T. *Carbohydr. Polym.* 2001, 41, 87-90.
[39] Di Renzo, F.; Valentin, R.; Boissiere, M.; Tourrette, A.; Sparapano, G.; Molviger, K. at al. *Chem. Mat.* 2005, 17, 4693-4699.
[40] Dubinin, M.M.; Serpinskii, V.V.; Dokl AN SSSR, 1981, 258, 1151-1154.
[41] Van den Berg, C., Bruin, S. In Water activity: influences on food quality; Rockland, L. B.; Stewart, F. Ed.; Academic Press: New York, 1981, 147–177.

[42] Chalich, A.E.; Bairamov, D.F.; Gerasimov, V.K.; Chalich, A.A.; Fel'dshtein, M.M. Visokomol soed 2003, 45, 1856-1861.

[43] Kablan, T.; Clement, Y. bi Y.; Francoisea, K.A.; Mathiasb, O.K. *Acta Chim. Slov.* 2008, 55, 677–682.

[44] Krasilnikova, O.K.; Vartapetian, R.Sh. *Macromol. Symp.* 2005, 222, 181-185.

[45] Valentın, J.L.; Carretero-Gonzalez, J.; Mora-Barrantes, I.; Chasse, W. Saalwachter, K. *Macromol.* 2008, 41, 4717-4729.

[46] Andreeva, G.A.; Polyakov, P.S.; Dubinin, M.M.; Nikolaev, K.M.; Ustinov, E.A. Izv. AN SSSR Ser him 1981, 10, 2188-2192.

[47] Dubinin, M.M. *Carbon* 1983, 21, 359-366.

[48] Ng, J.Y.; Cheung, W.H.; McKay, G.; *J. Colloid Interface Sci.* 2002, 255, 64-74.

[49] Juang R.-S.; Shao, H.-J. Water Res 2002, 36, 2999-3008.

[50] Stoeckli, F.; Lopez-Ramon, V.M.; Moreno-Castilla, C. *Langmuir* 2001, 17, 3301-3306.

[51] Kharitonova, A.G.; Krasilnikova, O.K.; Vartapetyan, R.Sh; Bulanova, A.V. *Koloid Zh.* 2005, 67, 416-420.

[52] Rhazi, M.; Desbrieres, J.; Tolaimate, A. *Polymer* 2002, 43, 1267-1276.

[53] Yuan, Y.-C.; Zhang, M.-Q.; Rong, M.-Z. *Acta Chim. Sinica* 2005, 63, 1753-1758.

[54] Juang, R.S.; Wu, F.C.; Tseng, R.L. *Wat. Res.* 1999, 33, 2403-2409.

[55] Wu, F.C.; Tseng, R.L.; Juang, R.Sh. *Ind. Eng. Chem. Res.* 1999, 38, 270-275.

[56] Yoshizuka, K.; Lou, Z.; Inoue, K. *React. and Functional Polym.* 2000, 44, 47-54.

[57] Ho, Y.S.; McKay, G. *Trans Inst. Chem. Eng.* 1999, 77 B, 165-173.

[58] Masoud, M.S., El-Enein, S.A., Abed, I.A., Ali, A.E. *Journ. Coord. Chem.* 2002, 55, 153-178.

[59] Liver, A.B.P. Inorganic electronic spectroscopy. Elsevier. 1987, 2, 146-278.

[60] Yavney, D.B.W., Doedens, R.J. JACS, 1970, 92, 21, 6350-6352.

[61] Cao, R., Shi, Q., Sun, D., Hong, M., Bi, W., Zhao, Y. *Inorganic Chemistry* 2002, 41, 23, 6161-6168.

[62] Chiessi, E., Branca, M., Palleschi, A., Pispisa, B. *Inorg. Chem.* 1995, 34, 2600-2609.

In: Focus on Chitosan Research
Editors: Arthur N. Ferguson and Amy G. O'Neill
ISBN 978-1-61324-454-8

Chapter 2

CHITOSAN: A PROMISING BIOMATERIAL FOR TISSUE ENGINEERING AND HEMORRHAGE CONTROL

Nan Cheng[1], Xudong Cao[1] and Henry T. Peng[2*]

[1]. Department of Chemical and Biological Engineering,
University of Ottawa, Ottawa, Ontario, Canada
[2]. Defense Research and Development Canada,
Toronto, Canada

ABSTRACT

Chitosan, a natural polysaccharide, and its derivatives are very important biomaterials because of their distinct physicochemical and biological properties such as biocompatibility, biodegradability, non-toxicity, antimicrobial and hemostatic properties. In addition, the existence of reactive functional groups in the chitosan molecule also makes it possible to tailor the chitosan material for specific tissue engineering and hemorrhage control applications. In the first part of this chapter, we will focus on the properties of chitosan and various modifications to improve its properties to enhance drug delivery and tissue regeneration. The modification of chitosan will be introduced from two aspects: chemical and physical modifications, i.e. grafting with other small molecules or biomolecules and blending chitosan with other different biomaterials, respectively. The applications of chitosan, especially its derivatives in drug delivery and tissue regeneration targeting different organs, such as cartilage, bone, skin, blood vessel, nerve system and liver will be discussed. Current improvements of these different artificial substitutes (based on chitosan) in different organs will be reviewed by the authors and other researchers . In the second part of the chapter, we will review the development of chitosan-based biomaterials for hemorrhage control in trauma and surgical injuries. Both chemical and physical modifications of chitosan will be discussed. Finally, we will elaborate on the challenges and future direction of using chitosan as a biomaterial for tissue engineering and hemorrhage control.

[*] Corresponding author: Henry T. Peng.

1. INTRODUCTION

Chitosan is a linear copolymer of 1,4 β-linked 2-acetamido-2-deoxy-D-glucopyranose and 2-amino-2-deoxy-D-glucopyranose. It is normally derived from naturally abundant chitin via deacetylation by enzymatic or alkaline treatment; its chemical structure is shown in Figure 1 [1]. Chitosan has two important functional groups, the specific functional group $–NH_2$ at the C-2 position and non-specific -OH groups at the C-3 and C-6 positions, as both are often involved in most chitosan chemical modifications to create different chitosan derivatives for different biological applications [2]. These chemical modifications include acylation, N-phthaloylation, tosylation, alkylation, Schiff base formation, reductive alkylation, O-carboxymethylation, N-carboxylalkylation, silylation and graft copolymerization. The existence of $–NH_2$ groups is an important feature of chitosan when compared with cellulose, the most abundant naturally occurring polymeric material, as the $–NH_2$ groups of the chitosan molecule have been shown to be the most active sites for chemical reactions. In addition, the hydroxyl groups in chitosan can be used for esterification and etherification.

Chitosan can be degraded by enzymes. Lysozyme is the primary enzyme responsible for *in vivo* chitosan degradation through hydrolysis of acetylated residues on the molecular chain. The degradation rate of chitosan is affected by the degree of crystallinity and deacetylation [3].

Chitosan has a cationic nature because of the existence of primary amines. This property allows chitosan to form electrostatic interactions with negatively charged molecules like DNA and some proteins, which is very useful in drug delivery systems and protection of drugs *in vivo*. At the same time, chitosan can be combined with a variety of biomaterials such as alginate, collagen, hyaluronic acid (HA), calcium phosphate, polymethylmethacylate (PMMA) and poly-L-lactic acid (PLA) to inherit different properties from different materials and these kinds of chitosan derivatives showed potential use in drug delivery systems and tissue engineering scaffolds [4]. Another advantage of chitosan is that it can be molded to various forms such as powder, paste, film, fiber, porous scaffold, etc. to specific applications and to target organs [4].

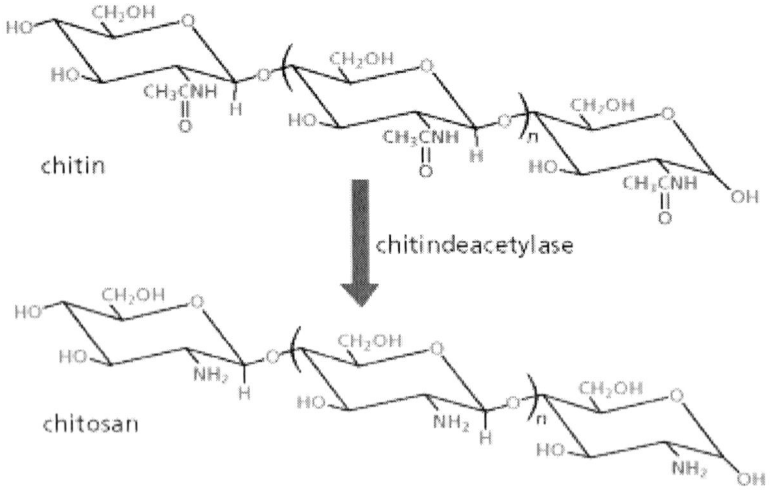

Figure 1. Chemical structures of chitin and chitosan.

The structure and properties of chitosan, such as the degree of deacetylation (DD), water solubility and molecular weight, determine its applications. The DD is an important parameter of chitosan. Many researchers have investigated the influence of DD on the efficiency of different drug delivery systems and implants. Taking gene delivery system as an example, Leong and his groups [5] studied the effect of DD on gene transfection. They used chitosan (CS) with different DD (90%, 70% and 62%) to achieve complete DNA/CS complexation and delivered the complex into a HEK293 (human embryonic kidney cell), a Hela (cervical cancer cell) and SW756 (HPV-18 positive cervical carcinoma cell) cells. According to the study, chitosan with the lowest DD required a larger amount to form a complete complex with the same amount of DNA. The gene expression is lower for chitosan with a lower DD due to the instability of the DNA/CS complex *in vitro* and *in vivo*. They believed that these results were due to the difference of the primary amines on chitosan by the various deacetylations. In addition, the solubility of chitosan relates to the deacetylation degree and the distribution of acetyl groups and primary amine groups in chitosan. It is a difficult parameter to control, but it is crucial for the determination of chitosan's properties and applications. Because of the harsh chemical conditions for many biomolecules and the stringent requirement of *in vivo* studies, highly deacetylated chitosan, which contains more primary amine groups and can be dissolved in a weak acidic solution, is the popular candidate for biological applications. Based on the different applications, there are specific requirements for the molecular weight, usually from 300 to 1,000 kDa [4].

Thanks to the above-mentioned properties, chitosan is believed to hold great potential as a biomaterial with broad applications. Furthermore, its gelation properties and high affinities to *in vivo* macromolecules have made chitosan and its derivatives ideal candidates for tissue engineering applications. This chapter focuses on its use as carriers of drug delivery, scaffolds for tissue regeneration and hemostatic agents for hemorrhage control.

2. Carriers of Drug Delivery

Due to the biocompatible and biodegradable properties, the interest in chitosan and its derivatives as carriers of drug delivery arises. The positive charge and the water solubility of chitosan and its derivatives are very important for the development of drug delivery systems. These properties enable the chitosan to form electrostatic interaction with negatively charged molecules and can be delivered in an aqueous environment, which is the requirement for *in vivo* circulation. In recent years, significant effort has been devoted to develop a biodegradable drug delivery system based on chitosan and its derivatives. Here we will introduce different systems using chemical modification and physical blends of chitosan.

2.1. Chemical Modified Chitosan

2.1.1. Surface Modified Chitosan as Vectors of Drug and Gene Delivery

Surface modification of chitosan usually happens with primary amine group on chitosan. By introducing different small molecules as the side chains of chitosan bulk chain, the optimization of drug loading efficiency, microsphere size, swelling properties and release control can be achieved. Based on specific applications and target organs, the methods, including the increase of solubility in neutral pH, the introduction of hydrophobic and hydrophilic molecules to render amphiphilicity, the creation of porous structure, and control of positive charges, etc. served different purposes.

The introduction of lactic acid (LA) is a good example of the application of increasing solubility in neutral conditions. LA grafted chitosan (LA-g-CS) was developed (Zhang et al. [6]) by the introduction of a D,L-lactic acid group on chitosan without a catalyst. It was prepared by dehydrating the solvent to cast a thin film of chitosan containing LAs. LA-g-CS nanoparticles were fabricated via a co-precipitation process by LA-g-CS in ammonium hydroxide to form coacervate drops. Depending on their results, the existence of LA renders the solubility of chitosan in neutral pH, and further increases drug encapsulation and drug release circulation. Unlike chitosan, LA-g-CS can be prepared in neutral pH and allows proteins or other pH-sensitive drugs incorporated in the complex with minimal or no denaturization. This is a very important improvement to encapsulate biosensitive drugs, most of which are proteins. LA-g-CS also prolonged the release of proteins according to their data. Using albumin as a model protein, LA-g-CS nanoparticles released 15% of protein over 4 weeks; in comparison, chitosan released 28% over 4 weeks. At the same time, the introduction of the lactyl segment on chitosan would not change the surface charge dramatically, which is considered a benefit to maintain a desired surface property that allows internalization of the drug to the negatively charged cells. Other small molecules, like methyl methacrylate, *N*-dimeth-ylaminoethyl methacrylate hydrochloride, and *N*-trimethylaminoethyl methacrylate chloride [7], have similar qualities as LA to increase the solubility of chitosan backbone in a wide range of pH. Because of the poor solubility of pure chitosan in neutral conditions, the effort to chemically modify chitosan and increase its water solubility is a vital subject for the delivery of biomolecules, especially proteins.

N-alkylated or acylated chitosan is a series of chitosan derivatives, which provide the chitosan backbone amphiphilic properties by the small molecules. The benefit of modifying chitosan with hydrophobic branches has been demonstrated by many researchers [8]. It was believed that the existence of hydrophobic branches would reduce hydration of the matrix and play an important role in the stabilization of the complex by hydrophobic interactions. The existence of hydrophobic branches opened a new perspective of chitosan on the application of the controlled release of hydrophobic drugs. It is also reported that the introduction of hydrophilic branches after the hydrophobic modification of chitosan will provide the formation of micelles [9]. Here, an example for increasing the solubility of water-insoluble drugs is amphiphilic N-octyl-O,N-carboxymethyl chitosan synthesized by Lin et al. [10]. The new amphiphilic derivative has a long alkyl group as hydrophobic moieties and carboxymethyl groups as the hydrophilic moieties. It can increase the loading efficiency of paciltaxel, a typical water-insoluble drug, up to 500 fold in water.

Carboxymethyl-hexanoyl chitosan, a amphiphilic chitosan, was developed by Liu and coworkers [11] to control the release of a water-soluble drug. The amphiphilic derivative was first modified by hydrophilic carboxymethylation to increase the flexibility of chitosan molecules in water followed by hydrophobic modification with hexanoyl groups to introduce amphiphilic properties. The carboxymethyl-hexanoyl chitosan exhibited spherical morphology and a core–shell configuration in aqueous solution, due to the hydrophobic interactions of hexanoyl groups in aqueous solution, and the negatively charged carboxymethyl groups (Figure 2). In a controlled release study of doxorubicin, an anticancer drug, researchers achieved a suitable control of drug release at 7 days. In short, the formation of amphiphilic derivatives of chitosan has the distinguished advantage of entrapping water-insoluble drugs, which is a major concern for these drugs in order to achieve a high efficiency of drug loading and sustainable drug release.

Reduction of positive charges of chitosan was a strategy usually applied in gene delivery systems. In the gene delivery system, the control of association and dissociation of the vector/DNA complex is more critical. Successful gene delivery and expression need the protection of vector from the degradation but it also needs the suitable release of genes after cell uptake of the complex. In general, a weak interaction of vector and gene is important for the gene delivery system. There are many methods which use chitosan to create weak interaction and stable complex. Thiolated chitosan (TCS) was prepared by Mohapatra et al. [12]. 33 kDa TCS was prepared by thioglycolic acid with chitosan. 1-ethyl-3-(3-dimethylaminopropyl) (EDC) was used to graft carboxylic groups in thioglycolic acid with primary amine groups in chitosan. A plasmid DNA containing a green fluorescent protein gene was applied to form TCS/DNA complex. By transfecting the complex with different ratios of TGS and DNA into HEK293 cells, researchers observed an improved transfection efficiency of TCS/DNA compared with CS/DNA. One possible explanation for the improvement was that TCS partially neutralized positive charges and loosens the tight DNA binding between pure chitosan with DNA. Alkylated chitosan is another example of the use of small molecules to reduce positive charge on chitosan for gene delivery study [13]. Alkylated chitosans were prepared by modifying chitosan with different chain lengths of alkyl bromide (Figure 3). It was shown that the alkyl chain weakens the electrostatic attraction between DNA and vectors. The transfection study in a C2C12 cell (mouse myoblast cell) showed that the transfection efficiency of plasmid DNA encoding CAT increased with the increase of alkyl side chains and the effect leveled off after C8. It is evident that the neutralization of positive charges on chitosan will benefit gene delivery in some cases.

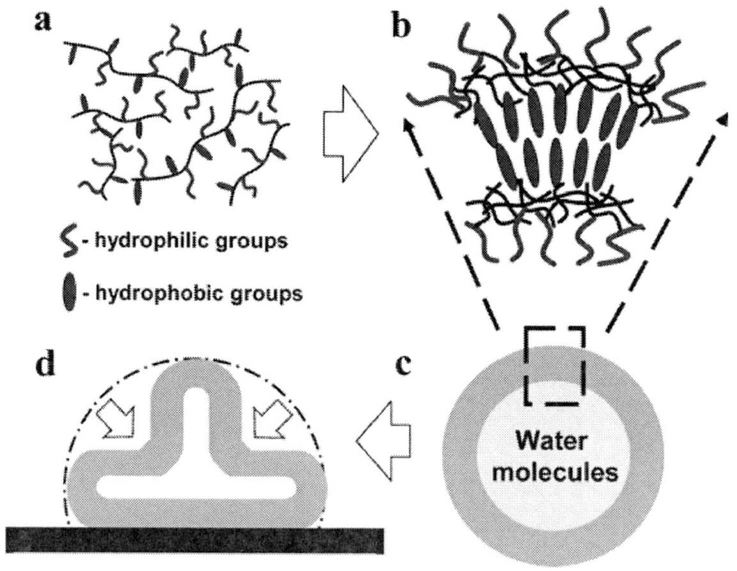

Figure 2. Scheme of formation process of carboxymethyl-hexanoyl chitosan nanocapsules.

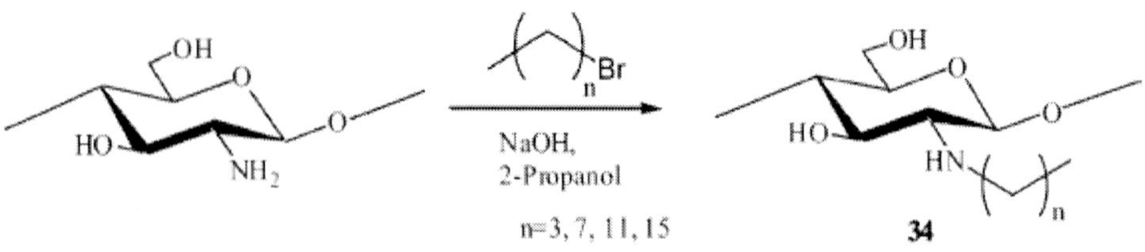

Figure 3. Synthesis of alkylated chitosan with different chain lengths of alkyl bromide.

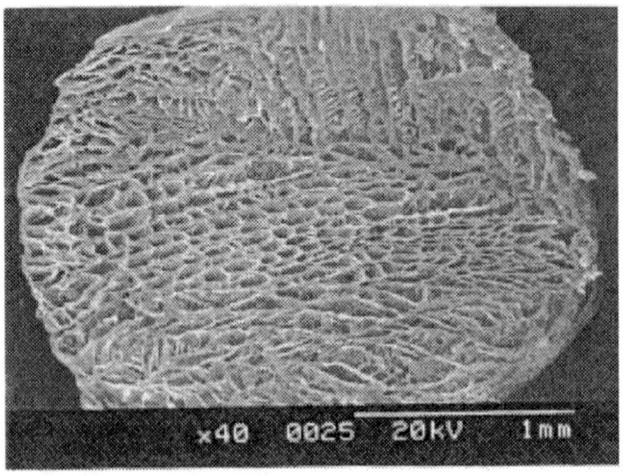

(a)

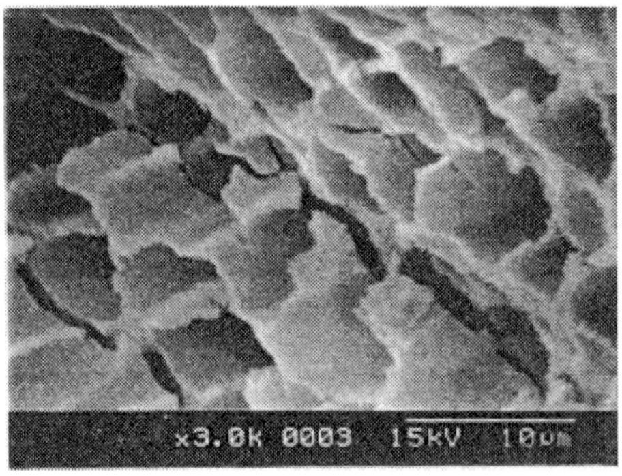

(b)

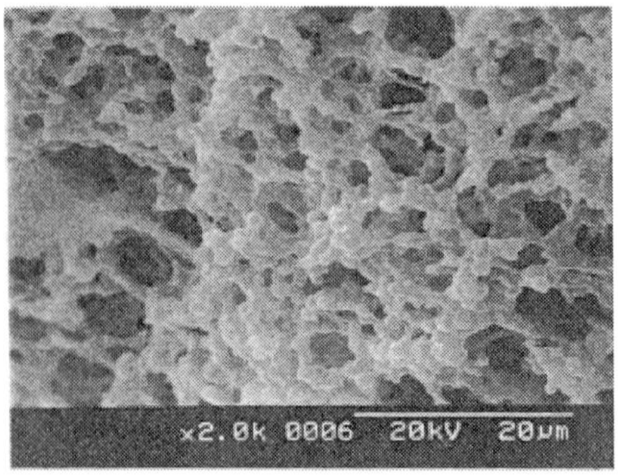

(c)

Figure 4. SEM images of phase-inversed chitosan.

Besides the chemical modification mentioned above, quaternized chitosan was used as a drug absorption enhancer to cross the intestinal epithelium due to its good mucoadhesive and permeability properties. It was usually used as insulin carriers. A Chinese research group prepared a derivative *N*-[(2-hydroxy-3-trimethylammonium) propyl] chitosan chloride by reacting chitosan with glycidyltrimethylammonium and using the Shirasu porous glass (SPG) membrane emulsification technique and the glutaraldehyde cross-linking method for uniform microspheres to control oral insulin administration. Both *in vitro* and *in vivo* studies showed a reduction in blood glucose levels [14]. The porous chitosan drug delivery system was also investigated by some researchers. The phase inversion technique was applied in the creation of porous chitosan and derivative for delivering a nonsteroidal anti-inflammatory drug, indomethacin [15]. In their study, using two stages of liquid-liquid and liquid-solid phase inversion from tripholyphosphate solution and the following chemical modifications with quaternary ammonium, porous chitosan derivatives containing aliphatic and aromatic acyl groups were generated (Figure 4). The drug was immobilized by either hydrophobic or electrostatic interaction and the chitosan derivatives showed the absorption of indomethacin and the potential use in a drug delivery system.

Small molecule-grafted chitosans were widely applied in drug delivery systems. The aim to induce small molecules was to optimize and create an efficient drug delivery vector by varying surface properties for different drugs such as charge and water solubility.

2.1.2. Macromolecule-Grafted Chitosans for Drug Delivery

Chitosan was usually combined and optimized by grafting with other macromolecules. These polymers, such as pluronic, poly (ethylene glycol) (PEG) families, render improved chitosan surface properties,

For example, PEG is the most popular synthetic polymer to graft with chitosan. PEG is a hydrophilic polymer with a low toxicity and biocompatibility. The general function of PEG is to enhance the water solubility of chitosan. Recently, PEG was used as a part of chitosan grafting, always combined with other hydrophobic molecules to create polymeric micelles. One example of these amphiphilic chitosans involving PEG is PEG conjugated N-octyl-O-sulfate chitosan [16]. The novel chitosan derivatives are with hydrophobic moieties of octyl and hydrophilic moieties of sulfate and polyethylene glycol monomethyl ether (mPEG) groups. Their *in vivo* studies with paclitaxel showed that the existence of PEG provides strong protein inhibition due to the hydrophilicity of PEG segment. The pharmacokinetic study in rats showed slower elimination in the initial stage of intravenous injection in PEG contained micelles. Higher targeting to the uterus was also seen in the tissue distribution study. So they believed that PEG side chains improved the efficiency of N-octyl-O-sulfate chitosan in targeting the ovaries and protected the drugs at the beginning of operations.

Another important polymer to graft with chitosan is pluronic families. Pluronic, PEO–*b*–PPO–*b*–PEO triblock copolymers, have been widely used in injectable drug delivery systems by micelle formation. Pluronic micelle could be formed in an aqueous solution above 25 C. Park et al. [17] developed Pluronic grafted chitosan copolymers by EDC reaction and after characterization, they concluded that Pluronic grafted chitosan was a reversible hydrogel and exhibited a lower critical solution temperature (LCST) at 34°C in aqueous solution. The proper LCST below body temperature and above room temperature make pluronic grafted chitosan a potential candidate as a thermo-sensitive drug control system with temperature as a stimulus. Later, Aminabhavi et al. [18] introduced a novel hydrogel microsphere of chitosan and pluronic F-127 using glutaraldehyde (GA) as a crosslinker to release 5-Fluorouracil, an anticancer drug. They investigated the effect of GA content and ratios of chitosan and pluronic F-127 on the morphology of the microspheres, drug loading and release. It was found that the release of drugs was related to the extent of crosslinking by GA and the amount of pluronic F-127. The morphology of these microspheres is smooth spherical shape (Figure 5).

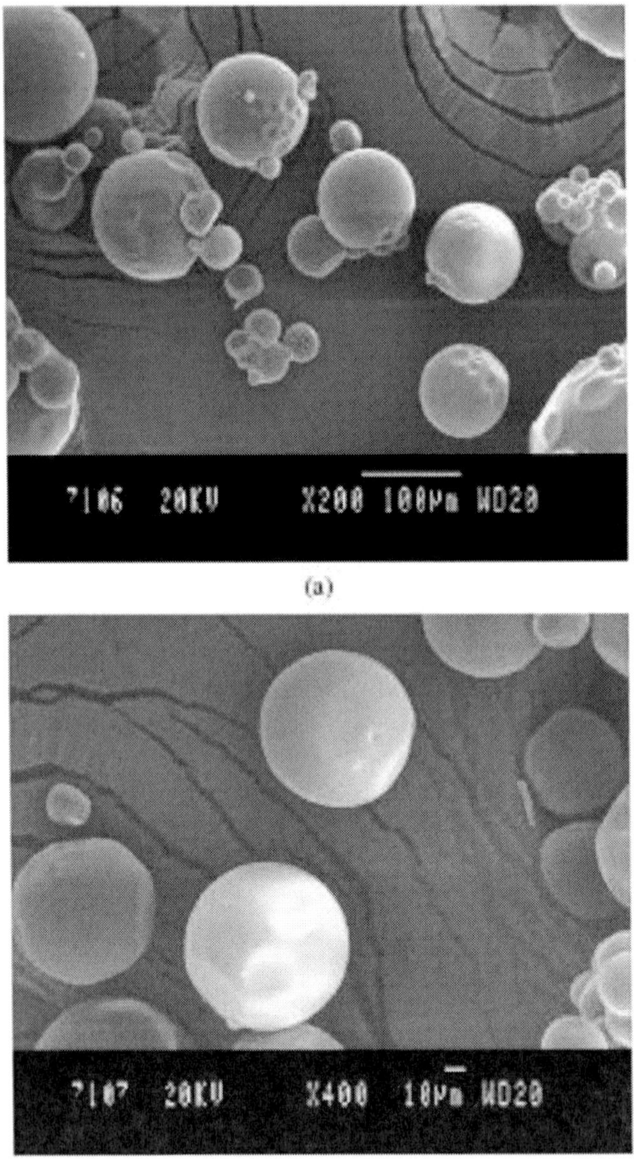

Figure 5. SEM images of 5-Fluorouracil loaded chitosan-co-pluronic F-127 microspheres with different magnification.

2.1.3. Chitosan Blended with other Macromolecules

Chitosan could also form polyelectrolyte (PEC) with many natural and synthetic materials. PECs of chitosan are different with covalent bonded chitosan derivatives. It is formed by ionic interaction as with ironically crosslinked networks. Chitosan PECs are formed between the positive charges of chitosan networks with negative charges of other molecules, usually macromolecules, in an aqueous solution [19-22]. Therefore, the preparation of a PEC with chitosan only requires a polyanionic polymer. The most commonly used polyanions to form a PEC with chitosan are polysaccharide with carboxylic groups such as alginate [23-25], chondroitin sulfate [26,27] and hyaluronic acid [28-31], proteins like collagen [32-34] and synthetic polymers like poly(L-lactide) [35]. As it is very important to maintain the equilibrium of positive and negative charges in PEC system, the change in pH, either acidic or basic changes, will cause the swelling and conformation changes. Also, the ionic strength and the temperature are important. The ionic interaction between the two polymers is not as strong as covalent binding, so the diffusion of drug molecules is dominated in the drug release mechanisms. But the difference of charge on the drug molecules will affect the diffusion efficiency.

The PEC of chitosan and alginate has been studied on the control release of nicardipine HCl, and both diffusion and relaxation of polymers are related to the drug release [36]. The chitosan/chondroitin sulfate PEC was studied as a colon delivery system and indomethacin was used as a model drug. In the study of drug release *in vitro*, the dependence of drug release on the concentration of chitosan was found. Also, the selective delivery of drugs to the colon could be achieved using this system [26,37].

As a carrier of a drug delivery system, chitosan has unique biodegradable and biocompatible advantages. Chitosan and its derivatives were proven to improve the loading and release of the poorly soluble drug. The chemical modification with either small or macro-molecules has potential application as a specific drug delivery system targeting different organs. The drug loading and controlled release is not only related to chitosan molecular weight, DD, but is also dependent on the charge density and the type of crosslinking. Various therapeutic agents such as anticancer, proteins, and anti-inflammatories have been studied with different chitosan derivatives.

3. Scaffolds for Tissue Regeneration

Tissue engineering is regarded as an ideal medical treatment to repair injured body parts and restore their functions by using regenerated tissues and artificial implants [38-40]. For regeneration of injured tissues, living cells, signal molecules and supporting materials are the fundamental elements. Basically, tissue engineering makes use of a polymer scaffold which is modified or fabricated to promote cell adhesion, migration, proliferation and differentiation[39]. The scaffolds serve as three-dimensional (3D) frameworks to support cell behaviors and functions. Chitosan and its derivatives are one of the most important biomaterials studied in tissue engineering, because they offer distinct physic-chemical and biological properties to help tissue regeneration. There are many benefits to using chitosan based scaffolds for cell study and tissue regeneration. These benefits include the relative rapid degradation by enzymes, non-toxicity, ease of development, existence of primary amines on the bulk surface, low inflammatory, and promotion of cell initial attachment and proliferation [4].

3.1. Surface Modified and Grafted Chitosan Scaffolds

The surface modification of chitosan for tissue engineering application is usually used to create biological recognition to cells. It is an important tool to design and define the cell-surface interaction. To introduce bioactive molecule on chitosan surfaces, many different methods have already been developed. They can be divided into three categories: chemical coupling, physical adsorption and ionic interaction [40,41]. The most commonly used chemical coupled bioactive molecules are cell adhesive peptide sequences, such as RGD (Arg-Gly-Asp) from fibronectin [42,43], YIGSR (Tyr-Ile-Gly-Ser-Arg) from laminin and IKVAV (Ile-Lys-Val-Ala-Val) also from laminin. Besides, in some cases, growth factors, like bone morphogenetic protein-2 (BMP-2), are also chemically coupled to the chitosan surface. But usually growth factors are physically entrapped into the chitosan scaffold to keep their bioactivity. BMP-2, fibroblast growth factor (FGF) and epidermal growth factor (EGF) [43] are generally encapsulated in the scaffold to promote cell adhesion, and proliferation [44]. Physical absorption or ionic interaction is a simple technique to achieve favorable properties and it is also a method used to minimize the damage to bioactive molecules [45,46].

Recently, more and more research interest was devoted to chitosan derivatives to serve as scaffolds for tissue regenerations. Either chemical modification of primary amine on chitosan or chitosan blending or grafting with other natural or synthetic materials. By choosing different side chains or blending materials, modifications will improve chitosan solubility and widen the field of its potential applications. Many preliminary studies have already shown that the chemical modification of chitosan will improve its potential and maintain its advantages like biocompatibility, biodegradability and anti-inflammatory [46,47]. Except chemical modified chitosan, blending and copolymers of chitosan with alginate [48], collagen [49], gelatin [50-52] and PLA [53] are all promising materials for tissue regeneration.

The chitosan scaffold was suggested as an attractive candidate as a bone regenerative material due to its proper biological and physical properties. It served as a temporary skeleton to help the regeneration of lost tissues. The advantages of chitosan scaffolds for bone regeneration include high and interconnected porous structure (Figure 6), osteoconductivity and ability to enhance bone formation [39]. Bone composes of organic composition like collagen type I, glycosaminoglycans (GAGs), proteoglycans and glycoprotein, and inorganic composition of hydroxyapatite crystal [39]. Chitosan is believed to support the function and expression of extracellular matrix (ECM) protein in human osteoblasts and chondrocytes[54]. Recently, differentiation of bone marrow stromal cells (BMSCs) on poly(lactide-co-glycolide) (PLGA)/chitosan scaffolds with neuron growth factor (NGF) were investigated, and differentiation toward osteoblasts with apparent deposition of calcium was found on the scaffold. With the introduction of NGF, mature neurons yielded [55]. The mechanical properties of the bone regenerative scaffolds are important to transmit mechanical force and form matrix mineralization [39]. Recently, the combination of chitosan with alginate [56] and PLGA [55] all showed improved mechanical properties. It suggested that the secondary composition to provide mechanical properties is an alternative strategy to improve bone regeneration.

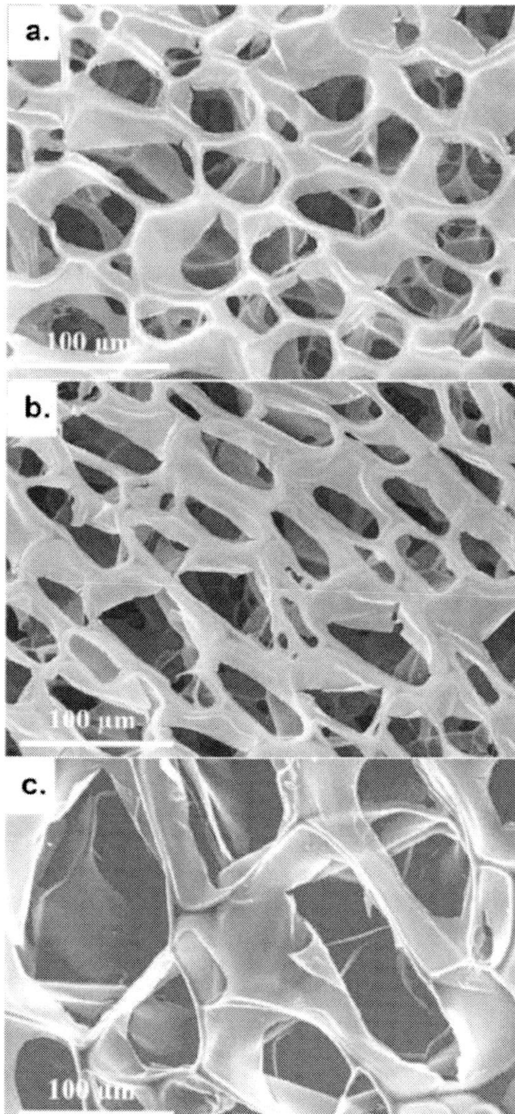

Figure 6. Chitosan scaffolds with porous structure: a 95% DD; b 88% DD; c 70% DD. (Adapted from www.maney.co.uk/journals/mst).

Cartilage regeneration requires the isolation of articular chondrocytes and their precursor cells for *in vitro* cultivation and then implantation into the joint [57]. Chitosan-alginate-hyaluronan complexes with RGD and chondrocytes could partially repair rabbit knee cartilage defects. Collagen-chitosan-GAG scaffold with transforming growth factor β1 (TGF-β1) was reported to promote cartilage regeneration. Chitosan-poly (N-isopropylacrylamide) (PNIPAAm) hydrogel could help mesenchymal stem cells (MSCs) differentiate to Chondrocytes [4,57].

Other tissue regenerations like nerve, blood vessel, and skin on chitosan and its derivates have also been studied. Chitosan was studied as a candidate material for nerve regeneration because of its distinct properties including biodegradability, biocompatibility, antibacterial activity and antitumor. Recently, there have been several studies using chitosan film, tube or fiber for peripheral nerve regeneration. Every structure of chitosan derivatives showed good capability of promoting cell adhesion, migration and proliferation [46,58]. Recent studies indicated that chitosan fiber with microporous structures could be used to deliver recombinant human vascular endothelial growth factor (rhVEGF) to induce the new vessels [59].

3.2. Chitosan Scaffolds with Microcapsules to Control Drug Release

Encapsulation of microsphere in polymeric scaffold is a strategy used to combine drug delivery into an artificial scaffold. Usually, it was used for the controlled release of growth factors. Several studies have been done with this strategy and most of them combined chitosan derivatives with PLGA microsphere. EGF encapsulated PLGA 50/50 microspheres were reported to be entrapped into a chitosan tube and the EGF bioactivity could last at least 14 days by the assessment of neurosphere forming bioassay [60]. TGF-β1 for cartilage regeneration was loaded into chitosan microspheres and then a three-composition collagen-chitosan-GAS scaffold was generated with encapsulated TGF-β1 [61]. The culture of rabbit articular chondrocytes showed an increase in the cell proliferation rate and GAG production in the scaffold with TGF-β1.

3.3. Chitosan Scaffolds with other Composition

Bioactive ceramic materials could be combined with the application of chitosan for bone repair and regeneration. The introduction of a bioceramic is due to its similar chemical composition to natural bone. However, to overcome its inherent hardness, which could cause the secondary damage to the bone, chitosan with its good biological properties was introduced in the system to mimic natural bone structure and composition. There was a study showing the development of the platelet-derived growth factor (PDGF) delivery system with a chitosan/tricalcium phosphate (TCP) sponge. By assessing the delivery system with a rat calvarial defect model, researchers found the carrier of chitosan/TCP enhanced bone regeneration, and in the space of degraded carriers, newly formed bone without connective tissue encapsulation was found [62]. Very recently, a nanocomposite of bioactive glass ceramic nanoparticles with chitosan scaffold was developed and demonstrated a potential scaffold for tissue engineering applications [63].

4. Chitosan for Hemorrhage Control

Chitosan has well-documented bioactivities, such as antibacterial [64] and hemostatic properties [65-67]. Chitosan-based injectable hydrogels have been reported [68] and used as a biological adhesive [69]. The hemostatic potential of chitosan has long been recognized [66,70], however, its hemostatic potency reported in the literature varies significantly which could be due to the differences in physiochemical properties [67].

In this section, we first review the literature on the studies of hemostatic properties of chitosan in different physical and chemical forms, followed by an overview of the mechanism of actions on the blood coagulation system. Our own work on thromboelastographic evaluation of chitosan is discussed as well.

Table 1. Comparison of different forms of chitosan in the study of its effects on blood coagulation

Physical form	Source	Mw and DD	Hemostatic efficacy	Reference
Solid film	Crab shell	80%	Reduced postsurgical bleeding after visceral and parietal peritoneal abrasion in rabbit	[74]
Solid films by solution casting		97%	Increased platelet adhesion	[75]
Heat-compressed spongy sheet with backing, e.g., HemCon™	Crab shell	>100,000, >85%	Reduced severe bleeding in animal models aand human trials	[76]
Sponge with a density of 0.053 g/cm	Shell of fly-larva	3.10×10^5, 68.5%	Achieved 100% hemostasis and survival rate in a rat hepatic hemorrhage model,	[77]
Power with a size of ~0.08 mm		97%	Reduced whole blood clottinging time	[75]
Powder (Celox™)	A propriety blend of different types of chitosan		Reduced severe bleeding in animal models	[78]
Nanoparticles in saline		90%	No effects as measured by thromboelastography	[79]
Microspheres (40-100 μm)	Crab shell	MW 1,380,000, 80%	In vitro assay	[80]
Crosslinked filaments	Crab shell	24×10^4, >99%	Induced blood coagula	[81]
Chitosan suspension in NaCl solution		1.86×10^6, 90.3%; 2.24×10^6, 71.8%; 3.80×10^5, 92.1%; 5.19×10^5, 64.4%	Promoted platelet adhesion	[71]
Solution (2%) in acetic acid	Prawn shell	126,000	Decreased whole-blood clottinging time by about 40% of normal	[66]
Solution (2 mg/ml) in 0.2% glacial acetic acid		800 to 1,500 k	Reduced 43% lingual bleeding times	[82]
Solution (2 mg/ml) in acetic acid		2000-3000 saccharide units	Stopped bleeding in cerebral cortical lesions in cats	[83]
Solution (1 wt%) in saline		1150	No effects as measured by thromboelastography	[84]
Solution (2 mg/ml) in 0.2% glacial acetic acid		2.24×10^6, 71.8%; 9.35×10^5, 66.4%; 5.19×10^6, 64.4%;	Erythrocyte aggregation and deformation No effects on coagulation factors	[71]

4.1. Physically Modified Chitosan

Chitosan may be manufactured in various physical forms including solutions; solid sheets/films, filament; powder, fiber, and hydrogel, all of which have shown hemostatic potential [67]. Unfortunately, the characteristics of chitosan were not completely described in many of the hemostatic studies as summarized in Table 1, which makes it difficult to define criteria for comparison. In addition to physical structures, the molecular weight and deacetylation degree may affect hemostasis [71]. For example, the modified whole blood clottingting time assay and the platelet coagulation test suggested that the oligochitosan obtained by sodium nitrite-induced depolymerization and methanol fractionation are much less procoagulant than their parental chitosan [72]. The source where chitosan was extracted also affects its properties [73]. On the other hand, the utilization and effectiveness of the hemostatic agents depend on the severity and site of bleeding. There is no universal animal model for comparing the efficacy of all promising hemostatic materials.

The use of chitin and chitosan for solid wound dressings has been well documented [85-87]. For example, chitin-chitosan dual-layer materials were formulated as wound dressings for treatment of mustard burns [85]. Only those with indication for hemorrhage control will be discussed herein.

The Chitosan solid dressing was normally prepared from its aqueous solution in 2% acetic acid by either a freeze- or air-drying method, followed by successive washings of water. It could be treated in an aqueous 20% ammonia solution and freeze dried again. The dressing showed similar hemostatic efficacy to a commercial collagen dressing in a rabbit cervical vein wound, but slower degradation and greater tissue response after subcutaneous implantation in rabbits [88]. The chitosan membrane was also found to reduce postsurgical bleeding after the abrasion of liver surface in the rabbit model [74].

A number of commercial products based on solid chitosan have been studied for hemostasis purpose: Clo-Sur PAD™ (Scion Cardio-Vascular, Miami, Florida) and Chito-Seal™ (Abbott Vascular Devices, Redwood City, CA). The former is composed entirely of chitosan, whereas the Chito-Seal™ pad is backed with cellulose coating. Chito-Seal™ has a tubular structure, and the diameters of the tubes are much larger (~15 μm in diameter). The in vitro thromboelastographic study showed raw chitosan and its medical products could have different hemostatic abilities as indicated by relative onset times for fibrin polymerization and clot stiffness in platelet-rich plasma [89,90]. In one clinical study, Clo-Sur PAD™ has been used to treat hemorrhage at arterial puncture sites for cardiac catheterization and found to reduce time to hemostasis, time to ambulation, and to decrease patient discomfort, however, a high technical failure rate of 19% requires further improvement before routine use can be recommended [91]. Another clinical study in this setting compared the Clo-Sur PAD™ with Chito-Seal™, and found that both pads reduced the average time to hemostasis by a small amount that did not translate into a reduction in overall bed rest time [92]. However, in a coagulopathic swine spleen-bleeding animal model, Clo-Sur PAD™ and Chito-Seal™ failed to control the bleeding [93].

The chitosan-based hemostatic products have also been compared with other commercial topical hemostats: poly-N-acetyl glucosamine (p-GlcNAc) based Syvek patch, collagen-based Actifoam®, cellulose-based Surgicel®, and fibrinogen/thrombin-based Bolheal® fibrin glue, in a swine model of splenic hemorrhage, as shown in Table 2, using different end points in terms of number of compressions and time to hemostasis percentage [93-96]. The thromboelastography (TEG) measurement was made in platelet-poor and platelet-rich plasma mixed with the chitosan medical products and raw materials (Vanson, Redmond, WA) in saline at 1 mg/mL and 0.2 M $CaCl_2$ solution using Haemoscope model 5000 and the data showed reduced R time by the Clo-Sur in the plasma compared with no material control [97]. This is in agreement with a further TEG study reported by Fischer et al. who demonstrated that a variety of glucosamine-based biopolymers including the marine-derived poly-N-acetyl glucosamine fiber, chitin, chitosan, could decrease R and increase MA in platelet-rich plasma [89]. These studies demonstrated that TEG can be used to compare *in vitro* hemostatic properties of biomaterials in different forms.

Although the hemostatic property of chitosan has been well known, its potential was not fully explored until recently when a novel process was invented to construct chitosan wound dressings with layered porous structures [76]. The process involved freeze-drying an aqueous solution of chitosan, and heating, compression and gamma –irradiation of the resulting sponge film. Optimal structures were open-porous consisting of uniform interconnected pores of about 50 μm in diameter or lamellar and hexagonal structures normal to the plane of cooling. These structures yielded a large specific surface area greater than 500 cm^2/g. Medical foam adhesive backing was then attached to the top surface of the film for easy handling and uniform application of pressure at a bleeding site. A product called HemCon™ (HemCon Medical Technologies Inc, Portland, OR) was developed based on the invention. Pusateri et al. demonstrated that the chitosan-based hemostatic dressing was capable of controlling severe parenchymal and large venous hemorrhage in a swine model of severe liver injury [98]. The chitosan dressing reduced blood loss (264 ml vs. 2,879ml, $p<0.001$) and increased survival (87.5% vs. 28.6%, $p=0.004$). The dressing is stable, rugged, light and relatively inexpensive, but is rigid and difficult to apply over a complex wound. In addition, the inconsistent performance of the dressing for hemorrhage control was noticed [99] perhaps due to batch-to-batch variation, a common problem seen in biopolymer production. On the other hand, physical parameters, such as degree of acetylation and molecular weight of chitosan, likely affected its hemostatic properties [72,73,100], and needs to be investigated. The dressing seemed to lose its tissue adhesiveness and eventually failed to maintain hemostasis for more than 2 h (time-limited efficacy) [99]. In a high-flow arterial wound model where a proximal arterial injury was created with 2.7-mm vascular punches in both femoral arteries of anesthetized swine, the HemCon™ maintained 100% and 84% hemostasis at 30 min and at 240 min, respectively, after application of one quarter of a standard production 10 ×10-cm chitosan-based dressing (HemCon™ dressing) with two folded 12-ply squares of 7.6 × 7.6-cm gauze and one roll of conforming gauze dressing compared with 21% and 7% by a 48-ply gauze dressing (48PG) made of two folded 12-ply squares of 7.6 × 7.6-cm gauze (Medline Industries, Inc., Mundelein, IL) and one-roll of conforming gauze dressing (Kerlix Lite; Kendall, Mansfield, MA) [101].

Advances in hemostatic materials have been made in the past few years. There is significant interest in using these hemorrhage control materials for battlefield injuries [102]. The chitosan-based HemCon™ has been evaluated in different severe animal bleeding models in comparison with other advanced hemostatic materials, such as QuikClot® and fibrin sealant dressing in lethal extremity arterial hemorrhages in swine [103,104], QuikClot® and its analogs in a swine model of lethal groin injury [105], On the other hand, clinical data have shown the efficacy of the product in pre-hospital settings for both military [106] and civilian uses [107]. The product is effective in hemorrhage control in various hospital procedures, such as hemodialysis [108], and oral surgery [109]. Furthermore, a new version of the product based on a dual-sided, flexible roll (Chitoflex™, HemCon Inc.) showed better performance in a lethal groin injury model of goats [110]. The product has also been used in as an effective adjunct for hemostasis in various surgical procedures, such as dacryocystorhinostomy surgery [111].

Animal and human studies have demonstrated the feasibility of using HemCon™ as an internal dressing to control hemorrhage and urinary leakage by sealing off the parenchymal wound following a laparoscopic partial nephrectomy (LPN) in a porcine model [112], laparoscopic repair of inferior vena caval injury [113], and pelvic battlefield trauma [114]. On the other hand, new hemostatic products have emerged and may soon replace the current chitosan dressing for hemorrhage control on the battlefield [115,116]

In addition to the films, a hemostatic product based on chitosan powders has been developed. Celox™ (MedTrade Products Ltd, Cheshire, UK) is a proprietary preparation of chitosan granules made from more than one type of chitosan with a large surface area [78]. It stops blood loss by forming a gel-like clot that sticks well to damaged tissues to plug the bleeding site as the material binds to the surface of red blood cells. The efficacy has been confirmed in swine models of a grade V hepatic injury [98], and a complex groin injury with transaction of the femoral vessels [78] or puncture of the extremity artery [101,116]. In addition, in a severe groin puncture injury model in swine, Celox™ showed superior efficacy in improving survival, hemostasis, and maintenance of mean arterial pressure than HemCon™ and Chitoflex™ [117]. Case reports have described its lifesaving use in patients undergoing cardiothoracic surgery where conventional techniques for hemostasis had failed [118]. However, in a lethal groin injury model in goat, the dual-sided, flexible, roll type outperformed the one-sided wafer and chitosan powder [110].

New chitosan dressings continue to emerge and be tested in severe hemorrhage of a hypothermic coagulopathic grade V liver injury model [119]. Compared with standard packing, the chitosan treatment led to significantly less blood loss, mean fluid resuscitative volume, and greater resuscitation mean arterial pressure. Hemostasis was achieved on average 5.2 minutes following modified chitosan and never achieved with the standard packing. At 1 hour post injury, all treatment animals survived compared with half of controls.

Table 2. Comparison of hemostatic biomaterials in swine models of splenic hemorrhage [93,94,96]

Commercial dressings	Average number of compression to hemostasis	Hemostasis achieved (%)
Syvek Patch	3	100
Clo-Sur Pad	10	0
Chito-Seal	9	25
Gauze	8	50
p-GlcNAc	1.4	91*
Actifoam	>3.6	25*
Surgicel	4.2	17*

* The ability to achieve hemostasis after two applications with the cycle of manual pressure for 20 sec followed by observation for 2 min.

Chitosan derived from a fly-larva shell was processed to a sponge for internal hemorrhage control [77]. The polymer was prepared by a modified gradual-base extraction associated with a freeze-drying method. In a rat hepatic hemorrhage model where a reproducible, 8x 8x5mm3 excision was created on the left lateral hepatic lobe, free bleeding was allowed for 10 s, the chitosan sponge was the most effective agent, controlling 90% of hemorrhage (9/10) during the first 1-min compression cycle and the rest 10% (1/10) during the second cycle. Both the gelatin sponge (GS) and the oxidized cellulose (OC) required significantly more compression cycles to achieve acute hemostasis ($p < 0.001$). GS even failed in achieving initial hemostasis twice. During a 4 h post treatment survival study, CS treatment led to 100% survival, while GS and OC treatment resulted in 25% (2/8) and 10% (1/10) mortality, respectively. Two GS treatments failed at about 65 and 95 min after operation, and one OC treatment failed at about 130 min postoperatively, both due to the same issue of re-bleeding. The 8-week follow-up study indicated that all other rats in each group were in good recovery, without hemorrhage or infection symptoms.

Chitosan filaments were formed by wet spinning chitosan solution to an aqueous 5% ammonia solution containing 40-43% ammonium sulfate, followed by washing and air drying [81]. When inserted into the lumen of a dog's peripheral veins, thick blood coagula appeared at 2 h.

Microcrystalline chitosan sealant was delivered via an arterial sheath at the completion of catheterization to improve hemostasis [120]. The sealant contained 18 wt% of calcium chloride chelated with 5.41% of microcrystalline chitosan with an average molecular weight of 4.4×10^4, a deacetylation degree of 74% and a water retention value of 1030%. The sealing agent significantly reduced manual compression time from 16.7 to 6.1 minutes to achieve hemostasis in heparinized dogs.

Liquid chitosan enhanced blood coagulation in vitro assays [66] and hemostasis in several animal studies involving bleeding from small vessels [82,83,121]. When added to whole blood, the chitosan solution brought down the whole-blood clotting time by about 40% of normal through its induced agglutination of erythrocytes due to the interaction of the positively charged chitosan with receptors containing muraminic acid residues on the cell surface. When used in a concentration of 2 mg/mL in 0.2% glacial acetic acid in lingual incisions, chitosan effectively decreased intraoral bleeding time by 43% compared to the control solution which consisted of 0.2% glacial acetic acid without chitosan in a heparinized rabbit model [82]. In cerebral cortical lesions in cats, chitosan solution (2 mg/ml) in glacial acetic acid was applied to the bleeding cortical defect and covered with a saline-soaked cottonoid. The hemostasis was achieved in an average of about 4 min with no toxic effects on the brain tissue [83].

Microcrystalline chitosan sealant was delivered via an arterial sheath at the completion of catheterization to improve hemostasis [120]. The sealant contained 18 wt% of calcium chloride chelated with 5.41% of microcrystalline chitosan with an average molecular weight of 4.4×10^4, a deacetylation degree of 74% and a water retention value of 1030%. The sealing agent significantly reduced manual compression time from 16.7 to 6.1 minutes to achieve hemostasis in heparinized dogs.

Chitosan films have been used as tissue adhesives in laser tissue welding for wound closure [122] and nerve repair [123].

4.2. Chitosan Coatings

Chitosan attached onto different substrates through physical or chemical bonds, leading to both pro- and anti-coagulant effects. For example, chitosan was immobilized onto polypropylene non-woven fabric [124]. The fabric was first grafted by poly(acrylic acid) using antenna-coupling microwave plasma and then was immobilized by chitosan using 1-ethyl-3-(3-dimethylaminopropyl)carbodiimide for activating -COOH functional groups on PAA to react with -NH_2 on chitosan. Blood clotting test showed increased platelet, red and white blood cells adhesion on the chitosan modified surface in comparison with non-treated and PAA-grafted surfaces. The adhesion is due to the interaction between positively charged chitosan and blood cells. Similar surface chemistries were used to immobilize chitosan onto thermoplastic polyurethane, which resulted in reduced coagulation activities [125]. This may be due to the difference in chitosan: surface-immobilized chitosan was partially N-carboxypropylated, compromising its interaction with platelets. So far, this approach has not made any significant impact on the development of hemostatic materials for hemorrhage.

A chitosan thin membrane (2-6 μm) coated on a segment of silk filament suture and inserted into the lumen of a dog's peripheral veins resulted in the formation of an intense thick blood coagulum on the material, but less or no blood coagulum formation after further treatment with acetic anhydride or n-hexanoic anhydride [126]. Chitosan coating on other polymer filaments (e.g., chitin, chitin-collagen, chitin-silk fibroin) was also thrombogenic and hemostatic [127].

Chitosan (Mw 100,000, degree of deacetylation 80%) was coated on poly(lactic acid) (PLLA) braided wires through a solution dipping method [128]. The coated wires showed increased blood coagulation due to chitosan-induced platelet adhesion and erythrocyte aggregation.

Chitosan (Mw 600 000, degree of deacetylation 80%) was coated on fibrous PLLA and bicomponent PLLA/poly(ethylene glycol) mats by immersion of the mats in the chitosan solution [129]. SEM micrographs confirmed the formation of a thin chitosan coating between some of the fibers with a 20 nm mean thickness of the film. In addition, chitosan on PLLA mats led to significant adhesion of erythrocytes, while a triple coating of chitosan on the mats caused agglutination, deformation, and aggregation of the erythrocytes and platelets, thus indicating high hemostatic activities of the obtained hybrid fibrous material. The obtained results are in good agreement with data on the thrombogenicity of chitosan-coated filament composites [126].

A chitosan layer was assembled on a quartz crystal through a layer-by-layer process with dextran exhibited procoagulant behavior against whole human blood when the substrate was immersed into blood for 30 min at 37°C [130]. It is noteworthy that the procoagulant activity was affected by the underlying layer of the assembly as the chitosan layer assembly with heparin showed anticoagulant activity.

Chemically grafted chitosan on fabrics (e.g., cotton gauze, print cloth) was prepared with a citric acid grafting onto cellulose by pad-dry-cure using 11.9% citric acid and 4% sodium hypophosphite, on the USP Type VII cotton gauze sponges; dried at 85°C for 5 min and cured at 160°C for 3 min [131]. The positive charge of the chitosan chains grafted on the fabric was expected to promote the aggregation of platelets locally, and promote gelling of blood to inhibit wicking into the fabric.

4.3. Chemically Modified Chitosan

Chitosan has been modified through its hydroxyl and amino groups for biomedical applications [132]; however, the chemical modification generally reduced hemostatic effects and thus had limited use for hemorrhage control.

Chitosan (Mw 800–1000 kDa, with an 80% degree of deacetylation) was modified to introduce azide and lactose moieties for use as a biological adhesive to stop bleeding and accelerate the wound healing process [69,133]. Specifically, the tissue adhesive was prepared from chitosan with the p-azidebenzoic acid and lactobionic acid moieties which were introduced through condensation reactions. The former moiety resulted in chitosan to become photocrosslinkable and the latter provided water solubility at neutral pH. In an animal model where the tails of anesthetized mice were cut off, 30 mg/ml of the photocrosslinkable chitosan aqueous solutions completely stopped the bleeding by UV-irradiation with finger pressure within 30 seconds, compared with 3 minutes of fibrin glue . (Beriplast P, Hoechst-Marion-Roussel, Tokyo, Japan) determined in a similar way after equal volumes of thrombin and fibrinogen solutions were mixed and applied onto the wounds. The photocrosslinkable chitosan was also used to reduce bleeding in endoscopic mucosal resection where 2.5% polymer solution in normal saline (0.3 mL) was injected into the submucosal layer of the rat stomach converted to a hydrogel (like a soft rubber) by 30-s ultraviolet irradiation [134]. In the Az-CH-LA–treated (with UV irradiation) group, blood loss was extremely low (about 7% of that seen in saline injected controls).

Chemical substitution of amino groups of chitosan has normally reduced its hemostatic properties. For example, sulfonated chitosan with varied degrees of deacetylation exhibited anticoagulant effects [135]. The principal mechanism for the anticoagulant activity is mediated through heparin cofactor II and is dependent on polysaccharide molecular weight [136]. N-acylatation of chitosan with various carboxylic anhydrides also decreased blood coagulation as indicated by increased time for the initiation and termination of coagulation and by reduced thrombus formation [137]. On the other hand, N,O-carboxymethylchitosan decreased whole blood clotting time and promoted platelet adhesion and aggregation in vitro [75].

Chitosan films were treated in saturated NO_2-glacial acetic acid solution, resulting in oxidization of the hydromethyl groups on the film surface to carboxyl groups and less coagulation of blood cells [138]. The reduced coagulation was ascribed to the change in surface charges from positive NH3 to negative COO- . However, the surface morphology should be considered as it became rough after the modification as shown by SEM.

Chitosan was N-acylated with butanoic, hexanoic and benzoic anhydride, and made into nanoparticles with a diameter of approximately 350 nm [79]. To evaluate their blood compatibility for intravenous injection to detoxify drugs, TEG was conducted as the only coagulation assay in the study where the acylated nanoparticles were suspended in a 30- μL saline solution and mixed with 30-μL $CaCl_2$ solution and 240 μL citrated human blood in a 3000C TEG® (Haemoscope Co, Niles, IL, USA) cup at 37°C. No significant changes in R, K, α and MA were observed compared to saline-treated control, illustrating good blood compatibility, which was attributed to the balance in the hydrophobic and hydrophilic properties. Other TEG parameters (e.g., LY30) could be included to understand any fibrinolytic effects.

Lee et al. used TEG to assess effects of water-soluble oligochitosan and its derivatives on blood coagulation [84]. Oligochitosan was modified with benzenesulfonyl chloride or dinitrobenzenesulfonyl chloride, which may be injected intravenously for drug detoxification of amitriptyline, a widely prescribed antidepressant frequently used for committing suicide. The electron deficient aromatic rings on the oligochitosan derivatives would allow selective and rapid adsorption of the drug through π-π complexation. Apparently, the blood compatibility is essential for the application and thus needs to be tested. Specifically, aliquots of 30-μL 1 wt% sample solution in saline, 30-μL $CaCl_2$ solution and 240-μL blood were mixed in a TEG cup at 37°C and the coagulation was measured using a model 3000C TEG machine (Haemoscope Co, Skokie, IL, USA). Four principle parameters: R, K, α and MA were analyzed by TEG, illustrating no significant difference in the coagulation of blood treated respectively with saline solution, initial and modified oligochitosan at the concentration of 1.0 and 0.1 wt% in the saline solution. This is in agreement with animal studies where the oligochitosan derivatives did not manifest acute cardiotoxicity and weight gain or loss for 4 weeks, when intravenously injected into rats via a tail vein. The study illustrated the capability of TEG itself as a reliable tool for evaluation of biomaterials' hemocompatibility.

Crosslinked chitosan microspheres (40–100 μm) were prepared using methods of emulsification and ethanol coagulant and subsequently crosslinked by formaldehyde and glutaraldehyde and treated with H_2O_2, which could shorten the clotting time and induce the adhesion and activation of platelets evaluated by dynamic blood clotting, platelet adhesion and activation, erythrocyte adhesion, hemolysis, and protein absorption assays [80].

4.4. Chitosan Composites

Chitosan has been combined with a variety of other biomaterials ranging from polymers to ceramics [139]. There are a few studies involving hemorrhage control. A chitosan-based dressing with improved hemostatic properties was prepared by incorporating polyphosphate [140]. The negatively charged phosphate polymer is a procoagulant agent present in dense granules of human platelets [141] and can form a polyelectrolyte complex with chitosan [142]. The hemostatic efficacy was only demonstrated by in vitro coagulation assays.

TraumaStat (OreMedix, Lebanon, OR) is a new product that combines chitosan with silica and polyethylene to form a dressing as conformable as gauze, with a large surface area (110 m^2/g) [143]. Silica acts as a potent activator of the intrinsic clotting cascade; chitosan forms a mucoadhesive component sealing the damaged tissue and holding the silica in place for close contact with blood; and polyethylene provides structure integrity and conformability. The large surface area would allow better interaction with blood. In swine femoral vessel transection models, TraumaStat outperformed standard gauze and chitosan-based HemCon [143] and Chitoflex (HemCon, Inc.) dressings [144].

By cross-linking , a chitosan/dextran gel was obtained and significantly improved hemostasis after endoscopic sinus surgery in a sheep model of chronic rhinosinusitis, with an average time to hemostasis of 4.09 min compared with 6.57 min in control (no treatment) [145].

4.5. Hemostatic Mechanisms

A number of studies have focused on understanding the hemostatic mechanisms of chitosan. Various mechanisms may contribute to its hemostatic properties and be postulated to be via vasoconstriction and the rapid mobilization of red blood cells, clotting factors, and platelets to the site of the injury, but the exact hemostatic mechanism of chitosan is still under study [146]. In one study chitosan (Mw=80,000 degree>80%) particles with 2.8 and 6.2 mm were suspended in a phosphate-buffered solution (PBS, pH 7.2) at a concentration of 30 mg/ml reduced blood coagulation time and enhanced platelet aggregation [146]. Another study showed that chitosan (Mw=50,000, deacetylation degree >90%) significantly promoted platelet adhesion and aggregation through enhancing the expression of platelet glycoprotein IIb/IIIa complex on platelet membrane surfaces and increasing the intracellular calcium level in platelets [147].

In wound, damaged vessel wall or ruptured atherosclerotic plaque, platelet adherence to exposed subendothelial tissue such as collagen, is a critical step in hemostasis or thrombosis. The initial adhesion can generate intracellular signals responsible for the activation of GPIIb/IIIa and release of thromboxane A_2, which further promotes platelet spreading, thus strengthening the stability of adhesion. Chitosan significantly enhances both platelet adhesion and aggregation, which may account for the interaction of platelets with damaged tissues and promoting the wound healing effect of chitosan [147].

The in vitro study also evaluated the effects of chitosan (deacetylation degree 97%) and its derivatives (N-acetylated chitosan (PNAC) 56%, N,O-carboxymethylchitosan (NOCC), N-sulfated chitosan, N-(2-hydroxy) propyl-3-trimethylammonium chitosan chloride) on human blood coagulation and platelet activation [75]. It was revealed that chitosan, NOCC significantly decreased whole blood clotting time with respect to that of the pure whole blood (a blank control), while PNAC and NOCC significantly lowered plasma recalcification time. In the platelet adhesion and activation studies of chitosan, PNAC and NOCC, NOCC attracted and activated the platelets most effectively, probably because of its carboxyl functional group and moderate hydrophilic surface.

Blood protein adsorption on approximately 10 nm thick chitosan film prepared by glutaraldehyde crosslinking of chitosan on aminopropyltriethoxysilane (APTES) coated silicon showed a weak activation of the intrinsic pathway of coagulation, but large amounts of fibrinogen and other plasma proteins bound to the chitosan, resulting in platelet binding and activation via GPIIb/IIIa to the adsorbed fibrinogen, a fact that may explain the previously observed procoagulant behavior of chitosan [148].

Scanning and transmission electron microscopy revealed that red blood cells bound to chitosan particles and became cup-shaped to a lesser extent than β-poly-N-acetyl glucosamine derived from microalgal sources [149].

4.6. Thromboelastographic Study of Chitosan

To develop an effective hemostatic agent for the treatment of hemorrhage, the leading cause of death on the battlefield and the second leading cause of civilian trauma death [150,151], we evaluated the hemostatic effects of different types of chitosan using a computerized TEG® Hemostasis System 5000 (Haemoscope Corporation, Niles, IL). Specifically, a powder material was mixed with 340-μL citrated human blood from healthy volunteers in a sampling cup and blood coagulation was initiated by adding 20 μL of 0.2 M calcium chloride. To account for intrinsic differences in blood drawn from different donors and/or at different times to exclude extraneous effects, the percent difference for the TEG® variables (e.g., onset of clot formation R, time to a specific level of clot strength K, rapidity of fibrin cross-linking α, and maximal clot strength MA) between a biomaterial-dosed sample and a blank from the same blood draw was calculated to quantitatively determine the biomaterial's effects on blood coagulation. The sign (+/-) denotes either positive or negative effects on blood coagulation. For example, negative changes in R, K and positive changes in α, MA indicate an enhancement of clot formation and strength. The magnitude suggests the extent of the effects. Although Figure 7 shows the certain hemostatic potency of the material, the percentage difference in the TEG® parameters from blank was not profound and increased with its amount from 0.5 to 2.0 mg, which is consistent with others' TEG results [116,152]. This may be because the material does not set off the normal cascade and its mode of action relies on adhesion to blood vessel, which cannot be measured in the TEG.

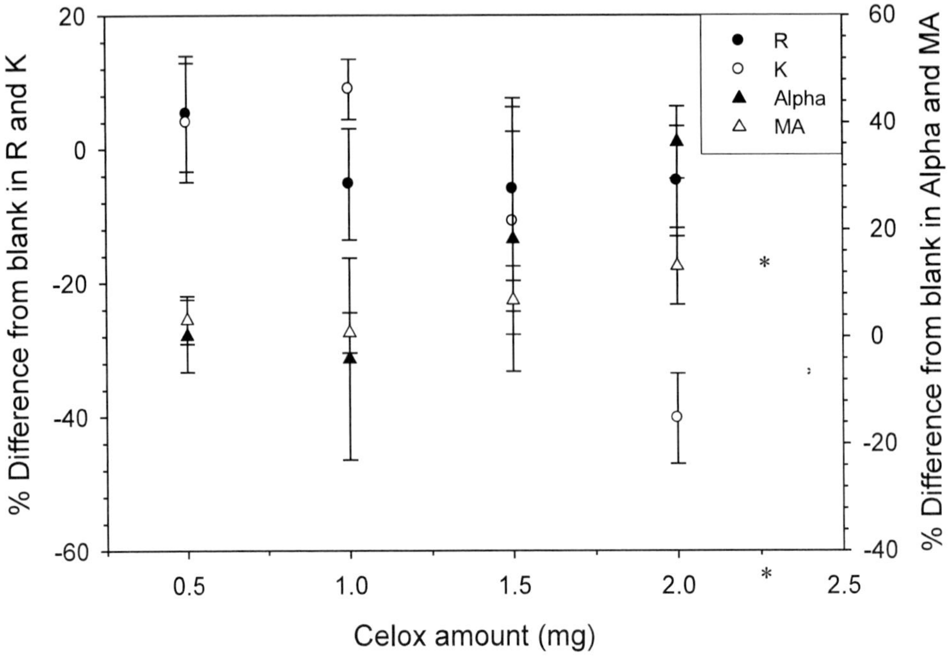

Figure 7. Plot of TEG® variables versus the amount of Celox mixed with blood. Data are expressed as mean ± SD (n=3 except n=5 for 0.5 mg). The powder material was mixed with 340-μL citrated human blood from healthy volunteers in a sampling cup and blood coagulation was initiated by adding 20 μL of 0.2 M calcium chloride. The percent difference for the TEG® variables between a material-dosed sample and a blank from the same blood draw was calculated to quantitatively determine the material's effects on blood coagulation. * Significant difference from 0.5 mg, respectively ($p<0.05$).

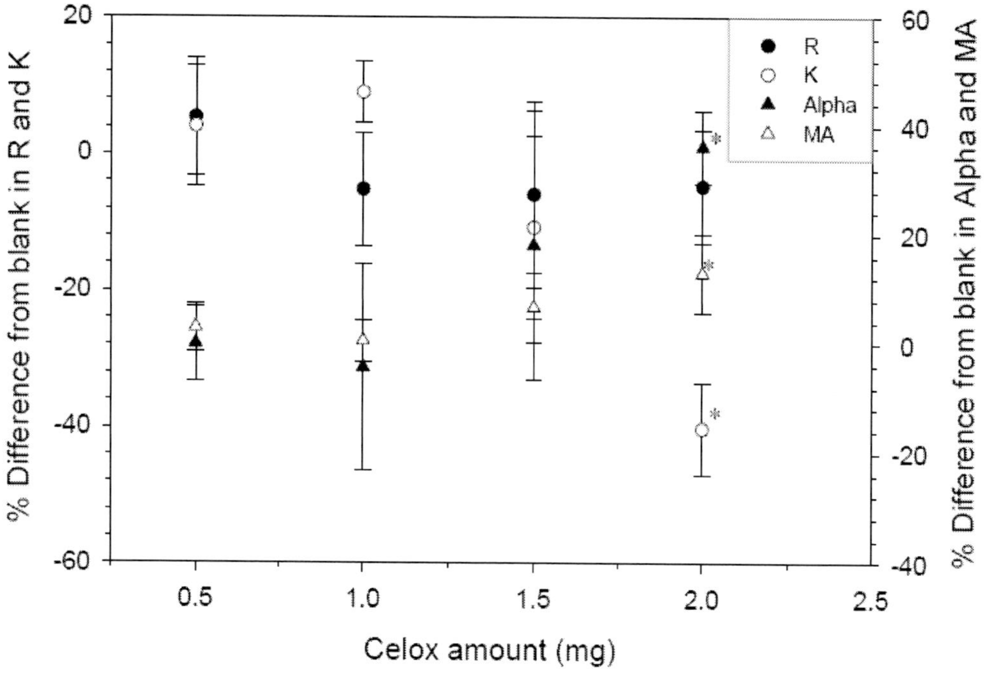

Figure 8. Effects of polymer and ceramic solid materials (1 mg) on blood coagulation. Data are expressed as mean ± SD (n=3). *, + Significant difference from blank and Celox, respectively ($p<0.05$).

We also used the TEG method to study the hemostatic properties of chitosan from different natural resources which would lead to different structures and properties [153]. Three types of chitosan were evaluated in comparison with Celox™ and QuikClot® in our study (Figure 8). They are all medical grades with one from squid and the other two from crustacean shells. Squid chitosan was derived from giant squid cartilage with a deacetylation degree higher than 90% and molecular weight higher than 1,000,000 (Arabio Co., Ltd., Guui-Dong, Gwangin-Gu, Korea). It has a predominant parallel (i.e., β) structure, which may show higher reactivity as a result of weak intermolecular forces, compared to the one in an α form [154]. In contrast, the other chitosans were originated from crustacean shells with less than 7% acetyl groups in either a random or block distribution (BioSyntech Canada Inc., Laval, QC, Canada), and both have an α structure.

Compared to the commercial hemostatic materials in powder forms, chitosan biomaterials were as effective as Celox as indicated by the TEG® parameters, but less effective than QuikClot, which is in agreement with their hemostatic mechanism mainly by tissue adhesion. As far as the onset of clot formation (R) is concerned, the block chitosan outperformed Celox, but not in terms of clot polymerization (α) and strength (MA). No significant differences were observed among the chitosan agents while block chitosan appears more effective than squid chitosan, likely due to their differences in chitosan source and physical structure. Furthermore, as implied by reduced R time (approximately 20% decrease), all chitosan materials exhibited hemostatic activities, perhaps due to their induced formation of the coagulum of red blood cells and platelet aggregation as reported by many investigators

[67]. Our study confirms that TEG is a useful tool to screen different hemostatic biomaterials prior to *in vivo* testing for the development of new agents for hemorrhage control.

5. Conclusions and Future Directions

Chitosan, almost the only cationic polysaccharide in nature, is nontoxic, biodegradable, antibacterial, antitumor and hemostatic. These special properties make it an important candidate for various biological applications. Due to the poor water solubility of pure chitosan, physical and chemical modifications of chitosan play important roles in these applications. The modifications may improve poor water solubility in neutral pH and mechanical properties of chitosan and widen its applications for tissue engineering and drug delivery system. The combination with other biomaterials and biomolecules is also a promising approach to utilization of both advantages from chitosan and the secondary or even third biomaterials. At the same time, the majority of studies carried out so far are only investigating *in vitro* conditions, so *in vivo* studies of many existing chitosan derivatives need to be carried out and seriously addressed. Appropriate sterilization needs to be established prior to clinical use [155].

A number of chitosan-based hemostatic agents may be applied topically to both external and internal tissue injuries. These are available in different forms (liquid, coating, powder, granule, fiber, film, hydrogel, sponge). The solid form, especially the sheet dressing, appears to be more effective in stopping severe bleeding given easy application with manual pressure, but may be less suitable for non-compressible internal bleeding. In contrast, chitosan in liquid and powder forms may be conformable to irregular wound surfaces for close contact with the origin of bleeding, especially for a small wound surface, and can be applied several times to the same area without concern for disrupting previous hemostatic areas. Chemical modifications of chitosan tend to reduce its hemostatic activities. There are currently no universal biomaterials that suit all applications due to different requirements for physical and handling properties as a result of different wound characteristics. New chitosan-based biomaterials continue to emerge. We believe that *in vitro* studies are important for further development of these materials for *in vivo* applications. The efficacy of some applications has been established, but some are awaiting further evaluations. Some may be viewed as an alternative, but some as complementary agents to improve surgical outcomes. Costs, side effects and efficacy must be considered when choosing a product for use in a specific surgical procedure.

References

[1] Karrer, P., and Hofmann, A. (1929). Polysaccharide xxxix. On the enzymatic breakdown of chitin and chitosan I. *Helvetica Chimica Acta,* 12, 616-637

[2] Rinaudo, M. (2006). Chitin and chitosan: Properties and applications. *Progress in Polymer Science,* 31, 603-632

[3] Shigemasa, Y., Saito, K., Sashiwa, H., and Saimoto, H. (1994). Enzymatic degradation of chitins and partially deacetylated chitins. *Int. J. Biol. Macromol.*, 16, 43-49

[4] Shi, C. M., Zhu, Y., Ran, X. Z., Wang, M., Su, Y. P., and Cheng, T. M. (2006). Therapeutic potential of chitosan and its derivatives in regenerative medicine. *Journal of Surgical Research*, 133, 185-192

[5] Kiang, T., Wen, H., Lim, H. W., and Leong, K. W. (2004). The effect of the degree of chitosan deacetylation on the efficiency of gene transfection. *Biomaterials,* 25, 5293-5301

[6] Bhattarai, N., Ramay, H. R., Chou, S. H., and Zhang, M. Q. (2006). Chitosan and lactic acid-grafted chitosan nanoparticles as carriers for prolonged drug delivery. *International Journal of Nanomedicine,* 1, 181-187

[7] Qian, F., Cui, F. Y., Ding, J. Y., Tang, C., and Yin, C. H. (2006). Chitosan graft copolymer nanoparticles for oral protein drug delivery: Preparation and characterization. *Biomacromolecules,* 7, 2722-2727

[8] Holme, K. R., and Hall, L. D. (1991). Chitosan derivatives bearing c-10-alkyl glycoside branches - a temperature-induced gelling polysaccharide. *Macromolecules*, 24, 3828-3833

[9] Prabaharan, M. (2008). Chitosan derivatives as promising materials for controlled drug delivery. *Journal of Biomaterials Applications*, 23, 5-36

[10] Huo, M. R., Zhou, H. P., Zhang, Y., and Lu, L. (2007). Synthesis and characterization of novel amphiphilic chitosan derivatives and its solubilizing abilities for water-insoluble drugs. Chem. *J. Chin. Univ.-Chin.,* 28, 1995-1999

[11] Liu, K. H., Chen, S. Y., Liu, D. M., and Liu, T. Y. (2008). Self-assembled hollow nanocapsule from amphiphatic carboxymethyl-hexanoyl chitosan as drug carrier. *Macromolecules*, 41, 6511-6516

[12] Lee, D., Zhang, W., Shirley, S. A., Kong, X., Hellermann, G. R., Lockey, R. F., and Mohapatra, S. S. (2007). Thiolated chitosan/DNA nanocomplexes exhibit enhanced and sustained gene delivery. *Pharmaceutical Research,* 24, 157-167

[13] Liu, W. G., Zhang, X., Sun, S. J., Sun, G. J., De Yao, K., Liang, D. C., Guo, G., and Zhang, J. Y. (2003). N-alkylated chitosan as a potential nonviral vector for gene transfection. *Bioconjugate Chemistry*, 14, 782-789

[14] Holle, L., Song, W., Holle, E., Wei, Y.-Z., Wagner, T., and Yu, X.-Z. (2003). A matrix metalloproteinase 2 cleavable melittin/advidin conjugate specifically targets tumor cells in vitro and in vivo. *Int. J. Oncol.* , 22, 93-98

[15] Mi, F. L., Shyu, S. S., Chen, C. T., and Lai, J. Y. (2002). Adsorption of indomethacin onto chemically modified chitosan beads. *Polymer*, 43, 757-765

[16] Qu, G. W., Yao, Z., Zhang, C., Wu, X. L., and Ping, Q. E. (2009). Peg conjugated n-octyl-o-sulfate chitosan micelles for delivery of paclitaxel: In vitro characterization and in vivo evaluation. *European Journal of Pharmaceutical Sciences,* 37, 98-105

[17] Chung, H. J., Go, D. H., Bae, J. W., Jung, I. K., Lee, J. W., and Park, K. D. (2005). Synthesis and characterization of pluronic((r)) grafted chitosan copolymer as a novel injectable biomaterial. *Curr. Appl. Phys.*, 5, 485-488

[18] Rokhade, A. P., Shelke, N. B., Patil, S. A., and Aminabhavi, T. M. (2007). Novel hydrogel microspheres of chitosan and pluronic f-127 for controlled release of 5-fluorouracil. *J. Microencapsul.*, 24, 274-288

[19] Liu, Z. H., Jiao, Y. P., Wang, Y. F., Zhou, C. R., and Zhang, Z. Y. (2008). Polysaccharides-based nanoparticles as drug delivery systems. *Advanced Drug Delivery Reviews*, 60, 1650-1662

[20] Krayukhina, M. A., Samoilova, N. A., and Yamskov, I. A. (2008). Polyelectrolyte complexes of chitosan: Formation, properties, and applications. *Uspekhi Khimii,* 77, 854-869

[21] Lankalapalli, S., and Kolapalli, V. R. M. (2009). Polyelectrolyte complexes: A review of their applicability in drug delivery technology. *Indian Journal of Pharmaceutical Sciences,* 71, 481-487

[22] Shu, S. J., Sun, C. Y., Zhang, X. G., Wu, Z. M., Wang, Z., and Li, C. X. Hollow and degradable polyelectrolyte nanocapsules for protein drug delivery. *Acta Biomaterialia,* 6, 210-217

[23] Ke, G. Z., Xu, W. L., and Yu, W. D. Preparation and properties of drug-loaded chitosan-sodium alginate complex membrane. *International Journal of Polymeric Materials,* 59, 184-191

[24] Elzatahry, A. A., Eldin, M. S. M., Soliman, E. A., and Hassan, E. A. (2009). Evaluation of alginate-chitosan bioadhesive beads as a drug delivery system for the controlled release of theophylline. *Journal of Applied Polymer Science,* 111, 2452-2459

[25] Sarmento, B., Ribeiro, A., Veiga, F., Sampaio, P., Neufeld, R., and Ferreira, D. (2007). Alginate/chitosan nanoparticles are effective for oral insulin delivery. *Pharmaceutical Research*, 24, 2198-2206

[26] Amrutkar, J. R., and Gattani, S. G. (2009). Chitosan-chondroitin sulfate based matrix tablets for colon specific delivery of indomethacin. *Aaps Pharmscitech*, 10, 670-677

[27] Chen, W. B., Wang, L. F., Chen, J. S., and Fan, S. Y. (2005). Characterization of polyelectrolyte complexes between chondroitin sulfate and chitosan in the solid state. *Journal of Biomedical Materials Research* Part A, 75A, 128-137

[28] Wu, J., Wang, X. F., Keum, J. K., Zhou, H. W., Gelfer, M., Avila-Orta, C. A., Pan, H., Chen, W. L., Chiao, S. M., Hsiao, B. S., and Chu, B. (2007). Water soluble complexes of chitosan-g-mpeg and hyaluronic acid. *Journal of Biomedical Materials Research* Part A, 80A, 800-812

[29] Jiang, H. L., Wang, Y. J., Huang, Q., Li, Y., Xu, C. N., Zhu, K. J., and Chen, W. L. (2005). Biodegradable hyaluronic acid/n-carboxyethyl chitosan/protein ternary complexes as implantable carriers for controlled protein release. *Macromolecular Bioscience,* 5, 1226-1233

[30] Kim, S. J., Lee, K. J., and Kim, S. I. (2004). Swelling behavior of polyelectrolyte complex hydrogels composed of chitosan and hyaluronic acid. *Journal of Applied Polymer Science*, 93, 1097-1101

[31] Kim, S. J., Shin, S. R., Lee, K. B., Park, Y. D., and Kim, S. I. (2004). Synthesis and characteristics of polyelectrolyte complexes composed of chitosan and hyaluronic acid. *Journal of Applied Polymer Science,* 91, 2908-2913

[32] Taravel, M. N., and Domard, A. (1995). Collagen and its interaction with chitosan .2. Influence of the physicochemical characteristics of collagen. *Biomaterials*, 16, 865-871

[33] Tonhi, E., and Plepis, A. M. D. (2002). Preparation and characterization of collagen-chitosan blends. *Quimica Nova,* 25, 943-948

[34] Zhang, Q. Q., Liu, L. R., Ren, L., and Wang, F. J. (1997). Preparation and characterization of collagen-chitosan composites. *Journal of Applied Polymer Science,* 64, 2127-2130

[35] Shu, S. J., Zhang, X. G., Teng, D. Y., Wang, Z., and Li, C. X. (2009). Polyelectrolyte nanoparticles based on water-soluble chitosan-poly (l-aspartic acid)-polyethylene glycol

for controlled protein release. *Carbohydr. Res.*, 344, 1197-1204
[36] Takka, S., and Acarturk, F. (1999). Calcium alginate microparticles for oral administration: I: Effect of sodium alginate type on drug release and drug entrapment efficiency. *J. Microencapsul.,* 16, 275-290
[37] Kofuji, K., Ito, T., Murata, Y., and Kawashima, S. (2000). The controlled release of a drug from biodegradable chitosan gel beads. *Chemical and Pharmaceutical Bulletin,* 48, 579-581
[38] Langer, R., and Vacanti, J. P. (1993). Tissue engineering. *Science*, 260, 920-926
[39] Thein-Han, W. W., Kitiyanant, Y., and Misra, R. D. K. (2008). Chitosan as scaffold matrix for tissue engineering. *Materials Science and Technology,* 24, 1062-1075
[40] Jiang, T., Kumbar, S. G., Nair, L. S., and Laurencin, C. T. (2008). Biologically active chitosan systems for tissue engineering and regenerative medicine. *Current Topics in Medicinal Chemistry,* 8, 354-364
[41] Karakecli, A. G., Satriano, C., Gumusderelioglu, M., and Marletta, G. (2008). Enhancement of fibroblastic proliferation on chitosan surfaces by immobilized epidermal growth factor. *Acta Biomaterialia,* 4, 989-996
[42] Park, K. M., Joung, Y. K., Park, K. D., Lee, S. Y., and Lee, M. C. (2008). Rgd-conjugated chitosan-pluronic hydrogels as a cell supported scaffold for articular cartilage regeneration. *Macromolecular Research,* 16, 517-523
[43] Tigli, R. S., and Gumusderelioglu, M. (2008). Evaluation of rgd- or egf-immobilized chitosan scaffolds for chondrogenic activity. *International Journal of Biological Macromolecules,* 43, 121-128
[44] Kirsebom, H., Aguilar, M. R., Roman, J. S., Fernandez, M., Prieto, M. A., and Bondar, B. (2007). Macroporous scaffolds based on chitosan and bioactive molecules. *J. Bioact. Compat. Polym.,* 22, 621-636
[45] Lopez-Lacomba, J. L., Garcia-Cantalejo, J. M., Casado, J. V. S., Abarrategi, A., Magana, V. C., and Ramos, V. (2006). Use of rhbmp-2 activated chitosan films to improve osseointegration. *Biomacromolecules,* 7, 792-798
[46] Itoh, S., Yamaguchi, I., Suzuki, M., Ichinose, S., Takakuda, K., Kobayashi, H., Shinomiya, K., and Tanaka, J. (2003). Hydroxyapatite-coated tendon chitosan tubes with adsorbed laminin peptides facilitate nerve regeneration in vivo. *Brain Research,* 993, 111-123
[47] Xu, H. X., Yan, Y. H., and Li, S. P. (2009). Chitosan-l- lactic acid scaffold for the regeneration of peripheral nerve and its ngf release properties. *Journal of Wuhan University of Technology-Materials Science Edition*, 24, 961-964
[48] Liao, I. C., Wan, A. C. A., Yim, E. K. F., and Leong, K. W. (2005). Controlled release from fibers of polyelectrolyte complexes. *Journal of Controlled Release*, 104, 347-358
[49] Zhang, Y. F., Cheng, X. R., Wang, J. W., Wang, Y. N., Shi, B., Huang, C., Yang, X. C., and Liu, T. J. (2006). Novel chitosan/collagen scaffold containing transforming growth factor-beta 1 DNA for periodontal tissue engineering. *Biochemical and Biophysical Research Communications*, 344, 362-369
[50] Machado, C. B., Ventura, J. M. G., Lemos, A. F., Ferreira, J. M. F., Leite, M. F., and Goes, A. M. (2007). 3d chitosan-gelatin-chondroitin porous scaffold improves osteogenic differentiation of mesenchymal stem cells. *Biomedical Materials,* 2, 124-131
[51] Guo, T., Zhao, J. N., Chang, J. B., Ding, Z., Hong, H., Chen, J. N., and Zhang, J. F.

(2006). Porous chitosan-gelatin scaffold containing plasmid DNA encoding transforming growth factor-beta 1 for chondrocytes proliferation. *Biomaterials,* 27, 1095-1103

[52] Liu, H. F., Yao, F. L., Zhou, Y., Yao, K. D., Mei, D. R., Cui, L., and Cao, Y. L. (2005). Porous poly (dl-lactic acid) modified chitosan-gelatin scaffolds for tissue engineering. *Journal of Biomaterials Applications*, 19, 303-322

[53] Lee, J. Y., Nam, S. H., Im, S. Y., Park, Y. J., Lee, Y. M., Seol, Y. J., Chung, C. P., and Lee, S. J. (2002). Enhanced bone formation by controlled growth factor delivery from chitosan-based biomaterials. *Journal of Controlled Release,* 78, 187-197

[54] Lahiji, A., Sohrabi, A., Hungerford, D. S., and Frondoza, C. G. (2000). Chitosan supports the expression of extracellular matrix proteins in human osteoblasts and chondrocytes. *Journal of Biomedical Materials Research,* 51, 586-595

[55] Kuo, Y.-C., Yeh, C.-F., and Yang, J.-T. (2009). Differentiation of bone marrow stromal cells in poly(lactide-co-glycolide)/chitosan scaffolds. *Biomaterials,* 30, 6604-6613

[56] Li, Z. S., Ramay, H. R., Hauch, K. D., Xiao, D. M., and Zhang, M. Q. (2005). Chitosan-alginate hybrid scaffolds for bone tissue engineering. *Biomaterials,* 26, 3919-3928

[57] Kim, I. Y., Seo, S. J., Moon, H. S., Yoo, M. K., Park, I. Y., Kim, B. C., and Cho, C. S. (2008). Chitosan and its derivatives for tissue engineering applications. *Biotechnology Advances,* 26, 1-21

[58] Cheng, M. Y., Deng, J. U., Yang, F., Gong, Y. D., Zhao, N. M., and Zhang, X. F. (2003). Study on physical properties and nerve cell affinity of composite films from chitosan and gelatin solutions. *Biomaterials,* 24, 2871-2880

[59] Linn, T., Erb, D., Schneider, D., Kidszun, A., Elcin, A. E., Bretzel, R. G., and Elcin, Y. M. (2003). Polymers for induction of revascularization in the rat fascial flap: Application of vascular endothelial growth factor and pancreatic islet cells. *Cell Transplantation,* 12, 769-778

[60] Goraltchouk, A., Scanga, V., Morshead, C. M., and Shoichet, M. S. (2006). Incorporation of protein-eluting microspheres into biodegradable nerve guidance channels for controlled release. *Journal of Controlled Release*, 110, 400-407

[61] Lee, J. E., Kim, K. E., Kwon, I. C., Ahn, H. J., Lee, S. H., Cho, H. C., Kim, H. J., Seong, S. C., and Lee, M. C. (2004). Effects of the controlled-released tgf-beta 1 from chitosan microspheres on chondrocytes cultured in a collagen/chitosan/ glycosaminoglycan scaffold. *Biomaterials,* 25, 4163-4173

[62] Lee, Y. M., Park, Y. J., Lee, S. J., Ku, Y., Han, S. B., Klokkevold, P. R., and Chung, C. P. (2000). The bone regenerative effect of platelet-derived growth factor-bb delivered with a chitosan/tricalcium phosphate sponge carrier. *J. Periodont.*, 71, 418-424

[63] Peter, M., Binulal, N. S., Soumya, S., Nair, S. V., Furuike, T., Tamura, H., and Jayakumar, R. Nanocomposite scaffolds of bioactive glass ceramic nanoparticles disseminated chitosan matrix for tissue engineering applications. *Carbohydrate Polymers,* 79, 284-289

[64] Rabea, E. I., Badawy, M. E.-T., Stevens, C. V., Smagghe, G., and Steurbaut, W. (2003). Chitosan as antimicrobial agent: Applications and mode of action. Biomacromolecules, 4, 1457-1465

[65] Okamotoa, Y., Yanoa, R., Miyatakea, K., Tomohirob, I., Shigemasac, Y., and Minami, S. (2003). Effects of chitin and chitosan on blood coagulation. *Carbohydr. Polym*, 53, 337-342

[66] Rao, S. B., and Sharma, C. P. (1997). Use of chitosan as a biomaterial: Studies on its safety and hemostatic potential. *J. Biomed. Mater. Res.*, 34, 21-28
[67] Whang, H. S., Kirsch, W., Zhu, Y. H., Yang, C. Z., and Hudson, S. M. (2005). Hemostatic agents derived from chitin and chitosan. *J. Macromol. Sci., Polym. Rev.*, 45, 309--323
[68] Bergera, J., Reista, M., Mayera, J. M., Feltb, O., Peppasc, N. A., and Gurny, R. (2004). Structure and interactions in covalently and ionically crosslinked chitosan hydrogels for biomedical applications. *Eur. J. Pharm. Biopharm*, 57, 19-34
[69] Ono, K., Saito, Y., Yura, H., Ishikawa, K., Kurita, A., Akaike, T., and Ishihara, M. (2000). Photocrosslinkable chitosan as a biological adhesive. *J. Biomed. Mater. Res.,* 49, 289-295
[70] Arand, A. G., and Sawaya, R. (1986). Intraoperative chemical hemostasis in neurosurgery. *Neurosurgery*, 18, 223-233
[71] Yang, J., Tian, F., Wang, Z., Wang, Q., Zeng, Y. J., and Chen, S. Q. (2008). Effect of chitosan molecular weight and deacetylation degree on hemostasis. *Journal of Biomedical Materials Research* - Part B Applied Biomaterials, 84, 131-137
[72] Lin, C.-W., and Lin, J.-C. (2003). Characterization and blood coagulation evaluation of the water-soluble chitooligosaccharides prepared by a facile fractionation method. *Biomacromolecules,* 4, 1691-1697
[73] Shepherd, R., Reader, S., and Falshaw, A. (1997). Chitosan functional properties. *Glycoconjugate J.*, 14, 535-542
[74] Fukasawa, M., Abe, H., Masaoka, T., Orita, H., Horikawa, H., Campeau, J. D., and Washio, M. (1992). The hemostatic effect of deacetylated chitin membrane on peritoneal injury in rabbit model. *Surg. Today,* 22, 333-338
[75] Janvikul, W., Uppanan, P., Thavornyutikarn, B., Krewraing, J., and Prateepasen, R. (2006). In vitro comparative hemostatic studies of chitin, chitosan, and their derivatives. *J. Appl. Polym. Sci.*, 102, 445-451
[76] McCarthy, S., Gregory, K., Wiesmann, W., and Campbell, T. (2008). Wound dressing and method for controlling severe life-threatening bleeding. *US Patent* 7371403,
[77] Gu, R., Sun, W., Zhou, H., Wu, Z., Meng, Z., Zhu, X., Tang, Q., Dong, J., and Dou, G. (2010). The performance of a fly-larva shell-derived chitosan sponge as an absorbable surgical hemostatic agent. *Biomaterials*, 31, 1270-1277
[78] Kozen, B. G., Kircher, S. J., Henao, J., Godinez, F. S., and Johnson, A. S. (2008). An alternative hemostatic dressing: Comparison of celox, hemcon, and quikclot. *Acad. Emerg. Med.*, 15, 74-81
[79] Lee, D.-W., Powers, K., and Baney, R. (2004). Physicochemical properties and blood compatibility of acylated chitosan nanoparticles. *Carbohydr. Polym*, 58, 371-377
[80] Wang, Q. Z., Chen, X. G., Li, Z. X., Wang, S., Liu, C. S., Meng, X. H., Liu, C. G., Lv, Y. H., and Yu, L. J. (2008). Preparation and blood coagulation evaluation of chitosan microspheres. *J. Mater. Sci. Mater. Med.*, 19, 1371-1377
[81] Hirano, S., Zhang, M., Nakagawa, M., and Miyata, T. (2000). Wet spun chitosan-collagen fibers, their chemical n-modifications, and blood compatibility. *Biomaterials,* 21, 997-1003
[82] Klokkevold, P. R., Fukayama, H., Sung, E. C., and Bertolami, C. N. (1999). The effect of chitosan (poly-n-acetyl glucosamine) on lingual hemostasis in heparinized rabbits. *J. Oral Maxillofac. Surg.*, 57, 49-52

[83] Brandenberg, G., Leibrock, L. G., and Shuman, R. (1984). Chitosan: A new topical hemostatic agent for diffuse capillary bleeding in brain tissue. *Neurosurgery*, 15, 9-13

[84] Lee, D.-W., and Baney, R. (2004). Oligochitosan derivatives bearing electron-deficient aromatic rings for adsorption of amitriptyline: Implications for drug detoxification. *Biomacromolecules*, 5, 1310-1315

[85] Loke, W. K., Lau, S. K., Yong, L. L., Khor, E., and Sum, C. K. (2000). Wound dressing with sustained anti-microbial capability. *J. Biomed. Mater. Res.,* 53, 8-17

[86] Peh, K., Khan, T., and Ch'ng, H. (2000). Mechanical, bioadhesive strength and biological evaluations of chitosan films for wound dressing. *Journal of pharmacy and pharmaceutical sciences,* 3, 303-311

[87] Paul, W., and Sharma, C. (2004). Chitosan and alginate wound dressings: A short review. *Trends Biomater. Artif. Organs.*, 18, 18-23

[88] Wang, X., Yan, Y., and Zhang, R. (2006). A comparison of chitosan and collagen sponges as hemostatic dressings. *J. Bioact. Compat. Polym.*, 21, 39-54

[89] Fischer, T., Bode, A., Demcheva, M., and Vournakis, J. (2007). Hemostatic properties of glucosamine-based materials. *J. Biomed. Mater. Res.,* 80A, 167-174

[90] Fischer, T. H., Bode, A. P., Demcheva, M., and Vournakis, J. N. (2007). Hemostatic properties of glucosamine-based materials. *Journal of Biomedical Materials Research* - Part A, 80, 167-174

[91] Mlekusch, W., Dick, P., Haumer, M., Sabeti, S., Minar, E., and Schillinger, M. (2006). Arterial puncture site management after percutaneous transluminal procedures using a hemostatic wound dressing (clo-sur p.A.D.) versus conventional manual compression: A randomized controlled trial. *Journal of Endovascular Therapy*, 13, 23-31

[92] Nguyen, N., Hasan, S., Caufield, L., Ling, F. S., and Narins, C. R. (2007). Randomized controlled trial of topical hemostasis pad use for achieving vascular hemostasis following percutaneous coronary intervention. *Catheter. Cardiovasc. Interv.*, 69, 801-807

[93] Fischer, T. H., Connolly, R., Thatte, H. S., and Schwaitzberg, S. S. (2004). Comparison of structural and hemostatic properties of the poly-n-acetyl glucosamine syvek patch with products containing chitosan. *Microsc. Res. Tech.*, 63, 168-174

[94] Chan, M. W., Schwaitzberg, S. D., Demcheva, M., Vournakis, J., Finkielsztein, S., and Connolly, R. J. (2000). Comparison of poly-n-acetyl glucosamine (p-glcnac) with absorbable collagen (actifoam), and fibrin sealant (bolheal) for achieving hemostasis in a swine model of splenic hemorrhage. *Journal of Trauma* - Injury, Infection and Critical Care, 48, 454-458

[95] Schwaitzberg, S. D., Chan, M. W., Cole, D. J., Read, M., Nichols, T., Bellinger, D., and Connolly, R. J. (2004). Comparison of poly-n-acetyl glucosamine with commercially available topical hemostats for achieving hemostasis in coagulopathic models of splenic hemorrhage. *J. Trauma*, 57, S29-S32

[96] Cole, D. J., Connolly, R. J., Chan, M. W., Schwaitzberg, S. D., Byrne, T. K., Adams, D. B., Baron, P. L., O'Brien, P. H., Metcalf, J. S., Demcheva, M., and Vournakis, J. (1999). A pilot study evaluating the efficacy of a fully acetylated poly-n- acetyl glucosamine membrane formulation as a topical hemostatic agent. *Surgery*, 126, 510-517

[97] Valeri, C. R., Srey, R., Tilahun, D., and Ragno, G. (2004). In vitro effects of poly-N-acetyl glucosamine on the activation of platelets in platelet-rich plasma with and

without red blood cells. *J. Trauma,* 57, S22-S25
[98] Pusateri, A. E., McCarthy, S. J., Gregory, K. W., Harris, R. A., Cardenas, L., McManus, A. T., and Goodwin Jr, C. W. (2003). Effect of a chitosan-based hemostatic dressing on blood loss and survival in a model of severe venous hemorrhage and hepatic injury in swine. *J. Trauma,* 54, 177-182
[99] Kheirabadi, B. S., Acheson, E. M., Deguzman, R., Sondeen, J. L., Ryan, K. L., Delgado, A., Dick Jr, E. J., Holcomb, J. B., and Alam, H. B. (2005). Hemostatic efficacy of two advanced dressings in an aortic hemorrhage model in swine. *J. Trauma*, 59, 25-35
[100] Amaral, I., Sampaio, P., and Barbosa, M. (2006). Three-dimensional culture of human osteoblastic cells in chitosan sponges: The effect of the degree of acetylation. *J. Biomed. Mater. Res.,* 76A, 335-346
[101] Gustafson, S. B., Fulkerson, P., Bildfell, R., Aguilera, L., and Hazzard, T. M. (2007). Chitosan dressing provides hemostasis in swine femoral arterial injury model. *Prehosp. Emerg. Care*, 11, 172-178
[102] Neuffer, M. C., McDivitt, J., Rose, D., King, K., Cioonan, C. C., and Vayer, J. S. (2004). Hemostatic dressings for the first responder: A review. *Mil. Med.*, 169, 716-720
[103] Acheson, E. M., Kheirabadi, B. S., Deguzman, R., Dick, E. J., Jr., and Holcomb, J. B. (2005). Comparison of hemorrhage control agents applied to lethal extremity arterial hemorrhages in swine. *J. Trauma,* 59, 865-874
[104] Ward, K. R., Tiba, M. H., Holbert, W. H., Blocher, C. R., Draucker, G. T., Proffitt, E. K., Bowlin, G. L., Ivatury, R. R., and Diegelmann, R. F. (2007). Comparison of a new hemostatic agent to current combat hemostatic agents in a swine model of lethal extremity arterial hemorrhage. *J. Trauma*, 63, 276-284
[105] Ahuja, N., Ostomel, T. A., Rhee, P., Stucky, G. D., Conran, R., Chen, Z., Al-Mubarak, G. A., Velmahos, G., Demoya, M., and Alam, H. B. (2006). Testing of modified zeolite hemostatic dressings in a large animal model of lethal groin injury. *J. Trauma*, 61, 1312-1320
[106] Wedmore, I., McManus, J. G., Pusateri, A. E., and Holcomb, J. B. (2006). A special report on the chitosan-based hemostatic dressing: Experience in current combat operations. *J. Trauma,* 60, 655-658
[107] Brown, M. A., Daya, M. R., and Worley, J. A. (2009). Experience with chitosan dressings in a civilian ems system *J. Emerg. Med.*, 37, 1-7
[108] Bachtell, N., Goodell, T., Grunkemeier, G., Jin, R., and Gregory, K. (2006). Treatment of dialysis access puncture wound bleeding with chitosan dressings. *Dialysis and Transplantation,* 35, 672-681
[109] Malmquist, J. P., Clemens, S. C., Oien, H. J., and Wilson, S. L. (2008). Hemostasis of oral surgery wounds with the hemcon dental dressing *J. Oral Maxillofac. Surg.*, 66, 1177-1183
[110] Sohn, V. Y., Eckert, M. J., Martin, M. J., Arthurs, Z. M., Perry, J. R., Beekley, A., Rubel, E. J., Adams, R. P., Bickett, G. L., and Rush Jr., R. M. (2009). Efficacy of three topical hemostatic agents applied by medics in a lethal groin injury model *J. Surg. Res.*, 154, 258-261
[111] Dailey, R. A. M. D. F. A. C. S., Chavez, M. R. M. D., and Choi, D. P. D. (2009). Use of a chitosan-based hemostatic dressing in dacryocystorhinostomy. SO - *Ophthalmic Plastic and Reconstructive Surgery*, 25, 350-353

[112] Xie, H., Khajanchee, Y. S., Teach, J. S., and Shaffer, B. S. (2008). Use of a chitosan-based hemostatic dressing in laparoscopic partial nephrectomy. *J. Biomed. Mater. Res.* Part B: Appl. Biomater., 85, 267-271

[113] Xie, H., Teach, J. S., Burke, A. P., Lucchesi, L. D., Wu, P.-C., and Sarao, R. C. (2009). Laparoscopic repair of inferior vena caval injury using a chitosan-based hemostatic dressing. *The American Journal of Surgery,* 197, 510-514

[114] Morrison, J. J., Mountain, A. J. C., Galbraith, K. A., and Clasper, J. C. (2010). Penetrating pelvic battlefield trauma: Internal use of chitosan-based haemostatic dressings. *Injury,* 41, 239-241

[115] Kheirabadi, B. S., Scherer, M. R., Estep, J. S., Dubick, M. A., and Holcomb, J. B. (2009). Determination of efficacy of new hemostatic dressings in a model of extremity arterial hemorrhage in swine. *Journal of Trauma* - Injury, Infection and Critical Care, 67, 450-459

[116] Kheirabadi, B. S., Edens, J. W., Terrazas, I. B., Estep, J. S., Klemcke, H. G., Dubick, M. A., and Holcomb, J. B. (2009). Comparison of new hemostatic granules/powders with currently deployed hemostatic products in a lethal model of extremity arterial hemorrhage in swine. *J. Trauma,* 66, 316-328

[117] Arnaud, F., Teranishi, K., Tomori, T., Carr, W., and McCarron, R. (2009). Comparison of 10 hemostatic dressings in a groin puncture model in swine. *J. Vasc. Surg.*, 50, 632-639

[118] Millner, R. W. J., Lockhart, A. S., Bird, H., and Alexiou, C. (2009). A new hemostatic agent: Initial life-saving experience with celox (chitosan) in cardiothoracic surgery. *Ann. Thorac. Surg.*, 87, e13-14

[119] Bochicchio, G., Kilbourne, M., Kuehn, R., Keledjian, K., Hess, J., and Scalea, T. (2009). Use of a modified chitosan dressing in a hypothermic coagulopathic grade v liver injury model. *Am. J. Surg.,* 198, 617-622

[120] Hoekstra, A., Struszczyk, H., and Kivekäs, O. (1998). Percutaneous microcrystalline chitosan application for sealing arterial puncture sites. Biomaterials, 19, 1467-1471

[121] Klokkevold, P. R., Subar, P., Fukayama, H., and Bertolami, C. N. (1992). Effect of chitosan on lingual hemostasis in rabbits with platelet dysfunction induced by epoprostenol. *J. Oral Maxillofac. Surg.*, 50, 41-45

[122] Garcia, P., Mines, M. J., Bower, K. S., Hill, J., Menon, J., Tremblay, E., and Smith, B. (2009). Robotic laser tissue welding of sclera using chitosan films. *Lasers Surg. Med.,* 41, 60-67

[123] Lauto, A., Stoodley, M., Marcel, H., Avolio, A., Sarris, M., McKenzie, G., Sampson, D. D., and Foster, L. J. R. (2007). In vitro and in vivo tissue repair with laser-activated chitosan adhesive. *Lasers Surg. Med.*, 39, 19-27

[124] Tyan, Y. C., Liao, J. D., and Lin, S. P. (2003). Surface properties and in vitro analyses of immobilized chitosan onto polypropylene non-woven fabric surface using antenna-coupling microwave plasma. *J. Mater. Sci. Mater. Med.*, 14, 775-781

[125] Lin, W. C., Tseng, C. H., and Yang, M. C. (2005). In-vitro hemocompatibility evaluation of a thermoplastic polyurethane membrane with surface-immobilized water-soluble chitosan and heparin. *Macromol. Biosci.*, 5, 1013-1021

[126] Hirano, S., and Noishiki, Y. (1985). The blood compatibility of chitosan and n-acylchitosans. *J. Biomed. Mater. Res.*, 19, 413-417

[127] Hirano, S., Zhang, M., Yamane, H., Kimura, Y., Matsukawa, K., and Nakagawa, M.

(2003). Chemical modification and some aligned composites of chitosan in a filament state. *Macromol. Biosci.*, 3, 620-628

[128] Hu, W., Huang, Z. M., Meng, S. Y., and He, C. L. (2009). Fabrication and characterization of chitosan coated braided plla wire using aligned electrospun fibers. *J. Mater. Sci. Mater. Med.*, 20, 2275-2284

[129] Spasova, M., Paneva, D., Manolova, N., Radenkov, P., and Rashkov, I. (2008). Electrospun chitosan-coated fibers of poly(l-lactide) and poly(l-lactide)/poly(ethylene glycol): Preparation and characterization. *Macromol. Biosci.*, 8, 153-162

[130] Serizawa, T., Yamaguchi, M., and Akashi, M. (2002). Alternating bioactivity of polymeric layer-by-layer assemblies: Anticoagulation vs procoagulation of human blood. *Biomacromolecules,* 3, 724-731

[131] Edwards, J. V., Howley, P., Prevost, N., Condon, B., Arnold, J., and Diegelmann, R. (2009). Positively and negatively charged ionic modifications to cellulose assessed as cotton-based protease-lowering and hemostatic wound agents. *Cellulose,* 16, 911-921

[132] Sashiwa, H., and Aiba, S. I. (2004). Chemically modified chitin and chitosan as biomaterials. *Progress in Polymer Science* (Oxford), 29, 887-908

[133] Ishihara, M., Nakanishi, K., Ono, K., Sato, M., Kikuchi, M., Saito, Y., Yura, H., Matsui, T., Hattori, H., Uenoyama, M., and Kurita, A. (2002). Photocrosslinkable chitosan as a dressing for wound occlusion and accelerator in healing process. *Biomaterials,* 23, 833-840

[134] Hayashi, T., Matsuyama, T., Hanada, K., Nakanishi, K., Uenoyama, M., Fujita, M., Ishihara, M., Kikuchi, M., Ikeda, T., and Tajiri, H. (2004). Usefulness of photocrosslinkable chitosan for endoscopic cancer treatment in alimentary tract. *Journal of Biomedical Materials Research* - Part B Applied Biomaterials, 71, 367-372

[135] Hirano, S., Tanaka, Y., Hasegawa, M., Tobetto, K., and Nishioka, A. (1985). Effect of sulfated derivatives of chitosan on some blood coagulant factors. *Carbohydr. Res.*, 137, 205-215

[136] Suwan, J., Zhang, Z., Li, B., Vongchan, P., Meepowpan, P., Zhang, F., Mousa, S. A., Mousa, S., Premanode, B., Kongtawelert, P., and Linhardt, R. J. (2009). Sulfonation of papain-treated chitosan and its mechanism for anticoagulant activity. *Carbohydr. Res.,* 344, 1190-1196

[137] Lee, K. Y., Ha, W. S., and Park, W. H. (1995). Blood compatibility and biodegradability of partially n-acylated chitosan derivatives. *Biomaterials*, 16, 1211-1216

[138] Yang, Y., Zhou, Y., Chuo, H., Wang, S., and Yu, J. (2007). Blood compatibility and mechanical properties of oxidized-chitosan films. *J. Appl. Polym. Sci.*, 106, 372-377

[139] Bergera, J., Reista, M., Mayera, J. M., Feltb, O., and Gurny, R. (2004). Structure and interactions in chitosan hydrogels formed by complexation or aggregation for biomedical applications. *Eur. J. Pharm. Biopharm*, 57, 35-52

[140] Ong, S. Y., Wu, J., Moochhala, S. M., Tan, M. H., and Lu, J. (2008). Development of a chitosan-based wound dressing with improved hemostatic and antimicrobial properties. *Biomaterials,* 29, 4323-4332

[141] Smith, S. A., and Morrissey, J. H. (2008). Polyphosphate as a general procoagulant agent. *J. Thromb Haemost*, 6, 1750–1756

[142] Mi, F. L., Shyu, S. S., Wong, T. B., Jang, S. F., Lee, S. T., and Lu, K. T. (1999). Chitosan-polyelectrolyte complexation for the preparation of gel beads and controlled

release of anticancer drug. Ii. Effect of ph-dependent ionic crosslinking or interpolymer complex using tripolyphosphate or polyphosphate as reagent. *Journal of Applied Polymer Science*, 74, 1093-1107

[143] Englehart, M. S., Cho, S. D., Tieu, B. H., Morris, M. S., Underwood, S. J., Karahan, A., Muller, P. J., Differding, J. A., Farrell, D. H., and Schreiber, M. A. (2008). A novel highly porous silica and chitosan-based hemostatic dressing is superior to hemcon and gauze sponges. *J. Trauma*, 65, 884-892

[144] Sambasivan, C. N., Cho, S. D., Zink, K. A., Differding, J. A., and Schreiber, M. A. (2009). A highly porous silica and chitosan-based hemostatic dressing is superior in controlling hemorrhage in a severe groin injury model in swine. *Am. J. Surg.*, 197, 576-580

[145] Valentine, R., Athanasiadis, T., Moratti, S., Robinson, S., and Wormald, P. J. (2009). The efficacy of a novel chitosan gel on hemostasis after endoscopic sinus surgery in a sheep model of chronic rhinosinusitis. *American Journal of Rhinology and Allergy*, 23, 71-75

[146] Okamoto, Y., Yano, R., Miyatake, K., Tomohiro, I., Shigemasa, Y., and Minami, S. (2003). Effects of chitin and chitosan on blood coagulation. *Carbohydr. Polym.*, 53, 337-342

[147] Chou, T. C., Fu, E., Wu, C. J., and Yeh, J. H. (2003). Chitosan enhances platelet adhesion and aggregation. *Biochem. Biophys. Res. Commun.*, 302, 480-483

[148] Benesch, J., and Tengvall, P. (2002). Blood protein adsorption onto chitosan. *Biomaterials*, 23, 2561-2568

[149] Smith, C. J., Vournakis, J. N., Demcheva, M., and Fischer, T. H. (2008). Differential effect of materials for surface hemostasis on red blood cell morphology. *Microsc. Res. Tech.,* 71, 721-729

[150] Sauaia, A., Moore, F. A., Moore, E. E., Moser, K. S., Brennan, R. R., Read, R. A., and Pons, P. T. (1995). Epidemiology of trauma deaths: A reassessment. *J. Trauma*, 38, 185-193

[151] Champion, H., Bellamy, R., Roberts, C., and Leppaniemi, A. (2003). A profile of combat injury. *J. Trauma*, 54, S13–S19

[152] Ostomel, T., Stoimenov, P., Holden, P., Alam, H., and Stucky, G. (2006). Host-guest composites for induced hemostasis and therapeutic healing in traumatic injuries. *J. Thromb. Thrombolysis,* 22, 55-67

[153] Shepherd, R., Reader, S., and Falshaw, A. (1997). Chitosan functional properties *Glycoconjugate J,* 14, 535-542

[154] Shimojoh, M., Fukushima, K., and Kurita, K. (1998). Low-molecular-weight chitosans derived from β-chitin: Preparation, molecular characteristics and aggregation activity. *Carbohydr. Polym, 35*, 223-231

[155] Rao, S. B., and Sharma, C. P. (1995). Sterilization of chitosan: Implications. *J. Biomater. Appl.,* 10, 136-143

In: Focus on Chitosan Research ISBN 978-1-61324-454-8
Editors: Arthur N. Ferguson and Amy G. O'Neill

Chapter 3

THE APPLICATIONS OF CHITOSAN WITH DIFFERENT MODIFICATIONS IN TISSUE ENGINEERING

Gan Wang[1,2], Qiang Ao[3], Yandao Gong[1] and Xiufang Zhang[1,*]

[1]. State Key Laboratory of Biomembrane and Membrane Biotechnology, School of Life Sciences, Tsinghua University, Beijing, China
[2]. Department of Chemistry and Biology, College of Science, National University of Defense Technology, Changsha, China
[3]. Institute of Neurological Disorders, Yuquan Hospital, Tsinghua University, Beijing, China

ABSTRACT

Chitosan, a cationic polymer, is the N-deacetylated derivative of chitin, which is abundant in nature. Chitosan has drawn much attention because of its good biocompatibility, biodegradability, low toxicity, and ability to be fabricated into various forms in tissue engineering. Because of the interesting biological properties, the utilizations of chitosan have been sufficiently developed. To make a breakthrough for the exploiture, modification will be a key point. Here we described several modifications including gamma irradiation, blending with PEG, PHB, gelatin, butyl bromide, collagen, polycation materials, nano-hydroxyapatite, coupled with GRGDS peptide, and carboxymethylation. Then the effect of the different modifications on chitosan cell affinity and biocompatibility has been discussed. Tissue engineering has emerged as promising field to treat the loss or malfunction of a tissue. A scaffold in tissue engineering plays an important role in accommodating and stimulating new tissue growth. Chitosan is one the widely used natural polymers in tissue engineering. Here we also discussed the applications of chitosan with different forms in tissue engineering including cartilage tissue, bone tissue, nerve tissue, vascular tissue, liver tissue, and skin tissue. All the descriptions have shown that the application of chitosan has a good prospect.

* Corresponding author. School of Life Sciences, Tsinghua University, Beijing 100084, China. Tel.: +86-10-62783261; fax: +86-10-62794214; E-mail address: zxf-dbs@mail.tsinghua.edu.cn.

1. Introduction

Chitosan [β-(1, 4)-2-amino-2-deoxy-D-glucan], a cationic polymer, is the N-deacetylated derivative of chitin, which is an abundant natural polysaccharide found in the exoskeleton of crustacean shells, shellfish like shrimp, and crabs. Chitosan can also be found in the cell wall of certain groups of fungi, particularly zygomycetes. Chitosan has drawn much attention because of its good biocompatibility, biodegradability, low toxicity, and ability to be fabricated into various forms in tissue engineering, such as film [1], porous scaffold [2], hydrogel [3], and tube [4-6]. Because of the interesting biological properties, the utilizations of chitosan have been sufficiently developed. At present, commercial use of chitosan has been carried out partly as non-modified form. In order to make a breakthrough for the exploiture, its modification will be a key point, because it is capable to introduce a variety of functional groups to improve the properties of chitosan. Until now, numerous works have been reported on the modifications of chitosan, including our works. In this chapter, we would like to describe different modifications of chitosan and the effect on cell affinity and biocompatibility. Then we discussed the applications of chitosan with different forms in tissue engineering including cartilage tissue, bone tissue, nervous tissue, vascular tissue, liver tissue, skin tissue, and so on.

2. Modifications of Chitosan

Due to its abundant active groups on the main chain, such as amino and hydroxyl groups, chitosan can be easily modified. Many derivatives can be prepared by chemical and physical modification of chitosan, bringing diverse new biological properties to chitosan. Many techniques for modification of chitosan have been attempted, including irradiation by γ-ray, blending, grafting, and crosslinking, and a great deal of progress has resulted: the capability of materials has been enhanced to a large extent, chitosan has been endowed with new characteristics, and thus the application of chitosan has been extended.

2.1. Physical Modifications

One defect of chitosan is its inferior mechanical properties, such as tenacity. Therefore, chitosan must to be modified to improve its mechanical properties. Yang *et al.* [7] reported the effect of gamma radiation on the mechanical properties and biocompatibility of chitosan membranes. The chitosan solution was injected into the wells of four 4-well tissue culture cluster dishes. Those four dishes were dried at 50 °C for about 11 h, and then irradiated by the gamma rays. The radiation doses for the four dishes were 12 kGy, 14 kGy, 16 kGy, and 18 kGy, respectively. They found that the rupture intensity was increased by 30% while the rupture strain was doubled by gamma irradiation in the range of 14-18 kGy. Fetal mouse cerebral cortex cell culture showed that the chitosan membranes with gamma ray irradiation retained their good biocompatibility. In the specified dose range, the gamma irradiation induced improved mechanical properties while keeping the excellent biocompatibility of the chitosan membrane.

Any clinical used biomaterial must have excellent properties in bulk as well as on the surface. Since it is the biomaterial surface that first comes into contact with the living tissue when the biomaterial is planted in the body, the initial response of the body to the biomaterial depends on its surface properties. An effective approach for developing a clinically applicable biomaterial is to modify the surface of a material that already has excellent biofunctionality and bulk properties [8]. Polymer blending is one of the most effective methods for providing new, desirable polymeric materials for particular applications. Chitosan is expected to be useful in the development of composite materials by blending with other polymers because it has many functional properties [9]. Blends are often used to improve tensile properties and to provide a stronger structural component as separation media that supports the active polymer. The physical properties of a polymer can also be altered by introducing a second polymer that improves the properties of the original polymer in certain aspects, such as hydrophobicity, lowered melt temperature, raised glass transition temperature, etc. [10]. Polyethylene glycol (PEG) is a faculty of water soluble polymers with many different molecular weights that exhibits useful properties such as protein resistance, low toxicity and immunogenicity. Zhang *et al.* [11] investigated the relationship between the PEG content and the biocompatibility and mechanical properties of blended membranes by blending with different PEGs. It was found that the mechanical properties of chitosan membrane could be improved slightly by the proper amount of PEG, but this improvement was not significant and additional PEG caused the properties to deteriorate. The addition of PEG significantly improved the biocompatibility of the materials by the improvement of the protein adsorption and cell adhesion, growth and proliferation. PEG of the proper concentration helped to maintain the natural structure of the protein adsorbed on the surface of the materials. The difference in molecular weights of PEG had no significant influence on the mechanical properties and biocompatibility.

Poly(3-hydroxybutyrate) (PHB) is a kind of polyester synthesized by some bacteria. PHB has attracted much attention for biomedical applications because of its excellent biocompatibility and biodegradability [12-14]. Blending chitosan with PHB will result in a completely biodegradable material with more desirable physical and biological properties. Cao *et al.* [15] developed an inexpensive method to prepare a novel blended material of chitosan and PHB using an emulsion blending technique. Chitosan/PHB blend films were successfully prepared. The results showed that PHB was uniformly entrapped as microspheres in chitosan matrices, which gave the blended films a rough surface. With the increase in PHB, the film surface roughness increased and the swelling capability decreased. In the wet state, the blended films exhibited a lower elastic modulus, a higher elongation-at-break and a higher tensile strength compared with chitosan films. NIH 3T3 fibroblasts cell culture experiments revealed that the blended films had better cytocompatibility than chitosan films.

Gelatin is a protein made soluble by hydrolysis of collagen derived from the skin, white connective tissue, and bones of animal. Gelatin contains carboxyl groups on its chain backbones and has the potential to mix with chitosan [16]. Moreover, gelatin is also a biodegradable polymer with many attractive properties, such as excellent biocompatibility, nonantigenicity, plasticity and adhesiveness, and it is widely used in biomedical and pharmaceutical fields [17]. Cheng *et al.* [18] created a series of chitosan-gelatin composite films by varying the ratio of components. They investigated the physical properties of the composite films and biological activity of the complex. The results indicated that free water content of the composite films increased with increasing gelatin content. In the wet state, composite film exhibited a lower modulus and a higher percentage of elongation-at-break compared with chitosan film. The composite film with 60wt% gelatin had the highest percentage of elongation-at-break, while its modulus was lowered to one-fifth that of chitosan film. The adsorption amount of fibronectin on composite film was much higher than on chitosan film. PC12 cells grown on composite films differentiated more rapidly and extend longer neurites than on chitosan films.

Poly-L-lysine is a well-known coating reagent used for supporting attachment and growth of neurons. Cheng *et al.* [19] prepared a series of chitosan–poly-L-lysine composite films by varying the ratio of components and studied the cell affinity of the composite materials with different content of poly-L-lysine. It was found that poly-L-lysine-blended chitosan significantly improved nerve cell affinity as indicated by increased attachment, differentiation, and growth of PC12 cells. Composite film with 3 wt% poly-L-lysine was an even better material in nerve cell affinity than collagen. Cheng *et al.* [1] reported that compared with chitosan blended with poly-L-lysine, chitosan blended with collagen and albumin also both improved the PC12 cells affinity by increasing attachment, differentiation and growth of PC12 cells. Fetal mouse cerebral cortex cells grew better on composite films than on chitosan. Chitosan blended with poly-L-lysine exhibited the best nerve cell affinity among these three types of composite material. Gong *et al.* [20] also found that both chitosan coated with poly-L-lysine and chitosan-poly-L-lysine mixture had excellent nerve cell affinity with the ability to promote nerve cell to grow and function normally. Zheng *et al.* [21] first investigated three important surface properties (surface nanotopography, chemistry, and wettability) of chitosan/poly-L-lysine composite films and their effects on cell behavior of MC3T3-E1 osteoblast-like cells. It was found that pure chitosan film had the surface consisting of tightly packed, grain-like particles. When blended with 0.1% poly-L-lysine, the surface changed and displayed a topography composed of granules. Interestingly, when blended with 0.25% and 0.5% poly-L-lysine, the surface topography was changed to be fiber dominant. As the concentration of poly-L-lysine increased to 1%, 3%, and 5%, however, particle-dominant topographies appeared again and displayed larger particle size and greater average height than those of pure chitosan film. MC3T3-E1 cells strongly responded to surface topography of composite films. On fiber-dominant surface, cells exhibited a significantly higher level of adhesion, proliferation, osteocalcin gene expression, and mineralization than on particle- or granule-dominant surface. Compared with surface topography, surface chemistry, and wettability were not always closely correlated with MC3T3-E1 cell behavior.

Polycations, as well-known coating reagents, are often used for supporting cell attachment and growth [19, 22]. Zheng *et al.* [23-25] used three polycations (poly-L-lysine, polyethyleneimine, and poly-L-ornithine) to modify chitosan with blending method. Surface nanotopography, chemistry, and wettability of three chitosan/polycation composite films were investigated. The cytocompatibility of these composite films was also systematically evaluated with the culture of MC3T3-E1 osteoblast-like cells, PC12 cells, and Schwann cells. The results demonstrated that chitosan/poly-L-lysine composite film possessed high cytocompatibility with osteoblasts due to its ECM-like surface topography and proper surface chemistry and wettability. PC12 cells strongly responded to surface topography as well as surface chemistry and wettability. Of all the composite films, chitosan/poly-L-lysine films with nanofiber-dominant surface significantly supported the outgrowth and differentiation of PC12 cells. On three different chitosan/polycation composite films, Schwann cells were connected to each other and exhibited greater proliferation, compared to the chitosan control.

Poly-L-lysine and poly-L-ornithine-modified substrates exhibited *in vitro* biocompatibility with Schwann cells and PC12 cells, which indicated that these substrates could serve as suitable substrates for peripheral nerve regeneration.

2.2. Chemical Modifications

The degree of deacetylation (DD) is an important parameter of chitosan, which influences not only the physicochemical characteristics [26, 27] but also the biological properties [28, 29] of chitosan materials. Cao *et al.* [30] prepared six kinds of chitosan samples with similar molecular weight but various DD in a range from 70.1 to 95.6% by the method of Mima [31] and measured their crystallinities, swelling properties, mechanical properties, and protein adsorption capacities. Then the authors investigated Schwann cells spreading and proliferation behaviors on these films. The results indicated that DD had a marked effect on the physicochemical properties and Schwann cell affinity of chitosan films. Higher DD chitosan films showed a greater crystallinity, a higher elastic modulus and tensile strength and a lower swelling index than those with lower DD. Chitosan films with DD higher than 90% presented better substrata for the spreading and proliferation of Schwann cells. These results suggested that chitosan with DD higher than 90% would be considered as a promising material for application in peripheral nerve regeneration.

Cross-linking is a general method to prolong the degradation of biomaterials. Cross-linking makes chitosan less susceptible to degradation by enzymes [32]. However, some cross-linking agents are generally cytotoxic that may impair the biocompatibility of the cross-linked biomaterials [33]. Both the biodegradation rate and the biocompatibility of the cross-linked chitosan conduits should be taken into consideration in the clinical application. In Cao *et al.* study [34], three kinds of cross-linking agents, including hexamethylene diisocyanate (HDI), epichlorohydrin (ECH) and glutaraldehyde (GA), were selected to cross-link chitosan films and the effects of cross-linking agents on the properties of the cross-linked films were examined. The cross-linked films' swelling properties, mechanical properties and biodegradability were evaluated. In addition, primary rat Schwann cells were cultured on these crosslinked films and the cell spread and proliferation behavior were investigated. They found that after cross-linking, the swelling index and the degradation rate of all the chitosan films decreased while their hydrophilicity and elastic modulus increased. However, the films crosslinked with ECH and GA were not better substrata for the growth of Schwann cells in comparison with CHI films. In contrast, the HDI cross-linked films enhanced the spread and proliferation of Schwann cells compared with CHI films. Therefore, surface cross-linking with HDI is a promising way to modify chitosan nerve repair conduits.

Degradability is a critical material property to many applications in tissue engineering [35]. The degradation of chitosan, however, is very slow and poorly controlled [36]. Chitosan degradation occurs via a slow and unpredictable dissolution process both *in vitro* and *in vivo*, likely due to its insolubility in physiological pH [37]. This is a significant drawback, as one naturally desires to fit the degradation rate of polymeric scaffolds to the rate of new tissue formation [38]. To accelerate the degradation rate of chitosan, Lu *et al.* [39] developed a technique to induce carboxyl groups within chitosan via a carboxymethylation reaction. Degradation of 1-ethyl-3(3-dimethylaminopropyl) carbodiimide hydrochloride (EDC) crosslinked carboxymethyl chitosan (CM-chitosan) was investigated by measuring the weight loss over time. Culture studies of Neuro-2a cells were also performed on these degradable CM-chitosan films. They found that EDC cross-linked CM-chitosan *in vitro* degraded much faster compared to chitosan, with the mechanical properties retained after carboxymethylation and cross-linking. After cross-linking, the hydrophilicity and elastic modulus of the CM-chitosan films decreased. The constructed matrix degraded to 30% in weight within 8 weeks of incubation in lysozyme solution. In addition, the EDC cross-linked CM-chitosan films enhanced the spread and provided a good proliferation substratum of Neuro-2a cells. However, the biocompatibility of CM-chitosan is likely to decrease at high degrees of carboxymethylation and cross-linking. An alternative approach to regulate degradation involves controlling the molecular weight distribution (MWD) of the polymer chains used to form scaffolds. The use of a bimodal MWD (i.e., a mixture of high MW polymer and lower MW polymer) allows one to control the degradation rate of cross-linked CM-chitosan scaffolds in an independent manner [40] and may allow one to maintain mechanically stable scaffolds with a controlled rate of degradation. Lu *et al.* [41] also investigated whether the degradation rate of chitosan could be controlled by using a combination of carboxymethylation of the chitosan prior to scaffold formation and utilization of a bimodal MWD during scaffold formation. The results demonstrated that carboxymethylation and bimodal MWD were successfully combined to regulate chitosan degradation. Controlling the MWD of properly tailored chitosan allows one to regulate the mechanical properties and degradation of chitosan in a sophisticated manner, while maintaining favorable cell interactions.

Li *et al.* [42] improved the mechanical properties and strengthened the biological properties of chitosan by modifying chitosan with butyl bromide. N-butyl chitosan with different substitution degrees was prepared by solid–liquid heterogeneous reaction. This study thoroughly analyzed the physical properties and MC3T3-E1 cell affinity of N-butyl chitosan. FTIR, free amino percentage analysis, and X-ray diffraction results confirmed that the substitution reaction occurred primarily on NH_2 moieties. The static contact angles experiment and the swelling ratio experiment showed the increased hydrophobicity of N-butyl chitosan. Compared with chitosan films, N-butyl chitosan films exhibited a higher elastic modulus, which is advantageous in bone tissue engineering. N-Butyl chitosan films were more favorable for MC3T3-E1 cell attachment and proliferation, and MTT assay indicated the cell viability of N-butyl chitosan films was greater than that of chitosan film. The results obtained from ELISA indicated the adsorption amount of fibronectin on butyl chitosan films was much higher than on chitosan films.

A number of extracellular matrix (ECM) proteins such as fibronectin, vitronectin, and laminin that contain the cell-binding domain RGD (Arg-Gly-Asp) have been shown to play a critical role in cell behavior, because they regulate gene expression by signal transduction set in motion by cell adhesion to the biomaterial [43, 44]. The GRGDS sequence is a very intriguing member of RGD family. $\alpha_V\beta_3$, $\alpha_V\beta_5$, and $\alpha_{IIb}\beta_3$ are the integrins most reported to be involved in bone function [45]. The GRGDS peptide has been shown to have comparable affinity to $\alpha_V\beta_3$ and $\alpha_{IIb}\beta_3$ at an intermediate level and may be useful if no particular integrin is to be addressed for cell adhesion [46]. It is also reported that GRGDS plays an essential role in adhesion, remodeling, and osseointegration at the interface between a biomaterial and bone [47]. The combination of the GRGDS sequence and chitosan may be beneficial to the culture of osteoblasts. Li *et al.* [48] investigated the immobilization of the GRGDS sequence on chitosan surface by the imide bond-forming reaction through a short four-carbon arm, as well as the modulation of MC3T3-E1 cell behavior on GRGDS-coupled chitosan. The results demonstrated that chitosan films could be successfully modified with GRGDS peptides, and the concentration of the immobilized GRGDS on the surface of chitosan was measured to be on the order of 10^{-9} mol/cm^2. GRGDS immobilization enhanced attachment and proliferation of MC3T3-E1 cells on chitosan, which has been shown to be dependent on peptide concentration. Competitive inhibition of MC3T3-E1 cell attachment with soluble GRGDS peptides indicated that MC3T3-E1 cell adhesion was specifically mediated by GRGDS-sensitive cell adhesion receptor. The cytoskeletal organization of MC3T3-E1 cells was highly affected by the immobilization of GRGDS peptide on chitosan. The migration distance of MC3T3-E1 cells on the GRGDS-coupled surface was greater than the distance observed on unmodified chitosan. In addition, the migration of osteoblasts was enhanced with increasing peptide concentration. The GRGDS-coupled chitosan surface accelerated differentiation and mineralization of the MC3T3-E1 cells as evidenced by expression of Alkaline Phosphatase (ALP) and calcium deposition. The ALP activity and calcium deposition were also affected by peptide concentration. Ho et al. [49] improved the chitosan scaffold for more cell attachment by immobilizing RGDS peptide sequence on it. When they cultured rat osteosarcoma cells on such modified scaffolds, they found that more cells attached on the scaffolds than unmodified scaffolds and formed bone like tissue. These results indicated that optimization of peptide on the surface of chitosan presented here may be an effective method for modulating osteoblast function in scaffold-based bone tissue engineering.

Hydroxyapatite (HA) is one of the major constituents of the inorganic component in human hard tissues (bones and teeth), and it is one of the most common biomaterials studied in bone tissue engineering because of its good biocompatibility [50, 51]. It can form a direct chemical bond with surrounding tissues and is osteoconductive, nontoxic, noninflammatory, and nonimmunogenic [52-55]. However, the migration of the nano-HA particles from the implanted site into surrounding tissues might cause damage to healthy tissue [56]. To find a resolution, composites of nano-HA and polymers were researched to find a material that retained the good properties of nano-HA and prevented the nano-HA particles from migrating. A composite of HA and chitosan therefore is expected to be a good biomaterial. Kong *et al.* [57] used a chemical wet method to produce a homogenous nano-HA/chitosan composite scaffold with porous structure. Chitosan and HA were combined homogenously through the in situ synthesis of nano-HA using the wet chemical method. The porous structure of the composite scaffolds was made by the lyophilization. The spongy scaffolds showed good porosity and some cells could grow in the pores of these 3-D scaffolds. On the pore walls of the scaffolds, the nano-HA particles were inlaid in the chitosan surface like islands. The composite scaffolds showed better biocompatibility than chitosan scaffolds. Cells grown on the composite scaffolds were in a better state and had a higher proliferation.

The first report on the modification of chitosan with sugars has been carried out by Hall and Yalpani [58, 59]. They synthesized sugar-bound chitosan by reductive N-alkylation using $NaCNBH_3$ and unmodified sugar or sugar-aldehyde derivative. At that moment, the sugar-bound chitosans were mainly investigated on the rheological study. Since the specific recognition of cell, virus, and bacteria by sugars has been discovered, this modification has generally been used to introduce cell specific sugars into chitosan. Morimoto *et al.* [60-63] reported the synthesis of sugar-bound chitosans such as D- and L-fucose and their specific interaction with lection or cells. Kato *et al.* [64] also prepared lactosaminated N-succinyl-chitosan and its fluorescein thiocarbanyl derivative as a liver-specific drug carrier in mice through asialoglycoprotein receptor.

N-phthaloylation of chitosan is an efficient procedure to improve its organosolubility. Moreover, N-phthaloylchitosan is a key intermediate to prepare regioselective derivatives of chitosan or to attach various functional groups in organic solvents. Liu *et al.* [65] reported the unique procedure by microwave irradiation to prepare N-phthaloylchitosan. Trimethylsililation (TMS) of chitosan may be another way of solubilization as suggested from the silylation of polysaccharides such as cellulose, dextran, and chitin. Kurita *et al.* [66] reported the efficient way for the TMS of chitosan. The TMS-chitosan was almost soluble in pyridine and swelled considerably in common organic solvents such as acetone, THF, and DMSO. Though TMS-chitosan was hydrolytically stable in neutral and weakly alkaline media, it was deprotected readily with acid. Polymeric materials having specific properties such as photoactivity, in particular the ability to participate in energy of electron transfer process, have become increasingly important in recent years. Wu *et al.* [67] prepared photosensitive chitosan containing anthracene chromophore. This photosensitive chitosan could lead to the application of the environmentally friendly photocatalytic system operating with visible light to conduct an efficient degradation of wide range of pollutants.

3. The Applications of Chitosan in Tissue Engineering

Tissue engineering, as stated by Langer and Vacanti, is "an interdisciplinary field that applies the principles of engineering and life sciences toward the development of biological substitutes that restore, maintain, or improve tissue function or a whole organ" [68]. It is the use of a combination of cells, engineering and materials methods, and suitable biochemical and physio-chemical factors to improve or replace biological functions. Tissue engineering is an emerging multidisciplinary field involving biology, medicine, and engineering that is likely to revolutionize the ways we improve the health and quality of life for millions of people worldwide by restoring, maintaining, or enhancing tissue and organ function. The foundation of tissue engineering for either therapeutic or diagnostic applications is the ability to exploit living cells in a variety of ways.

Tissue engineering aims to create medical devices that, once implanted, will replace or enhance tissue function that has been impaired by disease, injury, or age. These devices are typically created by seeding cells in a biomaterial scaffold. Cells attach to the scaffolds and reorganize them to form functional tissue by proliferating, synthesizing extracellular matrix, and migrating along the scaffold. This reorganization can begin to occur outside previously of the body in a bioreactor and then continue after implantation into a patient. As a key element in tissue engineering, scaffold plays a very important role in the growth of tissues. The scaffolds are often made of polymers designed to degrade slowly and safely in the body, disappearing as the cells make their renovations. They can be extremely complex chemically: an ideal scaffold must facilitate cell attachment without provoking an immune response, permit the diffusion of nutrients from the blood, and, at least initially, mimic the mechanical properties of the tissue. Additionally, the scaffold can be constructed with or designed to release factors that can beneficially manipulate the behavior of the cells in the device.

Chitosan is one of the widely used natural polymers in tissue engineering due to its biocompatibility, biodegradability, low toxicity, antimicrobial properties, ease of modification, and easy fabrication into various forms such as film, fiber, porous scaffold, microsphere, hydrogel, and tube.

3.1. Cartilage and Bone Tissue

Cartilage is a stiff yet flexible connective tissue found in many areas in the bodies of humans and other animals, including the joints between bones, the rib cage, the ear, the nose, the elbow, the knee, the ankle, the bronchial tubes and the intervertebral discs. It is not as hard and rigid as bone but is stiffer and less flexible than muscle. Cartilage is composed of specialized cells called chondrocytes that produce a large amount of extracellular matrix composed of collagen fibers, abundant ground substance rich in proteoglycan, and elastin fibers. Bones are rigid organs that form part of the endoskeleton of vertebrates. They function to move, support, and protect the various organs of the body, produce red and white blood cells and store minerals. Bone tissue is a type of dense connective tissue. Both bones and cartilages help in supporting and locomotion of the body. Chitosan has been widely used as scaffold material for regeneration of cartilage and bone tissues.

The microcarrier (MC) system recently has gained much attention in cartilage tissue engineering [69, 70]. An important advantage of this system is that cell-seeded MCs can be delivered directly to the place that needs repair [71]. MCs and expanded cells can be injected or administered directly, thus eliminating the need for reseeding the retrieved cells into a scaffold delivery system. Lu *et al.* [72] reported the nanofibrous collagen-coated CM-chitosan microcarriers (CMC-MCs) for chondrocyte culture. They found that naturally structured collagen on the polymer MCs played an important role in cell adhesion and cell growth. *In vitro* chondrocyte culture revealed better cell attachment, proliferation, and differentiation on the CMC-MCs immobilized with self-assembled collagen nanofibers. Cells were observed to grow into a tissue-like structure after 7 days of culture. Immunofluorescence and RT-PCR analysis confirmed that these MCs supported chondrocytes to express and produce cartilage ECM components. CMC-MCs prepared by a modified phase separation method combined with temperature controlled freeze-extraction could be expected to find important application as injectable scaffolds for cartilage tissue engineering. Nettles *et al.* [73] reported one of the preliminary experiments to demonstrate chitosan as scaffolds for cartilage tissue engineering. They concluded that chitosan scaffolds may be a useful alternative to synthetic cell scaffolds for cartilage tissue engineering. Hoemann *et al.* [74] used chitosan in solution form with chondrocytes to regenerate damaged cartilage. It was observed that the chitosan self-gelling solution could preserve chondrocyte viability and phenotype after injection and solidification, serve as a scaffolding to help build tissue with mechanical properties, and reside in articular defects in live animal joints for at least 1 week after implantation. Sechriest *et al.* [75] modified chitosan with chondroitin sulfate A and used it for supporting chondrogenesis. Chondrocytes cultured on this scaffold synthesized higher amounts of collagen II and exhibited limited cell division, characteristics of differentiated chondrocytes. Xia *et al.* [76] isolated autologous chondrocytes isolated from pig's auricular cartilage and seeded them onto the chitosan-gelatin scaffold. They engineered elastic cartilages at the porcine abdomen subcutaneous tissue. After 16 weeks of implantation, the engineered elastic cartilages had acquired not only normal histological and biochemical properties, but also normal mechanical properties. Lee *et al.* [77] developed a three-dimensional chitosan scaffold in combination with transforming growth factor-beta 1 (TGF-beta 1) loaded chitosan microspheres and evaluated the effect of the TGF-beta 1 release on the chondrogenic potential of rabbit chondrocytes in the scaffolds. They found that the scaffolds showed higher chondrocytes attachment levels than the chitosan scaffolds, regardless of the TGF-beta 1 microspheres. Both the proliferation rate and glycosaminoglycan production of the chitosan scaffolds incorporating TGF-beta 1 microspheres were significantly higher that of the control scaffolds without microspheres 10 days after incubation. Hsu *et al.* [78] evaluated the chitosan-alginate-hyaluronate complexes modified by an RGD-containing protein as tissue engineering scaffolds for cartilage regeneration. One month after the chondrocyte scaffolds were implanted into rabbit knee cartilage defects, partial repair was observed.

The regeneration of bone is one of the major difficulties in clinical surgery because many factors including trauma, tumor, and bone diseases such as osteitis and osteomyelitis can cause bone defects. To restore the structure and function of bone, many solutions have been used in therapy and research, including autografts, allografts, xenografts, and other artificial substitutes [79-81]. Autografts need secondary surgery to procure donor bone from the patient's own body and the amount of donor bone is limited [82, 83]. Allogenic bone substitutes and xenografts bear the risk of infections and immune response [84]. Artificial bone made of metals and bioceramics does not behave like true bone and it is hard to inosculate with surrounding tissues. HA/chitosan composite has been used in teeth and bones due to its chemical and crystallographic similarity to the carbonated apatite. Kong *et al.* [85] examined bioactivity of chitosan/nano-HA and chitosan scaffolds in an *in vitro* biomimetic process. The chitosan/nano-HA composite scaffolds showed better biomineral activity than chitosan scaffolds. In the composite scaffolds, nano-HA particles provided nuclei in the mineralization process. As a result more apatite formed on the composite scaffolds than on the chitosan scaffolds. The addition of nano-HA influenced the composition of the apatite layer. The results also suggested that nano-HA could enhance the coating of apatite layer on biomaterials, which could be used to produce apatite polymer composite scaffolds. Furthermore, it was shown that the structure of the apatite layer also influenced the viability and differentiation of preosteoblast cells. The apatite formation and cells viability reflected the properties of the substrate: the chitosan/nano-HA composite showed a higher bioactivity for bone tissue engineering. Kong *et al.* [86] also designed a novel chitosan/nano-HA composite scaffold for bone-graft applications. This multilayer scaffolds had both enhanced mechanical properties and increased pore size compared with the uniform porous scaffolds made with normal lyophilization. In addition, a guided bone regeneration membrane was combined with these scaffolds to prevent fibrous tissue growth into the bone defect site. Cell culture assessment suggested that the multilayer scaffold was more compatible than the uniform porous scaffold for cell ingrowth. *In vivo* New Zealand White rabbit assessment demonstrated that the multilayer scaffolds prevented the formation of fibrous tissue in the defect site and their central larger pores ensured the nutrition supply. The biomimetic multilayer scaffolds showed good biocompatibility and osteoconductivity *in vivo*. Zhao *et al.* [87] prepared a biodegradable HA/chitosan-gelatin network composite of similar composition to that of normal human bone as a three-dimensional biomimetic scaffold for bone tissue engineering. The osteoblasts cultured on the scaffold synthesized collagen I and proteoglycan like substrate and also formed bone like tissue during the culture period.

Lahiji *et al.* [88] cultured osteoblasts and chondrocytes on chitosan coated cover slips and analyzed the expression of collagens and aggrecans. Both types of cells on chitosan exhibited similar morphology as *in vivo* and viability greater than 90% even after seven days of culture. Seol *et al.* [89] showed the bone formation in chitosan sponges *in vitro*. They grew rat calvarial osteoblasts in porous chitosan sponges and monitored for cell density, ALP activity, and calcium deposition up to 56 days culture. Histological results corroborated bone formation with the sponges had occurred. Calcium phosphate incorporation into chitosan has been found to increase the strength of the matrices. Lee *et al.* [90] cultured fetal rat calvarial osteoblastic cells on the chitosan-tricalcium phosphate scaffold. The scaffolds supported the proliferation of osteoblastic cells as well as their differentiation as indicated by high ALP activities and deposition of mineralized matrices by the cells. Alginate has been also used to modify chitosan for bone tissue regeneration. One of the advantages of using alginate with chitosan is that unlike chitosan scaffolds which can only be fabricated from acidic solutions, the chitosan-alginate scaffolds can be prepared in acidic, basic, or neutral solution. This unique attribute provides a favorable environment for incorporating proteins with less risk of denaturation for sustained release *in vivo*. Studies showed that osteoblasts readily attached to the chitosan-alginate scaffold, proliferated well, and deposited calcified matrix. Park *et al.* [91] used chitosan-alginate with mesenchymal stem cells/bone morphogenetic protein-2 (BMP-2) as injectable material to generate new bone successfully. The composites were injected into the subcutaneous space on the dorsum of nude mice to investigate new bone tissue formation and were examined histologically over a 12-week period. The results showed that composites injected into the mouse were able to stimulate new bone formation.

3.2. Nervous Tissue

Nervous tissue is one of four major classes of vertebrate tissue. Nervous tissue is the main component of the nervous system—the brain, spinal cord, and nerves—which regulates and controls body functions. It is composed of neurons, which transmit impulses, and the neuroglia, which assist propagation of the nerve impulse as well as provide nutrients to the neuron. Neural tissue engineering is primarily a search for strategies to eliminate inflammation and fibrosis upon implantation of foreign substances. Often foreign substances in the form of grafts and scaffolds are implanted to promote nerve regeneration and to repair damage caused to nerves of both the peripheral and central nervous systems by an injury. Chitosan can be used as channel or conduit material to bridge the gaps between the two broken ends of a nerve.

Zhang *et al.* [92] used a designed mold shown in Figure 1 to prepare the hollow chitosan conduits. Chitosan acetic acid solution was injected gently into a homemade mold using a syringe. The mold included a stainless steel outer tube, a cone-shaped pedestal, a coping, and a mandrel that consisted of a stainless steel rod smocked with a Teflon tube. The cone-shaped pedestal was fixed on the outer tube. When the height of the chitosan solution in the mold was near the top opening of the tube, a coping was fixed onto the tube and the mandrel was inserted into the chitosan solution through the center port of the coping.

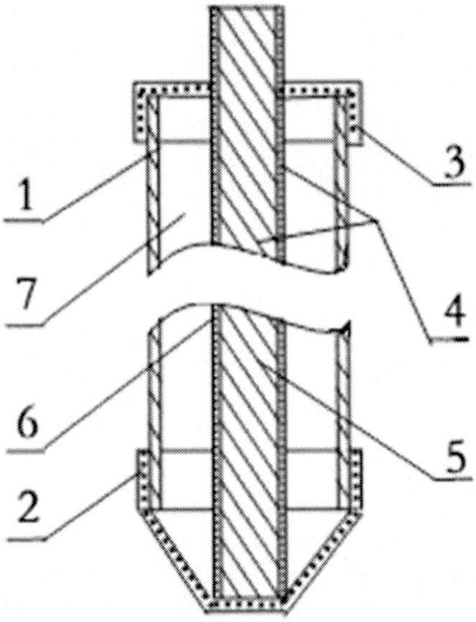

Figure 1. Schematic drawing of the optimally designed mold used in preparing hollow chitosan conduits. (1) stainless steel tube; (2) cone-shaped pedestal; (3) coping; (4) mandrel, including 5 and 6; (5) stainless steel rod; (6) Teflon tube; and (7) chitosan solution.

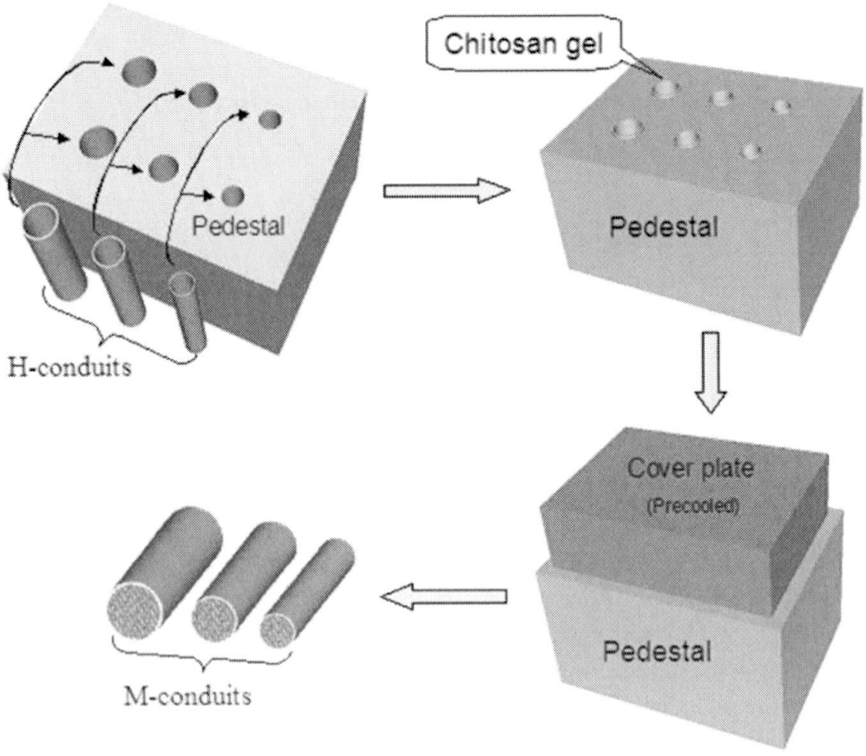

Figure 2. Schematic diagram for procedure of fabricating multimicrotubule chitosan nerve conduits with various diameters.

The mandrel was inserted into the bottom of the cone-shaped pedestal to ensure that the mandrel was located in the center of the tube and the wall of the final chitosan tube was even. The mold with chitosan solution was placed in liquid nitrogen or a freezer for 12 h to thoroughly freeze the chitosan solution. Then the frozen mold was placed at room temperature for several minutes and the pedestal and coping were removed. After 10 min or so, the outer layer of frozen chitosan solution thawed, and the outer tube was removed by extrusion. The frozen chitosan tube combined with the mandrel was dried in a freeze-dryer. The main drying temperature was -40°C and the main drying pressure was 12 Pa. After 12-15 h, the freeze-drying course of the chitosan solution was completed. Then the samples were immersed into 2% (w/v) sodium hydroxide (NaOH) solution and equilibrated for 10-20 min to eliminate the remaining acetic acid. Samples were rinsed several times using deionized water until the rinsing solution was neutral, and then equilibrated in 0.2 mol/L phosphate buffered saline (pH = 7.4) for 30 min. The surface water of the samples was absorbed by filter paper and dried at room temperature or freeze-dried, and then the mandrel was removed to produce a porous hollow chitosan conduit. As discussed before, EDC cross-linked CM-chitosan had improved biodegradability. Wang *et al.* [4] used CM-chitosan tubes prepared by the above method to achieve a successful peripheral nerve regeneration in a 10 mm rat sciatic nerve defect model. *In vitro* cell culture indicated that the CM-chitosan films had no cytotoxicity to Schwann cells. *In vivo* studies showed that CM-chitosan tubes had better nerve regeneration performance than the chitosan tube and demonstrated equivalence to nerve autografts, which is the current ''gold'' standard, as assessed by histomorphometry analysis.

On the basis of hollow chitosan conduits, several studies have been taken to fabricate different structural chitosan conduits. Ao *et al.* [93] used a uniquely designed mold (Figure 2) to produce multimicrotubule chitosan conduits (M-conduits). The mold composed of a styrofoam insulating pedestal with several holes and a stainless steel cover plate was used to produce M-conduits. The holes in the pedestal had various diameters and depths. In brief, the fabrication process was as follows. Corresponding hollow chitosan conduits (H-conduits) were inserted upright into the holes in the styrofoam pedestal, and filled with chitosan acetic acid solution, then rapidly covered with a precooled stainless steel cover plate, and placed in a freezer (-20 to -80°C). The styrofoam insulating pedestal enclosing the tubes could reduce the heat transfer through the side wall, and heat transfer occurred mainly through the top cover plate. Gradual phase separation then occurred uniaxially in the presence of a unidirectional temperature gradient from the top end to the lower end of the chitosan conduits. The microtubule diameters were controlled by adjusting the polymer concentration and cooling temperature, which is with increasing of solution concentration and decreasing of cooling temperature, the pore sizes became smaller, which was consistent with others' results. The phase-separated polymer/solvent samples were pulled out from the pedestal, placed in a cooled container, and dried in the freeze-dryer. The freeze-dried samples were deacidified and rinsed as aforementioned, then dried at room temperature. Thus M-conduits with various diameters and depths were prepared. The axially oriented microtubules extend through the full length of the conduit and the structures were homogeneous. These conduits contained longitudinally aligned microtubules for guiding the outgrowing nerve fibers, and an outer membrane for preventing ingrowth of fibrous tissue into the nerve gap. *In vitro* characterization demonstrated that the mold-based multimicrotubule chitosan conduits had suitable mechanical strength, microtubule diameter distribution, porosity, swelling, biodegradability, and nerve cell affinity, and so they hold promise in applications as nerve tissue engineering scaffolds. Wang *et al.* [94] cultured neural stem cells (NSCs) on M-conduits in DMEM/F12 medium with 10% fetal bovine serum (FBS). As a result of contact guidance by microtubule, a certain part of NSCs elongated along a preferred axis showing a morphology indication of neurons or neuronal progenitors. The immunofluorescence staining image showed the cells cultured on M-conduits had β-tubulin III positive immunoreactivity. M-conduits demonstrated a significant promotion in neuronal differentiation and an acceptable proliferation behavior. The differentiation and proliferation properties of NSCs cultured on M-conduits might be exploited in neural tissue engineering.

Wang *et al.* [95] describe a method of preparation of chitosan nerve conduits that introduced braided chitosan fibers to reinforce the porous structure. In brief, chitosan yarns were first utilized to prepare fabric tubes with various inner diameters and wall thicknesses using a flexible industrial knitting process. A segment of the knitted chitosan tube was cut off the fabric tube and a mandrel of proper size was inserted into it. Then, they were dipped into 2% (w/v) chitosan solution for 1 min. After being taken out of the chitosan solution, they were placed vertically at -80°C for 12 h, and subsequently lyophilized in a freeze-dryer. The freeze-dried samples were immersed into 2% aqueous solution of sodium hydroxide and equilibrated for 30 min to neutralize the acetic acid, then rinsed with deionized water several times until the rinsing solution was neutral and finally dried at room temperature. The mandrel was removed, and the porous chitosan fiber-based hollow tubes were obtained. *In vitro* direct contact test with neuronal cells and *in vivo* Wistar rat biocompatibility study indicated that the porous chitosan-based fiber-reinforced conduit were non-cytotoxic and tissue compatible, and therefore, represented a promising candidate conduit for peripheral nerve regeneration.

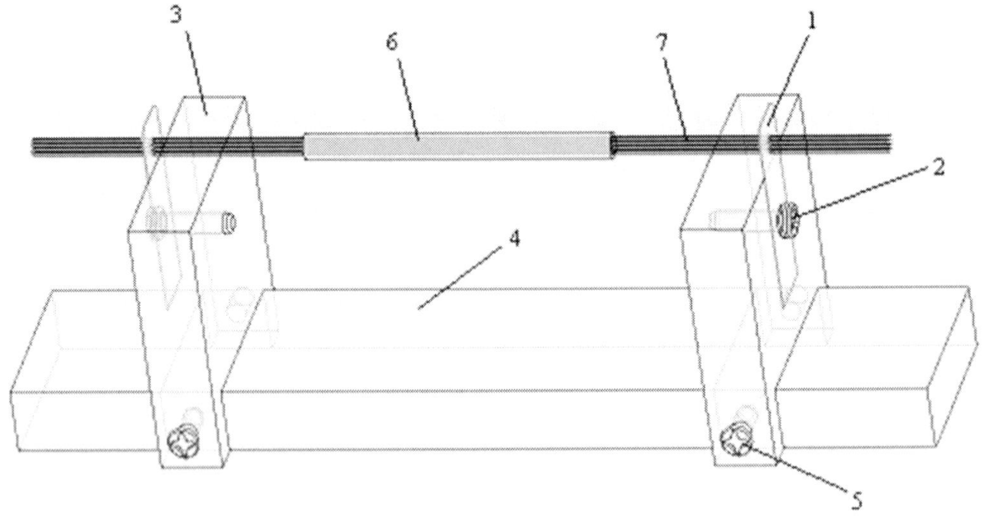

Figure 3. The optimally designed mold for the manufacture of guidance scaffolds: (1) fixing patches with fixing holes whose distribution was computer-aided designed and perforated under computer control, (2) setscrews to fix the patches to the moveable blocks, (3) moveable blocks to define the length of the scaffolds, (4) the pedestal with a smooth surface for moveable blocks to move freely, (5) screws to fix the moveable blocks with the pedestal, (6) prefabricated fiber-based chitosan hollow tube, and (7) the acupuncture needles.

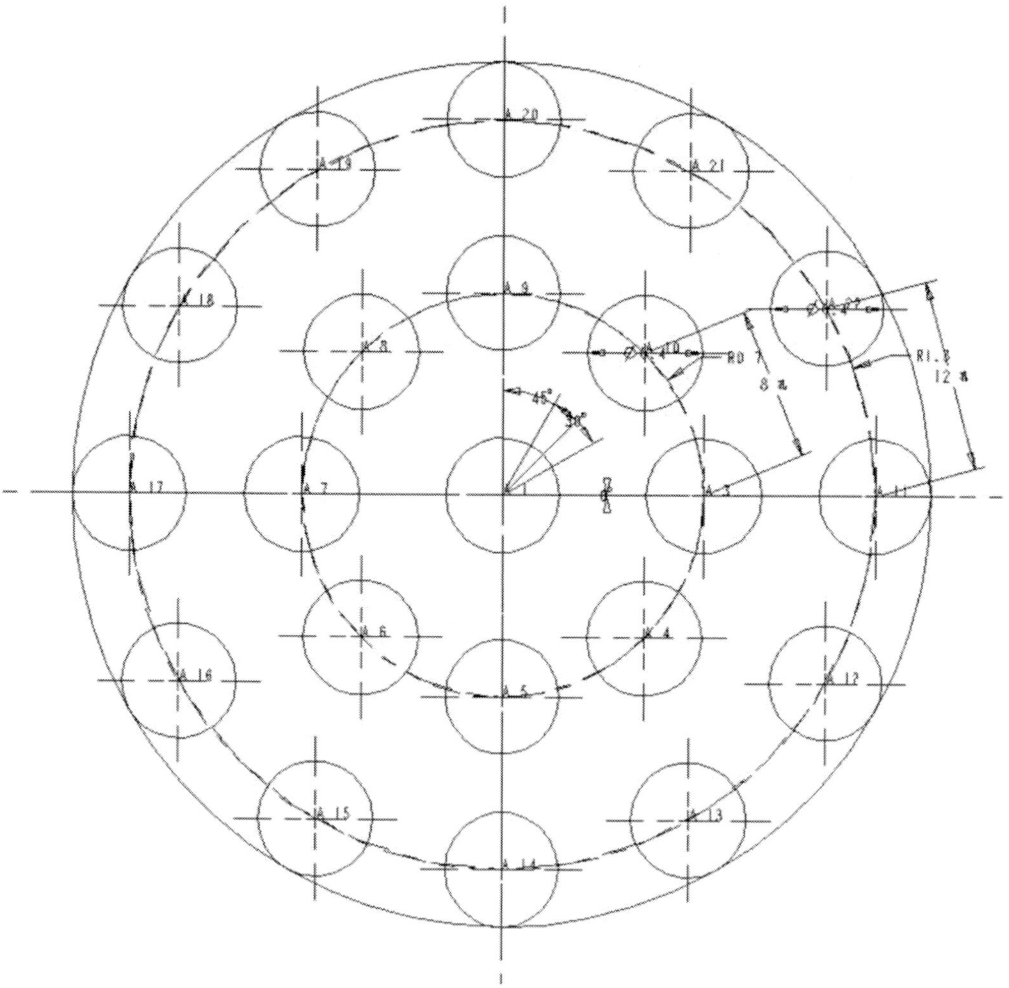

Figure 4. Representative scheme of the hole distribution and organization on the fixing patches. In this scheme, the largest outer circle diameter was 3.0 mm, which matched the inner diameter of the outer wall of the guidance scaffold. The two dashed concentric circles indicate the alignment of the holes. The 21 small circles (400 μm in diameter) represent the position of the holes that define the distribution of the macrochannels, while the crosses represent the center of the circles. As shown in the scheme, the distance between small circles could be changed flexibly under computer control by varying the angle from the inner circle to the outer ring of circles.

Wang *et al.* [96] also described a new process for producing chitosan nerve guidance scaffolds with desirable mechanical properties and a controllable inner structure for guiding neurite extension. As shown in Figure 3, the optimally designed mold was composed of acupuncture needles, two fixing patches, two moveable blocks, and a pedestal. The acupuncture needles were used as mandrels to define the longitudinally aligned macrochannels and secured by two fixing patches through the fixing holes at either end. The patches were fastened to the moveable blocks and the moveable blocks were located perpendicularly on the pedestal and could move along the pedestal freely to define the length of the scaffold (commonly from 10 to 150 mm). The most essential parts of the mold were the fixing holes, which determined the positions of the mandrels. As shown in Figure 4, the number, diameter, and distribution of the fixing holes were computer-aided designed and well controlled in a two dimensional (2D) coordinate system in accordance with the parameters selected for the acupuncture needles. Chitosan nerve guidance scaffolds were produced using the computer-aided designed mold and a phase separation technique. A segment of a prefabricated chitosan hollow tube was set between the two fixing patches. Then, under a microscope, each acupuncture needle was threaded through the fixing hole in one patch, through the chitosan tube, and then through the corresponding hole in the other patch. The needles were parallel to each other. Then, 2% (w/v) chitosan solution was slowly injected into the tube to fill the whole space between the needles. The moveable blocks were moved together, ensuring the two fixing patches were in close contact with the ends of the tube. Then, the whole assembly was placed at -20°C for 12 h, and subsequently lyophilized, deacidified, rinsed, and dried as mentioned earlier. Finally, the needles were slowly rotated and carefully removed from the structure to reveal the axially oriented macrochannels. The macroscopic appearance of the scaffold was shown in Figure 5. The cross-sectional geometry of the scaffolds and the channels was approximately circular with slight irregularities and the channels were well aligned within the inner diameter of the outer tube wall. The nine sequential slices had almost the same transverse morphology, which indicated that the scaffold had a continuous macrochanneled structure. Representative micro-CT images of the chitosan porous guidance scaffolds are shown in Figure 6. As expected, micro-CT imaging confirmed that the macrochannels defined by acupuncture needles extended through the whole length of the scaffold along the longitudinal axis. The porous knitted outer wall of the scaffolds produced in this study improved the scaffold mechanical properties. The radially interconnected micropores increased the surface area for cell adherence and can facilitate nutrient diffusion and cell transfer to the scaffold's interior. The longitudinally aligned macrochannels are expected to provide templates for cell migration and axonal elongation, and ultimately to support nerve regeneration in a natural fascicle form.

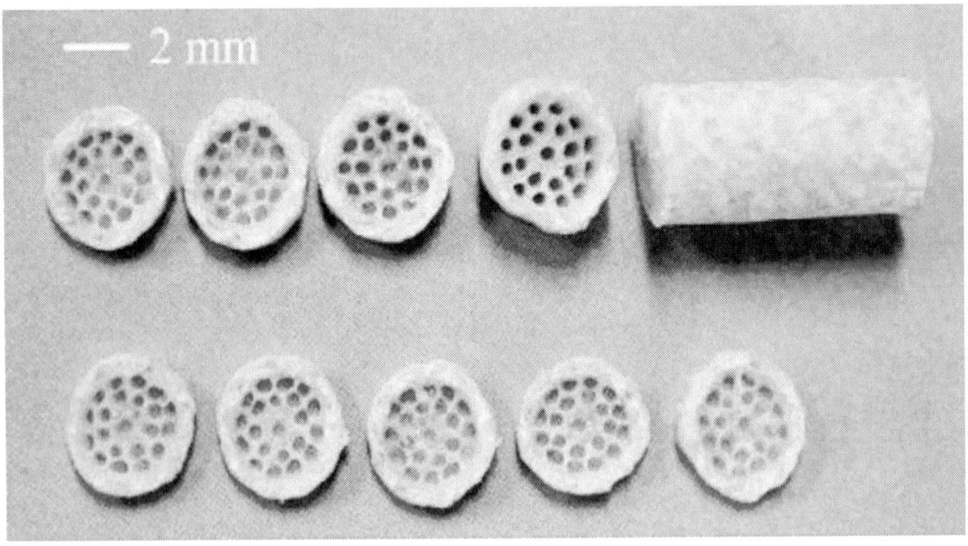

Figure 5. Optical images of guidance scaffolds.

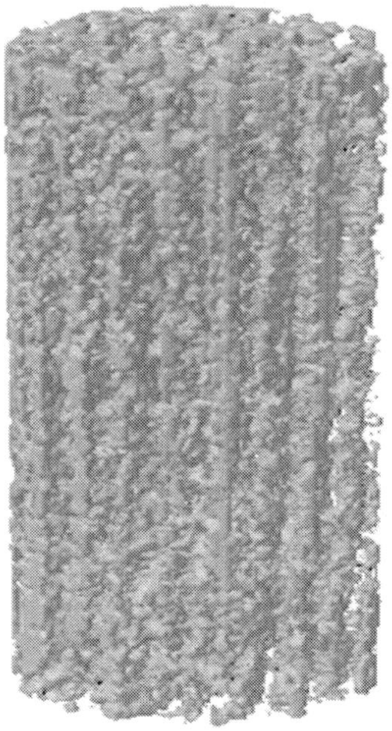

Figure 6. Representative micro-CT images of the guidance scaffolds. The macrochannels defined by acupuncture needles are visible as axially oriented extending through the whole length of the scaffold (top to bottom).

Rosales-Cortes *et al.* [97] studied the immunological response against chitosan. The results showed that when the regeneration of the axotomized sciatic nerve was induced through tubulization with chitosan, it did not induce any immunologic activation or depression in the dog. Hu *et al.* [98] observed the impact of bridging the sciatic nerve defect in different lengths with chitosan membranous tube on peripheral nerve regeneration in rats. They found that small gaps were regenerated more efficiently than the longer ones. Chitosan conduits loaded with neuroactive components have been used for nerve regeneration. Chavez-Delgado *et al.* [99] designed chitosan prostheses containing progesterone to fill in a gap in the rabbit facial nerve. Their results showed that neurosteroids allowed guided nerve regeneration more efficiently. Rosales-Cortes *et al.* [100] used chitosan prosthesis loaded with progesterone to induce and stimulate the regeneration of the sciatic nerve in dogs. Suzuki *et al.* [101] developed tendon chitosan tubes having the ability to bind peptides covalently, and examined the effectiveness of laminin peptides coupled to the tubular walls on sciatic nerve regeneration in rats. Histological findings suggested that the immobilized laminin as well as laminin oligopeptides might effectively assist nerve tissue extension. Wei *et al.* [102] prepared chitosan collagen films for nerve regeneration. In rat model with sciatic nerve defect, chitosan-collagen films were sutured into channel to bridge 5 mm, 10 mm nerve defects. The results showed that in 5 mm nerve defects, the quality of nerve regeneration was similar to that of the end to end anastomosis group. For 10 mm nerve defects, nerve regeneration was inferior to that of the end to end anastomosis group.

3.3. Vascular Tissue

The blood vessels are the part of the circulatory system that transport blood throughout the body. There are three major types of blood vessels: the arteries, which carry the blood away from the heart, the capillaries, which enable the actual exchange of water and chemicals between the blood and the tissues; and the veins, which carry blood from the capillaries back towards the heart. Blood vessels play an important role in virtually every medical condition. Vascular transplantation has been commonly used for the treatment of vascular diseases. The search for vascular substitute materials has been a half-century endeavor [103]. The emergence of tissue engineering opens new possibilities in reconstructive vascular grafting. Scaffolds made of different materials, such as synthetic polymers, natural materials, and decellularized xenogenous tissues, have been utilized in blood vessel tissue engineering. Recently, chitosan has been used as a scaffold of tissue engineered blood vessels.

Zhang *et al.* [104] developed a novel chitosan-based scaffold for blood vessel tissue engineering with a sandwich-like multilayered structure. The inner and outer surface of knitted chitosan tubes were mantled by chitosan/gelatin complex solution. Porous chitosan/gelatin complex were formed at the inner and outer surfaces of the scaffold by thermally induced phase-separation technique. Using that method, they were able to engineer mechanically robust vessels derived from chitosan, which was compatible with vascular smooth muscle cells. The scaffold had potential for immediate application after heparin modification, or co-culturing with vascular smooth muscle cells and endothelial cells to cultivate a blood vessel *in vitro*. Cuy *et al.* [105] explored chitosan as a promising material for heart valve tissue engineering applications. They evaluated the adhesive properties of the heart valve endothelial cells and attempted to enhance their growth on chitosan. The endothelial cells form the smooth internal lining of vascular organs, known as endothelium which ensures the friction free flow of blood. The cells also secrete growth regulators and vasoactive substances. The heart valve endothelial cells were cultured on chitosan and chitosan modified with collagen type IV. Composite chitosan/collagen type IV films were shown to improve heart valve endothelial cells growth and morphology over chitosan alone. Chupa *et al.* [106] investigated the potential of glycosaminoglycans (GAG)-chitosan and dextran sulfate (DS)-chitosan complex materials for controlling the proliferation of vascular endothelial cells and smooth muscle cells. The results indicated that while chitosan alone supported cell attachment and growth, GAG-chitosan materials inhibited spreading and proliferation of vascular endothelial cells and smooth muscle cells *in vitro*. In contrast, DS-chitosan surfaces supported proliferation of both cell types. *In vivo*, heparin-chitosan and DS-chitosan scaffolds stimulated cell proliferation and the formation of a thick layer of dense granulation tissue. These results indicated that the GAG-chitosan materials could be used to modulate the proliferation of vascular cells both *in vitro* and *in vivo*. Chung *et al.* [107, 108] photochemically grafted cell adhesive peptide GRGD on chitosan surface and showed the increase in cell adhesion and proliferation of human umbilical vein endothelial cells. Zhu *et al.* [109] immobilized chitosan and other molecules like gelatin and collagen on poly (L-lactic acid) membranes for increasing endothelium regeneration. Human umbilical vein endothelial cells *in vitro* proved that the cell proliferation rate and cell activity of biomacromolecule-immobilized PLLAs were improved compared with control PLLA. Meng *et al.* [110] designed a biomacromolecular layer-by-layer coating process of chitosan/heparin onto a coronary stent for the acceleration of the re-endothelialization and healing process after coronary stent deployment. The results of *in vitro* culturing of porcine iliac artery endothelial cells supported the rationale that the combination of chitosan and heparin could bring both endothelial cell compatibility and haemocompatibility to the stent surface. A porcine coronary injury model and arteriovenous shunt model were used for the further evaluation of the application of this kind of surface-modified stainless steel stent *in vivo*. The final results proved that this facile coating approach could significantly promote re-endothelialization and was safer compared with bare metal stents for its much improved anticoagulation property. Zhu *et al.* [111] fabricated novel human-like collagen (HLC) /chitosan tubular scaffolds by cross-linking and freeze-drying process. Human venous fibroblasts were expanded and seeded onto the scaffolds. Furthermore, the scaffolds were implanted into rabbits' livers to evaluate their biocompatibility. The results indicated that HLC/chitosan tubular scaffolds (1) exhibited interconnected porous structure; (2) achieved the desirable levels of pliability (elastic up to 30% strain) and stress of 300 +/- 16 kPa; (3) were capable of enhancing cell adhesion and proliferation and ECM secretion; (4) showed superior biocompatibility. This study suggested the feasibility of HLC/chitosan composite as a promising candidate scaffold for blood vessel tissue engineering.

3.4. Liver Tissue

Liver is a vital organ present in vertebrates and some other animals. It has a wide range of functions, including detoxification, decomposition of red blood cells, glycogen storage, protein synthesis, and production of biochemicals necessary for digestion. The liver is necessary for survival; there is currently no way to compensate for the absence of liver function. It lies below the diaphragm in the thoracic region of the abdomen. It also performs and regulates a wide variety of high-volume biochemical reactions requiring highly specialized tissues, including the synthesis and breakdown of small and complex molecules, many of which are necessary for normal vital functions. Live transplantation is the only available successful treatment for end stage liver disease but the shortage of donors, high cost and clinical complexities are the major limitations due to which very few people undergo transplantation. Liver tissue engineering is supposed to represent an edge over transplantation. Hepatocytes make up 70-80% of the cytoplasmic mass of the liver. In contrast to direct cell injection, tissue engineering approaches to hepatocyte transplantation focus on delivering cells on scaffolds which create an appropriate environment for liver function. These scaffolds, typically fabricated from polymers, serve as a three dimensional support to promote cellular attachment and organization, a substrate from which to deliver factors promoting growth and/or differentiation of the cells, a barrier between the implanted cells and the patient's immune system, or any combination of these roles. Chitosan has been used either alone or in combination with other polymers as scaffold for liver tissue engineering.

Kawase *et al.* [112] showed that non crosslinked chitosan gel was too fragile and could not be used for cell culture. They crosslinked the free amino groups of chitosan gel with glutaraldehyde to increase its strength. Rat hepatocytes cultured on crosslinked chitosan gels showed that the cells could stably attach to the surface and exhibited the same morphology *in vivo*. They also retained higher urea synthesis activity, a liver-specific function, than those on the collagen-coated surface. The results indicated that chitosan was a promising material as a scaffold for hepatocyte attachment, which would be applied to an effective bioartificial liver support system. Lin *et al.* [113] prepared N-caproyl chitosan nanoparticles surface modified with glycyrrhizin. The results indicated that N-caproyl chitosan nanoparticle surface modified with glycyrrhizin was a stable and effective drug delivery vehicle for hepatocyte targeting. Wang *et al.* [114] synthesized a new chitosan (CS) derivative, galactosylated chitosan (GC), and used it to prepare norcantharidin-associated GC nanoparticles (NCTD-GC NPs) by taking advantage of the ionic cross-linkage between the molecules of the anti-hepatocarcinoma medicine NCTD and of the GC as carrier. *In vitro*, and compared to CS-based NCTD-CS NPs, NCTD-GC NPs demonstrated satisfactory compatibility with hepatoma cells and strong cytotoxicity against hepatocellular carcinoma cells. *In vivo* antitumor activity of NCTD-GC NPs was evaluated in mice bearing H22 liver tumors. NCTD-GC NPs displayed tumor inhibition effect in mice, better than either the free NCTD or the NCTD-CS NPs. As a hepatocyte-targeting carrier, GC NPs were potentially promising for clinical applications. She *et al.* [115] proposed a model of the self-assembly and examined hepatocytes attachment and inflammatory response for the silk fibroin/chitosan (SFCS) scaffold. The results showed that compared with the polylactic glycolic acid scaffold, the SFCS scaffold further facilitated the attachment of hepatocytes. Keeping the good histocompatibility and combining the advantages of both fibroin and chitosan, the SFCS scaffold could be a prominent candidate for soft tissue engineering, for example, in the liver. Feng *et al.* [116] developed a novel natural nanofibrous scaffold with surface-galactose ligands to enhance the bioactivity and mechanical stability of primary hepatocytes in culture. The nanofibrous scaffold was fabricated by electrospinning a natural material, galactosylated chitosan (GC), into nanofibers with an average diameter of approximately 160 nm. The GC nanofibrous scaffolds displayed slow degradation and suitable mechanical properties as an ECM for hepatocytes according to the evaluation of disintegration and Young's modulus testing. The results of morphology characterization, double-staining fluorescence assay and function detection showed that hepatocytes cultured on GC nanofibrous scaffold formed stably immobilized 3D flat aggregates and exhibited superior cell bioactivity with higher levels of liver-specific function maintenance in terms of albumin secretion, urea synthesis and cytochrome P-450 enzyme than 3D spheroid aggregates formed on GC films. These spheroid aggregates could be detached easily during culture period from the flat GC films. They suggested such GC-based nanofibrous scaffolds could be useful for various applications such as bioartificial liver-assist devices and tissue engineering for liver regeneration as primary hepatocytes culture substrates. Chu *et al.* [117] designed a chitosan nanofiber scaffold via electrospinning to enhance cell attachment and promote liver functions of hepatocytes cultured in bioreactors. Data showed that hepatocytes on chitosan nanofiber scaffold exhibited better viability and tighter cell-substrate contact than cells on regular chitosan film. In addition, urea synthesis, albumin secretion and cytochrome P450 activity of hepatocytes on chitosan nanofiber scaffold were all 1.5 to 2 folds higher than the controls. Glycogen synthesis was also increased as compared with the controls. These results suggested the potential application of this chitosan nanofiber scaffold as a suitable substratum for hepatocyte culturing in bioreactors. Jiankang *et al.* [118] developed a novel fabrication process to create chitosan-gelatin hybrid scaffolds with well-organized architectures and highly porous structures by combining rapid prototyping, microreplication and freeze-drying techniques. A hepatocyte culture experiment was conducted to evaluate the efficiency of the well-defined chitosan-gelatin scaffold in facilitating hepatocyte growth in the inner layer of the scaffold *in vitro*. Scanning electron microscopy and histological analysis showed that hepatocytes could form large colonies in the predefined hepatic chambers, and these cavities could the completely filled with hepatocytes during 7 day culture. Albumin secretion and urea synthesis further indicated that the well-organized scaffolds were more suitable for hepatocyte culture. Jiang *et al.* [119] prepared galactosylated chitosan-graft-polyethylenimine (GC-g-PEI) copolymer by an imine reaction between periodate-oxidized GC and low-molecular-weight PEI. GC-g-PEI showed good DNA-binding ability and superior protection of DNA from nuclease attack and had low cytotoxicity compared to PEI 25K. GC-g-PEI/DNA complexes showed higher transfection efficiency than PEI 25K in both HepG2 and HeLa cell lines. Transfection efficiency into HepG2, which has asialoglycoprotein receptors, was higher than that into HeLa, which does not. GC-g-PEI/DNA complexes also transfected liver cells *in vivo* after intraperitoneal administration more effectively than PEI 25K. These results suggested that GC-g-PEI can be used in gene therapy to improve transfection efficiency and hepatocyte specificity *in vitro* and *in vivo*. In Seo *et al.* [120] study, as a synthetic extracellular matrix (ECM) for hepatocytes, a highly porous hydrogel (sponge-like) scaffold, 150-200 μm pore size in diameter, was fabricated with alginate (AL), galactosylated chitosan (GC), and heparin through electrostatic interaction. The optimal concentration of GC and heparin in AL/GC/heparin sponges to perform the best liver-specific function was 1 and 6 wt% to AL contents, respectively, where albumin secretion was maintained with maximal rates. Coculture of hepatocytes in AL/GC/heparin sponges with NIH3T3 in a transwell insert resulted in significant increase of liver-specific functions, such as improved albumin secretion rates, ammonia elimination rates, and ethoxyresorufin-O-deethylase activity by cytochrome P4501A1 compared to those in hepatocyte monoculture. The results suggested that hepatocytes as stable spheroids enhance liver-specific functions in AL/GC/heparin sponges, providing a new synthetic ECM to design bioartificial liver devices.

3.5. Skin Tissue

The skin is a soft outer covering of an animal, in particular a vertebrate. In mammals, the skin is the largest organ of the integumentary system made up of multiple layers of ectodermal tissue, and guards the underlying muscles, bones, ligaments and internal organs. Because it interfaces with the environment, skin plays a key role in protecting (the body) against pathogens and excessive water loss. Its other functions are insulation, temperature regulation, sensation, and the protection of vitamin B folates. Severely damaged skin will try to heal by forming scar tissue. This is often discolored and depigmented. The loss of this largest organ is incompatible with sustained life. In reconstructive surgery or burn treatment, substitution of the skin is often necessary. Skin tissue engineering emerged as an experimental regenerative therapy motivated primarily by the critical need for early permanent coverage of extensive burn injuries in patients with insufficient sources of autologous skin for grafting. It applies the principles and methods of both engineering and life sciences toward the development of substitutes to restore and maintain skin structure and function. Chitosan has been proved to be good material for skin tissue engineering.

Ma *et al.* [121] designed the bilayer structure of chitosan film and sponge as a scaffold of skin tissue engineering and a dermis substitute. The dry thickness of the film layer was 19.6 μm and that of the sponge layer was controlled at 60-80 μm. Human neofetal dermal fibroblasts were seeded in the chitosan sponge layer and cultured for 4 weeks. They found that the cells could grow and proliferate well. Sun *et al.* [122] fabricated three-dimensional collagen-chitosan scaffolds with type I collagen and chitosan through freeze drying and glutaraldehyde cross-linking. They investigated the microstructure of the scaffolds as well as the fibroblasts proliferation, cytokine secretion and cell cycle. Flow cytometry analysis indicated that cells in the scaffolds proliferated steadily. IL-6 concentration measurement by the ELISA test suggested that the scaffolds could promote secretion of the fibroblasts cytokine. These results showed that the fibroblasts and the scaffolds interact well with each other, and the fibroblasts had better proliferation ability and biological activity in the scaffolds than in monolayer culture. The scaffolds were a promising candidate for skin tissue repair and regeneration with enhanced biostability and good cytocompatibility. Murakami *et al.* [123] developed a hydrogel sheet composed of a blended powder of alginate, chitin/chitosan and fucoidan (ACF-HS) to create a moist environment for rapid wound healing. Full-thickness skin defects were made on the backs of rats and mitomycin C solution was applied onto the wound for 10 min in order to prepare healing-impaired wounds. After thoroughly washing out the mitomycin C, ACF-HS was applied to the healing-impaired wounds. Histological examination demonstrated significantly advanced granulation tissue and capillary formation in the healing-impaired wounds treated with ACF-HS on day 7, as compared to those treated with calcium alginate fiber and those left untreated. Chen *et al.* [124] prepared a biodegradable chitosan membrane with an asymmetric structure as a novel skin substitute. Chitosan was cross-linked with genipin and then frozen and lyophilized to yield a porous asymmetric membrane (CG membrane). Nanoscale collagen I particles were injected into the CG membrane to form an asymmetric CGC membrane. The results revealed that after 7 d of dynamic culture, many of the adhered cells exhibited a flat morphology and well spread on the surface of CGC membrane treated with 0.125 wt % of genipin. In animal studies, the CGC membrane seeded with fibroblasts and grown *in vitro* for 7 d was more effective than both gauze and commercial wound dressing, Suile, in healing wounds. An *in vivo* histological assessment indicated that covering the wound with the asymmetric CGC membrane resulted in its epithelialization and reconstruction. CGC membrane, thus, had great potential in skin tissue engineering. Sezer *et al.* [125] prepared fucoidan-chitosan hydrogels and investigated their treatment efficiency on dermal burns. Dermal burns was formed on seven adult male New Zealand white rabbits and the optimum gel formulation applied on the wounds. Control and treatment group biopsy samples were taken on days 7, 14 and 21 and each burn wound site was evaluated histopathologically. No edema was seen in tested groups except control after 3 d treatment. After 7 d treatment, fibroplasia and scar were fixed on wounds treated with fucoidan-chitosan gel and fucoidan solution. The best regeneration on dermal papillary formation and the fastest closure of the wounds were observed in fucoidan-chitosan hydrogels after 14 d treatment. Altman *et al.* [126] evaluated the potential of a silk fibroin-chitosan (SFCS) scaffold serving as a delivery vehicle for human adipose-derived stem cells (ASCs) in a murine soft tissue injury model. Their data showed that the extent of wound closure was significantly enhanced in the ASC-SFCS group versus SFCS and no-graft controls at postoperative day 8. Microvessel density at wound bed biopsy sites from 2 weeks postoperative was significantly higher in the ASC-SFCS group versus SFCS alone. Engrafted stem cells were positive for the fibroblastic marker heat shock protein 47, smooth muscle actin, and von Willebrand factor at both 2 and 4 weeks. GFP-positive stem cells were also found to differentiate into epidermal epithelial cells at 4 weeks postoperative. They concluded that human adipose-derived stem cells seeded on a silk fibroin-chitosan scaffold enhanced wound healing and showed differentiation into fibrovascular, endothelial, and epithelial components of restored tissue. Peng *et al.* [127] found that a composite nano-TiO2-chitosan with collagen artificial skin (NTCAS) showed promising characteristics of moderate water absorptivity, fine thickness, low density, and favorable time-dependent biodegradability. In wounded subjects, NTCAS showed a steady level of TNF-alpha, while the IL-6 level reached a peak on Day 7, although significantly lower than in the control and Duoderm groups. In the animal model, NTCAS showed better and faster recovery than the other groups, which could be attributed to the unique bactericidal effect of nano-TiO2 and immune-enhancing effect of chitosan. Zhou *et al.* [128] successfully prepared biocompatible carboxyethyl chitosan/poly(vinyl alcohol) (CECS/PVA) nanofibers by electrospinning of aqueous CECS/PVA solution. The potential use of the CECS/PVA electrospun fiber mats as scaffolding materials for skin regeneration was evaluated *in vitro* using mouse fibroblasts (L929) as reference cell lines. Indirect cytotoxicity assessment of the fiber mats indicated that the CECS/PVA electrospun mat was nontoxic to the L929 cell. Cell culture results showed that fibrous mats were good in promoting the cell attachment and proliferation. This novel electrospun matrix would be used as potential wound dressing for skin regeneration. Boucard *et al.* [129] processed bio-inspired bi-layered physical hydrogels only constituted of chitosan and water in the treatment of full-thickness burn injuries to assess whether this material was totally accepted by the host organism and allowed *in vivo* skin reconstruction of limited area third-degree burns. Veterinary experiments were performed on pig's skins and biopsies at days 9, 17, 22, 100 and 293, were analyzed by histology and immuno-histochemistry. All the results showed that chitosan materials were well tolerated and promoted a good tissue regeneration. They induced inflammatory cells migration and angiogenetic activity favouring a high vascularisation of the neo-tissue. At day 22, type I and IV collagens were synthesized under the granulation tissue and the formation of the dermal-epidermal junction was observed. After 100 days, the new tissue was quite similar to a native skin, especially by its aesthetic aspect and its great flexibility.

4. Conclusion

Chitosan is one of the most promising biomaterials in tissue engineering because it offers a distinct set of advantageous physico-chemical and biological properties that qualify them for a variety of tissue regeneration. All the above descriptions have shown that chitosan and its derivatives have a good application prospect. However, still a lot of challenges are required to identify the modifications of chitosan or composites, which can be used efficiently in various tissue engineering applications.

References

[1] Cheng M, Cao W, Gao Y, Gong Y, Zhao N, Zhang X. Studies on nerve cell affinity of biodegradable modified chitosan films. *J. Biomater Sci. Polym.* Ed 2003;14:1155-1167.

[2] Li Z, Ramay HR, Hauch KD, Xiao D, Zhang M. Chitosan-alginate hybrid scaffolds for bone tissue engineering. *Biomaterials* 2005;26:3919-3928.

[3] Fukuda J, Khademhosseini A, Yeo Y, Yang X, Yeh J, Eng G., et al. Micromolding of photocrosslinkable chitosan hydrogel for spheroid microarray and co-cultures. *Biomaterials* 2006;27:5259-5267.
[4] Wang G, Lu G., Ao Q, Gong YD, Zhang XF. Preparation of cross-linked carboxymethyl chitosan for repairing sciatic nerve injury in rats. *Biotechnol. Lett.* 2010;32:59-66.
[5] Itoh S, Yamaguchi I, Suzuki M, Ichinose S, Takakuda K, Kobayashi H, et al. Hydroxyapatite-coated tendon chitosan tubes with adsorbed laminin peptides facilitate nerve regeneration *in vivo*. *Brain Res*. 2003;993:111-123.
[6] Li X, Yang Z, Zhang A, Wang T, Chen W. Repair of thoracic spinal cord injury by chitosan tube implantation. *Biomaterials* 2009;30:1121-1132.
[7] Yang F, Li XH, Cheng MY, Gong YD, Zhao NM, Zhang XF, Yang Y. Performance Modification of Chitosan Membranes Induced by Gamma Irradiation. *J. Biomater Appl.* 2002;16:215-226.
[8] Ikada Y. Surface modification of polymers for medical applications. *Biomaterials* 1994; 15:725-736.
[9] Denuziere A, Ferrier D, Damour O, Domard A. Chitosan-chondroitin sulfate and chitosan-hyaluronate polyelectrolyte complexes: biological properties. *Biomaterials* 1998; 19:1275-1285.
[10] Rathke TD, Hudson SM. Review of chitin and chitosan as fiber and film formers. *J. Mater Sci*: Rev Macromolecular Chem Phys 1994;C34:375-437.
[11] Zhang M, Li XH, Gong YD, Zhao NM, Zhang XF. Properties and biocompatibility of chitosan films modified by blending with PEG. *Biomaterials* 2002;23:2641-2648.
[12] Yang X, Zhao K, Chen GQ. Effect of surface treatment on the biocompatibility of microbial polyhydroxyalkanoates. *Biomaterials* 2002;23:1391-1397.
[13] Chen LJ, Wang M. Production and evaluation of biodegradable composites based on PHB-PHV copolymer. *Biomaterials* 2002;23:2631-2639.
[14] Norma G, Chavati R, Rubén S, Juan F, Analía V, Julio S T. Characterization and application of poly(β-hydroxyalkanoates) family as composite biomaterials. *Polym Test* 2000;19:485-492.
[15] Cao W, Wang A, Jing D, Gong Y, Zhao N, Zhang X. Novel biodegradable films and scaffolds of chitosan blended with poly(3-hydroxybutyrate). *J. Biomater Sci. Polym* Ed 2005;16:1379-1394.
[16] Taravel MN, Domard A. Relation between the physicochemical characteristics of collagen and its interaction with chitosan: I. *Biomaterials* 1993;14:930-938.
[17] Peppas NA. In: Ratner B, editor, In biomaterials science. New York: Academic Press, 1996.p.63.
[18] Cheng M, Deng J, Yang F, Gong Y, Zhao N, Zhang X. Study on physical properties and nerve cell affinity of composite films from chitosan and gelatin solutions. *Biomaterials* 2003;24:2871-2880.
[19] Cheng M, Gong K, Li J, Gong Y, Zhao N, Zhang X. Surface modification and characterization of chitosan film blended with poly-L-lysine. *J. Biomater. Appl.* 2004; 19:59-75.
[20] Gong H, Zhong Y, Li J, Gong Y, Zhao N, Zhang X. Studies on nerve cell affinity of chitosan-derived materials. *J. Biomed. Mater Res*. 2000;52:285-295.

[21] Zheng Z, Zhang L, Kong L, Wang A, Gong Y, Zhang X. The behavior of MC3T3-E1 cells on chitosan/poly-L-lysine composite films: Effect of nanotopography, surface chemistry, and wettability. *J. Biomed Mater Res.* Part A 2009;89A:453-465.
[22] Hahn SK, Hoffman AS. Preparation and characterization of biocompatible polyelectrolyte complex multilayer of hyaluronic acid and poly-L-lysine. *Int. J. Biol. Macromol.* 2005;37:227-231.
[23] Zheng Z, Wei Y, Wang G, Gong Y, Zhang X. *In vitro* biocompatibility of three novel chitosan/polycation composite materials for nerve regeneration. *Neural Regeneration Research* 2008;3:837-842.
[24] Zheng Z, Wei Y, Wang G, Wang A, Ao Q, Gong Y, Zhang X. Surface properties of chitosan films modified with polycations and their effects on the behavior of PC12 cells. *J. Bioact. Compat. Pol.* 2009;24:63-82.
[25] Zheng Z, Wei Y, Wang G, Gong Y, Zhang X. Surface characterization and cytocompatibility of three chitosan/polycation composite membranes for guided bone regeneration. *J. Biomater Appl.* 2009;24:209-229.
[26] Chen RH, Hua D. Effect of N-Acetylation on the acidic solution stability and thermal and mechanical properties of membranes prepared from different chain flexibility chitosan. *J. Appl. Poly Sci.* 1996;61:749-754.
[27] Wan Y, Creber KAM, Peppley B, Bui VT. Ionic conductivity of chitosan membranes. *Polymer* 2003;44:1057-1065.
[28] Chatelet C, Damour O, Domard A. Influence of the degree of acetylation on some biological properties of chitosan films. *Biomaterials* 2001;22:261-268.
[29] Howling GI, Dettmar PW, Goddard PA, Hampson FC, Dornish M, Wood EJ. The effect of chitin and chitosan on the proliferation of human skin fibroblasts and keratinocytes *in vitro*. *Biomaterials* 2001;22:2259-2266.
[30] Cao W, Jing D, Li J, Gong Y, Zhao N, Zhang X. Effects of the Degree of deacetylation on the physicochemical properties and Schwann cell affinity of chitosan films. *J. Biomater. Appl.* 2005;20:157-177.
[31] Mima S, Miya M, Iwamoto R, Yoshikawa S. Highly deacetylated chitosan and its properties. *J. Appl. Polym. Sci.* 1983;28:1909-1917.
[32] Jameela SR, Jayakrishnan A. Glutaraldehyde cross-linked chitosan microspheres as a long acting biodegradable drug delivery vehicle: studies on the *in vitro* release of mitoxantrone and *in vivo* degradation of microspheres in rat muscle. *Biomaterials* 1995;16:769-775.
[33] Naimark WA, Pereira CA, Tsang K, Lee JM. HMDC crosslinking of bovine pericardial tissue: a potential role of the solvent environment in the design of bioprosthetic materials. *J. Mater Sci. Mater Med.* 1995;6:235-241.
[34] Cao W, Cheng M, Ao Q, Gong Y, Zhao N, Zhang X. Physical, mechanical and degradation properties, and Schwann cell affinity of cross-linked chitosan films. *J. Biomater. Sci. Polymer* Edn 2005;16:791-807.
[35] Lu L, Zhu X, Valenzuela RG, Currier BL, Yaszemski MJ. Biodegradable polymer scaffolds for cartilage tissue engineering. *Clinical Orthopaedics* 2001;391:251-270.
[36] Tomihata K, Ikada Y. *In vitro* and *in vivo* degradation of films of chitin and its deacetylated derivatives. *Biomaterials* 1997;18:567-575.
[37] Xie Y, Liu X, Chen Q. Synthesis and characterization of water-soluble chitosan derivate and its antibacterial activity. *Carbohydrate Polymers* 2007;69:142-147.

[38] Mi F, Tan Y, Liang H, Sung H. *In vivo* biocompatibility and degradability of a novel injectable chitosan based implant. *Biomaterials* 2002;23:181-191.

[39] Lu G, Kong L, Sheng B, Wang G, Gong Y, Zhang X. Degradation of covalently cross-linked carboxymethyl chitosan and its potential application for peripheral nerve regeneration. *European Polymer Journal* 2007;43:3807-3818.

[40] Kong HJ, Lee KY, Mooney DJ. Decoupling the dependence of rheological/mechanical properties of hydrogels from solid concentration. *Polymer* 2002;43:6239-6246.

[41] Lu G, Sheng B, Wang G, Wei Y, Gong Y, Zhang X. Controlling the degradation of covalently cross-linked carboxymethyl chitosan utilizing bimodal molecular weight distribution. *J. Biomater. Appl.* 2009;23:435-451.

[42] Li J, Gong Y, Zhao N, Zhang X. Preparation of N-butyl chitosan and study of its physical and biological properties. *J. Appl. Polym. Sci.* 2005;98:1016-1024.

[43] Anselme K. Osteoblast adhesion on biomaterials. *Biomaterials* 2000;21:667-681.

[44] Ruoslahti E. RGD and other recognition sequences for integrins. *Annu. Rev. Cell Dev. Biol.* 1996;12:697-715.

[45] Schaffner P, Dard MM. Structure and function of RGD peptides involved in bone biology. *Cell Mol. Life Sci.* 2003;60:119-132.

[46] Hersel U, Dahmen C, Kessler H. RGD modified polymers: biomaterials for stimulated cell adhesion and beyond. *Biomaterials* 2003;24:4385-4415.

[47] Yang XB, Roach HI, Clarke NMP, Howdle SM, Quirk R, Shakesheff KM, Oreffo ROC. Human osteoprogenitor growth and differentiation on synthetic biodegradable structures after surface modification. *Bone* 2001;29:523-531.

[48] Li J, Yun H, Gong Y, Zhao N, Zhang X. Investigation of MC3T3-E1 cell behavior on the surface of GRGDS-coupled chitosan. *Biomacromolecules* 2006;7:1112-1123.

[49] Ho MH, Wang DM, Hsieh HJ, Liu HC, Hsien TY, Lai JY, Hou LT. Preparation and characterization of RGD-immobilized chitosan scaffolds. *Biomaterials* 2005;25:3197-3206.

[50] Ishihara K, Arai H, Nakabayashi N, Morita S, Furuya K. Adhesive one cement containing hydroxyapatite particle as bone compatible filler. *J. Biomed Mater Res.* 1992; 26:937-945.

[51] Holmes RE, Bucholz RW, Mooney V. Porous hydroxyapatite as a bone-graft substitute in metaphyseal defects. *J. Bone Joint Surg.* 1986;68:904-911.

[52] Jarcho M. Calcium phosphate ceramics as hard tissue prosthetics. *Clin. Orthop.* 1981; 157:259-278.

[53] Ogiso M. Reassessment of long-term use of dense HA as dental implant: case report. *J. Biomed Mater. Res.* 1998;43:318-320.

[54] Hench LL. Bioceramics: from concept to clinic. *J. Am. Ceram. Soc.* 1991;74(7):1487-1510.

[55] Zhang R, Ma PX. Poly(L-hydroxyl acids)/hydroxyapatite porous composites for bone-tissue engineering. I. Preparation and morphology. *J. Biomed. Mater. Res.* 1999; 44: 446-455.

[56] Miyamato Y, Ishikawa KI, Takechi M, Toh T, Yuasa T, Nagayama M, Suzuki K. Basic properties of calcium phosphate cement containing atelocollagen in its liquid or powder phases. *Biomaterials* 1998;19:707-715.

[57] Kong L, Gao Y, Cao W, Gong Y, Zhao N, Zhang X. Preparation and characterization of nano-hydroxyapatite/chitosan composite scaffolds. *J. Biomed Mater Res.* Part A 2005;75A:275-282.
[58] Hall LD, Yalpani M. Formation of branched-chain, soluble polysaccharides from chitosan. *J. Chem. Soc. Chem. Commun.* 1980;1153-1154.
[59] Yalpani M, Hall LD. Some chemical and analytical aspects of polysaccharide modifications. III. Formation of branched-chain, soluble chitosan derivatives. *Macromolecules* 1984;17:272-281.
[60] Morimoto M, Saimoto H, Shigemasa Y. Control of functions of chitin and chitosan by chemical modification. Trends in Glycosci Glycotechn. 2002;14:205-222.
[61] Morimoto M, Saimoto H, Usui H, Okamoto Y, Minami S, Shigemasa Y. Biological activities of carbohydrate-branched chitosan derivates. *Biomacromolecules* 2001;2: 1133-1136.
[62] Li X, Tsushima Y, Morimoto M, Saimoto H, Okamoto Y, Minami S, Shigemasa Y. Biological activity of chitosan-sugar hybrids: specific interaction with lectin. *Polym. Adv. Technol.* 2000;11:176-179.
[63] Li X, Tsushima Y, Morimoto M, Saimoto H, Okamoto Y, Minami S, Shigemasa Y. Synthesis of chitosan-sugar hybrid and evaluation of its bioactivity. *Polym. Adv. Technol.* 1999;10:455-458.
[64] Kato Y, Onishi H, Machida Y. Biological characteristics of lactosaminated N-succinyl-chitosan as a liver-specific drug carrier in mice. *J. Control Release* 2001;70:295-307.
[65] Liu L, Li Y, Li Y, Fang YE. N-phthaloylation of chitosan by microwave irradiation. *Carbohydr. Polym.* 2004;57:97-100.
[66] Kurita K, Hirakawa M, Kikuchi S, Yamanaka H, Yang J. Trimethylsilylation of chitosan and some properties of the product. *Carbohydr. Polym.* 2004;56:333-337.
[67] Wu SZ, Zeng F, Zhu HP, Tong Z. Energy and electron transfers in photosensitive chitosan. *J. Am. Chem. Soc.* 2005;127:2048-2049.
[68] Langer R, Vacanti JP. Tissue engineering. *Science* 1993;260:920-926.
[69] Miyabayashi T, Clemmons RM, Farese JP, Uhl EW. Three-dimensional culture of feline articular chondrocytes in alginate MCs. *J. Vet. Med. Sci.* 2006;68:1239-1242.
[70] Malda J, Frondoza CG. MCs in the engineering of cartilage and bone. *Trends Biotechnol.* 2006;24:299-304.
[71] Yoon JR, Lee JS, Kim HJ, Lim HW, Lim HC, Park JH, et al. The use of poly(lactic–co–glycolic acid) MCs as injectable cell carriers for cartilage regeneration in rabbit knees. *J. Biomater. Sci. Polym.* Ed 2006;17:925-939.
[72] Lu G, Sheng B, Wei Y, Wang G, Zhang L, Ao Q, Gong Y, Zhang X. Collagen nanofiber-covered porous biodegradable carboxymethyl chitosan microcarriers for tissue engineering cartilage. *Eur. Polym. J.* 2008;44:2820-2829.
[73] Nettles DL, Elder SH, Gilbert JA. Potential use of chitosan as a cell scaffold material for cartilage tissue engineering. *Tissue Eng.* 2002;8:1009-1016.
[74] Hoemann CD, Sun J, Legare A, McKee MD, Buschmann MD. Tissue engineering of cartilage using an injectable and adhesive chitosan-based cell-delivery vehicle. *Osteoarthr. Cartilage* 2005;13:318-329.
[75] Sechriest VF, Miao YJ, Niyibizi C, Westerhausen-Larson A, Matthew HW, Evans CH, Fu FH, Suh Jk. GAG-augmented polysaccharide hydrogel: a novel biocompatible and

biodegradable material to support chondrogenesis. *J. Biomed. Mater Res*. 2000;49:534-541.

[76] Xia W, Liu W, Cui L, Liu Y, Zhong W, Liu D, Wu J, Chua K, Cao Y. Tissue engineering of cartilage with the use of chitosan-gelatin complex scaffolds. *J. Biomed. Mater. Res.* Part B 2004;71B:373-380.

[77] Lee JE, Kim SE, Kwon IC, Ahn HJ, Cho H, Lee SH, Kim HJ, Seong SC, Lee MC. Effects of a chitosan scaffold containing TGF-beta1 encapsulated chitosan microspheres on *in vitro* chondrocyte culture. *Artif Organs* 2004;28:829-839.

[78] Hsu SH, Whu SW, Hsieh SC, Tsai CL, Chen DC, Tan TS. Evaluation of chitosan-alginate-hyaluronate complexes modified by an RGD-containing protein as tissue-engineering scaffolds for cartilage regeneration. *Artif Organs* 2004;28:693-703.

[79] Hidalgo DA, Rekow A. A review of 60 consecutive fibula free flap mandible reconstruction. *Plast Reconstr. Surg*. 1995;96:585-596.

[80] Goldberg VM, Stevenson S. Natural history of autografts and allografts. *Clin. Orthop.* 1987;225:7-16.

[81] Costantino PD, Friedman CD. Synthetic bone graft substitutes. *Otolaryngol. Clin. North Am.* 1994;27:1037-1074.

[82] Block JE, Poser J. Does xenogeneic demineralized bone matrix have clinical utility as a bone graft substitute? *Med. Hypotheses* 1995;45:27-32.

[83] Sasso RC, Williams JI, Dimasi N, Meyer PR Jr. Postoperative drains at the donor site of iliac-crest bone grafts. A prospective, randomized study of morbidity at the donor site in patients who had a traumatic injury of the spine. *J. Bone Joint Surg. Am.* 1998;80: 631-635.

[84] Kokubo T, Kim HM, Kawashita M. Novel bioactive materials with different mechanical properties. *Biomaterials* 2003;24:2161-2175.

[85] Kong L, Gao Y, Lu G, Gong Y, Zhao N, Zhang X. A study on the bioactivity of chitosan/nano-hydroxyapatite composite scaffolds for bone tissue engineering. *Eur. Polym. J.* 2006;42:3171-3179.

[86] Kong L, Ao Q, Wang A, Gong K, Wang X, Lu G, Gong Y, Zhao N, Zhang X. Preparation and characterization of a multilayer biomimetic scaffold for bone tissue engineering. *J. Biomater. Appl.* 2007;22:223-239.

[87] Zhao F, Yin Y, Lu WW, Leong JC, Zhang W, Zhang J, Zhang M, Yao K. Preparation and histological evaluation of biomimetic three-dimensional hydroxyapatite/chitosan-gelatin network composite scaffolds. *Biomaterials* 2002;23:3227-3234.

[88] Lahiji A, Sohrabi A, Hungerford DS, Frondoza CG. Chitosan supports the expression of extra cellular matrix proteins in human osteoblasts and chondrocytes. *J. Biomed. Mater. Res.* 2000;51:586-595.

[89] Seol YJ, Lee J, Park YJ, Lee YM, Ku Y, Rhyu IC, Lee SJ, Han SB, Chung CP. Chitosan sponges as tissue engineering scaffolds for bone formation. *Biotechnol. Lett.* 2004; 26: 1037-1041.

[90] Lee YM, Park YJ, Lee SJ, Ku Y, Han SB, Choi SM, Klokkevold PR, Chung CP. Tissue engineering bone formation using chitosan/tricalcium phosphate songes. *J. Periodontol.* 2000;71:410-417.

[91] Park DJ, Choi BH, Zhu SJ, Huh JY, Kim BY, Lee SH. Injectable bone using chitosan-alginate gel/mesenchymal stem cells/BMP-2 composites. *J. Cranio-Maxill Surg.* 2005; 33: 50-54.

[92] Zhang XF, Cao WL, Gong YD, Gao Y, Li JM. A method for the preparation of porous chitosan tube. *China Patent* 2004 No. 02149086.4.
[93] Ao Q, Wang A, Cao W, Zhang L, Kong L, He Q, Gong Y, Zhang X. Manufacture of multimicrotubule chitosan nerve conduits with novel molds and characterization *in vitro*. *J. Biomed. Res*. Part A 2006;77A:11-18.
[94] Wang G, Ao Q, Gong K, Wang A, Zheng L, Gong Y, Zhang X. The effect of topology of chitosan biomaterials on the differentiation and proliferation of neural stem cells. *Acta Biomater*. 2010; 6(9):3630-3639.
[95] Wang A, Ao Q, Wei Y, Gong K, Liu X, Zhao N, Gong Y, Zhang X. Physical properties and biocompatibility of a porous chitosan-based fiber-reinforced conduit for nerve regeneration. *Biotechnol. Lett.* 2007;29:1697-1702.
[96] Wang A, Ao Q, Cao W, Yu M, He Q, Kong L, Zhang L, Gong Y, Zhang X. Porous chitosan tubular scaffolds with knitted outer wall and controllable inner structure for nerve tissue engineering. *J. Biomed. Mater Res*. Part A 2006;79A:36-46.
[97] Rosales-Cortes M, Peregrina-Sandoval J, Banuelos-Pineda J, Sarabia-Estrada R, Gomez-Rodiles CC, Albarran-Rodriguez E, Zaitseva GP, Pita-Lopez ML. Immunological study of a chitosan prosthesis in the sciatic nerve regeneration of the axotomized dog. *J. Biomater. Appl*. 2003;18:15-23.
[98] Hu Y, Liu YL, Dong YL, Gao XC, Bu HF, Yin ZS. An experimental study of repairing nerve defect in different lengths with chitosan membranous tube. *Chinese J. Clinical Rehab.* 2004;8:2564-2565.
[99] Chavez-Delgado ME, Mora-Galindo J, Gomez-Pinedo U, Feria-Velasco A, Castro-Castaneda S, Lopez-Dellamary Toral FA, Luquin-De Anda S,Garcia-Segura Lm, Garcia-Estrada J. Facial nerve regeneration through progesterone-loaded chitosan prosthesis A preliminary report. *J. Biomed. Mater. Res.* Part B 2003;67B:702-711.
[100] Rosales-Cortes M, Peregrina-Sandoval J, Banuelos-Pineda J, Castellanos-Martinez EE, Gomez-Pinedo UA, Albarran-Rodriguez E. Regeneration of the axotomized sciatic nerve in dogs using the tubulization technique with chitosan biomaterial preloaded with progesterone. *Rev. Neurol*. 2003;36:1137-1141.
[101] Suzuki M, Itoh S, Yamaguchi I, Takakuda K, Kobayashi H, Shinomiya K, Tanaka J. Tendon chitosan tubes covalently coupled with synthesized laminin peptides facilitate nerve regeneration *in vivo*. *J. Neurosci. Res*. 2003;72:646-659.
[102] Wei X, Lao J, Gu YD. Bridging peripheral nerve defect with chitosan-collagen film. *Chinese J. Traumatol*. 2003;6:131-134.
[103] Hess F. History of (micro) vascular surgery and the development of small-caliber blood vessel prostheses (with some notes on patency rates and re-endothelialization). *Microsurgery* 1985;6:59-69.
[104] Zhang L, Ao Q, Wang A, Lu G, Kong L, Gong Y, Zhao N, Zhang X. A sandwich tubular scaffold derived from chitosan for blood vessel tissue engineering. *J. Biomed. Mater. Res*. Part A 2006;77A:277-284.
[105] Cuy JL, Beckstead BL, Brown CD, Hoffman AS, Giachelli CM. Adhesive protein interactions with chitosan: consequences for valve endothelial cell growth on tissue-engineering materials. *J. Biomed. Mater. Res*. Part A 2003;67A:538-547.
[106] Chupa JM, Foster AM, Sumner SR, Madihally SV, Matthew HW. Vascular cell responses to polysaccharide materials: *in vitro* and *in vivo* evaluations. *Biomaterials* 2000; 21:2315-2322.

[107] Chung TW, Lu YF, Wang SS, Lin YS, Chu SH. Growth of human endothelial cells on photochemically grafted Gly-Arg-Gly-Asp (GRGD) chitosans. *Biomaterials* 2002; 23:4803-4809.

[108] Chung TW, Lu YF, Wang HY, Chen WP, Wang SS, Lin YS, Chu SH. Growth of human endothelial cells on different concentrations of Gly-Arg-Gly-Asp grafted chitosan surface. *Artif Organs* 2003;27:155-161.

[109] Zhu Y, Gao C, Liu X, He T, Shen J. Immobilization of biomacromolecules onto aminolyzed poly(L-lactic acid) toward acceleration of endothelium regeneration. *Tissue Eng.* 2004;10:53-61.

[110] Meng S, Liu Z, Shen L, Guo Z, Chou LL, Zhong W, Du Q, Ge J. The effect of a layer-by-layer chitosan-heparin coating on the endothelialization and coagulation properties of a coronary stent system. *Biomaterials* 2009;30:2276-2283.

[111] Zhu C, Fan D, Duan Z, Xue W, Shang L, Chen F, Luo Y. Initial investigation of novel human-like collagen/chitosan scaffold for vascular tissue engineering. *J. Biomed. Mater. Res.* Part A 2009;89A:829-840.

[112] Kawase M, Michibayashi N, Nakashima Y, Kurikawa N, Yagi K, Mizoguchi T. Application of glutaraldehyde crosslinked chitosan as a scaffold for hepatocytes attachment. *Biol. Pharma Bull.* 1997;20:708-710.

[113] Lin A, Chen J, Liu Y, Deng S, Wu Z, Huang Y, Ping Q. Preparation and evaluation of N-caproyl chitosan nanoparticles surface modified with glycyrrhizin for hepatocyte targeting. *Drug Dev. Ind. Pharm.* 2009;35:1348-1355.

[114] Wang Q, Zhang L, Hu W, Hu ZH, Bei YY, Xu JY, Wang WJ, Zhang XN, Zhang Q. Norcantharidin-associated galactosylated chitosan nanoparticles for hepatocyte-targeted delivery. *Nanomedicine* doi:10.1016/j.nano.2009.07.006.

[115] She Z, Liu W, Feng Q. Self-assembly model, hepatocytes attachment and inflammatory response for silk fibroin/chitosan scaffolds. *Biomed. Mater.* doi: 10.1088/1748-6041/4/4/045014.

[116] Feng ZQ, Chu X, Huang NP, Wang T, Wang Y, Shi X, Ding Y, Gu ZZ. The effect of nanofibrous galactosylated chitosan scaffolds on the formation of rat primary hepatocyte aggregates and the maintenance of liver function. *Biomaterials* 2009;30:2753-2563.

[117] Chu XH, Shi XL, Feng ZQ, Gu ZZ, Ding YT. Chitosan nanofiber scaffold enhances hepatocyte adhesion and function. *Biotechnol. Lett.* 2009;31:347-352.

[118] Jiankang H, Dichen L, Yaxiong L, Bo Y, Hanxiang Z, Qin L, Bingheng L, Yi L. Preparation of chitosan-gelatin hybrid scaffolds with well-organized microstructures for hepatic tissue engineering. *Acta Biomater.* 2009;5:453-461.

[119] Jiang HL, Kwon JT, Kim YK, Kim EM, Arote R, Jeong HJ, Nah JW, Choi YJ, Akaike T, Cho MH, Cho CS. Galactosylated chitosan-graft-polyethylenimine as a gene carrier for hepatocyte targeting. *Gene Ther.* 2007;14:1389-1398.

[120] Seo SJ, Choi YJ, Akaike T, Higuchi A, Cho CS. Alginate/galactosylated chitosan/heparin scaffold as a new synthetic extracellular matrix for hepatocytes. *Tissue Eng.* 2006;12:33-44.

[121] Ma J, Wang H, He B, Chen J. A preliminary *in vitro* study on the fabrication and tissue engineering applications of a novel chitosan bilayer material as a scaffold of human neofetal dermal fibroblasts. *Biomaterials* 2001;22:331-336.

[122] Sun LP, Wang S, Zhang ZW, Wang XY, Zhang QQ. Biological evaluation of collagen-chitosan scaffolds for dermis tissue engineering. *Biomed. Mater.* doi: 10.1088/1748-6041/4/5/055008.

[123] Murakami K, Aoki H, Nakamura S, Nakamura S, Takikawa M, Hanzawa M, Kishimoto S, Hattori H, Tanaka Y, Kiyosawa T, Sato Y, Ishihara M. Hydrogel blends of chitin/chitosan, fucoidan and alginate as healing-impaired wound dressings. *Biomaterials* 2010;31:83-90.

[124] Chen KY, Liao WJ, Kuo SM, Tsai FJ, Chen YS, Huang CY, Yao CH. Asymmetric chitosan membrane containing collagen I nanospheres for skin tissue engineering. *Biomacromolecules* 2009;10:1642-1649.

[125] Sezer AD, Cevher E, Hatıpoğlu F, Oğurtan Z, Baş AL, Akbuğa J. Preparation of fucoidan-chitosan hydrogel and its application as burn healing accelerator on rabbits. *Biol. Pharm. Bull.* 2008;31:2326-2333.

[126] Altman AM, Yan Y, Matthias N, Bai X, Rios C, Mathur AB, Song YH, Alt EU. IFATS collection: Human adipose-derived stem cells seeded on a silk fibroin-chitosan scaffold enhance wound repair in a murine soft tissue injury model. *Stem Cells* 2009;27:250-258.

[127] Peng CC, Yang MH, Chiu WT, Chiu CH, Yang CS, Chen YW, Chen KC, Peng RY. Composite nano-titanium oxide-chitosan artificial skin exhibits strong wound-healing effect-an approach with anti-inflammatory and bactericidal kinetics. *Macromol. Biosci.* 2008;8:316-327.

[128] Zhou Y, Yang D, Chen X, Xu Q, Lu F, Nie J. Electrospun water-soluble carboxyethyl chitosan/poly(vinyl alcohol) nanofibrous membrane as potential wound dressing for skin regeneration. *Biomacromolecules* 2008;9:349-354.

[129] Boucard N, Viton C, Agay D, Mari E, Roger T, Chancerelle Y, Domard A. The use of physical hydrogels of chitosan for skin regeneration following third-degree burns. *Biomaterials* 2007;28:3478-3488.

In: Focus on Chitosan Research
Editors: Arthur N. Ferguson and Amy G. O'Neill
ISBN 978-1-61324-454-8

Chapter 4

ENHANCED BIOCHEMICAL EFFICACY OF OLIGOCHITOSANS AND PARTIALLY DEPOLYMERIZED CHITOSANS

Riccardo A. A. Muzzarelli*
Emeritus Professor of Enzymology, University of Ancona,
Ancona, Italy

ABSTRACT

It is being realized that the performances of chitosans depend in most cases on their essential characteristic properties, such as the degree of acetylation and the average molecular weight. Crystalline polymorphic form, degree of crystallinity, polydispersity of molecular weight, zeta potential and other properties are also important. The selection of a certain chitosan for a particular application has consequences in terms of degree of efficacy, reproducibility of the results and suitability for scaling up in various areas such as pre-clinical tests, adoption for pharmaceutical formulations, environment protection. The requirements for reproducible quality of tailor-made chitosans become even more stringent when they are chemically or enzymatically modified.

Macrophages, fibroblasts and other human cells respond to the presence of chitosans in significantly different ways depending on their DA and MW, in terms of phenotypes, secretory repertoire, efficacy of their receptors, and final capacity to demonstrate superior biochemical significance compared to other polysaccharides. This is particularly true in wound healing, bone defect repair, control over inflammation, exertion of immunostimulation, efficacy in drug delivery, anti-cholesterolemic capacity.

Chitosans are also antioxidants and antistress agents, capable to scavenge free radicals in a mammalian organism: this action too is controlled by DA and MW, chelating activity, and capacity to interact with biochemicals like nitric oxide, superoxide dismutase and glutathione peroxidase. Chitosans of proper chain size lend themselves to be targeted to a tumor, where they suppress the proliferation of tumor cells or at least reduce their viability; moreover they exert anti-metastatic action.

Chitosans are also active on the fungal and bacterial cells, and the selection of the proper size of the chitosans permits to optimize the results against pathogens, in

* E-mail: muzzarelli.raa@gmail.com. www.chitin.

consideration of the multiplicity of mechanisms of action of oligochitosans and partially hydrolyzed chitosans, using a combination of approaches, including in vitro assays, killing kinetics, cellular leakage measurement, membrane potential estimation, and electron microscopy, in addition to transcriptional response analysis.

This review includes concise but detailed well-assessed protocols for the preparation of oligochitosans and partially hydrolyzed chitosans, mainly by using sodium nitrite, sonication, hydrogen peroxide, electrolysis, and hydrolytic unspecific enzymes.

1. Introduction

The chitins and chitosans of various origins and some of their derivatives obtained by chemical or enzymatic means are today protagonists in the research fields of wound healing, tissue engineering, gene therapy and collection of metal ions, thanks to their outstanding properties. Low molecular weight and oligomeric chitosans have been reported to provide health benefits such as immunity regulation, anti-tumor activity, liver protection, blood lipids lowering, anti-diabetic, anti-oxidant and anti-obesity capacities. The preparation protocols and the analytical methods were reviewed by Yin et al. 2009.

Basic information on these polysaccharides, relevant to the title topic, can be found in books and review articles [Muzzarelli, 1977, 2009a,b,c,d; 2010; Jollès and Muzzarelli, 1999; Muzzarelli and Muzzarelli, 2002; Muzzarelli et al., 1989, 1990, 1999; Varlamov et al., 2003; Kumar et al., 2004; Kurita, 2005; Rinaudo, 2006a,b; Mourya and Inamdar, 2008; Keong and Halim, 2009; Hollister, 2005; Uragami and Tokura, 2006].

2. Deacetylation Patterns and Nanostructures

The microstructure of chitosan, a linear sequence of glucosamine and N-acety-lglucosamine units, depends on the preparation conditions. Knowledge of the structural differences between chitosan preparations is important in determining the properties of chitosan and essential for structure-activity analysis where biological systems are concerned. Determination of the pattern of acetylation of chitosan samples (pattern of acetylation parameter) by C-13 NMR spectroscopy hitherto required depolymerization of the chitosans, but this step is not necessary for determining the pattern of acetylation of chitosan samples having < 41 kDa [Kumirska et al., 2009].

As part of an effort to develop a biodegradable nerve guidance channel based on chitin/chitosan, Freier et al. (2005) conducted a systematic in vitro study on the biodegradation and neural cell compatibility of chitosan and N-acetylated chitosan films conditioned with ammonia in methanol + water and extensively washed with water in order to eliminate the acetate ion. The in vitro degradation (pH 7.4, 37 °C) in the presence of 1.5 microg/ml lysozyme showed a progressive mass loss to greater than 50% within 4 weeks for films with 30-70% acetylation. In contrast, the degradation of samples with very low or high acetylation was minimal over the 4-week period. All chitosan-based films showed neural cell adhesion after 2 days of culture, but cell viability decreased with increasing degree of acetylation. Chitosan (0.5% DA) provided cell viability, ca. 8 times higher than that of 11% DA chitosan. Both chitosans supported more numerous and longer neurites than the other chitosans. Thus chitosan amine content should be tuned for optimal biodegradation and cell compatibility [Freier et al., 2005].

A colloidal solution was obtained from the deacetylation of chitin whiskers under alkaline conditions at 150°C under nitrogen and 2.45 GHz by using a microwave technique in only 1/7 of the time of the conventional method. FTIR and H1 NMR techniques confirmed that the DA was less than 10% within 3 hr. The wide-angle X-ray diffraction spectrum clearly showed that the highly crystalline chitin whiskers changed to amorphous chitosan. SEM micrographs showed the aggregation of branched nanofibers, whereas the TEM micrographs revealed the scaffold morphology. Figure 6B (a) in the original article confirms that the chitin whiskers have an aspect ratio of 16, which corresponds to that noted by Nair and Dufresne. Figures 6B (b) and (c) show that chitosan, after being treated with 40% and 50% alkaline solutions, is in whisker form; however, the morphology changes to a scaffold when a 60% alkaline solution is used for 6 h. In the past, the same authors reported a unique nanoscaffold chitosan, obtained from traditional alkaline treatment of chitin whiskers in glassware; here, they also find a similar result, when the alkaline treatment is carried out in a microwave reactor [Lertwattanaseri et al., 2009].

In several fields of surgery, the treatment of complicated tissue defects is an unsolved clinical problem. In particular, the use of tissue scaffolds has been limited by poor revascularization and integration. As a follow-up of a recent series of articles (see for instance Kozen et al., 2007), Scherer et al. developed a scaffold-like membrane with bioactive properties and tested the biologic effects in vitro and in vivo in diabetic wound healing. For this purpose, cells-nanofibers interactions were tested in vitro by cell metabolism and migration assays. In vivo, full thickness wounds in diabetic mice (n = 15 per group) were treated either with chitosan, with a cellulosic control material, or were left untreated. Wound healing kinetics, including wound re-epithelialization and wound contraction as well as microscopic metrics such as tissue growth, cell proliferation (Ki67), angiogenesis (PECAM-1), cell migration (MAP-kinase), and keratinocyte migration (p 63) were monitored over a period of 28 days. Messenger RNA levels related to migration (uPAR), angiogenesis (VEGF), inflammatory response (IL-1 beta), and extracellular matrix remodeling (MMP3 and 9) were measured in wound tissues. Results showed that the fibers stimulated cell metabolism and the in vitro migratory activity of endothelial cells and fibroblasts; the membranes accelerated wound closure mainly by re-epithelialization and increased keratinocyte migration (7.5-fold), granulation tissue formation (2.8-fold), cell proliferation (4-fold), and vascularization (2.7-fold) compared with control wounds. Expression of markers of angiogenesis (VEGF), cell migration (uPAR) and ECM remodeling (MMP3, MMP9) were up- regulated [Scherer et al., 2009].

Xiong et al. demonstrated the effects of deacetylated chitohexaose (hexamer) on tumor angiogenesis and its mechanism of action. Five fractions from dimer to hexamer were separated by a linear gradient solution of HCl on a cation-exchange resin, and then HCl was removed from the fractions with the aid of a charcoal column. The purity of the five fractions was analyzed by HPLC and the molecular masses were analyzed by MALDI-TOFMS. The hexamer had no toxic effect on normal ECV304 cells, but could inhibit the proliferation and migration of tumor-induced ECV304 cells in a dose-dependent manner. The mechanism was demonstrated through the detection of mRNA expression of VEGF, MMP-9, TIMP-1, TIMP-2 and uPA by RT-PCR, which showed that the hexamer down-regulated the VEGF and uPA mRNA expressions in ECV304 cells, but up-regulated the TIMP-1 mRNA expression [Xiong et al. 2009].

3. Modified Oligomers

The literature on modified chitosans and their partially depolymerized forms is abundant. The recent advances in enhancing the water solubility of chitosan as well as in bringing chitosan into organic solutions are also boosting the number of derivatives. For example, an efficient and chemoselective procedure for preparing highly organosoluble 3,6-di-O-tert-butyldimethylsilyl chitosans and oligomers permits to bring chitosan into organic solution for further reaction. The selective modification of the oligochitosans with 0.50 DA was achieved by using said silyl in chloride form as the reagent in combination with DMF/imidazole. The silyl-protected polymers displayed excellent solubility in a number of common organic solvents. The silylated chitosans showed unprecedented solubility in organic solvents. The combination of high organosolubility and stability of the TBDMS group in a wide range of conditions makes these derivatives valuable intermediates for the preparation of various N-derivatives as important materials for pharmaceutical and other applications: the 3,6-di-O-TBDMS-chitosan and oligochitosans were reacted with acetic anhydride and deprotected to obtain the N-acetyl derivatives [Runarsson et al., 2008].

Chung et al. improved the solubility of chitosan with the aid of the Maillard reaction with 1% chitosan and 2% reducing sugar (glucose or glucosamine) dissolved in 0.2 M acetic acid, adjusted to pH 6.0, and incubated at either 50 or 70 °C for 1-7 days. The solubility of modified chitosan derivatives was significantly greater than that of plain chitosan. The solubility of chitosan-glucosamine was higher than that of chitosan-glucose, and the chitosan-glucosamine derivative remained soluble at pH 10. Rheology revealed that the apparent viscosity of the water-soluble chitosan derivatives in aqueous solution depended upon system conditions such as pH, ionic strength, and temperature. The measured apparent viscosity decreased as all system conditions increased. As calculated by the Arrhenius equation, the activation energy of the derivatives in aqueous solution generally decreased upon increasing the extent of the Maillard reaction [Chung et al., 2006]. Similarly, the chitosan-maltose derivative remained soluble when the pH approached 10. Among chitosan-saccharide derivatives, the solubility of chitosan-fructose derivative was highest at 17.1 g/l. Considering yield, solubility and pH stability, the chitosan-glucosamine derivative was deemed the optimal water-soluble derivative. Compared with the acid-soluble chitosan, the chitosan-glucosamine derivative exhibited high chelating capacity for Zn^{2+}, Fe^{2+} and Cu^{2+} ions. Relatively high antibacterial activity against *Escherichia coli* and *Staphylococcus aureus* was noted for the chitosan-glucosamine derivative as compared with native chitosan. The water-soluble chitosan produced using the Maillard reaction may become a promising commercial substitute [Chung et al., 2005, 2006].

Chitooligomers easily turn brown during their shelf life. The factors influencing the browning of chitooligomers were investigated by Zeng et al. (2007) who attributed the browning to the structural change of the chitooligomers. Water-solubility, thermal stability and moisture-adsorption of chitooligomers decreased with the increase of browning. The time, temperature, pH, moisture, oxygen and reductant all had effect on the browning. The optimal preservation condition for the chitooligomers is low temperature and humidity, pH below 4 or above 10, and absence of oxygen.

4. Macrophages and other Human Cells

Macrophages are essential elements of innate immunity that control inflammatory reactions, healing processes, tissue homeostasis and immune tolerance by the co-ordinated release and clearance of soluble mediators. According to Kzhyshkowska et al., scavenger receptors (SRs) constitute a major class of receptors which direct endocytosed material for degradation. SR-mediated uptake can result in both pro-inflammatory and tolerogenic programming of macrophages. While effects of SRs on the level of signal transduction are well documented, the effect of endocytosis on the regulated secretion, in particular lysosomal secretion in macrophages remains elusive. The SR stabilin-1 shuttles between endosomal compartment and biosynthetic compartment and transports newly synthesized stabilin-interacting chitinase-like protein to the lysosomal secretory pathway. This sorting function of stabilin-1 is mediated by GGAs, clathrin adaptors responsible for the mannose-6-phosphate receptor-mediated transport of lysosomal hydrolases. Moreover, stabilin-1 internalizes hormone placental lactogen, transports it to the trans-Golgi network-associated transcytosis. Thus stabilin-1 is the only known SR that links endocytic clearance, intracellular sorting and trancytosis. A novel level of regulation for the secretory repertoire of macrophages was proposed in terms of cross-talk of uptake and release at the level of vesicular trafficking [Kzhyshkowska et al., 2009].

Factors affecting anti-inflammatory effect of oligochitosans (COSs) in lipopolysaccharide-stimulated RAW264.7 macrophage cells were investigated by Lee SH et al. The oligochitosans (10% DA) inhibited the NO secretion better than those with 50% DA, and the 5-10 kDa fraction of the former showed the highest inhibition activity. Said fraction (10% DA, 5-10 kDa) also inhibited the LPS-stimulated production of PGE_2, TNF-alpha and IL-6, as well as the expression of iNOS, COX-2, TNF-alpha, and IL-6, thus revealing itself as the most effective anti-inflammatory oligochitosan acting via down-regulation of transcriptional and translational expression levels of TNF-alpha, IL-6 and iNOS and COX-2 [Lee SH et al., 2009].

The neurotoxicity of glutamate and pyroglutamate is known since long [Choi, 1985, 1988; Coyle et al., 1993; McGeer et al., 1984], but this matter has been disregarded by those who have promoted the use of glutamate. Zhou et al. reported the protective effect of chitosan oligomers (MW 800) against glutamate-induced neurotoxicity in cultured hippocampal neurons. The cell viability assessments, together with Hoechst 33342 staining and flow cytometry for cell apoptosis analysis, indicated that cell apoptosis induced by 125 microM glutamate in cultured hippocampal neurons was attenuated in a concentration-dependent manner by pretreatment with oligomers. After measurement with Fluo 4-AM, the chitosan oligomers were found to depress glutamate-induced elevation in intracellular calcium concentration. The enzymatic assay indicated that they antagonized glutamate-evoked activation of caspase-3. These results collectively suggest that chitosan oligomers protect cultured hippocampal neurons from glutamate-induced cell damage by interfering with the elevation of calcium, and inhibiting caspase-3 activity [Zhou et al., 2008].

Yang et al. have investigated the in vitro neuronal differentiation of PC-12 cells. The morphologic observation and assessment using the specific reagent of tetrazolium salt WST-8 indicated that neurite outgrowths from PC-12 cells and their viability were enhanced by treatment of oligochitosans. The real-time quantitative RT-PCR and Western blot analysis showed that oligochitosan could upregulate the expression of neurofilament-H mRNA or protein and N-cadherin protein in PC-12 cells. The maximum effect of 0.1 mg/ml oligochitosan was obtained after 2 week culture. Thus oligochitosans possess good nerve cell affinity by supporting nerve cell adhesion and promoting neuronal differentiation and neurite outgrowth [Yang et al., 2009].

Gong et al. investigated the possible benefits of treatment with oligochitosans on nerve regeneration after crush injuries to peripheral nerves. The rabbits with the crushed common peroneal nerve were treated by daily intravenous injection of 1.5 or 3 mg/kg body weight of oligochitosans or identical volume of saline (as the control) for a 6-week period. The results showed that the compound muscle action potentials, the number of regenerated myelinated nerve fibers, the thickness of regenerated myelin sheaths, and the cross-sectional area of tibialis posterior muscle fibers were significantly improved in the nerves that received oligochitosans treatment in a dose-dependent way. Oligochitosans accelerated peripheral nerve regeneration after injury [Gong et al., 2009].

Chitosan is being used as a wound-healing accelerator in veterinary medicine. Chitosan enhances the functions of inflammatory cells such as polymorphonuclear leukocytes (PMN) (phagocytosis, production of osteopontin and leukotriene B4), macrophages (phagocytosis, production of interleukin (IL)-1, transforming growth factor beta1 and platelet derived growth factor), and fibroblasts (production of IL-8). As a result, chitosan promotes granulation and organization, therefore chitosan is beneficial for the large open wounds of animals. However, there are some reported complications of chitosan application. Firstly, chitosan causes lethal pneumonia in dogs which are given a high dose of chitosan, but this effect is restricted to dogs only. Secondly, intratumor injection of chitosan on tumor-bearing mice increased the rate of metastasis and tumor growth. Therefore, it was recommended to consider these effects of chitosan prior to drug delivery [Ueno et al., 2001].

The aim of Fernandes et al. was to study the effect of two partially depolymerized chitosans differing in molecular weight, i.e. <3 and <5 kDa, at concentrations in the range 5.0-0.005 mg/mL, on human red blood cells. The interactions of these two samples with RBC membrane proteins and with hemoglobin were assessed, as well as the RBC morphology and surface structure, analyzed by optical microscopy and atomic force microscopy. In the presence of either chitosan, no significant hemolysis was observed; however, at concentrations $>$ 0.1 mg/mL, changes in membrane binding hemoglobin were observed. Membrane protein changes were also observed with increasing oligochitosan concentration, including a reduction in both alpha- and beta-spectrin and in band 3 protein, and the development of three new protein bands: peroxiredoxin 2, calmodulin, and hemoglobin chains. Morphologic evaluation indicated that at high concentrations said chitosans interact with RBCs, leading to RBC adhesion, aggregation, or both. An increase in the roughness of the RBC surface with increasing concentration was observed. Overall, these findings suggest that the two partially depolymerized chitosans damage RBCs depending on their MW and concentration, and significant damage resulted from either a higher MW or a concentration greater than 0.1 mg/mL [Fernandes et al., 2008].

The use of tripolyphosphate permits to get chitosan in nanoparticles dispersed in a colloidal aqueous medium. By using a rational experimental design, Nasti et al. have studied the influence of a number of factors (pH, concentrations, ratios of components, different methods of mixing) in the preparation of chitosan/tripolyphosphate nanoparticles and in their coating with hyaluronic acid, aiming at the minimisation of size polydispersity, the maximisation of zeta potential and long-term stability, and at the control over average nanoparticle size. Three optimised nanoparticles have been developed (two uncoated and one hyaluronate-coated) and their toxicity on fibroblasts and macrophages has been evaluated: experiments showed the beneficial character of hyaluronate-coating in the reduction of toxicity (IC_{50} raised from 0.7-0.8 mg/mL to 1.8 mg/mL) and suggested that the uncoated chitosan/TPP nanoparticles had toxic effects following internalisation rather than membrane disruption [Nasti et al., 2009]. The importance of this work resides in particular on the detection of cytotoxicity of tripolyphosphate, an issue that has been so far neglected by the numerous authors that are proposing nanoparticles made of chitosan and tripolyphosphate.

5. Antioxidants and Antistress

The antioxidant potency of chitin derivative-glucosamine hydrochloride was investigated employing various established in vitro systems, such as superoxide, O_2*-, hydroxyl OH* radical scavenging, reducing power, and ferrous ion chelating potency. In the work by Xing et al., glucosamine hydrochloride had pronounced scavenging effect on superoxide radical. For example, the scavenging activity of glucosamine hydrochloride for O_2*- was 83.74 parts per thousand at 0.8 mg/mL. Second, the OH* scavenging activity of glucosamine hydrochloride was also strong and was about 54.89% at 3.2 mg/mL. Third, the reducing power of glucosamine hydrochloride was more pronounced. The reducing power of glucosamine hydrochloride was 0.632 at 0.75 mg/mL. However, ferrous ion-chelating potency was soft. Furthermore, ferrous ion-chelating potency, the radical scavenging rate, and the reducing power of glucosamine hydrochloride increased with their increasing concentration. The multiple antioxidant activity of glucosamine hydrochloride was evident as it showed considerable reducing power, superoxide/hydroxyl-radical scavenging ability. These in vitro results suggest the possibility that glucosamine hydrochloride used to alleviate oxidative stress, as an ingredient in health or functional food [Xing et al., 2006].

The antioxidant activity of various chitosan and chitin oligomers (supplied by Seikagaku, Tokyo) was tested in vitro with aminoguanidine, pyridoxamine, and Trolox as reference compounds. Hydroxylation of benzoate to salicylate by H_2O_2 in the presence of Cu^{2+} was effectively inhibited by chitobiose, chitotriose, aminoguanidine, pyridoxamine, and Trolox (their IC_{50} values were 18, 80, 85, 10, and 95 microM, respectively), whereas glucosamine, di-N-acetylchitobiose and tri-N-acetylchitotriose did not show any inhibitory activity. Chitobiose and chitotriose were more potent than the 3 reference compounds in scavenging hydroxyl radicals produced by photolysis of zinc oxide: IC_{50} values of the 2 oligomers were 30 and 5.5 microM, respectively. Such a scavenging activity of these 2 chitooligomers was also shown by the use of another system, a mixture of Fe^{3+}/EDTA/ascorbate/ H_2O_2, for producing hydroxyl radicals. Only chitobiose and Trolox, of the 10 compounds tested, had the ability to scavenge superoxide radicals generated by a non-enzymatic system using phenazine methosulfate and NADH. Keeping in mind that chitobiose and chitotriose are appreciably absorbed in the intestine of rats, both of them would act as effective antioxidants in vivo when orally ingested [Chen et al., 2003].

The inhibition of protein oxidation by reactive oxygen species would be beneficial to living organisms exposed to oxidative stress, because oxidized proteins are associated with many diseases and can propagate damages. Anraku et al. measured the ability of 2.8 kDa chitosan, D-glucosamine and N-acetyl glucosamine to protect human serum albumin from oxidation by peroxyl radicals derived from 2,2'-azobis(2-amidinopropane) dihydrochloride and N-centered radicals from 1,1'-diphenyl-2-picrylhydrazyl and from 2,2'-azinobis(3-ethylbenzothiazoline-6-sulfonic acid). Comparison with the antioxidant action of vitamin C showed that, on a molar basis, chitosan was equally effective in preventing formation of carbonyl and hydroperoxide groups in human serum albumin exposed to peroxyl radicals. it was also a potent inhibitor of conformational changes in the protein, assessed by spectrophotometry and intrinsic fluorescence. D-Glucosamine was much less effective and N-acetyl glucosamine was not an antioxidant. Protection of the albumin from peroxyl radicals was achieved by scavenging the peroxyl radical. Chitosan was also a good scavenger of N-centered radicals, with glucosamine and N-acetyl glucosamine much less effective. The results suggest that administration of low molecular weight chitosans may inhibit neutrophil activation and oxidation of serum albumin commonly observed in patients undergoing hemodialysis, resulting in reduction of oxidative stress associated with uremia [Anraku et al., 2008].

Treatment with partially depolymerized chitosan (20 kDa; DD 95 %) for 4 weeks produced a significant decrease in levels of plasma glucose, atherogenic index and led to increase in high density lipoprotein cholesterol in normal volunteers, as reported by Anraku et al. Chitosan treatment also lowered the ratio of oxidized to reduced albumin and increased total plasma antioxidant activity. There was good correlation between the latter and oxidized albumin ratio. The oxidized albumin ratio represents a potentially useful marker of oxidative stress. In vitro, albumin carbonyls and hydroperoxides were significantly decreased in a time-dependent manner in the presence of chitosan, compared with controls ($p < 0.05$). Chitosan also reduced two stable radicals in a dose- and time-dependent manner. Therefore chitosan has a direct antioxidant activity in systemic circulation by lowering the indices of oxidative stress in both in vitro and in vivo studies [Anraku et al., 2009].

Liu HT et al. investigated the protective effects of oligochitosans against hydrogen peroxide-induced oxidative damage on human umbilical vein endothelial cells ECV304. After 24 h pre-incubation with oligochitosans (25-200 microg/ml), the viability loss in ECV304 cells induced by H_2O_2 (300 microM) for 12 h was markedly restored in a concentration-dependent manner as measured by MTT assay. This effect was accompanied by a marked decrease in intracellular reactive oxygen species (by measuring intensity of DCFH fluorescence). Oligochitosans also exerted preventive effects on suppressing the production of lipid peroxidation such as malondialdehyde, restoring activities of endogenous antioxidants including superoxide dismutase and glutathione peroxidase, along with the capacity of increasing levels of nitric oxide and nitric oxide synthase. In addition, pre-incubation of oligochitosans with ECV304 cells for 24 h resulted in the reduction of apoptosis and the induction of cell cycle arrest in G(1)/S + M phase as assayed quantitatively by Annexin V-fluorescein isothiocyanate apoptosis detection kit using flow cytometry. These findings suggest that oligochitosans can effectively protect ECV304 cells against oxidative stress by

H_2O_2, which might be of importance in the treatment of cardiovascular diseases [Liu HT et al., 2009].

Chen SK et al. evaluated the effect of the functional groups sulfate, amine, and hydroxyl and their ionized forms on in vitro antioxidant capacities of low-molecular-weight polysaccharides prepared from agar (LMAG), chitosan (LMCH), and starch (LMST), and elucidated their structure / activity relationship. Ascorbic acid and ethylenediaminetetraacetic acid were used as positive controls. The in vitro antioxidant capacities of LMAG and LMCH were higher than that of LMST in the DPPH radical, superoxide radical, hydrogen peroxide, and nitric oxide radical scavenging and ferrous metal-chelating capacities. The different scavenging capacities may be due to the combined effects of the different sizes of the electron-cloud density and the different accessibility between free radical and LMPS, which, in turn, depends upon the different hydrophobicity of the constituent sugars [Chen SK et al., 2009].

Three N-carboxymethyl chitosan oligosaccharides with different degree of N-carboxymethylation (0.28, 0.41, and 0.54) were prepared by controlling of the amount of glyoxylic acid in the etherification process of chitosan oligosaccharide. Their antioxidant activities were evaluated by the scavenging of 1,1-diphenyl-2-picrylhrazyl radical (DPPH) radical, superoxide anion and determination of reducing power. With increasing substitution degree, the scavenging activity of N-CM-oligochitosans against DPPH radical decreased, and reducing power increased. As for superoxide anion scavenging, the order of efficacy for the degrees of N-carboxymethylation is 0.41 > 0.54 > 0.28. The difference may be related to the different radical scavenging mechanisms and donor effect of substituting carboxymethyl group [Sun et al., 2008].

6. Antitumor Efficacy

Hosomi et al. synthesized various oligosaccharides containing galactose and one glucosamine (or N-acetylglucosamine) units with β1–4, α1–6 and β1–6 glycosidic bonds; they were Galβ1–4GlcNH$_2$, Galα1–6GlcNH$_2$, Galα1–6GlcNAc, Galβ1–6GlcNH$_2$, Galβ1–4Galβ1–4GlcNH$_2$ and Galβ1–4Galβ1–4GlcNAc. Of these, Galα1–6GlcNH2 (MelNH$_2$) and glucosamine (GlcNH$_2$) had a suppressive effect on the proliferation of human leukemia K562 cells, but none of the other saccharides tested containing GlcNAc showed this effect; on the other hand, the proliferation of the human fibroblasts was suppressed by GlcNH$_2$ only. Upon adding Galα1–6GlcNH$_2$ or glucosamine to the culture, the number of K562 cells decreased strikingly within 72 h due to cell death. Furthermore, all of the cells were stained with Galα1–6GlcNH-FITC (MelNH-FITC) but neither the control cells nor the cells incubated with glucosamine were stained. The suppressive activity of MelNH$_2$ against the K562 cell proliferation was higher than that of GlcNH$_2$ (at 5 mM each) [Hosomi et al., 2009].

In an attempt to contribute to the understanding of the biochemical aspects involving recognition, differently charged (quaternary amino groups and sulfate groups) oligochitosan derivatives were synthesized and their anticancer activities were studied using three cancer cell lines, HeLa, Hep3B and SW480. Neutral red and MTT cell viability studies revealed that highly charged oligomers could significantly reduce cancer cell viability, regardless to the positive or negative charge. Fluorescence microscopy observations and DNA fragmentation studies confirmed that the anticancer effect of these highly charged oligomers were due to necrosis. However, the biochemistry for anticancer activity of strongly charged oligochitosans compared to their slightly charged counterparts remains unclear [Huang et al., 2009].

For the evaluation of the depression of tumor growth, the anti-metastatic potency and related pathways of oligomers extracted from fungi, oligomers having mainly greater than 3 but less than 8 units were prepared by continuous hydrolysis of fungal-derived chitosan with bromelin. The activity of metastasis related protein (matrix metalloproteinase-9, MMP-9) was inhibited by oligomers in Lewis lung carcinoma cells (LLC)- In a LLC-bearing mouse tumor growth and lung metastasis model, oligomers inhibited tumor growth and the number of lung colonies in LLC-bearing mice as well as the lung metastasis, and the survival time of the LLC-mice was prolonged. These results support the depression of tumor growth and the anti-metastatic potency of oligomers in cancer prevention [Shen et al., 2009].

In vitro and in vivo, oligomers significantly inhibited human hepatocellular carcinoma (HepG2) cell proliferation, reduced the percentage of S-phase and decreased DNA synthesis rate in oligomer-treated HepG2 cells. Expressions of cell cycle-related genes were analyzed and the results indicated that p21 was up-regulated, while PCNA, cyclin A and cdk-2 were down-regulated [Shen et al. 2009]. A series of chitooligomers (1.7 - 3.8 kDa) prepared by Tian et al. with the aid of hydrogen peroxide and selective precipitation in ethanol solutions were effective on the proliferation of L02 hepatocytes: at the initial stage of culture there was an inhibitory effect on proliferation, but the cultures recovered in cell proliferation and exhibited promotion effect in the following days. There was no significant effect of the chitooligomers on the functions of albumin secretion and urea synthesis of the hepatocytes [Tian et al., 2010].

Doxorubicin conjugated stearic acid-g-chitosan oligosaccharide polymeric micelles (DOX-CSO-SA) was synthesized by Hu FQ et al., via cis-aconityl bond between the anticancer drug doxorubicin (DOX) and stearic acid grafted chitosan oligosaccharide (CSO-SA). The CSO-SA micelles demonstrated fast internalization into tumor cells, and those with substitution degree 6.47% were used to synthesize DOX-CSO-SA. The critical micelle concentration was about 0.14 mg/ml. The micelles with 1 mg/ml CSO-SA had 32.7 nm diameter with a narrow size distribution and 51.5 mV surface potential. After conjugating with doxorubicin, the micellar size increased and the zeta potential decreased. The release rate of DOX from DOX-CSO-SA micelles increased significantly with the reduction of the pH from 7.2 to 5.0. The micelles effectively suppressed the tumor growth and reduced the toxicity typical of the commercial doxorubicin hydrochloride [Hu, FQ et al., 2009].

7. Antibacterial Activity

Raafat et al. investigated the antimicrobial mode of action of chitosan using a combination of approaches, including in vitro assays, killing kinetics, cellular leakage measurement, membrane potential estimation, and electron microscopy, in addition to transcriptional response analysis. The chitosan solution (DD 75-85 %, 50-190 kDa, 1% [wt/vol] in 1% acetic acid) was sterilized by autoclaving at 121°C for 20 min and stored at 4°C. The consequence of this thermal treatment on the MW remains unclear because the authors report the meaningless MW value 243.17 g/mol in their Table 2; in any case the sample was polydispersed. Chitosan exhibited a dose-dependent inhibitory effect on the growth of *Staphylococcus simulans*. A simultaneous permeabilization of the cell membrane to small cellular components, coupled to a significant membrane depolarization, was detected. Treatment of *S. simulans* with chitosan (20 microg/ml) resulted in a gradual leakage of UV-absorbing substances (likely nucleotide and coenzyme pools) from bacterial cells, followed by a plateau for up to 2 h. Optical density measurements of the treated culture revealed 50% reduction in culture density after 2 h, which was attributed to aggregation and flocculation of the cells in the presence of chitosan; attempts to detect protein leakage failed. Perturbation of membrane integrity, by antimicrobials for instance, leads to membrane depolarization and ultimately bacterial cell death. Raafat et al. found a substantial reduction in bacterial membrane potential when they monitored the distribution of the small lipophilic [^{3}H]TPP$^+$ ions between cells of *S. simulans* and the suspending medium in response to treatment with chitosan. A concomitant interference with cell wall biosynthesis was not observed. Chitosan treatment of *S. simulans* did not give rise to cell wall lysis; the cell membrane remained intact. The control cells showed intact plasma membranes of high electron density and an outer cell wall of medium electron density which was more or less uniform along the entire cell perimeter; sites of cell division were also evident. Cells treated with chitosan for 5 min showed irregular structures protruding from the cell surface, which might be chitosan deposits still attached to the negatively charged surface polymers. Interestingly, it seemed that the cell membrane became locally detached from the cell wall, giving rise to "vacuole-like" structures underneath the cell wall, possibly resulting from ion and water efflux and decreased internal pressure. Nonetheless, the membrane was well discernible in all sections and there was no evidence for cell wall lysis. As a point of difference from carboxybutyl chitosan which demonstrated an irregularly structured and frayed cell wall in carboxybutyl chitosan-treated microorganisms [Muzzarelli et al., 1990), the plain chitosan used by Raafat et al. seems to be less prone to seriously damage the wall, in agreement with the rationale that the bulkyness of the carboxybutyl groups has a more effective disturbing capacity on the structure.

The analysis of the transcriptional response data revealed that chitosan leads to multiple changes in the expression profiles of *Staphylococcus aureus* SG511 genes involved in the regulation of stress and autolysis, as well as genes associated with energy metabolism. Chitosan treatment reduced the bacterial growth rate, and this was clearly reflected in genetic expression profiles through the down-regulation of macromolecular biosynthesis, including genes involved in RNA and protein synthesis (14 ribosomal protein genes), as well as in the metabolism of carbohydrates, amino acids, nucleotides, and nucleic acids (six genes), lipids, and coenzymes (A long list of up-regulated and down-regulated genes is in Table 4 of original). To evaluate the possible involvement of teichoic acids of *S. aureus* in chitosan's antimicrobial activity and to analyze their role in chitosan susceptibility, Raafat et al. tested *S. aureus* strain SA113 together with four of its mutants lacking one or more genes involved in teichoic acid biosynthesis. Moreover, there is no evidence that chitosan is broken down by extracellular staphylococcal enzymes into active smaller fragments which might pass through the cell wall more easily. In addition, dialyzed chitosan was fully antimicrobial, suggesting that large molecules are responsible for its activity. Their main conclusion was that, when plain chitosan is used, membrane-bound energy generation pathways are affected, probably due to impairment of the proper functional organization of the electron transport chain, thus interfering with proper oxygen reduction and forcing the cells to shift to anaerobic energy production. They speculated that binding of chitosan to teichoic acids, coupled with a potential extraction of membrane lipids (predominantly lipoteichoic acid) results in a sequence of events, ultimately leading to bacterial death [Raafat et al., 2008].

The bactericidal activity of chitosan acetate solution against *E. coli* and *S. aureus* is well assessed. The integrity of the cell membranes of both species, and the permeability of the outer and inner membranes of *E. coli* were investigated by determining the release of materials that absorb at 260 nm, the release of cytoplasmic beta-galactosidase activity, and the changes in the fluorescence of cells treated with 1-N-phenylnaphthylamine: the latter test showed that the fluorescence reached a maximum within 10 minutes, thus confirming previous results [Helander et al., 2001], and that the chitosan was responsible of the rapid increase of the permeability of the bacterial cell membranes, and ultimately disrupted them severely, with the consequent release of cellular contents. This damage was started by the reaction between amino groups of chitosan and phosphate groups of phospholipid components of the cell membranes [Liu H et al., 2004].

These results on *E. coli* were confirmed most recently by Li XF et al. 2010: their plots of optical density at 610 nm *vs* time of contact with each of 3 chitosans showed the impressively higher bactericidal activity of the 50 kDa chitosan particularly at 4-6 h (o.d. 0.3 instead of 0.65-0.75 for chitosans of higher and lower MW, and control). Such trend was confirmed by the data on the release of cellular materials absorbing at 260 nm (K^+, PO_4^{3-}, DNA, RNA) and the loss of intracellular enzymes LDH and gamma-GT: accordingly the 50 kDa chitosan was more effective in damaging the membranes of *E. coli* than the 3 kDa chitosan. Interestingly, the residual concentrations of the 3, 50 and 1000 kDa chitosans after 30 min contact with *E. coli* were lowered by 4.90, 17.30 and 1.82 %, respectively [Li XF et al., 2010].

The antibacterial activity of membranes made of variously acetylated chitosans was investigated by Takahashi et al. in order to produce a high-performance membrane for technical use in separation processes. The antibacterial activity of powdered chitosan membrane was evaluated on the chitosan samples after drying and milling in terms of minimal inhibitory concentration (MIC). The MIC for *E. coli* was almost 200 (mg chitosan/ml bacterial suspension), and for *S. aureus* it was 40. *S. aureus* was more strongly inhibited by chitosan than the Gram-negative *E. coli*: in fact the inhibitory effect was recognized as a bactericidal effect. Antibacterial activity depended on the shape and the specific surface area of the powdered chitosan membrane. The influence of the deacetylation degree (DD) of the chitosan on inhibiting the growth of *S. aureus* was investigated by two methods: a conductimetric assay, and incubation using a mannitol salt agar medium in a Petri dish in the middle of which a square piece of membrane was positioned: the number of colonies in the agar layer over the membrane was much smaller than that in the peripheral agar (color photographs in the original). By both methods, chitosan with a higher DD successfully inhibited growth of *S. aureus*. The trend was clearer as DD increased from 83.9% to 92.2%. The sample with the highest DD was especially successful in inhibiting the growth of *S. aureus*. [Takahashi et al., 2008].

Atomic force microscopy (AFM) imaging was used to obtain high-resolution images of the effect of chitosans on the bacterial morphology [Eaton et al., 2008]. The AFM measurements were correlated with viable cell numbers, which show that the two species reacted differently to the high- and low-molecular-weight chitosan derivatives. The images obtained revealed not only the antibacterial effects, but also the response strategies used by the bacteria; cell wall collapse and morphological change, reflected cell death, whereas clustering of bacteria appeared to be associated with cell survival. In the case of the low-MW oligochitosan, the *E. coli* responded to the treatment by clustering probably due to the ionic interaction between oligochitosan and cell wall (since oligochitosan was protonated under the pH conditions used) or due to the production of EPS. This presumably protected the bacteria in the interior of the clusters from the action of the oligochitosan, preventing further bacterial population reduction after 4 h of treatment time. However, for the high-MW chitosan the polymer prevented this clustering mechanism; hence, many more isolated bacteria were observed, and also more evidence was seen of cell death (collapsed rods). The chitosan coating of the cells interfered with the communication between the cells, further hampering their defenses. In addition, nanoindentation experiments revealed mechanical changes in the bacterial cell wall induced by the treatment. The nanoindentation results suggested that despite little modification observed in the Gram-positive bacteria in morphological studies, cell wall damage had indeed occurred, since cell wall stiffness was reduced after oligochitosan treatment.

It is instructive to compare the data obtained from the AFM experiments to those obtained by standard cell-counting studies. The cell-viability studies on the two bacteria, using oligochitosan and chitosan, indicate that the trend in the results differs for the two organisms, and also for the two MW values. In the case of the oligochitosan they acted very quickly to reduce the number of viable cells of *E. coli*, by almost three orders of magnitude within the first 4 h. However, between 4 and 24 h the number of viable cells showed no further reduction. In the case of the chitosan, however, while the initial reduction in cell viability was slower, over the 24 h a greater reduction was achieved. According to the cell-viability studies, the action of the chitosan was slower, but the antibacterial effect was maintained for longer, resulting in a greater effect in the long term. It was confirmed that the antimicrobial effect is strongly dependent on the type of target microorganism and the MW of chitosan. Using a confocal laser scanning microscope showed that chitosan oligomers actually penetrate *E. coli*, suggesting that its antibacterial activity seemed to be mainly caused by the inhibition of the DNA transcription [Liu XF et al., 2001].

The aim of Ji QX et al. was to evaluate the applicability of chitosan to the dental field, along with N-[1-hydroxy3-(trimethylammonium)propyl] chitosan chloride (HTCC), prepared by reacting chitosan with glycidyltrimethylammonium chloride (GTMAC). The chitosan used in this work declaredly had MW 3.800 kDa (an unreliable value), and DD 86%. Thus little can be said, notwithstanding the indication of a MW reduction to 1080 kDa (also unreliable). Chitosan and HTCC were characterized by infrared and H^{1}NMR spectroscopy. Four oral strains (*Porphyromonas gingivalis* ATCC33277, *Prevotella intermedia* ATCC 25611, *Actinobacillus actinomycetemcomitans* Y4, and *Streptococcus mutans* Ingbritt C) were susceptible to chitosan and HTCC with minimum inhibitory concentrations (MICs) ranging from 0.25 to 2.5 mg/mL. The in vitro 3-(4,5-dimethyl-2-thizolyl)-2,5-diphenyl-2H-tetrazolium bromide (MTT) assay determined that chitosan at 2000, 1000, 100, and 50 microg/mL could stimulate the proliferation of human periodontal ligament cells. Instead, HTCC inhibited the proliferation at those concentrations, but accelerated the proliferation of ligament cells at 0.3-10 microg/mL. Transmission electron microscopy observations revealed that the ultra-structure of ligament cells was altered by HTCC treatment at 1000 microg/mL. In fact, TEM observations revealed morphological changes after cells were incubated with different concentrations of chitosan and HTCC. The quaternary nature of HTCC at the higher concentrations provided a strongly cationic environment and a strong electrostatic attraction with the negatively charged groups on the surface of ligament cells, which resulted in the alteration of the cells, thus causing a serious cytotoxic effect. One can deduce that the better antibacterial activity of HTCC is related to the cytotoxicity, although more research should be carried out on this point. Also, when HTCC is associated with other composites and forms a new structure such as a film, gel, microcapsule, microsphere or nanoparticle, reevaluation of biocompatibility will be necessary for planning applications of chitosan and HTCC in the dental field [Ji, QX et al., 2009].

Electrokinetic properties of complexes of chitosan with lipopolysaccharides from *E. coli* O55:B5, *Yersinia pseudo tuberculosis* 1B 598, and *Proteus vulgaris* 025 (48/57) and their size distribution were investigated by Davydova et al., using zeta-potential assay and quasi-elastic light scattering. The interaction of LPS from different microorganisms with chitosan at the same w/w ratio of components (1:1) resulted in the formation of complexes in which the negative charge of LPS was neutralized (LPS from E. coli) or overcompensated (*Y. pseudotuberculosis* and *P. vulgaris*). The size change of the endotoxin aggregates during binding with chitosan was observed. The binding constants of chitosan with LPSs were determined by a method based on the anionic dye Orange II. The LPS from *E. coli* possess higher affinity for chitosan in comparison with the two others samples of endotoxin [Davydova et al., 2008].

Considering the increasing microbial resistance to conventional antibiotics, essential oils have shown to be an important option against these pathogens. For sustained stability and prolonged release of essential oils from pharmaceutical formulations, some authors have studied the association of chitosan to them. The review by Pedro et al. (2009) disserts about the application of chitosan and essential oils in oral cavity care, pointing out that their association may be an interesting option in the food area as well.

8. Various Approaches to Depolymerization

Recent studies of chitosan have increased the interest in its conversion to oligochitosans because these compounds are water-soluble and have potential use in several biomedical applications. Furthermore, such oligomers may be more advantageous than chitosans because of their much higher absorption profiles at the intestinal level, which permit their facilitated access to systemic circulation and potential distribution throughout the entire human body.

8.1. Sodium Nitrite

Chitosans of low Mw (23 and 38 kDa) were obtained with the aid of sodium nitrite by Vila et al. after a depolymerization process of a 70 kDa chitosan. The reaction between chitosan and nitrite is very fast due to the liberation of gaseous products, i.e. nitrogen oxides, and therefore it may be uncontrollable, unless precautions are adopted. To avoid complete degradation of chitosan, those authors used 0.1 % nitrite solution to be added in various ratios to 0.2 g/l chitosan solution. The partially depolymerized chitosans and the original chitosan were then reacted with tripolyphosphate that spontaneously forms nanoparticles of chitosan tripolyphosphate salt, particularly useful because they have affinity for proteins and lend themselves to incorporation of the latter. Tetanus toxoid, used as a model antigen, was entrapped within chitosan nanoparticles. Tetanus toxoid-loaded nanoparticles were first characterized for their size, electrical charge, loading efficiency and in vitro release of antigenically active toxoid. Irrespective of the chitosan MW, the nanoparticles were in the 350 nm size range, and exhibited a positive electrical charge (+40 mV) and associated Tetanus toxoid quite efficiently (loading efficiency: 50-60%). In vitro release studies showed an initial burst followed by an extended release. The nanoparticles were then administered intranasally to conscious mice in order to study their feasibility as vaccine carriers. Chitosan nanoparticles were also labeled with FITC-BSA and their interaction with the rat nasal mucosa examined by confocal laser scanning microscopy. Tetanus toxoid-loaded nanoparticles elicited an increasing and long-lasting humoral immune response (IgG concentrations) as compared to the fluid vaccine. Similarly, the mucosal response (IgA levels) at 6 months post-administration of Tetanus toxoid-loaded chitosan nanoparticles was significantly higher than that obtained for the fluid vaccine. The chitosan tripolyphosphate nanoparticles can cross the nasal epithelia and, hence, transport the associated antigen. Interestingly, the ability of these nanoparticles to promote the access of the antigen to the immune system was not significantly affected by the MW. Indeed, high and long-lasting responses could be obtained using low MW chitosan. Furthermore, the response was not influenced by the 70-200 microg chitosan dose [Vila et al., 2004].

The deacetylation and depolymerization reactions of chitin/chitosan from three crustacean species (*Paralomis granulosa, Lithodes antarcticus and Palinurus vulgaris*) were evaluated under the same conditions. The average molecular weight and the mole fraction of N-acetylated units were the parameters studied in the resulting chitosans. The N-deacetylation follows a pseudo-first order kinetics and the apparent rate constants are very similar. The depolymerization process actuated with the aid of diluted potassium nitrite at 35°C follows a pseudo-first order kinetics for the three species, but in the first 9 min *P. vulgaris* shows a slightly lower depolymerization rate. In terms of ash contents, crystallinity and the physicochemical characteristics of chitin, the chitosans show different quality [Galed et al., 2005], thus demonstrating that the chemical depolymerization is affected by the origin (and seasonal variations) of the raw material.

8.2. Sonication

High intensity ultrasound irradiation has received much attention as a rapid, environmentally friendly, and byproduct-free method. The mechanism, kinetics, and application of ultrasound in the degradation of various synthetic polymers have been widely investigated. Cleavage of polymer chains by sonication with frequencies ranging from 20 to 100 kHz has been attributed mainly to the action of shear forces formed due to the relative movement between solvent and polymer molecules during the collapse of cavitation bubbles and the formation of microjets. Thus, the underlying cause of degradation of a polymer by ultrasound is considered to be primarily of a mechanical nature. However, at frequencies higher than 100 kHz, free HO* radicals formed by ultrasound in an aqueous solution have a significant role in the polymer degradation.

The influence of sonication at 41 W/cm^2 during extraction of chitin from the shells of the North Atlantic shrimp *Pandalus borealis*, on chitin yield, purity, and crystallinity was investigated by Kjartansson et al.. Shells were peeled, washed, lyophilized, ground, and suspended for 4 h in 0.25 M HCl (1: 40) at 40 °C followed by ultrasonication at 41 W/cm^2 for 1, and 4 h. Demineralized shells were lyophilized, resuspended in 0.25 M NaOH (1: 40), and ultrasonicated at 41 W/cm^2 for 1 and 4 h to remove proteins. The purity of extracted chitin was determined from the total amount of glucosamine. The crystallinity index and size of crystals were calculated from wide-angle X-ray scattering measurements. Scanning electron microscope images were recorded to evaluate morphological changes in samples. The yield of chitin decreased from 16.5 to 11.4% after 1 h sonication, with concomitant increased concentrations of depolymerized materials in the wash water. Sonication did not enhance the removal of minerals, but enhanced the removal of proteins. The crystallinity index of chitin decreased from 87.6 to 79.1 and 78.5% after 1 and 4 h of sonication, yielding chitosans with crystallinity indices of 76.7, 79.5, and 74.8% after deacetylation, respectively. Fourier transform infrared spectroscopy scans indicated that the degree of acetylation of chitins was unaffected by sonication. Comparison of the extraction results of *P. borealis* with those from freshwater prawns indicated that more impurities were left in *P. borealis* chitin, suggesting that composition and structural arrangement of chitin in shells influence the efficiency of ultrasound-assisted extraction [Kjartansson et al. 2006).

Chain scission increased with an increase in power of ultrasound and solution temperature, but a decrease in chitosan concentration. The chemical structure and polydispersity of the original and the fragments were nearly identical. A model based on experimental data to describe the relationship between chain scission and experimental variables (power of ultrasound; irradiation time; reduced concentration, c[eta]; and solution temperature) was proposed. It was concluded that ultrasonic irradiation is a suitable method to perform partial depolymerization and to obtain moderate macromolecules from large ones. Crab-shell chitosan with degree of acetylation 9% in acetic solution was exposed to ultrasonic irradiation at different time intervals at 60 °C. The intrinsic viscosity decreased from 19.6 to 4.2 dL/g but the DA did not change.

The degradation kinetics of chitosan with a DA of 43% at pH values in the range of 1.0–4.7 and under sonication were followed by measuring the solution viscosity: the maximum alteration of the DA occurred at pH 1.0 and optimal pH for prevention of deacetylation process was at pH 2.0. It was observed that the reduction of intrinsic viscosity was faster at the beginning of sonication, but the rate decreased gradually; the decrease in intrinsic viscosity was greater for chitosans having higher DA. The polydispersity of the chromatograms did not increase when the molecular weight of the fragments decreased, suggesting that the larger macromolecules were preferentially fragmented and fragmentation did not occur near the end of macromolecules; the degradation of chitosan is not a random process and the polydispersity of the fragments depends on the sequence of N-acetyl glucosamine and glucosamine in the chains.

The viscosity of the 0.6% chitosan solution at 5 °C was ten times greater than that at 45 °C. A certain amount of energy is consumed in the process of disaggregation, leading to smaller value for a. Secondary processes may be taken into consideration at higher temperatures as follows: (1) low concentration of acid (dilute solution of acetic acid as a solvent for chitosan) at high temperatures may induce random degradation of chitosan; (2) thermal motion should have an impact on polymer conformation, where a compact conformation should increase the susceptibility of the polymer to ultrasonic action; and (3) ultrasonic irradiation was shown to dissociate water molecules, forming free radicals (hydrogen, H*, and hydroxyl, OH*), which can lead to the formation of hydrogen peroxide.

Low concentration of hydrogen peroxide induced random degradation of chitosan. However, the latter effects of temperature on the polymer degradation may play only a minor role. In conclusion, wave created from an ultrasound with a frequency of 20 kHz induced fragmentation in chitosan solution.

The rate of chitosan fragmentation in 0.1 M acetic acid increased with an increase in power of ultrasound and temperature the latter being a minor factor. The polydispersity of the fragments did not change by progressing in fragmentation process. The optimum conditions for preparation of smaller molecular weight fragments from large macromolecules were a low chitosan concentration, a high ultrasonic power, and a high solution temperature. The results of ^{1}HNMR spectroscopy and elemental analysis indicate that the chemical structure and the DA of chitosan were not altered by ultrasonic irradiation. The relationship between chain scission and the experimental variables was described by a proposed parametric model. This model can be used to estimate chain scission of chitosan in 0.1 M acetic acid under different experimental conditions. Ultrasonic irradiation can be employed for partial depolymerization and preparation of moderate macromolecules from large ones.

Kasaai reviewed the methods for the determination of the degree of acetylation DA and classified them into three categories: (1) spectroscopy (IR, ^{1}HNMR, ^{13}CNMR, ^{15}NNMR, and UV); (2) conventional (various types of titration, conductometry, potentiometry, ninhydrin assay, adsorption of free amino groups of chitosan by picric acid); (3) destructive (elemental analysis, acid or enzymatic hydrolysis of chitin/ chitosan and followed by the DA measurement by colorimetry or high performance liquid chromatography, pyrolysis-gas chromatography, and thermal analysis using differential scanning calorimetry) methods. These methods have been compared for their performances and limitations as well as their advantages and disadvantages. The use of IR and NMR spectroscopy methods provides a number of advantages. They do not need long-term procedures to prepare samples, and they provide information on the chemical structure. ^{1}HNMR and UV techniques are more sensitive than IR, ^{13}CNMR and ^{15}NNMR spectroscopy. The IR technique is mostly used for a qualitative evaluation and comparison studies. Conventional methods are not applicable for highly acetylated chitin. The results of the latter methods are affected by ionic strength of the solvent, pH, and temperature of solution. In destructive methods, longer times are needed for the measurements compared to spectroscopy and conventional methods, but they are applicable to the entire DA range [Kasaai, 2009; Kasaai et al., 2008].

The degradation of chitosan by high-intensity ultrasound was investigated by gel permeation chromatography coupled with static light scattering. The molecular weight, radius of gyration, and polydispersity of chitosan were reduced by ultrasound treatment, whereas chitosan remained in the same random coil conformation and the degree of acetylation did not change after sonication. The results demonstrate that (1) the degradation of chitosan by ultrasound is primarily driven by mechanical forces and the degradation mechanism can be-described by a random scission model; (2) the degradation rate is proportional to Mw; and (3) the degradation rate coefficient is affected by ultrasound intensity, solution temperature, polymer concentration, and ionic strength, whereas acid concentration has little effect. Additionally, the data indicate that the degradation rate coefficient is affected by the degree of acetylation of chitosan and independent of the initial molecular weight (Wu et al. 2008). Chitosan remains in a random coil conformation after the sonication regardless of sonication time. This suggests that the degradation is not free radical induced because free radical degradation would result in the formation of macromolecular free radicals, and the recombination of these macromolecular free radicals would likely lead to the formation of side chains and a conformational change. The UV spectra of chitosan before and after sonication further suggest that degradation was not induced by free radicals. HIU did not alter the UV spectrum of chitosan aqueous solutions significantly, in contrast to the degradation carried out by a 360 kHz ultrasound, where byproducts containing carbonyl groups were formed as evidenced by a new absorbance peak at 265 nm in the UV spectra. This further strengthens the argument that at low frequencies the degradation is mainly due to mechanical forces [Wu et al., 2008].

The effect of inorganic salts such as sodium chloride on the hydrolysis of chitosan in a microwave field was investigated by Xing et al. While it is known that microwave heating is a convenient way to obtain a wide range of products of different molecular weights only by changing the reaction time and/or the radiation power, the addition of some inorganic salts was shown to effectively accelerate the degradation of chitosan under microwave irradiation. The molecular weight of the degraded chitosan obtained by microwave irradiation was considerably lower than that obtained by traditional heating. Moreover, the molecular weight of degraded chitosan obtained by microwave-assisted irradiation under the conditions of added salt was considerably lower than that obtained by microwave irradiation without added salt. Furthermore, the effect of ionic strength of the added salts was not linked with the change of molecular weight. FTIR spectral analyses demonstrated that a significantly shorter time was required to obtain a satisfactory molecular weight by the microwave irradiation-assisted inorganic salt method compared to conventional technology [Xing et al., 2005].

The effects of 360-kHz ultrasound on aqueous solutions of chitosan and starch were studied by Czechowska-Biskup et al. In Ar-saturated 2×10^{-2} mol/dm^3 chitosan solutions, pH 3, at an ultrasound dose rate of 170 W/kg, the average sonochemical chain scission yield in the sonication time range of 0-90 min is ca. 8×10^{-11}) mol/J. This yield depended on polymer concentration, ultrasound power and gas used to saturate the solution. Scission is accompanied by side reactions leading to the formation of carbonyl groups. Ultrasound-induced chain scission of starch proceeds with lower yield [Czechowska-Biskup et al., 2005].

8.3. Hydrogen Peroxide

Hydrogen peroxide was used to depolymerizes chitosan to yield water-soluble chitosan: an important aspect of this chemical approach is that the reagent does not introduce in the medium any permanent chemical such as counterions. A mathematical model between degradation conditions (H_2O_2 concentration, time and temperature) and the recovery of water-soluble chitosan was constructed using response surface methodology. Each factor showed a significant effect on the recovery. The optimal conditions to obtain the highest recovery of water-soluble chitosan were 5.5% of H_2O_2, 3.5 h, and 42.8 °C; the recovery was 93.5%. By determination of inhibition zone diameter, water-soluble chitosan showed significantly ($P<0.05$) higher inhibition capabilities against *E. coli, B. subtilis* and *S. aureus* than crude chitosan [Du et al., 2009].

Tian et al. demonstrated by FTIR and H-1 NMR the breakage of 1,4-beta-D-glucoside bonds of chitosan, leading to depolymerization. X-ray quantitative analysis shows the crystallinity degree of chitosan was changed in the depolymerization. In the crystalline region the chitosan is depolymerized by a peel-it-off process while the amorphous portion is depolymerized by a penetrating pattern. Moreover, the depolymerization rate of chitosan depends on the deacetylation degree of chitosan, the concentration of hydrogen peroxide and the reaction temperature [Tian et al., 2004]. The existence of a synergistic effect on the degradation of chitosan by hydrogen peroxide under irradiation with ultraviolet light was demonstrated by means of viscometry [Wang et al., 2005].

8.4. Enzymatic Depolymerization

With time, new improvements were introduced to enzymatic production and presently it has been developed to a continuous production process. The biological activities reported for oligochitosan, such as antimicrobial, anticancer, antioxidant, and immunostimulant effects depend on their physico-chemical properties. Kim and Rajapakse (2005) have summarized different enzymatic preparation methods of oligochitosan and some of their reported biological activities, as follows: (1) Antimicrobial activity, antimicrobial mechanisms of oligomers; effect of positive charge and molecular weight on antibacterial activity; charge properties of bacterial cell wall; effect of oligomers on the inhibition of fungal growth. (2) Antiviral activity of oligomers; inhibitory activity on animal viruses; inhibitory activity on bacteriophages. (3) Antitumor activity. (4) Antioxidant and radical scavenging activities. (5) Fat lowering and hypocholesterolemic effects of oligomers. (6) Immunostimulant effects of oligomers. (7) Other biological activities. Therefore their review article, within certain limits, represents the status of the knowledge on these topics in 2005.

Su, XW et al. determined the effect of chitosan on three human enteric viral surrogates: murine norovirus 1 (MNV- 1), feline calicivirus F-9 (FCV-179), and (ssRNA) bacteriophage MS2 (MS2). Chitosan oligosaccharide lactate (5 kDa) and water-soluble chitosan (53 kDa) at concentrations of 1.4, 0.7, and 0.35% were incubated at 37 °C for 3 h with equal volumes of each virus at high (ca. 7 log PFU/ml) and low (ca. 5 log PFU/ml) titers. Chitosan effects on each treated virus were evaluated with standardized plaque assays in comparison to untreated virus controls. The water-soluble chitosan at 0.7% decreased the FCV-F9 titer by 2.83 log PFU/ml, and also decreased MS2 at high titer by 1.18 - 1.41 log PFU/ml. Chitosan treatments had no effect on MNV-1 at high titers. Chitosan oligosaccharide showed similar trends against the viruses, but to a lesser extent compared with that of water-soluble chitosan. When lower virus titers (5 log PFU/ml) were used, plaque reduction was observed for FCV-F9 and MS2, but not for MNV-1 [Su et al., 2009].

When four chitosans variously deacetylated were used as substrates of a commercial lipase, the latter exhibited higher activity toward chitosans which were 82.8% and 73.2% deacetylated. The optimal temperature of the lipase was 55 °C for all chitosans. The chitosan hydrolysis carried out at 37 °C yielded a larger quantity of oligochitosans than that at 55 °C when the reaction time was longer than 6 h, and the yield of 24 h hydrolysis at 37 °C was 93.8%. Products analysis demonstrated that the enzyme produced glucosamine and oligochitosans with degree of polymerization of 2-6 and above, and it acted on chitosan in both exo- and endo-hydrolytic manner [Lee DX et al., 2008].

With the goal of achieving operational stability, the production of chitosan oligosaccharides was made by continuous hydrolysis of chitosan in an enzyme membrane reactor. The continuous-flow stirred-tank reactor was equipped with an ultrafiltration membrane with a molecular weight cut-off of 2 kDa, and the hydrolysis was accomplished with chitosanase from *Bacillus pumilus*. After optimization of the reaction parameters, such as the amount of enzyme, the yield of the target oligosaccharides produced in the membrane bioreactor with free chitosanase reached 52% on the basis of the fed concentration of chitosan. An immobilized chitosanase prepared by the multipoint attachment method was used to improve the operational stability of the membrane bioreactor. Under the optimized conditions, pentamer and hexamer were steadily produced at 2.3 g/L (46% yield) for a month: the half-life of the productivity of the reactor was estimated to be 50 days [Kuroiwa et al., 2009].

Chitosan was depolymerized either by HCl hydrolysis or enzymatic degradation with the commercial preparation Pectinex Ultra Spl. The oligochitosans released by both methods were selectively precipitated in methanol solutions and characterized using MALDI-TOF mass spectrometry. The enzymatic method yielded shorter fragments with a higher proportion of fully deacetylated chitooligomers. Conversely, acid hydrolysis of the starting chitosan resulted in fragments with degree of polymerization up to sixteen and more monoacetylated residues than with the enzymatic procedure [Cabrera et al., 2005]. This article is one of several making use of unspecific enzymes in the controlled depolymerization of chitosan.

8.5. Electrochemical Depolymerization

Chitosan (DD 89.8%; 479 kDa) was dissolved (0.3%) in 0.2 mol/L acetic acid + 0.3 mol/L sodium acetate solution. The electrolysis was performed with $Ti/TiO_2–RuO_2$ as anode, stainless steel as cathode and the spacing of 20 mm between the two electrodes. A DC potentiostat was used as the power supply; the solution was stirred constantly at 60 °C. After depolymerization, the solution was neutralized with 2 mol/L NaOH to pH 8–9, excess ethanol was added, and the products were collected by filtration and overnight drying in a vacuum at 60 °C.

When the chitosan was exposed to electrolysis [160 mA/cm^2], its MW decreased from 491 to 33 kDa within 60 min. When the current density was varied from 40 to 160 mA/cm^2 and the corresponding current potential was varied from 4.2 to 10.5 V, the MW of chitosan decreased to 159, 91, 59 and 33 kDa in 60 min, (67.6%, 81.4%, 88.0% and 93.3%, respectively), i.e. the increase of current density led to more extended depolymerization. Moreover, a linear relationship existed between the 1/MW and the reaction time, due to a random scission of chitosan chains. To confirm the electrochemical depolymerization of chitosan, GPC was used for qualitative evaluation of the molecular weight distribution [Cai et al., 2010]. It is evident that the electrolysis is a quite effective method for the depolymerization of chitosan.

Conclusion

According to the most recent literature, oligochitosans and partially depolymerized chitosans confirm to be endowed with enhanced peculiar activities in comparison to high molecular weight chitosans. They include stimulation of human cells in vitro and in vivo, radical scavenging and stress prevention, anti-tumor and anti-metastasis activity. Their production is currently a facile process, but their storage requires certain precautions in order to avoid pigment formation and aggregation. They lend themselves to a number of chemical or enzymatic modifications that are somewhat simpler to carry out considering the much lower viscosity of their solutions in comparison with chitosans, and the solubility in organic media for oligochitosans.

The literature reveals discrepancies in the evaluation of average molecular weights and polydispersity, depending on the analytical methods used, that sometimes lead to unreliable data: therefore an inter-laboratory standard evaluation would be welcome, if reproducibility of performances is sought. Particularly in the case of antibacterial activity, it has been noted in various studies that certain molecular weights of chitosans are more efficient than others in depressing the growth of some bacterial strains or in killing them; thus, really good results will be consistently obtained if optimum molecular weight, polydispersity, zeta potential and accompanying ions are identified. The same can be said for the exploitation of chitosan as a DNA carrier for genetic therapy.

ACKNOWLEDGMENTS

The author is grateful to Dr. Marilena Falcone, Central Library, Polytechnic University, Ancona, Italy, for assistance in handling the bibliographic information, and to Mrs. Maria Weckx for the preparation of the typescript.

REFERENCES

Anraku, M; Fujii, T; Furutani, N; Kadowaki, D; Maruyama, T; Otagiri, M; Gebicki, JM; Tomida, H. Antioxidant effects of a dietary supplement: Reduction of indices of oxidative stress in normal subjects by water-soluble chitosan. *Food and Chemical Toxicology* 47 (1). 2009. 104-109.

Anraku, M; Kabashima, M; Namura, H; Maruyama, T; Otagiri, M; Gebicki, JM; Furutani, N; Tomida, H. Antioxidant protection of human serum albumin by chitosan. *International Journal of Biological Macromolecules* 43 (2). 2008. 159-164.

Cabrera, JC; Van Cutsem, P. Preparation of chitooligosaccharides with degree of polymerization higher than 6 by acid or enzymatic degradation of chitosan. *Biochemical Engineering Journal* 25 (2). 2005. 165-172.

Cai, Q; Gu, Z; Chen, Y; Han, W; Fu, T; Song, H; Li, F. Degradation of chitosan by an electrochemical process. *Carbohydrate Polymers* 79, 2010, 783-785.

Chen, AS; Taguchi, T; Sakai, K; Kikuchi, K; Wang, MW; Miwa, I. Antioxidant activities of chitobiose and chitotriose. *Biological and Pharmaceutical Bulletin* 26 (9). 2003. 1326-1330.

Chen, SK; Tsai, ML; Huang, JR; Chen, RH. In Vitro Antioxidant Activities of Low-Molecular-Weight Polysaccharides with Various Functional Groups. *Journal of Agricultural and Food Chemistry* 57 (7). 2009. 2699-2704.

Choi, D.W. Glutamate neurotoxicity and diseases of the nervous system. *Neuron* 1. 1988. 623–634.

Choi, D.W. Glutamate neurotoxicity in cortical cell culture is calcium dependent. *Neuroscience Letters* 58. 1985. 293–297.

Chung, YC; Kuo, CL; Chen, CC. Preparation and important functional properties of water-soluble chitosan produced through Maillard reaction. *Bioresource Technology* 96 (13). 2005. 1473-1482.

Chung, YC; Tsai, CF; Li, CF. Preparation and characterization of water-soluble chitosan produced by Maillard reaction. *Fisheries Science* 72 (5). 2006. 1096-1103.

Coyle, JT; Puttfarken, P. Oxidative stress, glutamate and neurodegenerative disorders. *Science* 262. 1993. 689-695.

Czechowska-Biskup, R; Rokita, B; Lotfy, S; Ulanski, P; Rosiak, JM. Degradation of chitosan and starch by 360-kHz ultrasound. *Carbohydrate Polymers* 60 (2). 2005. 175-184.

Davydova, VN; Bratskaya, SY; Gorbach, VI; Solov'eva, TF; Kaca, W; Yermak, IM. Comparative study of electrokinetic potentials and binding affinity of lipopolysaccharides-chitosan complexes. *Biophysical Chemistry* 136 (1). 2008. 1-6.

Du, YJ; Zhao, YQ; Dai, SC; Yang, B. Preparation of water-soluble chitosan from shrimp shell and its antibacterial activity. *Innovative Food Science and Emerging Technologies* 10 (1). 2009. 103-107.

Eaton, P; Fernandes, JC; Pereira, E; Pintado, ME; Malcata, FX. Atomic force microscopy study of the antibacterial effects of chitosans on Escherichia coli and Staphylococcus aureus. *Ultramicroscopy* 108 (10). 2008. 1128-1134.

Fernandes, JC; Eaton, P; Nascimento, H; Belo, L; Rocha, S; Vitorino, R; Amado, F; Gomes, J; Santos-Silva, A; Pintado, ME; Malcata, FX. Effects of chitooligosaccharides on human red blood cell morphology and membrane protein structure. *Biomacromolecules* 9 (12). 2008. 3346-3352.

Freier, T; Koh, HS; Kazazian, K; Shoichet, MS. Controlling cell adhesion and degradation of chitosan films by N-acetylation. *Biomaterials* 26 (29). 2005. 5872-5878.

Galed, G; Miralles, B; Panos, I; Santiago, A; Heras, A. N-Deacetylation and depolymerization reactions of chitin/chitosan: Influence of the source of chitin. *Carbohydrate Polymers* 62 (4). 2005. 316-320.

Gong, YP; Gong, LL; Gu, XS; Ding, F. Chitooligosaccharides promote peripheral nerve regeneration in a rabbit common peroneal nerve crush injury model. *Microsurgery* 29 (8). 2009. 650-656.

Helander, IM; Nurmiaho-Lassila, EL; Ahvenainen, R; Rhoades, J; Roller, S. Chitosan disrupts the barrier properties of the outer membrane of Gram-negative bacteria. *International Journal of Food Microbiology* 71 (2-3). 2001. 235-244.

Hollister, SJ. Porous scaffold design for tissue engineering. *Nature Materials* 4. 2005. 518–524.

Hosomi, O; Misawa, Y; Takeya, A; Matahira, Y; Sugahara, K; Kubohara, Y; Yamakura, F; Kudo, S. Novel oligosaccharide has suppressive activity against human leukemia cell proliferation. *Glycoconjugate Journal* 26 (2). 2009. 189-198.

Hu, FQ; Liu, LN; Du, YZ; Yuan, H. Synthesis and antitumor activity of doxorubicin conjugated stearic acid-g-chitosan oligosaccharide polymeric micelles. *Biomaterials* 30 (36). 2009. 6955-6963.

Huang, R; Mendis, E; Rajapakse, N; Kim, SK. Strong electronic charge as an important factor for anticancer activity of chitooligosaccharides. *Life Sciences* 78 (20). 2006. 2399-2408.

Ji, QX; Zhong, DY; Lu, R; Zhang, WQ; Deng, J; Chen, XG. In vitro evaluation of the biomedical properties of chitosan and quaternized chitosan for dental applications. *Carbohydrate Research* 344 (11). 2009. 1297-1302.

Jollès, P; Muzzarelli, RAA, eds. *Chitin and Chitinases*. Basel, CH. Birkhauser Verlag. 1999.

Kasaai, MR. Various Methods for Determination of the Degree of N-Acetylation of Chitin and Chitosan: A Review. *Journal of Agricultural and Food Chemistry* 57 (5). 2009. 1667-1676.

Kasaai, MR; Arul, J; Charlet, G. Fragmentation of chitosan by ultrasonic irradiation. *Ultrasonics Sonochemistry* 15 (6). 2008. 1001-1008.

Keong, LC; Halim, AS. In vitro models in biocompatibility assessment for biomedical grade chitosan derivatives in wound management. *International Journal of Molecular Sciences* 10. 2009. 1300-1313.

Kim, SK; Rajapakse, N. Enzymatic production and biological activities of chitosan oligosaccharides: A review. *Carbohydrate Polymers* 62 (4). 2005. 357-368.

Kjartansson, GT; Zivanovic, S; Kristbergsson, K; Weiss, J. Sonication-assisted extraction of chitin from North Atlantic shrimps (Pandalus borealis). *Journal of Agricultural and Food Chemistry* 54 (16). 2006. 5894-5902.

Kozen, B; Kircher, S; Heanao, J; Godinez, F; Johnson, A. An alternative field hemostatic agent. Comparison of a new chitosan granule dressing to existing chitosan wafer, zeolite and standard dressings, in a lethal hemorrhagic groin injury. *Annals of Emergency Medicine* 50 (3). 2007. S60-S61.

Kumar, MNVR; Muzzarelli, RAA; Muzzarelli, C; Sashiwa, H; Domb, AJ. Chitosan chemistry and pharmaceutical perspectives. *Chemical Reviews* 104. 2004. 6017-6084.

Kumirska, J; Weinhold, MX; Sauvageau, JCM; Thoming, J; Kaczynski, Z; Stepnowski, P. Determination of the pattern of acetylation of low-molecular-weight chitosan used in biomedical applications. *Journal of Pharmaceutical and Biomedical Analysis* 50 (4). 2009. 587-590.

Kurita, K. Chitin and chitosan: functional biopolymers from marine crustaceans. *Marine Biotechnology* 8. 2006. 203-226.

Kuroiwa, T; Izuta, H; Nabetani, H; Nakajima, M; Sato, S; Mukataka, S; Ichikawa, S. Selective and stable production of physiologically active chitosan oligosaccharides using an enzymatic membrane bioreactor. *Process Biochemistry* 44 (3). 2009. 283-287.

Kzhyshkowska, J; Krusella, L. Cross-talk between endocytic clearance and secretion in macrophages (Review). *Immunobiology* 214 (7). 2009. 576-593.

Lee, DX; Xia, WS; Zhang, JL. Enzymatic preparation of chitooligosaccharides by commercial lipase. *Food Chemistry* 111 (2). 2008. 291-295.

Lee, SH; Senevirathne, M; Ahn, CB; Kim, SK; Je, JY. Factors affecting anti-inflammatory effect of chitooligosaccharides in lipopolysaccharides-induced RAW264.7 macrophage cells. *Bioorganic and Medicinal Chemistry Letters* 19 (23). 2009. 6655-6658.

Lertwattanaseri, T; Ichikawa, N; Mizoguchi, T; Tanaka, Y; Chirachanchai, S. Microwave technique for efficient deacetylation of chitin nanowhiskers to a chitosan nanoscaffold. *Carbohydrate Research* 344 (3). 2009. 331-335.

Li XF; Feng, XQ; Yang S; FU GQ; Wang TP; Su ZX. Chitosan kills Escherichia coli through damage to the cell membrane. *Carbohydrate Polymers* 79, 2010, 493-499.

Liu, H; Du, YM; Wang, XH; Sun, LP. Chitosan kills bacteria through cell membrane damage. *International Journal of Food Microbiology* 95 (2). 2004. 147-155.

Liu, HT; Li, WM; Xu, G; Li, XY; Bai, XF; Wei, P; Yu, C; Du, YG. Chitosan oligosaccharides attenuate hydrogen peroxide-induced stress injury in human umbilical vein endothelial cells. *Pharmacological Research* 59 (3). 2009. 167-175.

Liu, XF; Guan, YL; Yang, DZ; Li, Z; Yao, KD. Antibacterial action of chitosan and carboxymethylated chitosan. *Journal of Applied Polymer Science* 79 (7). 2001. 1324-1335.

McGeer, EG; Singh, E. Neurotoxic effects of endogenous materials. *Experimental Neurology* 86. 1984. 410-413.

Mourya, VK; Inamdar, NN. Chitosan-modifications and applications: Opportunities galore. *Reactive and Functional Polymers* 68 (6). 2008. 1013-1051.

Muzzarelli, RAA. *Chitin*. Oxford, UK. Pergamon Press. 1977.

Muzzarelli, RAA. Aspects of chitin chemistry and enzymology. In Paoletti, M; Musumeci, S, eds. *Binomium chitin-chitinase: emerging issues*. Hauppauge, NY, USA. Nova Science Publishing Ltd. 2009.

Muzzarelli, RAA. Chitins and chitosans for the repair of wounded skin, nerve, cartilage and bone. *Carbohydrate Polymers* 76. 2009. 167–182.

Muzzarelli, RAA. Chitosans: new vectors for gene therapy. In *Carbohydrate polymers: development, properties and applications*. Hauppauge, NY, USA. Nova Science Publishing Ltd. 2010.

Muzzarelli, RAA. Genipin-chitosan hydrogels as biomedical and pharmaceutical aids. *Carbohydrate Polymers 77*. 2009. *1-9*.

Muzzarelli, RAA; Biagini, G; Pugnaloni, A; Filippini, O; Baldassarre, V; Castaldini, C; Rizzoli, C. Reconstruction of parodontal tissue with chitosan. *Biomaterials* 10. 1989. 598-603.

Muzzarelli, RAA; Muzzarelli, C. Chitin and chitosan hydrogels. In Philips, GO; Williams, PA, eds. *Handbook of Hydrocolloids*. Second edition. Cambridge, UK. Woodhead Publishing Ltd. 2009. Pp. 854-888.

Muzzarelli, RAA; Muzzarelli, C. Natural and artificial chitosan-inorganic composites. *Journal of Inorganic Biochemistry* 92. 2002. 89-94.

Muzzarelli, RAA; Stanic, V; Ramos, V. Enzymatic depolymerization of chitins and chitosans. In *Methods in Biotechnology: Carbohydrate Biotechnology Protocols*, C. Bucke, ed., Totowa, USA, Humana Press, 1999.

Muzzarelli, RAA; Tarsi, R; Filippini, O; Giovanetti, E; Biagini, G; Varaldo, PE. Antimicrobial properties of N-carboxybutyl chitosan. *Antimicrobial Agents and Chemotherapy* 34. 1990. 2019-2023.

Nasti, A; Zaki, NM; de Leonardis, P; Ungphaiboon, S; Sansongsak, P; Rimoli, MG; Tirelli, N. Chitosan/TPP and chitosan/TPP-hyaluronic acid nanoparticles: systematic optimisation of the preparative process and preliminary biological evaluation. *Pharmaceutical Research* 26 (8). 2009. 1918-1930.

Pedro, AS; Cabral-Albuquerque, E; Ferreira, D; Sarmento, B. Chitosan: An option for development of essential oil delivery systems for oral cavity care (Review). *Carbohydrate Polymers* 76 (4). 2009. 501-508.

Raafat, D; von Bargen, K; Haas, A; Sahl, HG. Insights into the mode of action of chitosan as an antibacterial compound. *Applied and Environmental Microbiology* 74 (12). 2008. 3764-3773.

Rinaudo, M. Characterization and properties of some polysaccharides used as biomaterials. *Macromolecular Symposia* 245. 2006. 549-557.

Rinaudo, M. Chitin and chitosan: properties and applications. *Progress in Polymer Science* 31. 2006. 603-632.

Runarsson, OV; Malainer, C; Holappa, J; Sigurdsson, ST; Masson, M. tert-Butyldimethylsilyl O-protected chitosan and chitooligosaccharides: useful precursors for N-modifications in common organic solvents. *Carbohydrate Research* 343 (15). 2008. 2576-2582.

Scherer, SS; Pietramaggiori, G; Matthews, J; Perry, S; Assmann, A; Carothers, A; Demcheva, M; Muise-Helmericks, RC; Seth, A; Vournakis, JN; Valeri, RC; Fischer, TH; Hechtman, HB; Orgill, DP. Poly-N-acetyl glucosamine nanofibers: a new bioactive material to enhance diabetic wound healing by cell migration and angiogenesis. *Annals of Surgery* 250 (2). 2009. 322-330.

Shen, KT; Chen, MH; Chan, HY; Jeng, JH; Wang, YJ. Inhibitory effects of chitooligosaccharides on tumor growth and metastasis. *Food and Chemical Toxicology* 47 (8). 2009. 1864-1871.

Su, XW; Zivanovic, S; D'Souza, DH. Effect of chitosan on the infectivity of murine norovirus, feline calicivirus, and bacteriophage MS2. *Journal of Food Protection* 72 (12). 2009. 2623-2628.

Sun, T; Yao, Q; Zhou, DX; Mao, F. Antioxidant activity of N-carboxymethyl chitosan oligosaccharides. *Bioorganic and Medicinal Chemistry Letters* 18 (21). 2008. 5774-5776.

Takahashi, T; Imai, M; Suzuki, I; Sawai, J. Growth inhibitory effect on bacteria of chitosan membranes regulated with deacetylation degree. *Biochemical Engineering Journal* 40 (3). 2008. 485-491.

Tian, F; Liu, Y; Hu, KA; Zhao, BY. Study of the depolymerization behavior of chitosan by hydrogen peroxide. *Carbohydrate Polymers* 57 (1). 2004. 31-37.

Tian, M; Chen, F; Ren, DW; Yu, XX; Zhang, XH; Zhong, R; Wan, CX. Preparation of a series of chitooligomers and their effect on hepatocytes. *Carbohydrate Polymers* 79 (1). 2010. 137-144.

Ueno, H; Mori, T; Fujinaga, T. Topical formulations and wound healing applications of chitosan (Review). *Advanced Drug Delivery Reviews* 52 (2). 2001. 105-115.

Uragami, T; Tokura, S, eds. *Material science of chitin and chitosan*. New York, NY, USA. Springer. 2006.

Varlamov, VP; Bykova, VM; Vikhoreva, GA; Lopatin, SA; Nemtsev, SV, eds. *Modern perspectives in chitin and chitosan studies*. Moscow, Russia. Vniro. 2003.

Vila, A; Sanchez, A; Janes, K; Behrens, I; Kissel, T; Jato, JLV; Alonso, MJ. Low molecular weight chitosan nanoparticles as new carriers for nasal vaccine delivery in mice. *European Journal of Pharmaceutics and Biopharmaceutics* 57 (1). 2004. 123-131.

Wang, SM; Huang, QZ; Wang, QS. Study on the synergetic degradation of chitosan with ultraviolet light and hydrogen peroxide. *Carbohydrate Research* 340 (6). 2005. 1143-1147.

Wu, T; Zivanovic, S; Hayes, DG; Weiss, J. Efficient reduction of chitosan molecular weight by high-intensity ultrasound: Underlying mechanism and effect of process parameters. *Journal of Agricultural and Food Chemistry* 56 (13). 2008. 5112-5119.

Xing, RG; Liu, S; Guo, ZY; Yu, HH; Li, CP; Ji, X; Feng, JH; Li, PC. The antioxidant activity of glucosamine hydrochloride in vitro. *Bioorganic and Medicinal Chemistry* 14 (6). 2006. 1706-1709.

Xing, RG; Liu, S; Yu, HH; Guo, ZY; Wang, PB; Li, CP; Li, Z; Li, PC. Salt-assisted acid hydrolysis of chitosan to oligomers under microwave irradiation. *Carbohydrate Research* 340 (13). 2005. 2150-2153.

Xiong, CN; Wu, HG; Wei, P; Pan, M; Tuo, YQ; Kusakabe, I; Du, YG. Potent angiogenic inhibition effects of deacetylated chitohexaose separated from chitooligosaccharides and its mechanism of action in vitro. *Carbohydrate Research* 344 (15). 2009. 1975-1983.

Yang, YM; Liu, M; Gu, Y; Lin, SY; Ding, F; Gu, XS. Effect of chito-oligosaccharide on neuronal differentiation of PC-12 cells. *Cell Biology International* 33 (3). 2009. 352-356.

Yin, H; Du, YG; Zhang, JZ. Low molecular weight and oligomeric chitosans and their bioactivities. *Current Topics In Medicinal Chemistry* 9 (16). 2009. 1546-1559.

Zeng, LT; Qin, CQ; Chi, WL; Wang, LS; Ku, ZJ; Li, W. Browning of chitooligomers and their optimum preservation. *Carbohydrate Polymers* 67 (4). 2007. 551-558.

Zhou SL; Yang, YM; Gu, XS; Ding, F. Chitooligosaccharides protect cultured hippocampal neurons against glutamate-induced neurotoxicity. *Neuroscience Letters* 444 (3). 2008. 270-274.

In: Focus on Chitosan Research
Editors: Arthur N. Ferguson and Amy G. O'Neill
ISBN 978-1-61324-454-8

Chapter 5

THERMAL RELAXATION PROPERTIES AND GLASS TRANSITION PHENOMENA IN CHITOSAN FILMS

J. Betzabe González-Campos[1,2], Evgen Prokhorov[1], G. Luna-Bárcenas[1], L. Chacón-García[2] and Rosa E. N. del Río-Torres[2]

[1]. Centro de Investigación y de Estudios Avanzados del Instituto Politécnico Nacional, Querétaro, México
[2]. Universidad Michoacana de San Nicolás de Hidalgo., Morelia, Michoacán, México

ABSTRACT

In this chapter thermal relaxation properties of chitosan neutralized and non-neutralized films will be reported as a function of water content using dynamical mechanical analysis and dielectric spectroscopy in the temperature range from 20 to 250°C. Three relaxation processes have been observed in different temperature and frequencies ranges. For the first time, the low frequency α-relaxation associated with the glass-rubber transition, which relate to a plasticizing effect of water, has been detected by this technique in both chitosan forms in the temperature range 20-70°C and for moisture contents between 0.5 to 10 wt %. On this basis, the glass transition temperature was estimated in the range 86-102°C which shifts to higher temperature with decreasing moisture content and the glass transition vanishes in a dry material. A second low frequency relaxation was observed from 80°C to the onset of thermal degradation (240°C) and identified as the sigma-relaxation often associated with the hopping motion of ions in the disordered structure of the biomaterial. This relaxation exhibits a normal Arrhenius-type temperature dependence with activation energy of 86-88 kJ/mol and it is independent of water content. The non-neutralized chitosan possess higher ion mobility than the neutralized one as determined by the frequency location of the σ-relaxation. A high frequency (10^4 -10^8 Hz) secondary β-relaxation in neutralized and non-neutralized chitosan, related to side group motions by means of the glucosidic linkage is observed in the temperature range 20-120°C and moisture contents less than 3 wt % with Arrhenius activation energy of 46.0-48.5 kJ/mol. In the films with higher moisture contents this relaxation is not well resolved due to a superposition of two relaxation processes: beta and beta wet relaxations that merge into one common β-relaxation process.

INTRODUCTION

The knowledge of thermal behavior of a polymer is of scientific and technological interest when temperature is a processing variable in device fabrication. Among several physical properties dependent on temperature, a glass transition can be used to characterize a property of a polymeric material. The glass transition temperature (T_g) constitutes the most important mechanical property for all polymers; since at this temperature the polymer goes from hard-glass like state to a rubber like state. This property is a specific feature for polymers amorphous domains and upon synthesis of a new polymer, glass transition temperature is among the first properties measured since the state parameters will exhibit an abrupt change within the range of the glass transition temperature. Hardness, volume, Young's modulus, percent elongation-to-break, viscosity, density, thermal expansion coefficient and heat capacity are some of the physical properties that undergo a drastic change at the glass transition temperature. So any instrument that can detect the thermal, mechanical or dielectric changes can theoretically be used to measure the T_g of a material. The glass transition itself is currently not well understood theoretically and the nature of the glass transition can be described as one of the most serious unsolved and challenging problems in condensed matter physics [1]

In the last four decades, it has been controversial as to whether chitin and its main derivative chitosan exhibit a glass transition temperature (T_g). In semi-crystalline polymers, such as chitin and chitosan, a glass transition temperature characteristic of the amorphous material can usually be detected. However, it is controversial as to whether polysaccharides exhibit a glass transition temperature (T_g) [2]. For chitosan, some authors using several techniques, that include differential scanning calorimetry (DSC) and dynamic mechanical thermal analysis (DMTA), have reported T_g values from 20 to 222°C [3-10], whereas others do not observe the glass transition [11-14]. DMA technique has been a common tool to analyze relaxational process in chitosan, however, these studies are performed at one low fixed frequency and the identification with T_g is tenuous, thus the nature of each relaxation process is also controversial and several different values have been assigned.

In polymers the glass transition phenomenon has been related to the dielectric α-relaxation processes through the Vogel-Fulcher-Tammann (VFT) equation [15]. Dielectric spectroscopy is a significant technique used to investigate the relaxation properties of materials. The main advantage of dielectric technique over DSC and DMA analysis that attempt to measure molecular dynamics is the extremely broad frequency range covered. Some authors have reported molecular dynamics analysis in chitosan by dielectric measurements with no evidence associated with a glass transition [11, 16, 17]. These studies reported the β-relaxation process related to local main chain motions at the high frequency side; and at higher temperature the so-called σ-relaxation produced by proton migration [16, 17]. It is well known that this biopolymer is highly hydrophilic and small amounts of water affect its molecular relaxations, this effect has been reported on wet chitosan since it exhibits an additional relaxation referred as the β_{wet}-relaxation [11, 16, 17]. Recently, molecular relaxations in chitin [18] and chitosan [2] have been investigated using dielectric spectroscopy, these results show evidence of the low frequency α-relaxation present in *dry annealed* films, this relaxation process is related to the glass transition and a glass transition temperature is assigned by fitting the experimental data to the Vogel-Fulcher-Tammann model.

On the other hand, the physical and chemical properties of this biopolymer can be significantly changed by the presence of small amounts of water [2, 16, 18, 19]; since chitosan has a strong affinity for water and therefore, may be readily hydrated forming macromolecules with rather disordered structures [20], a true understanding of hydration properties is essential for several applications in materials science, food industry and biotechnology [20]. Dielectric spectroscopy is a significant technique used to investigate the hydration properties of materials [2, 18, 19] and glass transition phenomena as well. That is why the objective of this chapter is to review the dielectric and DMA analysis to discern the nature of different molecular relaxation process and its relation ship to the α-relaxation process and the glass transition in hydrated and dry chitosan thin films.

Chitosan Glass Transition Temperature Controversy

Chitosan possess good film-forming properties. Numerous publications have reported on studies of films made of chitosan [21]. Among others, an important parameter on films-forming processing is the glass transition temperature (T_g). It is important not only in optimizing manufacturing processes, but in understanding the reliability implications of exposure of the products to thermo-mechanical stresses as well. The application of the glass transition concept can be also a valuable aid in edible films research, as T_g affects their mechanical and barrier (gas, eater vapor) properties. The T_g is a key factor for deciding the usefulness of a polymer.

Regarding chitosan glass transition temperature, it have been main object of study during the last years, owing to the fact that there are a wide variety of assigned T_g values as it is shown in Table I. Being a natural polymer, the source and method of extraction, cristallinity, molecular weight and deacetilation degree are some properties that can influence the T_g [20] and this could be the reason for the huge range of reported values.

The concept of T_g only applies to non-crystalline solids, which are mostly either glasses or rubbers. Non-crystalline materials are also known as amorphous materials. Amorphous materials are materials that do not have their atoms or molecules arranged on a lattice that repeats periodically in space. For amorphous solids, whether glasses, organic polymers, or even metals, T_g is the critical temperature that separates their glassy and rubbery behaviors. A small change in temperature T_g could result in pronounced changes in the mechanical, thermal and dielectric properties of amorphous materials.

Table I. Summary of the Chitosan glass transition temperature (T_g) reported in literature

Author	Year	Technique	Tg (°C)	Observations
Chitosan Glass Transition Temperature assigned				
Kaymin et al.[22]	1980	DIL and DMA	55	Dried samples at 100°C and annealed at 196°C
Ogura et al.[23]	1980	DMA	140	
Pizzoli et al.[11]	1991	DMA and DS	130	Water induced d relaxation
Kim et al.[24]	1994	DMA	150	
Ratto et al.[3]	1995	DSC	34	Water content near to 0%
Arvanitoyannis et al. [10]	1997	DETA	110	
Guan et al. [25]	1998	DSC and DMA	90 and 105	
Dufresne et al.[26]	1999	DMA	No Tg	
Sakurai et al.[7]	2000	DSC and DMA	203	Measured after heating up to 180°C
Toffey et al.[27]	2001	DMA and DSC	60-93	DSC didn't show glass transition
Kittur et al. [12]	2002	DSC	No Tg	
Lazaridou et al. [21]	2002	DMA	95	DMA measurements demonstrate the plasticizing effect of water giving values near to 77°C for water content ca. 0.1 weight fraction
Zohuriaan et al. [28]	2004	DSC	No Tg	
Dong et al. [5]	2004	DIL, DMA, TSC and DSC	140-150	They found two more relaxations at 85 and 197°C. Tg values do not depend on deacetilation degree
Wu et al. [29]	2004	DMA	153	They report α-relaxation at 15-22°C
Mucha et al.[4]	2005	DMA	156-170	Tg values depends on deacetilation degree
Liu et al [6]	2006	DSC	92	Reported value from first scan
Shieh et al. [30]	2006	DMA	150	One relaxation at 25°C attributed to the glass to rubber transition of water-plasticized chitosan
Quijada-Garrido et al.[31]	2007	DMA and DSC	85-99	Tg value depends on frequency. DSC showed no glass transition.
Rao et al [8]	2008	DSC	222	Only one scan was performed
Mayachiew et al. [9]	2008	DSC	193 and 197	Tg value depends on drying technique.

Dilatometry (DIL).
Dynamic Mechanical Analysis (DMA).
Dielectric Spectroscopy (DS).
Dynamic Electrical Thermal Analysis (DETA).
Thermally Stimulated Current Spectroscopy (TSC).
Differential Scanning Calorimetry (DSC).

In Differential Scanning Calorimetry (DSC) analysis, the glass transition is defined as a change in the heat capacity as the polymer matrix goes from the glassy state to the rubbery state. This is a second order endothermic transition that requires heat to go through the transition, so in the DSC the transition appears as a step transition and not as a peak such as in a melting transition. DSC is the classic and "official" way to determine T_g even though in some cases there are polymeric materials that do not exhibit a sharp T_g by DSC, this had been the case of chitin and chitosan as well as cellulose [24, 29]. On the other hand, in Thermal Mechanical Analysis (TMA) this transition is associated with a change in the free molecular volume and defines the glass transition in terms of the change in the coefficient of thermal expansion (CTE) as the polymer goes from glass to rubber state. Each of these techniques measures a different result of the change from glass to rubber. The DSC is measuring a heat effect, whereas the TMA is measuring a physical effect i.e. the CTE. Both techniques assume that the effect happens over a narrow range of a few degrees in temperature. If the glass transition is very broad it may not be seen with either approach.

From the practical point of view, fundamental information on the processability of polymers is usually obtained through thermal analysis; it provides information of the main polymers transitions related to melting and glass-to-rubber transition to the crystalline and amorphous phases, respectively. In addition to the well established calorimetric techniques, experimental methods capable of revealing the motional phenomena occurring in the solid state have attracted increasing attention. In amorphous polymers, α-relaxation, as determined by Dielectric Spectroscopy (DS) and Dynamic Mechanical Analysis (DMA) corresponds to the glass transition and reflects motions of fairly long chain segments in the amorphous domains of the polymer involving long range motions. Relaxations at lower temperatures (labeled β, γ, δ...) are generally due to short range motions related to local movements of the main chain, or rotations and vibrations of terminal groups or other side chains. DS and DMA are well-established techniques for the measurement of thermal transitions including the glass transition; they are especially available in detecting T_g of a sample that cannot be observed by normal calorimetric measurements. For example, T_g of polymers having crosslinked network structure [32]. These two techniques were used in the present work for chitosan thermal analysis.

THE GLASS TRANSITION AND THE α-RELAXATION PROCESS

When studying a polymer like chitosan on a large frequency/time scale, its response under a dielectric or dynamic stimulus could exhibits several relaxations. More over, the peaks could be usually broad and sometimes a superposed processes [2, 18, 33]. Molecular processes cover a broad frequency range and they are associated with the length scale of the conformational mobility in the polymeric chain. Relaxation processes in polymeric materials involves several relaxation process that go from secondary process that are very local motions (β, δ, γ, etc) to primary relaxation process due to segmental mobility involving co-operativity (α-process), or even relaxation processes involving large or complete polymeric segments [34]. The relaxation rate, shape of the loss peak and relaxation strength depend on the motion associated to a given relaxation process. In general, the same relaxation/retardation processes are responsible for the mechanical and dielectric dispersion observed in polar materials [34].

The α-relaxation is related to the glass transition and in general they are not well understood, and the real microscopic description of the relaxation remains being a current problem of polymer science [35]. However it is well accepted that the dynamics of the glass transition is associated with the segmental motion of chains being cooperative in nature [35], which means that a specific segment moves together with its environment. For most amorphous polymers in the α-process the viscosity and consequently the relaxation time increase drastically as the temperature decreases.

Therefore molecular dynamics is characterized by a distribution of relaxation times and by a non-Debye response. A strong temperature dependence presenting departure from linearity or non-Arrhenius thermal activation is a well description of the glass transition process, due to the abrupt increase in relaxation time as temperature decreases developing a curvature near T_g. This dependence can be well described by the Vogel- Fulcher-Tammann (VFT) equation [34, 35]:

$$\log f_{p\alpha} = \log f_{\alpha,\infty} - \frac{A}{T - T_o} \qquad (1)$$

where $f_{\alpha,\infty}$= 10^{10}-10^{13}Hz and A are constants, and T_0 is the so called ideal glass transition or Vogel temperature, which is generally 30-70°K below T_g [15, 35].

In general the α-process is well defined in the frequency domain and shows a relatively broad and asymmetric peak. Several functions like the Cole-Cole and Cole-Davison are able to describe broad symmetric and asymmetric peaks in the frequency domain and the most general one is the model function of Havriliak and Negami (HN function) [35]. In order to investigate the nature of chitosan relaxation processes, the fitting of the complex permittivity in *dry annealed* and *dry* neutralized and non-neutralized samples were carried out using the well-known HN empirical model [15, 22]: $\varepsilon^* - \varepsilon_\infty = \frac{(\varepsilon_s - \varepsilon_\infty)}{\left[1 + (j\omega\tau)^\alpha\right]^\beta}$, where ε_s-ε_∞ and ε_∞ are the dielectric relaxation strength and the dielectric constant at the high frequency limit, respectively. The exponents α and β introduce a symmetric and asymmetric broadening of the relaxation. This general equation includes especial cases of relaxations processes for Debye (α and β=1), Cole-Cole (α=1) and Davison-Cole (β=1).

Dielectric Relaxation Processess in Polysaccharides

To investigate molecular motions by Dielectric Spectroscopy (DS) both, the repeating unit of a polymer and the attached side groups must own a permanent dipolar moment to look into dipolar fluctuations in broad temperature and frequency ranges, this requirement is well accomplished by chitosan. Dielectric spectroscopy of polysaccharides has been considered controversial by many scientists, up to now [16]. Different molecular groups of a repeated unit of a polymer are separated by dielectric relaxation spectroscopy with respect to the rate of its orientational dynamics [16].

Einfeldt et al [16] developed an exhaustive study of dielectric spectra of a great variety of polysaccharides including dry and wet cellulose-based materials, different wet starches and also derivatives of cellulose and starch. They observed different modes of relaxation processes in the sub-T_g range assigned as γ, δ, β_{wet}, β and σ relaxation processes involving side group motion, water effect and hoping motion of ions (these relaxation processes are resumed and extensively described elsewhere [16]). However they do not draw any discussion about the primary α-relaxation associated to the glass transition temperature since no evidence was found for a dynamics with VFT temperature dependence; the typical feature for the glass transition dynamics.

DS studies on chitosan are scarce and ambiguous. Even though the scientific and technologic importance of this polymer only a few papers exist concerning to its molecular dynamics. Pizzoli et al [11] showed low- and high-temperature relaxations in dry chitosan (they called "dry chitosan" to vacuum annealed up to 180°C samples). Below room temperature, they observed a frequency dependent secondary relaxation ascribed to local motions of small molecular units and assigned as γ relaxation. Whereas at high temperature (above 140°C) they suggested the occurrence of a second relaxation, concluding that their results bring no evidence in favor of the glass transition attribution. On the other hand, Viciosa et al [17] reported two main Arrhenian type relaxations processes in neutralized and non-neutralized chitosan; process I found below 0°C owing the characteristics of a secondary relaxation process related with local chain dynamics, and process II at higher temperatures correlated with dc conductivity.

Viciosa et al [33] also reported the β-wet process located at temperatures below 0°C in wet samples, vanishing after heating to 150°C. Finally, Einfeldt et al [16] found a β-relaxation in β-chitin and chitosan in the low temperature range with activation energies of 44.7 and 47.8 kJ/mol respectively. They also performed high temperature studies on chitin and chitosan [36], however, they do provided activation energies values for the σ-relaxation in these biopolymers.

Dynamic Mechanical Analysis (DMA) in Chitosan

Dynamic mechanical analysis (DMA) is a thermal analysis technique that measures the properties of materials while they are deformed under periodic stress [37]. Since polymers are viscoelastic materials, i.e. they simultaneously exhibit solid-like and liquid-like properties; they are by definition time-dependent. They exhibit the properties of a glass (high modulus) at low temperatures and those of a rubber (low modulus) at higher temperatures. The scanning of temperature during a DMA experiments provides information about the glass transition or α-relaxation, so the T_g can be measured by DMA. DMA can also be used to investigate the frequency dependent nature of the transition; since T_g has a strong dependence on frequency DMA can also resolve secondary transitions, like β, γ, and δ transitions, in many materials that the DSC technique is not sensitive enough to pick up.

As mentiooned above, in chitosan a wide variety of T_g values have been reported over the past years, basic studies were conduced on the molecular motion and thermal relaxation behavior of chitosan using the DMA technique, some of them taking into account moisture content influence on their physical properties. Ogura et al [23] showed the dynamical mechanical behavior of dry chitosan, two loss peak assigned to the γ-relaxation and α-relaxation where observed. They proposed a T_g for dry chitosan around 140°C. Guan et al. [25] found two peaks in the DMA spectra of neutralized chitosan films; at ca. -55°C and 105°C, ascribe to the β and α-relaxations respectively. Ratto et al. [3] showed a low amplitude transition centered at 90°C assigned to local chain motion. Meanwhile, Toffey et al. [27] assigned a T_g= 60-93°C depending of the acid used to form the film, they also showed an additional low temperature relaxation at -10°C designed it as the β-relaxation.

Lazaridou et al. [21] observed the moisture content plasticizing effect on the glass transition temperature of chitosan and estimated a T_g ca. 95°C for dry chitosan. While Yu-Bey Wu et al. [29] observed two tan δ peaks; one attributed to the β-relaxation at 15-22°C and another at 153°C designed as the α-relaxation. As well as Quijada-Garrido et al. [31] that reported a β-relaxation at -30 to -12°C and an α-peak at 85-99°C depending on frequency. Ahn et al. [38] assigned 161°C as the α-relaxation peak and a peak around 60°C to what they describe as a water-induced β-relaxation. Whereas, Sakurai et al. [7] estimate the T_g to be 203°C by the tan δ curve after a first heating it up to 180°C, even though the onset of chitosan thermal degradation is around 170°C [39]. Neto et al. [20] observed a main event at 50°C, in the abstract they ascribe this relaxation to the glass transition; however in the manuscript they do not dare to say that it is the glass transition temperature.

Dong et al [5] by four different techniques including DMA, assigned a T_g value of 140-150°C, they observed three more peaks at ca. 85°C, 197°C and 200°C attributed to a water-induced relaxation, an unknown α transition and decomposition respectively. M. Mucha et al [4] recognized a broad tan δ peak at 156-170°C as theα-relaxation at T_g, they also found three more events assigned as T_1=-21°C, T_2=24°C and T_3=43°C that move to higher temperatures at T_1=-12°C, T_2=32°C and T_3=45°C in preheated samples. T_1 was assigned as β-relaxation, associated with local motions of side groups and T_2 and T_3 to strongest complex relaxations occurring at ambient temperature ascribed to structural reorganization of packing of chitosan molecules due to an increase of residual water mobility.

Shieh et al. [30] found two peaks, one at 25°C and another near to 150°C, the occurrence of the low-temperature relaxation was attributed to the glass-to rubber transition of the water plasticized chitosan, and the relaxation peak near 150°C to the glass transition of large-scale cooperative molecular motions in the amorphous phase.

Even though the wide variety of T_g values reported, there are some authors that found no glass transition evidence in chitosan by DMA measurements [11-14, 39-41].

Materials and Experimental Methods

Neutralized and Non-Neutralized Chitosan Films Preparation

Chitosan medium molecular weight (Mw= 150,000 g/gmol) of 76 and 82 % of degree of deacetylation (DD) reported by the supplier and calculated according to [42], was purchased from Sigma-Aldrich. Acetic acid from J.T. Baker was used as received without further purification. Chitosan films were obtained by dissolving 1 wt % of chitosan in a 1 wt % aqueous acetic acid solution with subsequent stirring to promote dissolution. Chitosan films were prepared by the solvent cast method by pouring the solution into a plastic Petri dish and allowing the solvent to evaporate at 60°C. Chitosan films prepared from acetic acid solution have the amino side group protonated (NH_3^+groups); therefore the films need to be neutralized. The films were immerse into a 0.1 M NaOH solution during 30 min and washed with distilled water until neutral pH, a subsequent dried step in furnace at 130°C for 14 hr was needed. The results of both *neutralized* as well as *non-neutralized* chitosan films are shown.

Electrode Preparation for Dielectric Measurements

A thin layer of gold was vacuum-deposited onto both film sides to serve as electrodes using a device (Plasma Sciences Inc.) with a gold target (Purity 99.99%) and Argon as gas carrier. With a gas pressure set to 30 *m*Torr and voltage set to 0.2 *k*V. Sputtering time was 4 minutes onto each side. Rectangular small pieces (5 mm × 4 mm) of these films were prepared for dielectric measurements. The contact area and thickness were measured with a digital calibrator (Mitutoyo) and a micrometer (Mitutoyo), respectively.

Infrared Measurements

Chemical analysis and degree of deacetylation calculation of chitosan (CTS) and CTS acetate/SN composites was performed by Fourier-Transform Infrared spectroscopy (FTIR) on a Perkin Elmer spectrophotometer model Spectrum GX, using an ATR accessory in the range 4000-650 cm-1, resolution was set to 4 cm-1 and the spectra shown are an average of 32 scans. Chitosan samples prepared in the forms of potassium bromide (KBr) disk and film were studied.

Thermal Measurements

Free water content was determined by thermogravimetric analysis (TGA). TGA curves were obtained using a Mettler Toledo apparatus, model TGA/SDTA 851e, using a sample mass of ca. 3 mg and an aluminum sample holder under argon atmosphere with a flow rate of 75 mL/min. Heating rate was set to 10°C/min. For dielectric measurements, three types of films have been studied: (1) *Wet* chitosan, samples prepared under normal ambient conditions (25% humidity, at 20°C); (2) *Dry* chitosan, samples heated at 120°C for 24 h then cooled at ambient conditions for 10-15 min prior to measurement; and (3) *Dry annealed* chitosan, *dry* samples with an additional annealing at 120°C for 1 hour in the experimental impedance cell followed by cooling to 30°C under vacuum. After this conditioning, dielectric measurements are taken up to 250°C. The annealing time of 1 hour was chosen because further annealing did not change resistance of samples.

Regarding DMA measurements, *wet* samples corresponds to samples under normal ambient conditions, *dry* samples were heated at 75°C before DMA measurements and *dry annealed* samples were heated at 150°C before DMA measurements.

Modulated differential scanning calorimetry (MDSC) measurements were performed in a Q100 TA Instruments calorimeter, using a sample mass of ca. 3-4 mg. Heating rate was set to 3°C/min with a modulation amplitude and frequency of ± 1°C and 1/60 s^{-1} respectively. Standard aluminum pans were used within a nitrogen atmosphere flowing at a rate of 80μL/min. An empty pan was used as reference. First and second scans were analyzed.

Dielectric Measurements

Dielectric measurements in the frequency range from 10^{-1} to10^{6} Hz were carried out using Solartron 1260 impedance gain-phase analyzer with 1294 Impedance interface and in the frequency range 40Hz-100 MHz using an Agilent Precision Impedance Analyzer 4294A. The amplitude of the measuring signal was 100 mV. The Agilent 4294A was calibrated for fixture compensation according to [43] when connecting a direct with home-made vacuum cell to the impedance analyzer port. It includes: open, short and load compensation with the characteristic impedance of 100 Ω used. Three samples: *wet*, *dry* and *dry annealed* previously defined were studied. To ensure entire water removal for the *dry annealed* samples, heating for 1 h at 120°C is done prior to dielectric measurements.

An in-house impedance two contacts vacuum cell was used in conjunction with a Watlow's Series 982 microprocessor with ramping temperature controller for all dielectric measurements from 20°C to 250°C, which was programmed to produce a constant heating rate of 3°C/min between certain measurements temperature. Each sample was kept for 3 min at each temperature to ensure thermal equilibrium.

Dynamic Mechanical Analysis (DMA)

Dynamic mechanical analysis was made with a Ta Instruments Dynamical Mechanical Analyzer model RSA III. The heating rate was 5°C/min at 0.1 HZ frequency and an initial strain 0.1% .

Experimental Data Pre-Treatment

Figure 1 shows the complex Cole-Cole plot for *dry annealed* chitosan films. It exhibit characteristic semicircles at high frequencies and a quasi-linear response at low frequencies, the linear response at low frequencies can be associated with interfacial polarization in the bulk films and/or surface and metal contact effects [18, 44]. To analyze the dielectric relaxation of chitosan films is necessary to understand the nature of the low frequency part of dielectric spectrum.

To test the influence of gold contact on the chitosan dielectric spectra, a non-symmetrical contact array was prepared as reported elsewhere [2]. Window insert of Figure 1 shows dielectric measurements on wet Au-chitosan-Au sample with different applied bias voltage at 23°C. The bias increase leads to a reduction of barrier resistance that changes the low frequency contribution of dielectric spectra (which occur at large values of Z') as shown in window insert of Figure 1. As can be seen, a deviation from a semi-circle is seen at both temperatures and the response depends on the bias voltage. This behavior is a good indication of contact polarization effects related to the (partial) blocking of charge carriers at the film/electrode interface [45]. This low frequency part of the electrical response is easily influenced by imperfect contact between the metal electrode and the sample.

In addition to this electrode polarization, interfacial polarization effects are observed in the high temperature range (>120°C) for all chitosan films. This effect manifests as a "bulge" on the semicircle. Figure 1 illustrates this effect at 125, 135 and 145 °C. According to the classical model the appearance of the interfacial polarization in dielectric spectrum can be observed as appearance of additional semicircle [2, 18]. This is a common form of a discontinuity occurring in an inhomogeneous solid dielectric associated with internal interfaces; it is well-known as the Maxwell-Wagner-Sillars (MWS) polarization. In polysaccharides and biopolymers [16, 17] the interfacial polarization (MWS polarization) was observed in low frequency and high temperature ranges. In the low frequency range both contact and interfacial polarizations were observed in all samples. These polarizations have to be carefully considered since it is important to take into account only a so-called depressed semicircle that does not include contact and interfacial polarization effects. For most polysaccharides a commonly used plot of admittance versus frequency does not reveal the appearance of the extra semicircle related to interfacial polarization [18]. It is noteworthy that data treatment proposed in this study allows one to identify and separate these two processes (contact and interfacial polarization effects) and consequently before fitting dielectric data to models, this low frequency data needs to be discarded because a model-based analysis can be misleading if appropriate contact and interfacial polarizations are not considered.

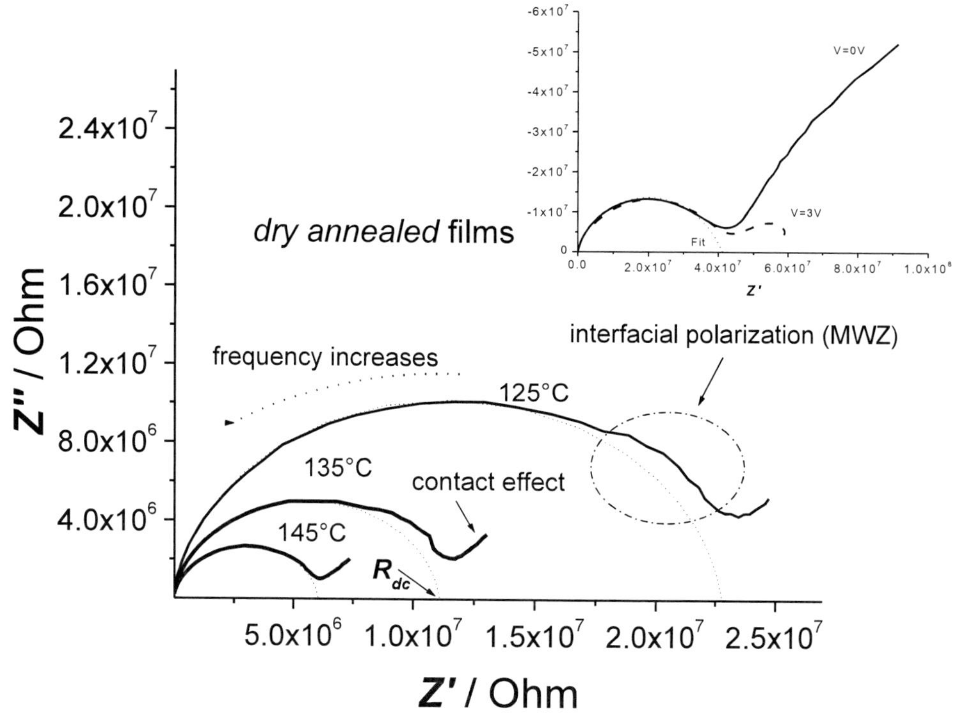

Figure 1. Complex dielectric spectrums of *dry annealed* chitosan films. Window insert: Dielectric spectra at different bias applied voltage. Contact effect is observed when applied voltage is changed from 0V to 3V in the quasi linear response (low frequency side), in a film tested at 25°C.

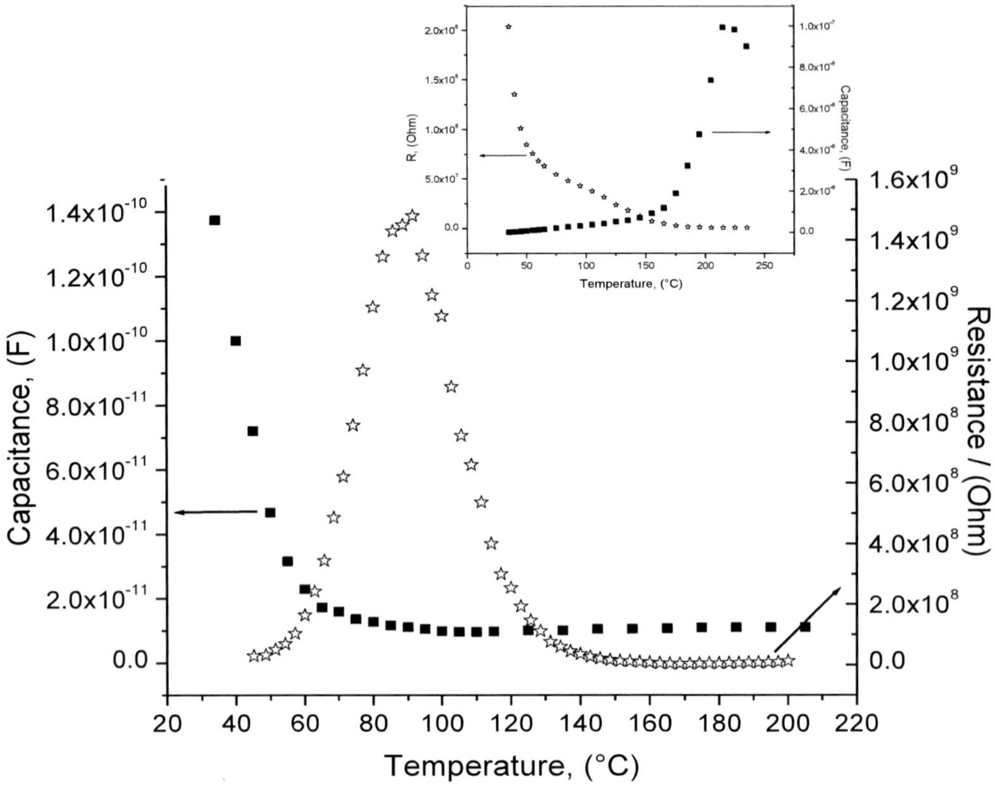

a.

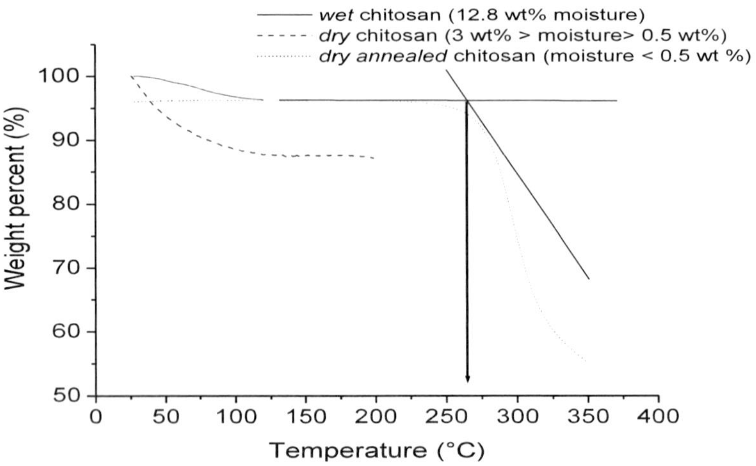

b.

Figure 2. a) Resistance (open stars) and capacitance (solid squares) as a function of temperature for wet film. Thermogravimetric response is also shown. Note that as temperature increases resistance increases due to water evaporation and finally decreases. Window inset shows the opposite behavior for *dry* films after water evaporation by annealing. b) Thermogravimetric analysis of solvent-cast chitosan for *wet*, *dry* and *dry annealed* films.

Moisture Content Effects on Chitosan Molecular Relaxations and *dc* Conductivity Correction

Chitosan is a hydrophilic polysaccharide and it is well known that moisture content has a significant influence on its physical properties [16, 18]. A true understanding of hydration properties is essential for several practical applications in materials science, food industry, biotechnology, etc [46]. Chitosan moisture content is affected by the number of ionic groups in the material as well as their nature. Hydroxyl and amine groups present in chitosan are important binding sites for water molecules. The glass transition phenomenon could be affected by moisture content, since it can work as a plasticizer [2]. Plasticization occurs in the amorphous region only, such that the degree of hydration is quoted as moisture content in the amorphous region.

Figure 2a shows the dependence of *dc resistance* and *capacitance versus temperature* for *wet* chitosan (12.8 wt% moisture content). It can be observed that in the temperature range 20-70°C the resistance increases and capacitance decrease. This behavior is ascribed to water modification effect on the relaxation mechanism of the matrix, since water has a lower resistance and higher dielectric constant. When water is present, biopolymer's resistance decreases; if temperature increases water evaporates and resistance increases (weight loss about 12.8 wt% in the TGA measurements is registered, see Figure 2b). When water content is below 3 wt% (*dry* and *dry annealed* samples of Figure 2b), the real dielectric behavior of chitosan is revealed. Films with near zero moisture content exhibit higher resistances at room temperature and lower resistance as temperature increases as it is shown in window insert of Figure 2a (*dry annealed* film). In summary, to obtain the dielectric behavior of pure chitosan without moisture influence, it is necessary to evaporate it by heat treatment and an additional annealing at 120°C, otherwise its evaporation mask electrical properties of the biopolymer.

For *dry* films, we emulated dielectric measuring conditions for TGA measurements: after overnight annealing in an oven at 120°C, these films were handled in ambient conditions for 10-15 min for TGA measurements (time to wait samples for TGA ≈ time for sample handling prior to dielectric measurements). The weight loss, and therefore moisture content is ca. 3.0%, indicates that chitosan reabsorbs water readily during the 10-15 min handling from the oven to the vacuum cell. So a second heat treatment in the vacuum cell prior to dielectric measurements is needed to obtain nearly *zero* weight percent of moisture. These films were reheated at 120°C in the impedance vacuum cell for 1 hour (*dry annealed* films moisture content ≈ 0.05 wt %).

Besides the variations of dielectric properties of a hydrated material due to the polar water molecules themselves, there is a second one due to the modification of the various polarization and relaxation mechanisms of the matrix material itself by water [19]. Furthermore, in the low-frequency region of measurements there is a third contribution often ignored in work dealing with high-frequency measurements, arising from the influence of moisture on conductivity and conductivity effects. The increase of electrical conductivity of the sample is the major effect present in wet samples, dielectric response is often masked by conductivity, and it superposes the dielectric processes in the loss spectra, this effect demands a conductivity correction of the dielectric loss spectra [15] because it strongly affects the modified loss factor ε''. In this case it can be expressed as:

$$\varepsilon'' = \varepsilon''_{\exp} - \frac{\sigma_{dc}}{\omega \varepsilon_0} \qquad (2)$$

where $\varepsilon''_{\exp}$ is the experimental loss factor value, σ_{dc} ($\sigma_{dc} = d/(R_{dc} \times S)$) is direct current conductivity, *d* and *S* are thickness and area of sample respectively, $\omega=2\pi f$ (*f* is frequency), and ε_0 is the permittivity of vacuum. As a general rule for polymers, σ_{dc} is determined from fitting of real component of the complex conductivity ($\sigma_{dc}=\sigma_0 f^n$, where σ_0 and n are fitting parameters) measured in the low frequency range where a plateau is expected to appear [36]. However, in our samples this plateau is not resolved, as a consequence polarization and contact effects cannot be discerned and the correct *dc* conductivity cannot be calculated by this method, so we use an alternative procedure to circumvent this problem. In the *Z"* versus *Z'* plot, the values of *dc* resistance R_{dc}, and the corresponding conductivity $\sigma_{dc}=d/(R_{dc}\times S)$, have been obtained from the extrapolation of the high frequency semicircle to the *Z'* axis as shown in Figures 1. On the other hand, Figure 3 shows experimental data before and after *dc* conductivity correction for *dry annealed* neutralized chitosan, it can be seen the high effectivity of the method previously described to discard conductivity effects; since it reduces significantly the conductivity contribution in the low frequency side of the spectra and at the same time allows disclosing the low frequency relaxations.

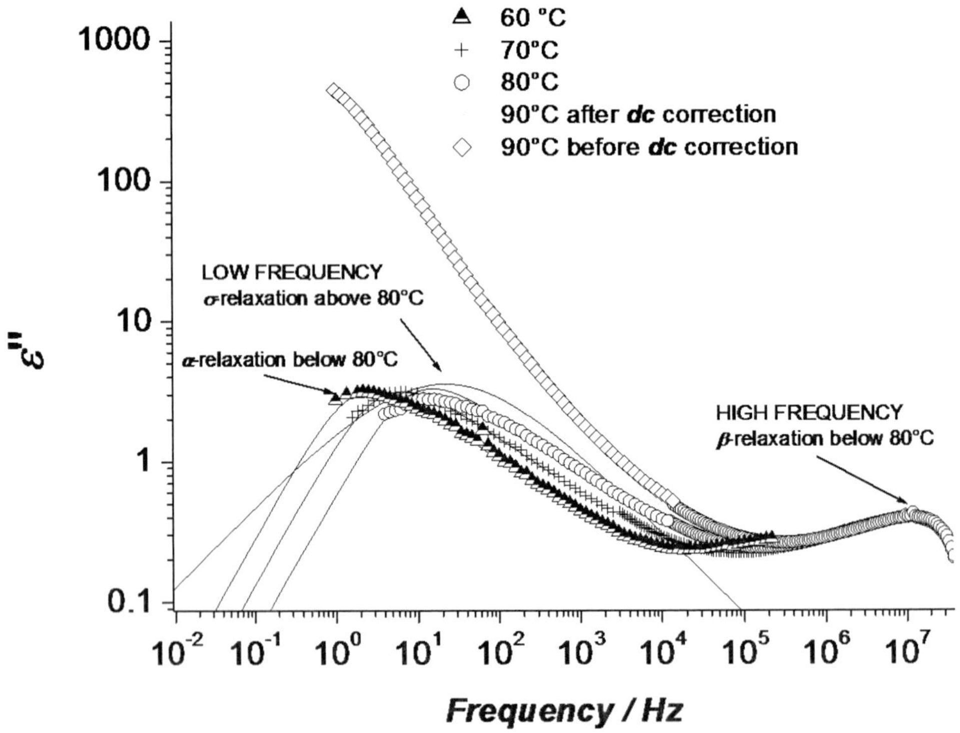

Figure 3. Dielectric spectra of neutralized chitosan. Note: low frequency relaxations are disclosed after *dc* conductivity correction. Low and high frequency relaxations can be fitted independently.

The primary α-relaxation associated with the glass transition, should appear below the kHz region [19]. After *dc* correction two different relaxations can be identified as seen in Figure 3 in the low frequency range (10^0-10^3 Hz) and one more above 10^5 Hz (high frequency side). This high frequency relaxation was previously reported as the secondary β-relaxation observed in the 10^4-10^8 Hz frequency range. The two relaxation processes are well defined and separated in the frequency range, and they can be investigated and correlated as the low and high-frequency relaxation processes independently. In the low frequency side and above 70°C a low frequency relaxation with Arrhenius temperature dependence (as will be shown later) is revealed, and below 70°C a relaxation with a non-Arrhenius temperature dependence that we designate as the primary or α-relaxation is disclosed. The same behavior was observed in non-neutralized films (nor shown here).

RESULTS AND DISCUSSION

Dielectric Spectroscopy Analysis

Low Frequency Relaxations

Once contact and interfacial polarization were discarded and *dc* correction was performed as described in previous section; high and low frequency relaxations were identified. The R_{dc}, and the corresponding *dc* conductivity σ_{dc} ($\sigma_{dc}= d/(R_{dc} \text{ x } S)$, where d is the film thickness and S the contact area) have been obtained from the intersection of the high frequency semicircle and the real-part axis on the impedance plane as it was shown in previous section and in Figure 1. The *log₁₀ dc conductivity versus 1000/T* dependence is shown in Figure 4. A similar behavior (moisture content-wise) in neutralized and non-neutralized films is shown, the non-neutralized chitosan films (open circles Figure 4) own higher conductivity values due to higher mobility and/or number of conducting species (NH_3^+ groups) [17].

The analysis of the conductivity plot shows two clear relaxations separated by temperature, so they can be analyzed in two temperature ranges. The "*low-temperature relaxation*" from 20 to 70°C and the "*high-temperature relaxation*" disclosed from 70°C to the onset of degradation ≈ 210°C.

Figure 4 shows conductivity results for the two chitosan used in this case: with 76 DD% and 82 DD% and obtained from shrimp and crab shell respectively. The same behavior described above is disclosed in both samples; below 70°C the characteristic α-relaxation behavior is present, and the high-temperature Arrhenius type relaxation is revealed above 70°C until the beginning of thermal degradation about 210°C. The Vogel temperature calculated from the fitting of our experimental data to the VFT relationship (equation 1) seems to be independent of the DD% and the raw material used for chitosan synthesis, since the same Vogel temperature (T_0 ≈269 K) from the VFT fit to the α-relaxation (explained later) is obtained in both chitosan samples (see Figure 4).

Regarding the influence of DD% on the glass transition temperature, M. Mucha et al [4] studied four different samples with 59, 67, 78 and 86 DD%, they stated that the temperature which they designed as the T_g of chitosan decreases from 167 to 156 °C with increasing DD% of chitosan. While Dong et al [5] found no influence of DD% on T_g for chitosan with 46, 64, 71, 91 and 100 DD %. They stated that the glass transition should not be influence by DD because the α-relaxation belongs to the motion of segments in the main chain and if this would happen, this relaxation process should be present due to side group (i.e., acetamino or amino group) motion.

Chitosan with 82 DD% exhibits lower conductivity values compared to that with 76 DD% (see Figure 4), this difference is ascribe to the neutralization process, since the sample corresponding to 76 DD% is in the non-neutralized form (experimental section for details). According to [17], this higher conductivity values indicates higher mobility due to the presence of more conductive species in the form of NH_3^+ groups. As can be seen, both chitosan forms show the same relaxation behavior in the whole temperature-range. Therefore the influence on thermal relaxation of the degree of deacetylation and the raw material used for its synthesis can be rule out for further analysis. The following molecular relaxations analysis presented here includes neutralized and non-neutralized chitosan with 82 DD%.

Low Temperature and Low Frequency Relaxation

Let us focus on the 20-70°C temperature range, this relaxation is strongly affected by moisture content; a decrease in conductivity as temperature increases is clearly shown in *wet* samples (moisture content > 3 wt% calculated by TGA analysis), this effect is related to water evaporation and vanishes above 100°C disguising the real dielectric properties of chitosan above this temperature. For moisture contents below 3 wt% and above 0.05 wt% (*dry* films), a non-Arrhenius behavior emerges while for moisture content below 0.05 wt% (*dry annealed* samples) this low temperature relaxation vanishes after the second heat treatment.

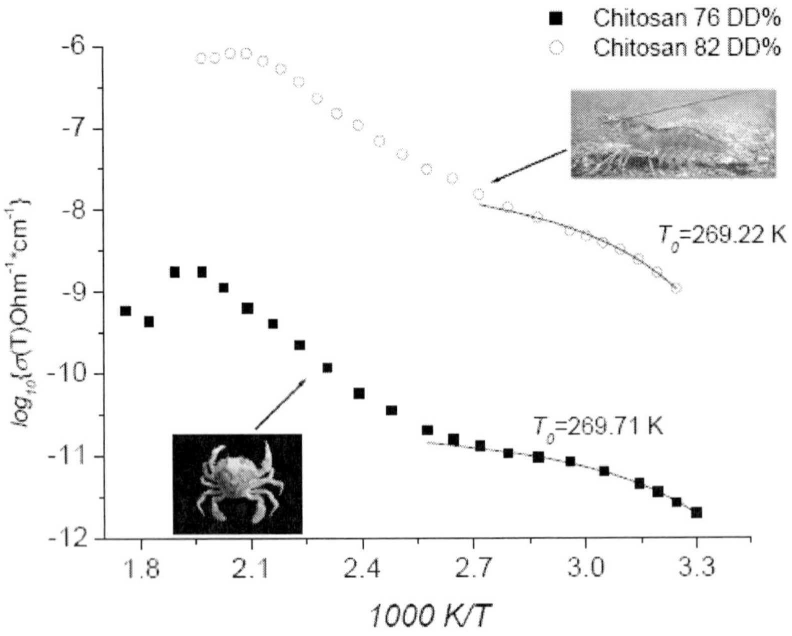

Figure 4. Temperature dependence of conductivity for chitosan (neutralized - solid squares and non neutralized - open squares). Note: the same Vogel temperature is obtained for both chitosan films with different DD%.

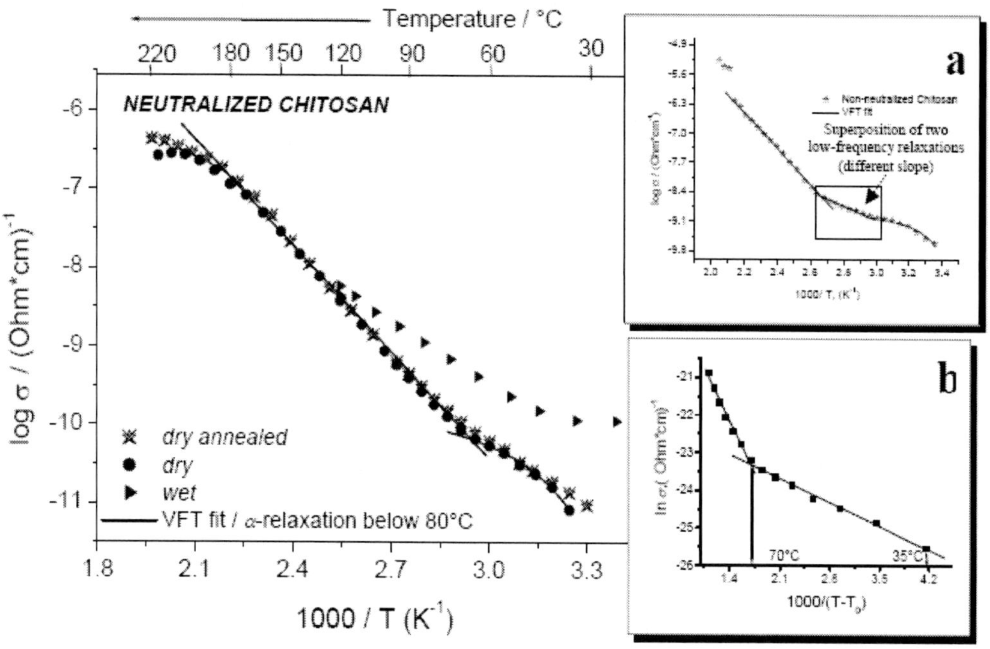

Figure 5. $\log_{10}\sigma$ *versus 1000/T* for *wet*, *dry* and *dry annealed* neutralized chitosan. Thermal relaxations are strongly affected by free-moisture content, α-relaxation is observed in the 20-70°C temperature range. Window insert: non-neutralized chitosan, a superposition of two low-frequency relaxations is observed.

This non-linear dependence disclosed in the 20-70°C is the typical trend of the α-relaxation behavior related to the dynamic glass transition. The temperature dependence of conductivity and relaxation time calculated from HN fitting (Figures 5 and 6) are well described by the VFT equation $\sigma = \sigma_0 \exp[-DT_0/(T-T_0)]$ and $\tau = \tau_0 \exp[DT_0/(T-T_0)]$ [15]. There is an excellent agreement between the Vogel temperature calculated from conductivity and relaxation time measurements independently (T_0 =284°K see Table II). Note that non-neutralized films exhibit higher conductivity than the neutralized ones, which is consistent with the excess number of protons in the non-neutralized films (NH_3^+ groups are present). Also, the T_0 value is independent of the presence of NH_3^+ groups.

The plot of *log* conductivity *versus* $1/(T-T_0)$ describes a straight line with negative slope in the 35-70°C temperature range. This linearity is a good indication of the α-relaxation; therefore all experimental data that fulfill this condition (Window insert of Figure 5b) belong to the α-relaxation process range (35 to 70°C).

The plot of relaxation time for the non-Arrhenius dependence present in the low frequency range below 70°C (Figure 6) as well as for secondary relaxations with Arrhenius behavior was obtained using the empirical Havriliak and Negami (HN) relation. For α-chitin (chitosan precursor) as well as for various cellulose-based materials, starches and non-polymeric glass forming liquids, it was found that the activation energies obtained from the *dc*-conductivity and the dielectric relaxation measurements are well correlated [36, 47, 48]. In glass-forming liquids, at high temperatures, both *dc*-conductivity and relaxation time show an Arrhenius behavior with the same activation energy. Below a cross over temperature (T_c) a Vogel-Fulcher-Tammann (VFT) behavior was observed [48].

Table II Parameter values for the VFT model in the range 20-70°C. The VFT model is generally used to describe α-relaxation. Both *dc* conductivity and relaxation time calculations are shown

α-relaxation			
VFTH Parameters:	τ_0 or σ_0	D	T_0(K)
NEUTRALIZED CHITOSAN			
Relaxation time	3.3×10^{-2}	0.32	284.3
dc Conductivity	3.1×10^{-10}	0.30	284.0
NON-NEUTRALIZED CHITOSAN			
Relaxation time	4.2×10^{-3}	0.43	283.9
dc conductivity	1.7×10^{-9}	0.10	283.9

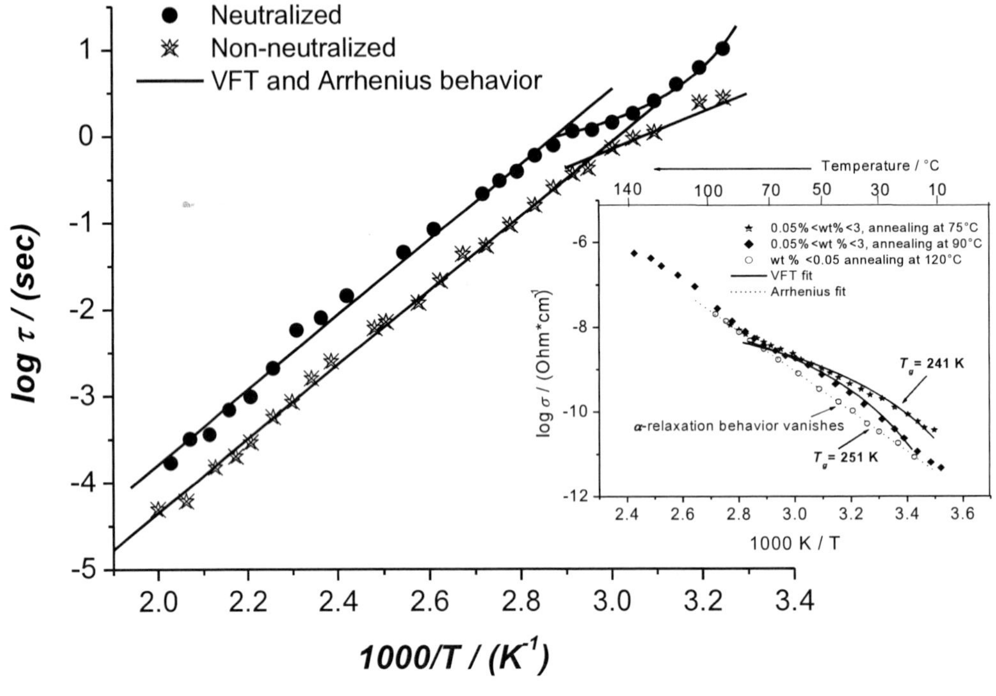

Figure 6. *log* relaxation time *versus 1000/T.* Lower relaxation time due to higher mobility conferred by NH_3^+ groups is demoted in non-neutralized films. Window insert: the α-relaxation is plasticized by water; T_0 shifts to higher values as moisture content decreases.

The two different temperature dependencies described above for glass-forming liquids are present in chitosan, and the similarity between *dc* conductivity and relaxation time for the two low frequency relaxations is clearly observed in Figures 5 and 6. Both dependences (*dc* conductivity (σ_{dc}) and relaxation time (τ) versus *1/T* plots) show the same features: an Arrhenius type relaxation will yield a straight line above 70°C (the crossover temperature (T_c) that separates the non-linear process from the linear process) whereas a non-Arrhenius relaxation will manifest as a curved line that suggests a VFT type or glass transition below 70°C in *dry* samples. For *wet* samples the decrease of conductivity as the temperature is increased from 20 to 70°C is likely due to the motion of water-polymer complex since water could be modifying the relaxation mechanism of the matrix material. And finally the vanishing of the non-Arrhenius behavior for moisture contents lower than 0.05 wt% labeled as *dry annealed* samples.

The Vogel temperature is the apparent activation temperature of the α-relaxation [15]. As proposed by a number of authors for different polymer systems using dielectric spectroscopy [15, 18, 49, 50] T_0 is usually 30-70K lower than T_g [20, 44]. In the present study, we take T_g= T_0+50 K and estimated a T_g value for neutralized and non-neutralized chitosan of 61°C (for moisture and heat treatment conditions explained before). The strong effect of moisture content on the α-relaxation process is also evidenced in window insert of Figure 6, since for *dry* samples the glass transition temperature shifts to higher values as moisture content decreases; it is possible to identify this effect when different annealing treatments at 75, 90 and 120°C are performed during measurements on the same sample. This shifting of the glass transition temperature to higher values as moisture contents decrease, points out a plasticizing effect of water on chitosan glass transition. Unfortunately, the accurate moisture content in the 0.05 to 3 wt% range according to the annealing treatment at different temperatures is difficult to determine since all the measurements are performed in a vacuum cell. However, it is noteworthy that moisture contents between 0.05 and 3 wt% are needed to distinguish the α-relaxation process by dielectric measurements before it vanishes at 0.05 wt% moisture content with the 120°C annealing treatment (see Figures 5, 6 and window insert of Figure 6).

Chitosan has functional groups like hydroxyls, amines and amides, which can act as hydrogen bond acceptors or donors. For this reason chitosan can be bonded or linked with hydrogen bond donors or acceptor compounds like water [31]. In the case of the glass transition temperature, plasticization occurs in the amorphous region only, such that the degree of hydration is quoted as moisture content in the amorphous region [51]. According to [52], the water sorption mechanism is composed of two main steps: water sorption on polymer sites and water clustering surrounding the first absorbed water molecules. Our results show that the α-relaxation process is strongly affected by the moisture content in the films as it is shown in Figures 5 and 6; while at percents higher than 3 wt% this process cannot be distinguished because of the free water effect, at lower percents (<3 wt% freezable bond water) a glass transition temperature can be assigned by the motion of a water-polymer complex in amorphous regions, a glass transition temperature can be assigned depending upon moisture content limited to be between 0.05 and 3 wt%. According to [31], the glass transition must be interpreted as torsional oscillations between two glucosamine rings across glucosidic oxygens and a cooperative hydrogen bonds reordering. Water sorption in hydrophilic polymers is usually a nonideal process leading to plasticization [52]. Water leads to an increase in the amounts of hydrogen bonds producing an increase in cooperative motion. Water hydrogen bonds between chitosan chains and increases *free volume*. When this happens chains can slide past each other more easily, so the time scale of the cooperative motion matches that of the experiment and the glass transition can be detected.

Nevertheless, if sample moisture is minimized (<0.05 wt %), chains are able to interact with each other giving rise to a denser packing. Thereby the mobility of polymer chains decrease and the glass transition phenomenon is not easily detected. Because of the absence of water, the glass transition temperature could be shifting to temperatures above 70°C (more rigid backbone). A similar effect than on Figures 5 and 6 (20-70°C temperature range) at different moisture contents was observed in the hydrogel poly(hydroxyethyl acrylate) (PHEA)

[19]. The results reported in this work [19] suggest that at h $\leq$ 0.21$\approx$ 17.3% water (where h is defined as grams of water per gram of dry sample) conductivity is governed by the motion of the polymeric chains, whereas at h $\geq$ 0.29 ($\approx$22.5 % of water), conductivity occurs through a separate water phase. At the highest moisture contents of 0.29 and 0.46 ($\approx$ 31.5 % of water) the *dc* conductivity dependence changes from VFT type to Arrhenius type [19].

When water is minimized (0.05 wt % of water, Figure 6) and plasticization is absent, the glass transition temperature could be shifting to higher temperatures. At this point (above 70°C), another molecular relaxation takes place and the α-relaxation is weaker and cannot be observed because of the *high temperature* relaxation effect, this is clearly observed in *dry annealed* chitosan below 70°C (Figures 5 and 6) by a change of the VFT behavior to a linear one with a different slope than that for the *high temperature* process. This "new" slope is a behavior halfway between the non-linear VFT behavior and the Arrhenius *high temperature* one, since under certain conditions of minimum moisture, the *high temperature* relaxation process is observed in the whole temperature range before the onset of thermal degradation as it is shown in Figures 5 and 6. For this *high-temperature* relaxation, non-neutralized chitosan films seem to be more sensitive to water, since in the *dry* state presents a particular behavior in the 70-100°C temperature range, which is recognized as a superposition of two relaxation processes, giving raise a different slope value as it is shown in window inset of Figure 5a.

As mentioned before, there exists a great controversy about the glass transition temperature (T_g) of chitosan; while some authors show no evidence of a glass transition by DSC, DMTA and dielectric measurements [11-14, 26, 28], others report a wide variety of values (see Table I). Some authors assigned values very close to chemical degradation [7-9], others seem to correspond to water elimination rather than a glass transition [6]. Some reported values that lie in the temperature range proposed in the present work [3, 22, 27]. The controversy and particularly the discrepancy in the glass transition temperature of chitosan may be related to an inefficient elimination of water, a heat treatment near the degradation temperature of chitosan, the film preparation technique or neutralization process. Regarding the deacetilation degree (%DD), our dielectric results showed no DD-effect on the T_g in agreement with the report of Dong et al. [5]. However, without doubt, the moisture content is a key factor that determines whether or not the glass transition temperature in chitosan can be observed since as it can be seen, the presence of water drastically affects the chitosan backbone mobility, especially in the α-relaxation region that corresponds to the cooperative motion and as a result, moisture content is probably the main cause of the wide glass transition temperatures range reported in literature.

High Temperature and Low Frequency Relaxation

The "*high temperature*" was defined in the temperature from 70°C to the onset of degradation $\approx$210°C (Figures 5 and 6). It is well described by the Arrhenius model and it was detected in both neutralized and non-neutralized chitosan. The slope of these curve represent the activation energy of each process. Both, the temperature dependence of *dc conductivity* as well as that for *relaxation time* are Arrhenius-type $\sigma = \sigma_0 \exp(-E_{a\sigma} / RT)$ and $\tau = \tau_0 \exp(E_{a\tau} / RT)$). For this relaxation, non-neutralized chitosan films seem to be more sensitive to water, since in the *dry* state presents a particular behavior in the 70-100°C temperature range, which is recognized as a superposition of two relaxation processes, giving raise a different slope value (see window insert of Figure 5). This relaxation has Havriliak and Negami fitting parameters α =0.72 ± 0.08 and β = 1.0, these parameters are temperature independent in both chitosan forms in agreement with previous studies [17]. It can be seen that the experimental data could be fitted with the model involving less parameters than the HN model in agreement with other authors [52, 53].

Our activation energy calculations ($Ea\sigma$ 80-88.2 kJ/mol for *conductivity* and $Ea\tau$=80-89 kJ/mol for *relaxation time*) are in agreement with previous reports for neutralized and non neutralized chitosan [17], and for polysaccharides [16]. The activation energy values for both films are quite close. This process so-called σ-relaxation has been widely studied and is associated with the hopping motion of ions in the disordered structure of the biomaterial [16]. Einfeldt et al. [16] observed this relaxation process in the high temperature range (>80°C), however, in our case, on minimum moisture conditions this relaxation process discloses in the whole temperature range until the onset of thermal degradation making difficult the α-relaxation detection. For the case of non-neutralized films, moisture has a strong effect on the α-relaxation. This is shown in the *dry* non-neutralized samples in the 70-100°C temperature range with moisture content in the range of 0.05-3.5 wt% (see window insert Figure 5a). Note that a lower slope (lower activation energy) ascribed to the superposition of α-and σ-relaxations describes this "new" Arrhenius-type relaxation for non-neutralized films. Activation energy is ca. $E_{a\sigma}$ (conductivity) ≈ 28.8 and $E_{a\tau}$ (time relaxation) ≈38.34 kJ/mol which is much smaller than that of neutralized films with similar moisture content (~80-85 KJ/mol,). The activation energy value of the relaxation time for the σ-relaxation process in the *dry* samples shows that water exerts a greater effect on the non-neutralized chitosan form; this is because of its superior ability to form hydrogen bonds providing a lower activation barrier for motion of ions. In this chitosan form, the σ-relaxation process is shifted to slightly higher frequencies compared to neutralized chitosan (not shown), this entails lower relaxation times (Figure 6). The ion mobility in non-neutralized chitosan is facilitated by NH_3^+ groups providing it higher conductivity.

Pizzoli et al. [11] by dynamic mechanical analysis and dielectric measurements observed a "high-temperature relaxation" near 140°C; the calculated activation energy was ~100 kJ/mol, and this value is in agreement with the activation energy of the σ-relaxation mentioned before. They did not interpret it as a glass transition and suggested that this relaxation arises from a molecular motion having a less co-operative character than the glass-to-rubber transition. Some authors have attributed this relaxation to the glass transition [4, 5], nonetheless it seems to be ion motion that yields this peak in dynamic mechanical spectra and not the glass transition, because the activation energy for segmental mobility should be greater.

Finally, in the low frequency range, a change from positive slope in the conductivity (Figure 5) to negative is disclose at 210°C and above denoting the onset of degradation, at this temperature the dependence of *resistance* and *capacitance versus temperature* also experiment a change in the slope (Figure 2a). TGA (Figure 2b) and DSC (not shown) measurements confirm this degradation.

High Frequency Relaxation

The well known high-frequency β-relaxation is identified in the 10^4-10^8 Hz frequency region for *dry* and *dry annealed* neutralized and non-neutralized films, however, in *wet* (12.8

wt % of moisture content) samples this relaxation is not well resolved and model-fitting is complicated. This is the reason why we cannot identify the β_{wet} relaxation process documented by other authors ascribed to biopolymer-swollen water motion [11, 33]. For *wet* polysaccharides, Enfieldt, et al [16] reported a superposition of two relaxation processes: β and β_{wet} relaxations that merge into one common β-relaxation process as water is driven off. The temperature dependence of the relaxation time for this high frequency relaxation is found to be Arrhenius-type (not shown) with an excellent linear fitting of the *relaxation time* as a function of *1000/T*. Our calculated activation energy values are in excellent agreement with previous reports [16, 17, 33]. This β-relaxation is a common reported relaxation process in chitosan [11, 16, 17, 22] and it has been related to side group motions by means of the glucosidic linkage [16, 54]. This is the main relaxation process found in all pure polysaccharides in the low temperature range (-135 to + 20°C) and corresponds to local chain dynamics [16]. For chitosan, we have found this secondary relaxation in the high temperature range in agreement with other authors (higher than 20°C) [16, 33], however, it is expected to appear in the high temperature range at the very high frequency end since Einfeldt et al [16] found this high frequency relaxation at temperatures as high as 120°C.

Dynamic Mechanical Analysis (DMA)

Effect of Moisture Content on DMA Relaxations

Neutralized and Non-neutralized chitosan films moisture weight percent was determined by TGA measurements under dry air flow emulating the DMA measurements environment. Figure 7 shows a weight loss of 11% for *wet* samples, 5.5 % for *dry* ones and 1.2% for *dry annealed* films. Moisture content in *wet* samples under dry air flow conditions are in agreement with our measurements reported for dielectric analysis. Chitosan DMA analysis shows three clear events (see Figure 8a) in non neutralized-chitosan films under *wet* and *dry* conditions; they can be distinguished by the presence of three *tan* δ peaks at 74°C, 182°C and 290°C in *wet* samples and at 88°C, 182°C and 290°C in films *drying* at 150°C before DMA measurements. These three relaxations are observed in neutralized films as well, it is shown in Figure 8b.

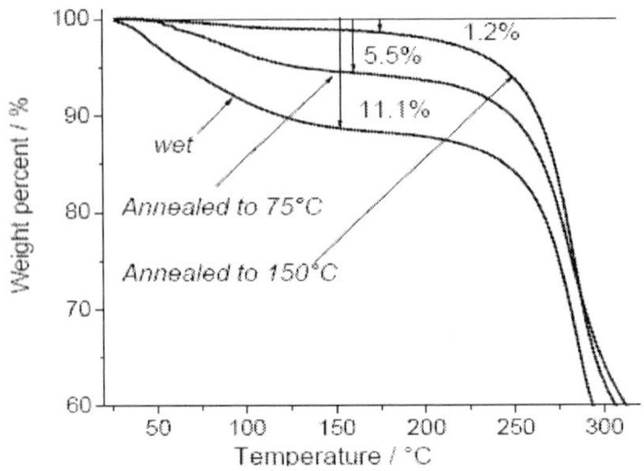

Figure 7. Thermogravimetric analysis for neutralized chitosan under air flow environment.

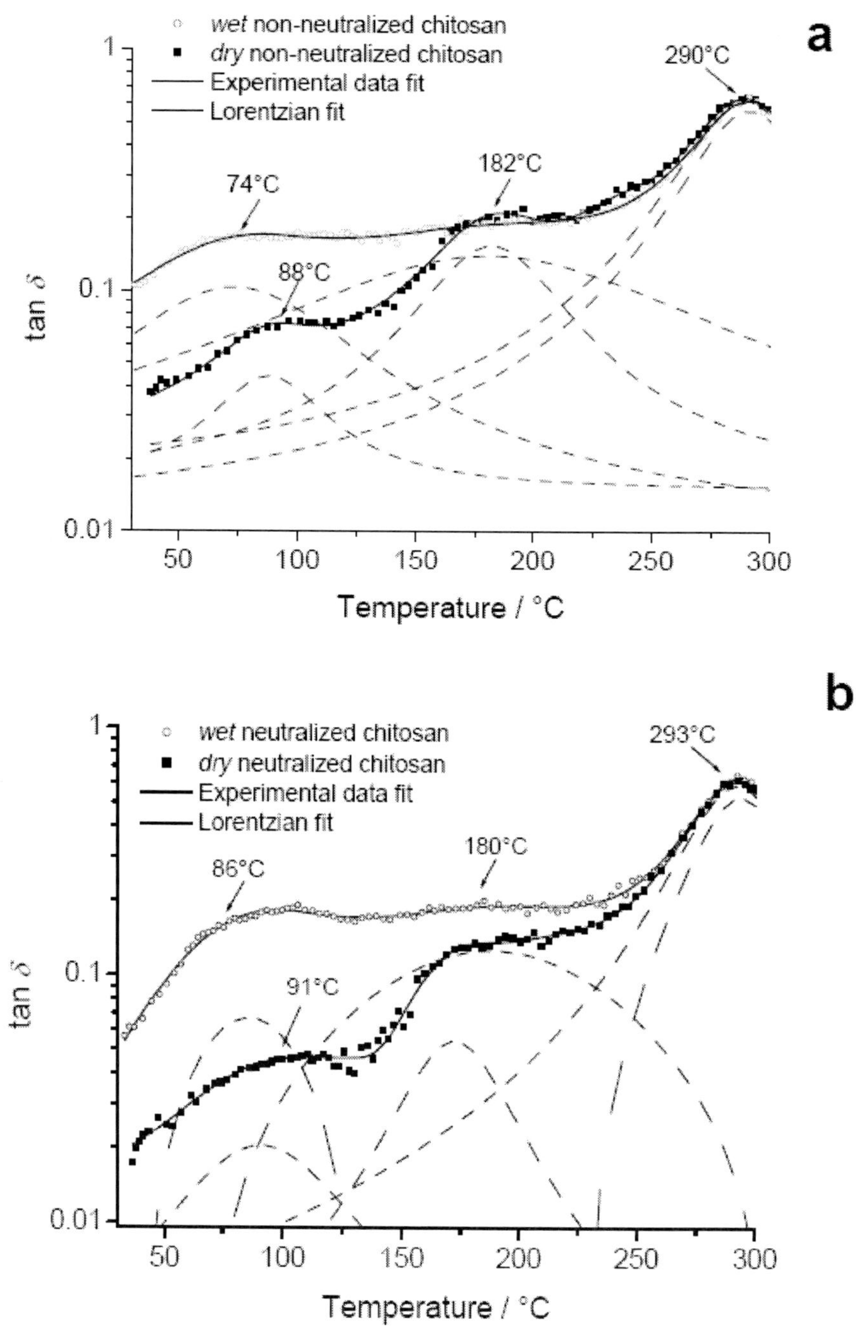

Figure 8. Viscoelastic properties for *wet* and *dry* (a) non-neutralized chitosan and (b) neutralized chitosan. Note: three main events are distinguished under *wet* conditions, storage and loss modules occur at lower temperatures when comparing with tan δ peaks.

Relaxation below 100°C

The molecular relaxation occurring below 100°C is highly affected by moisture content; the higher the moisture content, the lower the relaxation temperature, i.e., for *wet* samples the relaxation temperature shifts to lower temperatures in contrast to *dry* samples (see Figures 8 a and b). This phenomenon can be related to a plasticizing effect of water on the glass transition temperature of chitosan. The plasticizing effect is more clearly observed in Figure 9 since for *wet* samples the damping tan δ peak is present at 54°C whereas for samples dried at 75°C and at 150°C it shifts to 81°C and 89°C respectively. However, as well as it was shown by impedance measurements, this relaxation vanishes in neutralized chitosan films after the second heat treatment as it is shown in window insert of Figure 9. Quijada-Garrido et al. [31] by means of DMTA measurements assigned this relaxation process occurring below 100°C as the α-relaxation and hence as the glass transition temperature of chitosan, however, they do not provide a theoretically sound explanation about this statement and do not give additional data related to the effect of moisture content on this relaxation. They set the glass transition temperature of chitosan at 85°C, and observed a decrease in the temperature of the maximum of the α-relaxation with increasing glycerol content, suggesting a plasticizing effect on chitosan.

Lazaridou et al. [21] showed the thermomechanical behavior of chitosan and its blends with starch or pullulan. The location of the peak that they denoted as the α-relaxation for pure chitosan shifts to lower temperatures with increase of water content or addition of sorbitol due to plasticization of the polymer matrix. The glass transition temperature varies from 77°C at 10% of water to -23°C at about 30% of water and they estimated a glass transition temperature using the Gordon-Taylor equation for *dry* chitosan about 94.9°C. M. Mucha et al [4] showed two clear tan δ peaks occurring at 24°C and 43°C, moving to higher temperatures (32°C and 45°C respectively) in a second scan after heating at 130°C. Mucha et al. state that these peaks shift to higher temperatures because a chitosan film has lower amount of water, thus the system appears stiffer.

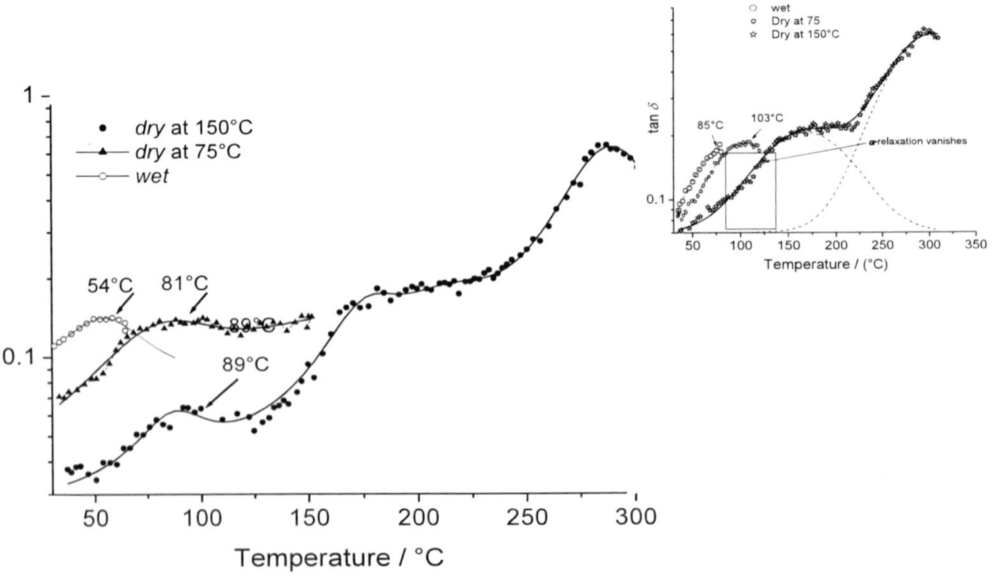

Figure 9. tan δ for non-neutralized chitosan. Note: the low temperature relaxation shifts to higher temperatures when moisture content decrease. Window insert: neutralized chitosan, as well as for dielectric analysis, α-relaxation vanishes after annealing.

These two peaks were recognized as a structural reorganization of packing of chitosan molecules due to an increase of residual water mobility, volume expansion and following change of hydrogen bond strength.

Toffey et al [27] reported a tan δ peak for wet samples ranging form 60°C to 93°C depending on the acidic media employed for chitosan dissolution, they used four acids: formic, acetic, propionic, and butyric acid and do not performed neutralization prior to DMA measurements. Likewise, in a different paper Toffey et al [27] observed tan δ transitions for chitosanium acetate located at progressively higher temperatures, indicating a shift of T_g to higher temperatures as the temperature range over the sample scanned increases. This rising temperature is usually attributed to the removal of residual moisture which plasticizes the material; nonetheless, in the case of chitosanium acetate, surprisingly, Toffey et al. [27] claim that this effect is due to conversion of chitosanium acetate to chitin instead of loss of residual moisture.

Relaxation above 100°C and below 200°C

The second tan δ peak located above 100°C (see Figures 8 and 9) had also been assigned as the α-relaxation of chitosan; Ogura et al [23] by means of the loss peak proposed a glass transition temperature around 140°C, however, they do not report moisture content on chitosan films. Dong et al [5] suggest a glass transition temperature for chitosan at 140°C-150°C, however, the DMA curves were obtained after the first run of heating up to 180°C and this temperature is high enough to induce thermal degradation in chitosan [39, 55], they also obtained a first peak at around 85°C ascribed to a water-induced relaxation. While M. Mucha et al [4] denoted a tan δ peak occurring at 156-170°C as the α-relaxation at T_g of chitosan according to deacetylation degree, however they do not give a theoretical statement to support this issue. Likewise, Y.B.Wu et al [29] designed the glass transition temperature of chitosan at 153°C by means of a tan δ peak.

Relaxation above 200°C

Surprisingly, the third peak disclosed above 200°C, has also been assigned has the glass transition temperature of chitosan. Sakurai et al [7] carried out DMA measurements in two cycles of heating: the first heating up to 180°C and the second heating up to 250°C after holding at 180°C for 5 min. In the first heating one large peak near 153°C is disclosed, however, the tan δ curve drastically changed in the second heating and a large shoulder appears at about 205°C, while the peak near 153°C disappears. They claim that the transition occurring at 205°C is related with the glass transition of chitosan, however, as it was shown above by TGA and dielectric measurements and supporting with literature reports [39], there is enough evidence to state that after the first heating up to 180°C chitosan is degradated. That is why the relaxation at 153°C disappears; therefore the process occurring at 205°C should be ascribed to the beginning of thermal degradation of the biopolymer and cannot be related with the glass-rubber transition.

Dynamic Mechanical Analysis (DMA) and Dielectric Spectroscopy Measurements Connection

As it was shown above, the understanding of dynamic mechanical analysis (DMA) experimental data and therefore the designation of the relaxation nature of each process disclosed in chitosan is a controversial issue. Without additional evidence it is difficult to identify the nature of each relaxation processes. In this case, DMA measurements support the dielectric spectroscopy results previously described.

Relaxation below 100°C

The "low frequency-low temperature" relaxation (20-70°C) obtained by dielectric spectroscopy as well as the relaxation disclosed below 100°C by DMA measurements is highly affected by moisture content, it suggests a plasticizing effect of water on the biopolymer. Dielectric results showed a non-Arrhenius behavior of this relaxation process that can be well described by the Vogel-Fulcher Tammann (VFT) equation, therefore, the first peak (dependent on moisture content) obtained below 100°C is related with the α-relaxation and consequently to the glass transition temperature of chitosan. The difference between the T_g value calculated by dielectric measurements and obtained by DMA measurements has been previously reported [31] for many polymers. This in part because dielectric spectroscopy is sensitive to fluctuations of dipole moments and mechanical relaxation monitors the fluctuations of internal stresses. It is noteworthy that in dielectric as well as DMA measurements, this relaxation is detected not any more after the second annealing.

Relaxation between 100°C and 200°C

The second peak occurring above 100°C and below 200°C, can be related to the σ-relaxation obtained by dielectric measurements. The activation energy disclosed by this relaxation process is in agreement with previous results reported for chitosan [16, 17], this process is associated with the hopping motion of ions in the disordered structure of the biomaterial.

Relaxation above 200°C

A change in the slope by dielectric measurements at 240°C has shown the beginning of chitosan thermal degradation; this statement is supported by DSC and TGA measurements as well as literature previous reports [39]. Therefore, the third peak obtained above 200°C is related with thermal degradation of chitosan.

CONCLUSIONS

The molecular dynamics of neutralized and non-neutralized chitosan was studied by dielectric spectroscopy and supported by dynamic mechanical analysis. The low frequency α-relaxation associated to the glass transition can be detected by dielectric spectroscopy once a pre-treatment of experimental data is performed. This relaxation process seems to be independent of the chitosan form evaluated (neutralized and non-neutralized) and is strongly influenced by moisture content, a glass transition can be calculated depending upon moisture content.

A plasticizing effect on chitosan α-relaxation is observed by dielectric spectroscopy and is supported by DMA analysis. For moisture contents less than 0.05 wt% the glass transition is difficult to observe due to a superposition of the α and σ-relaxation process. This work propose a successful method to monitor the plasticizing effect of water on chitosan glass transition by dielectric and DMA measurements, the same methodology is applied to TGA measurements in order to obtained the moisture weight percent with good agreement.

The well known σ-relaxation often associated with proton mobility is also observed in chitosan (neutralized and non-neutralized) from 80°C to the onset of degradation. On minimum moisture content conditions, this relaxation process could be detected in the whole temperature range before the onset of thermal degradation. It is strongly affected by moisture content for *dry* samples, by water effects, the activation energy shifts to lower values when compared to *dry annealed* samples. The non-neutralized chitosan showed an easier mobility in this ion motion process. This relaxation process exhibits a normal Arrhenius type temperature dependence with activation energy of 80-90 kJ/mol.

Finally, the high frequency secondary *β*-relaxation is also observed with Arrhenius activation energy of 46-48 kJ/mol.

REFERENCES

[1] Anderson, P.W. *Science,* 1995, 1615-1616.

[2] González-Campos, J. B.; Prokhorov, E.; Luna-Bárcenas, G.; Fonseca-García, A.; Sanchez. I. C. *J. Polym. Sci.* Part B:Polym Phys. 2009, 47, 2259-2271.

[3] Ratto J.; Hatakuyama T.; Blumstein R. B. *Polymer.* 1995, 427, 2915-2919.

[4] Mucha, M.; Pawlak, A. *Thermochim. Acta* 2005, 427, 69-76.

[5] Dong, Y.; Ruan, Y.; Wang, H.; Zhao, Y.; Danxia, B. *J. Appl. Sci.* 2004, 93, 1553-1558.

[6] Liu, Y.; Zhang, R.; Zhang, J.; Zhou, W.; Li, S. *Iranian Polym. J.* 2006, 15, 935-942.

[7] Sakurai, K.; Maegawa, T.; Takahashi, T. *Polymer.* 2000, 41, 7051-7056.

[8] Rao, V.; Johns, J. *J. Therm. Anal. Cal.* 2008, 92, 801-806.

[9] Mayachiew, P.; Devahastin S. *Drying Technol.* 2008, 26, 176-185.

[10] Arvanitoyannis, I. S.; Kolokuris, I.; Nakayama, A.; Yamamoto, N.; Aiba, S. *Carbohydr. Polym.* 1997, 34, 9-19.

[11] Pizzoli, M.; Ceccorulli, G.; Scandola, M. Carbohydr Res 1991, 222, 205-213.

[12] Kittur, F. S.; Harish P.; K.V.; Udaya, S. K.; Tharanathan, R. N. *Carbohydr. Polym.* 2002, 49, 185-193.

[13] Mano, J. F. *Macromol. Biosci.* 2008, 8, 69-76.

[14] Arvanitoyannis, I. S.; Nakayama, A.; Aiba S. *Carbohydr. Polym.* 1998, 37, 371-382.

[15] Raju GG. Dielectrics in Electrical Fields. Marcel Dekker Inc.: New York, 2003. pp. 138, 157, 248, 259.

[16] Einfeldt, J.; Meiβner, D.; Kwasniewski, A. *Prog. Polym. Sci.* 2001, 26, 1419–72.

[17] Viciosa, M. T.; Dionisio, M.; Silva, R.M.; Mano, J.F. *Biomacromolecules.* 2004, 5, 2073-2078.

[18] González-Campos, J.B.; Prokhorov, E.; Luna-Bárcenas, G.; Mendoza-Galván, A.; Sanchez, I. C.; Nuño-Donlucas, S. M.; García-Gaitan, B.; Kovalenko Y. *J. Polym. Sci.* Part B: Polym Phys. 2009, 932-943.

[19] Pissis, P. Electromagnetic Aquametry; Water in Polymers and Biopolymers by Dielectric Techniques; Springer: Berlin Heidelberg, 2005.

[20] Neto, C. G. T.; Giacometti, J. A.; Job, A. E.; Ferreira, F. C.; Fonseca, J. L. C.; Pereira, M. R. *Carbohydr. Polym.* 2005, 62, 97-103.

[21] Lazaridou, A.; Biliaderis, C. G. *Carbohydr. Polym.* 2002, 48, 179-90.

[22] Kaymin, I. F.; Ozolinya, G. A.; Plisko, Y. A. *Polym. Science* 1980, 22, 171-177.

[23] Ogura, K.; Kanamoto, T.; Itch, M; Miyashiro, H.; Tanaka, K. *Polym. Bullentin.* 1980, 2, 301-304.

[24] Kim, S. S.; Kim, S. J.; Moon, Y. D.; Lee, Y. M. *Polymer.* 199, 35, 3212-3216.

[25] Guan, Y.; Liu, X.; Zhang, Y.; Yao, K. *J. Appl. Polym. Sci.* 1998, 67, 1965-1972.

[26] Dufresne, A.; Cavaillé, J. Y.; Dupeyre, D.; Garcia-Ramirez, M.; Romero, *J. Polymer.* 1999, 40, 1657-1666.

[27] Toffey, A.; Glasser, W. G. *Cellulose.* 2001, 8, 35-47.

[28] Zohuriaan, M. J.; Shokrolahi, F. *Polym. Test.* 2004, 23, 575-579.

[29] Wu, Y. B.; Yu, S. H.; Mi, F. L.; Wu, C. W.; Shyu, S. S.; Peng, C. K.; Chao, A. C. *Carbohydr. Polym.* 2004, 57, 435-440.

[30] Shieh, Y. T.; Yang, Y. F. Euro Polym J. 2006, 42, 3162-3170.

[31] Quijada-Garrido, I.; Oglesias-González, V.; Mazón-Arechederra, J.M.; Barrales-Rienda, J.M. *Carbohydr. Polym.* 2007, 68, 173-186.

[32] Lee, Y. M.; Kim, S. H.; Kim, S. *J. Polymer.* 1996, 37, 5897-5905.

[33] Viciosa, M.T.; Dionisio, M.; Mano, J.F. *Biopolymers.* 2005, 81, 149-59.

[34] Dionísio, M.; Alves, N.M.; Mano, J.F. Review: Molecular dynamics in polymeric systems. e-polymers. 2004, 44. http://www.cqfb.fct.unl.pt/drs/docs/mano_030704.pdf

[35] Runt, J. P.; Fitzgerald, J. J. Dielectric Spectroscopy of Polymeric Materials; fundamentals and applications. American Chemical Society, Washington, DC 1997. pp 88, 89.

[36] Einfeldt, J.; Meiβner, D.; Kwasniewski, A. *J. Non-Cryst Solids* 2003, 320, 40-55.

[37] Menard, K. P. Dynamic mechanical analysis: a practical introduction. CRC Press, 2nd edition. Boca Raton Fl, USA 2008. p 2.

[38] Ahn, J. S.; Choi, H. K.; Cho, C. S. *Biomaterial.* 2001, 22, 923-928.

[39] Machado, A.A. S.; Martins, V. C. A.; Plepis, M.G. *J. Therm. Anal. Cal.* 2002, 67, 491-498.

[40] Yang, J. M.; Su, W. Y.; Leu, T. L. Yang, M. C. J Membr Sci. 2004, 236, 39-51.

[41] Ikejima, T.; Inoue, Y. *Carbohydr. Polym.* 2000, 41, 351-356.

[42] Mirzadeh, H.; Yaghobi, N.; Amanpour, S.; Ahmadi, H.; Mohagheghi, M. A.; Hormozi, F. *Iranian Polym. J.* 2002, 11, 63-68.

[43] Yanagawa, K. 4294A Precision Impedance Analyzer Operation Manual. *Agilent Technologies*. Japan 2000.

[44] Wintle, H. J. Conduction processes in polymers. In: Engineering dielectrics. Vol IIA, Electrical properties of solid insulating materials: Molecular structure and electrical behavior. Bartnikas, R., Eichorn, R. M., Eds.: ASTM Especial Technical Publication 783: Philadelphia, PA. 1983.

[45] F. Kremer, A. Schonhals (Eds.) Broadband Dielectric Spectroscopy, Springer-Verlag, Berlin 2003.

[46] Rowland, S.P. Water in polymers. American Chemical Society, Washington D.C. 1980; Vol 1.

[47] Stickel, F.; Fischer, E. W.; Richert, R. *J. Chem. Phys*. 1996, 104, 2043-2055.

[48] Cutroni, M.; Mandanici, A. *J. Chem. Phys*. 2001, 114, 7118-7123.

[49] Schönhals, A.; Kremer, F.; Hofmann, A.; Fischer, E.W. *Physical Review Letters*. 1993, 70, 3459-3462.

[50] Garcia, F.; Garcia-Bernabe, A.; Compan, V.; Diaz-Calleja, R.; Guzman, J.; Riande, E.; *J. Polym. Sci. B Polym. Phys.* 2001, 39, 286-299.
[51] Hodge, R.M.; Bastow,T.J. G.; Edward,H.; Simon, G.P.; Hill, A. *J. Macromolecules* 1996, 29, 8137-8143.
[52] Despond, S. ; Espuche, E.; Cartier, N.; Domard, A. *J. Polym. Sci.* Part B: Polym. Phys. 2005, 43, 48-58.
[53] Yagihara, S.; Yamada, M.; Asano, M.; Kanai, Y.; Shinyashiki, N.; Máximo, S.; Ngai, K. L. J. *Non-Cryst Solids* 1998, 235-237, 412-415.
[54] Dos Santos Jr, D. S.; Goulet, P. J. G.; Pieczonka, N. P.W.; Oliveira Jr, O. L.; Aroca, R. F. *Langmuir* 2004, 20, 10273-10277
[55] Balau, L.; Lisa, G.; Popa, M. I.; Tura, V.; Melnig, V. *Cent. Eur. J. Chem.* 2004, 2, 638-647.

In: Focus on Chitosan Research
Editors: Arthur N. Ferguson and Amy G. O'Neill
ISBN 978-1-61324-454-8

Chapter 6

BIOLOGICAL ACTIVITIES OF CHITIN, CHITOSAN AND RESPECTIVE OLIGOMERS

***F.reni K. Tavaria*[1], *João Fernandes* [1], *Alice Santos-Silva,*[2,3], *Sara Baptista da Silva*[2], *Bruno Sarmento* [2,4,*] *and Manuela Pintado*[1]**

[1]. Center of Biotechnology and Fine Chemistry, Biotechnology School of Portuguese Catholic University,
Rua Dr. António Bernardino de Almeida, 4200-072 Porto Portugal
[2]. Faculty of Pharmacy, University of Porto
Rua Anibal Cunha 164, 4050-047 Porto Portugal
[3]. Instituto de Biologia Molecular e Celular (IBMC) da Universidade do Porto,
Rua do Campo Alegre, Porto, Portugal
[4]. Department of Pharmaceutical Sciences, Instituto Superior de Ciências da Saúde-Norte,
Rua Central de Gandra 1317, 4585-116 Gandra Portugal

ABSTRACT

Chitin and chitosan have been receiving great attention as a novel functional material for their excellent biological properties such as biodegradation, immunological, antioxidant and antibacterial activities. However, its use has been scarcely developed due to its high molecular weight, which causes poor water solubility and high viscosity of the solutions. Compared with ordinary chitosan, chitosan oligomers have much improved water-solubility and some special biological functions due to the high absorption rate at the intestinal level, thus permitting its entrance in blood circulation and distribution all over the body. The potential application of chitin and chitosan, and respective oligomers, is multidimensional and it has been found in food and nutrition, biotechnology, material science, drug delivery, agriculture and environmental protection, and recently gene therapy too. In the realm of this chapter, we will focus on some biological activities attributed to those molecules, such as, antimicrobial and antifungal activities, wound healing, antioxidant, anti-inflammatory, anti-carcinogenic and anti-coagulant.

* Corresponding author. Tel.: +351 222 078949; fax: +351 222 003977. E-mail address: bruno.sarmento@ff.up.pt.

ABBREVIATION

Deacetylation degree	DD
Molecular weight	MW
Chitosan oligosaccharides	COS

1. INTRODUCTION

1.1. Chitin and Chitosan

The name 'chitin' was firstly used by Bradconnot in 1811. It is derived from the Greek word 'chiton', meaning tunic or 'coat of mail', since it was first identified from the exoskeleton (shell) of insects and crustaceans. Chitin is the second most abundant natural polysaccharide on Earth following cellulose. It is found mainly in crustaceans, molluscs, marine diatoms, insects, algae, fungi and yeasts. The processing industry of shellfish generates important waste with relevant impact on environmental pollution. Nevertheless, only a small percentage is used as animal feed or for chitin isolation. Its annual synthesis in marine and freshwater ecosystems is estimated at about 1600 and 600 million tons, respectively (Cauchie, 1997). Sources of chitin include shellfish, oyster and squid, harvested in quantities of about 29.9, 1.4 and 0.7 million tons a year (Synowiecki and Al-Khateeb, 2000). The chitin content in crustaceans normally ranges from 2 to 12% of the whole body mass and the composition of crustacean shells is 30-40% protein, 30-50% mineral salts and 13-42% chitin (Johnson and Peniston, 1982).

Chitin is a polysaccharide composed of β-(1->4)-linked *N*-acetyl-D-glucosamine (GlcNAc) residues with an acetamide group at the C2 position. Partial deacetylation of chitin lead to chitosan that is a copolymer of glucosamine (β(1–4)-linked 2-amino-2-deoxy-D-glucose) and N-acetylglucosamine (2-acetamido-2-deoxy-D-glucose).

Chitosan is, in fact, a collective name representing a family of de-*N*-acetylated chitins deacetylated to different degrees. Generically, the term chitosan has been applied when the extent of deacetylation is above 70% and the term chitin is used when the extent of deacetylation is insignificant, or below 20% (Baldrick, 2010). Chitosan polymers may present different molecular weights (MW) (50–2000 kDa), viscosity and degree of deacetylation (DD) (40–98%).

Recently, the commercial value of chitin has increased due to the beneficial properties associated to its soluble derivatives, applied essentially in the fields of chemistry, biotechnology, agriculture, food processing, medicine, dentistry, veterinary, environmental protection and paper or textile production. Both chitin and chitosan exhibit valuable biological activities, which have made these polysaccharides increasingly popular. Typical activities include antitumor, anticarcinogenic, immunoadjuvant, hypolipidemic, hemostatic, promotion of wound healing, prebiotic by enhancement of probiotic bacteria growth (e.g. *Lactobacillus bifidus*) and antimicrobial (Synowiecki and Al-Khateeb, 2003). Besides this, other positive aspects include the fact that they are derived from a natural source, biologically reproducible, biodegradable, biocompatible, non-toxic, biologically functional and changeable in molecular structure. Chitin and chitosan are structurally similar to heparin, chondroitin sulphate and hyaluronic acid, which are biologically important mucopolysaccharides in all mamals. Nevertheless, chitosan is almost the only cationic polysaccharide in nature (Sashiwa and Aiba, 2004), rendering unique properties in regard to biomedical applications. These important features further enhance the stimulus for new investigation by devising methodology to allow for applications in several commercial products. Therefore, new applications of chitin, chitosan and their derivatives can be expected in the future.

However, one of the main constrains is that chitosan is generally insoluble under physiological conditions due to its strong hydrogen-bonding network. Therefore, to improve solubility, several derivatives have been synthesized by chemical modification. Some of these include modifications with poly(ethylene glycol) (PEG) or addition of other chemical groups resulting in various chitosan derivatives, as listed in Table I. These derivatives have provided a platform for sustained release formulations at a controlled rate, prolonged residence time, improved patient compliance by reducing dosing frequency, enhanced bioavailability and a significant improvement in therapeutic efficacy (Chopra et al., 2006).

Table I. Chitin and chitosan derivatives

Chitin derivatives	Chitosan derivatives	References
Chitosan	*O* and *N*-carboxymethylchitosans	Rinaudo, 2006; Synowiecki and Al-Khateeb, 2003
Carboxymethylchitin	Chitosan 6-*O*-sufate	Rinaudo, 2006; Synowiecki and Al-Khateeb, 2003
Fluorinated chitin	N-methylene phosphonic chitosans	Chow and Khor, 2001; Rinaudo, 2006
N- and *O*-sulfated chitin	Trimethylchitosan ammonium	Tokura et al., 1992; Synowiecki and Al-Khateeb, 2003
(diethylamino)ethyl-chitin	Carbohydrate branched chitosans	Kurita et al., 1990; Rinaudo, 2006
Phosphoryl chitin	Chitosan-grafted copolymers	Andrew et al., 1998; Rinaudo, 2006
Mercaptochitin	Alkylated chitosans	Yoshino et al., 1992; Rinaudo, 2006
Chitin carbamates	Cyclodextrin-linked chitosans	Vincendon, 1992; Rinaudo, 2006
Oligochitins	Chitosan oligosaccharides (COS)	Rinaudo, 2006; Tharanathan and Kittur, 2003; Kim and Rajapakse, 2005

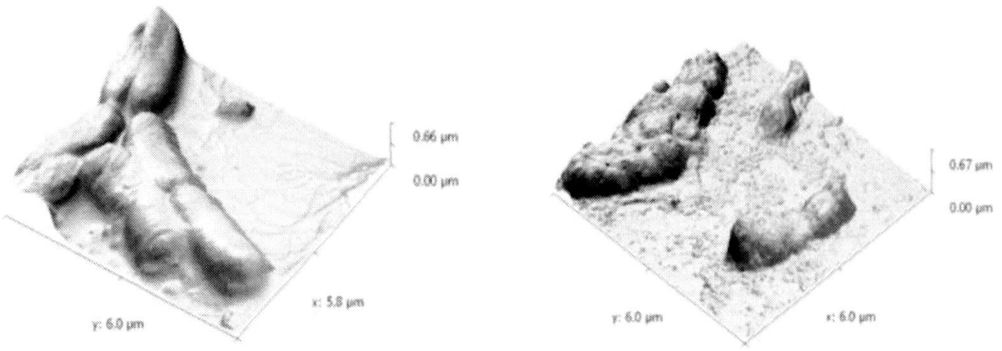

Fernandes et al., 2009.

Figure 1. Gram-positive bacterium *Bacillus cereus* imaged by atomic force microscopy, before (left) and after (right) contact with COS at 0.25 %(w/v).

1.2. Chitosan Oligosacharides

Chitosan oligosaccharides (COS), correspond to the hydrolyzed products of chitosan, which are readily soluble in water due to their shorter chain lengths and free amino group in *D*-glucosamine units (Jeon et al., 2000). Its lower viscosity and greater solubility at neutral pH have attracted the interest of many researchers to use chitosan in its oligosaccharide form. Besides these properties, COS also exhibit relevant health benefits such as lowering blood cholesterol levels, lowering high blood pressure, providing protection against infection, controlling arthritis and enhancing antitumor properties (Kim and Rajapakse, 2005). Unlike high MW chitosan, COS are easily absorbed through the intestine, quickly get into the blood allowing systemic biological effects in the organism. These biological activities are greatly affected by its MW, degree of polymerization, DD, charge distribution and nature of the chemical modification (Muzzarelli, 1994). In view of all this, COS have attracted a great interest in several industries, especially attributed to its antimicrobial, antioxidant and nutritional quality enhancer properties (Shahidi et al., 1999).

2. Anti-Microbial Activity

Chitosan and COS have shown antimicrobial activity against a wide range of target organisms, including Gram-positive (Figure 1) and -negative bacteria, yeasts and molds. Its antimicrobial activity, usually assessed by determination of Minimum Inhibitory Concentration (MIC), has been reported for several microorganisms according different chitosan MW and DD (Table II, III and IV).

The antimicrobial effect varies considerably with the molecular structure – both degree of polymerization and level of deacetylation affect independently the antimicrobial activity of chitosan, though it has been suggested that the influence of the MW on the antimicrobial activity is greater than the influence of the DD. Positively charged amino group located at C-2 position of the glucosamine monomer interacts with the negatively charged carboxylic acid group of macromolecules present on the cell surface and form polyelectrolyte complexes; this facilitates the binding of chitin/chitosan and its derivatives to the cell wall. The number of these amino groups has proven to play a major role in antibacterial activity, suggesting that chitosan has higher activity than that found in chitin. Chitosan with a higher DD results in greater adsorbed amounts of chitosan (Chung et al., 2004). The size and conformation appears to be fundamental on the effectiveness of chitosans, controlling the mechanism of action as described below. The mobility, attraction and ionic interaction of small chains are easier than bigger ones, which allow the adoption of an extended conformation and an effective binding to the cell surface (Goy et al., 2009); target organism - chitosan and its derivatives have been proven more effective upon Gram-negative bacteria than Gram-positive bacteria. The cell surface is a key factor to establish the amount of adsorbed chitosan, implying that the antibacterial mode of action is dependent upon the target microorganism. Evidence has been provided to demonstrate the relationship between the antibacterial activity of chitosan and the surface characteristics of cell wall, namely the polarity and hydrophilicity of cell wall. Negatively charged surfaces, induce the adsorption of higher amount of chitin/chitosan to the cell surface. More adsorbed chitosan evidently results in greater changes in the structure and in the permeability of the cell wall membrane (Kim and Rajapakse, 2005); and environment in which it is applied — chitosan exhibits antibacterial activity mainly in an acidic medium, which is usually attributed to the poor solubility of chitosan above pH 6.4.

Table II. Minimum inhibitory concentrations (MIC) of different chitosan and COS upon various Gram-negative bacteria

Target bacterium	DD (%)	MW (kDa)	MIC (%)	Ref.
Escherichia coli	85	1.2	0.10	Fernandes et al., 2008
Escherichia coli	85	6	0.06	Gerasimenko et al., 2003
Escherichia coli	69		0.01	Chen et al., 1998
Escherichia coli O-157	89	5–10	0.12	Jeon et al., 2001
Salmonella typhi	89	5-10	0.12	
Salmonella enterica	75-85	~190	>0.10	Marques et al., 2008
Salmonella typhimurium	69		>0.2	Chen et al., 1998
Klebsiella pneumoniae	85	1.2	0.10	Fernandes et al., 2010b
Pseudomonas aeruginosa	50-90	5–10	0.25	Kim andRajapakse, 2005
Pseudomonas aeruginosa	85	1.2	0.20	Fernandes et al., 2010b
Pseudomonas fluorescens	94	43	0.008	Devlieghere et al, 2004
Aeromonas hydrophila	69		0.05-0.20	Chen et al., 1998
Enterobacter aeromonas	94	43	0.006	Devlieghere et al, 2004
Vibrio cholera	69		0.02	Chen et al., 1998
Vibrio parahaemolyticus	69		0.01	Chen et al., 1998
Vibrio parahaemolyticus	75	1–10	0.4	Park et al., 2004

Table III. Minimum inhibitory concentrations (MIC) of different chitosan and COS upon various Gram-positive bacteria

Target bacterium	DD (%)	MW (kDa)	MIC (%)	Ref.
Bacillus cereus	75–90	1–10	0.125	Kim andRajapakse, 2005
Bacillus cereus	85	470	0.05	No et al., 2002
Bacilus subtilis	89	685	0.06	Jeon et al., 2001
Clostridium perfringens	90	600	>0.25	Chen et al, 2002
Listeria monocytogenes	94	43	0.02	Devlieghere et al, 2004
Listeria monocytogenes	85	26	0.25	No et al., 2002
Staphylococcus aureus	80-85	628	0.10	Fernandes et al., 2008
Staphylococcus aureus	50–90	1–10	0.125	Kim andRajapakse, 2005
Staphylococcus aureus	89	685	0.06	Jeon et al., 2001
Staphylococcus epidermis	89	685	0.06	Jeon et al., 2001
Staphylococcus epidermis	80-85	628	0.10	Fernandes et al., 2010b
Staphylococcus epidermis	75–90	5–10	0.063	Kim andRajapakse, 2005
Streptococcus thermophilus	85	600	0.005	Ausar et al., 2002
Propionibacterium freudenreichii	85	600	0.005	Ausar et al., 2002
Streptococcus faecalis	89	685	0.03	Jeon et al., 2001
Streptococcus mutans	89	685	0.008	Jeon et al., 2001
Lactobacillus plantarum	85	12	0.06	Gerasimenko et al., 2003
Bifidobacterium bifidum	85	12	0.0005	Gerasimenko et al., 2003
Lactobacillus casei	89	685	0.03	Jeon et al., 2001
Lactobacillus brevis	85	1106	0.05	No et al., 2002
Bacillus megaterium	85	470	0.05	No et al., 2002
Lactobacillus bulgaricus	89	685	<0.03	Jeon et al., 2001
Lactobacillus bulgaricus	85	600	0.005	No et al., 2002

As consequence, lower pH results in a higher antibacterial activity due to a more positively charged polymer with stronger affinity for cells (higher number of cationized amines available to interact), in addition to the 'hurdle effect' of inflicting acid stress on the target organisms. On the contrary, at neutral pH, COS are easily soluble, and guarantee antibacterial activity (Tsai and Su, 1999 No et al., 2003). Since different types of chitosans and particular environmental conditions are used in literature reports, some discrepancies in antimicrobial activity are found, which in some cases may lead to contradictory conclusions.

Although the information about antibacterial activity of chitosan and COS is still limited, the mostly accepted mechanism of action upon bacteria explains that the physiological pH in the cell is around neutral, which makes chitosan water-insoluble molecules to precipitate, and stack on the microbial cell surface. Therefore, an impermeable layer around the cell is formed, blocking the channels, which are crucial for living cells. The formation of this layer around the cell prevent the transport of essential solutes (causing internal osmotic imbalances) and may also destabilize the cell wall beyond repair, thus inducing severe leakage of intracellular electrolytes such as potassium ions and other low MW constituents (e.g. proteins, nucleic acids, glucose, and lactate dehydrogenase) and ultimately cell death (Sudharshan et al., 1992; Andres et al., 2007). When COS are dissolved at neutral pHs, no precipitate is formed and they may eventually penetrate the microbial cell (Figure 2) and alter permeability characteristics of the membrane and further prevent the entry of materials or cause leakage of cell constituents that finally leads to death (Liu et al., 2001; Kim and Rajapakse, 2005). The differential action exhibited by chitosan and its oligomers is therefore primarily dependent on the degree of polymerisation of the molecules. However, it was established a lower limit for COS activity, since it was demonstrated that COS with ca. 5-6 residues or lower possess little or no activity. Another proposed mechanism is the interaction of COS with bacterial DNA, which leads to the inhibition of the mRNA and protein synthesis, via the integration of COS into the nuclei of the microorganisms (Liu et al., 2001). This mechanism of action has been reported mainly in Gram-negative bacteria, where the thin layer of peptidoglycan on the cell wall facilitates to get through (Zheng and Zhu, 2003).

A third mechanism proposed for chitosan and COS involves metals chelation, suppression of spore elements and binding to essential nutrients required for microbial growth. It is unquestionable that chitosan molecules surrounding bacteria might complex metals and blockage some essential nutrients to flow, contributing to cell death. It is well known that chitosan has excellent metal-binding capacities, where the amine groups are responsible for the uptake of metal cations by chelation. In general, such mechanism proved to be more efficient at high pH, because positive ions are bounded to chitosan, since the amine groups are unprotonated, and the electron pair on the amine nitrogen is available for donation to metal ions. Nevertheless, this is, evidently, not a crucial antimicrobial activity, given that the sites available for interaction are limited and the complexation reaches saturation according metal concentration (Wang et al., 2005).

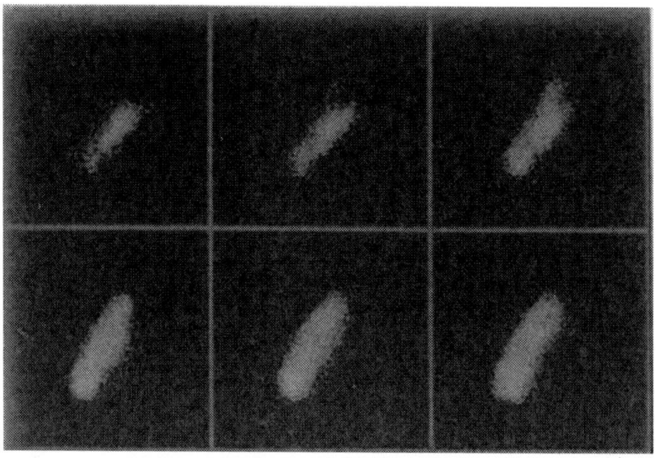

Liu et al., 2001.

Figure 2. Fluorescence micrographs of FITC-labeled COS (MW = 8kDa) accumulated in the Gram-negative bacterium *E. coli* cell, thus proving the penetration of these in the microbial cell.

Table IV. Minimum inhibitory concentrations (MIC) of different chitosan and COS upon various types of.fungi

Target organisms	DD (%)	MW (kDa)	MIC (%)	Ref.
Aspergillus fumigates	90	558	0.015	Lam and Diep, 2003
Aspergillus parasiticus	98	49.1	> 0.20	Tsai et al., 2002
Aspergillus niger	90	146	> 0.50	Seo et al., 2002
Fusarium oxysporum	76	285	0.50	Tsai et al., 2002
Colletotrichum gloeosporioides	93.2	1000	0.001	Jun-ang et al., 2009
Saccharomyces cerevisiae	-	930	0.05	Gil et al., 2004
Candida albicans	90	146	0.10	Seo et al., 2002
Candida albicans	>80	400-500	0.60	Balicka-Ramisz et al., 2005
Candida albicans	80-85	628	0.15	Fernandes et al., 2010b
Microsporum canis	>80	400-500	0.11	Balicka-Ramisz et al., 2005
Trichophyton mentagrophytes	>80	400-500	0.22	Balicka-Ramisz et al., 2005

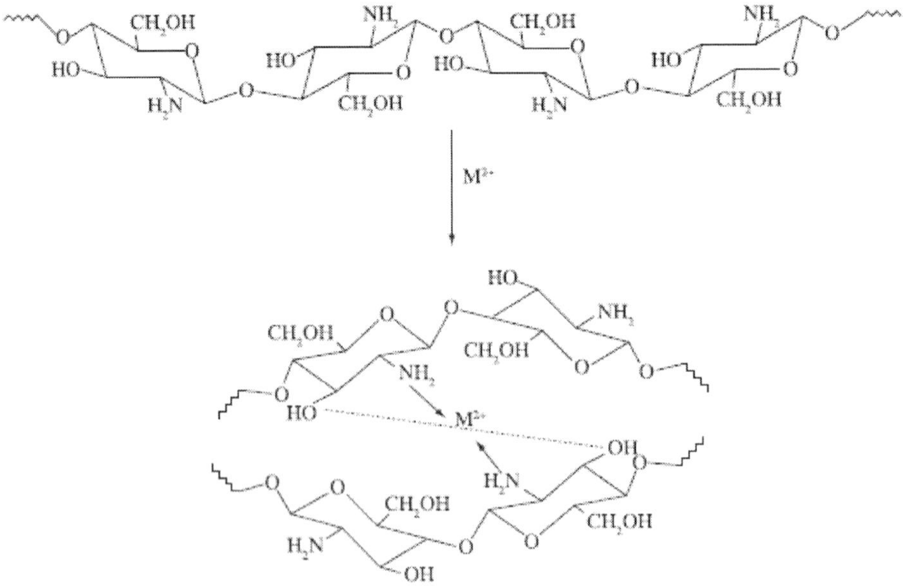

Figure 3. Metal-chitosan complexation model according to Wang et al. (2005).

Chitosan also exhibits antifungal activity upon molds and yeasts (Table IV). This activity is assumed to be fungistatic rather than fungicidal.

Generally, chitosan has been reported as being very effective in inhibiting spore germination, germ tube elongation and radial growth. The antifungic mechanism of chitosan involves cell wall morphogenesis with chitosan molecules interfering directly upon fungal growth, similarly to the effects observed in bacteria cells. The inhibition mechanism of COS against fungi is also similar to that of bacteria explained above. Microscopic observation showed that COS diffuse inside hyphae interfering on the enzymes activity responsible for the fungus growth. The damaging efficiency of chitosan upon fungal cell walls is also dependant on the concentration, DD and pH of the surrounding environment (Goy et al., 2009).

Table V. Different methodologies used to assess the antioxidant capacity of chitosan and its derivatives

Assay	Oxidative agent	MW (kDa)	DD (%)	Protection (%)	Concentration (% (w/v))	Ref.
Reduction of an artificial free radical						
TEAC	ABTS radical	2.8	50	100	0.5	Tomida et al., 2009
		<3	80-85	~70	0.1	Fernandes et al., 2010a
		107	80-85	~50	0.1	
DPPH Radical Scavenging	DPPH radical	6-7	76.54	92.37	0.25	Huang et al., 2006
		1-5	90	94.13	0.5	Je et al., 2004
Hydroxyl radical scavenging	OH^-	1-5	90	95	0.0075	Prashanth et al., 2007
		9	-	~75	0.325	Xing et al., 2005
		760	-	~10	0.325	
Metal chelating	Cu^{2+}	9	.	~37.5	0.025	Xing et al., 2005
		760	.	~12.5	0.025	
	Fe^{2+}	6-7	76.54	~80	0.1	
Carbon-centered radical scavenging	AAPH	6-7	76.54	75.03	0.125	Huang et al., 2006
Superoxide anion scavenging	O^{2-}	1-5	90	64.73	0.1	Prashanth et al., 2007
		9	-	~90	0.16	Xing et al., 2005
		760	-	~30	0.16	
Biomolecules protection/Biological systems						
Lipid peroxidation	Linoleic acid lipid radicals	400	91.8	45.5	~4x10^{-6}	Li et al., 2002
		4.6	91.8	49.2	~4x10^{-6}	
Protein oxidation	AAPH	2.8	50	~30	0.5	Tomida et al., 2009
DNA	$Fe^{2+} + H_2O_2$	1-5	90	>80	0.0001	Prashanth et al., 2007
Erythrocytes	AAPH	<3	80-85	24.9	0.0005	Fernandes et al., 2010a
Phages	H_2O_2	<3	80-85	90	0.005	
Human embryonic hepatocytes	H_2O_2	<2	90	80	0.2	Xu et al., 2009

3. Anti-Oxidant Activity

Scavengers of free radicals are considered preventive antioxidants and the presence of these compounds can affect the oxidative sequence at different biological levels. Chitosan and COS have been identified as potent radical scavengers by several authors (Kim and Rajapakse, 2005; Xiong et al., 2007). On the contrary, chitin and its oligomers have been described as ineffective, since this activity has been proved to be mainly dependent on the DD and strongly affected by chitosan MW. Even though the precise mechanism of radical scavenging activity of COS is not clear; it was reported that amino and hydroxyl groups attached to C-2, C-3 and C-6 positions of the pyranose ring react with unstable free radicals to form stable macromolecule radicals.

As seen above for antimicrobial activity, some discrepancies in literature may also be found for antioxidant activity of these molecules. This inconsistency is not only related with the variables associated to chitosan molecules (source, molecule structure and environment conditions), but also with the inexistence of a simple universal method by which antioxidant activities can be measured accurately and quantitatively (Je et al., 2004; Fernandes et al., 2010a). Numerous methodologies (Table V) have been used by different authors to demonstrate the significant scavenging capacity of chitosans against different radical species. In some cases the antioxidant activity was even better than those pertaining to reference antioxidants (e.g. aminoguanidine, pyridoxamine and trolox) (Chen et al., 2003).

Most of the methods are based on the ability of chitosan/COS to reduce a stable artificial free radical, namely scavenging of DPPH or ABTS radicals, carbon centered radical scavenging, hydroxyl or superoxide radical assays, and metal ion chelating (Figure 3). However, biological assays involving lipid peroxidation and protein oxidation inhibition, as well as DNA, erythrocytes and phages protection or even *in vivo* tests have been reported. Although the results obtained by these different methods demonstrated the important effect of MW of chitosan, all concluded that as higher the DD, higher is the scavenging activity.

4. Anti-Inflammatory Activity

Inflammation is a physiologic response to numerous stimuli such as infection, tissue trauma or different stressful factors. Generally, this process is considered an acute, transitory phenomenon and is associated with a systemic response called the acute-phase response (APR). One clear indication of this response is the increase in synthesis and secretion by the liver of several plasma proteins with decreases in others (Bayne and Gerwick, 2001) as can be seen in Figure 4. These acute-phase proteins are involved in a variety of defense-related activities such as, controlling the dispersal of infectious agents, repairing tissue damage, killing microbes and restoring a healthy state (Dautremepuits et al., 2004).

The strong immune stimulatory activity attributed to chitosan derivatives has been linked to the presence of *N*-acetyl-D-glucosamine residues (Porporatto et al., 2003). High MW chitosan upregulates production of IL-1, TNF-α, granulocyte macrophage colony stimulating factor (GM-CSF), nitric oxide (NO) and interleukin-6 (IL-6) in macrophages. Besides this, COS enhances TNF-α and IL-1β release and internalization in macrophages that is mediated by a lectin like receptor with mannose specificity (Feng et al., 2004).

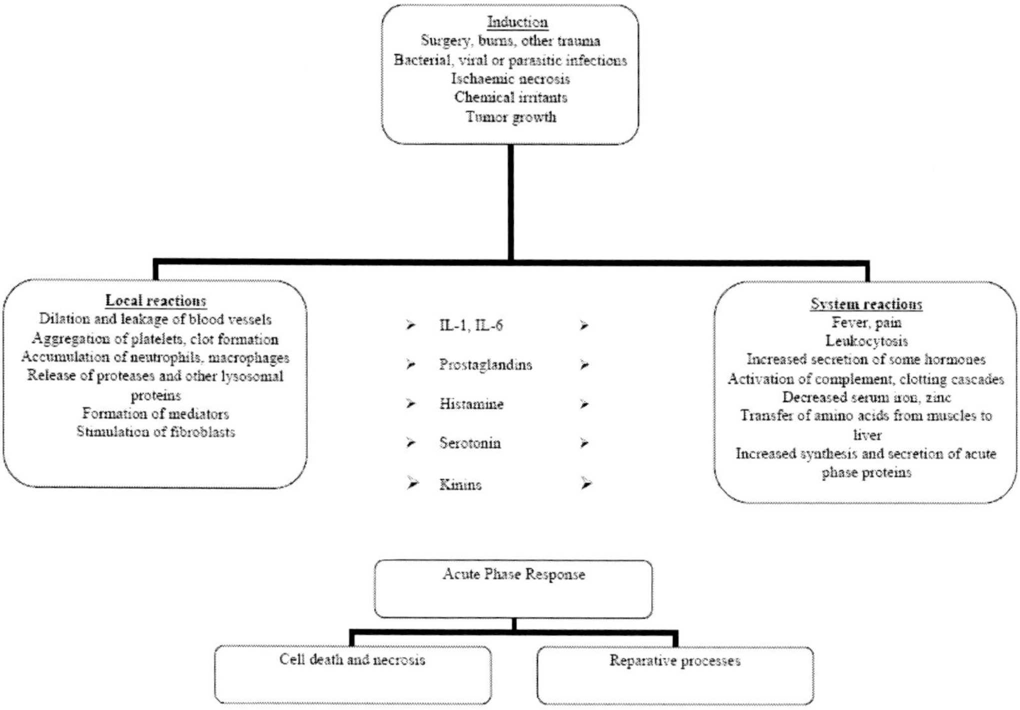

Figure 4. Diagram illustrating an acute phase response (adapted from Bayne and Gerwick, 2001) Macrophages are one of the most important immunocompetent cell types and they have a pivotal role in host defense and immuno-adjusting. Activated macrophages can release cytokines to defend against microbial infection and lyse tumor cells.

The anti-inflammatory effects of three types of chitin derivatives, namely phosphated chitin (P-chitin), phosphated-sulfated chitin (PS-chitin) and sulfated chitin (S-chitin) were investigated using a canine model of chitosan-induced pneumonia (Miyatake et al., 2003). P-chitin revealed to be the most effective anti-inflammatory agent followed by PS-chitin and S-chitin. Khanal et al. (2002) reported that P-chitin potentially recovered neutrophils deformability, which was lost by incubation with chitosan activated serum. Chitosan is known to activate the complement system (Minami et al., 1997), and C5a was found in the activated serum. C5a is a strong chemoattractant of neutrophils and responsible for their loss of deformability.

Chitosan also enhances the functionality of inflammatory cells such as PMN, macrophages and fibroblasts (Porporatto et al., 2003). Low MW chitosan binds to mannose receptors, which mediate the internalization of the chitosan particles. In fact, binding of *N*-acetyl-D-glucosamine to the specific receptors is thought to be a prerequisite for enhancing macrophage activation. Chitosan promotes the production of transforming growth factor β (TGF-β) and platelet-derived growth factor by human monocyte-derived macrophage (Ueno et al., 2001a) as well as production of interleukin (IL)-1 by mouse macrophages (Nishimura et al., 1986). It is known that a restricted pattern of activity of arginase and nitric oxide synthase is present in healing wounds. When studying the effect of chitosan on the arginine metabolic pathways of resident and inflammatory rat macrophages, Porporatto and co-workers (2003) found that low MW chitosan moderately activated the inducible nitric oxide synthase (iNOS) and arginase pathways in resident macrophages, whereas chitosan strongly enhanced arginase activity in inflammatory macrophages.

Tumor necrosis factor–α (TNF–α) is an inflammatory cytokine produced by the various cells including macrophages, lymphocytes, neutrophils and mast cells. It is a key mediator in the induction of apoptosis and development of humoral immune response; however, when present at high concentrations, its effects can be damaging such as inducing tissue injury and septic shock. Yoon and collaborators (2007) demonstrated that COS inhibited LPS-induced inflammatory effects in RAW 264.7 macrophage cells, suggesting that chitosan oligosaccharides may have an anti-inflammatory effect via stimulation of TNF-α. Oral administration of an antigen can result either in local and systemic priming or tolerance, being the basis of this dichotomy still poorly understood. The effectiveness of mucosal immune responses is influenced by factors such as nature of the antigen, dose, genetic background, antigen uptake and concentration of the immunologically relevant antigen that gains access to the internal milieu via the mucosa (Strobel and Mowat, 1998). Orally administered chitosan (MW ca. 80 kDa, 85% DD) in the absence of protein antigen, enhances the T helper cell type 2 (Th2)/Th3 microenvironment at the mucosal level by stimulating the production of regulatory cytokines (Porporatto et al., 2005). Chitosan (MW ca. 50 KDa, 85% DD) was also shown to be implicated in the modification of the uptake and/or distribution of the relevant antigen (type II collagen) promoting an anti-inflammatory effect early after feeding (Porporatto et al., 2004). Co-administered with protein antigens, chitosan modifies the uptake and/or the distribution of the relevant antigen and enhances the regulatory cytokines associated to the antigen-specific stimulation early after feeding (Porporatto et al., 2004). The results reported in these studies opened the possibility of using chitosan to modulate the immune response to orally administered antigens. In view of this, in a posterior research effort, Porporatto and co-workers (2008) demonstrated that chitosan enhanced the tolerance to an articular antigen with a decrease in the inflammatory response, showing consequent improvement in clinical signs.

5. Anti-Tumoral Activity

The antitumor activity is another feature observed by chitosan and its derivatives, dependent on their structural characteristics such as MW and DD (Kim and Rajapakse, 2005). In addition, immunostimulation property of COS is also thought to be responsible for antitumor activities, because chitin oligomers, which are not cationic, also exhibit antitumor activities (Tokoro et al., 1998). High MW chitosans (i.e. water insoluble chitosans) have been reported as being ineffective as antitumor agents (Ko et al., 1986; Kimura and Okuda, 1999), confirming the idea that anti-tumoral activity increases with the decrease of the MW (Table VI). The most accepted mechanism states that COS, water soluble chitosans and chitin oligomers inhibit the growth of tumor cells by exerting immunoenhancing effects and not by directly damaging tumor cells. It is suggest that the antitumor activity of water soluble chitosans and COS might be related, in part, to an enhancement of the proliferation of cytolytic lymphocytes – natural killer cells (Maeda and Kimura, 2004; Kim and Rajapakse, 2005).

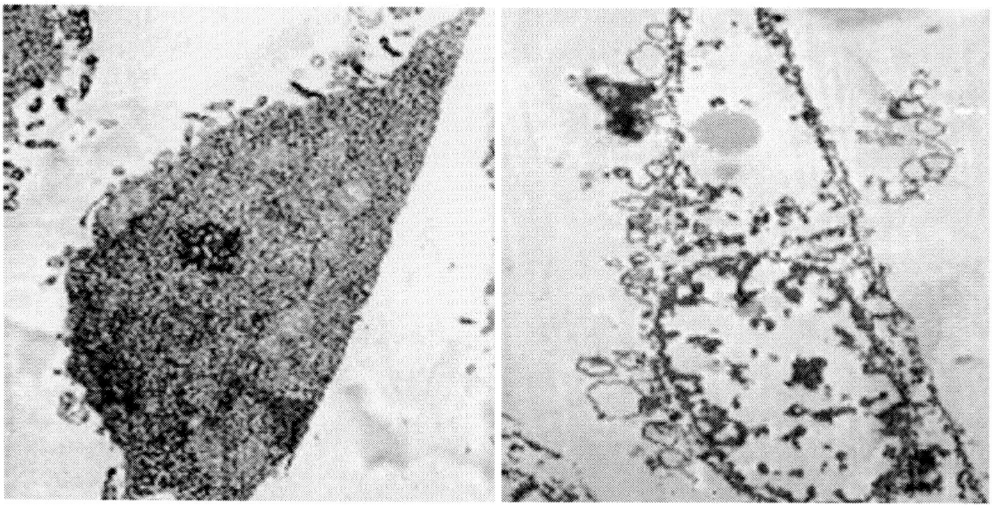

Qi et al., 2005.

Figure 5. Transmission electron microscopic photographs of human gastric carcinoma MGC803 cells cultured for 24 h in the absence (left) or presence of 100 μg/mL chitosan nanoparticles (right) *in vitro* (TEM, ×5 000).

Furthermore, it seems likely that COS and water soluble chitosans may induce the activation of macrophages through the production of cytokines such as IFN-γ, IL-12, and IL-18 (Maeda and Kimura, 2004). However, it has already been shown (Qi et al., 2005) that chitosan nanoparticles are able of directly damage tumor cells by provoking membrane disruption and cytoplasm leakage, inducing typical morphological features of necrosis (Figure 5).

Another proposed mechanism for the anti-tumoral effect was demonstrated using Ehrlich ascites tumor, HT-29 and CT26 colon carcinoma cells (Prashanth and Tharanathan, 2005; Hossain and Takahashi, 2008; Wang, 2008). It involves the induction of apoptosis on tumoral cells mainly by COS and in a lower extent by water soluble chitosans, as a function of nucleosomal DNA fragmentation. Still controversial are the effects of these compounds upon angiogenesis. Some reports suggest a potent angioinhibitory effect by COS and chitosans (Guzmán-Moralesa et al., 2009), correlated with a repression upon the expression of vascular endothelial growth factor (VEGF, most potent inducer of angiogenesis) or matrix metalloprotease-2 (Nam et al., 2007; Shen et al., 2009). However, others showed that chitosan particles had no effect on the release of angiogenic factors, and in some cases it even suggests that the mRNA expressions of VEGF and matrix metalloprotease-2 may be increased (Wang et al., 2008).

Although has been proved that water-insoluble chitosan do not exhibit direct activity against tumor cells, it has a relevant role in cancer treatment associated with chemotherapeutic drugs. Chitosan (ca. 600 kDa) has been shown to prevent the adverse reactions such as myelotoxicity, gastrointestinal toxicity, immunotoxicity, and body weight loss induced by the cancer chemotherapeutic drugs as 5-FU, cisplatin or doxorubicin, without reducing the antitumor activities of these drugs (Kimura and Okuda, 1999; Kimura and Onoyama, 2000; Maeda and Kimura, 2004). However, the main contribution in cancer treatment is explained by chitosan ability to act as an efficient and safe drug/gene carrier. In the form of micro- or nanoparticles or even as a gel or tablets, chitosan has been successfully used as clinical carrier for local drug/gene delivery system, owing primarily to its biocompatibility and biodegradability (Son et al., 2003; Ta et al., 2009; Gao et al., 2010). Several systems based in chitosan have been reported to improve the therapeutic efficiency of anti-cancer drugs at tumor sites, while concomitantly reduce associated serious side effects.

Medium MW COS ranging from 1.5 to 5.5 kDa could effectively inhibit the growth of sarcoma 180 solid (S180) or Uterine cervix carcinoma No. 14 (U14) tumor in BALB/c mice (Jeon and Kim, 2002). Chitosan nanoparticles were able to significantly inhibit the growth of Sarcoma-180 and mouse hepatoma H22 *in vivo* (Qi and Xu, 2006); dose and particle size effects were found to be relevant to the activity as well as the administration route.

Table VI. Effect of chitosan, chitin and respective oligomers upon the growth of different tumors in mice and different cancer cell lines

Chitosan / COS			Cell line / Tumor type	Inhibition (%)	Ref.
MW (kDa)	DD (%)	Dose (mg/Kg) / Inhibition dose (%)			
21	-	300	Mouse sarcoma-180	88	Maeda and Kimura, 2004
<1.2	-	300	Mouse sarcoma-180	91	
<1.5	100	500	MM 46 solid tumor	55	Suzuki et al., 1986
<1.5	100	300	Mouse sarcoma-180	93	
<1.5	100	10	Fibrosarcoma (Meth-A)	41	Tokoro et al., 1998
220	85	0.5	Mouse hepatoma H22	32	Qi and Xu, 2006
<10	-	/ 0.1%	Human leukemia (U937)	69-57	Wang, 2008
<10	-	/ 0.1%	Mouse rectal carcinoma (CT26)	69-57	
1.5-5.5	90	50	Uterine cervix tumor	73.6	Kim and Rajapakse, 2005
<2	84	0.05	Ehrlich ascites tumor	90	Prashanth and Tharanathan, 2005
780	81.2	/ 0.016%	293 cells	28	Zheng et al., 2006
780	81.2	/ 0.016%	HeLa	22	
426	90.3	/ 0.1%	HeLa human cervical carcinoma	50.8	Ai et al., 2008
1-3	85	/ 0.5%	Human breast carcinoma cells (MDA-MB-231)	30	Nam and Shon, 2009
24	-	/ 0.1%	Human hepatocellular carcinoma cell lines	51	
24	-	500	HepG2 tumor model in SCID mice	>50	Shen et al., 2009
<1.5	-	/ 0.1%	Human stomach adenocarcinoma cell lines	~60	
$(GlcNAc)_6$	0	10	Meth-A solid tumor	42	Tokoro et al., 1998
$(GlcNAc)_6$	0	1	Lewis lung carcinoma	20-30	Tsukada et al., 1990

Intravenous injection and smaller particle size led to higher tumor weight inhibition. Water-soluble COS, prepared by hydrolysis with hemicellulase and its hydrolysates were able to inhibit the growth of sarcoma 180 (S180) tumor cells in mice, with the maximum inhibitory rate of 64.2%, when administrated by intraperitoneal injection. Oral administration was also effective in the reduction of tumor weight reaching an inhibitory rate of 33.7% (Qin et al., 2002). Antitumor compounds are believed to exert effects on the immune system by stimulating leukocytes, cytotoxic T cells and natural killer cells (Tokoro et al., 1998). Their findings further suggest that chitosans with MW of 21 or 46 kDa may act as immunomodulators (inducing the activation of macrophages through the production of cytokines such as IFN-γ, IL-12 and IL-18) in the intestinal immune system due to enhancement of the cytotoxic activity of IELs. Furthermore, a study by Nishimura and co-workers (1986) revealed that partially deacetylated chitin and carboxymethyl chitin with an adequate degree of substitution were effective toward controlling tumor cells. Macrophages are one of the most important immuno-competent cell types present in mammalian tissues. Activated macrophages can kill tumor cells and pathogens either by direct contact or by releasing diffusible cytotoxic molecules, such as nitric oxide (NO), inter-leukin-1 (IL-1), tumor necrosis factor-alpha (TNF-α) and reactive oxygen intermediates (Higushi et al., 1990). Studies have demonstrated that high MW chitosan have a stimulatory effect on NO and TNF-α production (Jeong et al., 2000). Oligochitosans (average MW of 1000 Da) could significantly increase the activity of inducible nitric oxide synthase and consequently induce the synthesis of nitric oxide and tumor necrosis factor alpha (TNF-α) in macrophages as demonstrated by Yu et al. (2004). The antitumor activity of chitosan-copper complexes were evaluated by Zheng et al. (2006) who concluded that the complexes inhibited the proliferation of the two tumor cell lines, HeLa and 293. Heavy metals including copper are cytotoxic to cells at 10^{-3} mol/L, however, when diluted to 10^{-4}/L, the percentage of inhibition sharply decreased.

Nevertheless, some adverse effects were also reported as the result of high doses of chitosan: lethal pneumonia in dogs (Minami et al., 1996) and increased rate of metastasis and tumor growth in tumor-bearing mice. In acute toxicity tests the LD50 reported for chitosan was over 1.5 g/ kg orally in rats, over 10 g/kg subcutaneously in mice, 5.2 g/kg intraperitoneally in mice and 3.0 g/kg in rats (Ueno et al., 2001c). Despite this, the number of reported successful uses of chitosan in metastasis suppression and tumor reduction far overcome those of lethal overcomes (Nam and Shon, 2009).

6. Wound Healing Activity

N-acetylglucosamine is the monomeric unit of chitin, but also occurs in hyaluronic acid, an extracellular macromolecule that is implicated in wound repair. Therefore, chitin possesses favorable characteristics to promote rapid dermal regeneration and accelerate wound healing suitable for applications such as simple wound coverings or sophisticated artificial skin matrixes (Khor and Lim, 2003). Cho et al. (1999) compared the treatment of wounded skin with chitin, chitosan and water-soluble chitin powders and solutions and the water-soluble chitin solution was found to have the highest tensile strength and the fastest healing rate followed by the water-soluble chitin powder, chitin and finally, chitosan. Similarly, Ueno et al. (1999) demonstrated that chitosan incorporated in cotton enhanced wound healing by promoting infiltration of PMN cells at the wound site and then inducing active biodebridement (removal of damaged or dead tissue) by these cells (Morimoto et al., 2001), an essential process in wound healing. Chitosan-treated wounds showed histologically regeneration signs such as severe polymorphonuclear leukocyte infiltration with increased granulation, higher amounts of collagen and osteopontin at the wound site (Ueno et al., 2001b). Chitosan accelerates migration of PMN to the wound area, which secrete inflammatory mediators such as tumor necrosis factor (TNF)-α, interleukin (IL)-1, IL-8, IL-12, macrophage inflammatory protein (MIP)-1α and MIP-1β in addition to phagocytosis. Complement activation by chitin and chitosan was studied by Minami et al. (1998), who suggested that the alternative pathway was undertaken by these compounds, by activating complement components C3 and C5, but not C4. Later on, Suzuki and colleagues (2003) showed that C3 activation increased in a chain-length dependent manner in chitosan and non-water soluble COS. Binding of C3b to chitosan was observed. In 2000, the same group evaluated the influence of some physico-chemical properties on this activity, namely degree of deacetylation. They concluded that complement activation by chitosan-based mucopolysaccharides is more relevant for solid samples and compounds with heterogeneous acetylation, since they had a stable ability on complement activation when compared to those with homogeneous acetylation. Thus, chitin and chitosan may act directly to change C3 to C3i, or indirectly activating serum proteases.

Minagawa and co-workers (2007) additionally studied the effect of the MW of chitin/chitosan on wound healing using a linear incisional wound model in rats. Wound healing ability was higher for the chitosan group (D-glucosamine (GlcN), COS, than for the chitin-based compounds (N-acetyl-D-glucosamine (GlcNAc) chito-oligosaccharide (NACOS), and chitin). The same trend was reported for collagenase activity. The higher the deacetylation degree, the stronger the break strength becomes and more activated fibroblasts appear at the site. The use of a taurine-chitosan gel formulation in mice also yielded similar results (Değim et al., 2002), increasing wound tensile strength by decreasing malondialdehyde levels and increasing those of hydroxyproline.

In a study using chitosan dressed skin grafts, Stone and collaborators (2000) demonstrated that as a semi-permeable biological dressing, chitosan maintained a sterile wound exudate beneath a dry scab, preventing dehydration and contamination, thus improving conditions for healing. Furthermore, it facilitated wound re-epithelization and nerve regeneration within a vascular dermis, as well as an earlier return to normal skin color.

7. Anticoagulant Activity

The first step in the early wound healing process is hemostasis due to blood coagulation. Platelets are the most important component in the process. Furthermore, platelets release cytokines that enhance the healing process during blood coagulation. Chitin and chitosan enhance the release of the platelet derived growth factor-AB (PDGF-AB) and the transforming growth factor-β1 (TGF-β1) from the platelets, and this effect is more effective with chitosan than with chitin (Okamoto et al., 2003). On the other hand, chitin aggregates platelets more efficiently than chitosan, an effect which does not directly reflect blood coagulation.Chitosan tends to attract circulating plasma proteins, which adsorb to the material surface, due to its positive charge. This renders chitosan as highly thrombogenic in blood contacting applications, resulting in platelet adhesion and activation on the surface of the material lead to the thrombus formation and ultimately, device failure (Sagnella and Mai-Ngam, 2005).

The hemostatic activity of chitosan does not depend on any part of the blood coagulation cascade (Whang et al., 2005). Many studies indicate that the red blood cells readily activate coagulum formation induced by chitosan. Among these, the most significant observation is the '*in vitro*' study by Radhakrishnan et al. (1991), where they found a potent platelet aggregation induced by chitosan, which was proportional to the platelet concentration in the plasma. The mechanism for this phenomenon was later investigated by Chou et al. (2003), who demonstrated that chitosan is an effective inducer for rabbit platelet adhesion and aggregation. Red blood cells are thought to agglutinate as crosslinked, as they are bound together by chitosan polymer chains. Therefore, the hemostatic capacity of chitosan is dependent on MW, degree of deacetylation and other characteristics related to its polycationic properties interfering with the negatively charged molecules at the cell surface.

About the effect of chitosan MW on blood coagulation, Lee, back in 1974 already demonstrate that low molecular weight chitosans were unable to initiate the formation of firm coagulum. In 1994, Muzzarelli suggested that solid-state and high molecular weight chitosan exhibited hemostatic activity, whereas water-soluble depolymerized chitosan and its oligomers did not. The *in vivo* biological studies carried out by Hirano et al. (1985) showed that blood coagulum was formed on suture surfaces coated with fully deacetylated chitosan rather than on sutures impregnated with acetylated chitosan. In solid-state chitosan, macromolecular chains exist as a crystal state and can absorb platelets to induce coagulation, but it lacks the ability to aggregate erythrocytes. On the other hand, chitosan in acetic acid physiological saline solution is positively charged due to protonation of amidocyanogen and has the ability to induce erythrocyte aggregation as well as their deformation (Yang et al., 2008). Low deacetylated chitosan is more efficient in initiating hemostasis than middle or high deacetylated chitosan. Chitosan with 80% DD might have hemostatic activity through the modulation of fibrinolytic activity of peritoneal exudative macrophages (Fukasawa et al., 1992), while chitosan with a DD of 58% could stimulate the release of β-thromboglobulin (β-TG) and platelet factor 4 (Sugamori et al., 2000). Although the hemostatic mechanism of chitosan is still elusive up to date, it is a fact that chitosan acetic acid solution induces clotting (Malette, 1983) and erythrocyte deformation and this initiation process is independent of platelets or coagulation factors (Klokkevold et al., 1991, 1992).

Acetylated chitosan proved to be a strong activator of the alternative pathway of the complement and prevented deposition of anti-fibrinogen, which may explain its procoagulant activity (Benesch and Tengvall, 2002).

Chitin heparinoids, produced by sulfonation of marine crab shell chitosan, revealed strong anticoagulant activities, related to the inhibition of Factor Xa and thrombin activity in a similar mechanism as heparin (Vongchan et al., 2003). The presence of sulfated groups is an essential requirement for anticoagulant activity. In some cases, increasing the degree of sulfation was beneficial to the activity, however, other structural parameters such as, position of the sulfate groups, MW, configuration and conformation of the uronic acid and the backbone structure of the polysaccharide, all showed to interfere in the activity (Huang et al., 2003). This group also studied the influences of the acyl or quaternary groups on the anticoagulant activity of chitosan sulfates with respect to activated partial thromboplastin time (APTT), thrombin time (TT) and prothrombin time (PT). The propanoyl and hexanoyl groups increased APTT activity, and propanoyl groups also increased the TT anticoagulant activity slightly, while the *N,O*-quaternary chitosan sulfate showed only a slight TT coagulant activity.

Recently, a chitosan-based wound dressing with improved hemostatic properties has been developed by Ong et al. (2008). The refined chitosan dressing also incorporates polyphosphate (a procoagulant) and an antimicrobial (silver). This formulation accelerated blood clotting, increased platelet adhesion, generated thrombin faster and absorbed more blood than chitosan alone.

Water-soluble COS and highly deacetylated COS obtained by fractionation are much less procoagulant than their parental chitosan compound (Lin and Lin, 2003) and can therefore be utilized in biomedical applications in which blood coagulation is not desired.

8. Other Biological Activities

Hypertension is one of the major independent risk factors for arteriosclerosis, stroke, myocardial infarction and end stage renal disease. ACE is a dipeptidylcarboxypeptidase which catalyzes the formation of angiotensin II, a strong pressor, from angiotensin I and inactivates bradykinin, which has a depressor effect. Research work on chitosan and its oligomers has identified their potential to inhibit angiotensin converting enzyme (ACE), which plays an important role in blood pressure regulation in mammals. Park et al. (2003) reported that COS prepared from partially different deacetylated chitosans possess selective activities with the lowest acetylated variety showing the highest ACE inhibitory activity. Carboxylated COS, with a structural feature similar to captopril had the ACE inhibitory activity enhanced with increased substitution degree (Huang et al., 2005). This inhibition was found to be competitive via obligatory binding site of the enzyme. Other modifications of chitosan include the synthesis of aminoethyl chitooligosaccharide (AE-COS) which promoted ACE inhibitory effects (Ngo et al., 2008). Furthermore, administration of NaCl plus chitosan as a dietary salt to hypertensive rats successfully prevented hypertension (Park et al., 2009). This composition may be applied as a substitute table salt or as an ingredient in foods requiring salt addition, appropriate for patients recommended to reduce salt intake.

Other biological activities attributed to chitosan and its derivatives include its hypocholesteromic effects due to fat binding preventing their absorption through the gut (Kanauchi et al., 1995). Similarly, COS also reduced blood cholesterol levels and this is not dependent on their MWs (Kim and Rajapakse, 2005). However, the exact mechanism by which chitosan and COS lower blood cholesterol has not been completely elucidated so far. Some suggested mechanisms include ionic binding of COS with bile salts and bile acids, which inhibit micelle formation during lipid digestion (Remunan-Lopez et al., 1998). Another hypothesis is that chitosan and its oligomers can directly trap lipids and fatty acids (Tanaka et al., 1997). Sugano et al. (1988) have reported contraditory results by which fat binding and cholesterol lowering effect of chitosan cannot be explained by the ion binding hypothesis. Moreover, other evidences from animal studies suggest that the effective lowering of cholesterol level can be explained due to the ability of COS to increase the excretion of neutral sterol and indigestion of dietary fats. In addition, Sugano et al (1988) reported that almost all chitosan preparations could prevent increase in blood cholesterol at 5% dietary level.

9. CONCLUSION

Chitosan and COS have shown to possess antimicrobial activity against a wide range of target organisms, including Gram-positive and -negative bacteria, yeasts and molds, varying considerably with both MW and DD. Chitosan and COS have also been identified as potent radical scavengers, while chitin and its oligomers have been described as ineffective, since this activity has been proved to be mainly dependent on the degree of deacetylation and is strongly affected by the MW. A strong immune stimulatory activity is also attributed to chitinous derivatives; high MW chitosan upregulates production of IL-1, TNF-α, granulocyte macrophage colony stimulating factor (GM-CSF), nitric oxide (NO) and interleukin-6 (IL-6) in macrophages. Besides this, COS enhances TNF-α and IL-1β release and internalization in macrophages, while also improve the functionality of inflammatory cells such as PMN, macrophages and fibroblasts. The antitumor activity observed by chitosan and its derivatives is also dependent on structural characteristics such as molecular size and DD. Chitin oligomers, which are not cationic, also exhibited antitumor activities, due to their immunostimulatory properties. High MW chitosans have been reported as being ineffective as antitumor agents, confirming the idea that anti-tumoral activity increases with decreasing MW. Wound healing properties of chitosan are explained by the fact that it accelerates migration of PMN to the wound area, which secrete inflammatory mediators, promoting rapid dermal regeneration and accelerated wound healing, as does chitin. It is believed that these compounds use the alternative pathway of complement activation, which increases in a chain-length dependent manner in chitosan and non-water soluble COS. The hemostatic capacity of chitosan is dependent on MW, DD and other characteristics related to its polycationic properties. High MW chitosan exhibits hemostatic activity, whereas water-soluble depolymerized chitosan and its oligomers do not.

As can be seen from this chapter, many concerted biological activities can be attributed to chitin, chitosan and their respective derivatives. Although these activities are extensively reported in the scientific literature, data comparison and conclusion drawing is difficult, as unfortunately, most experiments do not convey complete information about the chemical characteristics of the chitin/chitosan used. In addition, the fact that these activities are affected by several factors at the same time further conceals the conclusions. The effort to complement these studies is much needed, so as to reach definite and reliable conclusions.

REFERENCES

Ai, H., Wang, F., Yang, Q., Zhu, F. and Lei, C. (2008). Preparation and biological activities of chitosan from the larvae of housefly, *Musca domestica. Carbohydrate Polymers, 72*, 419-423.

Andres, Y., Giraud, L., Gerente, C. and Cloirec, P. (2007). Antibacterial effects of chitosan powder: mechanisms of action. *Environmental Technology*, *28*, 1357-1363.

Andrew, C. A., Khor, E., and Hastings, G. W. (1998). The influence of anionic chitin derivatives on calcium phosphate crystallization. *Biomaterials*, *19*, 1309-1316.

Ausar, S. F., Passalacqua, N., Castagna, L. F., Bianco, I. D. and Beltramo, D. M. (2002). Growth of milk fermentative bacteria in the presence of chitosan for potential use in cheese making. *International Dairy Journal*, *12*, 899-906.

Baldrick, P. (2010). The safety of chitosan as a pharmaceutical excipient. *Regulatory Toxicology and Pharmacology*, *56*, 290–299.

Balicka-Ramisz, A., Wojtasz-Pajak, A., Pilarczyk, B., Ramisz, A. and Laurans, L. (2005). Antibacterial and antifungal activity of chitosan. *Proceedings of 12th International Congress on Animal Hygene, 2*, 406-408.

Bayne, C. J. and Gerwick, L. (2001). The acute phase response and innate immunity of fish. *Developmental and Comparative Immunology*, *25*, 725-743.

Benesch, J. and Tengvall, P. (2002). Blood protein adsorption onto chitosan. *Biomaterials*, *23*, 2561-2568.

Cauchie, H. M. (1997). An attempt to estimate crustacean chitin production in the hydrosphere. In A. Domard, G. A. F. Roberts, and K. M. Vårum (Eds.), *Advances in Chitin Science* (pp. 32-38). Lyon: Jacques Andre Publisher.

Chen, A. S., Taguchi, T., Sakai, K., KikuchiI, K., Wang, M. W. and Miwa, I. (2003). Antioxidant activities of chitobiose and chitotriose. *Biological and Pharmaceutical Bulletin, 26*, 1326-1330.

Chen, C., Liau, W. and Tsai, G. (1998). Antibacterial effects of N-sulfonated and N-sulfobenzoyl chitosan and application to oyster preservation. *Journal of Food Protection, 61*, 1124-1128.

Chen, Y. M., Chung, Y. C., Wang, L. W., Chen, K. T. and Li, S. Y. (2002). Antibacterial properties of chitosan in waterborne pathogen. *Journal of Environmental Science and Health, Part A*, *37*, 1379-1390.

Cho, Y. W., Cho, Y. N., Chung, S. H., Yoo, G. and Ko, S. W. (1999). Water-soluble chitin as a wound healing accelerator. *Biomaterials, 20,* 2139-2145.

Chopra. S., Mahdi, S., Kaur, J., Iqbal, Z., Talegaonkar, S. and Ahmad, F. J. (2006). Advances and potential applications of chitosan derivatives as mucoadhesive biomaterials in modern drug delivery. *Journal of Pharmacy and Pharmacology*, *58*, 1021-1032.

Chou, T., Fu, E., Wu, C. and Yeh, J. (2003). Chitosan enhances platelet adhesion and aggregation. *Biochemical and Biophysical Research Communications*, *302*, 480-483.

Chow, K. S. and Khor, E. (2001). New fluorinated chitin derivatives: synthesis, characterization and cytotoxicity assessment. *Carbohydrate Polymers*, *47*, 357-363.

Chung, Y. C., Su, Y. P., Chen, C. C., Jia, G., Wang, H. L., Wu, J. C. G. and Lin, J. G. (2004). Relationship between antibacterial activity of chitosan and surface characteristics of cell wall. *Acta Pharmacologica Sinica, 25*, 932-936.

Dautremepuits, C., Betoulle, S., Paris-Palacios, S. and Vernet, G. (2004). Immunology-related perturbations induced by copper and chitosan in carp (*Cyprinus carpio* L.). *Archives of Environmental Contamination and Toxicology*, *47*, 370-378.

Değim, Z., Çelebi, N., Sayan, H. Babül, A., Erdoğan, D. and Take, G. (2002). An investigation on skin wound healing in mice with a taurine-chitosan gel formulation. *Amino Acids, 22,* 187-198.

Devlieghere, F., Vermeulen, A. and Debevere, J. (2004). Chitosan: antimicrobial activity, interactions with food components and applicability as a coating on fruit and vegetables. *Food Microbiology*, *21*, 703-714.

Feng, J., Zhao, L. and Yu, Q. (2004). Receptor-mediated stimulatory effect of oligochitosan in macrophages. *Biochemical and Biophysical Research Communications*, *317*, 414-420.

Fernandes, J. C., Eaton, P., Gomes, A. M., Pintado, M. E. and Malcata, F. X. (2009). Study of the antibacterial effects of chitosans on *Bacillus cereus* (and its spores) by atomic force microscopy imaging and nanoindentation. *Ultramicroscopy, 109*, 854-860.

Fernandes, J. C., Eaton, P., Nascimento, H., Gião, M. S., Ramos, O. S., Belo, L., Santos-Silva, A., Pintado, M. E. and Malcata, F. X. (2010a). Antioxidant activity of chitooligosaccharides upon two biological systems: erythrocytes and bacteriophages, *Carbohydrate Polymers*, *79*, 1101-1106.

Fernandes, J. C., Tavaria, F. K., Soares, J. C., Ramos, O. S., Monteiro, J. M., Pintado, M. E. and Malcata, F. X. (2008). Antimicrobial effects of chitosans and chitooligosaccharides, upon *Staphylococcus aureus* and *Escherichia coli*, in food model systems. *Food Microbiology, 25*, 922-928.

Fernandes, J., Tavaria, F., Fonseca, S. C., Ramos, O. S., Pintado, M. E. and Malcata, F. (2010b). *In vitro* screening for anti-microbial activity of chitosans and chitooligosaccharides, aiming at potential uses in functional textiles. *Journal of Microbiology and Biotechnology, 20*, 311-318.

Fukasawa, M., Abe, H., Masaoka, T., Orita, H., Horikawa, H., Campeau, J. D., Washio, M. (1992). The hemostatic effect of deacylated chitin membrane on peritoneal injury in a rabbit model. *Surgery Today, 22*, 333–338.

Gao, J. Q., Zhao, Q. Q., Lv, T. F., Shuai, W. P., Zhou, J., Tang, G. P., Liang, W. Q., Tabata, Y. and Hu, Y. L. (2010). Gene-carried chitosan-linked-PEI induced high gene transfection efficiency with low toxicity and significant tumor-suppressive activity. *International Journal of Pharmaceutics*, *387*, 286-294.

Gerasimenko, D. V., Avdienko, I. D., Bannikova, G. E., Zueva, O. Y. and Varlamov, V. P. (2003). Antibacterial effects of water-soluble low molecular-weight chitosans on different microorganisms. Applied Biochemistry *and* Microbiology, *40*, 253-257.

Gil, G., Mónaco, S., Cerrutti, P. and Galvagno, M. (2004). Selective antimicrobial activity of chitosan on beer spoilage bacteria and brewing yeasts. *Biotechnology Letters, 26*, 569-574.

Goy, R. C., Britto, D. and Assis, O. B. G. (2009). A review of the antimicrobial activity of chitosan. *Polímeros, 19*, 241-247.

Guzmán-Moralesa, J., El-Gabalawyb, H., Phama, M. H., Tran-Khanha, N., McKeec, M. D., Wud, W., Centolae, M. and Hoemannaf, C. D. (2005). Effect of chitosan particles and dexamethasone on human bone marrow stromal cell osteogenesis and angiogenic factor secretion. *Bone*, *49*, 617-626.

Higuchi, M., Higashi, N., Taki, H. and Osawa, T. (1990). Cytolytic mechanisms of activated macrophages: tumor necrosis factor and L-arginine dependent mechanisms act synergistically as the major cytolytic mechanisms of activated macrophages. *Journal of Immunology*, *144*, 1245-1251.

Hirano, S. and Noishiki, Y. (1985). The blood compatibility of chitosan and N-acylchitosans. *Journal of Biomedical Materials Research*, *19*, 413-417.

Hossain, Z. and Takahashi, K. (2008). Induction of permeability and apoptosis in colon cancer cell line with chitosan. *Journal of Food and Drug Analysis, 16*, 1-8.

Huang, R., Du, Y., Yang, J. and Fan, L. (2003). Influence of functional groups on the in vitro anticoagulant activity of chitosan sulfate. *Carbohydrate Polymers, 338*, 483-489.

Huang, R., Mendis, E. and Kim, S.- K. (2005). Improvement of ACE inhibitory activity of chitooligosaccharides (COS) by carboxyl modification. *Bioorganic and Medicinal Chemistry, 13*, 3649-3655.

Huang, R., Rajapakse, N. and Kim, S.- K. (2006). Structural factors affecting radical scavenging activity of chitooligosaccharides (COS) and its derivatives. *Carbohydrate Polymers, 63*, 122-129.

Je, J. Y., Park, P. J. and Kim, S. K. (2004). Radical scavenging activity of hetero-chitooligosaccharides. *European Food Research and Technology, 219*, 60-65.

Jeon, Y. J. and Kim, S. K. (2002). Antitumor activity of chitosan oligosaccharides produced in an ultra filtration membrane reactor system. *Journal of Microbiology and Biotechnology, 12*, 503-507.

Jeon, Y. J., Park, P. J. and Kim, S.- K. (2001). Antimicrobial effect of chitooligosaccharides produced by bioreactor. *Carbohydrate Polymers, 44*, 71-76.

Jeon, Y. J., Shahidi, F. and Kim, S.- K. (2000). Preparation of chitin and chitosan oligomers and their applications in physiological functional foods. *Food Reviews International, 61*, 159-176.

Jeong, H.- J., Koo, H.- N., Oh, E.- Y., Chae, H.- J., Kim, H.- R. and Suh, S.- B. (2000). Nitric oxide production by high molecular weight water soluble chitosan via nuclear factor-κB activation. *International Journal of Immunopharmacology, 22*, 923-933.

Johnson, E. L. and Peniston, Q. P. (1982). Utilization of shellfish waste for chitin and chitosan production. In R. E. Martin, G. J. Flick, C. E. Hobard, and D. R. Ward (Eds.), *Chemistry and Biochemistry of Marine Food Products* (pp. 415-419). Westport, CT: A VI Publishers, Co.

Jun-ang, L., Aixian, J., Guoying, Z. and Yuanhao, H. (2009). Antifungal Activity of Chitosan against Colletotrichum gloeosporioides. *3rd International Conference on Bioinformatics and Biomedical Engineering*, DOI: 10.1109/ICBBE.2009.5162373.

Kanauchi, O., Deuchi, K., Imasato, Y., Shizukuishi, M. and Kobayashi, E. (1995). Mechanism for the inhibition of fat digestion by chitosan and for the synergistic effect of ascorbate. *Bioscience, Biotechnology, and Biochemistry, 59*, 786-790.

Khanal, D. R., Okamoto, Y., Miytake, K., Shinobu, T. Shigemasa, Y., Tokura, S. and Minami, S. (2002). Phosphated chitin (P-chitin) exerts protective effects by restoring the deformability of polymorphonuclear neutrophil (PMN) cells. *Carbohydrate Polymers, 48*, 305-311.

Kim, S.- K. and Kajapakse, N. (2005). Enzymatic production and biological activities of chitosan oligosaccharides (COS): A review. *Carbohydrate Polymers, 62*, 357-368.

Kimura, Y and Okuda, H. (1999). Prevention by chitosan *Japanese Journal of Cancer Research, 90*, 765-774.

Kimura, Y., Onoyama, M., Sera, T. and Okuda, H. (2000). Antitumour activity and side effects of combined treatment with chitosan and cisplatin in sarcoma 180-bearing mice. Journal *of Pharmacy and Pharmacology, 52*, 883-890.

Klokkevold, P. R., Lew, D. S., Ellis, D. G. and Bertolami, C. N. (1991). Effect of chitosan on lingual hemostasis in rabbits. *Journal of Oral and Maxillofacial Surgery, 49*, 858-863.

Klokkevold, P. R., Subar, P., Fukayama, H. and Bertolami, C. N. (1992). Effect of chitosan on lingual hemostasis in rabbits with platelet dysfunction induced by epoprostenol. *Journal of Oral and Maxillofacial Surgery, 50*, 41-45.

Ko, S., Takeshi, M., Yoshio, O., Akio, T., Shigeo, S. and Masuko, S. (1986). Antitumor effect of hexa-N-acetylchitohexaose and chitohexaose. *Carbohydrate Research, 151*, 403-408.

Kurita, K., Koyama, Y., Inoue, S. and Nishimura, S. (1990). ((Diethylamino) ethyl) chitins: preparation and properties of novel aminated chitin derivatives. *Macromolecules, 23*, 2865-2869.

Lam, N. D. and Diep, T. B. (2003). A preliminary study on radiation treatment of chitosan for enhancement of antifungal activity tested on fruit- spoiling strains. *Nuclear Science and Technology, 2*, 54-60.

Lee, V. F. P. (1974). Solution and shear properties of chitin and chitosan. Ph.D. Dissertation, University of Washington.

Li, W., Jiang, X., Xue, P. and Chen, S. (2002). Inhibitory effects of chitosan on superoxide anion radicals and lipid free radicals. *Chinese Science Bulletin, 47*, 887-889.

Lin, C.- W. and Lin, J.- C. (2003). Characterization and blood coagulation evaluation of the water-soluble chitooligosaccharide prepared by a facile fractionation method. *Biomacromolecules, 4*, 1691-1697.

Liu, X. F., Guan, Y. L., Yang, D. Z., Li, Z. and Yao, K. (2001). Antibacterial action of chitosan and carboxymethylated chitosan. *Journal of Applied Polymer Science, 79*, 1324-1335.

Maeda, Y. and Kimura, Y. (2004). Antitumor effects of various low molecular-weight chitosans are due to increased natural killer activity of intestinal intraepithelial lymphocytes in sarcoma 180–bearing mice. *Nutrition and Cancer, 134*, 945-950.

Malette, W. G. and Quigley, H. J. (1983). Method of achieving hemostasis. U. S. Patent 4394373.

Marques, A., Encarnação, M., Pedro, S. and Nunes, M. L. (2008). *In vitro* antimicrobial activity of garlic, oregano and chitosan against *Salmonella enterica*. World Journal of Microbiology *and* Biotechnology, *24*, 2357-2360.

Minagawa, T., Okamura, Y., Shigemasa, Y., Minami, S. and Okamoto, Y. (2007). Effects of molecular weight and deacetylation degree of chitin/chitosan on wound healing. *Carbohydrate Polymers, 67*, 640-644.

Minami, S., Masuda, M., Suzuki, H., Okamoto, Y., Matsuhashi, A., Kato, K. and Shigemasa, Y. (1997). Subcutaneous injected chitosan induces systemic activation in dogs. *Carbohydrate Polymers, 33*, 285-294.

Minami, S., Oh-oka, M., Okamoto, K., Miytake, A., Matsuhashi, Y., Shigemasa, Y. and Fukumoto, Y. (1996). Chitosan-inducing hemorrhagic pneumonia in dogs. *Carbohydrate Polymers, 32*, 115-122.

Minami, S., Suzuki, H., Okamoto, Y., Fujinaga, T. and Shigemasa, Y. (1998). Chitin and chitosan activate complement via the alternative pathway. *Carbohydrate Polymers, 36*, 151-155.

Miyatake, K., Okamoto, Y., Shigemasa, Y., Tokura, S. and Minami, S. (2003). Anti-inflammatory effect of chemically modified chitin. *Carbohydrate Polymers, 53*, 417-423.

Morimoto, M., Saimoto, H., Usui, H., Okamoto, Y., Minami, S. and Shigemasa, Y. (2001). Biological activities of carbohydrate-branched chitosan derivatives. *Biomacromolecules*, *2*, 1133-1136.

Muzzarelli, R. (1994). Significance of chitin-based medical items. In: S. Dumitriu (ed.), *Polymeric Biomaterials* (pp. 179-197). New York, NY: Marcel Dekker.

Nam, K.- S. and Shon, Y.- H. (2009). Suppression of metastasis of human breast cancer cells by chitosan oligosaccharides. *Journal of Microbiology and Biotechnology*, *19*, 629-633.

Ngo, D.- N., Qian, Z.- J., Je, J.- Y., Kim, M.- M. and Kim, S.- K. (2008). Aminoethyl chitooligosaccharides inhibit the activity of angiotensin converting enzyme. *Process Biochemistry*, *43*, 119-123.

Nishimura, S., Ishihara, C., Ukei, S., Tokura, S. and Azuma, I. (1986). Stimulation of cytokine production in mice using deacetylated chitin. *Vaccine*, *4*, 151-156.

Nishimura, S., Nishi, N., Tokura, S., Nishimura, K. and Azuma, I. 1986. Bioactive chitin derivatives. Activation of mouse-peritoneal macrophages by O-(carboxymethyl) chitins. *Carbohydrate Research 146*, 251-258.

No, H. K., Lee, S. H., Park, N. Y. and Meyers, S. P. (2003). Comparison of physicochemical, binding, and antibacterial properties of chitosans prepared without and with deproteinization process. *Journal of Agricultural Food Chemistry, 51*, 7659-7663.

No, H. K., Park, N. Y., Lee, S. H. and Meyers, S. P. (2002). Antibacterial activity of chitosans and chitosan oligomers with different molecular weights. *International Journal of Food Microbiology*, *74*, 65-72.

Okamoto, Y., Yano, R., Miyatake, K., Tomohiro, I., Shigemasa, Y. and Minami, S. (2003). Effects of chitin and chitosan on blood coagulation. *Carbohydrate Polymers, 53*, 337-342.

Ong, S.- Y., Wu, J., Moochhala, S. M., Tan, M.- H. and Lu, J. (2008). Development of a chitosan-based wound dressing with improved hemostatic and antimicrobial properties. *Biomaterials*, *29*, 4323-4332.

Park, P.- J., Je, J.- Y. and Kim, S.- K. (2003). Angiotensin I converting enzyme (ACE) inhibitory activity of hetero-chitooligosaccharides prepared from partially different deacetylated chitosans. *Journal of Agricultural and Food Chemistry*, *51*, 4930-4934.

Park, P.- J., Lee, H.K. and Kim, S.K. (2004). Preparation of hetero-chitooligosaccharides and their antimicrobial activity on *Vibrio parahaemolyticus. Journal of Microbiology and Biotechnology, 14*, 41-47.

Park, S.- H., Dutta, N. K., Baek, M.- W., Kim, D.- J., Na, Y.- R., Seok, S.- H., Lee, B.- H., Cho, J.- E., Cho, G.- S. and Park, J.- H. (2009). NaCl plus chitosan as a dietary salt to prevent the development of hypertension in spontaneously hypertensive rats. Journal *of Veterinary Science*, *10*, 141-146.

Porporatto, C., Bianco, I. D. and Correa, S. G. (2005). Local and systemic activity of the polysaccharide chitosan at lymphoid tissues after oral administration. *Journal of Leukocyte Biology*, *78*, 62-69.

Porporatto, C., Bianco, I. D., Cabanillas, A. M. and Correa, S. G. (2004). Early events associated to the oral co-administration of type II collagen and chitosan: induction of anti-inflammatory cytokines. *International Immunology*, *16*, 433-441.

Porporatto, C., Bianco, I. D., Riera, C. M. and Correa, S. G. (2003). Chitosan induces different L-arginine metabolic pathways in resting and inflammatory macrophages. *Biochemical and Biophysical Research Communications*, *304*, 266-272.

Porporatto, C., Canali, M. M. Bianco, I. D. and Correa, S. G. (2008). The biocompatible polysaccharide chitosan enhances the oral tolerance to type II collagen. *Clinical and Experimental Immunology*, *155*, 79-87.

Prashanth, K. V. H. and Tharanathan, R. N. (2005). Depolymerized products of chitosan as potent inhibitors of tumor-induced angiogenesis. *Biochimica et Biophysica Acta (BBA) - General Subjects, 1722*, 22-29.

Prashanth, K. V. H., Dharmesh, S. M., Rao, K. S. J., Tharanathan, R. N. (2007). Free radical-induced chitosan depolymerized products protect calf thymus DNA from oxidative damage. *Carbohydrate Research, 342*, 190-195.

Qi, L. and Xu, Z. (2006). *In vivo* antitumor activity of chitosan nanoparticles. *Bioorganic and Medicinal Chemistry Letters*, *16*, 4243-4245.

Qi, L. F., Xu, Z. R., Li, Y., Jiang, X. and Han, X. Y. (2005). *In vitro* effects of chitosan nanoparticles on proliferation of human gastric carcinoma cell line MGC803 cells. *World Journal of Gastroenterology*, *11*, 5136-5141.

Qin, C., Du, Y., Xiao, L., Li, Z. and Gao, X. (2002). Enzymic preparation of water-soluble chitosan and their antitumor activity. *International Journal Biological Macromolecules*, *31*, 111-117.

Radhakrishnan, V. V., Vijayan, M. S., Sambasivan, M., Jamaluddin, M. and Rao, S. B. (1991). Potential hemostasis application of chitosan in neurosurgical procedures: an experimental study. *Biomedicine: Journal of the Indian Association of Biomedical Research*, *2*, 3-6.

Remunan-Lopez, C., Portero, A., Vila-Jato, J. L. and Alonso, M. J. (1998). Design and evaluation of chitosan/ethylcellulose mucoadhesive bilayered devices for buccal drug delivery. *Journal of Controlled Release*, *55*, 143-152.

Rinaudo, M. (2006). Chitin and chitosan: Properties and applications. *Progress in Polymer Science*, *31*, 603-632.

Sagnella, S. and Mai-Ngam, K. (2005). Chitosan based surfactant polymers designed to improve blood compatibility on biomaterials. *Colloids and Surfaces B: Biointerfaces*, *42*, 147-155.

Sashiwa, H. and Aiba, S.- I. (2004). Chemically modified chitin and chitosan as biomaterials. *Progress in Polymer Science*, *29*, 887-908.

Seo, S. B., Ryu, C. S., Ahn, G. W., Kim, H. B., Jo, B. K., Kim, S. H., Lee, J. D. and Kajiuchi, T. (2002). Development of a natural preservative system using the mixture of chitosan-*Inula helenium L.* extract. International Journal of Cosmetic Science, *24*, 195-206.

Shahidi, F., Vidana Arachchi, J. K. and Jeon, Y. J. (1999). Food applications of chitin and chitosans. *Trends in Food Science and Technology*, *10*, 37-51.

Shen, K. T., Chen, M. H., Chan, H. Y., Jeng, J. H. and Wang, Y. J. (2009). Inhibitory effects of chitooligosaccharides on tumor growth and metastasis. *Food and Chemical Toxicology, 47*, 1864-1871.

Son, Y. J., Jang, J. S., Cho, Y. W., Chung, H., Park, R. W., Kwon, I. C., Kim, I. S., Park, J. Y., Seo, S. B., Park, C. R. and Jeong, S. Y. (2003). Biodistribution and anti-tumor efficacy of doxorubicin loaded glycol-chitosan nanoaggregates by EPR effect. *Journal of Controlled Release*, *91*, 135-145.

Stone, C. A., Wright, T. C., Powell, R. and Devaraj, V. S. (2000). Healing at skin graft donor sites dressed with chitosan. *British Journal of Plastic Surgery*, *53*, 601-606.

Strobel, S. and Mowat, A. M. (1998). Immune response to dietary antigens: oral tolerance. *Immunology Today*, *19*, 173-181.

Sudharshan, N. R., Hoover, D. G. and Knorr, D. (1992). Antibacterial action of chitosan. *Food Biotechnology, 6*, 257-272.

Sugamori, T., Iwase, H., Maeda, M., Inoue, Y. and Kurosawa, H. (2000). Local hemostatic effects of microcrystalline partially deacetylated chitin hydrochloride. *Journal of Biomedical Materials Research*, *49*, 225-232.

Sugano, M., Watanabe, S., Kishi, A., Izume, M. and Ohtakara, A. (1988). Hypocholesterolemic action of chitosans with different viscosity in rats. *Lipids*, *23*, 187-191.

Suzuki, K., Mikami, T., Okawa, Y., Tokoro, A., Suzuki, S. and Suzuki, M. (1986). Antitumor effect of hexa-N-acetylchitohexaose and chitohexaose. *Carbohydrate Research, 151*, 403-408.

Suzuki, Y., Miyatake, K., Okamoto, Y., Muraki, E. and Minami, S. (2003). Influence of the chain length of chitosan on complement activation. *Carbohydrate Polymers*, *54*, 465-469.

Suzuki, Y., Okamoto, Y., Morimoto, M., Sashiwa, H., Saimoto, H., Tanioka, S.- I., Shigemasa, Y. and Minami, S. (2000). Influence of physico-chemical properties of chitin and chitosan on complement activation. *Carbohydrate Polymers*, *42*, 307-310.

Synowiecki, J. and Al-Khateeb, N. A. (2000). The recovery of protein hydrolysate during enzymatic isolation of chitin from shrimp *Crangon crangon* processing discards. *Food Chemistry*, *68*, 147-152.

Synowiecki, J. and Al-Khateeb, N. A. (2003). Production, properties, and some new applications of chitin and its derivatives. *Critical Reviews of Food Science and Nutrition*, *43*, 145-171.

Ta, H. T., Dass, C. R., Larson, I., Choong, P. F. M. and Dunstan, D. E. (2009). A chitosan hydrogel delivery system for osteosarcoma gene therapy with pigment epithelium-derived factor combined with chemotherapy. *Biomaterials*, *30*, 4815-4823.

Tanaka, Y., Tanioka, S.- I., Tanaka, M., Tanigawa, T., Kitamura, Y. and Minami, S., Okamoto, Y., Miyashita, M. and Nanno, M. (1997). Effects of chitin and chitosan particles on BALB/c mice by oral and parenteral administration. *Biomaterials*, *18*, 591-595.

Tharanathan, R. N. and Kittur, F. S. (2003). Chitin — The undisrupted biomolecule of great potential. *Critical Reviews in Food Science and Nutrition*, *43*, 61-87.

Tokoro, A., Tatewaki, N., Suzuki, K., Mikami, T., Suzuki, S. and Suzuki, M. (1998). Growth inhibitory effect of hexa-*N*-acetylchitohexaose and chitohexaos and Meth-A solid tumor. *Chemical and Pharmaceutical Bulletin*, *36*, 784-790.

Tokura, S., Saiki, I., Murata, J., Makabe, T., Tsuta, Y., and Azuma, I. (1992). Inhibition of tumor-induced angiogenesis by sulphated chitin derivatives. In C. J. Brine, P. A. Sandford, and J. P. Zikakis (Eds.), *Advances in chitin and chitosan* (pp. 87-95). London and New York: Elsevier.

Tomida, H., Fujii, T., Furutani, N., Michihara, A., Yasufuku, T., Akasaki, K., Maruyama, T., Otagiri, M., Gebicki, J. M. and Anraku, M. (2009). Antioxidant properties of some different molecular weight chitosans. *Carbohydrate Research*, *344*, 1690-1696.

Tsai, G. J. and Su, W. H. (1999). Antibacterial activity of shrimp chitosan against *Escherichia coli*. *Journal of Food Protection, 62*, 239-43.

Tsai, G. J., Su, W. H., Chen, H. C. and Pan, C. L. (2002). Antimicrobial activity of shrimp chitin and chitosan from different treatments and applications of fish preservation. *Fisheries Science, 68*, 170-177.

Tsukada, K., Matsumoto, T., Aizawa, K., Tokoro, A., Naruse, R., Suzuki, S. and Suzuki, M. (1990). Antimetastatic and growth-inhibitory effects of N-acetylchitohexaose in mice bearing Lewis lung carcinoma. *Japanese Journal of Cancer Research, 81*, 259-65.

Ueno, H., Mori, T. and Fijinaga, T. (2001c). Topical formulations and wound healing applications of chitosan. *Advanced Drug Delivery Reviews*, *52*, 105-115.

Ueno, H., Murakami, M., Okumura, M., Kadosawa, T., Uede, T. and Fijinaga, T. (2001b). Chitosan accelerates the production of osteopontin from polymorphonuclear leukocytes. *Biomaterials*, *22*, 1667-1673.

Ueno, H., Nakamura, F., Murakami, M., Okumura, M., Kadosawa, T. and Fijinaga, T. (2001a). Evaluation effects of chitosan for the extra-cellular matrix production by fibroblasts and the growth factors production by macrophages. *Biomaterials*, *22*, 2125-2130.

Ueno, H., Yamada, H., Tanaka, I., Kaba, N., Matsuura, M., Okumura, M., Kadosawa, T. and Fuginaga, T. (1999). Accelerating effects of chitosan for healing at early phase of experimental open wound in dogs. *Biomaterials*, *20*, 1407-1414.

Vincendon, M. (1992). Chitin carbamates. In C. J. Brine, P. A. Sandford, and J. P. Zikakis (Eds.), *Advances in chitin and chitosan* (pp. 556-564). London and New York: Elsevier.

Vongchan, P., Sajomsang, W., Kasinrerk, W., Subyen, D. and Kongtaweiert, P. (2003). Anticoagulant activities of the chitosan polysulfate synthesized from marine crab shell by semi-heterogenous conditions. *ScienceAsia*, *29*, 115-120.

Wang, J., Chen, Y., Ding, Y., Shi, G. and Wan, C. (2008). Research of the degradation products of chitosan's angiogenic function. *Applied Surface Science, 255*, 260-262.

Wang, S. L. (2008). Reclamation of chitinous materials by bromelain for the preparation of antitumor and antifungal materials. *Bioresource Technology, 99*, 4386-4393.

Wang, X., Du, Y., Fan, L., Liu, H. and Hu, Y. (2005). Chitosan- metal complexes as antimicrobial agent: synthesis, characterization and structure-activity study. *Polymer Bulletin, 55*,105-113.

Whang, H. S., Kirsch, W., Zhu, Y. H., Yang, C. Z. and Hudson, S. M. (2005). Hemostatic agents derived from chitin and chitosan. *Polymer Reviews*, *45*, 309-323.

Xing, R., Liu, S., Guo, Z., Yu, H., Wang, P., Li, C., Li, Z. and Li, P. (2005). Relevance of molecular weight of chitosan and its derivatives and their antioxidant activities *in vitro*. *Bioorganic and Medicinal Chemistry*, *13*, 1573-1577.

Xiong, S. L., Li, A. L., Jin, Z. Y. and Chen, M. (2007). Effects of oral chondroitin sulfate on lipid and antioxidant metabolisms in rats fed a high-fat diet. *Journal of Food Biochemistry, 31*, 356-369.

Xu, Q., Ma, P., Yu, W., Tan, C., Liu, H., Xiong, C., Qiao, Y. and Du, Y. (2009). Chitooligosaccharides protect human embryonic hepatocytes against oxidative stress induced by hydrogen peroxide. *Marine Biotechnology*, DOI: 10.1007/s10126-009-9222-1.

Yang, J., Tian, F., Wang, Z., Wang, Q., Zeng, Y.- J. and Chen, S.- Q. (2008). Effect of chitosan molecular weight and deacetylation degree on hemostasis. *Journal of Biomedical Materials Research Part B: Applied Biomaterials*, *84*, 131-137.

Yoon, H. J., Moon, M. E., Park, H. S., Im, S. Y. and Kim, Y. H. (2007). Chitosan oligosaccharide (COS) inhibites LPS-induced inflammatory effects in RAW 264.7 macrophage cells. *Biochemical and Biophysical Research Communications*, *358*, 954-959.

Yoshino, H., Ishii, S., Nishimura, S., and Kurita, K. (1992). Preparation and characterization of mercapto-chitin derivatives. In C. J. Brine, P. A. Sandford, and J. P. Zikakis (Eds.), *Advances in chitin and chitosan* (pp. 565-570). London and New York: Elsevier.

Yu, Z., Zhao, L. and Ke, H. (2004). Potential role of nuclear factor-κappaB in the induction of nitric oxide and tumor necrosis factor-alpha by oligochitosan in macrophages. *International Immunopharmacology*, *4*, 193-200.

Zheng, L. Y. and Zhu, J. F. (2003). Study on antimicrobial activity of chitosan with different molecular weights. *Carbohydrate Polymers*, *54*, 527-530.

Zheng, Y., Yi, Y., Qi, Y., Wang, Y., Zhang, W. and Du, M. (2006). Preparation of chitosan-copper complexes and their antitumor activity. *Bioorganic and Medicinal Chemistry Letters*, *16*, 4127-4129.

In: Focus on Chitosan Research
Editors: Arthur N. Ferguson and Amy G. O'Neill
ISBN 978-1-61324-454-8

Chapter 7

THE POTENTIAL OF CHITOSAN IN DRUG DELIVERY SYSTEMS

Sara Baptista da Silva[1], João Fernandes[2], Freni Tavira[2], Manuela Pintado[2] and Bruno Sarmento[1,*]

[1.] Department of Pharmaceutical Technology,
Faculty of Pharmacy, University of Porto
[2.] Center of Biotechnology and Fine Chemistry,
Biotechnology School of Portuguese Catholic University

ABSTRACT

Chitosan has been exploited in promising drug delivery systems to release drugs in the target place, at the appropriate time and dosage, improving their bioavailability and, consequently, the therapeutic efficiency. On top of that, also allows the protection, control and release of biotech drugs such as peptides, genes, vaccines, antigens as well as synthetic drugs, which are highly limited to cross biological barriers and reach the target site, without collateral damage. Besides being easily manipulated, chitosan has important features, which make it unique. Its hydrosolubility and positive charge enable negative interactions with charged polymers, macromolecules, and polyanions when in contact with aqueous environment. Its exceptional biological properties such as mucoadhesiveness, permeation-enhancing effect across biological surfaces, biocompatibility and non-toxicity/non-antigenic make it useful for transmucosal drug delivery, improving absorption of the paracellular route. However, a better and future application of chitosan as therapeutic agent clearly depends on the design of appropriate carriers. As revised in this chapter, chitosan has been used in preparing films, beads, intragastric floating tablets, microparticles and nanoparticles for applications in the pharmaceutical field, surgeries and bone restructuring. Controlled drug delivery technology represents one of the borders of science, which involves a multidisciplinary scientific approach, contributing to human health care. Chitosan, with all its chemical and biological multipotential emerged as a promising solution for different conditions and medical applications, overcoming common problems of other drug carriers. As also

* Corresponding author. Tel.: +351 222 078949; fax: +351 222 003977. E-mail address: bruno.sarmento@ff.up.pt.

described in this chapter, several clinical trials involving chitosan-based delivery systems are on the roll to explore in clinical the advantages of this exceptional material.

Keywords: *Chitosan; controlled drug delivery; gene delivery; solutions; films; tablets, hydrogels; microparticles; nanoparticles; medical applications;*

1. Introduction

Pharmaceutical sciences have being moved forward to develop new therapeutics based on both biomolecules and synthetic compounds. Nevertheless, most of these new active compounds are unstable and must be protected from degradation in the physiological environment, besides the lower partition coefficients and poor absorption that constrain the transport across biological barriers [1]. Thus, the efficacy of most drugs clearly depends on the design of appropriate carriers for their delivery, protection, control and release [2]. Among the different approaches explored so far, colloidal carriers are particularly interesting, especially those made of mucoadhesive polymers to assure their epithelium permanence [3]. For this application, chitosan has come to be of particularly interesting for the association and delivery of labile macromolecular compounds [4]. Chitosan carriers have an exceptional potential for drug delivery, especially for mucosal, since these systems are stable in contact with physiological fluids and barriers, control drug release and protect against adverse conditions like mucosal enzymes and biological protective fluids. Due to mucoadhesion, combination of the particle size, particle surface chemistries, charge and the unique absorption enhancing properties, the chitosan potential in the medical field is widely promising. This innovative drug carrier is expected to develop and improve stability, bioavailability and therapeutic efficacy of several conditions, without compromising the safety performance of the drug and target conditions. The accomplishment combination of natural or synthetic compounds made of chitosan the key to success in the local or systemic administration of therapeutic agents. The different formulations such as films, tablets, hydrogels, micro and nanosytems are expected to optimize, characterize and select the drug performance, improved properties, greater stability and great potential for specific applications. Pharmaceutically, the systems are expected to increase the capacity, to maintain the drug activity during the preparation process, to optimize the release of the compound from the carrier system, and to ensure a good control of the physicochemical properties and stability of the compounds. Pharmacokinetics and toxicological relevance of chitosan systems are guaranteed in *in vitro* model systems in molecular, subcellular and cellular levels, as well as the therapeutic efficacy and safety *in vivo* should be proven.

2. Chitosan Proprieties and Biomedical Potential

Chitosan is an alkaline-deacetylated chitin biodegradable polysaccharide commonly derived from the exoskeletons shells of crustaceans and insects [5]. Its extraction is a cost-effective way to valorise seafood wastes and it is produced with different deacetylation degrees (DD) [6-7], and molecular weights (MW) [8]. Chitin is synthesized by an enormous number of living organisms and considering the amount of chitin produced annually in the world, it is the most abundant polymer after cellulose [6]. Chitin is obtained on an industrial scale from shrimp and crustaceans in general, however, the pupae of silkworms are also an alternative source of chitin and, consequently, of chitosan [9]. Another sources include the production of chitosan from fungi by fermentation [10], from chrysalides (a by-product from the silk industry) [9], or from sources of chitin (e.g. β-chitin) obtained from squid pens [4]. However, the most commonly obtained form of chitosan is the α-chitosan from crustacean, which represents approximately 70% of the organic compounds in such shells [4].

For the chitin isolation from the raw materials and subsequent chitosan production, the original sources in solid form are washed with water, desiccated at room temperature and cut into small pieces. Demineralization is carried out at room temperature using hydrochloric acid [9]. The de-proteinization should also be performed.

There are small difference in the chemical structure of chitin and chitosan, but these differences are extremely important when drug delivery is thought [11]. Considering the similar structures of chitin and chitosan with cellulose, both are made by linear β-(1-4)-linked monosaccharides. However, the functional groups connected to the second carbon in the repeating units differ from one to another [12]. Chitin is a linear homopolymer composed of β -(1,4)-linked *N*-acetyl-glucosamine units [11], while chitosan is a linear co-polymer polysaccharide consisting of β (1–4)-linked 2-amino-2-deoxy-D-glucose (D-glucosamine) and 2-acetamido-2-deoxy-D-glucose (N-acetyl-D-glucosamine) units, as depicted in figure 1 [4]. Chitin is insoluble in water and most common organic solvents used in pharmaceutical technology and consequently not useful in the development of drug delivery systems [13]. In contrast, chitosan is a rather insoluble in water and organic solvents, but soluble in dilute aqueous acidic solution (pH < 6.5), which can convert the glucosamine units into a soluble protonated amine form [11]. It is readily soluble in dilute organic acids such as acetic, citric and malic acid as well as hydrochloric acid [14]. This positive charge of chitosan is very useful since enables negative interaction with polyanions [13]. Chitosan is usually characterize in terms of MW, which commonly ranges from about 10 to 1000 kDa, and DD, in the range of 50% to 95% [15]. The biodegradability and biological properties of chitosan is frequently dependent on the relative proportions of N-acetyl-D-glucosamine and D-glucosamine residues [4] as well as on the MW.

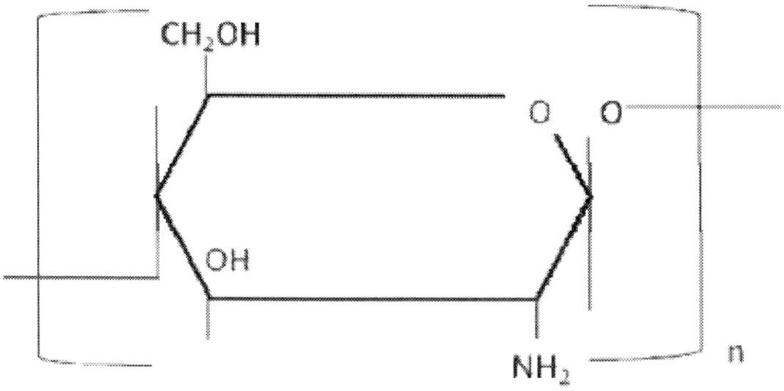

Figure 1. Structure and molecular formula of chitosan.

Table 1. Principal properties of chitosan in relation to its use in biomedical applications

Potential Biomedical applications	Principal characteristics
Surgical sutures	Biocompatible
Dental implants	Biodegradable
Artificial skin	Nontoxic, biological tolerance
Bone rebuilding	Non-antigenic
Corneal contact lenses	Mucoadhesiveness
Time-control release drugs for animals and humans	Renewable
Encapsulating material	Film forming
Tissue engineering	Hydrating agent
Cell culture	Hydrolyzed by lyzosyme
Nerve regeneration	Efficient against bacteria, viruses, fungi
	Anti-tumor
	Anti-Hypercholesterolemia

Chitosan has biological properties, unique and exceptional, that promote its use as a drug carrier [2], particularly to enhance transiently the permeability of mucosal barriers [13], increasing the effect on cell permeability [16], biocompatibility and biodegradability [17], being simultaneously non-toxic, non-antigenic [18] and mucoadhesive [3, 17]. Previous reports also indicated that chitosan possess various biological activities, such as antitumor effects [19], anti-hypercholesterolemia [20] and antimicrobial activity against several pathogen and spoilage bacteria that has already been widely demonstrated [21]. Currently, chitosan is widely used as a supporting material for tissue engineering applications, cell culture and nerve regeneration [22].

Table 1, summarizes the main potential biomedical properties of chitosan and associated characteristics. The better understanding of this unique cationic polymer invites new and several applications.

3. Chitosan Based Delivery Systems

Controlled-release technology emerged during the 1980s with a remarkable and increasing importance. This pharmaceutical technology that allows the predictable and reproducible release of a drug into a specific environment over an extended period of time, creating an optimal response, with minimum side-effects and prolonged efficacy, is a borderline of science [23].

The cationic nature of chitosan has been conveniently exploited for the development of new drug delivery systems and actually, a variety of chitosan-based delivery vehicles have been described for pharmaceutical field as well as for other medical applications [2]. In addition, the further chemical modification of chitosan is a powerful tool to control the interaction of the polymer with drugs, to enhance the load capability and to tailor the release profile of the particles [2]. Hence, chemically modified chitosan improves its bulk properties for the preparation of sustained drug release systems [2], which enhance its versatility in the biomedical and biotechnological fields [24].

Chitosan can be employed to formulate a variety of pharmaceutical dosage forms, namely solutions, hydrogels, films, tablets, microparticles, and nanoparticles. It is also a protagonist in other advanced fields, like non-viral vector for DNA and gene delivery [24-25]. Drug delivery applications include oral, nasal, parenteral and transdermal administration, implants and gene delivery. The transmucosal administration of drugs is being largely exploited by chitosan-based systems [7].

3.1. Chitosan Solutions

The study of chitosan solutions and related properties has much merit considering the multi-potential of the polymer and its many applications in the biomedical field. The preparation of the aqueous solution is also a mandatory step before obtaining any type of secondary materials such as films, gels, sponges, fibers, or particles. Despite this, most of the properties such as flocculation, adsorption and biological activities are accomplished in aqueous solution; the interactions between molecules of chitosan and metal ions, proteins, cells and bacteria have been widely studied in solutions [26]. The gross conformation of chitosan in solution may be spherical shape, random coil, and rod shape, which is manipulated by two sets of parameters: structure parameters, such as MW and DD and solution parameters, such as ionic strength, solvent, temperature and pH. Molecular weight can induce a conformation transitions — the conformations of small MW chitosans are stiffer and more extended than those of higher MW. In general, an increase of ionic strength makes the molecule contract [26]. These parameters are obtained from the plots of log intrinsic viscosity, sedimentation coefficient, diffusion coefficient, radius of gyration and log MW of the polymer [26]. The most widely used methods to characterize the conformation transition are capillary viscosity, analytical ultracentrifugation, static and dynamic light scattering [26]. Commonly chitosan solutions are easily prepared dissolving chitosan in organic acid like acetic or lactic acid.

Chitosan solutions have demonstrated most of their applications in the mucosal absorption, mainly due to the transiently ability of chitosan molecules to open tight junction of epithelial cells. Globally, it can be concluded that chitosan is able to enhance the paracellular route of absorption by tight junction disruption, inducing the translocation of tight junction proteins from the membrane to the cytoskeleton [27]. Regarding the effect of chitosans salts on drug intestinal permeability *in vitro*, there is general agreement that these polymers are potent absorption enhancers for poorly absorbed drugs such as atenolol and peptides [27]. Studies with buserelin, 9-deglycinamide, 8-arginine vasopressin and insulin have evidenced a strong increase in the transport of these drugs in the presence of chitosan glutamate and chitosan hydrochloride (acidic environment) [27]. Studies using chitosan glutamate and western blotting of Caco-2 cells fractions, observed translocation of tight junction proteins (ZO-1 protein) from the membrane to the cytoskeleton in response to treatment with chitosan [28].

Yu and his collaborators developed a chitosan solution for nasal insulin delivery [29]. They reported that chitosan concentration, osmolarity, medium and absorption enhancing ability of chitosan solution have significant effect on nasal insulin absorption. This may represent a new administration of insulin, safer, convenient for diabetic patients and with good applicability.

Recent research has also exploited the ability of chitosan solutions to control the spread of cytokine, and improve the immunoadjuvant properties in vaccine applications [30]. Sustained, local delivery of immunomodulatory cytokines in chitosan solution is under investigation for its ability to enhance vaccine and anti-tumor responses, both pre-clinically and clinically. In short, the study concluded that the chitosan solution maintained a deposit of bioactive antigen and cytokines, induced a cell expansion of the lymph nodes, including an increase in dendritic cells and antigen presenting cells. The resulting increase in the capacity of antigen presentation was exploited to improve both the humoral and cellular responses to the vaccine. The ability of chitosan to form deposits of antigens and cytokines, in addition to its safety profile and inherent versatility makes it a promising platform for the delivery of vaccines and cytokines [30].

3.1. Films

Use of porous biomaterials in the film development attracts scientists in particular concerning the biomedicine applications [22]. In the field of tissue engineering, porous polymer scaffolds can provide a framework for the seeded cells until they are well organized into a functioning tissue, especially in bone regenerative therapy [22]. In the field of surgical, the porous biomaterial is usually used as wound dressing to absorb wound fluid and promote healing [22]. On the other side, films, erodible and non-erodible inserts, rods and shields are the most logical delivery systems aimed at remaining for a long resident time in contact with physiologic surfaces like skin and the mucosal [5].

Biofilms and coatings, by themselves or acting as drug carriers have been particularly considered worldwide [8]. Porous membrane materials based on chitosan, adds advantages of biocompatibility, barrier properties, non-toxicity, non-polluting and low cost [8, 22]. Besides biological activities, chitosan has attracted notable interest due to its good mechanical and oxygen barrier properties, becoming a promising edible film component [19] applicable to pharmaceutical and food industries. The functional properties of chitosan-based films can be improved by combining them with other hydrocolloids [31]. The cationic properties of chitosan offer good opportunities to take advantages of electron interactions with numerous compounds (mainly macromolecules) and incorporate specific properties into the material [19]. Examples are chitosan blends with hyaluronic acid, alginate, rice, collagen and oleic acid [12]. Several aspects of the chitosan film-forming properties have been described, including the effect of solvent, concentration, storage effects, and the influence of important structural characteristics such as the MW of chitosan and its DD [32].

To produce chitosan films, usually chitosan is dissolved in acetic acid or hydrochloric acid, and then the solvent is removed by drying and final dry film collected. Another methodologies involves the precipitation of chitosan by common coagulating agents (e.g., sodium hydroxide), and finally made into film [12]. Other preparation techniques (casting, extrusion, etc.) and drying conditions may have influence on film properties. Mechanical film properties can be modified by controlling chain polymer orientation [19]. The relevant applications of these chitosan films include bone cell adhesion and growth, blood compatibility and cell adhesion [12].

Li-Fang et al [33] developed the application of chitosan films as coatings to oral dosage forms, and evaluated the directed solid dosage forms to the large intestine, where the component of the polysaccharide chitosan, incorporated into the blend film, would be digested. This digestion allowed the direct delivery of the drug present in the dosage form to the colon [33]. Since chitosan was reported to have intragastric floating characteristics and prolonged retention of the dosage form in the stomach, novel citrate cross-linked chitosan film was prepared by dipping chitosan film into citrate solution in order to may be useful for site-specific drug delivery in the stomach [34]. The results indicated that the citrate-chitosan film was useful in drug delivery such as for the site-specific drug controlled release in stomach.

Another study demonstrated the use and potential of chitosan films for production of mono and bilayer chitosan films containing corticosteroids as a drug carrier and controlled release in the eye [5]. Corticosteroids are most common use in eye drops for inflammation following eye surgery, such as after cataract surgery and corneal operations and others. Mono and bilayer corticosteroids chitosan films were successfully obtained and results suggested that these films are potential sustained release carriers for corticosteroids. The monolayer corticosteroids chitosan film might be a promising ocular delivery carrier for corticosteroids in few hours and bilayer corticosteroids-chitosan film in weeks. The importance and application of chitosan films also cover the local drug delivery system of antibiotics [35]. Local antibiotic delivery is another area of study designed to provide alternative methods of treatment to clinicians for compromised wound sites where avascular zones can prevent the delivery of antibiotics to the infected tissue. A recent study demonstrated that incorporating antibiotics in chitosan films could provide alternative methods of treating musculoskeletal infections [35].

Novel chitosan based polyelectrolyte complexes were developed and optimized in order to obtain films possessing the optimal functional properties (flexibility, resistance, water vapour transmission rate and bioadhesion) to be applied on skin [36]. The development was based on the combination of chitosan and two polyacrylic acid polymers with different crosslinkers and crosslinking densities. The optimized film, including adhesive property, has shown very good properties for application in the skin and represents a very promising formulation for further incorporation of drugs for topical and transdermal administration.

3.2. Tablets

Various studies with chitosan regarding controlled release delivery systems have been conducted for oral dosage forms, from film coated pellets, tablets or capsules to more sophisticated and complicated delivery systems such as osmotically driven systems, systems controlled by ion exchange mechanism, systems using three dimensional printing technology and systems using electrostatic deposition technology [37]. The most common controlled delivery system have been tablets and granules because of its effectiveness, low cost, ease of manufacturing and prolonged delivery time period, where the drug is uniformly dissolved or dispersed throughout the polymer [37]. The tableting process is associated with relatively high pressure in order to form suitable compacts. However, not only the tableting excipients are deformed during the process of tablet formation, but also the tablet itself. This can lead to total or partial damage to such materials, namely lost of biological activity of proteins and enzymes, polymorphic transformation of excipients or damage of the coating material. Most recently, different excipients were tested in order to prevent such damages. Amongst others, polysaccharides like chitosans and carrageenans have shown to be advantageous because of their elastic tableting behaviour [38]. Several reports have been published on the use of chitosan as tablet excipients. It was applied as a vehicle for sustained release tablets, a direct compressible diluent, a tablet disintegrant and a tablet binder [14]. As a diluent, chitosan was used for preparation of direct compressed tablets [39-40] where drug release was controlled. Studies using chitosans as directly compressible tablet excipients showed the potential of chitosans for use in modified release drug delivery systems without the need for additional adjuvants [39]. Chitosan also showed higher binder efficiency than other tablet binders such as methylcellulose and sodium carboxymethylcellulose [41] and used as a binder for colon specific drug delivery tablets with slow drug release compared with other polysaccharides or synthetic polymers [42]. Chitosan was utilized as tablets disintegrant [41] and showed bioadhesive properties in mixture with sodium alginate and in the form of thiolated chitosan derivative with slow drug release for intra-oral drug delivery tablets [43]. Furthermore, the solubilizing and amorphizing properties of low MW chitosan toward naproxin made it an optimal carrier for developing fast release oral tablet [44]. Depending mainly on ionic interaction, chitosan was also used for the preparation of tablets matrix to control drug release [45-46].

When used in a matrix-type tablet formulation, chitosan forms a gel-barrier in an acid environment that can modulate or constrain drug release. Furthermore at acidic pHs amines of chitosan are protonated and can therefore interact with oppositely charged drug ions, serving as excipient for modified release drug delivery systems [40].

Chitosan were studied as excipients in the preparation of prolonged theophylline tablet. These tablets showed higher drug bioavailability than of the commercial ones, which becomes a new potential formulation to respiratory problems [41].

The biological potential of chitosan adds also clear benefits to the tablet formulation and process. In a recent research five different polysaccharides with potential antioxidant activity for extended-release matrix tablets were compared [47]. The results suggest that chitosan would be potentially useful in an extended-release tablet with the higher antioxidant activity, able to catch the most diverse and natural oxidative species, usually involved in different pathologies.

A new study concerning vaginal infections and inflammations were evaluated using chitosan tablets [48]. Topical administration of the antibacterial metronidazole represents the most common therapy in the treatment of bacterial vaginosis caused by *Trychomonas vaginalis*. The formulations generally available for such therapy are creams, gels, vaginal lavages and vaginal suppositories. In this study, a new dosage form, containing metronidazole was developed with the aim to realize vaginal mucoadhesive tablets by including bioadhesive polymers as chitosan. This kind of delivery systems suitable for formulating metronidazole for topical application represents a good alternative to traditional dosage forms for vaginal topical administration in the treatment of infections or inflammations. These solutions overlap the limitations of conventional therapies that are not suitable to assure drug permanence on the vaginal mucosa surface for adequate time assuring the complete elimination of bacteria and pathology eradication.

Nonetheless, all applications of chitosan as tablet excipients were not in its derivative forms. However, attempts have been made to improve chitosan property by developing derivative salts. Chitosan derivatives such as glutamate, aspartate and hydrochloride salts have been used for colon-specific drug delivery and to enhance the delivery of therapeutic peptide across intestinal epithelia [14].

3.3. Hydrogels

Hydrogels are networks of hydrophilic polymers that can absorb large quantities of water without dissolution. Due to their physical properties resembling human tissue and its excellent tissue compatibility, hydrogels have been extensively studied for biomedical applications. They can be used as soft contact lenses [49], tissue engineering scaffolds [50], drugs vehicles, controlled-release system [51]. In addition, hydrogels have the potential for further healing [52]. They can absorb excess wound exudates, protect the wound from secondary infection and effectively promote the healing process by providing an environment for moist wound healing [53]. Even can also be removed without causing trauma to the wound [53].

Several models of hidrogels have been studied, including chitosan, poly (vinyl alcohol) (PVA) that is a water-soluble polyhydroxy polymer and alginate. However, chitosan has been widely exploited in hydrogel formulation and in practical applications because of its easy manipulation, excellent chemical resistance, physical properties, biodegradability and low price [53]. This polymer is used to produce hydrogels with well-known properties that are used for delivery of proteins and synthetic drugs [20]. Since this compound is also polyelectrolyte its ionic form produces complexes through hydrogen bonding or electrostatic interactions. Besides this, another interesting property of chitosan is its ability to gel in contact with specific polyanions. This gelation process is due to the formation of inter and intramolecular bonding mediated by these polyanions [2]. In the last decade, different chitosan hydrogels were produced for drug delivery in micro or nano-scale using the polyelectrolyte complexation technique [24, 25]. There are many factors that affect the relevant properties of the capsules of chitosan [4], in particular the composition, MW and DD of chitosan [20]. Several methods have been developed in which the particle size of chitosan hydrogel and its related properties are quite distinct, according to the method of preparation and the reaction conditions that are employed. One of the major factors that may influence the final properties is the method of preparing hydrogel. Moreover, few attempts were made to correlate statistically the reaction conditions with the final properties of chitosan hydrogel. Liu and his collaborators [54] evaluated the influence of chitosan MW and its concentration, along with pH, upon the swelling behavior of microcapsules of chitosan-alginate, and postulated that all factors have an effect on the behavior of the hydrogel swelling. The alginate-chitosan hydrogels are commonly prepared by ionic complexation using alginate as a gel core [55] and then characterized by the vibration modes of their main groups using fourier transform infrared spectroscopy. Other hydrogels formulation procedures can be performed by UV crosslinking. In this method, lactose moieties are introduced into chitosan to obtain much better water-soluble chitosan at neutral pH, and photoreactive azide groups are added to provide the ability to form a gel through crosslinking azide groups with amino groups [56]. This photocrosslinkable chitosan is then exposed to UV irradiation to form an insoluble and adhesive hydrogel within 60 s. Hydrogel has the consistency of transparent and soft rubber [56].

The crosslinking can also be performed by high temperature. It is based on the neutralization of a chitosan solution with a polyol counterionic dibase salt such as β-glycerophosphate. Chitosan/glycerophosphate is a thermosensitive solution, which is liquid at room temperature and solidifies into a white hydrogel at body temperature [56].

In addition, crosslinking can be achieved by high pH, employing the pH-sensitive property of chitosan solutions at low pH. Once injected into the body, these polymer solutions face different environmental pH conditions and form gels [56].

The chitosan hydrogel formulation can also be made by, freezing, thawing or chemical methods. Irradiation has the advantages of easy control of processing, without adding initiators or cross-linkers that can be harmful and difficult to remove, and also has the option of combining the hydrogel formation and sterilization in one technological step. The main disadvantage of hydrogels prepared by irradiation is its poor mechanical strength. However, hydrogels prepared by freeze-thaw for example, of aqueous solutions of PVA has good mechanical strength, are stable at room temperature, and does not require initiators or cross-linkers. The main disadvantage of this type of hydrogel is its opaque appearance and limited expansion capability [53].

The hydrogel yields are evaluated through the weight difference, placing the washed hydrogels into pre-weighed flasks and then into a stove at 50 °C until dryness. The chitosan hydrogel particles can then be visualized and characterized by particle size and size distribution using an inverted optical microscope [55] as well as, by solubility, X-ray diffraction, thermal analysis, and solvent uptake [57].

An ophthalmic delivery system with improved mechanical and mucoadhesive properties that could provide prolonged retention time for the treatment of ocular diseases were evaluated considering chitosan hydrogel formulation. For this, an *in situ* forming gel was developed by the combination of a thermosetting polymer, poly (ethylene oxide)–poly (propylene oxide)–poly (ethylene oxide) with chitosan. Therefore, the final formulation presented adequate mechanical and sensorial properties and remained in contact with the eye surface for a prolonged time. In conclusion, the *in situ* forming gel comprised of poloxamer/chitosan is a promising tool for the topical treatment of ocular diseases [58].

To overlap the limitation of topical delivery of antimicrobial agents and to prolong active drug concentrations in the oral cavity, it was designed a hydrogel formulation containing chitosan for delivery of chlorhexidine gluconate to the oral cavity [59]. Chitosan prolongs the adhesion time of oral gels and drug release also inhibiting the adhesion of *Candida albicans* to human buccal cells since it has antifungal activity [59]. The antifungal agent, chlorhexidine gluconate also induce the reduction of *Candida albicans* adhesion to oral mucosal cells.

The preparation and characterization of thiol-modified chitosan, which formed crosslinked hydrogels, was also described to characterize *in vitro* release kinetics of insulin encapsulated in different chitosan MW hydrogel and evaluated for their potential use as a scaffold for the culture of NIH 3T3 cells [60]. The results demonstrated that insulin is not immobilized locally within the gel network. Since the incorporation into the gel has no impact on insulin stability it may be assumed that the chitosan thermogelling system is an attractive delivery system for peptides and proteins.

The main goal of other study was to developed a chitosan bioadhesive gel for nasal delivery of insulin [61]. The proposed gel formulation could be useful preparation for controlled delivery of insulin through the nasal route and may represent an alternative treatment to diabetes.

Mucoadhesive chitosan lactate gels were developed intended for the controlled release of lactic acid onto vaginal mucosa [62]. The conclusions finding makes it reasonable to envisage a complete release of lactate from the tested formulations in vaginal environment.

Chitosan hydrogels were also studied as an alternative to conventional therapies of cancer [56]. A recent research reports an *in situ* gelling chitosan-based hydrogel system that sustains the release of a potential anti-cancer gene (pigment epithelium-derived factor) to the tumor site. A significant reduction of the primary osteosarcoma in a clinically relevant orthotopic model was measured. The combination of plasmid treatment and chemotherapy together with the use of this delivery system led to the highest suppression of tumor growth without side effects. The results obtained from this study demonstrate the potential application of a hydrogel system as an anti-cancer drug delivery for successful chemo-gene therapy [63].

Another notorious study focuses on the current use of injectable to form *in situ* chitosan hydrogels in cancer treatment [56]. Formulation protocols for *in situ* hydrogel systems, their cytotoxic properties, loading and *in vitro* release of drugs, their effect on cell growth *in vitro*, inhibition of tumor growth *in vivo* using mouse models, and future directions to enhance this technology were discussed. In conclusion, chitosan gelling systems due to their antibacterial, biocompatible, biodegradable and mucoadhesive properties are a potential carrier for various cancer treatments.

These hydrogels may also be useful to detect the localized growth of cells [56], which can also be directed to innovative methods of diagnosis

3.4. Microparticles

Microparticls are defined as multiparticulate delivery systems, usually spherical with size varying from 1 to 1000 μm, containing a core active substance [23]. The terms microcapsules and microspheres are often used synonymously. Spheres and spherical particles are also used for a large size and rigid morphology [23].

The use of microparticles-based therapy allows drug release to be carefully tailored to the specific treatment site through the choice and formulation of various drug–polymer combinations. The total dose of medication and the kinetics of release are the variables, which can be manipulated to achieve the desired result. Using innovative microencapsulation technologies, and by varying the copolymer ratio, MW of the polymer among other parameters, microparticles can be developed into an optimal drug delivery system, which will provide the desired release profile [11].

Different preparation of microparticles can be considered, mainly physical methods, chemical crosslinked methods and miscellaneous. Physical methodologies involve ionotropic gelation, emulsification and ionotropic gelation, modified emulsification and ionotropic gelation, floating hollow chitosan microspheres by ionic interaction with sodium dioctyl sulfosuccinate, coacervation and complex-coacervation [11]. Crosslinking with other chemicals are used for emulsion crosslinking method, multiple emulsion method, precipitation–chemical crosslinking and crosslinking with a naturally occurring agent [11]. Miscellaneous methods are thermal crosslinking, solvent evaporation method, spray drying, interfacial acylation, coating by chitosan solution and reacetylated chitosan microsphere [11]. Chitosan with different MW and concentration, degradation rate of chitosan particles and drug concentration interfere on microparticle properties [2].

Chitosan microparticles are used to provide controlled release of many drugs and to improve the bioavailability of degradable substances such as protein or enhance the uptake of hydrophilic substances across the epithelial layers. These microparticles are being investigated both for parenteral and oral drug delivery [11]. Having in mind bio/mucoadhesive properties of natural biopolymers, chitosan microparticules systems have potential for colon targeting. In order to achieve localization and prolonged residence time in the colon, matrices should have optimal particle size, between 4 and 15 μm [64]. Carrier systems in that size range are able to attach more efficiently to the mucus layer and accumulate in the affected region without the need for macrophage uptake. This novel formulation will offer efficient treatment of colon inflammatory diseases like ulcerative colitis and Chron's disease [64], increasing therapeutic concentration (at the site of inflammation) and activity and minimizing side effects that occur by conventional systemic absorption. Chitosan microparticulate carrier systems are also efficient in the treatment of inflammatory bowel diseases [64]. Budesonide is one of the most used drug substances in the treatment of active inflammatory bowel diseases. Chitosan microparticles loaded with budesonide were produced using novel one step spray-drying procedure. Coated microparticles were suitable candidates for oral delivery of budesonide with controlled release properties for local treatment of inflammatory bowel diseases.

Chitosan microparticles also represents a promising polymer in nasal peptide delivery [65] prolonging the residence time of nasal drug delivery systems at the site of drug absorption. Additionally, chitosan improves the absorption of peptides by opening transiently the tight junctions, as it was explain in section (3.1).

Oral administration of the nonsteroidal anti-estrogen tamoxifen is the treatment of choice for metastatic estrogen receptor-positive breast cancer. Chitosan microparticles were developed for tamoxifen delivery into the lymphatic system [66], improving tamoxifen oral bioavailability and decreasing its side effects. These data underlines other potential therapies to this serious cancer condition.

It was also reported the importance of chitosan microparticles in the purification of imunoglobulin G from human plasma by affinity chromatography using linoleic acid attached chitosan microparticles [67]. It was concluded that the microparticles allowed just one-step purification of imunoglobulin G from human plasma.

Chitosan and its derivative N-trimethyl chitosan chloride, given as microparticles associated to the non-toxic mucosal adjuvant LTK63, were evaluated for intranasal immunization with the group C meningococcal conjugated vaccine. The bactericidal activity measured in serum of mice immunized intranasally with the conjugated vaccine formulated with this delivery system and the LT mutant was superior to the activity in serum of mice immunized sub-cutaneous. Importantly, intranasal but not parenteral immunization, induced bactericidal antibodies at the nasal level, when formulated with both delivery system and adjuvant [68].

In another study, it was evaluated the ability of chitosan microparticles to enhance both the systemic and local immune responses against diphtheria toxoid after oral and nasal administration in mice. Significant systemic humoral immune responses were also found after nasal vaccination with diphtheria toxoid associated to chitosan microparticles. Diphtheria toxoid associated to chitosan microparticles results in protective systemic and local immune response against this toxoid after oral vaccination and in significant enhancement of imunoglobulin G production after nasal administration. Hence, these *in vivo* experiments demonstrate that chitosan microparticles are very promising mucosal vaccine delivery systems [69]. Other similar studies considered the chitosan microparticles as encapsulating agent of large amounts of antigens such as ovalbumin, or tetanus toxoid [70]. Besides chitosan particles are a promising candidate for mucosal vaccine delivery, mucosal vaccination not only reduces costs and increases patient compliance, but also limits the invasion of pathogens through mucosal sites.

3.6. Nanoparticles

Nanosystems as well as previously mentioned microparticles, are able to deliver drugs to the right place, at appropriate times and at the right dosage, also improving their bioavailability, efficiency and reducing citotoxicity associated to other systemic drugs carriers, actually becomes one of the most attractive areas of research in drug delivery [3]. These nanocarriers are submicron particles containing entrapped drugs intended for enteral or parenteral administration, which may prevent or minimize the drug degradation and metabolism as well as cellular efflux, extending the shelf-life [25]. Nanoparticles have a special role in targeted drug delivery in the sense that they have all the advantages of liposomes including the size property, but unlike liposomes, nanoparticles have a long shelf-life [71]. Some researchers have also observed that the number of nanoparticles, which cross the intestinal epithelium is greater than that of the microparticles and hydrophilic nanoparticles generally have longer resident time in blood then microspheres [71]. With their easy accessibility in the body, nanoparticles can also be transported via the circulation to different body sites. These particulate delivery systems have been shown to enhance the immune response following mucosal application. Nanoparticles have been made of safe materials, including synthetic bio-degradable polymers, natural biopolymers, lipids and polysaccharides and have the potential for overcoming important mucosal barriers, such as the intestinal, nasal and ocular barriers [25]. Chitosan based nanoparticles shows great potential for delivering macromolecular therapeutics (in particular drugs and genes) and control the complete release of the drugs in their native forms across biological barriers [16]. Important advantages of these nanoparticles include their rapid preparation under extremely mild conditions and also their ability to incorporate bioactive compounds [20]. Chitosan nanoparticles topically applied into the eye has been proven to increase the residence time of drugs in the precorneal area due to their adhesive properties and, therefore, could prolong the penetration of drugs into the intraocular structures [3]. An important research shows that a chitosan derivative can be use to prepare norcantharidin-associated nanoparticles by taking advantage of the ionic cross-linkage between the drug molecule and of the chitosan carrier [72]. As a hepatocyte-targeting carrier, chitosan nanoparticles are potentially promising for clinical applications.

The potential of nanoparticles as a vaccine delivery has also been demonstrated in several studies [73-75]. Nanoparticle-mediated gene delivery is an alternative to viral gene delivery. Nanoparticles offer the potential for safe, targeted, and efficient gene delivery in a variety of organs [76].

Borges et al [77], recently described a delivery system that is composed of a nanoparticulated chitosan core to which the hepatitis B surface antigen (AgHBs) was adsorbed and subsequently coated with sodium alginate. The enhancement of the immune response observed with the antigen-loaded nanoparticles demonstrated that chitosan is a promising platform for parenteral delivery of this antigen, since it resulted in a mixed Th1/Th2 type immune response.

Chitosan nanoparticles has also been produced as a carrier system for the nasal delivery of a monovalent influenza subunit vaccine [78]. The intranasal administered antigen-chitosan nanoparticles induced higher immune responses compared to the other intranasal antigen formulations, and these responses were enhanced by intranasal booster vaccinations. Moreover, among the tested formulations only intranasal administered antigen-containing chitosan nanoparticles induced significant Imunoglobulin A levels in nasal washes of all mice tested, demonstrating that chitosan nanoparticles are a potent new delivery system for intranasal administered influenza antigens.

Another study indicate that chitosan nanoparticles are a good carrier for DNA vaccines against tuberculosis by pulmonary delivery, which may provide an advantageous delivery route compared to intramuscular immunization, due increasing higher immunogenicity [79].

Chitosan nanoparticles have been also introduced as a useful carrier for peptide oral delivery, because they can protect these compounds from degradation. Insulin, like other peptides, has low therapeutic activity when administered orally due to degradation by proteolytic enzymes [80].

Sarmento et al [81] developed chitosan nanoparticles as drug carrier of insulin. Insulin was entrapped in different polyanion/chitosan nanoparticulate systems with high efficiency, to study morphologic and physical properties of resulting nanoparticulate complexes and to investigate insulin release behaviour under gastrointestinal conditions [81]. These nanoparticulate complexes appear to possess good properties for oral protein delivery, particularly those containing dextran sulfate/chitosan polyelectrolytes, which provided highest insulin association efficiency and retention of insulin in gastric simulated conditions. However alginate/chitosan nanoparticles also appear as promising in oral delivery system for insulin and potentially for other therapeutical proteins [82]. Moreover, it was demonstrated that blood glucose levels of diabetic rats can be effectively controlled by insulin-loaded chitosan nanoparticle administration, following either single or multiple oral administration. In addition, the hypoglycemic effect was observed for more than 24 h [83].

4. Clinical Trials

There are many definitions of clinical trials, generally are studies of biomedical and health research related to human beings following a pre-defined protocol [84]. These tests can only take place when quality and safety of the test are guaranteed by the Health Authority or the Ethics Committee recognized by the country, which will run the clinical trial. The randomized controlled trial is commonly accepted as the gold standard research method for evaluating health care interventions [85]. In any clinical trial it is desirable not only to achieve similar numbers of patients in each treatment group but also to ensure that patient groups are similar with respect to prognostic factors such as age or stage of disease [85].

Nowadays, many studies have come and explore the multi-potential of chitosan as a carrier for drug delivery and release [84]. Some relevant examples are described.

A. Efficacy and Safety of Hep-40 Chitosan for Mild to Moderately Elevated Cholesterol

This study is now completed and approved [86], which follows the discovery of lipid-lowering effect of chitosan by binding to fatty acids and cholesterol in the gastrointestinal tract restricting their absorption (Table 2). The study is designed to determine if HEP-40 chitosan (Enzymatic Polychitosamine Hydrolysate - 40kDa), a short-chained chitosan with a MW of 40 kDa, is safe and effective in lowering density lipoprotein LDL-cholesterol levels in patients who have not been previously treated with lipid-lowering agents and who have cholesterol levels that are mild to moderately above the levels recommended by the National Cholesterol Education Program Adult Treatment Panel III (NCEP ATP III) guidelines.

The primary objective was to evaluate the clinical benefit of administering HEP-40 chitosan at different doses and at different dosing regimens compared with placebo. Clinical benefit is defined as the reduction in LDL-cholesterol after 4 weeks of active treatment. This is a multi-centre, randomized, double-blind, placebo-controlled study. Following a 4-week Pre-Randomization Phase, where patients are instructed to maintain a stable diet, and are randomized to one of the following study groups for a 12-week Active Treatment Phase. Patients diagnosed with borderline, mild or moderate hypercholesterolemia, defined as LDL-C levels between 2.0 mmol/L and 4.5 mmol/L with a stable diet and willing to continue on the dietary regimen recommended by their physician (NCEP Step 1 Diet) for the duration of the study are considered. There are several excluded criteria in this trial such as: known cardiac disease; high risk of developing coronary artery disease; any condition affecting a major organ system, such as liver or kidney disease or malignancy; uncontrolled diabetes mellitus; HIV or hepatitis B or C positive; concurrent use of corticosteroids; and allergy or intolerance to crustaceans and/or seafood products. Success results were the percent change in serum LDL-C between the baseline and 4-week visit compared to placebo; percent change in serum LDL-C from baseline to 8- and 12-weeks of treatment compared to placebo; percent change in serum total cholesterol from baseline to 12 weeks of treatment compared to placebo; percent change in serum high-density lipoprotein (HDL-C) from baseline to 12 weeks of treatment compared to placebo; percent change in serum triglycerides from baseline to 12 weeks of treatment compared to placebo; at least safety and tolerability over the 12-week active treatment period, as determined by treatment-emergent adverse events.

B. Safety and Tolerability of Chitosa-n-acetylcysteine Eye Drops in Healthy Young Volunteers

Is a study that is not yet open for participant recruitment, it is estimated for a completion date – December 2010 [87]. Considering that the "dry eye syndrome" DES is a highly prevalent ocular disease, in particular in the elderly population and that current therapy for patients suffering from DES is the use of topically administered lubricants, no "ideal" formulation has yet been found. Recently, Croma Pharma has introduced chitosan-N-acetylcysteine eye drops, designed for treatment of symptoms related to “dry eye syndrome", Table 2. The new formulation comprises N-acetylcysteine, which has been used in ophthalmology because of its mucolytic properties for several years. Based on theoretical considerations, one can hypothesize that the new chitosan derivative may prove an increased adhesion to mucins of the ocular surface and may therefore be particularly beneficial in reducing the symptoms associated with DES.

Men and women aged between 18 and 45 years healthy with normal physical examination/laboratory values, medical history and with normal ophthalmic findings, (ametropia < 6 dpt) were considered. Exclusion criteria are: patients who participated in a clinical trial in the 3 weeks preceding the study; abuse of alcoholic beverages; symptoms of a clinically relevant illness in the 3 weeks before the first study day; ametropia of 6 or more dpt and Pregnancy. Success results are the differences finding between the treated and the non-treated eye with Chitosan- N- Acetylcysteine eye drops.

C. Trial of a Novel Chitosan Hemostatic Sealent in the Management of Complicated Epistaxis

This study is enrolling participants by invitation only and it is aimed to be a prospective clinical trial to investigate the efficacy of a chitosan-coated nasal packing (ChitoFlex® used in conjunction with the HemCon Nasal Plug) in the management of difficult spontaneous epistaxis and to evaluate its healing effect on nasal mucosa [88]. Furthermore, this study will help to determine if there are any non-desirable effects that chitosan may have on the nasal cavity, such as the production of fibrosis and foreign body reaction. Current therapies include epistaxis nasal packing, chemical or electric cauterization, and arterial ligation or embolization. There are many products available on the nasal market and are used to stop or prevent bleeding nose that have also antimicrobial properties to be effective in bleeding control, which have to be well tolerated. Currently, there is no perfect package for nose, but some are closer to the ideal than others. These nasal packs can be classified as non-absorbable and absorbable nasal pack. ChitoFlex ® is made of chitosan, is a new hemostatic product marketed and is used as a dressing for wounds. Since the chitosan has a positive charge it attracts red blood cells, which have a negative charge. Red blood cells create a seal over the wound because they are attracted to the bandage, forming a very tight and coherent seal. The ChitoFlex ®-dressing has been demonstrated as an antibacterial barrier in laboratory testing upon a variety of Gram-positive and - negative bacterial organisms. Men and women with more over than 18 years and with Epistaxis despite nasal packing or rebleeding after removal of the packing were considered. Are not part of the study patients unable or unwilling to provide informed consent and with prior diagnosis of disease or medical condition affecting the ability of blood to clot (e.g., hemophilia.). Success will be defined as achieving active control of bleeding before patient leaves the physician's office. This study should also help to determine the benefits acquired by the use of this new product in terms of presence/absence of post-packing tissue scarring. This will be accessed by endoscopic examination of the nasal cavity following removal of HemCon material.

D. Chitosan Dressings to Facilitate Safe Effective Debridement of Chronic Wounds and Minimize Wound Bacterial Re-Colonization

This study is not yet open for participant recruitment, it is expected to be completed at August 2010 [89]. The purpose of the study is to determine if the use of HemCon chitosan-based dressings is effective to facilitate safe, effective debridement of chronic wounds in the operating room and inpatient ward settings and to minimize bacterial re-colonization of wounds (Table 2). With the study it is supposed to demonstrate that debridements performed using HemCon dressings at the bedside can be performed safely without excessive bleeding in eligible patients. It is also a purpose to compare the levels of bacterial load between debrided wounds treated with HemCon dressings and debrided wounds treated with gauze and saline dressings at 2 days and 5 days after debridement. Other following objectives will be also achieved by the study: to determine the cost efficacy, if any, between wounds debrided at the bedside with HemCon dressings and wounds debrided in the operating room setting; to determine the cost efficacy, if any, between wounds debrided in the operating room where hemostasis is achieved with a HemCon dressing and between debrided wounds where hemostasis is achieved with traditional cauterization methods and to compare comfort levels in patients treated with HemCon dressings as compared to traditional gauze dressings. Men and women older than 18 years and able to provide written informed consent with a wound on their body with an eschar and/or significant slough present will be considered. Female must not be pregnant, subjects must be willing to consent to a blood transfusion in the rare event that a transfusion is necessary and patients must be hospitalized for at least 5 days to enable controlled dressing changes by the service. Cannot be part of the study pre-debridement hemoglobin level < 7.0 g/dL candidates that refuse blood transfusions or unable to provide written informed consent, subjects with sensitivity to chitosan or the gauze dressings used in this study, or any local anesthetic needed for a debridement, subjects who are in the intensive care unit and who have grossly infected wounds that may reasonably be expected to require multiple debridement procedures prior to clearance of bacteria and nonviable tissue from the wound and for wounds situated in the lower extremity.

Table 2, recap the several addressed conditions that have been focus of the chitosan clinical trials, with correspondent alternative and tested treatment followed by the study. These examples of chitosan based therapy suggest its multiple and diverse potential and allow a real sense of current biomedical importance.

Table 2. Chitosan clinical trials with associated condition, suggested treatment including chitosan and phase of the study

Trial	Condition	Product	Phase
A	Hypercholesterolemia	Device: HEP-40 chitosan	II
B	Dry Eye Syndroms	Device: Chitosan- N- Acetylcysteine eye drops	I
C	Epistaxis	Device: 2009-I-Epistaxis-1	
D	Wound Debridement	Device: HemCon Dressings and HemCon ChitoGauze; chitosan-based Device: Gauze and saline dressings	Pre-clinical tests

5. Conclusion

The physical and chemical properties of chitosan, such as inter- and intramolecular hydrogen bonding and the cationic charge in acidic medium with its several bioactivities such as nontoxicity, biocompatibility, biodegradability and transmucosal absorption allows chitosan to become an excellent candidate for drug carrier and excipient for controlled release systems.

Chitosan is a promising candidate for oral cavity drug delivery, for the enhancement of absorption of drugs using buccal delivery system, and to assure longer residence in the stomach allowing stomach-specific drug delivery. In the intestine, chitosan acts as an absorption enhancer by increasing the residence time of dosage forms at mucosal sites, inhibiting proteolytic enzymes, and increasing the permeability of protein and peptide drugs across mucosal membranes. A number of *in vitro* and *in vivo* studies showed its importance as a suitable material for efficient nonviral gene and DNA vaccine delivery.

The versatility and biodegradability of this polymer permits its easily manipulation in various forms and its derivatives can be also naturally digested by lysozomal enzymes, in vivo. Besides solutions, a number of chitosan-based colloidal systems have been explored for bioactive molecules carriers, like films, tablets, hydrogel, micro and nanoparticle. In these formulations chitosan can reach the target sites with low collateral damages, toxicity and higher efficiency.

The emergence of innovative medicines demands effective strategies to enhance drug permeation and bioavailability. A better understanding of the mechanisms of action of these novel drugs vehicles will provide their further optimization, opening exciting opportunities to improve the administration of macromolecules, such as proteins, peptides, genes, vaccines, antigens and synthetic compounds, which allows the treatment of several diseases.

Therefore, useful knowledge for future adaptation is provided for pharmaceutical industry, medical assistance and human health care.

References

[1] Garcia-Fuentes, M., Csaba, N. and Alonso, M. J., Nanostructured Chitosan Carriers for Oral Protein and Peptide Delivery. *Protein Delivery,* 2007,

[2] Prabaharan, M. and Mano, J. F., Chitosan-Based Particles as Controlled Drug Delivery Systems. *Drug Delivery,* 2005, 12, 41–57.

[3] Campos, A., Diebold, Y., Carvalho, E., Sánchez, A. and Alonso, M. J., Chitosan Nanoparticles as New Ocular Drug Delivery Systems: in Vitro Stability, in Vivo Fate, and Cellular Toxicity. *Pharmaceutical Research,* 2004, 21, 803-810.

[4] George, M. and Abraham, E., Polyionic hydrocolloids for the intestinal delivery of protein drugs: Alginate and chitosan - a review. *Journal of Controlled Release,* 2006, 114, 1-14.

[5] Rodrigues, L. B., Leite, H. F., Yoshida, M. I., Saliba, J. B., Junior, A. S. C. and Faraco, A. A. G., In vitro release and characterization of chitosan films as dexamethasone carrier. *International Journal of Pharmaceutics,* 2009, 368, 1-6.

[6] Prow, T. W., Bhutto, I., Grebe, R., Uno, K., Merges, C., Mcleod, D. S. and Lutty, G. A., Nanoparticle-delivered biosensor for reactive oxygen species in diabetes. *Vision Research,* 2008, 48, 478-485.

[7] Rinaudo, M., Chitin and chitosan: Properties and applications. *Progress in Polymer Science,* 2006, 31, 603-632.

[8] Vásconez, M. B., Flores, S. K., Campos, C. A., Alvarado, J. and Gerschenson, L. N., Antimicrobial activity and physical properties of chitosan–tapiocastarch based edible films and coatings. *Food Research International,* 2009, 42, 762-769.

[9] Paulino, A. T., Simionato, J. I., Garcia, J. C. and Nozaki, J., Characterization of chitosan and chitin produced from silkworm crysalides. *Carbohydrate Polymers,* 2006, 64, 98-103.

[10] Wu, T., Zivanovic, S., Draughon, A. and Sams, C. E., Chitin and ChitosansValue-Added Products from Mushroom Waste. *Journal of Agriculture and Food Chemistry,* 2004, 52, 7905-7910.

[11] Sinha, V. R., Singla, A. K., Wadhawan, S., Kaushik, R., Kumria, R., Bansal, K. and S., D., *Chitosan microspheres as a potential carrier for drugs. International Journal of Pharmaceutics,* 2004, 274, 1-33.

[12] Shih, C.-M., Shieh, Y.-T. and Twu, Y.-K., Preparation and characterization of cellulose/chitosan blend films. *Carbohydrate Polymers,* 2009, 78, 169-174.

[13] Alonso, M. J. and Sanchéz, A., The potential of chitosan in ocular grug delivery. *Journal of Pharmacy and Pharmacology,* 2003, 55, 1451-1463.

[14] Nunthanid, J., Laungtana-anan, M., Sriamornsak, P., Limmatvapirat, S., Puttipipatkhachorn, S., Lim, L. Y. and Khor, E., Characterization of chitosan acetate as a binder for sustained release tablets. *Journal of Controlled Release,* 2004, 99, 15-26.

[15] Dyer, A. M., Hinchcliffe, M., Watts, P., Castile, J., Jabbal-Gill, I., Nankervis, R., Smith, A. and Illum, L., Nasal Delivery of Insulin Using Novel Chitosan Based Formulations: A Comparative Study in Two Animal Models between Simple Chitosan Formulations and Chitosan Nanoparticles. *Pharmaceutical Research* 2002, 19,

[16] Shi, Y. and Huang, G., Recent Developments of Biodegradable and Biocompatible Materials Based Micro/Nanoparticles for Delivering Macromolecular Therapeutics. *Critical Reviews in Therapeutic Drug Carrier Systems,* 2009, 26, 29–84.

[17] Aranaz, I., Mengíbar, M., Harris, R., Paños, I., Miralles, B., Acosta, N., Galed, G. and Heras, Á., Functional Characterization of Chitin and Chitosan. *Current Chemical Biology,* 2009, 3, 203-230.

[18] Sousa, F., Guebitz, G. M. and Kokol, V., Antimicrobial and antioxidant properties of chitosan enzymatically functionalized with flavonoids. *Process Biochemistry,* 2009, 44, 749-756.

[19] Garcıa, M. A., Pinotti, A., Martino, M. and Zaritzky, N., Electrically treated composite FILMS based on chitosan and methylcellulose blends. *Food Hydrocolloids,* 2009, 23 722-728.

[20] Sashiwa, H. and Aiba, S.-I., Chemically modified chitin and chitosan as biomaterials. *Progress in Polymer Science,* 2004, 29, 887-908.

[21] Fernandez-Saiz, P., Lagaron, J. M. and Ocio, M. J., Optimization of the Film-Forming and Storage Conditions of Chitosan as an Antimicrobial Agent. *Journal of Agriculture and Food Chemistry* 2009, 57, 3298–3307.

[22] Zhao, Q. S., Ji, Q. X., Xing, K., Li, X. Y., Liu, C. S. and Chen, X. G., Preparation and characteristics of novel porous hydrogel films based on chitosan and glycerophosphate. *Carbohydrate Polymers,* 2009, 76, 410-416.

[23] Kumar, M. N. V. R., A review of chitin and chitosan applications. *Reactive and Functional Polymers,* 2000, 46, 1-27.

[24] Muzzarelli, R. A. A., Genipin-crosslinked chitosan hydrogels as biomedical and pharmaceutical aids. *Carbohydrate Polymers,* 2009, 77, 1-9.

[25] Motwani, S. K., Chopra, S., Talegaonkar, S., Kohli, K., Ahmad, F. J. and Khar, R. K., Chitosan–sodium alginate nanoparticles as submicroscopic reservoirs for ocular delivery: Formulation, optimisation and in vitro characterisation. *European Journal of Pharmaceutics and Biopharmaceutics,* 2008, 68, 513–525.

[26] Qun, G. and Ajun, W., Effects of molecular weight, degree of acetylation and ionic strength on surface tension of chitosan in dilute solution. *Carbohydrate Polymers,* 2006, 64, 29-36.

[27] Cano-Cebrián, M. J., Zornoza, T., Granero, L. and Polache, A., Intestinal Absorption Enhancement Via the Paracellular Route by Fatty Acids, Chitosans and Others: A Target for Drug Delivery. *Current Drug Delivery,* 2005, 2, 9-22.

[28] Smith, J., Wood, E. and Dornish, M., Effect of Chitosan on Epithelial Cell Tight Junctions *Pharmaceutical Research,* 2004, 21, 43-49.

[29] Yu, S., Zhao, Y., Wu, F., Zhang, X., Lü, W., Zhang, H. and Zhang, Q., Nasal insulin delivery in the chitosan solution: in vitro and in vivo studies. *International Journal of Pharmaceutics,* 2004, 281, 11–23.

[30] Zaharoff, D. A., Rogers, C. J., Hance, K. W., Schlom, J. and Greiner, J. W., Chitosan solution enhances the immunoadjuvant properties of GM-CSF. *Vaccine,* 2007, 25, 8673-8686.

[31] Vargas, M., Albors, A., Chiralt, A. and Gonzalez-Martinez, C., Characterization of chitosan–oleic acid composite films. *Food Hydrocolloids,* 2009, 23, 536-547.

[32] Ferreira, C. O., Nunes, C. A., Delgadillo, I. and Lopes-da-Silva, J. A., Characterization of chitosan–whey protein films at acid pH. *Food Research International,* 2009, 42, 807-813.

[33] Li-Fang, F., Wei, H., Yong-Zhen, C., Bai, X., Qing, D., Feng, W., Min, Q. and De-Ying, C., Studies of chitosan/Kollicoat SR 30D film-coated tablets for colonic drug delivery. *International Journal of Pharmaceutics,* 2009, 375, 8-15.

[34] Shu, X. Z., Zhu, K. J. and Song, W., Novel pH-sensitive citrate cross-linked chitosan film for drug controlled release. *International Journal of Pharmaceutics,* 2001, 212, 19-28.

[35] Noel, S. P., Courtney, H., Bumgardner, J. D. and Haggard, W. O., A Potential Local Drug Delivery System for Antibiotics. *Clinical Orthopaedics and Related Research,* 2008, 466, 1377-1382.

[36] Silva, C. L., Pereira, J. C., Ramalho, A., Pais, A. A. C. C. and Sousa, J. J. S., Films based on chitosan polyelectrolyte complexes for skin drug delivery: Development and characterization. *Journal of Membrane Science,* 2008, 320, 268-279.

[37] Abdul, S. and Poddar, S. S., A flexible technology for modified release of drugs: multi layered tablets. *Journal of Controlled Release,* 2004, 97, 393- 405.

[38] Schmid, W. and Picker-Freyer, K. M., Tableting and tablet properties of alginates: Characterisation and potential for Soft Tableting. *European Journal of Pharmaceutics and Biopharmaceutics,* 2009, 72, 165-172.

[39] Rege, P. R., Shukla, D. J. and Block, L. H., Chitinosans as tableting excipients for modified release delivery systems. *International Journal of Pharmaceutics,* 1999, 181 49-60.

[40] Rege, P. R., Garmise, R. J. and Block, L. H., Spray-dried chitinosans Part II: in vitro drug release from tablets made from spray-dried chitinosans. *International Journal of Pharmaceutics,* 2003, 252, 53-59.

[41] Alsarra, I. A., El-Bagory, I. and Bayoni, M. A., Chitosan and sodium sulfate as excipients in the preparation of prolonged release theophylline tablets. *Drug development and industrial pharmacy,* 2005, 1, 385-395.

[42] Sinha, V. R. and Kumria, R., Binders for colon specific drug delivery: an in vitro evaluation. *International Journal of Pharmaceutics* 2002, 249, 23-31.

[43] Roldo, M., Hornof, M., Caliceti, P. and Bernkop-Schnurch, A., Mucoadhesive thiolated chitosan as platforms for oral controlled drug delivery: synthesis and ion vitro evaluation. *European Journal of Pharmaceutics and Biopharmaceutics,* 2004, 57, 115-121.

[44] Mura, P., Zerrouk, N., Mennini, N., Maestrelli, F. and Chemtob, C., Development and characterization of naproxen-chitosan solid system with improved drug dissolution properties. *European Journal of Pharmaceutical Sciences* 2003, 19, 67-75.

[45] Tapia, C., Costa, E., Terraza, C., Munita, A. M. and Azdani-Pedram, M., Study of the prolonged release of theophylline from polymeric matrices based on grafted chitosan with acrylamide. *Pharmazie,* 2002, 57, 744-749.

[46] Bernkop-Schnurch, A., Schuhbauer, H., Clausen, A. E. and Hanel, R., Development of a sustained release dosage form for alpha-lip acid. I. Design and in vitro evaluation. *Drug Development and Industrial Pharmacy* 2004, 30, 27-34.

[47] Tomida, H., Yasufuku, T., Fujii, T., Kondo, Y., Kai, T. and Anraku, M., Polysaccharides as potential antioxidative compounds for extended-release matrix tablets. *Carbohydrate Research,* 2010, 345 82-86.

[48] Perioli, L., Ambrogia, V., Paganoa, C., Scuota, S. and Rossi, C., FG90 chitosan as a new polymer for metronidazole mucoadhesive tablets for vaginal administration. *International Journal of Pharmaceutics,* 2009, 377, 120-127.

[49] Opdahl, A., Kim, S. H., Koffas, T. S., Marmo, C. and Somorjai, G. A., Surface mechanical properties of pHEMA contact lenses: Viscoelastic and adhesive property changes on exposure to controlled humidity *Journal of Biomedical Materials Research Part A,* 2003, 67A, 350-356.

[50] Nowak, A. P., Breedveld, V., Pakstis, L., Ozbas, B., Pine, D. J. and Pochan, D. e. a., Rapidly recovering hydrogel scaffolds from selfassembling diblock copolypeptide amphiphiles. *Nature,* 2002, 417(6887), 424-428.

[51] Qiu, Y. and Park, K., Environment-sensitive hydrogels for drug delivery. *Advanced Drug Delivery Reviews,* 2001, 53, 321-339.

[52] Sen, M. and Avci, E. N., Radiation synthesis of poly(N-vinyl-2-pyrrolidone)- kappa-carrageenan hydrogels and their use in wound dressing applications. I. Preliminary laboratory tests. *Biomedical Materials Research Part A,* 2005, 74A, 187-196.

[53] Yang, X., Liu, Q., Chen, X., Yu, F. and Zhu, Z., Investigation of PVA/ws-chitosan hydrogels prepared by combined c-irradiation and freeze-thawing. *Carbohydrate Polymers,* 2008, 73, 401-408.

[54] Liu, X., Xu, W., Liu, Q., Yu, W., Fu, Y., Xiong, X., Ma, X. and Yuan, Q., Swelling behaviour of alginate-chitosan microcapsules prepared by external gelation or internal gelation technology. *Carbohydrate Polymers,* 2004, 56, 459-464.

[55] Abreu, F. O. M. S., Bianchini, C., Forte, M. M. C. and Kist, T. B. L., Influence of the composition and preparation method on the morphology and swelling behavior of alginate–chitosan hydrogels. *Carbohydrate Polymers,* 2008, 74, 283-289.

[56] Ta, H. T., Dass, C. R. and Dunstan, D. E., Injectable chitosan hydrogels for localised cancer therapy. *Journal of Controlled Release,* 2008, 126, 205-216.

[57] Kandile, N. G., Mohamed, M. I., Zaky, H. T., Nasr, A. S. and Abdel-Bary, E. M., Synthesis and properties of chitosan hydrogels modified with heterocycles. *Carbohydrate Polymers,* 2009, 75, 580-585.

[58] Gratieri, T., Gelfuso, G. M., Rocha, E. M., Sarmento, V. H., Freitas, O. d. and Lopez, R. F. V., A poloxamer/chitosan in situ forming gel with prolonged retention time for ocular delivery. *European Journal of Pharmaceutics and Biopharmaceutics,* 2010, In press,

[59] Senel, S., Ikinci, G., Kas, S., Yousefi-Rad, A., Sargon, M. F. and Hıncal, A. A., Chitosan films and hydrogels of chlorhexidine gluconate for oral mucosal delivery. *International Journal of Pharmaceutics,* 2000, 193 197-203.

[60] Wu, Z. M., Zhangb, X. G., Zhengb, C., Li, C. X., Zhanga, S. M., Donga, R. N. and Yua, D. M., Disulfide-crosslinked chitosan hydrogel for cell viability and controlled protein release. *European Journal of Pharmaceutical Sciences,* 2009, 37, 198-206.

[61] Varshosaz, J., Sadrai, H. and Heidari, A., Nasal delivery of insulin using bioadhesive chitosan gel. *Drug delivery,* 2006, 13, 31-38.

[62] Bonferoni, M. C., Giunchedi, P., Scalia, S., Rossi, S., Sandri, G. and Caramella, C., Chitosan Gels for the Vaginal Delivery of Lactic Acid: Relevance of Formulation Parameters to Mucoadhesion and Release Mechanisms. *AAPS PharmSciTech,* 2006, 7,

[63] Ta, H. T., Dass, C. R., Larson, I., Choong, P. F. M. and Dunstan, D. E., A chitosan hydrogel delivery system for osteosarcoma gene therapy with pigment epithelium-derived factor combined with chemotherapy. *Biomaterials,* 2009, 30, 4815-4823.

[64] Crcarevska, M. S., Dodov, M. G. and Goracinova, K., Chitosan coated Ca–alginate microparticles loaded with budesonide for delivery to the inflamed colonic mucosa. *European Journal of Pharmaceutics and Biopharmaceutics,* 2008, 68, 565-578.

[65] Krauland, A. H., Guggi, D. and Bernkop-Schn¨urch, A., Thiolated chitosan microparticles: A vehicle for nasal peptide drug delivery. *International Journal of Pharmaceutics,* 2006, 307, 270-277.

[66] Coppi, G. and Iannuccelli, V., Alginate/chitosan microparticles for tamoxifen delivery to the lymphatic system. *International Journal of Pharmaceutics,* 2009, 367, 127-132.

[67] Uyguna, D. A., Uyguna, M., Karagözlera, A., Öztürka, N., Akgöla, S. and Denizli, A., A novel support for antibody purification: Fatty acid attached chitosan beads. *Colloids and Surfaces B: Biointerfaces,* 2009, 70, 266-270.

[68] Baudner, B. C., Verhoef, J. C., Giuliani, M. M., Peppoloni, S., Rappuoli, R., Giudice, G. D. and Junginger, H. E., Protective immune responses to meningococcal C conjugate vaccine after intranasal immunization of mice with the LTK63 mutant plus chitosan or

trimethyl chitosan chloride as novel delivery platform. *Journal of drug targeting* 2005, 13, 489-498.

[69] Lubben, I. M. V. d., Kersten, G., Fretz, M. M., Beuvery, C., Verhoef, C. and Junginger, H. E., Chitosan microparticles for mucosal vaccination against diphtheria: oral and nasal efficacy studies in mice. *Vaccine,* 2003, 21, 1400-8.

[70] Van Der Lubben, I. M., Verhoef, J. C., Borchard, G. and Junginger, H. E., Chitosan for mucosal vaccination. *Advanced Drug Delivery Reviews,* 2001, 52, 139-144.

[71] Wu, Y., Yang, W., Wang, C., Hu, J. and Fu, S., Chitosan nanoparticles as a novel delivery system for ammonium glycyrrhizinate. *International Journal of Pharmaceutics,* 2005, 295, 235-245.

[72] Wang, Q., Zhang, L., Hu, W., Hu, Z.-H., Bei, Y.-Y., Xu, J.-Y., Wang, W.-J., Zhang, X.-N. and Zhang, Q., Norcantharidin-associated galactosylated chitosan nanoparticles for hepatocyte-targeted delivery. *Nanomedicine: Nanotechnology, Biology, and Medicine* (in press),

[73] Amidi, M., Romeijn, S. G., Borchard, G., Junginger, H. E., Hennink, W. E. and Jiskoot, W., Preparation and characterization of protein-loaded N-trimethyl chitosan nanoparticles as nasal delivery system. *Journal of Controlled Release,* 2006, 111, 107-116.

[74] Amidi, M., Romeijn, S. G., Verhoef, J. C., Junginger, H. E., Bungener, L., Huckriede, A., Crommelin, D. J. A. and Jiskoot, W., N-trimethyl chitosan (TMC) nanoparticles loaded with influenza subunit antigen for intranasal vaccination: biological properties an immunogenicity in a mouse model. *Vaccine* 2007, 25, 144-153.

[75] Sayın, B., Somavarapu, S., Li, X. W., Thanou, M., Sesardic, D., Alpar, H. O. and S¸ enel, S., Mono-N-carboxymethyl chitosan (MCC) and N-trimethyl chitosan (TMC) nanoparticles for non-invasive vaccine delivery. *International Journal of Pharmaceutics,* 2008, 363, 139-148.

[76] Prow, T. W., Bhutto, I., Kim, S. Y., Grebe, R., Merges, C., McLeod, D. S., Uno, K., Mennon, M., Rodriguez, L., Leong, K. and Lutty, G. A., Ocular nanoparticle toxicity and transfection of the retina and retinal pigment epithelium. *Nanomedicine: Nanotechnology, Biology, and Medicine,* 2008, 4, 340–349.

[77] Borges, O., Silva, M., Sousa, A. d., Borchard, G., Junginger, H. E. and Cordeiro-da-Silva, A., Alginate coated chitosan nanoparticles are an effective subcutaneous adjuvant for hepatitis B surface antigen. *International immunopharmacology,* 2008, 8, 1773-80.

[78] Amidi, M., Romeijn, S. G., Verhoef, C., Junginger, H. E., Bungener, L., Huckriede, A., Crommelin, D. J. A. and Jiskoot, W., N-trimethyl chitosan (TMC) nanoparticles loaded with influenza subunit antigen for intranasal vaccination: biological properties and immunogenicity in a mouse model. *Vaccine,* 2007, 25, 144-53.

[79] Bivas-Benita, M., Meijgaarden, K. E. V., Franken, K. L. M. C., Junginger, H. E., Borchard, G., Ottenhoff, T. H. M. and Geluk, A., Pulmonary delivery of chitosan-DNA nanoparticles enhances the immunogenicity of a DNA vaccine encoding HLA-A*0201-restricted T-cell epitopes of Mycobacterium tuberculosis. *Vaccine,* 2004, 22, 1609-15.

[80] Bayat, A., Larijani, B., Ahmadian, S., Junginger, H. E. and Rafiee-Tehrani, M., Preparation and characterization of insulin nanoparticles using chitosan and its quaternized derivatives. *Nanomedicine : nanotechnology, biology, and medicine,* 2008, 4, 115-20.

[81] Sarmento, B., Martins, S., Ribeiro, A., Veiga, F., Neufeld, R. and Ferreira, D., Development and Comparison of Different Nanoparticulate Polyelectrolyte Complexes as Insulin Carriers. *International Journal of Peptide Research and Therapeutics,* 2006, 12,

[82] Sarmento, B., Ribeiro, A., Veiga, F., Sampaio, P., Neufeld, R. and Ferreira, D., Alginate/Chitosan Nanoparticles are Effective for Oral Insulin Delivery. *Pharmaceutical Research,* 2007, 24,

[83] Sarmento, B., Ribeiro, A., Veiga, F., Ferreira, D. and Neufeld, R., Oral Bioavailability of Insulin Contained in Polysaccharide Nanoparticles. *Biomacromolecules,* 2007, 8, 3054-3060.

[84] Trials, C. *Clinical trials: A service of the U.S. National Institutes of Health.* [cited 2010 21 of January]; http://clinicaltrials.gov/ct2/results?term=chitosan.

[85] Scott, N. W., McPherson, G. C., Ramsay, C. R. and Campbell, M. K., The method of minimization for allocation to clinical trials: a review. *Controlled Clinical Trials,* 2002, 23, 662-674.

[86] Trials, C. *Efficacy and Safety of HEP-40 Chitosan for Mild to Moderately Elevated Cholesterol.* [cited 2010 21 of January]; http://clinicaltrials.gov/ct2/show/NCT00454831?term=chitosanandrank=3].

[87] Trials, C. *Safety and Tolerability of Chitosan-N-acetylcysteine Eye Drops in Healthy Young Volunteers.* [cited 2010 21 of January]; http://clinicaltrials.gov/ct2/show/NCT01015209?term=chitosanandrank=4].

[88] Trials, C. *Trial of a Novel Chitosan Hemostatic Sealant in the Management of Complicated Epistaxis.* [cited 2010 21 of January]; http://clinicaltrials.gov/ct2/show/NCT00863356?term=chitosanandrank=2].

[89] Trials, C. *Chitosan Dressings to Facilitate Safe Effective Debridement of Chronic Wounds and Minimize Wound Bacterial Re-colonization.* [cited 2010 21 of January]; http://clinicaltrials.gov/ct2/show/NCT01035944?term=chitosanandrank=1].

In: Focus on Chitosan Research
Editors: Arthur N. Ferguson and Amy G. O'Neill
ISBN 978-1-61324-454-8

Chapter 8

SOLID STATE SYNTHESIS AND MODIFICATION OF CHITOSAN

Tatiana A. Akopova, Alexander N. Zelenetskii and Alexander N. Ozerin*
Enikolopov Institute of Synthetic Polymer Materials, Russian Academy of Sciences, 70 Profsoyuznaya Str., 117393 Moscow, Russia

ABSTRACT

This chapter deals with a novel method of chemical modification of polysaccharides, in particular solid-state synthesis of chitosan and its composites with biocompatible synthetic polymers. The technique is based on a variety of chemical and physical transformations at conditions of plastic flow, which are realized in polymeric solids at pressure and shear deformation. The extrusion in the solid state is one of the most environmentally friendly methods of polysaccharide modification, since it does not require the presence of any catalysts, initiators, as well as organic solvents. It is desirable for biomedical applications, as it allows avoiding toxic remains over final composite materials. At the same time it provides numerous possibilities to circumvent many processing obstacles typical for preparation of natural polymers-based hybrid materials and to achieve better results compared to those of reactions in a melt or in a solution. Compared with traditional methods for design and manufacture of hybrid polysaccharide-based materials, solid-state reactive blending is a relatively simple, cost effective, and convenient way to improve the composite properties. In particular, grafting of polyester moieties onto chitosan chains was found to occur under selected deformation conditions. These polymeric materials demonstrate an amphiphilic behavior with dispersion propensity in organic solvents and could be promising for various biomedical applications. Microfibers containing chitosan up to 40% can be obtained by electrospinning of these dispersions. A uniform structure of the obtained polylactide/chitosan materials is confirmed by improved mechanical properties of films as compared with a model molten system with the same composition. Also it was shown recently that chitosan-g-poly(vinyl alcohol) copolymers can be produced by simultaneous alkaline

* Enikolopov Institute of Synthetic Polymer Materials, Russian Academy of Sciences, 70 Profsoyuznaya Str., 117393 Moscow, Russia.

solid-state deacetylation of chitin and poly(vinyl acetate) in an extruder. The main feature of these co-polymeric systems is solubility in neutral water at low and moderate temperatures. The obtained graft-copolymers possess excellent ability to stabilize nano-scale particles of both organic-inorganic origins. The dependence of chemical composition, morphologies and macroscopic properties on the ratios of co-extruded components as well as on the processing conditions is discussed.

1. INTRODUCTION

In this chapter we describe the structural transformations of chitin that take place during its solid-state processing in an extruder as well as related morphology, structural, mechanical, and relaxation properties of chitosan, polylactide-chitosan and poly(vinyl alcohol)-chitosan composites produced by the solid-state synthesis.

Chitin is a poly-N-acetylglucosamine and is the second after cellulose natural carbohydrate polymer in abundance. The unique sorption properties of chitosan, which is a highly deacetylated chitin, and the great abilities of its chemical modification, salt and complex formation determine its wide applications [1, 2]. Chitosan is metabolised by certain human enzymes, especially lysozyme, and is considered to be biodegradable. Due to its positive charges at physiological pH, chitosan is also bioadhesive, has bacteriostatic effects and promotes cell attachment and proliferation. In medical and pharmaceutical applications, chitosan and its derivatives are used as a component in hydrogels [3], polymeric membranes, porous scaffolds for controlled drug release and cell transplantation [4, 5]. Materials based on chitosan in both polymeric and oligomeric forms can be designed as injectable scaffolds, microcarriers, nanoparticles, micro- and nanofibers that are attractive materials for tissue engineering [6, 7]. It was demonstrated that the composite chitosan-based materials can mimic the structure of natural extracellular matrices and have the potential for cell cultivation in vitro with the idea of implanting them later on to promote tissue regeneration. Also it is shown that nano-scale assemblies of small molecules (e.g., a vesicle) or other nano-components (e.g., a protein or nanoparticle) can be connected to sites along a chitosan chain. These structures, in turn, can be connected into larger supramolecular assemblies and deposited at specific device addresses or formed as free-standing films [8]. Thus, chitosan may emerge as an integral material for soft matter (bio)fabrication.

Chemical derivatization of chitosan, including grafting with synthetic polymers, provides a powerful means to promote new biological activities and to modify its mechanical properties. The primary amino groups on the molecule are reactive and provide a mechanism for side group attachment using a variety of reaction conditions. This modification generates a material with lower stiffness and often altered solubility. It was demonstrated that the chitosan-based copolymers are good potential candidates for carriers of bioactive compounds and wound-dressing materials due to solubility at physiological pH and biocompatibility. Of course, the precise nature of changes in chemical and biological properties depends on the nature of the side group or grafted chain. For example, PEG-g-chitosan copolymers possess enhanced solubility in water, and a modification of chitosan with aliphatic acids, polyesters or peptides provides hydrophobic molecules compatibility.

The specific features of supramolecular structure of polysaccharides largely control reactivity of the functional groups in reactions that usually take place under heterogeneous conditions. To prepare highly substituted derivatives, preliminary activation of polysaccharides is required. Traditionally, a polysaccharide modification carries out in organic or alkaline aqueous media that, however, results in a great amount of sewage water. In particular, chitosan is usually prepared by treating of chitin with concentrated aqueous NaOH solutions (40–50% w/v) at 10–20 molar excess of alkali and at elevated temperatures. Moreover, the multistep process of deacetylation is required to prepare chitosan with low degree of acetylation (DA) [9].

Low compatibility of synthetic polymers with natural ones as well as impossibility for processing polysaccharides in the molten state strongly limits design and manufacture of hybrid polysaccharide-based polymeric systems. The yield of graft copolymers forming under melt processing of thermoplastic polymers with polysaccharides is insubstantial. Thus, working out new procedures for synthesis of polysaccharide derivatives and copolymers is one of the promising trends in modern chemistry of the polysaccharides.

Reactive mixing in the solid state of all compounds at conditions of plastic flow which are realized in polymeric solids at joint action of pressure and shear deformation by applying external mechanical energy seems to be one of the promising methods for modifying infusible and poorly soluble polymer substances like polysaccharides (such as cellulose, starch, chitin, and chitosan). The so-called "mechanochemical" approach has been successfully used for inorganic solids (in particular, for alloying, interaction of soft metals with ceramics, activation of minerals for catalysis, etc.), and for inorganic–organic composite materials, especially for nanocomposites [10-19]. As the final stage in development of transformations of solids towards technological level, numerous syntheses were performed on a laboratory and industrial scale using mills of various types. The organic reactions that take place under conditions of shear deformation, so-called "solvent-free reactions", are currently receiving a great deal of interest. Organic solid-state reactions have numerous advantages since they are infinitely high-concentration reactions and proceed much more efficiently and faster than solution reactions in many cases, as described by Toda and co-authors [20]. Solid-state reactions are important not only for their high efficiency and selectivity but also for their simplicity and cleanness. In other words, solid-state reactions may be regarded as more economical and ecologically favorable procedures in chemistry.

In Russia, systematic studies of organic reactions conducted under conditions of plastic flow were started in the early seventies. Since then hundreds of organic reactions were investigated and described in several reviews [21-27]. The following principal regularities were found in these studies: key role of plastic deformation in degree of conversion; a presence of critical pressure depending on chemical nature of the substance; a proof that the reactions occur exclusively in solid state (hot spots possibility is eliminated) and are almost independent of a temperature; a dependence of relaxation transitions and melting temperatures of polymers on pressure with coefficient of 0.15–0.25 K/MPa [21]. The latter determines that the polymers under deformation are mostly in a glassy state when pressure rises. However, elevated temperature is often required to increase deformability of the material (especially simultaneous deformability in the case of polymer blends).

Since organic substances, including polymers, have low thermal stability and large strain relaxation times as compared with inorganic substances, the most appropriate techniques for their solid-state modification are Bridgman anvils and extruders, specially designed for slow continuous deformation of solids. At the laboratory-scale, the Bridgman anvil is the most effective device to realize mechanical syntheses of organic compounds (pressure ranges 0.5–

10 GPa, angle of rotation up to thousands of degrees). Compression of substance in thin layers with anvils, made of high-strength steel, is identical to uniform compression. The lower anvil can rotate, with deformation of the substance proportional to the angle of rotation [28].

The grinding of the substances under conditions of joint action of high pressure and shear deformation is only one of the results of intense mechanical treatment, along with molecular-level mixing, accumulation of structural defects, formation of free radicals, double electric layers and excited electron states [25]. Plastic deformation eliminates the problems of reactive mixing and allows to reach quantitative yield of the reactions. The uniqueness of this method relies upon a variety of chemical (e.g.: grafting, polymeranalogous transformation, polymerization and polycondensation reactions) and physical transformations occurring at conditions of plastic flow.

We are engaged in the studies of the mechanisms and kinetics of the chemical reactions proceeding under the conditions of plastic flow in solid state of polymers, in particular polysaccharides, as well as low-molecular-weight organic compounds. This approach to design of new polysaccharide-based materials was realized with a semi-industrial co-rotating twin-screw extruder (ZE 40A×40D UTS, Berstorff, Germany). The extruder has a variable set of processing elements adapted to specific applications which provide a high shear strain and powerful dispersive action. The main working elements of the extruder perform compression and deformation of the material similarly to the Bridgman anvils (Figure 1). Effectiveness of deformation is determined by the position of these elements, their number and temperature in different zones. The extruder has a barrel length to diameter ratio of 40, a screw diameter of 40 mm and can be successfully used to transfer the processes to industrial scale. Productivity for solid-state processes: 10…30 kg/h; zone temperatures range: - 5…300 °C; screw rotation speed: 10…500 rpm.

The most attractive feature of this technology is that the entire modification process proceeds in the solid state, and does not require melting of components or any solvents as reaction medium. Thus, solid-state technique undoubtedly is one of the most environmentally friendly and promising methods of polysaccharide modification. We have shown previously that solid-state reactive blending (SSRB) in simple one-step procedure allowed to produce chitosan with high degree of deacetylation, its salts of different acids, including mixed salts, acyl-, carboxymethyl-substituted derivatives etc. [29-31]. Despite the solid state of initial components, the reactions of cellulose, chitin and chitosan with low-molecular-weight compounds under the plastic flow conditions yields the products that are relatively homogeneous with respect to their chemical properties [32, 33].

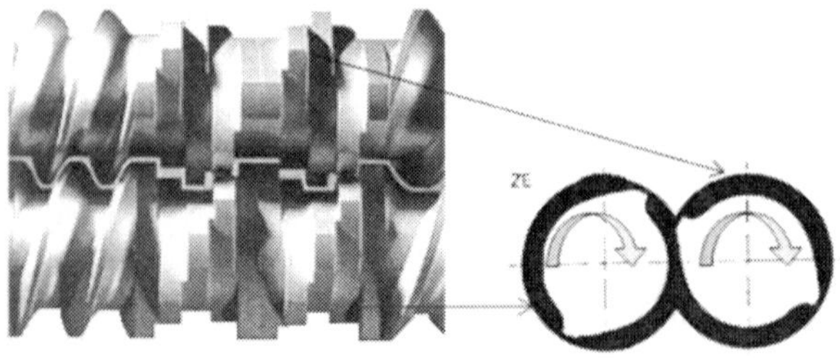

Figure 1. ZE-40 extruder processing elements.

It should be specially noted that the reactions under these conditions often proceed as such; this means no need in any initiators or catalysts.

Design of new efficient and ecologically safe processes for polysaccharide modification is one of the main purposes of our current studies. New approaches to development of polysaccharide-based composite materials are essential. By application of solid-state co-extrusion to mixtures of polysaccharides like chitin, chitosan, starch or cellulose with synthetic polymers or monomers a number of the polysaccharide-based new materials was obtained. Of particular interest are the ultra-fine blends of chitosan with polyethylene obtained in a presence of maleic anhydride. The films made of these products have improved plasticity as compared with those of materials produced by co-extrusion of neat homopolymers. The modified blends can be used for preparation of special sorbents and plastics [34].

Blending of synthetic polymers with polysaccharides which results in derivatization (in particular, grafting) of natural polymers is very promising way to reduce costs and to develop biodegradable plastics. In recent years, it has been found that SSRB (co-extrusion and particularly the Bridgman anvils technique) is capable of inducing grafting of synthetic polymers of different hydrophilicity onto polysaccharide chain, needless of any chemical reagents or solvents [35]. The latter is desirable for biomedical applications, for allows to avoid toxic remains over final composite materials. At the same time it provides numerous possibilities to circumvent many processing obstacles typical for preparation of natural-polymers-based hybrid materials and to achieve better synthetic results as compared with those of reactions in the melt or solution.

2. Solid-State Synthesis of Chitosan

The results [36-41] obtained while using various methods of polymer structure investigations (IR spectroscopy, NMR, X-ray scattering, microscopy, sorption techniques, etc.), showed that chitin, just as cellulose, is characterized by an ordered fibrillar structure, a developed system of intra- and intermolecular hydrogen bonds, a high degree of crystallinity, and polymorphism. The main objective of this chapter is to analyze the structural and chemical conversions in solid chitin-alkali mixtures during their processing in an extruder and to study the structure of the formed chitosan. The proposed approach has numerous advantages because allow to considerably decrease the process duration, consumption of reagents and, in such a way, to improve the economical and ecological parameters of the process. So, alkali excess required for almost complete deacetylation of chitin at solid-state synthesis is two-four times less than those usually reported in the literature [42]. In here we summarize the data related to solid-state alkaline deacetylation of chitin that were obtained in recent years.

We used α-chitin isolated from Far East crab shells (Vostok-Bor, Vladivostok, Russia; 600 kDa) as a starting material. In order to uniform the particle sizes, preliminary grinding of initial chitin flakes was performed in extruder at 5-10°C. Native chitin and chitin after treatment were analyzed by FTIR and WAXS methods. No essential distinctions were revealed by FTIR. Decreasing in molecular weight (40%) was observed as it usually takes place under mechanical activation of polysaccharides at these conditions. The ash and moisture content of the grinded chitin were measured and calculated: 0.50±0.2% ash; 5.6±0.2% water. Deformation under the action of shear and stress in absence of water at room temperature is shown to reduce the crystallinity of the neat initial chitin [43]. However, subsequent addition of water restores the degree of crystallinity so that it becomes equal to the value characteristic for the initial chitin. Extrusion processing of chitin at room temperature with the addition of water has virtually no effect on the crystallinity and degree of ordering of the chitin crystal lattice, a result which is similar to that obtained for simple dry grinding at an elevated (180°C) temperature.

Chitosan, produced by alkaline deacetylation of the grinded chitin, is represented by the extruded mixture of chitin and alkali at molar ratio of 1 : 5 (equal mass ratio) and at various treatment duration. Sodium hydroxide (98.5%, microprills with typical grain size distribution in diameter: 1.2 – 2.0 mm < 2%, 0.5 – 1.2 mm > 93%, and 0 – 0.5 mm < 5%) of analytical reagent grade were purchased from Merck (Germany) and was used without additional treatment. The studied chitosan samples included a set of chitosan samples isolated by washing with deionized water up to neutral reaction of the extrusion mixture (Chs-1, 2, 3). The compositions and a mean yield of chitosan after a purification procedure are listed in Table 1. The extruded mixture of chitin with potassium hydroxide at a similar to other samples molar ratio of the components (Chs4) was studied "as received" for an understanding of conversion mechanism. For comparison, we also studied Chs5 sample produced by a traditional deacetylation in an alkali solution under laboratory conditions.

A 2 kg batch of each trial was made and extruded. The optimal screw speed was 100 rpm. The solid mixture of chitin and alkali a little at a time fed into extruder, which was preliminary heated. The temperature of co-extrusion was 170-180°C in all cases of treatment.

Wide-angle X-ray scattering (WAXS) analysis of reactive blends was performed in order to reveal structural transformations that take place during the processing of solid chitin with alkali in an extruder. The data were recorded for the Chs4 sample in the transmission mode (CuKα radiation, wavelength $\lambda = 0.1542$ nm). It can be seen that reactive mixtures haven't a crystal phase of polymers and alkali after extrusion at conditions of chitin deacetylation, whereas a side reaction product – potassium acetate, is formed (Figure 2). These data leads us to conclusion that the alkaline deacetylation of chitin at the conditions of solid-state synthesis proceeds through a preliminary formation of the amorphous solid solution.

Table 1. Reactive mixtures and chitosan characteristics after purification

Sample Code	Reactive mixture (w:w)	Number of run and duration	Yield of chitosan (%)	Ash[a] (%)	Moisture[b] (%)
Chs1	Chitin/NaOH (1:1)	1 (20 min)	82	0.6	2.3
Chs2		2 (40 min)		0.5	2.2
Chs3		3 (60 min)		0.3	2.7

[a] The purification procedure was performed as follows: washing by deionized water a) by centrifugation 30/70 v/v – twice; b) by filtration up to neutral reaction 1/10 v/v on average; c) by electrodialysis (1-2 days) to remove the residual salts completely for sample characterization.

[b] Water suspensions of the refined chitosan samples were lyophilized at temperature –10…+30°C.

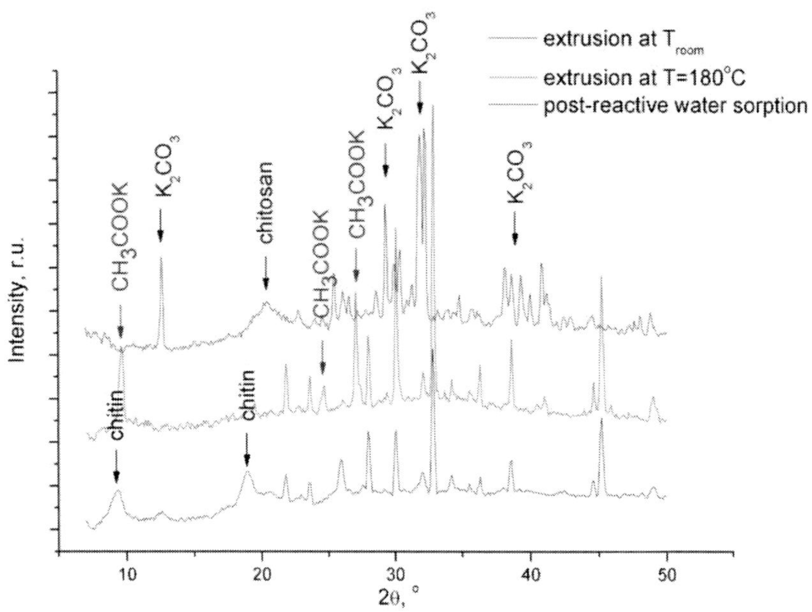

Figure 2. WAXS curves showing structural evolution of chitin/alkali reactive mixture.

Table 2. Summary of the results of chitosan samples characterization

Sample code	Solubility in 0.1 M HCl (%)	Viscosity average MW × 10^{-4} (g/mol)	DD (%)	
			titration	^{1}H NMR
Chs1	97.6	5.4	90.9	90.0
Chs2	97.8	4.0	96.0	95.0
Chs3	98.0	1.8	97.3	95.0

The structure of the obtained products (Chs1-3) was characterized by potentiometric titration data, GPC-MALS, NMR, FTIR spectroscopy analysis, WAXS, and viscosity measurements. Summary of the results on average molecular weight (MW) and degree of chitin deacetylation (DD) are presented in Table 2.

The viscosity average MW of chitosans was calculated by Mark-Houwink equation $[\eta] = K_m M^a$ where $K_m = 1.38 \times 10^{-4}$ and $a = 0.85$ [44]. The viscosity measurements in 0.3 M acetic acid – 0.2 M sodium acetate buffer at 25°C are adjusted well with GPC-MALS method in 0.1 M nitric acid (pH-value 2.6) for samples Chs1 and Chs2 (Table 3). The GPC-curves of both samples showed unimodal distribution with M_w/M_n lower than 2. Unlike that Chs3 sample has bimodal MW distribution that complicates evaluation of average MW by GPC method. The most probable reason of such a behavior can be explained by branching of low-molecular-weight chitosan during preparation and purification procedure due to reaction of aldehyde end groups with amine groups forming in course of co-extrusion (Figure 3).

At the same time, in all of the GPC-MALS profiles log MW of the eluted molecules decreased linearly as the elution volume increased and the curves are superposed. The latter confirm that chemical and topological structure of the polymers is identical.

Table 3. Molecular weight and ratio of Mw/Mn determined by MALS for the obtained chitosan samples

Sample code	M_w	M_n	M_w/M_n
Chs1	64,800	34,000	1.9
Chs2	36,000	18,200	2.0
Chs3	59,700	19,300	3.1

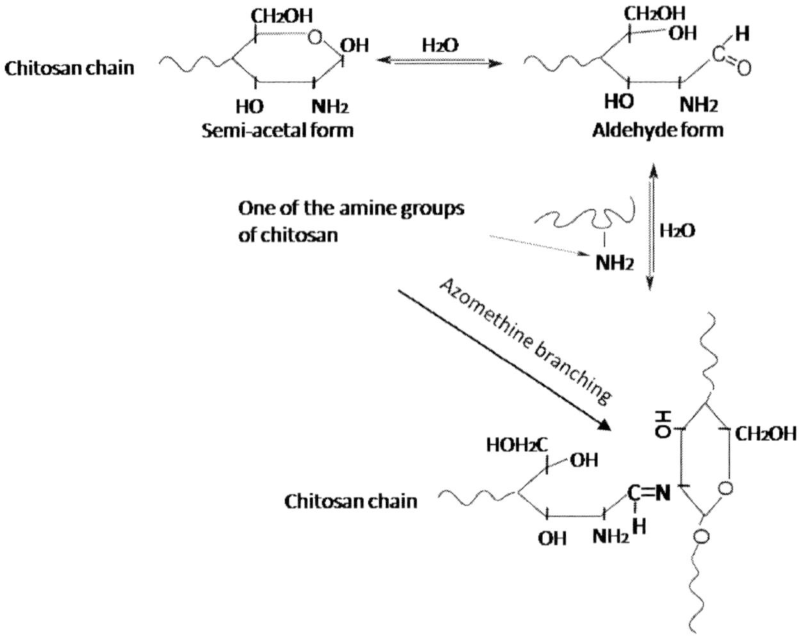

Figure 3. The Scheme of expected branching of low-molecular-weight chitosan during preparation and purification procedure.

^{1}H NMR specta (D_2O–DCl; reference TSP) of all samples adjusted well with the published spectra of an almost N-deacetylated polysaccharides with signals of H-3, H-4, H-5 of the sugar unit 4.0–4.5 (multiplet); H-6 at 3.6 (triplet); H-1 at 5.0 and 5.3 (duplets) which are assigned to the N-acetylated and N-deacetylated units, relatively; and signal at 2.4 (ppm) due to the protons of (–CH_3) groups of acetylated units. The spectrum of Chs2 sample is presented in Figure 4. The assignments were almost identical for the all samples under consideration. The integral ratio of the H-signals of N-acetylated units to those of glucosamine structures suggested that the degree of acetylation (DA) is not more than 0.05 per monosaccharide unit. DOSY NMR the analysis shows uniformity of the chitosan samples and an absence of impurity signals (Figure 5).

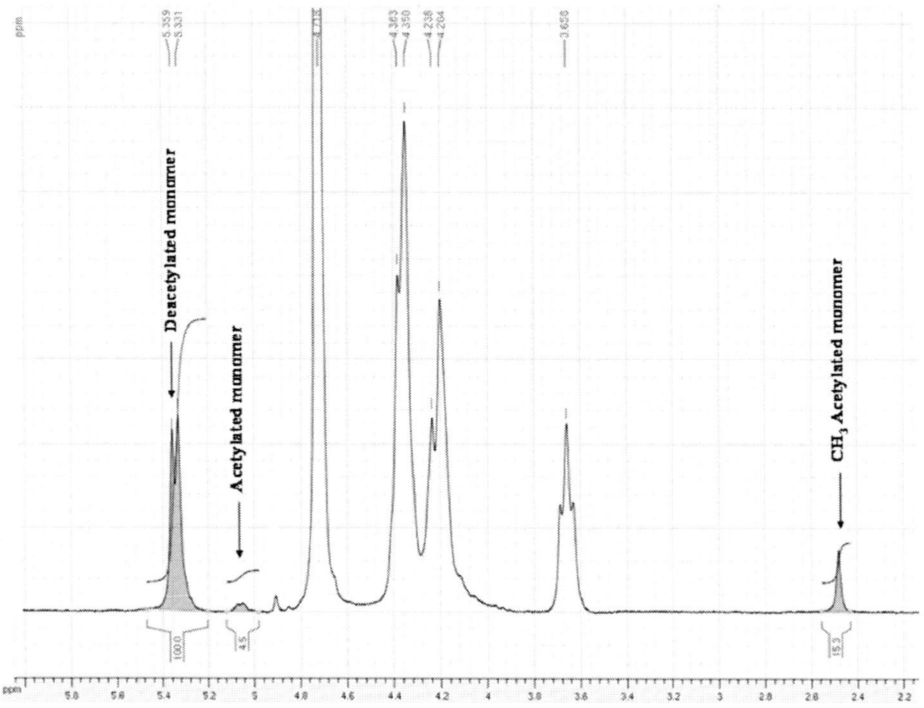

Figure 4. ^{1}H NMR spectrum of Chs2 in D_2O–DCl, reference TSP.

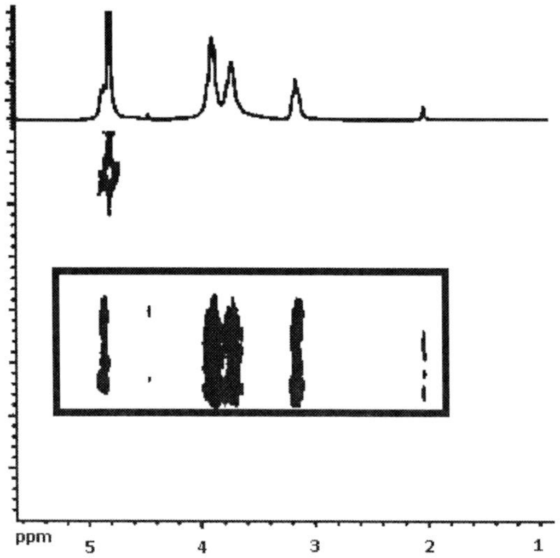

Figure 5. The 2D NMR spectrum (DOSY) of the Chs3 sample.

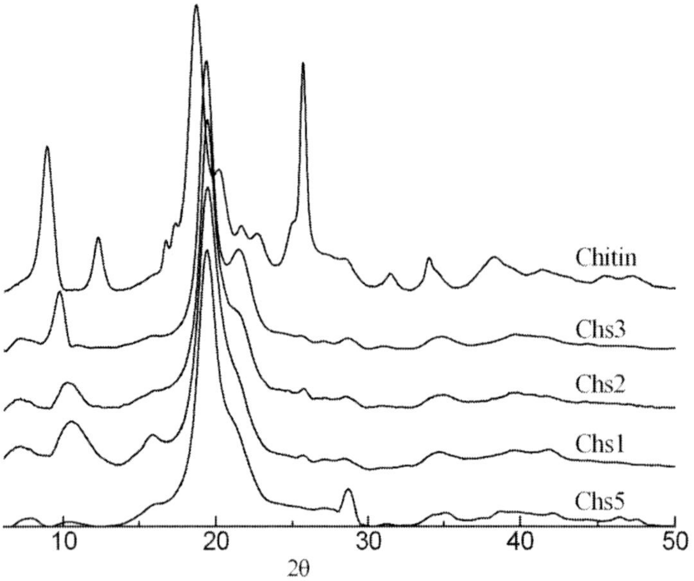

Figure 6. WAXS curves of the grinded initial chitin and the chitosan obtained by the solid-state synthesis: Chs1, Chs2, and Chs3 samples; chitosan prepared by traditional suspension method (Chs5).

Decreasing in MW confirmed by the viscosity measurements of the Chs3 sample is accompanied by formation of well-ordered structure with relatively higher crystallinity as compared to other two samples as well as to the Chs5 sample (DA 0.15) prepared by deacetylation in a liquid phase of NaOH (Figure 6). According to the literary data, both lower DA and lower MW can be caused it.

As a whole, solid-state synthesis using a twin-screw extruder is shown to yield a more amorphous product in comparison with chitosan produced by the traditional suspension method (at a comparable level of acetamide group conversion into amine groups). The degree of crystallinity of the chitosan obtained through solid-state synthesis doesn't exceed 25-30% [43]. The dimensions of both chitin and chitosan crystallites after solid-state treatment agree with the values obtained for the original chitin and for chitosan synthesized by the traditional suspension method. At the same time, the crystalline phases of these samples seem to be formed under different conditions.

FTIR analysis of the obtained chitosan samples (Figure 7) strongly confirmed the range of DA values obtained by titration and NMR methods. The IR-spectra were recorded in absorbance mode at a resolution of 4 cm^{-1} and processed with Win-IR software v.4, Bio-Rad Digilab Division. All the spectra were normalized by employing as internal standard the composite band of C–O stretching vibrations of pyranose ring at 1075 $см^{-1}$.

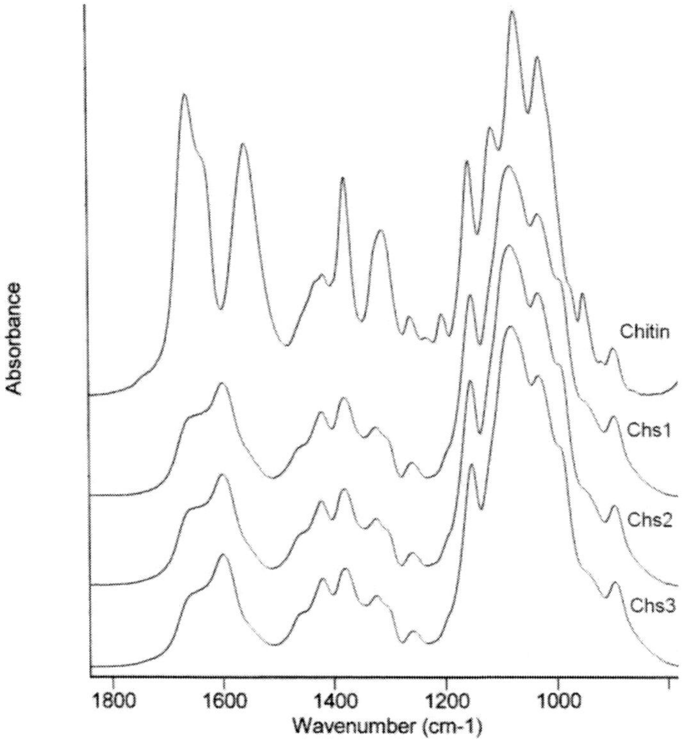

Figure 7. FTIR spectra of the grinded initial chitin and the obtained chitosan samples.

As is known, the doublet at 1663 and 1626 cm^{-1} in spectrum of initial chitin is assigned to $\nu_{C=O}$ and arises from two types of H-bonds in which the C=O groups are involved in α-chitin [45]. A band appearing at 1597 cm^{-1} with the decrease of the DA of initial chitin expected for primary amines. IR-spectra of the samples represent the IR-bands of the carbonyl stretching, $\nu_{C=O}$ (amide I, at 1655 cm^{-1}), and the NH bending, δ_{NH} (amide II, at 1550 cm^{-1}), which clearly change their intensities along a decreasing in DA.

Both FTIR and WAXS analysis showed that all obtained chitosan samples are homogeneous with respect to their chemical structure. It is clearly appeared that duration of the process is a key parameter on which deacetylation degree of chitin as well as MW of chitosan formed at conditions of solid-state synthesis are mainly depended.

Comparison of the surface with the bulk of the formed chitosan particles was performed with FTIR-ATR methods. The obtained data confirm that the solid-state synthesis, which does not require swelling prior to modification of the polymer, provides high deacetylation in the bulk of chitin particles due to the absence of diffusion restrictions in the solid-state co-extrusion process, unlike traditional synthesis of chitosan in aqueous concentrated NaOH solutions.

3. Grafting of Poly(Vinyl Alcohol) onto Chitosan

Poly(vinyl alcohol) (PVA) is a hydrolysis product of poly-(vinyl acetate) (PVAc) and is a biocompatible water-soluble synthetic polymer. Besides, it is recognized as one of the few synthetic polymers particularly biodegradable under both aerobic and anaerobic conditions [46, 47]. PVA is used as a basic material for a variety of biomedical applications, including contact lens material, skin and artificial cartilage replacement material, etc., because of its non-toxicity and desirable physical properties such as good ability to form films and fibers. Since chitosan is considered as fully biodegradable in human body polymer, its blends with PVA are attractive materials for numerous biomedical applications. These materials in a form of composite films, hydrogels, and scaffolds for cell attachments were prepared and characterized as biocompatible materials having desired properties for controlled drug delivery and tissue engineering [48–54]. The authors report a good compatibility of the components mainly due to hydrogen bonds which are formed between the functional groups of both polymers, results in an increase in mechanical properties of blend films. Among PVA/chitosan composites materials, temperature- and pH-responsive hydrogels and beads with porous cross-linked structure have been most widely studied. The physically cross-linked composites as well as those of materials prepared by using of different initiators and common cross-linking agents are essentially insoluble but good swelling in non-acidic aqueous solutions.

At the same time, the main disadvantage of the commonly used "chitosan-based" method for many biomedical applications, in particular as a component of micro- and nanocapsules for encapsulation of sensitive to pH-value bioactive compounds, relates to chitosan poor solubility at physiological pH. Graft copolymerization is a known way to materials with new properties. This approach is widely used in the case of polysaccharides as well. In this chapter an emphasis will be laid on possibilities of the SSRB technique to prepare chitisan-based copolymers with enhanced solubility and other properties to meet the need of biomedicine.

The solid-state synthesis of chitosan modified by grafting with PVA in order to prepare their water-soluble copolymers was performed. From the view point that the reactive blend formed in course of chitosan preparation is an amorphous solid solution, it seemed be very promising to conduct the blending of chitosan with PVA by employing as precursors chitin and PVAc. In other words, to conduct an alkaline deacetylation and blending of the polymers concurrently. In this case an efficient blending of the new formed PVA and chitosan as a result of their impossibilities to form a crystalline phase during simultaneous plastic deformation should be observed.

The dependence of chemical composition, morphologies and macroscopic properties on the ratios of co-extruded components as well as on the processing conditions was studied. PVAc (beaded, viscosity average MW 350,000) was added to reactive mixture in course of co-extrusion of chitin with sodium hydroxide after the first run (reactive blend for Chs1) at molar ratio of PVAc/Chs of 1.4 (CP1 sample) or 3 (CP2). For comparison, we also prepared CP3 sample while PVAc was added in 10% *i*-PrOH solution. All the mixtures were subjected to co-extrusion at 60°C. The obtained blend was inserted into Soxlett extraction apparatus for low-molecular-weight compounds removal. The samples were exposed to water/organic mixture (ethanol/ethyl acetate/water) extraction for 48 hours followed by electrodialysis (1-2 days) in aqueous solution (including dispersed phase). Then the water suspensions of the samples were freezed and lyophilized at temperature –10…+30°C up to moisture content 2-3%.

Formation of chitosan and PVA under the employed processing conditions was confirmed by fractionation of the reaction products with various solvents and subsequent examination of the fractions with the techniques of IR spectroscopy, WAXS, GPC, DSC and elemental analysis.

The results of fractionation of the products by a sequential dissolving in neutral water at ambient (fraction A) and elevated (80°C, fraction B) temperatures, and in dilute acetic acid (AcOH, fraction C) as well as formulation of the fractions are listed in Table 4.

It was revealed by FTIR and WAXD analysis that the DD of both chitin and PVAc in this process are almost complete (up to 95-98%). Since chitosan is insoluble in neutral water (in an olygomeric form as well), it can be clearly seen that the employed conditions of synthesis result in formation of extensively grafted co-polymeric systems.

To measure mechanical characteristics and to perform structural studies, films of different thicknesses were prepared by casting 5% copolymer aqueous solutions onto a glass substrate. The cast solutions were dried at 50°C for 2 h. The films containing chitosan as acetate were held in 1 M NaOH for 1 h, carefully washed by deionized water, and dried. The obtained films were mechanically strong, transparent, and homogeneous. All the samples were held at a constant relative humidity of 44% for 24 h prior to tests.

Both FTIR and elemental analyses were used to evaluate a content of chitosan in all the obtained fractions. Glucosamine content was calculated from the results of elemental analysis using C/N ratio for both pure and modified chitosan. Content of the glucosamine units by FTIR analysis was calculated from a ratio of the intensities of the bands assigned to chitosan -NH_2 bending at 1597 cm^{-1} and $-CH_2-$ bending in PVA at 850 cm^{-1}.

Figure 8 shows the FTIR spectra of both copolymeric (CP1-B) and blend films as well as the spectrum of CP1-A fraction with high content of chitosan. The samples for analyses were in a form of thin films (5-10 μm thickness). As the reference IR spectrum for evaluation of the content of chitosan in the obtained products, a model blend film containing 20%-wt of chitosan was used. The calculations showed that the CP1-A and CP1-B fractions contain 50 and 5 mol% of chitosan, respectively.

Table 4. The results of fractionation of CP samples and formulation of the fractions: A – soluble in neutral water at RT, B – soluble in neutral water at 80°C, C – soluble in dilute acetic acid

Sample code	Fraction[a]	Content of the fraction (%-wt)	% of chitosan (wt / mol)	
			FTIR	Elemental analysis
CP1	A	30	80/50	60/40
	B	30	17/5	19/6.5
	C	40	98/90	98/90
CP2	A	10	15/4.5	not available
	B	25	10/3	
	C	65	98/90	
CP3	A	20	30/10	not available
	B	30	15/4.5	
	C	50	98/90	

[a] Soluble in AcOH fractions (C) were precipitated by 1M NaOH under intense stirring up to solution pH of 8-9, then washed with water by filtration and dried.

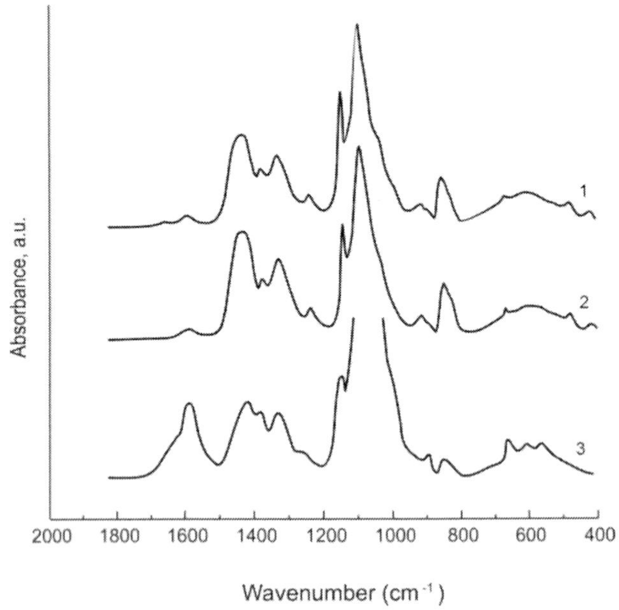

Figure 8. FTIR spectra of PVA/chitosan samples: (1) a reference blend film; (2) CP1-B fraction; (3) CP1-A fraction.

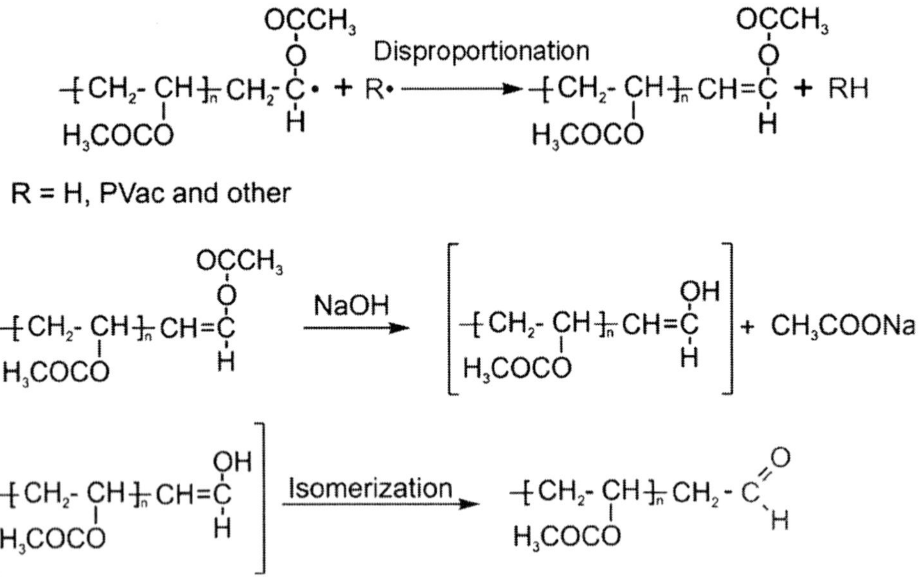

Figure 9. The Scheme of formation of the aldehyde end-groups in PVA.

The most probable results of the intense mechanical treatment of PVAc in the presence of alkali, along with its deacetylation, are a mechano-induced scission of polymer chains and formation of enol/aldehyde structures due to disproportionation/isomerization reactions (Figure 9).

Degradation of the formed PVA as a result of ketones/enol formation and the chain scission is other possible mechanism to obtain reactive end groups of the polymer. Indeed, it is a well-known pathway to creation a hydrogel structure of PVA using a physical method consisting of repeated freezing and thawing cycles of aqueous solutions of the polymer or by irradiation [55-57]. Cross-linking of the polymer under these conditions has to be present to avoid dissolution of the hydrophilic polymer chain segments into the aqueous phase.

The solid-state alkaline deacetilation of neat PVAc in extruder leads to formation of insoluble fractions up to 50 wt%. Surprisingly enough, while the reactive mixtures contain chitosan, about 97-98% of entire blend composition are comprised of soluble products. So, it leads to conclusion that a radical mechanism is not dominant at the employed conditions of SSRB but grafting of PVA chains onto chitosan can proceeds through the Schiff reaction of the formed aldehyde end groups of PVA with the amine groups. The proposed synthetic scheme for preparation of the PVA/chitosan copolymers is shown in Figure 10.

We suppose also that PVAc as a component with a flexible chain is acting as surfactant, thus assisting better deformability and better intermixing of all the components. The increase in PVAc concentration in the reactive mixture (CP2 sample) slightly influences a yield and composition of the fractions enriched with PVA or chitosan (CP1-B and CP1-C). Also these fractions can contain PVA and chitosan homopolymers, respectively. At the same time, the content of chitosan in CP2-A fraction is essentially lower (from 50 up to 4–5 mol%). The yield of this fraction also is decreased. Thus, an increase in plasticity of the systems under co-extrusion affects an efficiency of the mechanochemical processes. The close results are received at introduction of PVAc as a 10% solution in *i*-PrOH. This solvent affects plasticity of the reactive mixture not so much at a high content of chitosan but probably serves as a free radical scavenger preventing also an oxidation of PVA in alkaline medium.

WAXD analysis shows that X-ray diffraction pattern of CP1-B copolymer is similar to that of neat PVA. At the same time, as follows from Figure 11, the films made of this copolymer demonstrate much better mechanical properties than the PVA-homopolymer. These facts indicate that the distribution of the copolymer over the PVA-matrix is perfectly uniform. This is the very reason for improved plasticity of such materials.

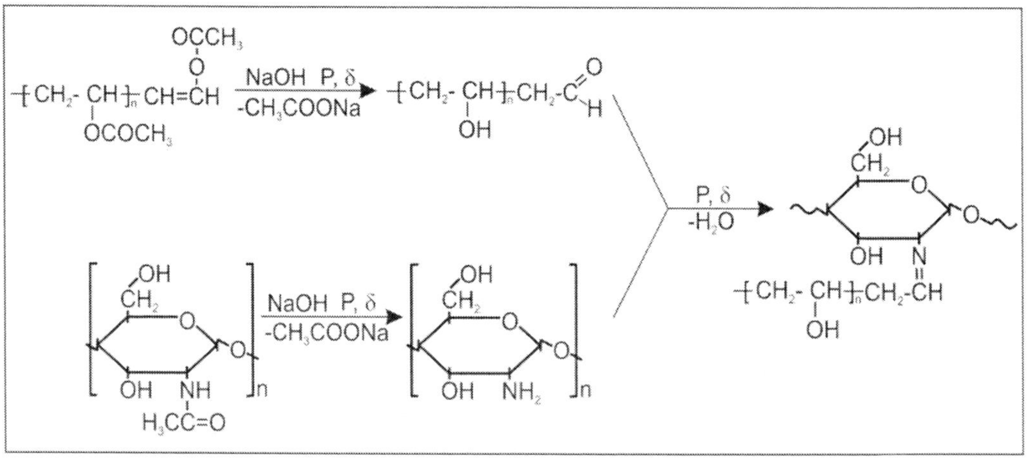

Figure 10. The proposed synthetic scheme for preparation of the PVA/chitosan copolymers.

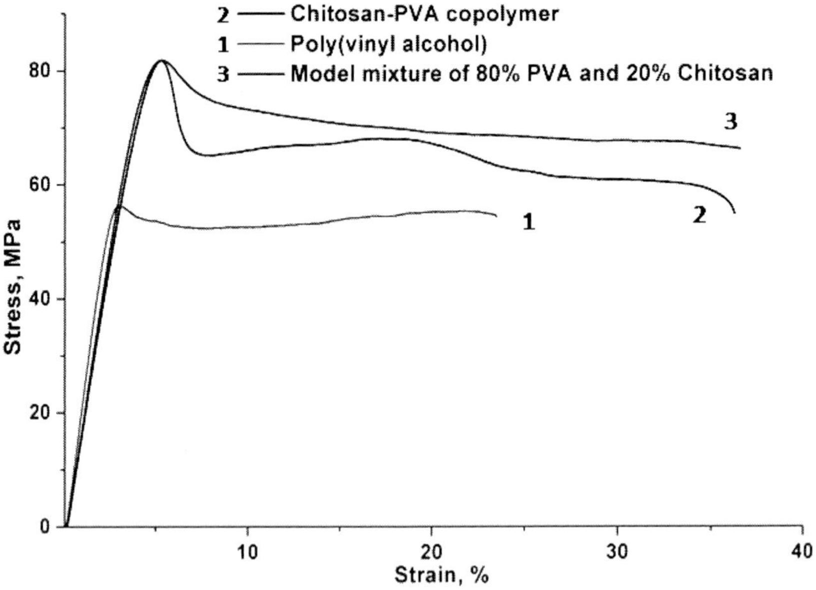

Figure 11. The curves of a strain dependence on stress for pure PVA films and the films of CP1-B and the model mixture.

Table 5. Tensile properties of PVA and PVA/chitosan films

Sample	Content of chitosan (wt-%)	Tensile strength (MPa)	Modulus of elasticity (GPa)	Strain at tensile strength (%)
PVA (Mowiol 66-100)	0	62	2.9	2.6
Blend film 1	10	98	2.2	4.6
Blend film 2	20	85	1.9	5.0
Blend film 3	30	83	1.8	5.5
Blend film 4	40	82	1.8	5.2
Blend film 5	80	60	1.5	4.6
Copolymer (17 wt-% of Chitosan)	15	81	2.3	4.6
Relative error		±10%	±5%	±10%

The results of comparative physical-mechanical tests of copolymeric and blend films prepared by solvent casting technique are listed in Table 5. To prepare model blend films, a 10% water solution of poly(vinyl alcohol) (Mowiol; 66-100; MW 100,000; a content of vinylacetate units is 1 %) and a 10% solution of chitosan in 2% acetic acid were used. It was, in particular, shown that the copolymeric films based on 17 wt% chitosan and 83 wt% PVA (CP1-B sample) demonstrate a similar in mechanical strength and plasticity properties as compared with the model blend films .

The films cast from graft-copolymer solutions possessed fairly good strength characteristics but were relatively brittle. To reduce the brittleness of such materials and to improve their elongation-at-break, it is possible to add a plasticizer (glycerol) if this additive does not affect the set of other characteristics of the material.

The SEM micrographs of the surface of all the PVA/chitosan samples indicate their homogeneous structure. The analysis of structure of CP1-B copolymer films with ACM (160 kHz) (Figure 12) shows a presence of heterogeneous regions of 100 nm which can be related to aggregates of rigid-chain polymer (chitosan).

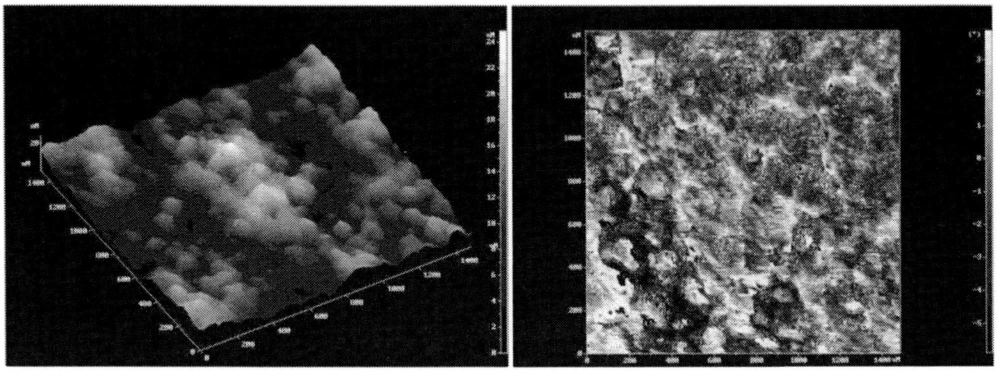

Figure 12. AFM images of a surface of the film made of CP1-B copolymer (left – topography, and right – phase contrast, 1.5×1.5 μm).

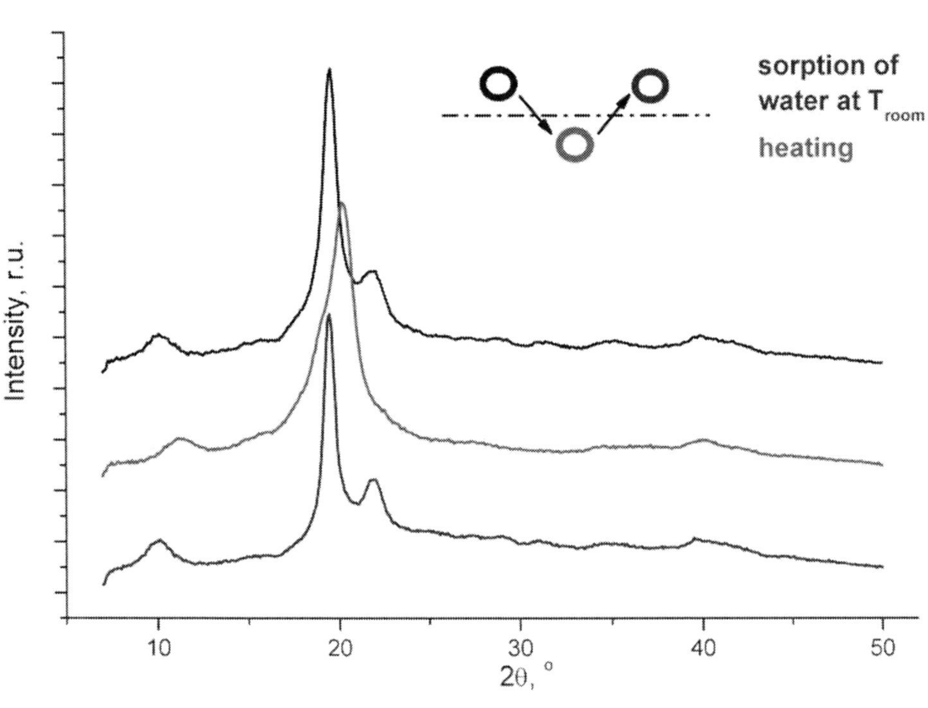

Figure 13. WAXS curves of the CP1-A copolymer (powder) during heat treatment at 120°C followed by sorption of water.

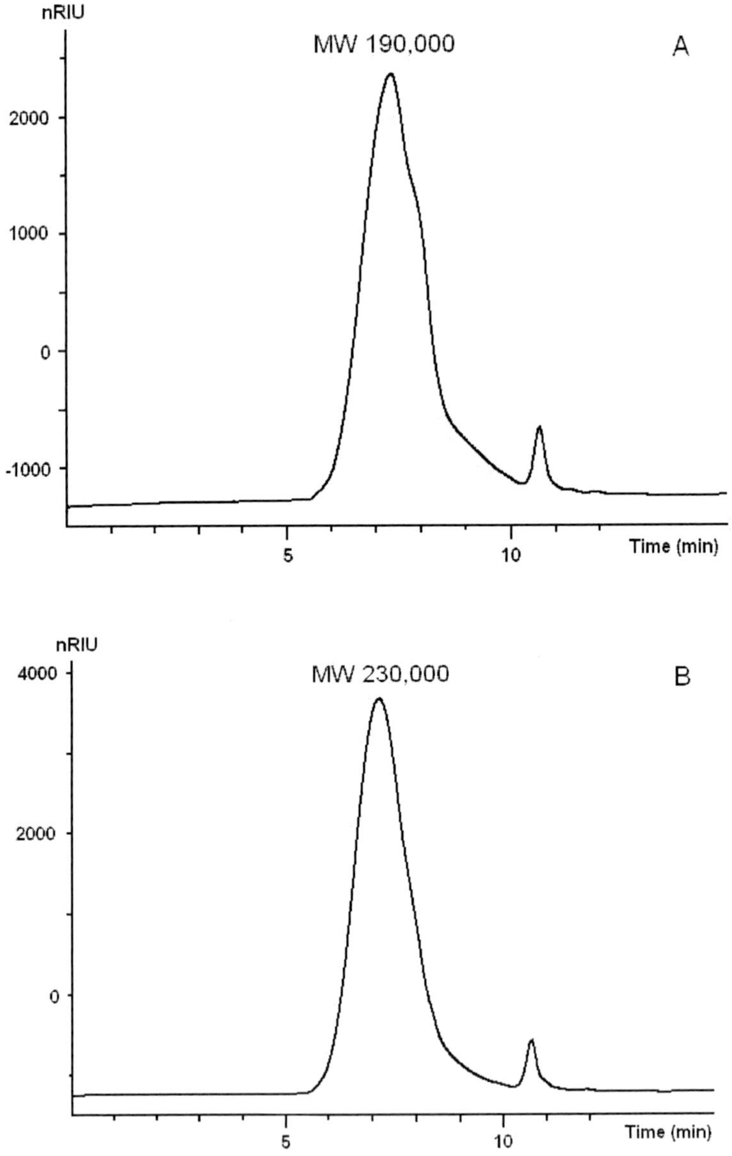

Figure 14. GPC curves of the CP1-A and CP1-B samples in (H_2O + 0.02% NaN_3).

The CP1-A copolymer of 80%-wt of chitosan shows WAXS curve typical for chitosan with high degree of crystallinity (Figure 13). Unlike that the blend films with approximately the same compositions were an almost amorphous. It is not surprising for graft-copolymers if length of grafted chains is long enough. A brief heat treatment at 120°C of CP1-A sample in order to remove water leads to dramatic changes. The peaks typical for crystalline structure of chitosan completely disappear and are replaced by a new peak of crystalline structure of PVA. All these changes are completely reversible, namely after water sorption the picture in all the details returns back to the initial one.

The reason for such a behavior most probably is related to easy swelling of highly hydrophilic PVA. It seems to be of logic to propose that an important feature of the system under consideration is that all the PVA is statistically distributed in the form of relatively short graft-chains. This very model explains the solubility of the copolymers in neutral water at ambient temperatures that is impossible for each of homopolymers taken separately.

The DSC melting endotherm recorded during the first heating run for the CP1-A sample has demonstrated the PVA melting peak at Tm = 218°C slightly shifted to lower temperatures as compared to that of neat PVA.

The average MW of the samples was determined by using GPC on Agilent 1200 (PLaquel – OH 40 column, 25°C, a flow-rate 1 of mL/min) pre-calibrated with dextran standards. Figure 14 shows the GPC curves of the CP1-A and CP1-B samples with (H_2O + 0.02% NaN_3) as the solvent/eluent. A 0.2% solution of the sample in eluent was used. The relative amounts of copolymers were monitored by means of an on-line refractive index (RI) detector. The calculated values of average MW of the soluble in neutral water fractions were 190,000 and 230,000 g/mol for CP1-A and CP1-B samples, respectively.

Interaction of the pure polymers under SSRB conditions was studied as well. Chitosan was refined from the reactive blend after solid-state deacetylation of chitin, then dried and blended in extruder with neat commercial PVA (Mowiol 40-88; average MW 205,000; a degree of hydrolysis of 86.7–88.7 mol) at equal molar ratio and various temperatures. The blending had no success in grafting in this case. The soluble at ambient temperature fraction was less than 5% in respect to entire composition and contained just a negligible quantity of amine groups in the solution. Thus, the proposed synthetic scheme by employing as the precursors PVAc and chitin seem to be much more efficient to prepare chitosan-g-PVA copolymers. To the best of our knowledge such systems were obtained first time.

These systems that can be formed as films, microfibers and microcapsules were demonstrated to be very promising materials for numerous biomedical applications, in particular, as carriers of bioactive compounds and wound dressings [58, 59]. The developed stable polyelectrolyte microcapsules of the desired size and membrane thickness based on the chitosan-g-PVA copolymers were successfully used for a long term cultivation of animal cells [60, 61]. In addition, the obtained copolymers unlike those of their composite mixtures possess an excellent ability to stabilize the dispersions of inorganic particles, in particular TiO_2 nano-dispersions which play important role as photocatalytic sensibilizers of a multitude of biological processes [62, 63]. Thus, these copolymers could be successfully employed to prepare new hybrid polymeric systems with entrapped nanofillers both organic and inorganic origins.

4. Solid-State Reactive Blending of Chitosan with Polyesters

A use of chitosan as drug delivery agent or a biomaterial that can support chondrogenesis has drawn a considerable attention in the applications as a component of cartilage tissue scaffolds [64, 65].

Composites and graft-copolymers of polysaccharides with polyesters, like chitosan and poly(lactide)s, which are biodegradable and biocompatible materials, are of great practical and theoretical importance. . Poly(lactide)s (PLA) and lactide copolymers are biodegradable, compostable, producible from renewable resources, and nontoxic to the human body and the environment [66]. PLA undergoes a scission in the body to lactic acid, a natural intermediate in metabolism. However, a lack of functional groups to promote cell attachment limits its usage in some biomedical fields, for example for tissue engineering. Nevertheless, PLA ability to form compatible blends with some other polymers can provide an improvement of physicochemical properties of biomaterials based on PLA [67].

The use of materials based on chitosan and PLA for tissue engineering and prolonged or controlled drug delivery has been reported [68-71]. It was shown that with improved toughness, controlled biodegradability, and chemical functionalities, such blends can be a good candidate for new biomaterials. Cytocompatibility of chitosan-g-PLA copolymers and chitosan/PLA mixed systems has been also demonstrated [72, 73]. In such compositions chitosan can interact more directly with cells whereas PLA provides both mechanical strength and stiffness to the biodegradable structure [74]. Moreover, due to its cationic nature, chitosan could act as a buffer to minimize the drop of pH resulting from the PLA degradation. It is found that this kind of complex materials evoke a minimal foreign body reaction.

The well-known low compatibility of polyesters with polysaccharides strongly limits design and manufacture of hybrid chitosan/PLA polymeric systems in the molten state of the polyester. High viscosity of the melts, especially at fabrication of composite materials with high chitosan content, requires glycerol and water as plasticizers. It results in change for the worse of adhesion as well as leads to hydrolysis of polyester matrix. The yield of graft copolymers forming under melt or solution processing is insubstantial. The grafting occurs on the very surface of the polysaccharide particles. The size of chitosan particles remains practically unchanged as compared with the initial ones. The studies of morphology of these blends confirmed that chitosan and polyester are phase-separated systems [75]. Moreover, in all cases of traditional methods of blending/grafting, it can be found that the choice of common solvents, which is closed related with morphology properties of the blends, is rather difficult.

All of the above mentioned lead to conclusion that the SSRB method has numerous advantages to produce the PLA/chitosan composite materials.

In this chapter we study the morphology as well as structural, mechanical, and relaxation properties of the blends of chitosan with semi-crystalline poly(L-lactide) (PLA) and amorphous poly(D,L-lactide-co-glycolide) (PLGA) obtained at various temperatures below melting point of PLA and glass transition of PLGA and at different molar ratio of components.

Polymers used were a commercial poly(L-lactide) (Sigma) with an average MW of 160,000; poly(D,L-lactide-co-glycolide) (Resomer, type 50:50, Boehringer Ingelheim) with MW of 52,000; and chitosan Chs1 sample having DA of 0.10 and average MW of 50,000 g/mol.

To prepare chitosan-polyester blends, chitosan powders were first mixed with polyester pellets in extruder at 50°C and at a screw rotation speed of 100 rpm. To create a final blend the mixtures were treated for a 10 min using a 300 g batch for each trial at various temperatures: 50, 100, 110, 120, and 130°C in the case of PLA; and at 60°C for PLGA-based system. The chitosan/PLA samples (CPL) at molar ratio of 1:3 (40/60 wt-%) and chitosan/PLGA samples (CPLG) at mass ratio of 60/40 were obtained.

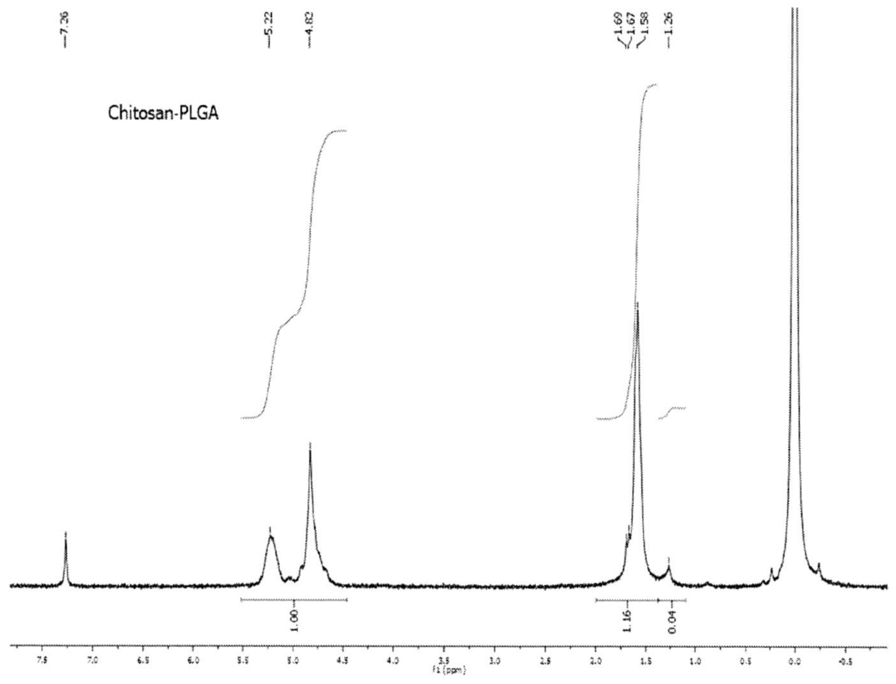

a.

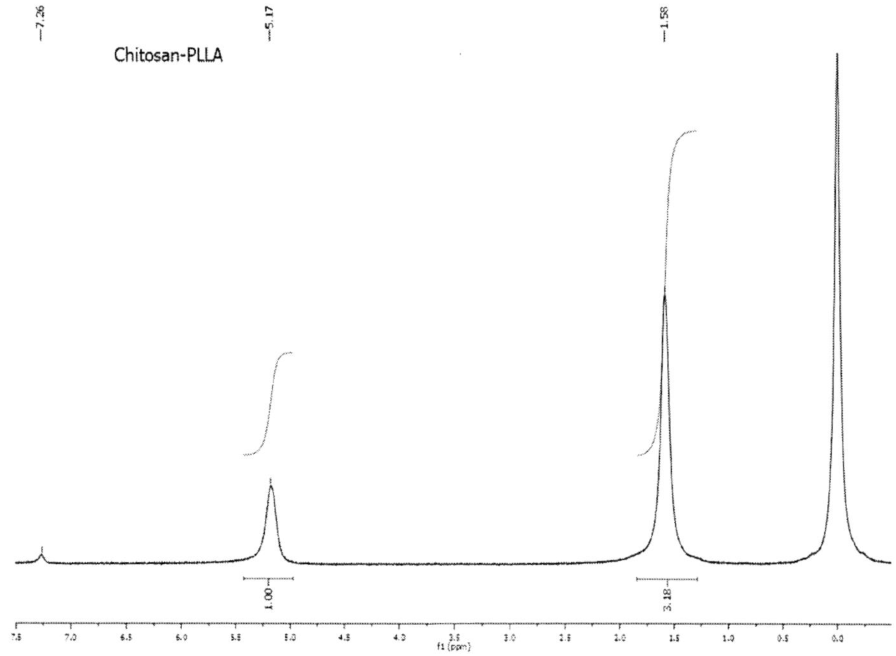

b.

Figure 15. [1]H.NMR spectra of (A) CPLG/60°C and (B) CPL/130°C samples.

In order to characterize and to process the blends, we have investigated their solubility in $CHCl_3$, the solvent being commonly used for polyester analysis. No one of the polymers gave an optically transparent solutions in the chlorinated solvent. Nevertheless, these both materials give a rise to stable dispersions whatever the concentration assessed. Dispersion capacity has been quantified by DLS when analyzing a mean light scattering intensity of these solutions at 90°. It was shown that these dispersions give rise to clear DLS autocorrelation curve thus confirming the formation of nano- microdispersions. The mean size derived from the deconvolution of these autocorrelation curves has revealed that the mean diameter of the CPLG dispersion is substantially lower (~ 200 nm) compared to the other polymer samples which have mean size well above 1 μm [76]. As a conclusion of this dissolution study, it can be stated that these polymeric materials demonstrate an amphiphilic behavior with dispersion propensity in organic solvents. Obviously, this property is totally different compared to what we would observe if we take a physical mixture of two parent materials. It is also worth to mention that the polyester characteristics clearly affect an ability to form nano-scale dispersion. Indeed, the CPLG sample appears as the more dispersible material in opposition to the CPL samples. This property is correlated with a difference in crystallinity and average MW of the parent polyesters.

In order to verify the solubility level of the obtained polymeric material sequences in this organic medium, we also determined their ^{1}H.NMR signal in $CDCl_3$ (30 mg/mL; TMS as an internal reference). As shown in Figure 15, main proton peaks in NMR spectra correspond mainly to the signals coming from the polyester sequences. The proton peaks in the spectra of CPL and CPLG blends are not resolved so well as in the spectra typically observed for homopolyesters. Thus, these NMR data confirm that the polysaccharide moieties are mainly internalized within a core of the nano-microdispersions generated within this organic solvent, and therefore they are not visible in this nuclear spectrometry mode.

Fractional analysis performed in acidic medium for CPL samples obtained at various temperatures (Table 6) strongly confirmed a presence of copolymer fraction in the obtained blends (at least 10%-wt). It can be seen also that the solutions contain a fraction which is not precipitated with 5% NH_4OH. The ninhydrin assay of solutions after precipitation showed a presence of NH_2-groups coming from modified chitosan in olygomeric form.

Table 6. Fractionation data for CPL samples in 4% formic acid

Sample code	Weight of sample[a] (g)	Insoluble in 4% HCOOH fraction (enriched by PLA) (g) / (%)	Soluble fraction[b] (enriched by chitosan) (g) / (%)	Σ weight of fractions (g)	Loss (%)
CPL/50°C	0.7368	0.5073 / 68.9	0.1996 / 27.1	0.7069	4.0
CPL/100°C	0.9754	0.6074 / 62.3	0.3340 / 34.2	0.9414	3.5
CPL/110°C	1.1824	0.7715 / 65.3	0.3609 / 30.5	1.1324	4.2
CPL/120°C	0.7240	0.4645 / 64.2	0.2164 / 29.9	0.6809	5.9
CPL/130°C	0.8293	0.4985 / 60.1	0.2314 / 27.9	0.7299	12.0

[a] Chitosan: PLA mass ratio is 40:60.

[b] Weight of precipitated with 5% NH_4OH fraction.

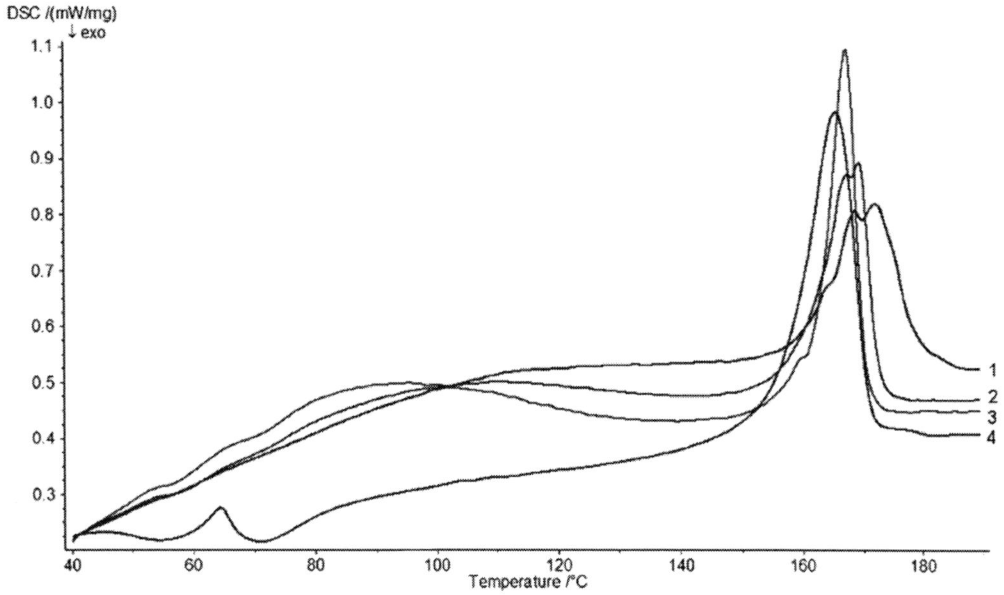

Figure 16. First heating DSC curves for (*1*) CPL/50°C, (*2*) CPL/100°C, (*3*) CPL/130°C, and (*4*) neat PLA samples.

The grafting can be a result of N-acylation of chitosan amine groups by polyester end carboxyl groups. Moreover, as it has been shown in our previous studies [30], ionic interaction between chitosan amine groups and carboxylic groups occurs easy under shear deformation conditions.

The DSC melting endotherms recorded during the first heating run for neat PLA polyester and corresponding CPL samples prepared by extrusion at various temperatures are shown overlaid in Figure 16. In the first heating run, neat PLA has demonstrated a glass transition at 61.2°C with $\Delta Cp\ (T_g) = 0.330$ J g^{-1}, and the melting peak at $T_m = 165.5$°C with $\Delta H = 40.6$ J g^{-1}.

It should be noted that chitosan does not reveal sharp endothermic transitions in the temperature region of 130-190°C, and hence all the endotherms shown in Figure 16 originate from PLA melting. It may be seen also that PLA melting peaks in the compositions shift to higher temperatures and acquire clear multiphase structure due to chitosan presence.

Noteworthy, DSC melting endotherm of neat PLA itself usually consists of two overlapping peaks that can be separated with the aid of curve-fitting routines. These peaks indicate to coexistence of at least two crystal phases differing in terms of their thermodynamical stability in the parent polymer. Commonly, an appearance and a co-existence of these more or less discrete phases in PLA is related to thermal prehistory of the sample.

Extrusion of PLA in the presence of chitosan, however, results in preparation of the blends whose melting endotherms clearly differ from those characteristics for neat PLA (*cf.* Figure 16). Figure 17, for instance, shows that the blend endotherm is considerably more complex as compared to that of neat PLA and comprises complementary peaks belonging to newly formed phases. The latter obviously indicate to occurrence of specific interaction of polylactide chains with chitosan moieties. It is of interest also that the changes in temperature of the extruder treatment have resulted in variations in the phase content of the PLA endotherm.

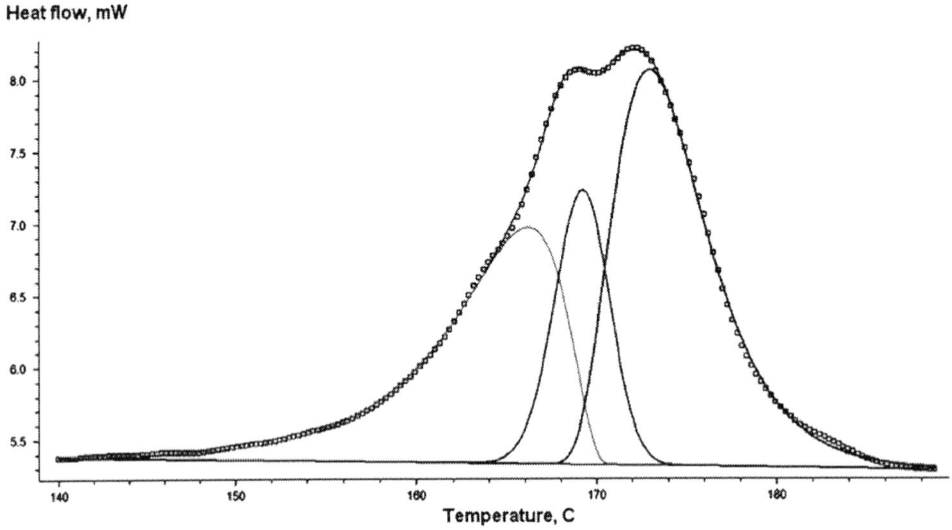

Figure 17. Separation into constituent phases of the complex melting endotherm for CPL/50°C sample.

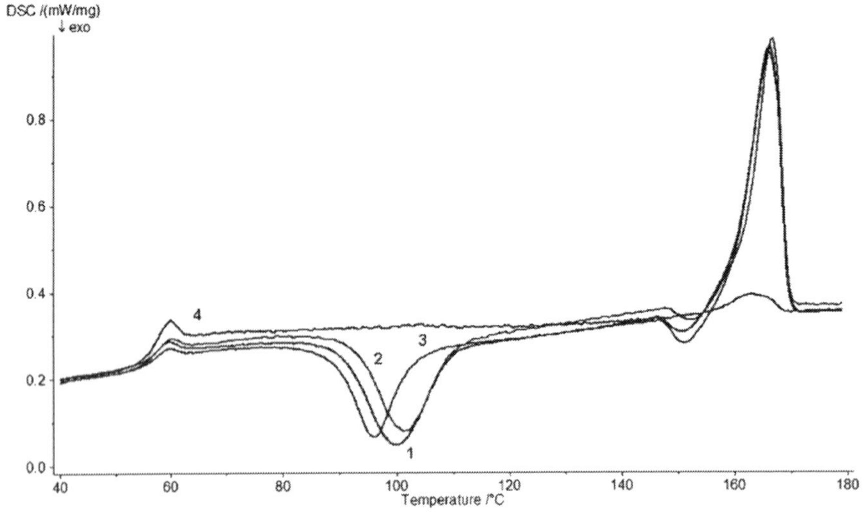

Figure 18. Second heating DSC curves for (*1*) CPL/50°C, (*2*) CPL/100°C, (*3*) CPL/130°C, and (*4*) neat PLA samples.

For all CPL blends melted in DSC and then cooled at 10 K min^{-1} rate, the strong exothermic peaks have been observed during the second DSC heating run in the temperature range of 80-120°C (Figure 18). These exoterms belong to so called "cold crystallization" of PLA chains. The fact that neat PLA has not revealed any cold crystallization peaks at the same conditions infers that chitosan exerts clear nucleating effect on PLA. This observation is similar to that one reported by Correlo et al. [71] who melt processed chitosan and PLA at 50 wt% of chitosan. In that case the PLA cold crystallization in the presence of chitosan was observed in the first heating run. They also render chitosan to be a nucleating agent for PLA. The authors noted, however, that differences could be associated with higher decrease of PLA molecular weight during blend processing.

The data presented above does not permit to determine, however, in what crystal form PLA is present in the blends and whether it forms a separate phase in blend compositions. In order to evaluate structural and relaxation properties of the obtained materials, WAXS and DMA analyses were performed. The films of CPL samples were prepared by press forming at 180°C. The mechanical properties of the films were measured as well.

Table 7. Tensile properties of the films of neat PLA and CPL samples obtained at 50 and 130°C

Sample	Tensile strength (MPa)	Strain at tensile strength (%)	Modulus of elasticity (GPa)
PLLA	48.0	2.4	2.3
CPL/50°C fast cooling	51.0	2.0	2.9
CPL/130°C fast cooling	18.3	0.7	2.7
CPL/130°C after annealing	21.3	0.7	3.0

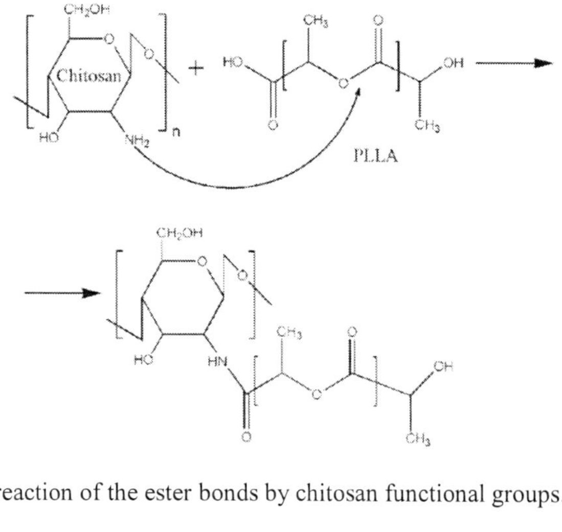

Figure 19. Aminolysis reaction of the ester bonds by chitosan functional groups.

The tensile strength, modulus of elasticity, and elongation are shown in Table 7. The tensile strength of the blends prepared by low-temperature extrusion is is close to the values characteristic for neat PLA. Employment of higher temperature during extrusion of CPL blends led to a decrease of the tensile strength. Under conditions of solid-state synthesis a decrease of PLA molecular weight and, thus, a decrease in the tensile strength of the films, can occur also due to aminolysis reactions and alcoholysis of the ester bonds by chitosan functional groups (shown in Fig 19). It provides polyester moieties grafting onto chitosan chain and formation of uniform structure of the chitosan/PLA materials. In the case of films prepared by slow cooling of the melt, PLA crystallizes and the film strength increases. Wide-angle XR scattering data for the slowly annealed and rapidly quenched hot-pressed films of CPL blend confirmthe differences in their crystallinity.

The prepared films were compared in terms of their morphology with films of a model mixture of the same composition (40 wt% of chitosan) prepared with the aid of microcompounder at 190°C and then cast from the melt. The size of chitosan particles in the latter case remained unchanged (about 100 μm) , and the samples visually appeared to be heterogeneous. The films prepared in this model experiment were highly brittle and unsuitable for evaluating their mechanical properties.

According to the DMAdata,

fast cooling of the composite films, promoting amorphization of PLA phases, has resulted to a decrease of the dynamic storage modulus E' of the sample (Figure 20). It is corresponding to the region of a glass-transition (at 60°C) and correlates with WAXS and DSC data for the CPL samples.The same relaxation process but not so clearly manifested is observed in the sample subjected to annealing. In both composite films the transition in the state of plastic flow is not completed. It is obvious that mechanical properties of compositions above T_g of PLA are determined exclusively by the chitosan matrix. Unlike hydrophilic systems based on chitosan and PVA, the given system is stable in terms of mechanical properties in the presence of the moisture [63].

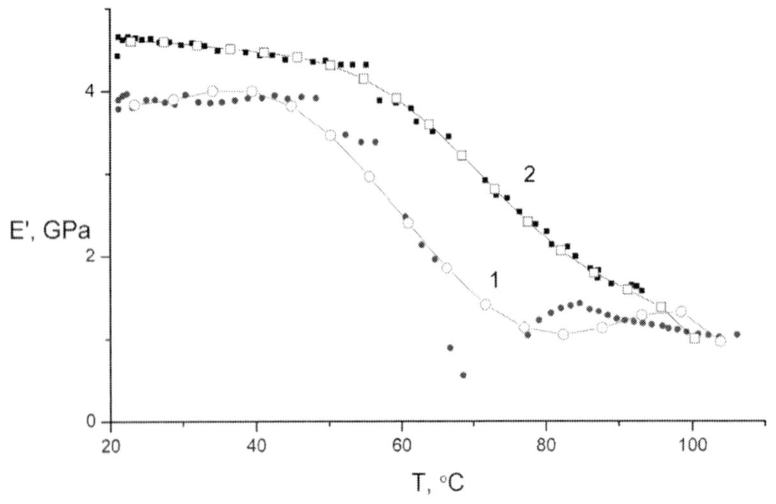

Figure 20. DMA curves of the dependence of storage modulus E' on temperature for (*1*) amorphous and (*2*) crystalline films of CPL/130°C sample.

Thus, the blending of chitosan/polyester mixtures under SSRB conditions leads to substantial changes of their structure. Strong interaction in the blends at the solid-state co-extrusion has resulted in facilitating a degree of chitosan particle dispersability. Microfibers from the colloidal (with particle size of 200-1000 nm) CPL dispersions in chloroform containing up to 40 wt% of chitosan were obtained by electrospinning process. This approach could be extended over a broader range of polyalcanoates (PHB, PCL and the like). These amphiphilic composites can be assuredly used to promote tissue regeneration.

5. Conclusion

This chapter has dealt with a novel method of chemical modification of chitin and chitosan in a simple one-step procedure under conditions of shear deformation. The most attractive feature of this method is that the entire modification process proceeds in the solid state, and thus does not require melting of components or any solvents as reaction medium. The structural transformations of chitin that take place during its solid-state processing in an extruder as well as related morphology, structural, mechanical, and relaxation properties of chitosan, polylactide-chitosan and poly(vinyl alcohol)-chitosan composites produced by the solid-state synthesis are studied.

The materials were processed with a semi-industrial co-rotating twin-screw extruder (ZE 40A×40D UTS, Berstorff, Germany) having a variable set of processing elements that provide a high shear strain and powerful dispersive action.

It is apparent that the alkaline deacetylation of chitin at the conditions of solid-state synthesis proceeds through a preliminary formation of the amorphous solid solution. The deacetylation degree of chitin achieved in this process is almost complete. The proposed approach has numerous advantages because it allows us to considerably decrease the process duration, consumption of reagents and, in such a way, to improve the economical and ecological parameters of the process. A summary of the obtained results showed that all obtained chitosan samples are homogeneous with respect to their chemical structure. Comparison of the surface with the bulk of the formed chitosan particles confirm that the solid-state synthesis, which does not require swelling prior to modification of the polymer, provides high deacetylation in the bulk of chitin particles due to the absence of diffusion restrictions in the solid-state synthesis. This approach to preparation of chitosan is also shown to yield a more amorphous product in comparison with chitosan produced by the traditional suspension method (at a comparable level of acetamide group conversion into amine groups). The dimensions of both chitin and chitosan crystallites after solid-state treatment agree with the values obtained for the original chitin and for chitosan synthesized by the traditional method. At the same time, the crystalline phases of these samples seem to be formed under different conditions. This result agrees with the general theory of mechanochemistry [25]. Processing of chitin and synthesis of chitosan in the solid state are accompanied not only by dispersion of the crystalline phase at the macroscopic level. In addition, the mechanically activated materials demonstrate the formation of labile nanoscale structures. Inside these structures, dislocationless mass transfer ensures deformation-induced molecular-level mixing, which is sufficient, in particular, for the solid-state deacetylation reaction with a quantitative yield.

At present the most commonly applied ways to control the affinity and to improve the mechanical properties of biopolymer-based compositions are blending with synthetic polymers of different hydrophilicity. It was shown in this chapter that the solid-state reactive blending results in modification of the chitosan properties (in particular, solubility) due to grafting of the synthetic polymers onto chitosan chains.

Of particular interest are the products of concurrent extrusion of chitin with poly(vinyl acetate) in the presence of NaOH, which under conditions of solid state shear deformation were transformed into the grafted system comprised of almost completely deacetylated chitosan and poly(vinyl alcohol). The main feature of these co-polymeric systems is solubility in water at neutral pH. These systems can form films, microfibers and microcapsules and due to their biocompatibility and enhanced solubility were demonstrated to be very promising materials for numerous bio-medical applications.

The obtained chitosan-polyester systems were found to possess amphiphilic properties and to form micelle-like stable ultra-fine dispersions in organic solvents. Obviously, this property totally differs from what we would observe in the case of a non-modified chitosan which would not dissociate in these solvents at all. A uniform structure of the obtained materials is confirmed by their enhanced mechanical properties as compared with a model molten system with the same composition. This approach to design new polysaccharide-based materials can be extended to a wide range of polysaccharides and polyesters. Such targeted modification leads to biodegradable and biocompatible materials. At the same time it provides numerous possibilities to circumvent many processing obstacles typical for preparation of natural-polymers-based hybrid materials. Thus, solid-state reactive polymer blending seems to be a very promising way to new polymeric compositions.

The authors express their gratitude to Dr. Leonid Vladimirov, Dr. David Lanson (Rhodia-CRTA Aubervilliers), and to all other colleagues for providing chitosan sample analyses and fruitful discussions. This research was financially supported by the Russian Foundation for Basic Research (RFBR, project no. 10-03-01022-a); by the grant of the President of the RF for the support of leading scientific schools (SS-4371.2010.3). Additional support provided by a French Corporation, RHODIA, is gratefully acknowledged.

REFERENCES

[1] Muzzarelli, R. A. A.; Muzzarelli, C. *J. Adv. Polym. Sci.* 2005, vol.186, 151-209.

[2] Gupta, K. C.; Ravi Kumar, M. N. V. *J. Macromol. Sci.-Reviews in Macromol. Chem. Phys.* 2000, vol. 40, 273-308.

[3] Berger, J.; Reist, M.; Mayer, J. M.; Felt, O.; Peppas, N. A.; Gurny, R. *Eur. J. Pharm. Biopharm.* 2004, vol. 57(1), 19–34.

[4] Yang, T. L.; Young, T. H. *J. Biomaterials* 2008, 29(16):2501-8.

[5] Chen, M. H.; Hsu, Y. H.; Lin, C. P.; Chen, Y. J.; Young, T. H. *J. Biomed Mater Res* A 2005, vol. 74(2), 254-62.

[6] Suh, J.-K.; Matthew, H. W. *J. Biomaterials* 2000, vol. 21, 2589-2598.

[7] Duan, B.; Wu, L.; Li, X.; Yuan, X.; Li, X.; Zhang, Y.; Yao, K. *J. Biomater Sci. Polym.* Ed. 2007, vol. 18(1), 95-115.

[8] Payne, G. F.; Raghavan, S. R. *J. Soft Matter* 2007, vol. 3, 521-527.

[9] Mima, S.; Miya, M.; Iwamoto, R.; Yoshikawa, S. *J. Appl. Polym. Sci.* 1983, vol. 28, 1909-1917.
[10] Lyakhov, N. Z.; Grigorieva T. F.; Barinova, A. P.; Vorsina, I. A. *Russian Chemical Reviews* 2010, vol. 79(3), 189-203.
[11] Bityagin, P. Yu. *Russian Chemical Reviews* 1994, vol. 63(12), 965-976.
[12] Heinicke, G. *Tribochemistry*; Hanser, Munich, 1984.
[13] Gielen, M.;Willen, R.; Wrackmeyer, B. *Solid-state organometallic chemistry, methods and applications*; Wiley, Chichester, 1999.
[14] Boldyrev, V. V. *Russian Chemical Reviews*, 2006, vol. 75, 177–189.
[15] Avvakumov, E.; Senna, M.; Kosova, N. *Soft mechanochemical synthesis: A Basis for New Chemical Technologies*; Boston, Kluwer Academic Publishers, 2001.
[16] Grigorieva, T. F.; Barinova, A. P.; Lyakhov, N. Z. *Journal of Nanoparticle Research* 2003, vol. 5, 439-453.
[17] Aylmore, M. G.; Lincoln, F. J; Cosgriff, J. E.; Deacon, G. B.; Gatehouse, B. M.; Sandoval, C. A.; Spiccia, L. *Eur. J. Solid State Inorg. Chem.* 1996, vol. 33, 109-119.
[18] Gaffet, E.; Bernard, F.; Niepce, J.-C.; Charlot, F.; Gras, C.; Le Caër, G.; Guichard, J.-L.; Delcroix, P.; Mocellin, A.; Tillement, O. (1999) *J. Mater Chem.* 1999, vol. 9, 305-314.
[19] Fernández-Bertran, J. F. *Pure Appl. Chem.* 1999, vol. 71, 581-586.
[20] Kaupp, G. In Organic Solid State Reactions; Toda, F.; Eds.; *Top Curr. Chem.* 254; Springer-Verlag: Berlin, 2005; vol. 254, 95–183.
[21] Zharov, A. A., in *High-Pressure Chemistry and Physics of Polymers*, ed. by A.L. Kovarskii; London, Tokyo: CRC Press, 1994; Ch. 7, 268.
[22] Enikilopov, N. S. *Russian Chemical Reviews* 1991, vol. 60(3), 283-287.
[23] Dubinskaya, A. M. *Russian Chemical Reviews* 1999, vol. 68(8), 637–652.
[24] Butyagin, P. Yu.; Dubinskaya, A. M.; Radtsig, V. A. *Russian Chemical Reviews* 1969, vol. 38(4), 290-305.
[25] Bityagin, P. Yu. *Colloid* J. 1999, vol. 61, 537-544.
[26] Prut, E. V.; Zelenetskii, A. N. *Russian Chemical Reviews* 2001, 70 (1), 65-79.
[27] Boldyreva, E. V. *J. Mol. Struct.* 2004, vol. 700, 151-155.
[28] Bridgman, P.W. *Phys. Rev.* 1935, vol. 48, 825-847.
[29] Akopova, T. A.; Rogovina, S. Z.; Vikhoreva, G. A.; Zelenetskii, S.N. *Polymer Science* (B) 1995, vol. 37, 528-531.
[30] Rogovina, S. Z.;, Vikhoreva, G. A.; Akopova T. A.; Gorbacheva I. N. *J. Appl. Polym. Sci.* 2000, vol.76, 616-622.
[31] Rogovina, S. Z.; Akopova T. A.; Vikhoreva, G. A. *Polym. Degrad. and Stability*, 2001, vol. 73, 557-560.
[32] Rogovina S. Z.; Akopova T. A. *Polymer Science, Ser.* C 1994, vol. 36, 487-494.
[33] Rogovina S. Z.; Vikhoreva G. A.; Akopova T. A. In "Chitosan in Pharmacy and Chemistry", Ed. R.A.A. Muzzarelli and C. Muzzarelli; Italy, ATEC, 2002; 449.
[34] Akopova, T. A.; Vladimirov, L. V., Zhorin, V. A.; Zelenetskii A. N. *Polymer Science*, Ser. B 2009, vol. 51, 124–134.
[35] Ozerin, A. N.; Zelenetskii, A. N.; Akopova, T. A.; et al., RF Patent № 2292354, 2007.
[36] Liang, C. Y.; Marchessault, R. H. *J. Polym. Sci.* 1960, vol. 43, 71-84.
[37] Pearson, F G.; Marchessault, R. H.; Liang, C. Y. *J. Polym. Sci.* 1960, vol. 43, 101-116.
[38] Gardner, K. H.; Blackwell, *J. Biopolymers* 1975, n. 14, 1581-1595.

[39] Domszy, J. G.; Roberts, G. A. F. *Makromol. Chem*. 1985, vol. 186, 1671-1677.

[40] Salmon, S.; Hudson, S. M. *J. Macromol. Sci.* C 1997, vol. 37, 199-276.

[41] Li, J.; Revol, J. F.; Marchessault, R. H. *J. Appl. Polym. Sci.* 1997, vol. 65, 373-380.

[42] Rogovina, S. Z.; Akopova, T. A.; Vikhoreva, G. A. *J. Appl. Polym. Sci.* 1998, vol.70, 927–933.

[43] Mogilevskaya, E. L.; Akopova, T. A.; Zelenetskii, A. N.; Ozerin, A. N. *Polymer Science,* Ser. A, 2006, vol. 48, 116–123.

[44] Gamzazade, A. I.; Shlimak, V. M.; Sklar, A. M.; Shticova, E. V.; Pavlova, S. A.; Rogojin, S. V. *Acta Polym.* 1985, vol. 36, 420–424.

[45] Duarte M. L. *Int. J. Biol. Macromol.* 2002, vol. 31, 1–8.

[46] Matsumura, S.; Tomizawa, N.; Toki, A.; Nishikawa, K.; Toshima, K. *Macromolecules* 1999, vol. 32, 7753–7761.

[47] Corti, A.; Solaro, R.; Chiellini, E. *Polym. Degrad. Stab*. 2002, vol. 75 (3), 447–458.

[48] Lee, S. Y.; Pereira, B. P.; Yusof, N.; Selvaratnam, L,; Yu, Z.; Abbas, A. A.; Kamarul, T. *Acta Biomater*. 2009, vol. 5, 1919–1925.

[49] Berger, J.; Reist, M.; Mayer, J. M.; Felt, O.; Peppas, N. A.; Gurny, R. *Eur. J. Pharm. Biopharm.* 2004, vol. 57, 19–34.

[50] Mucha, M.; Ludwiczak, S.; Kawinska, M. Carbohyd. Polym. 2005 vol. 62, 42–49.

[51] Kim, S. T.; Park, S. J.; Kim, S. *J. Reactive and Functional Polymers* 2003, vol. 55, 53–59.

[52] Srinivasa, P. C.; Ramesh, M. N.; Kumar, K. R.; Tharanathan, R. N. *Carbohydrate Polymers* 2003, vol. 53, 431–438.

[53] Mangala, E.; Kumar, T. S.; Baskar, S.; Rao, K. P. *Trends Biomat. Art. Organs*. 2003, vol. 17, 34-40.

[54] Wang, T.; Turhan, M.; Gunasekaran, S. *Polym. Int.* 2004, vol. 53, 911-918.

[55] Cascone, M.G.; Maltinti, S.; Barbani, N. *J. Mat. Sci.*: Materials in medicine 1999, vol. 10, 431-435.

[56] Zhang, S.-J.; Yu, H.-Q. *Water Research* 2004, vol. 38, 309–316.

[57] Kim, S.J.; Lee, K.J.; Kim, S.I.; Lee, K.B.; Park, Y.D. *J. Appl. Polym. Sci.* 2003, vol. 90, 86-90.

[58] Zelenetskii, A. N.; Akopova, T. A.; Kildeeva, N. R.; et al. *Russ. Chem. Bull. Int.* Ed., 2003, vol. 53, 2073- 2077.

[59] Akopova, T. A.; Markvicheva, E. A.; Vladimirov, L. V.; et al. *Ext. Abst. Eur. Polym.* Congress 2005, Moscow, Russia.

[60] Markvicheva, E.; Zaitseva-Zotova, D.; Akopova, T.; Zelenetskii, A. *Eur. J. Cell Biol.* 2007, vol. 86S1, 63.

[61] Zaytseva-Zotova, D.; Balysheva, V.; Tsoy, A.; Akopova, T.; Vladimirov, L.; Bolotine, L.; Goergen, J.-L.; Markvicheva, E. *Macromol. Bio Sci.,* submitted.

[62] Ozerin, A. N.; Zelenetskii, A. N.; Akopova, A. N.; et al. *Polymer Science*. Ser. A 2006, vol. 48, 638- 643.

[63] Ozerin, A. N.; Perov, N. S.; Zelenetskii, A. N.; Akopova, T. A.; et al. *Nanotechnologies in Russia* 2009, vol. 4, 331–339.

[64] Breyner, N.M.; Hell, R.C.; Carvalho, L.R.; Machado, C.B.; Peixoto Filho, I.N.; Valério, P.; Pereira, M.M.; Goes, A.M. *Cells Tissues Organs* 2009, E-pub ahead of print.

[65] Jin, R.; Moreira Teixeira, L.S.; Dijkstra, P.J.; Karperien, M.; van Blitterswijk, C.A.; Zhong, Z.Y., Feijen, *J. Biomaterials* 2009, vol. 13, 2544-2551.

[66] Simamora, P.; Chern, W. *J. Drugs Dermatol.* 2006, vol. 5(5), 436-440.
[67] Jiao, Y.P.; Cui, F.Z.; *Biomed. Mater.* 2007, vol. 2(4), R24-37.
[68] Cui Y.L., Qi A.D., Liu W.G., Wang X.H., Wang H., Ma D.M., Yao K.D., *Biomaterials* 2003, vol. 24, 3859-3868.
[69] Ding, Z.; Chen, J.; Sao, G.; Chang, J.; Zhang, J.; Kang, E.T. *Biomaterials* 2004, vol. 25, 1059-1067.
[70] Sébastien, F.; Stéphane, G.; Copinet, A.; Coma, V. *Carbohyd. Polym.* 2006, vol. 65 (2), 185-193.
[71] Correlo, V.M.; Boesel, L.F.; Bhattacharya, M.; Mano, J.F.; Neves, N.M.; Reis, R.L. *Macromolecular Mat. Eng.* 2005, vol. 290, 1157-1165.
[72] Yao, F.; Chen, W.; Wang, H.; Liu, H.; Yao, K.; Sun, P.; Lin, H. 2003, *Polymer*, vol. 44, 6435-6441.
[73] Bhattarai, N.; Ramay, H. R.; Chou, S.-H.; Zhang, M. *Int. J. Nanomedicine* 2006, vol. 1(2), 181–187.
[74] Prabaharan, M.; Rodriguez-Perez, M. A.; de Saja, J. A.; Mano, J. F. *J. Biomed. Mater. Res.* Part B: Appl. Biomater. 2007, vol. 81, 427-434.
[75] Correlo, V.M.; Boesel, L.F.; Bhattacharya, M.; Mano, J.F.; Neves, N.M.; Reis, R.L. *Mater. Sci. Eng.* A 2005, vol. 403 (1-2), 57-68.
[76] Demina, T.S.; Akopova, T.A.; Shchegolikhin, A.N.; Grandfils, Ch.; Markvicheva, E.A.; et al. *J. Biomater. Sci. Polymer* Edn, submitted.

Reviewed by G.E. Zaikov, N.M. Emanuel Institute of Biochemical Physics, Russian Academy of Sciences, 4 Kosygina Str., 119334 Moscow, Russia

In: Focus on Chitosan Research
Editors: Arthur N. Ferguson and Amy G. O'Neill
ISBN 978-1-61324-454-8

Chapter 9

CHITOSAN DERIVED SMART MATERIALS

Ashutosh Tiwari[1]*, Dohiko Terada[1], Chiaki Yoshikawa[2] and Hisatoshi Kobayashi[1]*

1. Biomaterials Center, National Institute for Materials Science
2. International Center for Materials Nanoarchitectonics
1-2-1, Sengen, Tsukuba 3050047, Ibaraki, Japan

ABSTRACT

Chitosan (CHIT), a non-toxic, biocompatible, and biodegradable natural polymer finds tremendous commercial application in its native and modified form. In the recent years, considerable efforts have been devoted to fabricate CHIT based smart materials with adopting attractive approaches, for examples, the ability to alter texture in a controlled mode via external stimulus including stress, electric or magnetic fields, temperature, moisture and pH. In the chapter, we have discussed the recent studies of the preparation of CHIT based smart materials from their various perspectives with put their special attention in the field of medical diagnostic and treatments. The modifications are illustrated basically on the stimuli-responsive CHIT as molecular device materials, biomimetic materials, hybrid-type composite materials, functionalised polymers, supermolecular systems, information- and energy-transfer materials, environmentally friendly materials, *etc.* at synergetic levels along their striking applications in both strategic and civil sectors.

* Corresponding author: E-mail: tiwari.ashutosh@nims.go.jp (A. Tiwari) and kobayashi.hisatoshi@nims.go.jp (H. Kobayashi); Tel: (+81) 29-860-4495; Fax: (+81) 29-859-2247.

1. INTRODUCTION

Chitosan (CHIT), a polycationic biopolymer is usually obtained from alkaline deacetylation of chitin, which is the main component of the exoskeleton of crustaceans such as shrimp, crab and lobster [1]. It is α (1→4) 2-amino 2-deoxy β-D-glucan, having structural characteristics similar to glycosaminoglycans (Figure 1). It is the characteristic polysaccharide of several important phyla, like arthropoda, annelida, mollusca, coelenterata and of many fungi, e.g. euascomycetes, zygomycetes, basidiomycetes and deuteromycetes. CHIT displays interesting characteristics including biodegradability, biocompatibility, chemical inertness, high mechanical strength, good film-forming properties and low cost [2, 3]. The physicochemical and biological properties of CHIT led to the recognition for number of its applications such as fabrication of artificial muscles [4, 5], wastewater treatment [6], functional membranes [7], food packaging [8-10], drug delivery systems [11, 12] and bio-sensors [13-15].

Moreover, CHIT has three reactive groups, i.e. primary (C-6) and secondary (C-3) hydroxyl groups on each repeat unit and the amino (C-2) group on each deacetylated unit. These reactive groups can be chemically modified for altering its physical and chemical properties. The typical reactions involving the hydroxyl groups are etherification and esterification. Selective *O*-substitution can be achieved by protecting the $-NH_2$ groups during the reaction. As *N*-protected CHIT derivatives, several schiff bases of CHIT and *N*-phthaloyl CHIT have been reported [16]. CHIT can also be modified by either cross-linking [17] or graft copolymerization [18-22]. In many cases, the chemically modified CHIT shows greater activity than the original polymer. In general, the smart properties of the natural CHIT are altered to a remarkable degree by the introduction of very small amount of substituent groups of either neutral or ionic types.

In the recent years, CHIT has drawn attention in the field of smart materials which offer a reversible and yet discontinuous molecular phase change in response to various external physicochemical factors not only because of their unique properties but also because of their potential for significant technological applications [1]. Chemical signals, such as pH, metabolites and ionic factors, will alter the molecular interactions between polymer chains or between polymer chain and ions present in the material system. The physical stimuli, such as temperature or electrical potential, may provide various energy sources for altering molecular interactions. These interactions will change the properties of polymer materials such as solubility, swelling behaviour, configurations of conformational change, redox (i.e., reduction-oxidation) state and crystalline or amorphous transition [23-27].

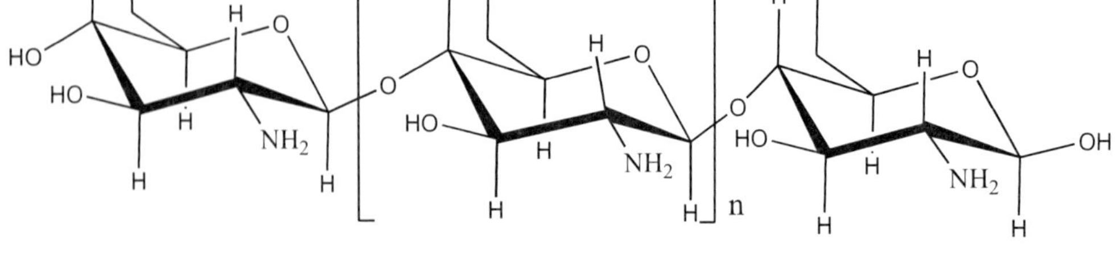

Figure 1. Chemical structure of CHIT.

Further, CHIT based smart materials have found potential applications in biomedical fields [28]. In this context, temperature- and pH-responsive hydrogels have been most widely studied because these two factors are crucial to the human body [29, 30]. Thermo-sensitive hydrogels are of great interest in therapeutic delivery and tissue engineering as injectable depot systems. In a positive thermo-reversible system, polymer solutions having an upper critical solution temperature (UCST) shrink by cooling below the UCST. While the polymer solutions having a lower critical solution temperature (LCST) contract by heating above the LCST. In a negative thermo-reversible system, polymers having LCST below human body temperature have a potential for injectable system applications. These negative thermo-reversible hydrogels are liquid at room temperature (20 – 25 °C) and undergo gelation when in contact with body fluids (36 – 37 °C), due to an increase in temperature. Different thermal setting gels have been described in the literature, including for example acrylic acid copolymers and *N*-isopropylacrylamide (NIPAAm) derivative [31-34]. Poly(*N*-isopropylacrylamide) (PNIPAAm) is well known to have a thermally-reversible property. It exhibits a LCST around 32 °C in aqueous solution [35]. PNIPAAm hydrogels having a chemically crosslinked structure were characterized by a thermo-sensitive nature in which they swell below the LCST and shrink above LCST in water.

Lately, a great attention has been paid to the development of stimuli-responsive CHIT based redox materials with the unique properties such as biocompatibility, biodegradability and biological functions, especially for the biomedical applications. They may be prepared by combining pH-responsive redox polymers with natural based polymeric component to form graft copolymers. A number of graft copolymers have been considered to be incorporated into pH-responsive material in this respect, among them polyaniline (PANI) has the greatest potentiality. The aim of this chapter is to focus on the different methods of preparation of CHIT based smart materials such as CHIT grafted/blended with PNIPAAm, PANI, PPy, etc. designed as stimuli-responsive polymeric materials for biomedical device applications. In this review, various types of strategies involved in the preparation of stimuli-responsive materials, polyelectrolyte complexes (PECs) and core-shell microgels based on CHIT and stimuli-responsive and their properties such as phase transition temperature, swelling behaviour, redox behaviour, mechanical strength, morphology and conductivity are discussed in details.

2. Preparation of Materials

2.1. By Graft Copolymerization

Graft copolymerization of vinyl monomers onto biopolymers using free radical polymerization has attracted the interest of many scientists in the last two decades. This technique enables the production of new polymeric smart materials using CHIT having desired properties and enlarges the field of the potential applications of them by choosing various types of side chains.

Several research articles have been reported on the graft copolymerization of various acrylic monomers onto polysaccharides containing hydroxyl groups using ceric ammonium nitrate (CAN) as an initiator [36, 37]. The most important feature of the oxidation with ceric ion is that it proceeds *via* a single electron transfer with the formation of free radicals on the reducing agent. Recently, NIPAAm monomer was grafted onto CHIT using CAN as shown in Figure 2 [38-40]. The resulted CHIT copolymer showed thermally reversible hydrogel, which exhibits a LCST around 32 °C in aqueous solutions. The percentage of grafting (%G) and the efficiency of grafting (%E) gradually increased with the concentration of NIPAAm up to 0.5 *M*, and then decreased due to the increase in the homopolymerization of NIPAAm. The solubility of CHIT was significantly reduced after grafting with NIPAAm. CHIT-*g*-NIPAAm hydrogels exhibited relatively high equilibrium water content (EWC), and their swelling ratios reached the equilibrium state within about 30 min. All the samples exhibited higher swelling ratios at pH 4 than at pH 7. At 35 °C, which is above the LCST of NIPAAm, the EWC of these hydrogels dramatically decreased compared with those at 25 °C. In addition, their pH-dependent swelling behaviours were more significantly at 25 than at 35 °C. From the plot of EWC vs. temperature as a function of pH, the phase transition of the hydrogels was clearly observed as the temperature increased in all the hydrogels. Also, the decrease of EWC induced by increasing temperature at pH 4 and 6 was more intensive than that at pH 7 and 9. Furthermore, CHIT-*g*-NIPAAm hydrogels showed a decrease of EWC as the temperature increased. The lower EWC value of hydrogel was found to be due to the decrease of the number of free amine groups in the CHIT backbone after grafting. A preliminary *in vitro* cell study showed nontoxic and biocompatible properties. These properties could be very useful in biomedical and pharmaceutical applications.

A temperature-sensitive graft copolymer was prepared by grafting NIPAAm onto a CHIT derivate whose amino groups were protected by phthaloyl groups using AIBN initiator [41]. The deprotection of the phthaloyl groups yields CHIT-*g*-PNIPAAm copolymers with free amino groups. The grafting degree is influenced by the monomer concentration, reaction temperature, initiator concentration and reaction time. In comparison with the native CHIT, CHIT-*g*-PNIPAAm had much better solubility in aqueous solutions in the range pH 1-12. The graft copolymer also exhibited thermal sensitivity. A thermosensitive hydrogel was synthesized by graft copolymerization of NIPAAm with carboxymethylCHIT using APS as an initiator [42]. The influence of the content of carboxymethylCHIT grafted on the properties of the resulted hydrogels was examined. In comparison with the conventional PNIPAAm hydrogels, the resulted hydrogels was found to be improved thermosensitive properties, including enlarged water content at room temperature and faster deswelling/swelling rate upon heating. Therefore, authors suggested that this thermosensitive and biodegradable hydrogel may have the potential applications in controlled drug delivery system and separation and purification of some biological materials such as proteins, enzymes and amylose.

Semi-inter penetrated networks (semi-IPNs) are defined as a composition in which one or more polymers are crosslinked, linear or branched. Verestiuc *et al.* prepared semi-IPNs by the free radical polymerization of NIPAAm onto CHIT using tetraethyleneglycoldiacrylate (TEGDA) as the crosslinking agent [43]. The influence of the degree of crosslinking and that of the ratio of CHIT to PNIPAAm on the pH/temperature induced phase transition behaviour and swelling characteristics of the hydrogel system were investigated. The proportion of CHIT that could be entrapped in the matrix was found to increase as the crosslinking density of the network increased; a maximum CHIT/NIPAAm ratio of 0.46 was obtained. In addition, as the CHIT content and crosslinking density increased, the phase transition temperature of the CHIT interpenetrated hydrogels was seen to become less well defined and to shift towards lower temperatures. The incorporation of CHIT into the structure induced pronounced pH sensitivity: the swelling degree of the hydrogel was seen to vary from ~100% at basic pH to over 2100% at acidic pH.

A full-IPN hydrogel was synthesized by chemical combination of methylene bis-acrylamide (MBAM) cross-linked PNIPAAm network with formaldehyde cross-linked CHIT network [44]. It was demonstrated that the properties of the gels, including the extractability of PNIPAAm within it, the phase transition behaviour, the swelling dynamics in aqueous phase, the swelling behaviour in ethanol/water mixtures and even the microstructure are quite different from those of the semi-IPN CHIT/PNIPAAm hydrogels. The LCST of the full-IPN hydrogel was at least 4-5 °C higher than that of the corresponding semi-IPN hydrogels. The difference of the two gels in the phase transition behaviour was attributed to the difference in their microstructures. The results showed that the semi-IPN hydrogels swell faster than the corresponding full-IPN hydrogels, and the swelling ratio for the semi-IPN hydrogels are almost independent of temperature. It was also noted that the swelling ratios of the full-IPN hydrogels kept almost constant as long as the concentration of ethanol was lower than 30% (v/v). Above that concentration, the swelling ratio started to decrease along with increasing ethanol concentration. When ethanol concentration was beyond 70% (v/v), the swelling ratio was kept constant again. IPN hydrogel of PNIPAAm/CHIT were prepared by free radical polymerization. During this reaction, PNIPAAm was crosslinked with bis(acrylamide) and CHIT was post-cross-linked with glutaraldehyde [45].

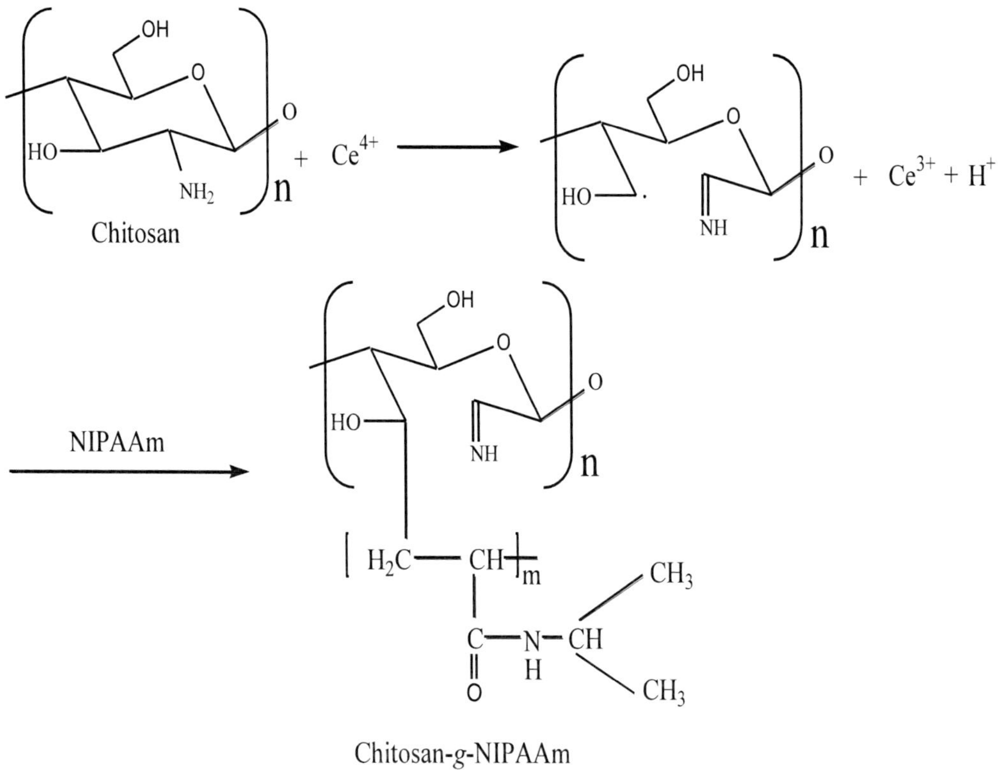

Figure 2. Reaction scheme for the preparation of CHIT-*g*-NIPAAm.

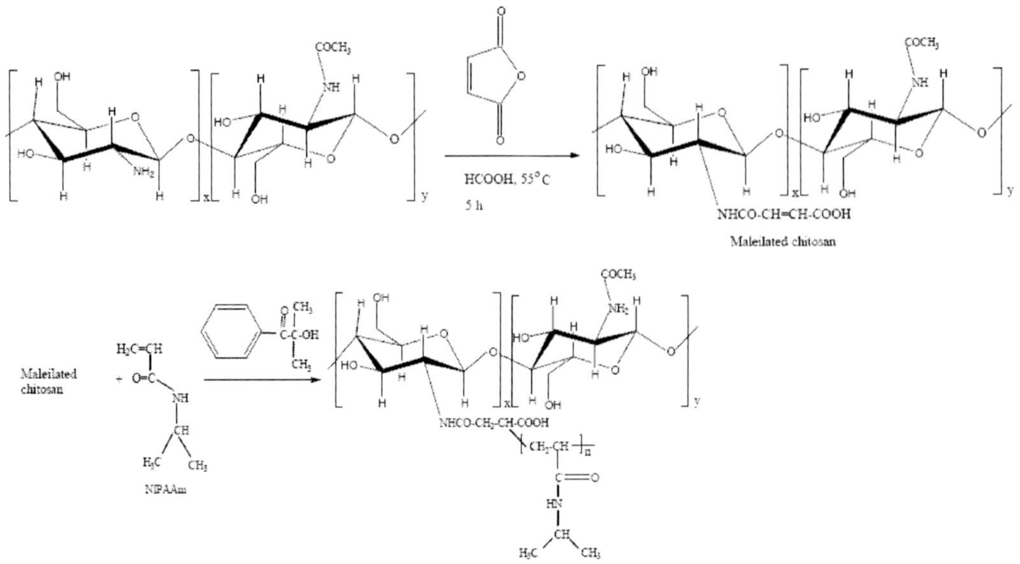

Figure 3. Preparation of maleilated CHIT and maleilated CHIT-*g*-NIPAAm.

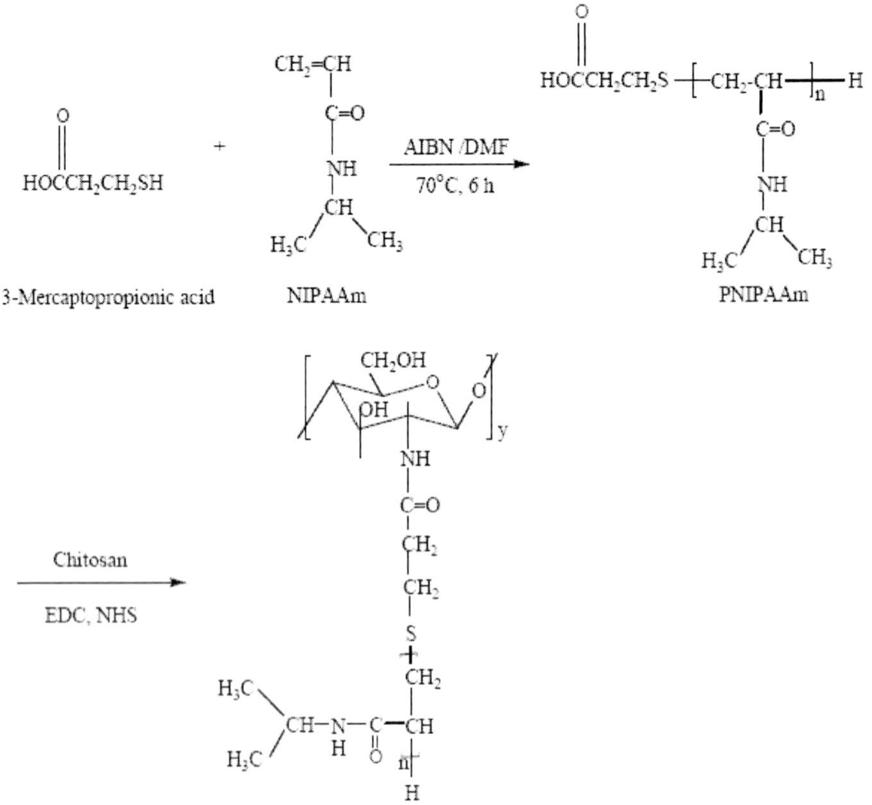

Figure 4. Preparation of carboxyl group-terminated PNIPAAm and comb-type graft hydrogel.

The selection of an adequate degree of post-crosslinking with glutaraldehyde allowed for preparing stimuli-sensitive hydrogel that are able to load large amounts of diclofenac and to control their release. An increase in the post-cross-linking degree of CHIT caused a decrease in the enthalpy of the transition, and in the absolute value of the transition heat capacity increment, as well as a broadening of the heat capacity peak. However, the transition enthalpy and the changes in heat capacity decreased upon increasing the cross-linking density of the CHIT network, owing to existence in the IPN of smaller microdomains.

Recently, a great interest has been made to graft natural polymers with vinyl monomers using radiation method. Graft copolymers based on a maleilated CHIT and NIPAAm were synthesized by UV radiation technique [46]. Here, CHIT was first modified with maleic anhydride to produce maleilated CHIT and then NIPAAm was grafted onto maleilated CHIT (Figure 3). The swelling ratio of maleilated CHIT-*g*-PNIPAAm depended on both pH value and temperature of the aqueous solution. The pH-dependent swelling behaviour was due to the co-existence of amino groups and carboxylic acid groups in the maleilated CHIT chain, for which the swelling ratio had the lowest values at pH 4 and 7. The temperature dependent swelling behaviour was derived from the grafted PNIPAAm component with which the swelling ratio started to decrease at 32 °C. The copolymer samples with higher grafting ratio showed the constant optical transmittance at low temperature and then started to decrease at 32 °C. The fact that the phase transition temperature was the same as the LCST of pure PNIPAAm indicates that precipitation of the grafted PNIPAAm phase was not affected by the CHIT component. This is because of its graft copolymer structure. The effect of grafting ratio on the optical transmittance was clearly observed, when the temperature was raised above 32 °C. As grafting ratio increased, the phase transition became distinct. The phase transition behaviour of sample with the greatest grafting ratio became similar to pure PNIPAAm. When the grafting ratio was very low, the decrease in optical transmittance at transition temperature was nearly indiscernible, and it exhibited almost the same optical transmittance behaviour as that of pure CHIT. Two-stage thermal degradation behaviour was observed for the copolymer samples, corresponding to the degradations of CHIT and grafted PNIPAAm chains, respectively.

Hydrogels based on PNIPAAm grafted CHIT were prepared via γ-radiation [47]. In this the grafting percentage and grafting efficiency increased with the increase of monomer concentration and total irradiation dose. The swelling ratios of CHIT-*g*-PNIPAAm hydrogels were increased with the increase of grafting percentage, which indicated that the swelling behaviour of the hydrogels depends on the amount of the grafted branches. The LCST of these hydrogels was found to be about 28 °C. The swelling ratios of the hydrogels decreased with the increase in pH value of the buffer solutions due to the pH sensitivity of CHIT in the buffer solution. Glycidyl methacrylated CHIT was synthesized and used to prepare dual-sensitive hydrogel with NIPAAm monomer via photopolymerization technology [48]. DCS and optical transmittance results indicated that hybrid hydrogel with low CHIT composition ratio exhibit obvious phase transition. Further, the hydrogels were used as drug carriers for controlled drug release study. It was observed that the pH value of external buffer solution had great effect on drug release rate due to the interaction between drug and hydrogel network. Moreover, hydrogel composition was also found to be influenced the drug release rate.

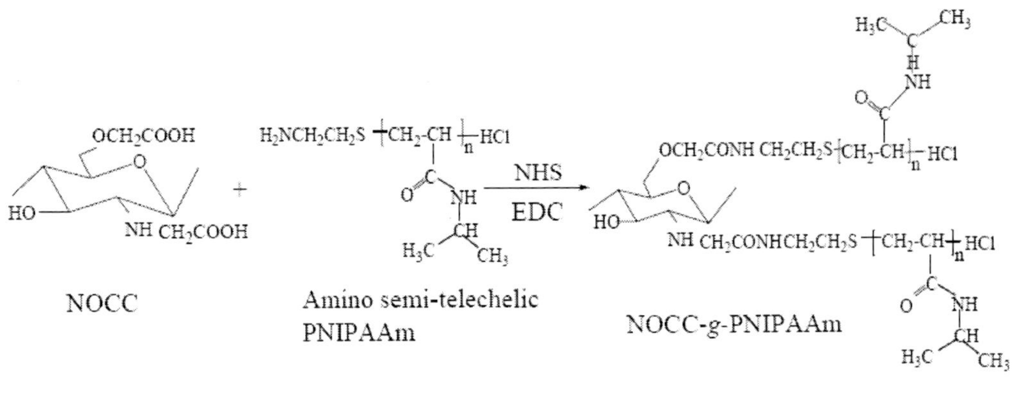

Figure 5. Reaction scheme for the preparation of NOCC-*g*-PNIPAAm.

The water soluble ethyl-3-(3-dimethylaminopropyl) carbodiimide (EDC) catalyzes the formation of amide bonds between carboxylic acids and amines by activating carboxyl groups to form an *O*-urea derivative. This derivative reacts readily with nucleophiles. Lee *et al.* synthesized comb-type graft hydrogels, composed of CHIT and semitelechelic PNIPAAm by using EDC [49]. Semitelechelic PNIPAAm with carboxyl end group was synthesized by radical polymerization using 3-mercaptopropionic acid as the chain-transfer agent, and then grafted onto CHIT (Figure 4). The comb-type hydrogels were prepared with two different graft yields and grafting regions, such as surface- and bulk-grafting, and then compared with a CHIT hydrogel. Results revealed that the introduction of the PNIPAAm side chain disturbed the ordered arrangement of the CHIT molecule, resulting in an increase in the EWC. In addition, the thermal stability of bulk graft hydrogel was relatively low and that of surface graft hydrogel was similar to that of CHIT. In the swelling/deswelling behaviours, comb-type graft hydrogels showed rapid temperature and pH sensitivity because of the free-ended PNIPAAm attached to the CHIT main chain and the CHIT amino group itself, respectively. The swelling ratio of the surface- and bulk-grafted hydrogels dramatically decreased at LCST of PNIPAAm (32 °C). The stepwise swelling behaviour confirmed that the swelling process was repeatable, in accordance with the temperature and pH changes.

Recently, thermo and pH-responsive hydrogel was prepared by reacting *N, O*-carboxymethyl CHIT (NOCC) with amino-telechelic PNIPAAm (PNIPAAm-NH_2) as shown in Figure 5 for drug delivery application [50]. The phase transition behaviour of the hydrogels was found to be dependent on the NOCC/PNIPAAm-NH_2 weight ratio and pH value of the medium. The hydrogels with a higher content of PNIPAAm showed a definite phase transition at 32 °C as it occurs in pure PNIPAAm. The swelling ratio of the beads was higher in pH 2.1 than pH 7.4. Moreover, the swelling ratio of the beads was decreased with the increase in PNIPAAm content of the beads. The release profile of entrapped ketoprofen from NOCC-*g*-PNIPAAm beads showed a slower and controlled release and found that the release behaviour was influenced by both the pH and temperature of the medium. It is expected that these smart hydrogels may be useful to develop drug delivery systems with improved drug loading capacity and controlled release behaviour. Thermo-sensitive *in situ* gel-forming properties of PNIPAAm-*g*-CHIT hydrogel and its potential utilization for ocular drug delivery was investigated [51]. This study demonstrated that PNIPAAm-*g*-CHIT may have possible utilization in improving the efficacy, bio-availability, and pharmacokinetic properties of water-soluble eye drugs such as timolol maleate. MTT assay in this research did not detect any cytotoxic effect of PNIPAAm-*g*-CHIT under the experimental conditions. Moreover, this research supports the possible role of thermosensitive polymer hydrogels in controlled release of therapeutic agents for treatment of glaucoma and other eye diseases.

Temperature-sensitive hydrogels composed of PNIPAAm with CHIT and CHIT/ hyaluronic acid were prepared in order to examine their physicochemical characteristics, *in vitro* drug release, and *in vivo* pharmacodynamics as reaction shown in Figure 6 [52]. The incorporation of CHIT and hyaluronic acid into PNIPAAm did not greatly affect the transition temperature as determined by UV-visible spectrophotometry, DSC, and viscometry. These hydrogels had gelation temperatures well below body temperature; thus, they readily became gels, making them ideally suited to function as injectable drug depots. The hysteresis of the reversed LCST was significantly lowered in CHIT and CHIT/hyaluronic acid hydrogels as compared to PNIPAAm by a 2 °C reduction in temperature. Based on SEM images, CHIT exhibited the densest entanglement of its crosslinked structure, followed by CHIT/hyaluronic acid and PNIPAAm. Each polymer system had distinct characteristics as a carrier for drugs. It was found that the pore size of these hydrogels and the affinity between the drug and hydrogel may have contributed to the main mechanisms determining drug release. It was also noted that the presence of hyaluronic acid in the systems helped to prevent the disintegration in the *in vitro* and *in vivo* environments. Based on the *in vitro* and *in vivo* results, it was concluded that such biocompatible delivery systems can facilitate the controlled release of nalbuphine and improve the duration of action after intravenous administration in rats, particularly in the case of the CHIT/hyaluronic acid hydrogel.

Trimethyl CHIT-*g*-PNIPAAm was synthesized with different grafting ratios [53]. Particles ranging from 200 to 900 nm were prepared by mixing the trimethyl CHIT-*g*-PNIPAAm copolymers or trimethyl CHIT with DNA, which were mainly controlled by the N/P ratio while less influenced by the temperature variation. Along with increase of the N/P ratio, the zeta potentials of these particles were found to be monotonously increased. At a given N/P ratio, the zeta potentials were almost same at 25 °C for the trimethyl CHIT-*g*-PNIPAAm copolymers regardless of their grafting ratios and the existence of serum proteins. However, at 37 °C the values were significantly decreased in a solution containing serum protein. Ethidium bromide competitive binding assay showed that trimethyl CHIT-*g*-PNIPAAm had stronger ability to combine with DNA at 40 °C when the PNIPAAm chains are in a collapsed state. *In vitro* culture of HEK293 cells demonstrates that the grafting of PNIPAAm will not affect cellular uptake of the trimethyl CHIT-*g*-PNIPAAm/DNA particles compared with trimethyl CHIT/DNA particles. The optimized gene transfection efficiency achieved by trimethyl CHIT-*g*-PNIPAAm was comparable to Lipofectamine™ 2000, while no obvious cytotoxicity was observed for this vector.

Recently, microgels with more complex structures, such as a multi-responsive core-shell, have received increasing attention because of the tuneable properties of the individual responsive components. Various types of core-shell microgels have been prepared from PNIPAAm grafted natural polymers. Leung *et al.* developed smart microgels that consist of well-defined temperature-sensitive cores with pH-sensitive shells based on PNIPAAm and CHIT [54, 55]. The unique core-shell nanostructures, which had narrow size distributions, exhibited tuneable responses to pH and temperature. Comparable hydrodynamic size expansions of microgel particles were observed when the pH was lowered from 7 to 3. A measurement of the zeta-potential of the microgels indicated that the positive potential increased with the decrease of pH. Therefore, the increase of particle size at a lower pH value was attributable to the increase of positive charges of amino groups and charge repulsion. The increase of CHIT hydrophilicity with lowering the pH of the solution had little influence on the volume phase transition temperature of the PNIPAAm.

The properties of crosslinked poly[CHIT–NIPAAm/methacrylic acid (MAA)–methyl methacrylic acid (MMA)] core-shell type copolymer particles were examined [56]. Here, the crosslinked copolymer of NIPAAm and CHIT was prepared as the core and the copolymer of MAA and MMA was prepared as the shell. The weight ratio of MAA/MMA and the concentration of shell monomers (MAA and MMA) in the feed of the reaction mixture had been changed to investigate their effects on the particle size, reaction rate, zeta-potential, specific surface area, and surface functional groups of the latex particles. The swelling and thermo-responsive behaviour of the film made from these core–shell latices were studied under different pH values of buffer solution. With an increase of the weight ratio of MAA/MMA or a decrease of the shell thickness of particles, the swelling ratio of the sample increased. In the pH 4, the swelling behaviour and the zeta potential of particles were influenced by the combination effect of shell thickness, NIPAAm, and CHIT. At pH 9, the swelling behaviour and the zeta potential were influenced by the combination effect of shell thickness, NIPAAm, and MAA.

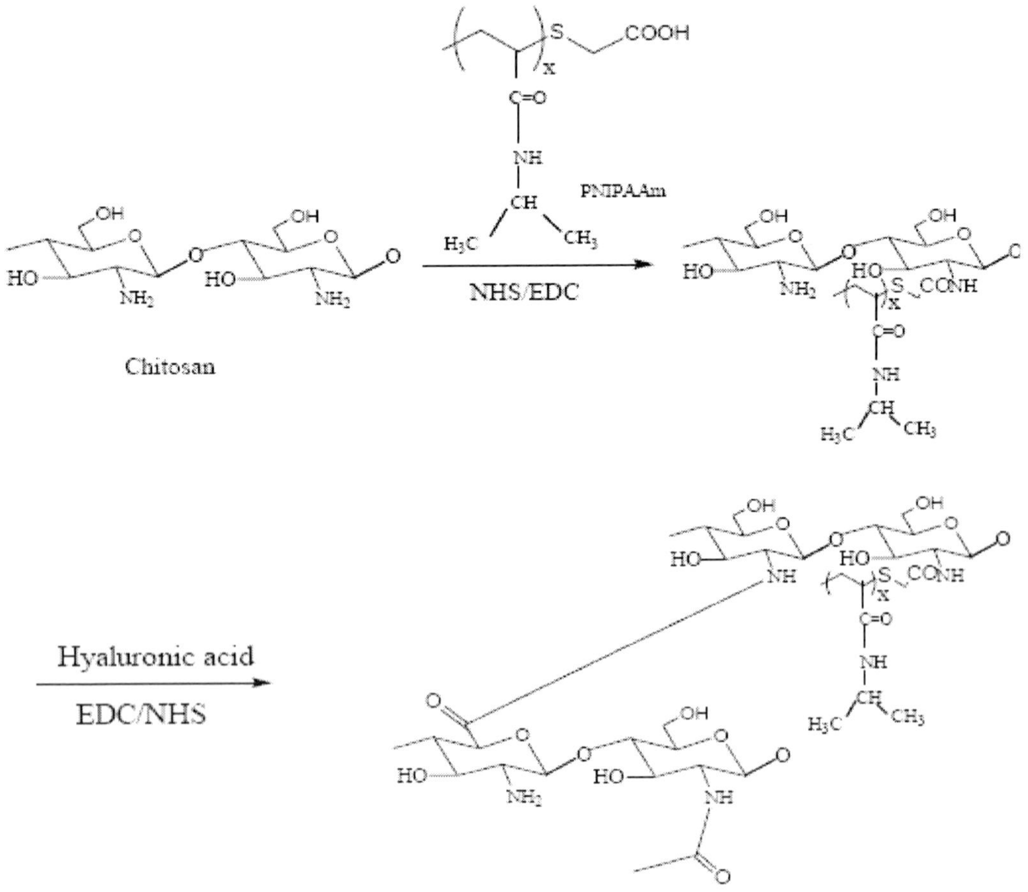

Figure 6. Reaction scheme for the synthesis of thermo-sensitive CHIT and CHIT/hyaluronic hydrogels.

CHIT-*g*-NIPAAm particles were synthesized by soapless emulsion copolymerization method [57]. An anionic initiator ammonium persulfate (APS) and a cationic initiator 2,2'-azobis(2-methylpropionamidine)dihydrochloride (AIBA) were used to initiate the reaction of copolymerization. The CHIT-*g*-NIPAAm copolymer synthesized by using APS showed a homogeneous morphology and exhibited the characteristic of a LCST at 32°C. The copolymer synthesized by using AIBA showed core–shell morphology, and the characteristic of LCST was insignificant. In another research, core-shell type chitosan/calcium phosphate composite fibers were prepared by a facile wet spinning method; the chitosan aqueous solution with PO_4^{3-} ions. The resulting fibers formed a unique core-shell structure, i.e., core of chitosan with shell of calcium phosphate. The mechanical property of the fibers have reinforced by initial concentration of chitosan solution [58]. Further, LCST of the CHIT-*g*-NIPAAm depended on the morphology of the copolymer particles. Below the LCST, the increase of either the CHIT/NIPAAm weight ratio or the crosslinking agent decreased the swelling ratio of the CHIT-*g*-NIPAAm copolymer. At temperature below the LCST, both PNIPAAm and CHIT had hydrophilic behaviour. The hydrophilic behaviour of PNIPAAm was more significant than that of CHIT; therefore, the swelling ratio of the CHIT-*g*-NIPAAm copolymer decreased with the increase of the CHIT/NIPAAm weight ratio. However, at temperature above LCST, PNIPAAm was hydrophobic and difficult to swell in the aqueous medium, so the hydrophilic behaviour of CHIT dominated the swelling ratio of the CHIT-*g*-NIPAAm copolymer. Therefore, the increase of the CHIT/NIPAAm weight ratio increased the swelling ratio of the copolymer. The swelling ratio of the copolymer was decreased with the increase in crosslinking density and pH value. Also, it exhibited a much less extend of swelling ratio variation in the same temperature range. In addition, with increasing the temperature above the LCST, the swelling ratio was decreased in contrast to those of homo-PNIPAAm beads. The IPN gel beads showed cyclic deswelling and swelling behaviours with a stepwise temperature change whereas the homo-PNIPAAm beads exhibited a very rapid deswelling kinetic profile in the early stage.

2.2. By Oxidative Copolymerization

CHIT is a non-conducting biomaterial; however, it has an excellent biocompatible property that favours the immobilization of biomolecules over its surface [59]. The surface immobilization and protection of the bio-sensing elements, i.e., the enzyme, antibody, DNA, etc. are one of the key challenges for fabricating a highly sensitive bio-sensors [60-63]. It was also reported that modifying CHIT with conducting polymers such as PANI improves its redox properties that influence the electron transfer kinetics during the course of bio-molecular detections [64].

In other hand, PANI has attracted considerable attention because of its good environmental stability, good redox reversibility and good electrical conductivity. These properties provide possible applications in battery electrodes [65, 66], electro chromic devices [67, 68], photoelectric cell [69, 70], light-emitting diode [71] and sensors [72, 73]. The general formula of PANI is as follows:

$$[(\text{-B-NH-B-NH})_y\ (\text{-B-N=Q=N-})_{1-y}]_n$$

where, B and Q denote the C_6H_4 rings in the benzenoid and quinoid forms, respectively [74-76]. The conducting state of PANI (i.e., emeraldine salt) can be achieved by the protonation of nitrogen in its 50% oxidized emeraldine state (y = 0.5), also called emeraldine base, using the chemical method as given by Chiang and Mac Diarmid in 1986 [77]. PANI exists in a variety of forms that differ in chemical and physical properties [78-81]. The most common green protonated emeraldine salt has conductivity on a semiconductor level of the order of 10^0 Scm^{-1}, many orders of magnitude higher than that of common polymers (>10^{-9} Scm^{-1}) but lower than that of typical metals (<10^4 Scm^{-1}). Protonated PANI (i.e., PANI-HCl) converts to a nonconducting blue emeraldine base, when treated with ammonium hydroxide (Figure 7).

The changes in physicochemical properties of PANI occurring in the response to various external stimuli are used in various applications [82, 83], for example, in organic electrodes, sensors, and actuators [84-86]. Other uses are based combination of electrical properties typical of semiconductors with materials parameters characteristic of polymers, like the development of 'plastic' microelectronics [87], electro chromic devices [88], tailor-made composite system [89, 90], and smart fabrics [91]. The establishment of the physical properties of PANI reflecting the conditional of preparations is thus of fundamental importance. Hence, PANI is a unique polymer whose variable conductivity can be controlled by the protonation of the imine sites on the main polymer chain [92]. Although pure PANI is not compatible for the biological specimens in a long time, its composites usually provide a sustainable matrix for the biological molecules [93]. It was reported that the sensitivity and selectivity of PANI to biological substances such as enzyme, antibody, DNA, mRNA, nucleic acids, etc. can be improved through copolymerization with polymers that have a variety of functional groups such as -OH, -COOH, -NH2, etc. [94-98]. Thus, CHIT-co-PANI has been synthesized for the development of electronic biomaterials as promising sensing probes for the detection of biomolecules. In the present chapter, we have reviewed preparation and bio-sensing properties of pH-responsive redox CHIT-co-PANI copolymers. The chemical reactions, morphology, redox and electrical properties and bio-sensing behaviours of CHIT-co-PANI copolymers are extensively descried.

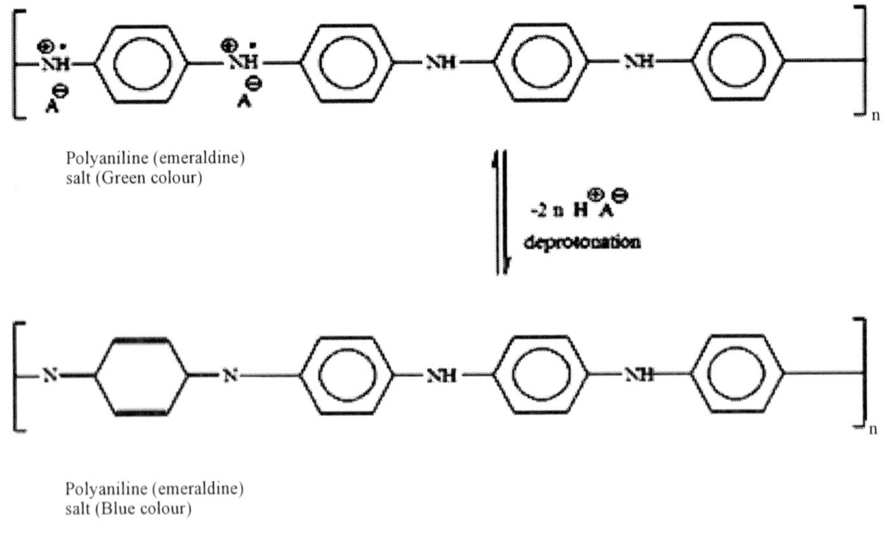

Figure 7. PANI (emeraldine salt) depronated with alkaline and formed PANI (emeraldine base).

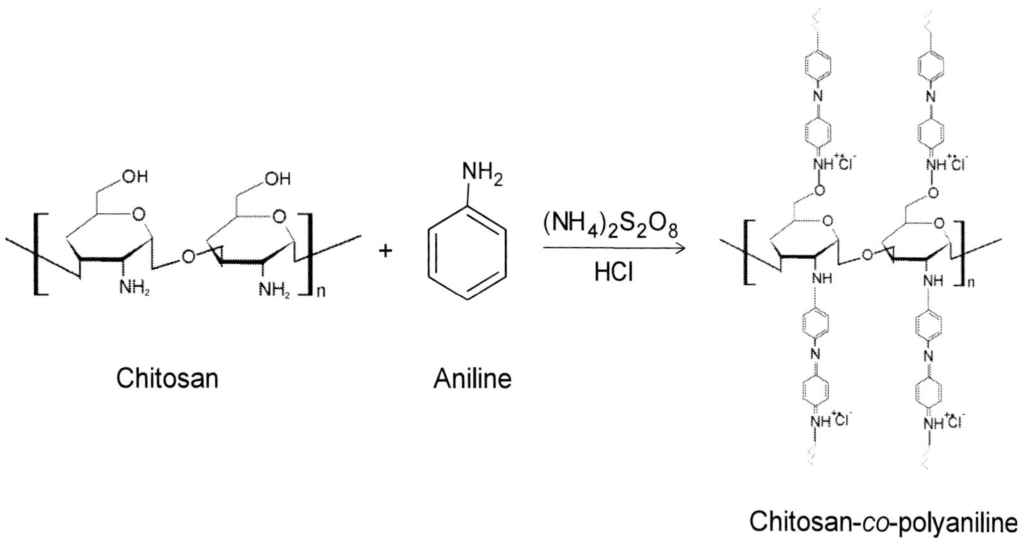

Figure 8. Synthesis of HCl doped CHIT-*co*-PANI using $(NH_4)_2S_2O_8$ as oxidizing agent in acidic medium.

A polymer comprising molecules with one or more species of block connected to the main chain as side chains, these side chains having constitutional or configurational features that differ from those in the main chain. In a graft copolymer, the distinguishing feature of the side chains is constitutional, i.e., the side chains comprise units derived from at least one species of monomer different from those which supply the units of the main chain. The graft copolymerization of PANI onto CHIT was reported using ammonium persulfate, $(NH_4)_2S_2O_8$/HCl as redox initiator as shown in Figure 8 [96]. It was observed that $(NH_4)_2S_2O_8$/HCl redox system can be efficiently used in the graft copolymerization of PANI onto CHIT. Silva *et al.* were synthesized pH-responsive PANI colloids by using CHIT as a steric stabilizer. The process was enzymatic polymerization using toluenesulfonic or camphorsulfonic acids as a doping agent [96].

2.3. By Electrochemical Polymerization

Electrochemical synthesis in organic chemistry is the synthesis of chemical in an electrochemical cell [100]. The main advantage of electrochemical synthesis over an ordinary redox reaction, is avoidance of the potential wasteful other half-reaction and the ability to precisely tune the required potential. Tiwari *et al.* have been reported electrochemical synthesis of CHIT-*co*-PANI in the acidic medium [101]. In a typical procedure, aniline and CHIT solution was mixed with HCl in an electrochemical cell and the CHIT-*co*-PANI was chronoamperometrically synthesized onto an ITO coated glass surface using a three-electrode assembly with ITO glass as working, platinum as counter, and Ag/AgCl as reference electrodes. The resulting CHIT-*co*-PANI/ITO electrode was washed with deionized water followed by a phosphate buffer saline solution of pH 7.0 in order to neutralize the electrode surface (Figure 9).

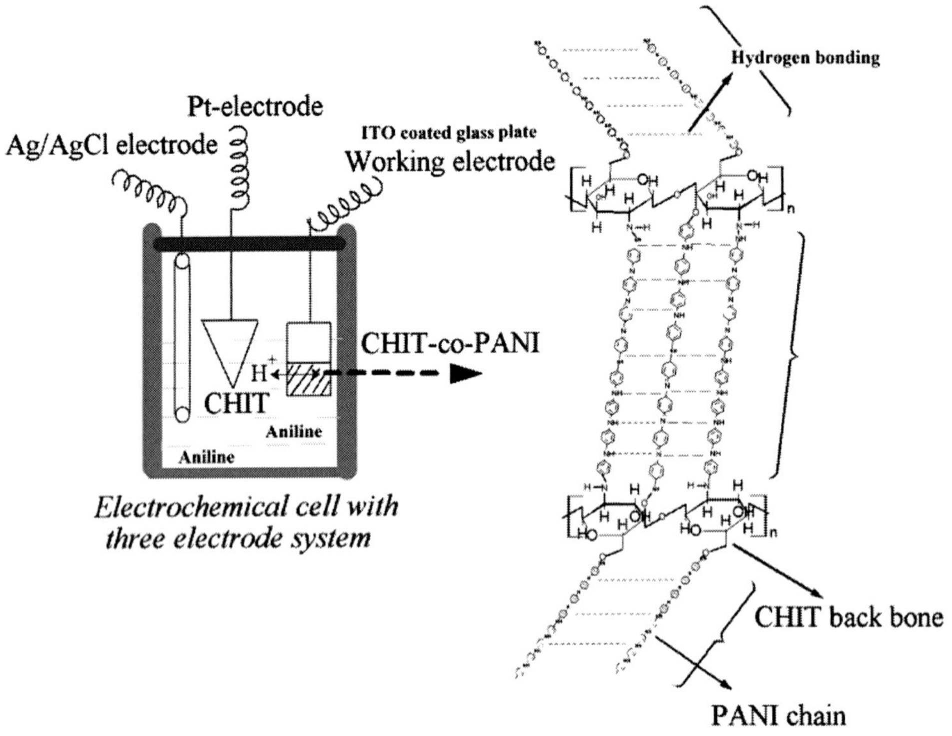

Figure 9. Electrochemical synthesis of CHIT-*co*-PANI onto indium-tin oxide (ITO) coated glass plate.

In the electrochemical copolymerization of CHIT-*co*-PANI, the aniline monomer initially became protonate with acid (i.e., H^+ of HCl) and propagated to form an intermediate called PANI radical cation [a].

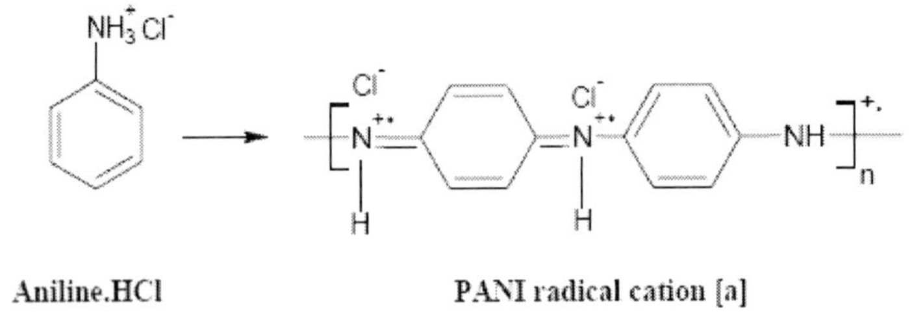

PANI radical cation simultaneously generated CHIT macro radicals [b] by the abstraction of hydrogen from the -OH and $-NH_2$ groups of the CHIT macromolecules.

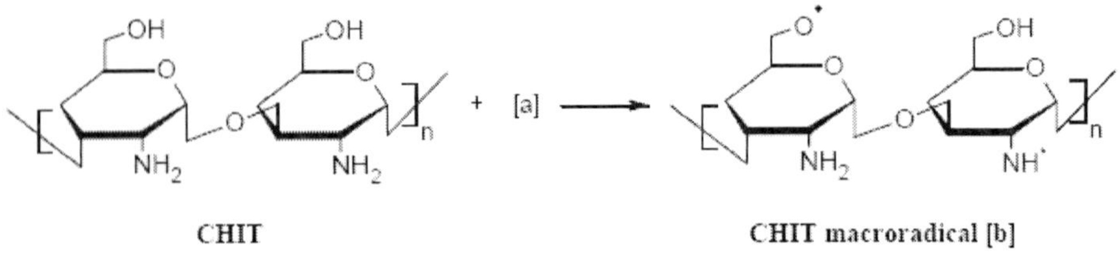

The PANI cation radicals and CHIT macro radicals then copolymerized and yielded CHIT-*co*-PANI.

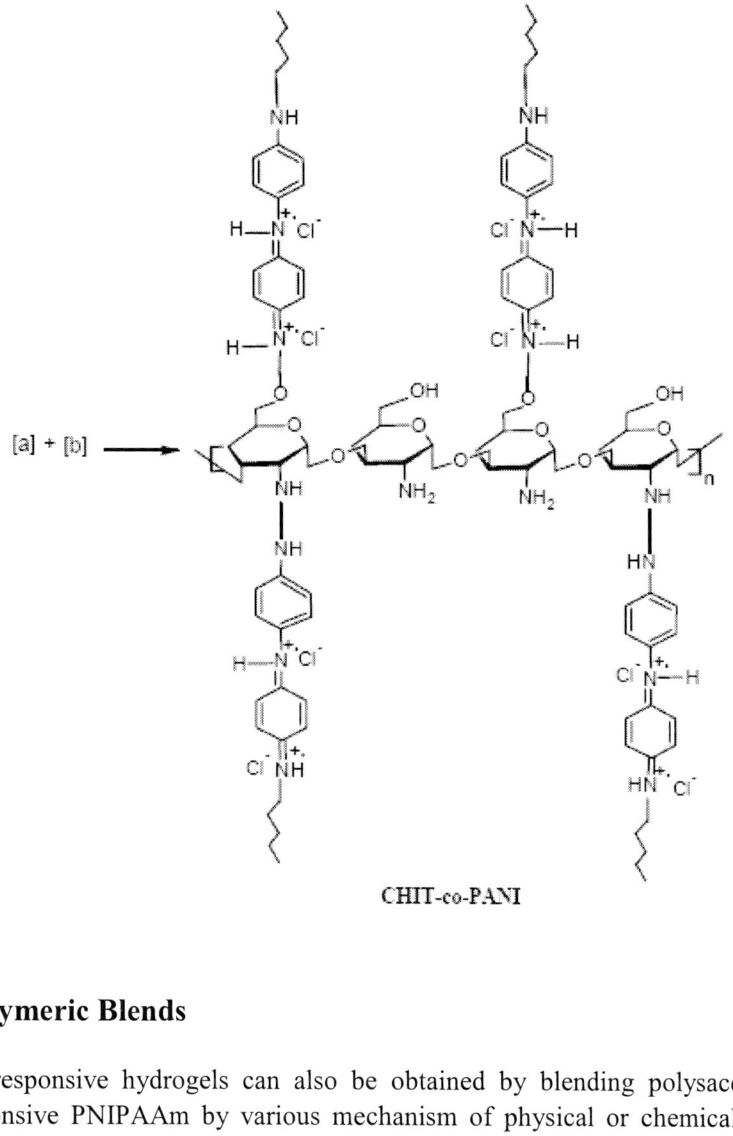

CHIT-co-PANI

2.4. By Polymeric Blends

Stimuli-responsive hydrogels can also be obtained by blending polysaccharides with thermo-responsive PNIPAAm by various mechanism of physical or chemical crosslinking such as hydrophobic association or covalent, ionic and hydrogen bonding. The ability to form a gel through the crosslinking of CHIT with PNIPAAm has been well documented.

A series of physically crosslinked hydrogels composed of maleilated CHIT and NIPAAm were prepared and characterized with respect to their swelling behaviours [102]. These hydrogels showed broad pH-sensitivities below 32 °C, but above that temperature exhibited sharp pH-dependent phase transition behaviours which depended on the content of maleilated CHIT in the hydrogels. The change of ratio of NIPAAm to maleilated CHIT was not affected LCST significantly, while it affected EWC at high pH region. The EWC of the hydrogels was strongly dependent on both pH and temperature. Since negatively charged carboxyl groups of maleilated CHIT were incorporated into the polymer network, the gel swelled at high pH region due to ionic repulsion of the carboxyl groups, and collapsed at low pH values because of protonation of carboxyl groups. The differences in EWC between low pH and high pH generally increased as the content of carboxyl groups in the hydrogels increased. Its transition pH shifted to a slightly higher pH in all hydrogels as the temperature increased since the gels hydrophobicity increased with temperature.

Many hydrogels are formed from water-soluble polymers by crosslinking them using crosslinking agents such as glutaraldehyde or polymerizing hydrophilic monomers in the presence of a crosslinker. Chemically crosslinked polymers seem to be one of the candidates to improve wet strength. CHIT can form gels with glutaraldehyde through the Schiff base reaction. The properties of glutaraldehyde crosslinked CHIT/PNIPAAm hydrogel were examined [103, 104]. Except the transparence property, the CHIT/glutaraldehyde, and PNIPAAm-containing CHIT/glutaraldehyde gels exhibited similar properties in all aspects examined.

The CHIT-PNIPAAm/glutaraldehyde gel was transparent below 30°C, whereas opaque above 32°C. IPN-hydrogel membrane of PNIPAAm/carboxymethyl CHIT was prepared, and the effects of the feed ratio of components, swelling medium and irradiation dose on the swelling and deswelling properties of the hydrogel was systematically studied [105]. The results showed that the introduction of carboxymethyl CHIT did not shift the LCST (at 32 °C), which was similar to the pure PNIPAAm. The lowest swelling ratio was found at pH 2. There was little influence of irradiation dose on the thermo- and pH-sensitivity of the IPN hydrogel, increasing dose only decreased the swelling ratio as shown in Figure 10. The PNIPAAm:carboxymethyl CHIT (1:4 w/w) hydrogel was not thermo-sensitive in distilled water, whereas it showed a discontinuous volume phase transition in pH 2 and a continuous one in pH 8 buffer.

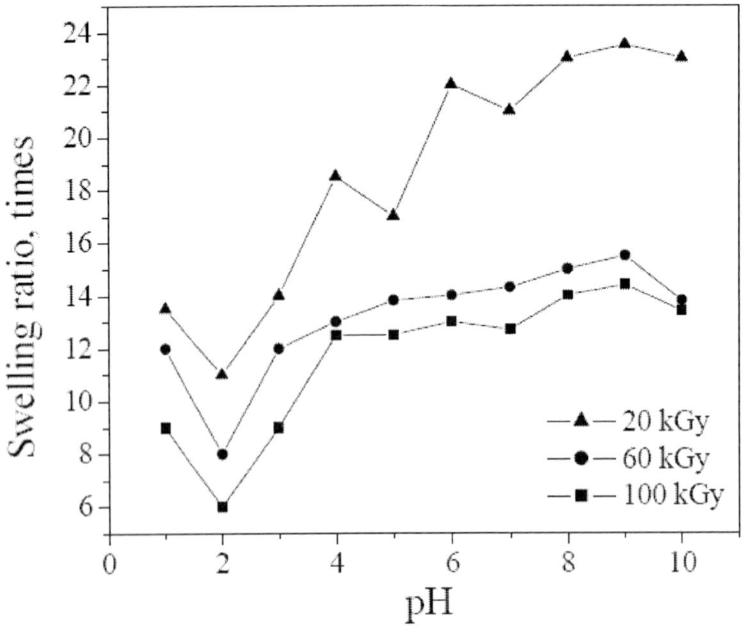

Figure 10. Effect of pH on swelling ratio of IPN-hydrogels at 15 °C with the comparison of different irradiation dose.

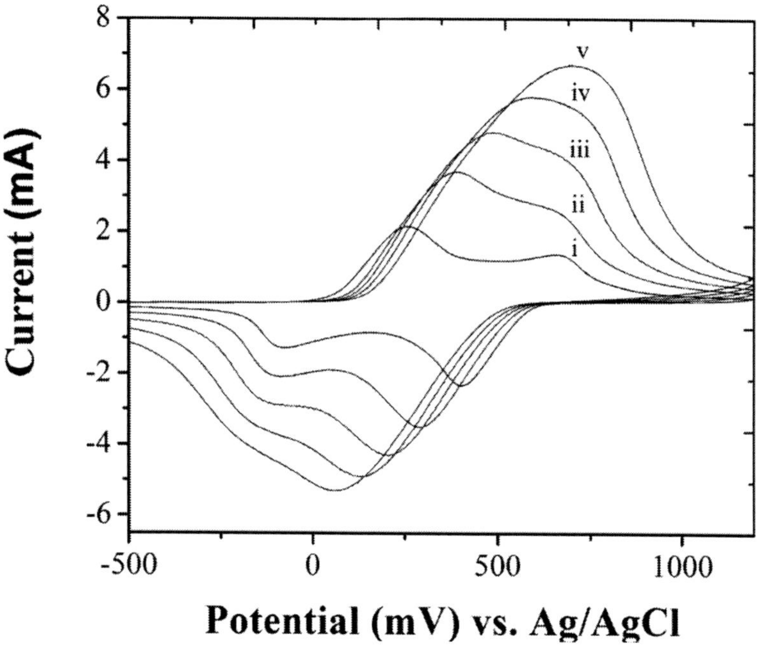

Figure 11. Scan rate and rate constant dependence of the CV curves. CV of the CHIT-*co*-PANI film using Ag/AgCl as reference electrode in 0.5M HCl at different scan rates: (i) at 10 mVs^{-1}; (ii) at 20 mVs^{-1}; (iii) at 30 mVs^{-1}; (iv) at 40 mVs^{-1}; and (v) at 50 mVs^{-1}. The oxidation peaks are shown at 250 and 710 mV and reduction peak is shown at 405 mV.

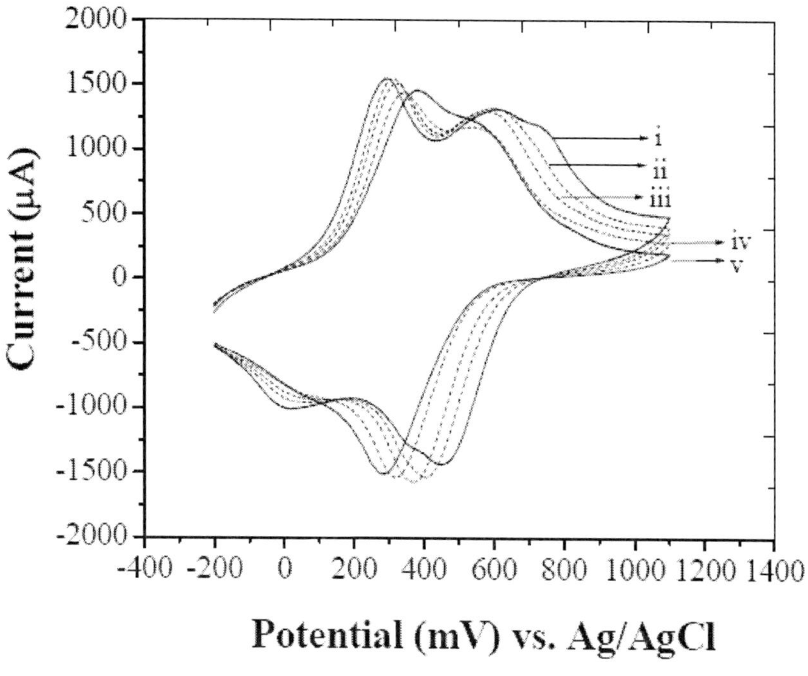

Figure 12. CV of the CHIT-*co*-PANI/$WO_3.nH_2O$ electrode vs. Ag/AgCl in 0.5M HCl at 10 mV/s scanning rates: (i) at 100 ppb; (ii) at 200 ppb; (iii) at 300 ppb; (iv) at 400 ppb; and (v) at 500 ppb.

3. Electrochemical Behaviour

PANI is a typically redox material that can be easily create electronic moment by producing electrons. In fact, the reversibility of the PANI is so well behaved and can be realized by the cyclic voltammetry (CV) study [101]. The acid doped analogues of PANI are an especially powerful oxidizing as well as reducing agent, due to inductive receiving/donation of electron. Figure 11 shows the anodic as well as cathodic current peaks of CHIT-*co*-PANI that influence the voltage scan rate. This is because of reversible electron transfer in between the protonated PANI and CHIT. With the increase in the scan rate, peak current was increased due to the reversible electron transfer reaction on the surface of CHIT-*co*-PANI.

The effect of doping on the CHIT-*co*-PANI can also be observed with the CV. Figure 12 shows the variation of current versus NO_3^- doping concentration with the CHIT-*co*-PANI profile. The redox current decreased with an increase in NO_3^- concentration.

A decrease in the redox current results from the immediate entrapping of NO_3^- within the CHIT-*co*-PANI electrode, which was produced from the NO_2 gas in the electrochemical cell. NO_3^- ions can be prepared by the reaction of NO_2 with H^+/H_2O and produces HNO_3, thus, an equimolar amount of NO_3^- ions via the following reaction schemes. The generally accepted overall reaction schemes between NO_2 and H_2O/O_2 are shown below ranging from equation (1) to (3):

$$4NO_2\,(g) + 2H_2O + O_2 \longrightarrow 4HNO_3 \quad \text{----------------} \quad (1)$$

$$HNO_3 \longrightarrow H^+ + NO_3^-\,(pK_a = -2) \quad \text{----------------} \quad (2)$$

$$HCl \longrightarrow H^+ + Cl^-\,(pK_a = -7) \quad \text{----------------} \quad (3)$$

The PANI copolymer chains in the CHIT-*co*-PANI electrode initially formed emerdeline chloride in the medium of HCl/H_2O. Once the NO_2 gas enters into the electrochemical cell, NO_3^- species are generated that form emeraldine nitrate with the PANI copolymer (Figure 13).

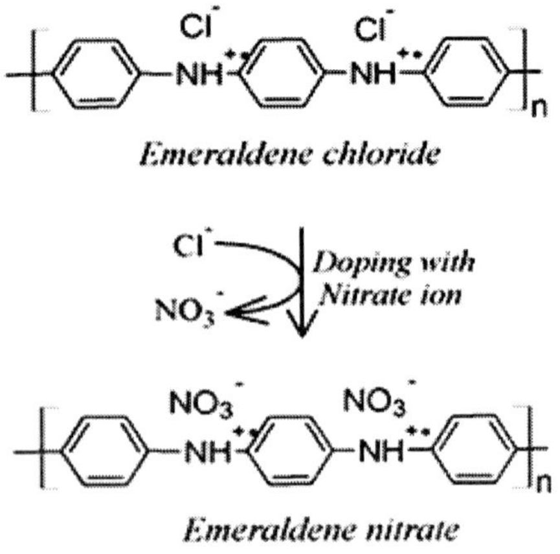

Figure 13. Chemical doping of the PANI chains in the CHIT-*co*-PANI copolymer electrode viz. from emerdeline chloride to emerdeline nitrate.

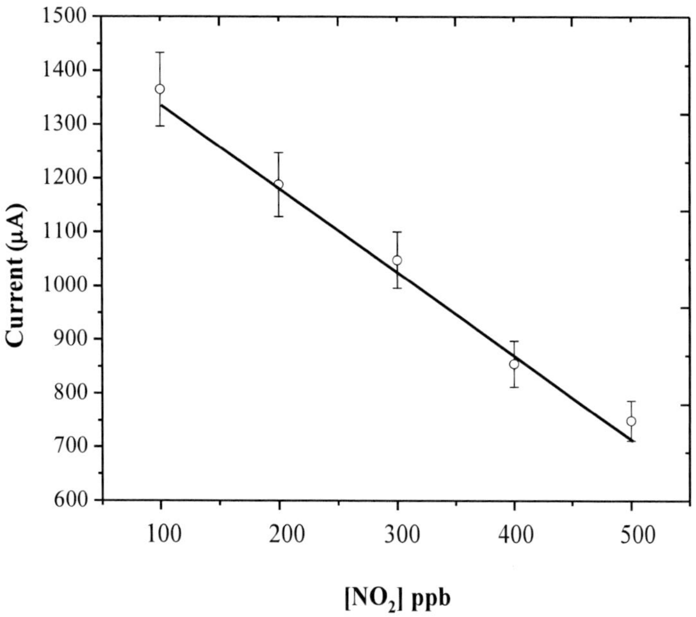

Figure 14. Effect of NO_2 concentration (i.e., a source of NO_3^-) on the current of the CHIT-*co*-PANI electrode at a fixed potential.

The electroactivity of emereldine chloride is higher than that of emeraldine nitrate due to the higher pK_a value of HCl compared with HNO_3. Therefore, a decrease in redox current was observed with an increase in the NO_2 concentration. As shown in Figure 14, the current of the CHIT-*co*-PANI electrode decreased linearly with the NO_2 concentration ranging from 100 to 500 ppb.

A recent work has been reported on the synthesis of PANI/CHIT composites and blends. Shin *et al.* have investigated synthesis methods for preparing CHIT/PANI-ES IPNs [106]. Semi IPNs based on PANI and crosslinked CHIT have been strongly depend on the synthesis conditions. In that study FTIR and NMR studies have also revealed the structure of the semi-IPNs adjusted to aqueous HCl solution (pH <1) product was similar to that observed for the doped PANI (emeraldine salt form).

PANI can be rapidly "switched" by the addition of acids and bases that protonate and unprotonate the base sites within the polymer. This leads to the dependency of the polymer states, and thus the reactions, upon the pH of the solution. In solutions of pH > 4, PANI loses it electroactivity entirely because the emeraldine salt (ES), the only conducting form of the polymer, is dedoped to form the insulator emeraldine base (EB). The same groups also investigated the effect of pH condition on the properties of a PANI/crosslinked CHIT semi IPN [107]. They found that the conductivity of the semi-IPNs increases with increasing PANI contents, adjusted to pH <1, formed a blended structure. The observed increase in the electrical conductivity caused by interaction of the components also reflects the charge transfer being associated with the acidic doping of PANI.

4. Conducting Property

Thanpitcha *et al.* prepared PANI/CHIT blend films by using a solution casting method and mechanical and electrical properties of blends were investigated in terms of blend composition and the doping conditions: acid type, acid concentration, and doping time [108]. Smooth, flexible, and mechanically robust blend films were obtained at PANI content lower than 50 wt%. To become electrically conductive, they doped PANI with HCl. The electrical conductivity of the doped films increases with increasing PANI content. However, high concentrations of HCl (2-6 M) and long doping times (15-24 h) led to a decrease in electrical conductivity. This result indicated the over protonation of PANI chains in the blend films. Moreover, increasing strength and using smaller anions for acid dopant induced higher electrical conductivity values of the blend films. The mechanical properties of the blend films were strongly affected by the doping treatment with HCl. The inferior mechanical properties of the blend films after doping were presumed to be due to hydrolysis of CHIT.

The comparatively poor mechanical strength of conducting polymer-based actuators has led researchers to investigate the possibility of obtaining composite materials and interpenetrating networks with other polymers, carbon nanotubes (CNT), etc. [109]. Combination of a conducting polymer and CHIT (hydrogel) is expected to lead to improved properties that can be utilized for designing actuator devices. A hydrogel can produce a large swelling strain, even though the strain offered by a conducting polymer is relatively low. However, the conducting polymer improves the property by providing a low operational voltage, a short diffusion path, and a fast response time. PANI/CHIT semi-IPN networks and dual-mode actuation behaviour in CHIT/PANI/CNT composite fibers was reported previously. [110]. Ismail *et al.* have fabricated a new type of electroactuating biopolymer hydrogel/PANI microfiber by wet spinning a CHIT solution, followed by the in situ chemical polymerization of aniline [111]. This novel biomaterial showed an enhanced chemical and electrochemical actuation in response to pH and an electrical stimulus. The fibers showed a reasonable electrical conductivity of 2.856×10^{-2} S/cm at room temperature. This novel semiconducting biomaterial showed enhanced electrochemical actuation, resulting in a strain of about 0.39%, and a chemical actuation corresponding to a strain of 6.73% upon switching the pH between pH = 0 and 1.

Yavuz *et al.* are the first group who reported substituted PANI/CHIT composites [112]. Composites were synthesized chemically and ammonium peroxydisulfate was used as oxidant and composites were characterized by measurements of conductivity, FTIR, UV-vis, SEM and TGA techniques. FTIR spectra of the composites revealed that there is a strong interaction between substituted PANIs and CHIT. Among the substituted PANI/CHIT composites synthesized, poly(*N*-ethylaniline)/CHIT (PNEANI/CHIT) has the highest conductivity with a value of 1.68×10^{-4} S/cm. The P2EANI/Ch composite exhibited higher thermal stability than the other composites. SEM images of the composites showed an agglomerated granular morphology of substituted PANI particles coated on the surface of CHIT (Figure 15).

Electrically conductive polymer biomaterials have been proven to be another promising alternative for developing new biodegradable conduits used for restoring the function of injured peripheral nerves or the generation of a nerve gap since the early 1990s by using electrical stimulation *in situ* [113-115]. Except for being effectively used for nerve regeneration [116, 117] these conductive biomaterials have also shown a capacity to facilitate the growth of other types of cells, such as endothelial cells [118], bone cells [119] and chromaffin cells [120]. In most cases, these conductive biomaterials are obtained in the form of blends or composites by using biodegradable polymers as matrices and an intrinsically conductive polymer, PPy, as a conductive component because PPy has an acceptable biocompatibility with mammalian cells [121].

Wan *et al.* worked on the fabrication of novel conductive poly(DL-lactide)/CHIT/polypyrrole (PPy) complex membranes [122]. Using poly(DL-lactide)/CHIT blends as matrices and PPy as a conductive component, several kinds of membranes with various compositions are prepared. A percolation threshold of PPy as low as 1.8 wt% is achieved for some membranes by controlling the CHIT proportion between 40 and 50 wt%. SEM images exhibit that the membranes with a low percolation threshold show a two phase structure which consists of poly(DL-lactide) and CHIT phases. Dielectric measurements indicate that there is limited miscibility between the poly(DL-lactide) and CHIT but PPy is nearly immiscible with the other two components. Based on the structural characteristics of the membranes, the PPy particles are suggested to be localized at the interface between two phases.

Yang and Lu [123] have demonstrated the one-step synthesis of AgCl/PPy core/shell nanostructures with controllable shell thickness in the presence of CHIT and thus hollow PPy nanoparticles was reported. AgCl, as a template, was formed during the initial polymerization by the interaction between metal cations and oxidizing anions and adsorbed with CHIT whereupon to induce the coating of PPy layers (Figure 16). The method shows the merits of easier preparation compared with the conventional stepwise process and controlled shell thickness could be obtained by adjusting the quantity of pyrrole added.

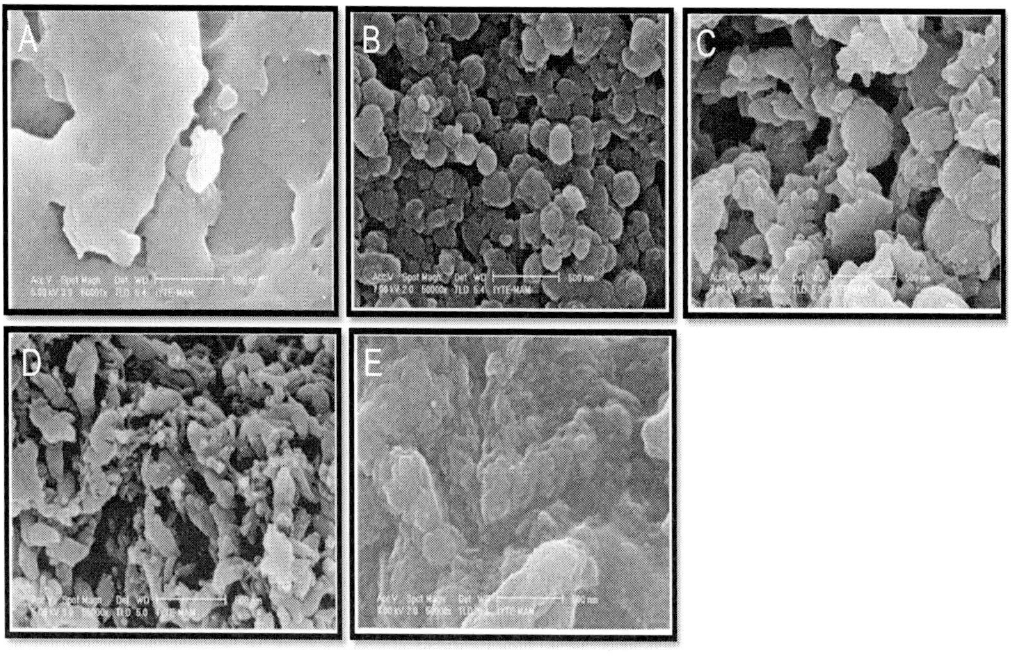

Figure 15. Scanning electron micrographs of the composites; (a) CHIT, (b) PANI/CHIT, (c) P2EANI/CHIT, (d) PNEANI/CHIT and (e) PNMANI/CHIT.

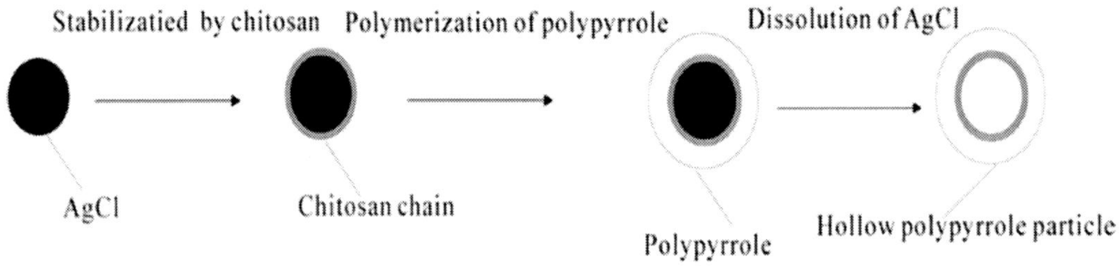

Figure 16. The possible mechanism for the formation of core/shell and hollow particles.

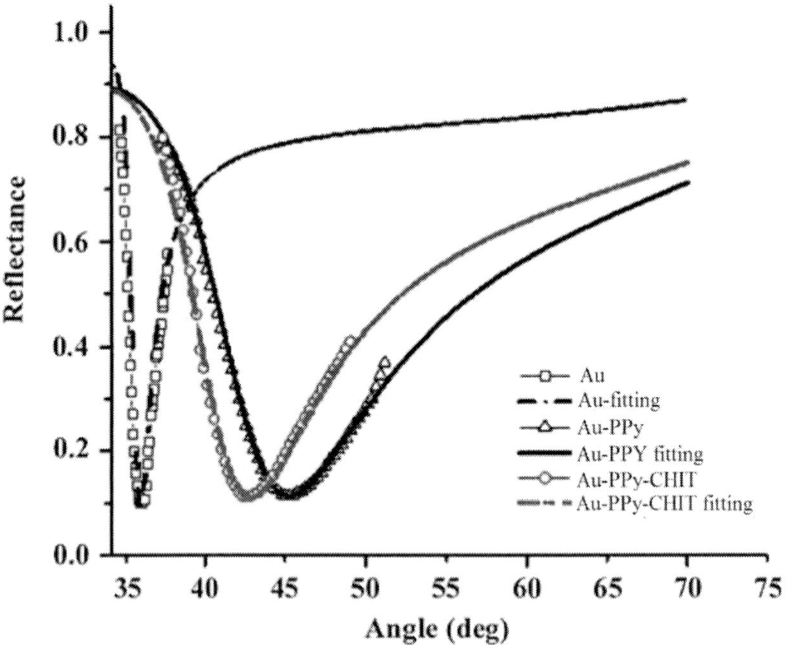

Figure 17. Reflectivity versus surface plasmon resonance angle curves for Au, Au–PPy, and Au–PPy–CHIT thin films.

Another study has been done by Abdi *et al.* [124], they prepared PPy-CHIT composite film and investigated the electrical and optical properties of PPy–CHIT composite. Their work focused on the effects of CHIT content on the electrical conductivity, photoacoustic effect [125], EMI-SE [126, 127], and dielectric constant of the resulting films. The refractive indexes of the samples were successfully measured by SPR technique (Figure 17).

Since the SPR angle shifted clearly to the right and produced a sharp peak of resonance angle, this technique is capable for using in sensitive optical sensors. It seems that the EMI shielding effectiveness of 33.9 dB for the PPy–CHIT composite films is very attractive in any electromagnetic interference (EMI) shielding applications where a minimum shielding effectiveness of 30 dB is required (Table 1).

Table 1. EMI shielding effectiveness of PPy-CHIT composition films with various concentrations of CHIT and PPy film without CHIT

CHIT conc. (%w/v)	*C* (Scm^{-1})	*Tr* (%)	*Ab* (%)	*Re* (%)	Total Atten	*SE* (dB)
0.3	39.3	0.75	16.05	83.2	99.25	21.2
0.5	49.4	0.20	8.09	91.7	99.79	26.9
0.7	69.1	0.04	7.56	92.4	99.96	33.9
0.9	42.1	0.29	11.91	87.8	99.71	25.3
1.1	33.5	1.50	18.12	80.4	98.52	18.3
PPy	35.2	1.01	16.21	82.8	99.01	19.8

C = conductivity; *Tr* = Transmittance; *Ab* = Absorbance; *Re* = Reflection; *SE* = Shielding effectiveness.

5. Bio-Sensor Applications

Bio-sensors are device for the detection of a bio-analyte that combines bio-chemical component with a physicochemical detector [128]. It consists of three elements: (1) the sensitive element i.e. biological material such as tissue, microorganisms, organelles, cell receptors, enzymes, antibodies, nucleic acids, etc. The sensitive elements can be created by biological engineering; (2) the transducer or the detector element that works in a physicochemical way; optical, piezoelectric, electrochemical, etc. and transforms the signal resulting from the interaction of the analyte with the biological element into another signal i.e., transducers that can be more easily measured and quantified; and (3) associated electronics or signal processors that is primarily responsible for the display of the results in a user-friendly way. The most widespread example of a commercial bio-sensor is the glucose bio-sensor, which uses the enzyme GOD to break blood glucose down. Electrochemical bio-sensors are normally based on enzymatic catalysis of a reaction that produces or consumes electrons (such enzymes are rightly called redox enzymes) [129, 130]. Schematic diagram of a bio-sensor is shown in Figure 18.

The layer-by-layer electrostatic self-assembly of CHIT-*co*-PANI was investigated for the fabrication of glucose bio-sensor using GOD as sensing elements [131]. The multilayer GOD/CHIT-*co*-PANI sensing electrode is expected to have a rapid response and a better sensitivity simultaneous with a higher response current as the layer number increases. It was observed that the steady-state current response time of GOD/CHIT-*co*-PANI is faster than that of GOD/CHIT under identical condition. Furthermore, the bio-sensor with GOD/CHIT-*co*-PANI has a greater response current than its counterpart. Due to the conductivity of PANI, CHIT-*co*-PANI has a faster transition of electron than CHIT. Accordingly, it can enhance the output current and reduce the time of response. The glucose bio-sensors fabricated using the GOD/CHIT-*co*-PANI multilayer film was found to be exhibited a faster response and a higher output current to glucose in the normal and diabetic level.

The performance of such bio-sensors usually depends on the physicochemical properties of the modified CHIT electrodes [132]. An amperometric bio-sensor of rutin in pharmaceutical formulations was fabricated by immobilization of gilo (*Solanum gilo*) crude extract on the chemically crosslinked CHIT with epichlorohydrin and glutaraldehyde matrix (Figure 19) [133].

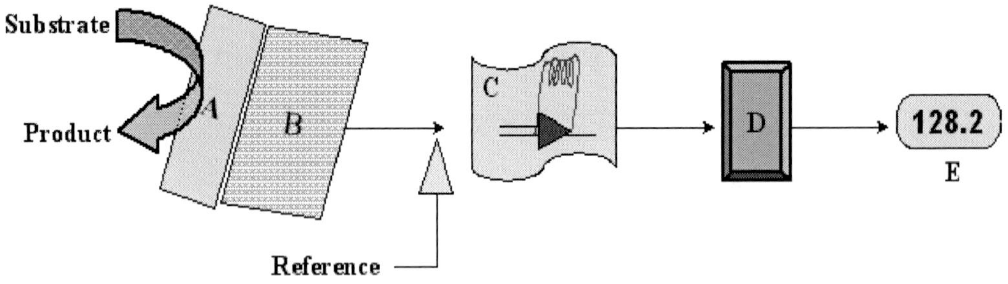

Figure 18. Schematic diagram showing the main components of a bio-sensor. The biocatalyst (A) converts the substrate to product. This reaction is determined by the transducer (B) which converts it to an electrical signal. The output from the transducer is amplified (C), processed (D) and displayed (E).

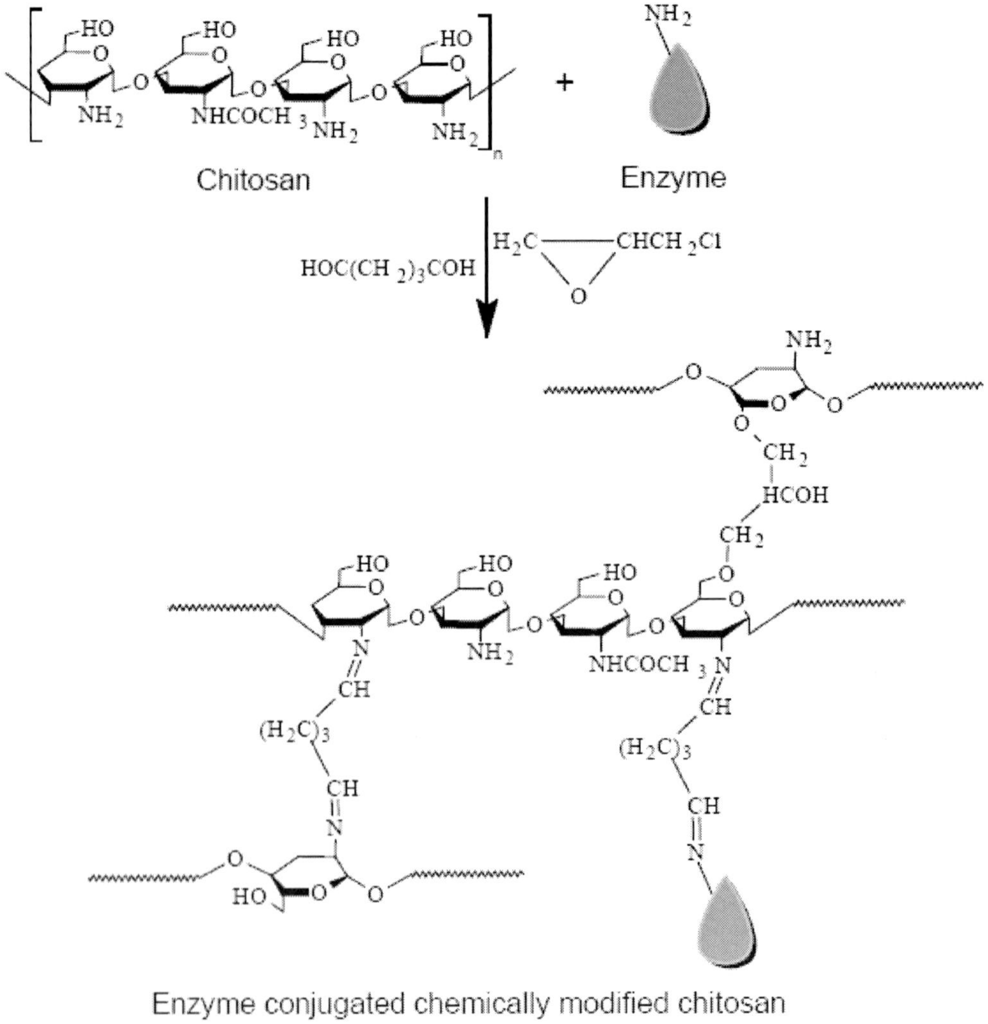

Figure 19. Reaction scheme for the preparation of rutin bio-sensor.

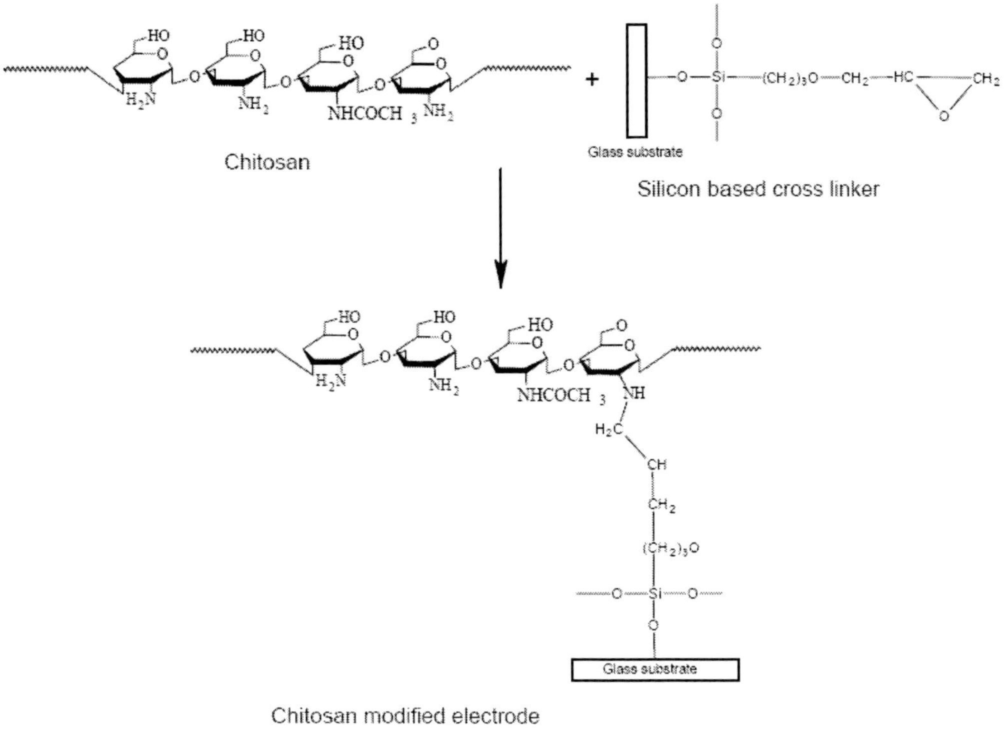

Figure 20. Preparation of rutin bio-sensor based on laccase immobilized CHIT microspheres.

Since rutin is a natural flavone derivative and act as a scavenger of various oxidizing species, the quantification of rutin is of considerable interest. In the bio-sensor, the gilo tissue acts as a source of peroxidase. The bioelectrode exhibited a linear response for rutin concentrations from 3.4×10^{-7} to 7.2×10^{-6} M and the recovery of rutin from the samples ranged from 96.2 to 102.4%. The detection and quantification limits were 2.0×10^{-8} and 6.3×10^{-8} M, respectively. The lifetime of this bio-sensor was found to be 8 months with ~500 uses.

Furthermore, Fernandes *et al.* fabricated rutin bio-sensor based on laccase immobilized on microspheres of CHIT crosslinked with tripolyphosphate in pharmaceutical formulations by square wave voltammetry (Figure 20) [134].

In the bio-sensor, laccase catalyzes the oxidation of rutin to the corresponding *o*-quinone, which is electrochemically reduced back to rutin at +0.35 V vs. Ag/AgCl. The resultant reduction current is used as the analytical response. The lifetime of this bio-sensor was found to be 320 days (at least 930 determinations). The results obtained for rutin in pharmaceutical formulations using the proposed bio-sensor were in agreement with those obtained with the standard method at the 95% confidence level.

Recently, ferrocene-branched CHIT derivatives are synthesized by reductive *N*-alkylation of CHIT with ferrocenecarboxaldehyde and used as a functionalized matrix to immobilize glucose oxidase on glassy carbon electrodes [135]. Ferrocene present in the electrode exhibit an excellent redox activity and establish efficient electrical communication between glucose oxidase and the electrodes for the oxidation of glucose. Gabrovska *et al.* reported urea bio-sensor using urease immobilized poly(acrylonitrile)-CHIT composite membranes [136]. It was found that the average size of the pore under a selective layer base poly(acrylonitrile) membrane is 7 μm, while the membrane coated with 0.25% CHIT shows a reduced pore size-small. In this study, urease was covalently immobilized onto poly(acrylonitrile)-CHIT composite membranes using glutaraldehyde. Both the amount of bound protein and relative activity of immobilized urease were measured. The highest activity was found to be as 94%.

Recently in another report, CHIT-*co*-PANI base immunosensor was investigated for the detection of ochratoxin-A [137]. The CHIT-*co*-PANI conducting biopolymer film was coated on ITO electrode using electrochemical polymerization process. Rabbit antibody (IgGs) was immobilization on CHIT-*co*-PANI, CHIT and PANI matrixes. Electrochemical impedance spectroscopy measurements showed low charge transfer resistance (R_{CT}) of CHIT-*co*-PANI and PANI. Ochratoxin-A interaction with IgGs had increased R_{CT} values and showed linear response up to 10 ng/mL ochratoxin-A concentration in electrolyte (Figure 21).

Relative change in R_{CT} was higher in CHIT-*co*-PANI due to higher proportion of carboxylic and hydroxyl functionalities at CHIT-*co*-PANI matrix surfaces. The absolute sensitivity of PANI, CHIT, and CHIT-*co*-PANI were found to be 16 ± 6, 22 ± 9 and 53 ± 8 Ω mL/ng, respectively. Due to the conductivity of PANI, CHIT/PANI showed low R_{CT} than CHIT. Therefore, bio-sensor could vary the output resistance and provide the electrical signal with respect to bioanalyte. The ochratoxin-A bio-sensor fabricated using the CHIT-*co*-PANI film was found to be exhibited a faster response and a higher output current to ochratoxin-A.

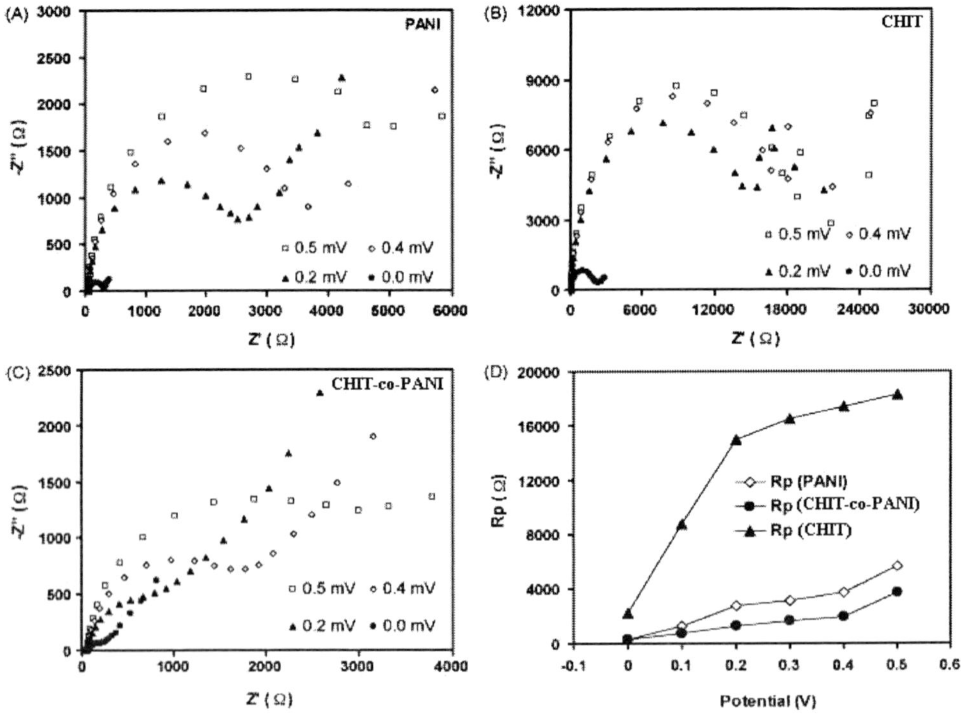

Figure 21. Nyquist plots at different potentials (A) PANI, (B) CHIT, (C) CHIT-*co*-PANI and (D) charge transfer resistance (R_{CT}) for PANI, CHIT-*co*-PANI and CHIT at different potentials.

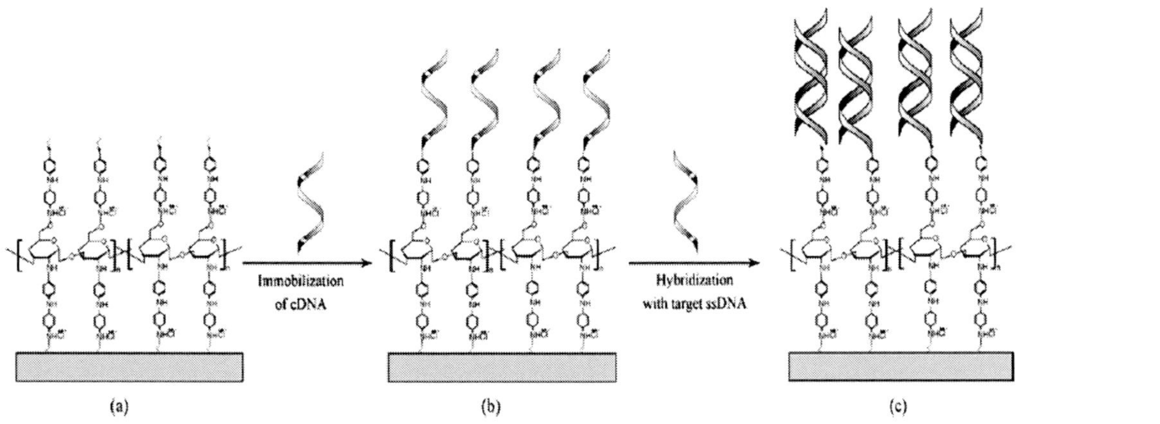

Figure 22. Schematic representation of preparation of CHIT-*co*-PANI/ITO electrode, immobilization of cDNA (BRCA1) over CHIT-*co*-PANI/ITO electrode, and hybridization of the ssDNA onto CHIT-*co*-PANI/ITO electrode.

DNA analysis plays an ever-increasing role in a number of areas related to human health such as diagnosis of infectious diseases, genetic mutations, drug discovery, forensics and food technology. Conventional methods for the analysis of specific gene sequences are based on either direct sequencing or DNA hybridization. The sequencing technology was invented by Maxam and Gilbert [138]. Worldwide, breast cancer is the second most common type of cancer and the second-leading cause of cancer death in women. Although considerable advancements have been made in detection and treatment in recent years, the majority of breast cancers are still detected in their last stage of advancement [139, 140]. Tiwari *et al.* suggested an electrochemical breast cancer bio-sensor based on a CHIT-*co*-PANI copolymer [141]. CHIT-*co*-PANI copolymer coated onto indium-tin-oxide (ITO) was fabricated by immobilizing the complementary DNA (cDNA) probe (42 bases long) associated with the breast cancer susceptible gene BRCA1 (Figure 22). Both the CHIT-*co*-PANI/ITO and the cDNA/Ch-*co*-PANI/ITO electrodes were characterized with Fourier transform infrared spectroscopy, atomic force microscopy, cyclic voltammetry (CV) and electrochemical impedance spectroscopy (EIS). The FTIR spectrum showed that there is an interaction between CHIT-PANI. And FTIR spectra confirmed the immobilization of cDNA onto the CHIT-*co*-PANI/ITO electrode. In the same work the selectivity of the bio-sensor was investigated by the hybridization of the cDNA immobilized CHIT-*co*-PANI/ITO bioelectrode with one-base-mismatch ssDNA sequences. After incubation with the mismatched ssDNA sequence, the bioelectrode showed a negligible decrease in the peak oxidation current at 0.32 V, indicating that the bio-sensor had good selectivity only for the ssDNA. The cDNA/CHIT-*co*-PANI/ITO bioelectrode was employed for the hybridization of the ssDNA, which had a detection limit of 0.05 fmol and showed an excellent sensitivity and reproducibility. The voltammetric sensitivity of the bio-sensor was 2.104 μA/fmol.

6. Conclusion

The combination of functional properties like thermal, electric, conducting, optical and biological with properties of CHIT has led to the development of a wide range of bio-functional materials which could produce a overabundance of biological compatible strategy for the development of eco-friendly technology. Engineering of CHIT as stimuli-responsive advanced bio-materials as biomimetic materials, hybrid-type composite materials, molecular device materials, functionalised polymers, supermolecular systems, information- and energy-transfer materials, environmentally friendly materials, *etc.* at various levels has establish its outstanding applications in a wider mode. The smart materials based on CHIT are prepared by graft copolymerization, oxidative copolymerization, electrochemical and blending with polymer and/or metals which would be capable to give fascinating direction for modern machinery.

Acknowledgments

The author, AT is owe his heartfelt gratitude to the National Institute for Materials Science, Japan for providing infrastructure facility and to JSPS, Japan for generous financial support under project P-09607 to carry out this work.

References

[1] Muzzarelli, R. In *CHIT.* Muzzarelli, R.; Ed.; Oxford: Pergamon Press, 1793; pp 144-176.
[2] Wang, B.; Zhang, J.; Cheng, G.; Dong, S. *Anal. Chim. Acta* 2000, *1-2*, 111- 118.
[3] Mucha, M.; Wankowicz, K.; Balcerzak, J. *e-Polymers* 2007*, 016*, 1-10.
[4] Kim, S.J.; Yoon, S.G.; Kim, I.Y.; Kim, S.I. *J. Appl. Polymer Sci.* 2004, *91*, 2876-2880.
[5] Kim, S.J.; Shin, S.R.; Kim, N.G.; Kim, S.I. *Smart Materials and Structures* 2004, *13*, 1036-1039.
[6] Crini, G. *Bioresource Technology* 2006, *9*, 1061-1085.
[7] Won, W.; Feng, X.; Lawless, D. *J. Membr. Sci.* 2002, *209*, 493–508.
[8] Arvanitoyannis, I.; Kulokuris, I.; Nakayama, A.; Yamamoto, N.; Aiba, S. *Carbohydr. Polymer* 1997, *34*, 9-19.
[9] Arvanitoyannis, I.; Nakayama, A.; Aiba, S. *Carbohydr. Polymer* 1998, *37*, 371-382.
[10] Arvanitoyannis, I. *J. Macromol. Sci.* 1999, *C39*, 205-271.
[11] Puttipipatkhachorn, S.; Nunthanid, J.; Yamamoto, K.; Peck, G. E. *J. Contr. Release* 2001, *75*, 143-153.
[12] Nunthanid, J.; Laungtana-anan, M.; Sriamornsak, P.; Limmatvapirat, S.; Puttipipatkhachorn, S.; Lim, L.Y.; Khor, E. *J. Contr. Release* 2004, *99*, 15-26.
[13] Tsai, Y.; Chen, S.; Liaw, H.W. *Sensor Actuator B Chem.* 2007, *125*, 474-481.
[14] Wu, B.Y.; Hou, S.H.; Yin, F.; Li, J.; Zhao, Z.; Huang, J.D.; Chen, Q. *Biosens. Bioelectron* 2007, *22*, 838-844.

[15] Wu, Z.; Feng, W.; Feng Y.; Liu Q.; Xu, X.; Sekino, T.; Fujii, A.; Ozaki, M. *Carbon* 2007, *45*, 1212-1218.
[16] Arnold, C.; Wu, J.R. *Methods in Enzymology*, 1988, *161*, *447-452.*
[17] Cao, Z.; Ge, H.; Lai, S. *Eur. Polym. J.* 2001, 37, 2141-2143.
[18] Kurita, K, *Prog. Polym. Sci.* 2001, *26*, 1921-1971.
[19] Yu L.; Li L.; Xiaofeng S.; Yue-e F. *Rad. Phys. Chem.* 2005, *74,* 297-301.
[20] Tao S.; Peixin X.; Qing L.; Jian X.; Wenming X, *Euro. Polym. J.* 2003, *39,*189-192.
[21] Guoqi, F.; Jichao, Z.; Hao, Y.; Li, L.; Binglin, H. *React. Funct. Polym.* 2007, *67,* 442-450.
[22] Prashanth, K. V. H.; Tharanathan, R. N. *Carbohydr. Polym.* 2003, *54, 343-351.*
[23] Dong, L. C.; Hoffman, A. S. *J. Controlled Release* 1991, *15*, 141 – 152.
[24] Hoffman, A. S.; Afrassiabi, A.; Dong, L. C. *J. Controlled Release* 1986, *4*, 213 – 222.
[25] Khare, A. R.; Peppas, N. A. *Biomaterials* 1995, *16*, 559 – 567.
[26] Ulbrich. K.; Subr, V.; Podperova, P.; Buresova, M. *J. Controlled Release* 1995, *34*, 155 – 165.
[27] Beltran, S.; Hooper, H. H.; Blanch, H. W.; Prausnitz, J. M. *J. Chem. Phy.* 1990, *92*, 2061 – 2066.
[28] Zhang, J.; Nicholas, A. *Macromolecules* 2000, *33*, 102 – 107.
[29] Nishi, S.; Kotaka, T. *Polym. J.* 1989, *21*, 393 – 402.
[30] Tanaka, Y.; Kagami, Y.; Matsuda, A.; Osada, Y. *Macromolecules* 1995, *28*, 2574 – 2576.
[31] Han, C. K.; Bae, Y. H. *Polymer* 1998, *39*, 2809 – 2814.
[32] Vernon, B.; Kim, S. W.; Bae, Y. H. *J. Biomater. Sci. Polymer Edn.* 1999, *10*, 183 – 198.
[33] Topp, M. D. C.; Dijkstra, P. J.; Talsma, H.; Feijen, J. *Macromolecules* 1997, *30*, 8518 – 8520.
[34] Ebara, M.; Aoyagi, T.; Sakai, K.; Okana, T. *Macromolecules* 2000, *33*, 8312 – 8316.
[35] Schild, H. G. *Prog. Polym. Sci.* 1992, *17*, 163 – 249.
[36] Nonaka, T.; Hashimoto, K.; Kurihara, S. *J. Appl. Polym. Sci.*1997, 66, 209 – 216.
[37] Goni, I.; Gurruchaga, M.; Valero, M.; Guzman, G. M. *J. Polym. Sci. Polym. Chem.* 1983, 21, 2573 – 2580.
[38] Chung, H. J.; Bae, J. W.; Park, H. D.; Lee, J. W.; Park, K. D. *Macromol. Symp.* 2005, *224*, 275 – 286.
[39] Lee, J. W.; Jung, M. C.; Park, H. D.; Park, K. D.; Ryu, G. H. *J. Biomater. Sci. Polymer Edn* 2004, *15*, 1065 – 1079.
[40] Kim, S. Y.; Cho, S. M.; Lee, Y. M.; Kim, S. J. *J. Appl. Polym. Sci.* 2000, *78*, 1381 – 1391.
[41] Mu, Q.; Fang, Y. *Carbohyd. Polym.* 2008, *72*, 308–314.
[42] Zhang, H.; Zhong, H.; Zhang, L.; Chen, S.; Zhao, Y.; Zhu, Y. *Carbohyd. Polym.* 2009, doi:10.1016/j.carbpol.2009.02.026.
[43] Verestiuc, L.; Ivanov, C.; Barbu, E.; Tsibouklis, J. *Int. J. Pharm.* 2004, *269*, 185 – 194.
[44] Wang, M.; Fang, Y.; Hu, D. *React. Funct. Polym.* 2001, *48*, 215 – 221.
[45] Lorenzo, C. A.; Concheiro, A.; Dubovik, A. S.; Grinberg, N. V.; Burova, T. V.; Grinberg, V. Y. *J. Controlled Release* 2005, *102*, 629 – 641.

[46] Don, T. M.; Chen, H. R. *Carbohyd. Polym.* 2005, *61*, 334 – 347.
[47] Kim, I. Y.; Kim, S. J.; Shin, M. S.; Lee, Y.M.; Shin, D. I.; Kim, S. I. *J. Appl. Polym. Sci.* 2002, *85*, 2661 – 2666.
[48] Han, J.; Wang, K.; Yang, D.; Nie. J. *Int. J. Bio. Macromol.* 2009, *44*, 229–235.
[49] Lee, S. B.; Ha, D. I.; Cho, S. K.; Kim, S. J.; Lee, Y. M. *J. Appl. Polym. Sci.* 2004, *92*, 2612 – 2620.
[50] Prabaharan, M.; Mano, J. F. *e-Polymers* 2007, *043*, 1 – 14.
[51] Cao, Y.; Zhang, C.; Shen, W.; Cheng, Z.; Yu, L.; Ping, Q. *J. Controlled Release* 2007, *120*, 186–194.
[52] Fang, J. Y.; Chen, J. P.; Leu, Y. L.; Hu, J. W. *Eur. J. Pharm. Biopharm.* 2008, *68*, 626–636.
[53] Mao, Z.; Ma, L.; Yan, J.; Yan, M.; Gao, C.; Shen, *J. Biomaterials* 2007, *28*, 4488–4500.
[54] Leung, M. F.; Zhu, J.; Harris, F. W.; Li, P. *Macromol. Rapid Commun.* 2004, *25*, 1819 – 1823.
[55] Leung, M. F.; Zhu, J.; Harris, F. W.; Li, P. *Macromol. Symp.* 2005, *226*, 177 – 185.
[56] Lin, C. L.; Chiu, W. Y.; Lee, C. F. *J. Colloid Interface Sci.* 2005, *290*, 397 – 405.
[57] Lee, C. F.; Wen, C. J.; Lin, C. L.; Chiu, W. Y. *J. Polym. Sci. Part A: Polym. Chem.* 2004, *42*, 3029 – 3037.
[58] Matsuda, A.; Ikoma, T.; Kobayashi, H.; Tanaka, *J. Mat. Sci. Eng. C-Biomimetic and Supramolecular Systems*. 2004, *24*, 723-728.
[59] Zhang, M.; Li, X.H.; Gong, Y.D.; Zhao, N.M.; Zhang, X.F. *Biomaterials* 2002, *23* 2641.
[60] Feng, K.J.; Yang, Y.H.; Wang, Z.J.; Jiang, J.H.; Shen, G.L.; Yu, R.Q. *Talanta* 2006, *70*, 561.
[61] Abbaspour, A.; Mehrgardi, M.A. *Anal. Chem.* 2004, *76*, 5690.
[62] Tonya, M.H.; Tarlov, M.J. *J. Am. Chem. Soc.* 1997, *119*, 8916.
[63] Tlili, C.; Korri-Youssoufi, H.; Ponsonnet, L.; Martelet, C.; Jaffrezic-Renault, N.J. *Talanta* 2005, *68*, 131.
[64] Xin, X.; Guang, R.; Juan, C.; Qiang, L.; Dong, L.; Qiang, C. *J. Mat. Sci.* 2006, 41, 3147.
[65] MacDiarmid, A.G.; Mu, S.L.; Somasiri, N.L.D.; Wu, W. *Mol. Cryst. Liq. Cryst.* 1985, *121*, 187.
[66] Novak, P.; Muller, K.; Santhanam, K.S.V.; Hass, O. *Chem. Rev.* 1997, *97*, 207.
[67] Kobayashi, T.; Yonevama, N.; Tamura, H. *J. Electroanal. Chem.* 1984, *177*, 281.
[68] Batich, C.D.; Laitinen, H.A.; Zhou, H.C. *J. Electroanal. Soc.* 1990, *137*, 883.
[69] Desilvestro, J.; Hass, O. *Chem. Commun.* 1985, 346.
[70] Dong, Y.H.; Mu, S.L. *Electrochim. Acta.* 1991, *36*, 2015.
[71] Karg, S.; Scott, J.C.; Salem, J.R.; Angelopoulos, M. *Synth. Met.* 1996, *80*, 111.
[72] Bartlett, P.N.; Whitaker, R.G. *Bio-sensor* 1988, *3*, 359.
[73] Yang, Y.F.; Mu, S.L. *J. Electroanal. Chem.* 1997, *71*, 432.
[74] Chiang, J.-C.; MacDiarmid, A.G. *Synth. Met.* 1986, *13*, 193.
[75] MacDiarmid, A.G.; Chiang, J.C.; Richter, A.F.; Epstein, A.J. *Synth. Met.* 1987, *18*, 285.
[76] MacDiarmid, A.G.; Epstein, A.J. *Faraday Discuss. Chem. Soc.* 1989, *88*, 317.

[77] Stejskal, J.; Kratochvil, P.; Jenkins, A.D. *Polymer* 1996, 37, 367.
[78] Trivedi, D.C. *In Handbook of Organic Conductive Molecules and Polymers*, Nalwa, H.S. (Ed.), Wiley, Chichester 1997, *2*, 505-7572.
[79] Gospodinova N.; Terlemezyan, L. *Prog. Polym. Sci.* 1998, *23*, 1443.
[80] Levi, B.G. *Phys, Today* 2000, *53*, 19.
[81] MecDiarmid, A.G *Angew. Chem, Int. Ed.* 2001, 40, 2581.
[82] Jin, Z.; Su, Y.; Duan, Y. *Sens. Actuators B* 2001, *72*, 75.
[83] Sotomayor, P.T.; Raimundo, I.M.; Zarbin, A.J.G.; Rohwedder, J.J.R.; Netto, G.O.; Alves, O.L. *Sens. Actuators B* 2001, *74*, 157.
[84] Kane-Maguire L.A.P.; Wallace, G.G. *Synth. Met.* 2001, *119*, 39.
[85] Hamers, R.J. *Nature* 2001, *412*, 489.
[86] Rosseinsky, D.R.; Mortimer, R.J. *Adv. Mater.* 2001, *13*, 783.
[87] Prokes, J.; Krivka, I.; Tobolkova, E.; Stejskal J. *Polym. Degrad. Stab.* 2000, *68*, 261.
[88] Elyashevich, G.K.; Terlemezyan, L.; Kuryndin, I.S. Lavrentyev, V.K.; Mokreva, P.; Yu, E.; Rosova, Yu, N. *Sazanov. Thermochim. Acta* 2001, *374*, 23.
[89] El-Sherif, M.A.; Yuan, J.; MacDiarmid, A.G.; *J. Intelligent Mater. Syst. Struct.* 2000, *11*, 407.
[90] Kuzmin, S.V.; Sáha, P.; Sudar, N.T.; Zakrevskii, V.A.; Sapurina, I.; Solosin, S.; Trchova, M.; Stejskal, J. *Thin Solid Films* 2008, *516*, 2181.
[91] Tai, H.; Jiang, Y.; Xie, G.; Yu, J.; Chen, X.; Ying, Z. *Sens. Actuators B* 2008, *129*, 319.
[92] Tiwari, A.; Sen, V.; Dhakate, S.R.; Mishra, A.P.; Singh, V. *Polym. Adv. Technol.* 2008, *19*, 909.
[93] Tiwari, A.; Singh, S.P. *J. Appl. Polym. Sci.* 2008, *108*, 1169.
[94] Tiwari, A. *J. Polym. Res.* 2008, *15*, 337.
[95] Tiwari, A. *J. Mcromol. Sci. A* 2007, *44*, 735.
[96] Tiwari, A.; Singh, V. *Exp. Polym. Let.* 2007, *1*, 308.
[97] Silva, R.C.; Escamilla, A.; Nicho, M.E.; Padron, G.; Perez, A.L.; Marin, E.A.; Moggio, I.; Garcia, J.R. *Eur. Polym. J.* 2007, 3471.
[98] Silva, R.C. *Langmuir* 2007, *23*, 8.
[99] Sperry, J.B.; Wright, D.L. *Chem. Soc. Rev.* 2006, *35*, 605.
[100] Steckhan, E. *Topics in current chemistry, Electrochemistry,* Springer, NY, USA, 1988, *3*.
[101] Tiwari, A.; Gong, S. *Electroanalysis* 2008, *20*, 1775.
[102] Shin, M. S.; Kang, H. S.; Park, T. G.; Yang, J. W. *Polym. Bull.* 2002, *47*, 451 – 456.
[103] Wang, M.; Qiang, J.; Fang, Y.; Hu, D.; Cui, Y.; Fu, X. *J. Polym. Sci. Part A: Polym. Chem.* 1999, *38*, 474 – 481.
[104] Goycoolea, F. M.; Heras, A.; Aranaz, I.; Galed, G.; Fernandez-Valle, M. E.; Monal, W. A. *Macromol. Biosci.* 2003, *3*, 612 – 619.
[105] Chen, J.; Sun, J.; Yang, L.; Zhang, Q.; Zhu, H.; Wu, H.; Hoffman, A. S.; Kaetsu, I. *Radiation Phy. Chem.* 2007, *76*, 1425–1429.
[106] Kim, S.J.; Shin, S.R.; Spinks, G.M.; Kim, I.Y.; Kim, S.I. *J Appl Polymer Sci* 2005, *96*, 867–873.
[107] Shin, S.R.; Park, S.J.; Yoon, S.G.; Spinks,G.M.; Kim, S.I.; Kim, S.J. *Synth. Met.* 2005, *154*, 213–216.

[108] Thanpitcha, T.; Sirivat, A.; Jamieson, A.M.; Rujiravanit, R. *Carbohydr Polymer* 2006, *64*, 560–568.
[109] Wan, Y.; Fang, Y.; Hu, Z.; Wu, Q. *Macromol. Rapid Commun.* 2006, *27*, 948–954.
[110] Spinks, G.M.; Shin, S.R.; Wallace, G.G.; Whitten, P.G.; Kim, I.Y.; Kim, S.I.; Kim, S.J. *Sens. Actuators B* 2007, *121*, 616–621.
[111] Ismail, Y.A.; Shin,S.R.; Shin, K.M.; Yoon, S.G.; Shon, K.; Kim, S.I.; Kim, S.J. *Sensor Actuator B Chem* 2008, *129*, 834–840.
[112] Yavuz, A.G.; Uygun, A.; Bhethanabotla, V.R. *Carbohydr Polymer*, 2009, *75*, 448-450.
[113] Wallace,G. G.; Kane-Mauire, L. A. P. *Adv. Mater.* 2002, *14*, 953-960.
[114] Valentini, R. F.; Vargo, T. G.; Gaedella Jr.,vJ.A.; Aebischer, P. *Biomaterials,* 1992, *13*, 183-190.
[115] Wong, J. Y.; Langer, R.; Ingber, D. E. *Mater. Res. Soc. Symp. Proc.* 1994, *331*, 141.
[116] Schmidt, C. E.; Shastri, V. R.; Vacanti, J. P.; Langer, R. *Proc. Natl. Acad. Sci. USA* 1997, *94*, 8948-8953.
[117] Kotwal,A.; Schmidt, C. E. *Biomaterials* 2001, *22*, 1055-1064.
[118] Garner, B.; Georgevich, A.; Hodgson, A. J.; Liu, L.; Wallace, G. G. J. *Biomed. Mater. Res.* 1999, *44*, 121-129.
[119] De Giglio, E. ; Sabbatini, L. ; Zambonin, P. G. *J. Biomater. Sci. Polym. Ed.* 1999, *10*, 845-856.
[120] Aoki, T.; Tanino, M.; Ogata, N.; Kumakura, K. *Biomaterials* 1996, *17*, 1971-1974.
[121] Wong, Y.; Langer, R.; Ingber, D. E.; *Proc. Natl. Acad. Sci. USA* 1994, *91*, 3201-3204.
[122] Wan, Y.; Fang, Y.; Hu, Z.; Wu, Q. *Macromol. Rapid Commun.* 2006, *27*, 948–954.
[123] Yang, X.; Lu, Y. *Polymer* 2005, *46*, 5324–5328.
[124] Abdi, M.M.; Kassim, A.; Mahmud, H. N. M. E.; Yunus, W.M.M. ; Talib, Z.A.; Sadrolhosseini, A.R. *J. Mater. Sci.* 2009, *44*, 3682–3686.
[125] Costa, A.C.R.; Siqueira, A.F. *J. Appl. Phys.* 1996, *80*, 5579-5582.
[126] Kim, M.S.; Kim, H.K.; Byun, S.W.; Jeong, S.H.; Hong, Y.K.; Joo, J.S.; Song, K.T.; Kim, J.K.; Lee, C.J.; Lee, J.Y. *Synth. Met.*, 2002, *126*, 233-239.
[127] Kim, H.K.; Kim, M.S.; Chun, S.Y.; Park, Y.H.; Jeon, B.S.; Lee, J.Y. *Mol. Cryst. Liq. Crys.t* 2003, *405*,161-169.
[128] Samjin, C.; Jaing, Z. *Sens. Actu. A* 2006, *128*, 317.
[129] Liu, X.; Shi, L.; Niu, W.; Li, H.; Xu, G. *Biosens. Bioelectr.* 2008, *23*, 1887.
[130] Tiwari, A.; Gong, S. *Electroanalysis* 2008, *20*, 2119.
[131] Xin, X.; Guang, R.; Juan, C.; Qiang, L.; Dong, L.; Qiang, C. *J. Mat. Sci.* 2006, *41*, 3147.
[132] Cho, W.J.; Huang, H.J. *Anal. Chem.* 1998, *70*, 3946.
[133] Oliveira, I.R.W.Z.; Fernandes, S.C.; Vieira, I.C. *Journal of Pharmaceutical and Biomedical Analysis* 2006, *41*, 366.
[134] S.C. Fernandes, de Oliveira, I.R.W.Z.; Filho, O.F.; Spinelli, A.; Vieira, I.C.; *Sens. Actuators B* 2008, *133*, 202.
[135] Yang, W.; Zhou, H.; Sun, C. *Macromolecular Rapid Communications* 2007, *28*, 265.
[136] Gabrovska, K.; Georgieva, A.; Godjevargova, T.; Stoilova, O.; Manolova, N. *Journal of Biotechnology* 2007, *129*, 674.
[137] Khan, R.; Dhayal, M. *Bio-sensors* 2009, *24*, 1700.

[138] Maxam AM.; Gilbert W. A new method for sequencing DNA. *Proc. Natl. Acad. Sci. USA* 1977, *74*, 560–564.

[139] Bouchardy, C.; Fioretta, G.; Verkooijen, H.M.; Vlastos, G.; Schaefer, P.; Delaloye, J.F.; Caspar, I.N.; Majno, S.B.; Wespi, Y.; Forni,M.; Chappuis,P.; Sappino, A.P.; Rapiti,E.; *Br. J. Cancer* 2007, *96*, 1743-1746.

[140] Sims, A.H.; Howell, A.; Howell, S.J.; Clarke, R.B. *Nat. Clin. Pract. Oncol.* 2007, *4*, 516-525.

[141] Tiwari, A.; Gong, S. *Talanta* 2009, *77*, 1217.

In: Focus on Chitosan Research
Editors: Arthur N. Ferguson and Amy G. O'Neill
ISBN 978-1-61324-454-8

Chapter 10

CHITOSAN COPOLYMERS AND IPNS FOR CONTROLLED DRUG RELEASE

P. P. Kundu[1]*, Vinay Sharma[1] and Kamlesh Kumari[2]

[1.] Deptartment of Polymer Science and Technology,
Calcutta University, Calcutta, India
[2.] Department of Chemical Technology,
Sant Longowal Institute of Engineering and Technology,
Longowal, India

ABSTRACT

Chitosan is a linear aminopolysaccharide obtained from the alkaline deacetylation of chitin. Chitosan is both biocompatible and biodegradable, making it an attractive material for use in drug delivery, gene delivery, cell culture, and tissue engineering applications. The achievement of predictable and reproducible release of an active agent into a specific environment over an extended period of time has significant advantage. A number of biodegradable polymers are potentially useful for this purpose, including synthetic as well as natural substances. Because chitosan is a swellable polymer with properties that can be tailored by varying the degree of deacetylation and molecular weight, it has been extensively investigated for use as a retardant polymer in matrix tablet formulations. The purpose of this review is to take a close look at the applications of chitosan IPNs and copolymers in controlled drug release. Based upon the present research and available products, some new and futuristic approaches in this area are thoroughly discussed.

* Corresponding author. E-mail: ppk923@yahoo.com.
1 Calcutta University, 92, A. P. C. Road, Calcutta-700 009, India.
2 Sant Longowal Institute of Engineering and Technology, Longowal-148106, India.

1. INTRODUCTION

1.1. Chitosan

Chitosan (poly [β-(1-4)-2-amino-2-deoxy-D-glucopyranose]), a non-toxic and biocompatible cationic polysaccharide, is produced by partial deacetylation of chitin derived from naturally occurring crustacean shells. Chitin is made up of a linear chain of acetylglucosamine groups while chitosan is obtained by removing enough acetyl groups (CH3-CO) for the molecule to be soluble in most diluted acids. This process is called deacetylation. The actual difference between chitin and chitosan is the acetyl content of the polymer. The molecular formula of chitosan (CHI) is $C_6H_{11}O_4N$ and its structure is shown in Figure 1. It is a linear polysaccharide composed of randomly distributed β-(1-4)-linked D-glucosamine (deacetylated unit) and N-acetyl-D-glucosamine (acetylated unit). The term chitosan embraces a series of polymers that vary in molecular weight (from approximately 10,000 to 1 million Dalton) and degree of deacetylation (in range of 50-95 %).

Since chitosan displays mucoadhesive properties [1], strong permeation enhancing capabilities for hydrophilic compounds and a safe toxicity profile, it has received considerable attention as a novel excipient in drug delivery system [2] and has been included in the European Pharmacopoeia since 2002. It has a number of commercial and possible biomedical uses [3-5]. Chitosan's has recently gained approval in the USA for use in bandages and other hemostatic agents [6]. The different theoretical mechanisms by which bioadhesion can be attained are shown in Table 1 [7].

The main commercial sources of chitin are the shellfish waste such as shrimps, lobster, crabs, squid pens, prawns and crawfish. Chitosan is found in nature, to a lesser extent than chitin, in the cell walls of fungi. Chitin is economically feasible and ecologically desirable because large amounts of shell wastes are available as a by-product of the seafood industry. Production of chitosan from these is inexpensive and easy. Commercially, chitosan is available in the form of dry flakes, solution and fine powder.

1.2. Interpenetrating Networks (IPNs)

An IPN is defined as an intimate combination of two polymersboth in network form, at least one of which is synthesized or crosslinked in the immediate presence of other [8-12]. Since its first synthesis in 1960 [13], this novel technique of synthesizing multi polymer systems helped in achieving tailor made blends with desired range of specific properties. A broad range of properties ranging from toughened elastomers to high impact plastics can be obtained when a glassy polymer is blended with a rubbery one. The ultimate properties depend on the relative proportions of the two components and also on the physical nature of the continuous phase [14-18].

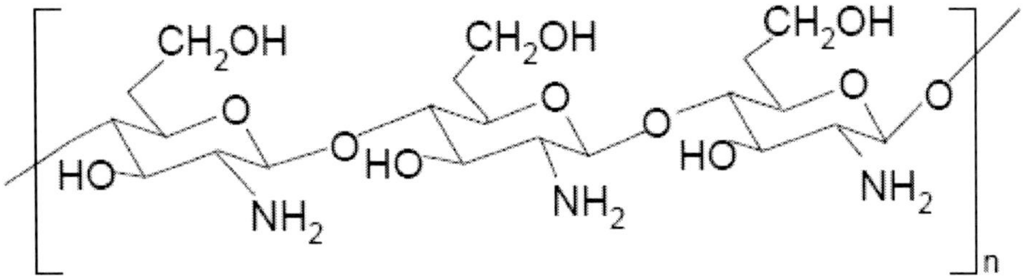

Figure 1. Structure of Chitosan.

Table 1. Different possible theoretical methods for bioadhesion of chitosan

Types	Method	Remarks
Electronic Theory	Attractive electrostatic forces between glycoprotein mucin network and the bioadhesive material.	Electron transfer occurs between the two forming a double layer of electric charge at the interface.
Adsorption Theory	Surface forces resulting in chemical bonding.	Strong primary forces: covalent bonds. Weak secondary forces: ionic bonds, hydrogen bonds and van der Walls' forces.
Wetting Theory	Ability of bioadhesive polymers to spread and develop intimate contact with the mucus membranes.	Spreading coefficients of polymers must be positive. The contact angle between polymer and cells must be near to zero.
Diffusion Theory	Physical entanglement of mucin strands and the flexible polymer chains. Interpenetration of mucin strands into the porous structure of the polymer substrate.	For maximum diffusion and best bioadhesive strength: solubility parameters of the bioadhesive polymer and the mucus glycoproteins must be similar.
Fracture Theory	Analyses the maximum tensile stress developed during the detachment of the buccal drug delivery system from the mucosal surfaces.	Does not require physical entanglement of bioadhesive polymer chains and mucin strands, hence appropriate to study the bioadhesion of hard polymers, which lack flexible chains.

1.3. Controlled Drug Release

Controlled drug release (CDR) occurs when a polymer, whether natural or synthetic, is judiciously combined with a drug or other active agent in such a way that the active agent is released from the material in a predesigned manner. The ideal drug release system [19] should be inert, biocompatible, mechanically strong, comfortable for the patient, capable of achieving high drug loading, safe from accidental release, simple to administer and remove, and easy to fabricate and sterilize. Providing control over drug delivery can be the crucial factor at times when traditional oral or injectable drug formulations cannot be used. These include situations requiring the slow release of water-soluble drugs, the fast release of low-solubility drugs, drug delivery to specific sites, drug delivery using nano particulate systems, delivery of two or more agents with the same formulation, and systems based on carriers that can dissolve or degrade and be readily eliminated. Basically, controlled release drug delivery systems serve two functions. The first, drug delivery, involves the transport of the drug to a particular part of the body. It may be accomplished in a number of ways: intravenously, transdermally, orally or vaginally [20].

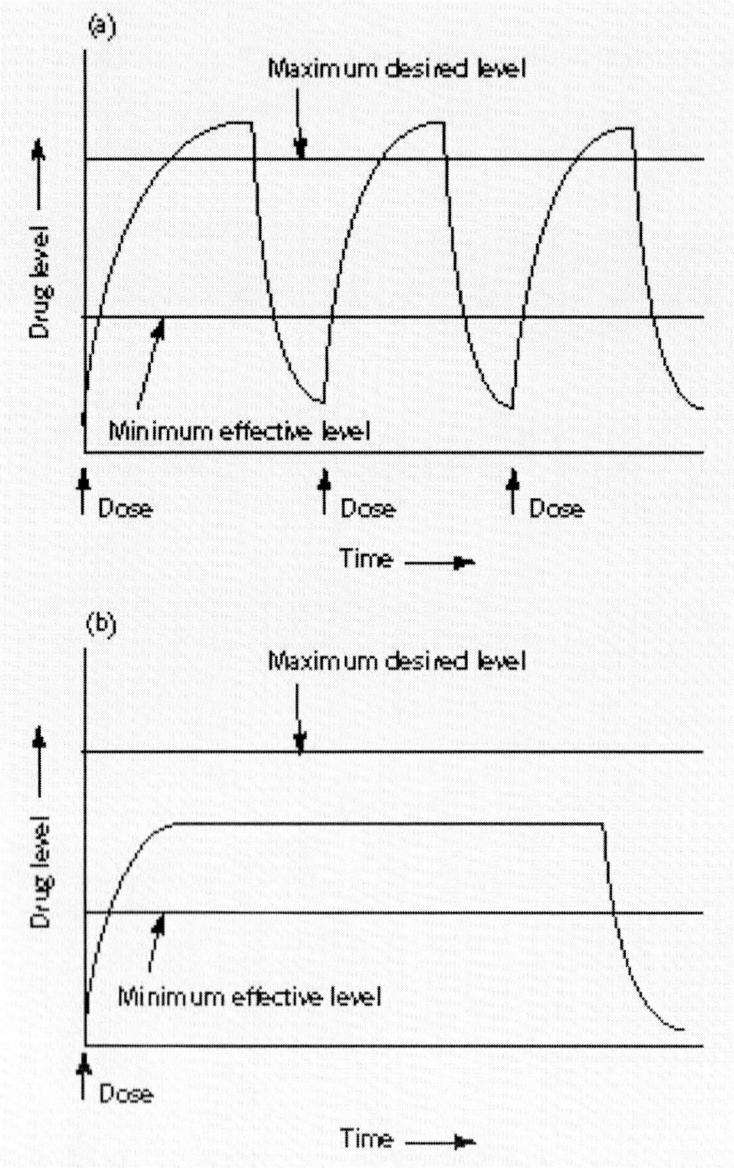

Reproduced with permission from Medical Plastics and Biomaterials, 1997, 34 © MPB.

Figure 2. Drug levels in blood with (a) traditional drug dosing and (b) controlled delivery dosing.

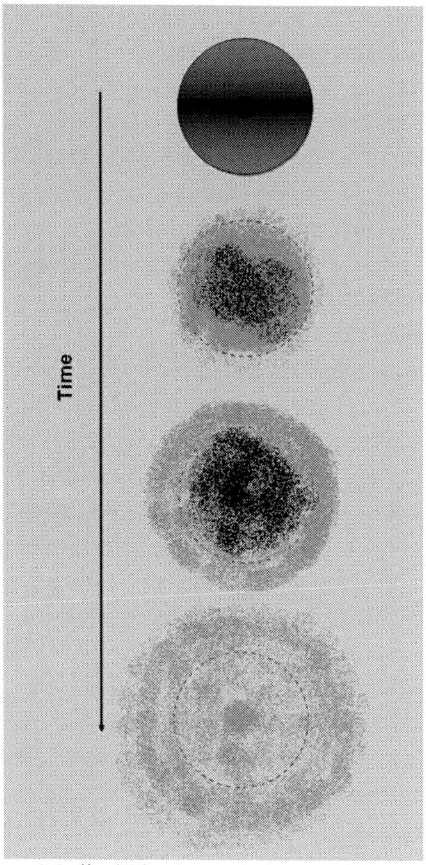

Reproduced with permission from Medical Plastics and Biomaterials, 1997, 34 © MPB.

Figure 3. Drug delivery from a typical matrix drug delivery system.

The second function, controlled release, describes the rate at which the drug is made available to the body once it has been delivered. The release of the active agent may be constant or cyclic over a long period, or it may be triggered by the environment or other external stimuli. In any case, the purpose behind controlled drug delivery is to achieve more effective therapies while eliminating the potential for both under and overdosing. Other advantages of using controlled delivery systems can include the maintenance of drug levels within a desired range, the need for fewer administrations, optimal use of the drug in question, and increased patient compliance. While these advantages can be significant, the potential disadvantages cannot be ignored: the possible toxicity or non-biocompatibility of the materials used, undesirable by-products of degradation, any surgery required to implant or remove the system, the chance of patient discomfort from the delivery device, and the higher cost of controlled-release systems compared with traditional pharmaceutical formulations [21].

The goal of many of the original controlled release systems was to achieve a delivery profile that would yield a high blood level of the drug over a long period of time [21]. With traditional tablets or injections, the drug level in the blood follows the profile shown in Figure 2a, in which the level rises after each administration of the drug and then decreases until the next administration. The key point with traditional drug administration is that the blood level of the agent should remain between a maximum value, which may represent a toxic level, and a minimum value, below which the drug is no longer effective. In controlled drug delivery, the drug level in the blood remains constant as shown in Figure 2b, between the desired maximum and minimum levels, for an extended period of time. Depending on the formulation and the application, this time may be anywhere from 24 hours to 5 years or even more.

In vitro release from biodegradable polymeric drug delivery systems has been shown to be influenced by many factors, such as 1) release medium composition (buffer system, ionic strength, pH, surfactants); 2) test conditions (temperature, agitation, frequent supernatant removal); 3) drug stability, solubility and adsorption to degrading matrix; and 4) formation of internal acidic microenvironment [22-28].

Basically, there are three primary mechanisms by which active agents can be released from a delivery system: diffusion, degradation, and swelling followed by diffusion [21]. Any or all of these mechanisms may occur in a given release system. Diffusion occurs when a drug or other active agent passes through the polymer that forms the controlled-release device. The diffusion can occur on a macroscopic scale, as through pores in the polymer matrix or on a molecular level, by passing between polymer chains. A polymer and active agent can be mixed to form a homogeneous system, also referred to as a matrix system. Matrix systems are those in which drug is distributed uniformly in inert polymeric matrix [29]. Diffusion occurs when the drug passes from the polymeric matrix into the external environment. The active agent travels through a swelled polymer matrix as a result of which the effective diameter of the matrix increases, as shown in Figure 3(a-d) [21]. The release rate normally decreases with this type of system, since the active agent has a progressively longer distance to travel and therefore requires a longer diffusion time to release.

In other methods, the drug can be administered either via implantable system *i.e.* orally or via transdermal system. The system shown in Figure 4a is representative of an implantable or oral reservoir delivery system, whereas the system shown in Figure 4b illustrates a transdermal drug delivery system, in which only one side of the device will be delivering the drug. For the systems shown in Figures 4a and 4b, the drug delivery rate can remain fairly constant. In these systems, a reservoir-whether solid drug, dilute solution, or highly concentrated drug solution within a polymer matrix is surrounded by a film or membrane of a rate-controlling material. The only structure effectively limiting the release of the drug is the polymer layer surrounding the reservoir. Since this polymer coating is essentially uniform and of a non changing thickness, the diffusion rate of the active agent can be kept fairly stable throughout the lifetime of the delivery system.

For the diffusion controlled systems described thus far, the drug delivery device is fundamentally stable in the biological environment and does not change its size either through swelling or degradation. In these systems, the combinations of polymer matrices and bio-active agents chosen must allow for the drug to diffuse through the pores or macromolecular structure of the polymer upon introduction of the delivery system into the biological environment without inducing any change in the polymer itself. It is also possible for a drug delivery system to be designed so that it is incapable of releasing its agent or agents until it is placed in an appropriate biological environment [21]. Swelling-controlled release systems are initially dry and, when placed in the body will absorb water or other body fluids and swell. The swelling increases the aqueous solvent content within the formulation as well as the polymer mesh size, enabling the drug to diffuse through the swollen network into the external environment [21].

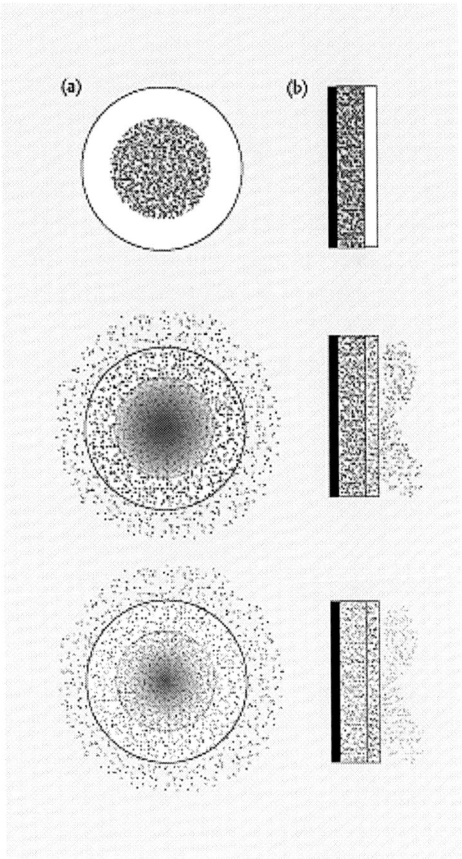

Reproduced with permission from Medical Plastics and Biomaterials, 1997, 34 © MPB.

Figure 4. Drug delivery from typical reservoir devices: (a) implantable or oral systems, and (b) transdermal systems.

2. Correlation between CDR and IPN/Copolymers

Drug delivery systems are essential components of controlled drugs release. The polymers have the advantage that they can be processed with various reproducible techniques in different well-defined combinations (graft, block and star copolymers, networks, and modified linear polymers) based on synthetic or natural polymers and mixtures. In the last decade, several kind of polymeric systems are proposed as drug carriers like capsules, liposomes, microparticles, nanoparticles, IPNs and copolymers [30]. These components must be biocompatible, biodegradable, and display a desired biodistribution providing a long-term availability of the therapeutic at specific target over time.

The use of copolymers allows precise control of the polymer's properties and quality of the polymers. The mechanical and physicochemical properties of such copolymers can be precisely tailored by varying the chemical composition. Polymer blends/IPNs are also attractive biomaterials because their properties can be easily adjusted by varying the ratio and nature of the components. This class of materials has considerable potential, owing in part to the large number of existing polymers that can be mixed. They also have the advantage of being easily produced by simple techniques. Blends or mixtures containing synthetic and/or natural polymers have been used in drug delivery. The use of IPNs has received a great deal of attention in recent years and different kinds of IPNs have been synthesized with different compositions. They are widely used in the development of new synthetic polymer scaffolds with improved biomechanical properties [30].

IPNs and copolymers (the modified forms of natural polymers) are capable of delivering drugs at constant rates over an extended period of time. In these systems, the rate of drug release is controlled by the balance between the drug diffusion across the concentration gradient and the polymer relaxation rate as a result of the diffusion controlled process due to swelling. The drug diffusion from the IPNs depends on the mesh size of the polymer network, which usually decreases with the increasing degree of crosslinker [31] The crosslinking also induces an increase in the crystallinity of the polymer chain as a result of which release rate decreases. The extent of grafting ratio also plays an important role in the drug release from the copolymer. With an increase in grafting, the drug release rate increases.

3. Chitosan IPNs and Copolymers

Due to its specific structure and properties, chitosan has found a number of applications in drug delivery including as an absorption enhancer for hydrophilic macromolecular drugs and as a gene delivery system [32-34]. However, the applications of chitosan in the biomedical field are limited, by its poor solubility in physiological media. Chitosan is soluble only in aqueous acidic solutions below pH 6.5, in which the primary amino groups of chitosan are protonated. To improve the poor water solubility of chitosan, several derivatives have been proposed. For example, the modification of chitosan by quaternization of the amino groups [35-37], N-carboxymethylation [38] and PEGylation [39,40] have been reported. Blending of chitosan with other polymers [41-43] and crosslinking are both convenient and effective methods of improving the physical and mechanical properties of chitosan for practical applications. Immunization studies carried out on rats using glutaraldehyde crosslinked chitosan spheres [44] showed promising tolerance by the living tissues of the rat muscles.

3.1. Chitosan Copolymers - Preparation and Use in CDR

Aminabhavi *et. al.* [45] prepared microspheres of polyacrylamide-grafted-chitosan crosslinked with glutaraldehyde. The detailed reaction is shown in Scheme 1. These copolymers were used to encapsulate indomethacin, a non-steroidal anti-inflammatory drug. The microspheres were produced by water/oil emulsion technique and encapsulation of indomethacin was carried out before crosslinking of the matrix. Microspheres were characterized for drug-entrapment efficiency, particle size, and water transport into the polymeric matrix as well as for drug-release kinetics. Dynamic swelling experiments suggested that, with an increase in crosslinking, the transport mechanism changed from Fickian to non-Fickian. The release of indomethacin depends upon the crosslinking degree of the network and also on the amount of drug loading. This was further supported by the calculation of drug-diffusion coefficients using the initial time approximation. The drug release in all the formulations followed a non-Fickian trend and the diffusion was relaxation-controlled.

Aminabhavi and Kumbar [31] synthesized modified chitosan microspheres and studied the effect of the grafting ratio on the controlled release of nifedipine (NFD) through microspheres. The grafting of acrylamide onto a chitosan backbone was carried out at three different acrylamide concentrations. The synthesis of the grafted polymer was achieved by $K_2S_2O_8$-induced free-radical polymerization. Microspheres of polyacrylamide-*g*-chitosan crosslinked with glutaraldehyde were prepared to encapsulate NFD, a calcium channel blocker and an antihypertensive drug. The microspheres of polyacrylamide-*g*-chitosan were produced by a water-oil emulsion technique with three different concentrations of glutaraldehyde as the crosslinking agent.

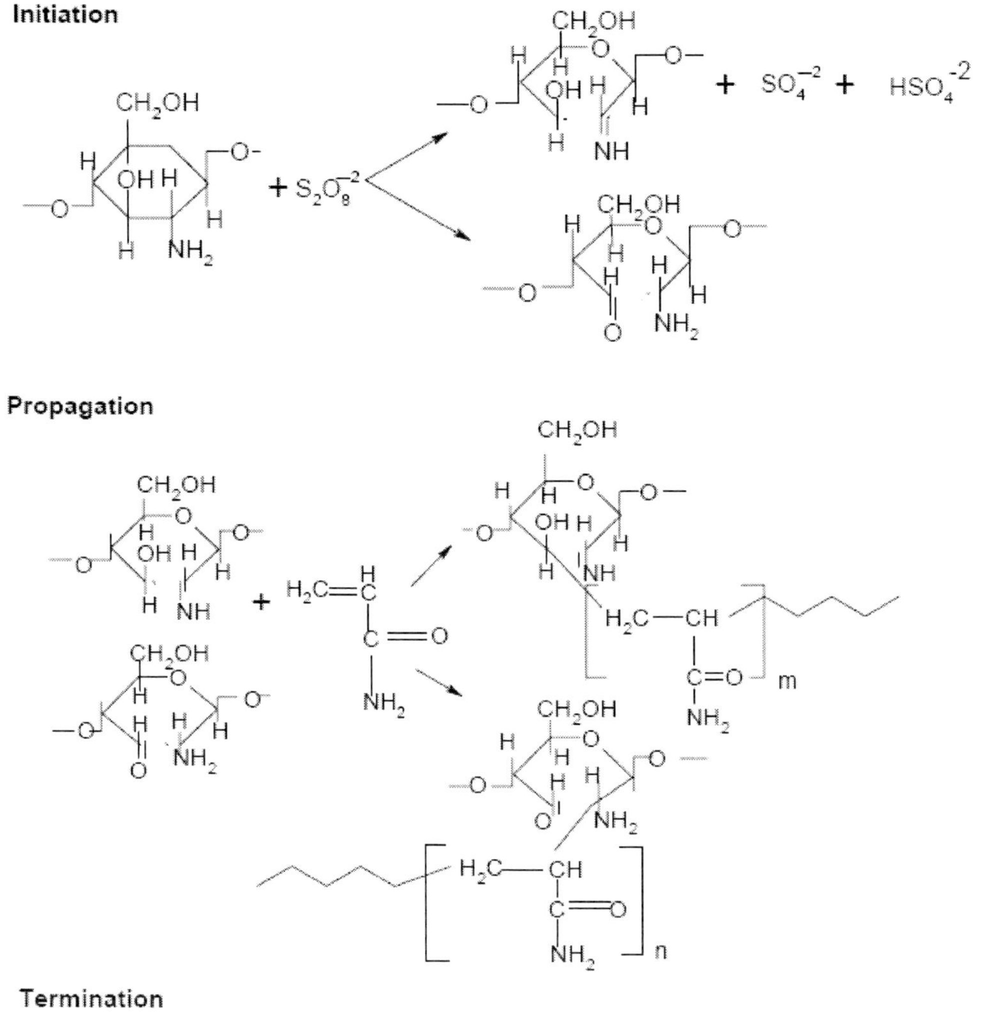

Reproduced with permission from Journal of Applied Polymer Science, 2003, 87, 1527 © John Wiley and Sons, Inc.

Scheme 1. Grafting and crosslinking reaction of PAAm on to chitosan.

The microspheres were characterized by the particle size; the water transports into these microspheres, as well as the equilibrium water uptake, were studied. Release of NFD depended on the crosslinking of the network and on the amount of drug loading. Calculations of drug diffusion coefficients with the initial time and later time approximation method further supported this. The drug release suggested that the time dependence of the NFD release followed zero-order kinetics.

Mao [46] synthesized the poly (ethylene glycol)-g-trimethyl chitosan block copolymers for the intranasal delivery of insulin. This block copolymer will increase both the solubility of chitosan in water, and the biocompatibility of trimethyl chitosan (TMC). A series of copolymers with different degrees of substitution were obtained by grafting the activated PEG of different molecular weight onto TMC via primary amino groups. Solubility experiments demonstrated that PEG-g-TMC (400) copolymers were completely water-soluble over the entire pH range of 1-14 regardless of the PEG molecular weight, even when the graft density was as low as 10%. Cytotoxicity of the copolymers and corresponding insulin complexes was studied using a methyl tetrajolium (MTT) assay.

All polymers exhibited a time- and dose-dependent cytotoxic response that increased with molecular weight. It was found that PEGylation with PEG, 5 kDa and 20 kDa significantly decreased the cytotoxicity of TMC 400 kDa, and complex formation with insulin led to further increases in concentration of the copolymers resulting in 50% inhibition of cell growth, i.e. IC_{50} value. Properties of PEG-g-TMC copolymer insulin complexes were further investigated to gain an insight into the effect of the copolymer composition on the complex properties. All complexes investigated had a particle size in the range of 150-300 nm and were positively charged. The ξ potentials decreased with increasing graft density. Association efficiency increased with PEG graft ratio, and extremely high association efficiency, that is, 96.22%, was obtained with PEG (5k)679-g-TMC(400) copolymer (679 indicates the average number of PEG chains per TMC 400 kDa macromolecule). Based on the good solubility of the copolymers and the specialty of polyethylene glycol, it was believed that these copolymers could be used to enhance the therapeutic and biotechnological potentials of other macromolecules.

To increase the antimicrobial activities of chitosan, Jung *et al.* [47] grafted two anionic soluble monomers, mono (2-methacryloyl oxyethyl) acid phosphate (MAP) and vinylsulfonic acid sodium salt (VSS) onto chitosan. The grafted copolymers had zwitterionic property. Antimicrobial activities of chitosan and graft copolymers depended largely on the amount and type of grafted chains as well as changes of pH, against *Candida albicans, Trichophyton rubrum,* and *Trichophyton violaceum.* The best antimicrobial activity among tested samples was at pH 5.75 with demonstrating strain selectivity against *Candida albicans* and *Trichophyton violaceum* due to the difference in affinity between cell wall of fungi and chitosan or its derivatives.

Lee *et al.* [48] prepared novel water-soluble thermosensitive chitosan copolymers by graft polymerization of N-isopropylacrylamide (NIPAAm) onto chitosan using cerium ammonium nitrate (CAN) as an initiator. They have investigated Sol-gel transition behavior by the cloud point measurement of the chitosan-g-NIPAAm aqueous solution. The gelling temperature was examined using the vial inversion method. The percentage of grafting and efficiency of grafting (%) were investigated according to concentrations of monomer and initiator. The maximum grafted chitosan copolymer was obtained with 0.4 M NIPAAm and $6x10^{-3}$ M CAN. Water-soluble chitosan-g-NIPAAm copolymers were prepared successfully and they formed thermally reversible hydrogels, which exhibit a lower critical solution temperature (LCST) around 32 °C in aqueous solutions. A preliminary in vitro cell study showed nontoxic and biocompatible properties. These results suggested that chitosan-g-NIPAAm copolymer could be very useful in biomedical and pharmaceutical applications as an injectable material for cell and drug delivery.

Lee and coworkers [49] synthesized Poly [*N*-isopropylacrylamide (NIPAAm)-chitosan] crosslinked copolymer particles by soapless emulsion copolymerization of NIPAAm and chitosan. An anionic initiator [ammonium persulfate (APS)] and a cationic initiator [2,2′-azobis(2-methylpropionamidine)dihydrochloride (AIBA)] were used to initiate the reaction of copolymerization. The chitosan-NIPAAm copolymer synthesized by using APS as the initiator showed a homogeneous morphology and exhibited the characteristic of a lower critical solution temperature (LCST). The copolymer synthesized by using AIBA as an initiator showed a core-shell morphology, and the characteristic of LCST was insignificant.

LCST of the chitosan-NIPAAm copolymer depended on the morphology of copolymer particles. Below the LCST, the increase of either the chitosan/NIPAAm weight ratio or the crosslinking agent decreased the swelling ratio of chitosan–NIPAAm copolymer. Regardless of the kind of initiator used to synthesize the copolymer, an increase of the pH value decreased the swelling ratio of the copolymer. In addition, the chitosan-NIPAAm copolymer particles were processed to form copolymer disks. The effect of various variables such as the chitosan/NIPAAm weight ratio, the concentration of crosslinking agent, and the pH values on the swelling ratio of chitosan-NIPAAm copolymer disks was investigated. Caffeine was used as the model drug to study the characteristics of drug loading of the chitosan-NIPAAm copolymer disks. Variables such as chitosan/NIPAAm weight ratio and the concentration of the crosslinking agent significantly influenced the behavior of caffeine loading. Two factors (pore size and swelling ratio) affected the behavior of caffeine release at 37 °C from the chitosan-NIPAAm copolymer disks.

A new biodegradable chitosan graft copolymer, chitosan-g-polycaprolactone, was synthesized, by Liu *et al.* [50] by the ring-opening graft copolymerization of ε-caprolactone onto phthaloyl-protected chitosan (PHCS) at the hydroxyl group in the presence of tin (II) 2-ethylhexanoate catalyst via a protection-graft-deprotection procedure. Toluene acted as a swelling agent to favor the graft reaction in the heterogeneous system. The grafting reactions were conducted with various PHCS/monomer/toluene feed ratios to obtain chitosan-g-polycaprolactone copolymers with various polycaprolactone contents. After deprotection, the phthaloyl group was removed and the amino group was regenerated. Thus, the obtained chitosan-g-polycaprolactone was an amphoteric hybrid with a large amount of free amino groups and hydrophobic polycaprolactone side chains. These copolymers exhibited new thermal transition and improved solubility in view of the intractable nature of the original chitosan.

3.2. Chitosan IPNs - Preparation and Use in CDR

Yao *et al.* [51] reported a procedure for the preparation of semi-IPN hydrogel based on glutaraldehyde-crosslinked chitosan with an interpenetrating polyether network. The pH sensitivity, swelling and release kinetics and structural changes of the gel in different pH solutions were studied [52,53]. They also [54,55] studied the dynamic water absorption characteristics, correlation between state of water and swelling kinetics of chitosan–polyether hydrogels by applying techniques such as DSC and positron annihilation lifetime spectroscopy. The hydrolysis of the gel was controlled by the amount of crosslinker added. Higher crosslink density of semi-IPNs resulted in a lower degree of swelling and slower hydrolysis [56]. Chlorhexidini acetas and cimetidine were used as model drugs for drug release studies. The swelling of gels was faster with high drug release at pH greater than 6 in comparison to that at pH greater than 6 [51,57].

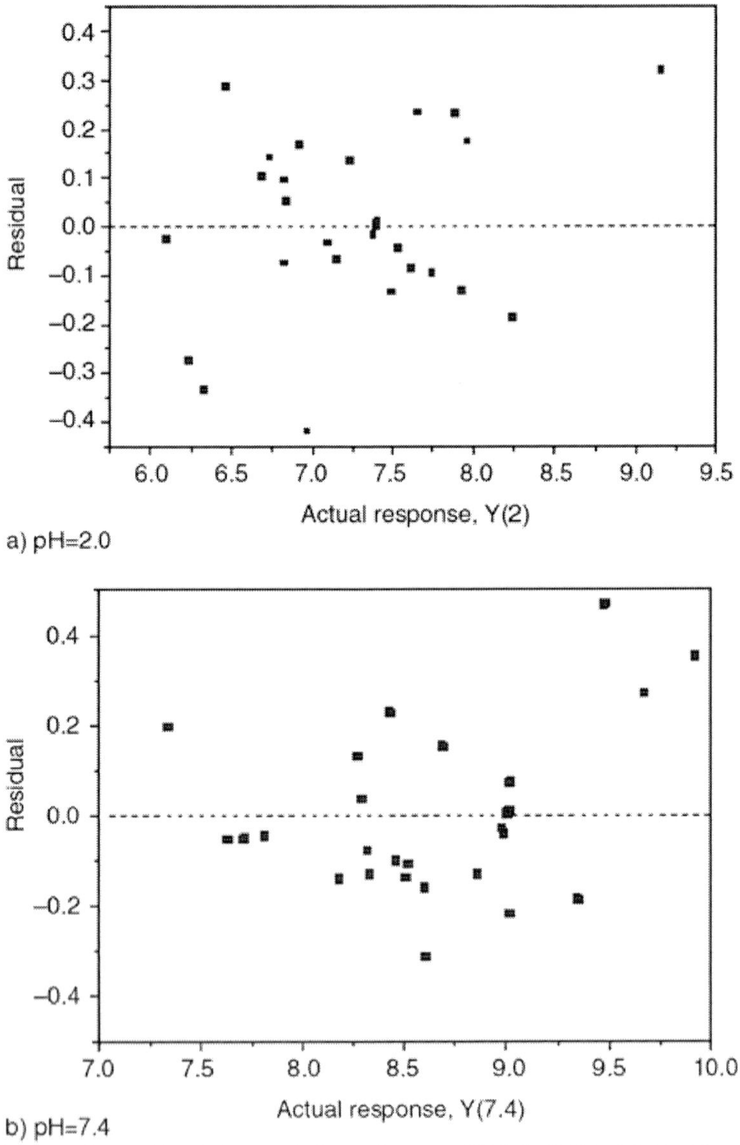

Figure 5. Response surface plots for drug release response a) Y(2), b) Y(7.4) for pH 2 and pH 7.4, respectively.

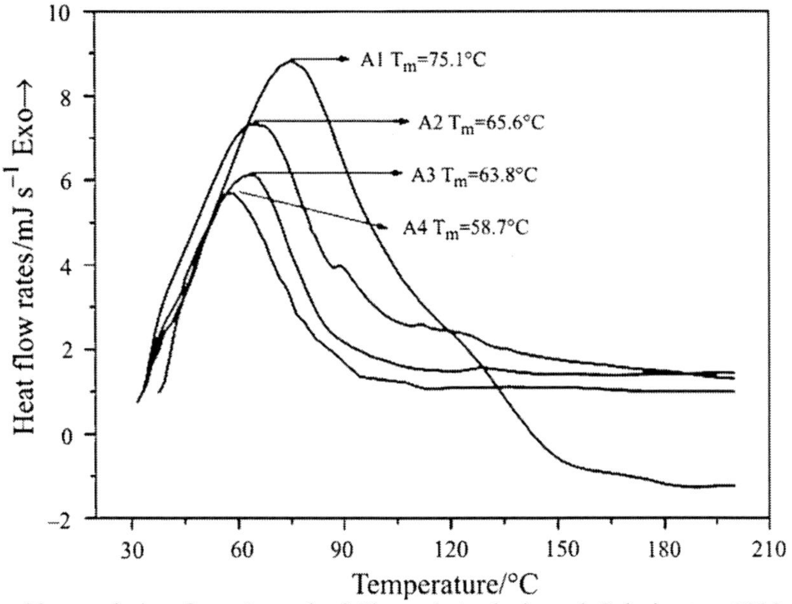

Reproduced with permission from Journal of Thermal Analysis and Calorimetry, 2009, 98, 472 © Akademiai Kiado, Budapest, Hungary.

Figure 6. DSC plots indicating the effect of concentration of chitosan on crosslinking of chitosan and alanine (mass % of chitosan in mixture: 100% (A1); 70% (A2); 50%, (A3); and 30% (A4)).

Kumari and coworkers studied the IPNs from chitosan-alanine [58-60], chitosan-glutamic acid [61,62] and chotsan-lysine [63] blends. In these studies, they used chlorpheniramine maleate (CPM) as model drug for the investigation of release profile and glutaraldehyde was used as crosslinking agent. In vitro studies have been carried on semi interpenetrating polymer network (IPN) beads of chitosan-alanine [58] as carrier for the controlled release of chlorpheniramine maleate (CPM) drug. A viscous solution of chitosan-alanine was prepared in 2% acetic acid solution, extruded as droplets by a syringe to NaOH-methanol solution and crosslinked using glutaraldehyde as a crosslinker. The swelling behavior of crosslinked beads in different pH solutions was measured at different time intervals. The swelling behavior was observed to be dependent on pH and degree of crosslinking. The structural and morphological studies of beads were carried out by using a scanning electron microscope. The drug release experiments of different drug loading capacity beads were performed in solutions of pH 2 and pH 7.4. The concentration of the released drug was evaluated using UV spectrophotometer. The results suggest that chitosan–alanine crosslinked beads are suitable for controlled release of drug.

Kumari *et al.* [59] investigated the effects of formulation variables on the release of drug and optimized the formulation of chitosan-alanine beads loaded with chlorpheniramine maleate (CPM) for controlled release using response surface methodology (RSM). Drug loaded beads of chitosan-alanine were prepared and crosslinked by using glutaraldehyde as crosslinker. The release behavior of drug was affected by preparation variables. A central composite design was used to evaluate and optimize the effect of preparation variables; chitosan concentration (X_1), percentage of crosslinker (X_2), concentration of drug (X_3) and release time (X_4) on the cumulative amount of drug release, Y in solutions of pH = 2.0 and pH = 7.4, respectively. The influence of each parameter was studied by factorial design analysis. Analysis of variance (ANOVA) was also used to evaluate the validity of the model. The statistical parameters reveal strong evidence that the constructed models for drug release in pH = 2 and pH = 7.4 are reliable. The residual plots showing homogeneity of data for drug release are shown in Figure 5. All the experimental data points are uniformly distributed around the mean of response variable showing the scatter of the residuals versus actual response values.

Kumari *et al.* [60] also studied the DSC cure kinetics of the chitosan-alanine beads. The curing of chitosan–alanine with glutaraldehyde as curing agent and using CPM as model drug was carried out with the help of differential scanning calorimeter (DSC). The effect of concentration of chitosan and percentage of crosslinker on the curing was studied at a rate of 5 °C/min. Cure kinetics were measured from 30 to 200 °C at four different heating rates (3, 5, 7 and 10 °C/min). It was observed that the crosslinking of chitosan-alanine was an exothermic process which resulted in a positive peak in the curves [Figure 6]. An increase in activation energy (E_α) was observed with extent of conversion. Typical non-isothermal DSC thermogram of chitosan–alanine samples along with curatives in the temperature range of 30–200 °C are shown in Figure 6. The DSC curves of curing of chitosan and alanine with glutaraldehyde reflect the curing process. The curing peaks in the curves are also superimposed with the dehydration peaks. As the same concentration of glutaraldehyde is present in these samples, the dehydration peak area will be equal. Thus, one can discuss comparative curing amongst these samples. It is observed that the curing starts from the beginning of the scanning in all of the samples. The curing of chitosan and alanine with glutaraldehyde is a fast reaction, as the process of curing starts immediately after the mixing of the constituents.

In case of sample A1, the curve line rises to a maximum peak temperature, 75.1 °C and the subsequent fall shows the completion of curing reaction. As the relative concentration of chitosan decreases, the height and width of curing peak decreases and the maximum peak temperature shifts towards a lower temperature for the given heating rate of 5 °C/min. The reason of this shifting of peak temperature may be due to the fact that as the concentration of chitosan decreases, the effective number of amino groups available to react with the aldehyde group of glutaraldehyde decreases. Due to the decrease in concentration of amino groups, the curing reaction completes at faster rate and thus, the curing peaks are observed at relatively lower temperature as compared to pure chitosan. In other words, as the concentration of chitosan increases, the number of unreacted amino groups increases. The curing reaction takes more time to complete, hence the rate of reaction slows down and forcing the curing peak to shift towards right, a higher end of temperature in the figure. With a decrease in the relative concentration of chitosan, the enthalpy of cure decreases, due to the limited extent of reaction between the amino group of chitosan and aldehyde group of crosslinker. As the area under the peak decreases with the shifting of curing towards lower temperatures, the enthalpy of curing process decreases with a decrease in chitosan concentration. It is observed that in the absence of alanine as in sample A1, the enthalpy is 1637.56 Jg^{-1} which decreases to 303.89 Jg^{-1}, for sample A4 having maximum concentration of alanine.

Kumari and Kundu also studied the cure kinetics [61] and release profile [62] of IPNs synthesized from chitosan and glutamic acid using glutaraldehyde as crosslinker. The curing of chitosan-glutamic acid with glutaraldehyde as curing agent in the presence of CPM was carried out with the help of DSC. The effect of concentration of chitosan and percentage of crosslinker on the curing of chitosan-glutamic acid was studied at a heating rate of 5 °C/min.

Cure kinetics were measured by the DSC in a range of 25 to 220 °C at four different heating rates (3, 5, 7, and 10 °C/min). The activation energy (E_α) was determined by Flynn, Wall, and Ozawa method for curing of the samples. An increase in activation energy (E_α) was observed with the extent of conversion. The area under the peak decreased with the shifting of curing temperature towards lower temperatures. This indicated a decrease in the enthalpy of curing with a decrease in chitosan concentration. It was observed that in the absence of glutamic acid, the enthalpy was 1637.56 Jg^{-1}, which decreased to 264.82 Jg^{-1} for sample having maximum concentration of glutamic acid.

In-vitro studies of chitosan and glutamic acid IPNs were also reported by Kumari and Kundu [62]. A viscous solution of chitosan-glutamic acid was prepared in 2% acetic acid solution, extruded as droplets through a syringe to alkali-methanol solution and the precipitated beads were crosslinked using glutaraldehyde solution. Swelling and drug release studies were carried out. Transport of release medium through the semi-IPN depended upon its pH and extent of crosslinking. The structural and morphological studies of beads were carried out by scanning electron microscopy (SEM).

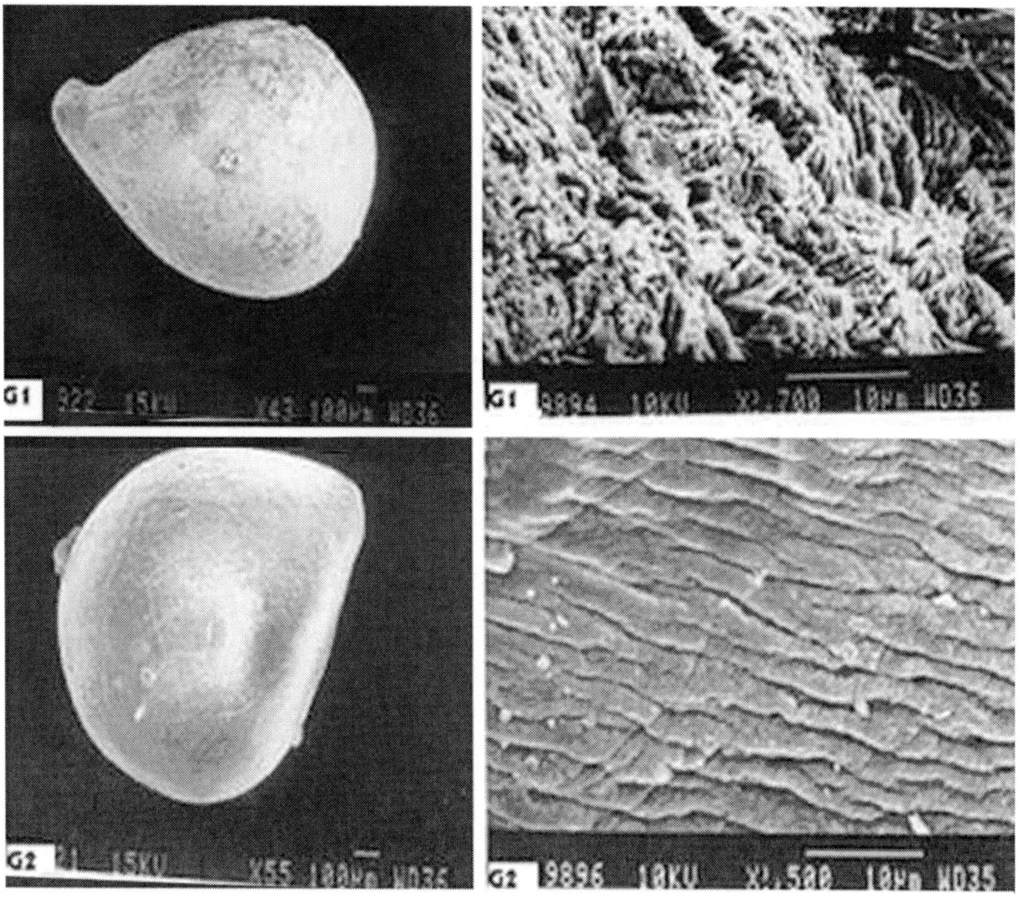

Figure 7. (a) SEM micrographs of the crosslinked beads (G1, G2) and (b) their morphology.

The SEM of two samples G1 and G2 is shown in Figure 7. Crosslinked chitosan–glutamic acid beads prepared by using different concentrations of glutaraldehyde had rough and rigid surfaces. It was observed that the shape of the beads, as shown in Figure 7a, was nearly spherical for samples G1 and G2. This was due to the different compositions of the crosslinked beads. As the concentration of chitosan decreased, the viscosity of the solution also decreased and it became easier to extrude the spherical beads through a syringe. The approximate size of the beads was in the range of 1-1·2 mm. From the morphology of the beads as shown in Figure 7b, the rubbery, fibrous and folded surfaces of the beads were observed. With higher concentration of crosslinker (G1), the chains come closer to each other and exhibit a regular, fibrous structure but with decreasing degree of crosslinker (G2), the structural morphology changes to layered and big fibrous bunches.

The swelling of beads and release rate of CPM in different pH solutions are studied. It is observed that the release of CPM is more in basic medium than in acidic medium due to higher degree of swelling. From these investigations, it is evident that the rate of swelling of matrix and release of drugs is dependent on the degree of crosslinking, weight ratio of chitosan-glutamic acid and the pH of solution. Therefore, by varying the weight ratio of chitosan-glutamic acid, concentration of glutaraldehyde and drug loading, the desired release rates can be achieved. Further, it also helps in optimizing drug entrapping capacity and its sustained release for an extended period of time. The larger surface area of beads as well as their ease of handling made them ideal agents of controlled release. According to the earlier studies, the amount and percentage of drug released in pH 7·4 was higher in case of chitosan-alanine system, whereas, it was slightly lower in pH 7·4 for the chitosan-glutamic acid system. This was due to the better compatibility of aliphatic end of CPM with glutamic acid because of the presence of long chain of hydrocarbon in it.

Kumari and Kundu [63] also investigated the semi-interpenetrating polymer network (semi-IPN) of chitosan and lysine with varying amounts of glutaraldehyde crosslinker for the control release of CPM. The cross-linked beads were dried by different drying techniques such as air-drying, oven-drying and freeze-drying. These semi-IPNs were characterized under a scanning electron microscope (SEM). Swelling studies of these beads were carried out in different pH (2.0 and 7.4) solutions. The effect of concentration of cross-linking agent and curing period on the swelling as well as on the drug release was studied. The results indicated that the size of matrix was largely dependent on the curing time of beads, concentration of glutaraldehyde and technique of drying. The freeze-dried beads exhibited a relatively higher percentage of swelling in the range of 66-89% as compared to oven-dried beads (53-74%) and air-dried beads (39-61%). The drug loaded beads which were cured for different time intervals followed by drying were tested for *in-vitro* release of CPM drug. The rate of drug release from freeze-dried beads was much faster than that from the oven dried and air-dried beads.

Lee *et al.* [64] reported a procedure for preparing semi-IPN polymer network hydrogels composed of β-chitosan and poly (ethylene glycol) diacrylate macromer (PEGM). The hydrogels were prepared by dissolving a mixture of PEGM and β-chitosan in aqueous acetic acid. The resulting mixture was then casted into films, followed by subsequent crosslinking with 2, 2-dimethoxy-2-phenylacetophenone as non-toxic photo initiator by UV irradiation. They studied the crystallinity, thermal and mechanical properties of the gels.

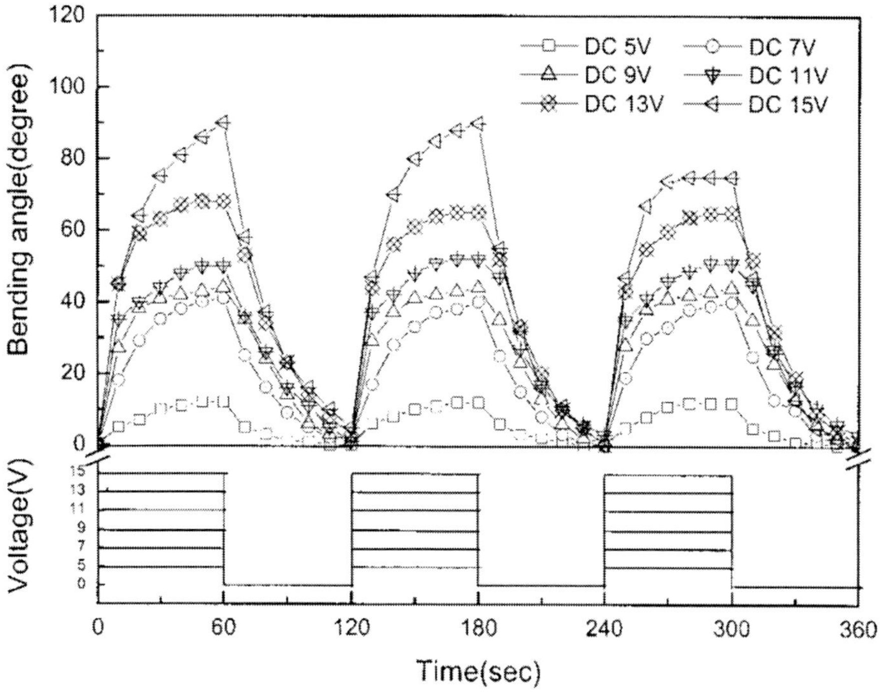

Reproduced with permission from Journal of Applied Polymer Science, 2002, 86, 2287 © John Wiley and Sons, Inc.

Figure 8. Variation of bending angle as a function of the applied voltage in aqueous NaCl at 0.2% by weight.

Kumar *et al.* [65,66] reported some advances in controlled-release formulations using gels of chitin and chitosan. Gupta *et al.* [67,68] investigated the swelling behavior, solubility, hydrolytic degradation, and loading capacity of semi-interpenetrating polymer network beads of chitosan-glycine (CHI-G) and chitosan-poly (ethylene glycol) (CHI-PEG), crosslinked with different concentrations of glutaraldehyde for controlled release of drugs. They observed that the swelling rates were directly proportional to the degree of crosslinking. In both the cases, the degree of swelling was very high in solution of pH 2.0 compared to that of pH 7.4, which was due to the inherent hydrophobicity of the chitosan beads dominating at high pH value, thus preventing faster swelling in neutral and alkaline media. Structural changes that take place during the process of swelling of crosslinked beads in solutions of different pH were supported by the IR and UV spectra recorded at different stages of the swelling. It has been observed from the dissolution profiles of CPM and isoniazid (INH) from the crosslinked chitosan beads at the various time intervals that the release of drugs from the beads in acidic medium was higher than that occurred in alkaline medium. In these investigations, PEG was used to enhance the swelling, solubility, and release properties by forming intermolecular crosslinks with chitosan hydroxyl groups.

Kim and coworkers [69] prepared IPN hydrogels composed of poly (vinyl alcohol) (PVA) and chitosan by UV irradiation method. Swelling behavior of these PVA/chitosan IPN hydrogel was studied by immersion of the gel in NaCl aqueous solutions at various concentrations. The IPN hydrogel exhibited high equilibrium water content (EWC) in the range 75-85%. The swelling ratio decreased with increasing concentration of aqueous NaCl solution. The stimuli response of the IPN hydrogel in electric fields was also investigated. When the PVA/chitosan IPN hydrogel in NaCl electrolyte solution was subjected to an electric field, the IPN bent toward the negative electrode. When the electric stimulus was removed, the IPN hydrogel returned to its original position [Figure 8]. The variation of bending angle of the IPN hydrogel as a function of the applied voltage in aqueous NaCl at 0.2% by weight is shown in Figure 8. The bending angle and the bending speed of the PVA/chitosan IPN increased with increasing applied voltage and concentration of NaCl aqueous solution. The PVA/chitosan IPN also showed stepwise bending behavior depending on the electric stimulus. The results of dielectric analysis (DEA) show that the IPN exhibited two T_gs, indicating the presence of phase separation in the IPN.

Kim *et al.* [70] also prepared PVA/chitosan IPNs of different compositions and studied the swelling kinetics by immersion of the hydrogel films in deionized water at various temperatures and in buffer solutions at various pH. The PVA/chitosan IPN hydrogels exhibited a swelling change in response to the external stimuli such as pH and temperature. The swelling ratio increased with an increase in the content of chitosan and was higher in acidic medium than in alkaline medium and the swelling exponent decreased with an increase in the molar ratio of hydrophilic groups of chitosan in IPNs. The overall swelling process was associated with anomalous diffusion due to polymer relaxation. The diffusion coefficient values increased with an increase in temperature and the content of chitosan.

Kim *et al.* [71] prepared PVA/Chitosan semi-IPNs by using acryloyl chloride crosslinked PVA and chitosan. Chitosan cannot be crosslinked with acryloyl chloride because the hydroxyl group of chitosan is poorly active. Therefore, it was not an AB-crosslinked copolymer but an IPN. The swelling ratio, free water and bound water contents of IPNs were measrsured. IPN hydrogels exhibited a relatively high swelling ratio in the range of 210-350% at 35 °C. The swelling ratio of PVA/ chitosan IPN hydrogels depended on pH and temperature. The swelling ratio increased with increasing molar ratio of hydrophilic groups of chitosan in IPNs. IPN with low free water contents exhibited the lowest equilibrium water content (EWC) value.

Saraydin and Ekici [72] prepared new IPN systems containing chitosan, poly (N-vinylpyrrolidone) and poly(acrylamide) polymers. IPNs were synthesized by radical polymerization of acrylamide monomers in presence of glutaraldehyde and N, N'-methylene-bis-acrylamide as crosslinkers and the other polymers. Glutaraldehyde was used in different concentrations to control the network porosity of IPNs. Spectroscopic and thermal analyses of these cylindrical shaped IPNs were carried out with fourier transform infrared spectroscopy, thermogravimetric, and thermomechanical analysis. Swelling studies of IPNs were carried out at pH 1.1 and pH 7.4 at 37 °C. The swelling and diffusion parameters of IPNs in these solutions were calculated. Amoxicillin as a bioactive species was entrapped to the IPNs during synthesis and in vitro release kinetics of IPNs was investigated. The experimental data of swelling and release studies suggested that the swelling and release process obeyed second-order kinetics. The release of entrapped amoxicillin from IPNs depended on the degree of crosslinking of the polymer and pH of the medium at body temperature. The amoxicillin release was higher in acidic as compared to alkaline medium.

Patel and Amiji [73] prepared and characterized freeze-dried chitosan-poly(ethylene oxide) (PEO) and chitosan-PEO semi-IPNs for site-specific antibiotic delivery in the stomach. The hydrogels were allowed to swell and release the antibiotics-amoxicillin and metronidazole in enzyme-free simulated gastric fluid (SGF, pH 1.2) and simulated intestinal fluid (SIF, pH 7.2) at 37 °C. Compared to the air-dried hydrogels, freeze-drying in the swollen state did significantly influence the swelling and the drug release properties. The porous matrix of the freeze-dried hydrogels was confirmed by SEM [Figure 9]. Figure 9 shows the SEM of freeze dried and air dried chitosan-PEO semi-IPN (sIPN). The top view of the freeze dried hydrogels (A) showed a highly porous surface with an approximate pore size of 8-10 μm in diameter. The surface of the air dried chitosan-PEO sIPN was completely non-porous (B).

The cross-sectional view of hydrogel beads showed open cell structure (C) for freeze dried and the air dried hydrogel did not have open cells. These hydrogels were able to swell faster than the air-dried hydrogels due to the rapid influx of SGF by capillary action. In addition to the porosity of the matrix, pH sensitivity due to the ionization of glucosamine residues in the acidic medium and the influence of high molecular weight of PEO in the freeze-dried chitosan-PEO semi-IPN were apparent on the swelling and drug release properties. After 1 hour in SGF, the swelling ratio (SR) of freeze-dried chitosan-PEO was 16:1. In contrast, the SR in SIF after 1 hour was only 8:60.

A B C D

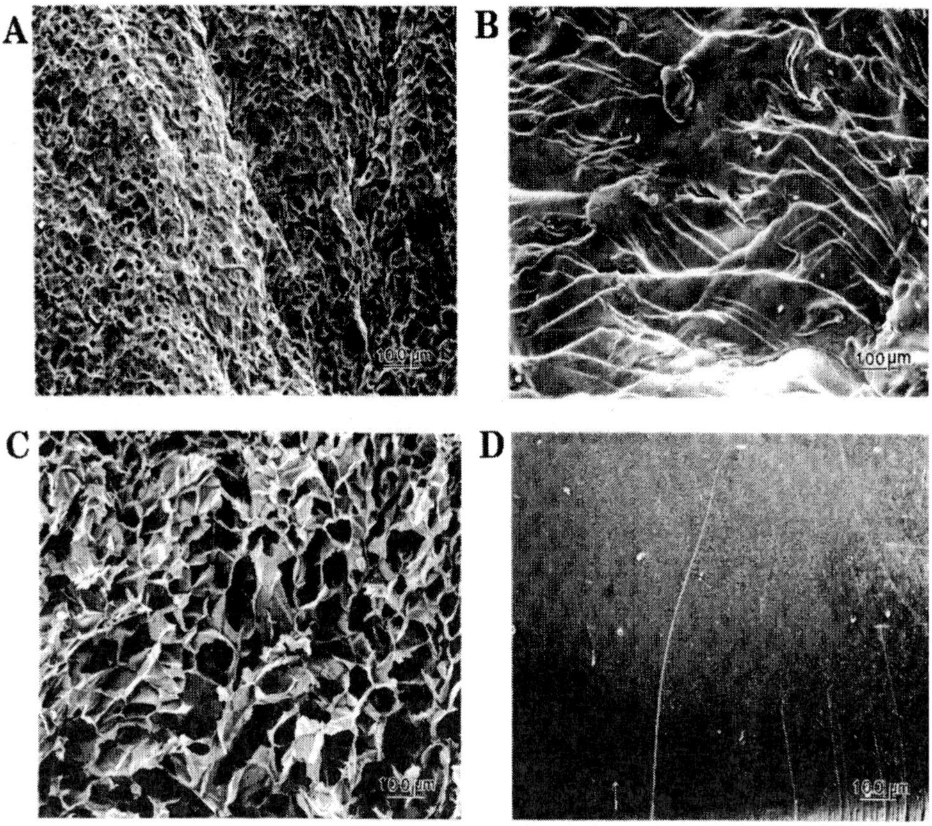

Figure 9. SEM of freeze dried and air-dried chitosan-PEO sIPN. The micrographs depict top view of freeze dried (A) and air dried (B) chitosan-PEO sIPN and cross-sectional view of freeze-dried (C) and air dried (D) chitosan-PEO sIPN. The original magnification was 100X and scale bar is equal to 100 μm.

The addition of high molecular weight PEO did influence the initial swelling and antibiotic release from the freeze-dried chitosan-PEO semi-IPN. More than 65% of the entrapped amoxicillin and 59% of metronidazole were released from the freeze-dried chitosan-PEO semi-IPN after 2 h in SGF. The results suggested that freeze-dried chitosan-PEO semi-IPN could be suitable for localized antibiotic delivery in the low pH environment of the gastric fluid for the treatment of H. *pylori* infection.

Similar studies were performed by Risbud *et al.* [74] for controlled release of antibiotic drug based on chitosan/polyvinyl pyrrolidone (PVP) by crosslinking chitosan-PVP blend with glutaraldehyde to form a semi-interpenetrating polymer network (semi-IPN). The freeze-drying process has generated matrices with high porosity compared to non-porous hydrogels obtained by the air-drying method. Freeze-dried hydrogels exhibited superior pH-dependent swelling properties over non porous air-dried hydrogels, which could be attributed to their porous nature. The increased swelling of hydrogels under acidic conditions was due to the protonation of the primary amino group on chitosan. Within 3 h, freeze-dried membranes released 73.2 and 51% of amoxicillin as against 31.68 and 27% by air-dried membranes in solutions of pH 1.0 and 2.0, respectively, and, thus, had superior drug-release properties compared with air-dried hydrogels. These results suggested that freeze-dried membranes could be useful for localized delivery of antibiotics in the acidic environment of the gastric fluid.

Chen *et al.* [75] developed a pH-sensitive hydrogel based controlled release system, composed of a water soluble chitosan derivative (*N,O*-carboxymethyl chitosan, NOCC) and alginate blended with genipin, to form a semi interpenetrating polymer networks (semi-IPNs). Genipin, which is used in herbal medicine and in the fabrication of food dyes, is a naturally occurring cross-linking agent. It is significantly less cytotoxic than glutaraldehyde and may provide a lesser extent of cross-linking to form a semi interpenetrating polymeric network (semi-IPN) within the developed hydrogel system. NOCC is a chitosan derivative having carboxymethyl substituent on some of both the amino and primary hydroxyl sites of the glucosamine units of the chitosan structure. These authors synthesized and characterized the swelling of NOCC/alginate-based hydrogels as a function of pH values. Release profiles of a model protein drug (bovine serum albumin, BSA) from test hydrogels were studied in simulated gastric and intestinal media. The percentage of decrease of free amino groups and cross-linking density for the NOCC/alginate hydrogel cross-linked with 0.75 mM genipin were 18% and 26 mol/m^3, respectively. At pH 1.2, the swelling ratio of the genipin-cross-linked NOCC/alginate hydrogel was limited (2.5) due to formation of hydrogen bonds between NOCC and alginate. At pH 7.4, the carboxylic acid groups on the genipin-cross-linked NOCC/alginate hydrogel became progressively ionized. In this case, the hydrogel swelled more significantly (6.5) due to a large swelling force created by the electrostatic repulsion between the ionized acid groups. The amount of BSA released at pH 1.2 was relatively low (20%), while that released at pH 7.4 increased significantly (80%).

Biomedical or pharmaceutical activity depends on how water molecules associate with the polymer. Water exists in polymer networks in three different physical states: free water which can freeze at the usual freezing point, intermediate water which freezes at a temperature lower than the usual freezing point and bound water which cannot freeze at the usual freezing point. Considering the importance of water state, Khalid *et al.* [76] synthesized a crosslinked chitosan reference gel and a chitosan-poly (ethylene oxide) semi-interpenetrating network (semi-IPN) and characterized the behavior of both the networks.

Swelling studies were performed on the two kinds of hydrogels by differential scanning calorimetry (DSC) at pH 1.2 and by the gravimetric method at pH 1.2 and pH 7.2. Both methods lead to similar results. The pH-dependent swelling properties were improved for the semi-IPN than in reference hydrogels. The results obtained by DSC measurements and thermogravimetric analysis techniques indicated that the semi-IPN contained more bound water than the reference gel probably due to the presence of the hydrophilic poly (ethylene oxide) chains. The semi-IPN displayed improved mechanical properties compared to the reference gel.

Ren and coworkers [77] prepared a semi-IPN hydrogel based on crosslinked chitosan with glutaraldehyde interpenetrating polyether polymer network (semi-IPN) and investigated its pH-sensitivity, swelling and release kinetics and structural changes of the gel in different pH solutions. The gel can be hydrolyzed in acid medium due to a higher swelling degree which results in the cleavage of imine bonds within the network. However, at pH ≥ 7, the swelling of the gel is limited and thus it is difficult to hydrolyse. The hydrolysis rate increases with decrease of ionic strength of the swelling solution. The amount of polyether and crosslinker added can be used to control the hydrolysis of the gels. To combine the antacid and antiulcer activity of chitosan with the pH-dependent swelling of the chitosan/polyether hydrogel, cimetidine, a histamine H_2 antagonist used to treat ulcers, was selected as a model drug. The drug release results at pH = 1.0 and 7.8 indicated that drug release from the gel depends on pH.

At the beginning, because of the swelling of gel at both pH = 1.0 and 7.8, drug released from the surface of the matrix and the release rate is controlled by diffusion. At pH = 1.0, the gel keeps on swelling and the drug is released gradually. Whereas due to deswelling of the gel at pH = 7.8 resulting from the transformation of amino cations within the network to amino groups and the decrease of the amount of drug left, the relaxation process begins to superimpose control on the drug release, which results in no drug being released continuously from the matrix. However, there is still drug release in pH = 7.8 buffer at the initial stage which may be due to a different controlled release mechanism in the swelling processes of the drug loaded matrix.

In addition, peptic ulcer in the gut is very common and their recurrence is also quite frequent, which is very often due to *Helicobacter pylori (H. pylori)* and hence very difficult to cure just by H_2 antagonists or proton pump inhibitors, without eradication of *H. pylori* by an efficient antibiotic. The hydrogels, having excellent swelling properties at acidic medium, is a good candidate for this dual therapy with site specificity, high sensitivity, and safety without the side effects of the triple therapy. Bhat *et al.* [78] prepared hydrogels by cross-linking chitosan and polyvinyl pyrrolidone blend with glutaraldehyde to form sIPN. The sIPN's were casted into membranes and vigorously tested for biocompatibility. Their swelling kinetics and pH-dependant swelling were also studied. Subsequently, ciprofloxacin and roxatidine were incorporated in the membranes. On studying the drug release patterns, it was found that there is a sustained release pattern, which could make them potential candidates for novel, dual therapy for the eradication of *H. pylori* and cure of peptic ulcer in the gut and prevention of gastric carcinoma.

Rao and coworkers [79] cultured human epidermoid carcinoma cells (HEp-2, Cincinnati) over a biodegradable collagen-chitosan scaffold. Glutaraldehyde was used as cross-linking agent for the development of scaffold. Various types of scaffolds were prepared using different proportionate mixtures of collagen and chitosan. These scaffolds were characterized by FT-IR, DSC and Thermo gravimetric analysis (TGA). Equilibrium swelling studies were carried out in phosphate buffer of physiological pH (7.4) to study its swelling characteristics at a slightly alkaline pH. The scaffold developed from 60 parts of collagen and 40 parts of chitosan showed optimum swelling property and was selected as the best scaffold for performing in vitro culture studies. *In-vitro* culture studies were carried out using HEp-2 cells, over the selected scaffold and its growth morphology was determined through optical photographs taken at different magnifications at various days of culture. The results suggest that the scaffolds prepared from collagen and chitosan can be utilized as a substrate to culture HEp-2 cells and can also be used as an in vitro model to test anti-cancerous drugs.

4. Conclusion

In recent years, controlled drug delivery formulations and the polymers used in these systems have become much more sophisticated, with the ability to do more than simply extend the effective release period for a particular drug. For example, current controlled-release systems can respond to changes in the biological environment (like pH, temperature etc.) and deliver or cease to deliver drugs based on these changes. In addition, materials have been developed that should lead to targeted delivery systems, in which a particular formulation can be directed to the specific cell, tissue, or site where the drug is to be delivered. Much of the development of novel materials in controlled drug delivery is focusing on the preparation and use of these responsive polymers with specifically designed macroscopic and microscopic structural and chemical features. Chitosan, an "intelligent" material, offers great promise by the virtue of its biocompatible, good absorption enhancing, controlled drug release properties, as well as bioadhesive nature. The degree of deacetylation and derivatization with various side chains can be a source of manipulation for specific drug-delivery applications. Systems like copolymers of chitosan with desirable hydrophilic/ hydrophobic interactions and IPNs have a wide scope in the field of controlled drug delivery.

Although there are many benefits of controlled drug delivery systems, the industry is still in its infancy and the emerging technologies offer new possibilities that scientists have only begun to explore. In the coming years a remarkable progress is expected, - a number of new therapies will be possible; we will be more able to target drugs to specific cells; and will learn to make drug release contingent on physiological stimuli. Also important is the potential savings that will accompany the better efficiency and bioavailability of medications. Thus, this article serves as a useful tool for the beginners as well as for the researchers actively involved in this fascinating area of IPNs and copolymers.

References

[1] Sinha, V.R.; Singla, A.K.; Wadhawan, S.; Kaushik, R.; Kumria, R.; Bansal, K.; Dhawan, S. Chitosan microspheres as a potential carrier for drugs. *International Journal of Pharmaceutics* 2004, 274, 1-33.

[2] Bernkop-Schnuch, A. Chitosan and its derivatives: potential excipients for peroral peptide delivery systems. *International Journal of Pharmaceutics* 2000, 194, 1-13.

[3] Khor, E.; Lim, L.Y. Implantable applications of chitin and chitosan. *Biomaterials* 2003, 24, 2339-2349.
[4] Sashiwa, H.; Aiba, S. Chemically modified chitin and chitosan as biomaterials. *Progress Polymer Science* 2004, 29, 887-908.
[5] Bitterman, R.J.; Forbes-Mckean, K.A. Methods of use of biomaterials and injectable implant containing biomaterial. 2008, USP 20080200430A1.
[6] Pusateri, A.E.; McCarthy, S.J.; Gregory, K.W.; Harris, R.A.; Cardenas, L.; McManus, A.T.; Goodwin, C.W. Effect of a chitosan-based hemostatic dressing on blood loss and survival in a model of severe venous hemorrhage and hepatic injury in swine. *The Journal of Trauma: Injury, Infection and Critical Care* 2003, 54, 177-182.
[7] Duchene, D.; Touchard, F.; Peppas, N.A. Pharmaceutical and medical aspects of bioadhesive systems for drug administration. *Drug Development and Industrial Pharmacy* 1988, 14, 283-318.
[8] Frisch,H.L.; Frisch, K.C.; Klempner, D. Advances in interpenetrating polymer networks. *Pure Applied Chemistry* 1981, 53, 1557-1566.
[9] Sperling,L.H; Interpenetrating polymer networks: An overview. In *Interpenetrating Polymer Networks*; Klempner, D.; Sperling, L.H.; Utracki, L.A.; Eds. American Chemical Society, 1994, pp 3-38.
[10] Sperling, L.H.; *Interpenetrating polymer networks and related materials*; Plenum Press: New York, 1981.
[11] Thomas, D.A.; Sperling, L.H. Interpenetrating polymer networks. In *Polymer Blends*; Paul, D. R., Newman, S., Eds.; Academic Press: New York, 1978; Vol. 2, pp 20−29.
[12] Miyazaki, S.; Ishii, K.; Nadai, T. The use of chitin and chitosan as drug carriers. *Chemical and Pharmaceutical Bulletin* 1981, 29, 3067-3069.
[13] Miller, J.R. Interpenetrating polymer networks. Styrene–divinylbenzene copolymers with two and three interpenetrating networks, and their sulphonates. *Journal of Chemical Society*, 1960, 1311-1318.
[14] Frisch, K.C.; Klempner, D.; Migdal, S. Polyurethane-polyacrylate interpenetrating polymer networks. I. *Journal of Polymer Science, Polymer Chemistry Edition* 1974, 12, 885-896.
[15] Frisch, K.C.; Klempner, D.; Migdal, S. Morphology of a polyurethane-polyacrylate interpenetrating polymer network. *Polymer Engineering and Science* 1974, 14, 76-78.
[16] Frisch, H.L.; Klempner, D.; Frisch, K.C. Recent advances in interpenetrating polymer networks. *Polymer Engineering and Science* 1982, 22, 1143-1152.
[17] Sperling, L.H. Recent advances in interpenetrating polymer networks. *Polymer Engineering and Science* 1985, 25, 517-520.
[18] Devia, M.N.; Manson, J.; Sperling, L.; Conde, A. Simultaneous interpenetrating networks based on castor oil polyesters and polystyrene. *Polymer Engineering and Science* 1978, 18, 200-203.
[19] Lou, L. Novel drug delivery systems. *Current Review of Pain* 1999, 3, 411-416.
[20] Shepherd, M.F. Chapter 7: Medication dosage forms and routes of administration. In *Manual of Pharmacy Technicians* Fred, L. (ed). American Society of Health-System Pharmacists, Bethedsda, MD. 2005, pp 113-127.
[21] Peppas, L.B. Polymers in controlled drug delivery. *Medical Plastics and Biomaterials Magazine* November 1997, pp 34.

[22] Peppas, N.A.; (ed), *Hydrogels in Medicine and Pharmacy*, CRC Press, Boca Raton, FL, 1986.

[23] Ha, C.S.; Gardella, J.A. Surface chemistry of biodegradable polymers for drug delivery systems. *Chemical Reviews* 2005, 105, 4205-4232.

[24] Kumari, A.; Yadav, S.K.; Yadav, S.C. Biodegradable polymeric nanoparticles based drug delivery systems. *Colloids and Surfaces B: Biointerfaces* 2010, 75, 1-18.

[25] Winzenburg, G.; Schmidt, C.; Fuchs, S.; Kissel, T. Biodegradable polymers and their potential use in parenteral veterinary drug delivery systems. *Advanced Drug Delivery Reviews* 2004, 56, 1453-1466.

[26] Edlund, U.; Albertsson, A.C. Degradable polymer microspheres for controlled drug delivery. *Advances in Polymer Science* 2002, 157, 67-112.

[27] Kshirsagar, N.A. Drug delivery systems. *Indian Journal of Pharmacology* 2000; 32, S54-S61.

[28] Onishi, H.; Machida, Y. *In vitro* and *in vivo* evaluation of microparticulate drug delivery systems composed of macromolecular prodrugs. *Molecules* 2008, 13, 2136-2155.

[29] Dash, A.K.; Cudworth II, G.C. Therapeutic applications of implantable drug delivery systems. *Journal of Pharmacological and Toxicological Methods* 1998, 40, 1-12.

[30] Kumar, M.N.V.R.; Kumar, N.; Domb, A.J.; Arora, M. Pharmaceutical polymeric controlled drug delivery systems. *Advances in Polymer Science*, 2002, 160, 45-117.

[31] Kumbar, S.G.; Aminabhavi, T.M. Synthesis and characterization of modified chitosan microspheres: Effect of the grafting ratio on the controlled release of nifedipine through microspheres. *Journal of Applied Polymer Science* 2003, 89, 2940-2949.

[32] Kim, S.Y.; Ha, J.C.; Lee, Y.M. Poly(ethylene oxide)-poly(propylene oxide)-poly(ethylene oxide)/poly(e-caprolactone) (PCL) amphiphilic block copolymeric nanospheres II. Thermo-responsive drug release behaviors. *Journal of Controlled Release* 2000, 65, 345-358.

[33] Yang, S.; Washington, C. Drug release from microparticulate systems. In *Microencapsulation: Methods and Industrial Applications*, Benita, S. (ed), 2nd Edition, CRC Press, Boca Raton, FL 2006, pp 183-211.

[34] Singla, A.K.; Chawla, M. Chitosan: some pharmaceutical and biological aspects–an update. *Journal of Pharmacy and Pharmacology* 2001, 53, 1047-1067

[35] Kim, C.H.; Choi, K.S. Synthesis and antibacterial activity of quaternized chitosan derivatives having different methylene spacers. *Journal of Industrial and Engineering Chemistry(Korean)* 2002, 8, 71-76

[36] Sieval, A.B.; Thanou, M.; Kotze, A.F.; Verhoef, J.C.; Brussee, J.; Junginger, H.E. Preparation and NMR characterization of highly substituted N-trimethyl chitosan chloride. *Carbohydrate Polymers* 1998, 36,157-165.

[37] Le, D. P.; Milas, M.; Rinando, M.; Desbrieres, J. Water soluble derivatives obtained by controlled chemical modifications of chitosan. *Carbohydrate Polymers* 1994, 24, 209-214.

[38] Muzzarelli, R.A.A.; Tanfani, F.; Emanuelli, M. N-(carboxymethylidene)chitosans and N-(carboxymethyl)chitosans: Novel chelating polyampholytes obtained from chitosan glyoxylate. *Carbohydrate Research* 1982, 107, 199-214.

[39] Saito, H.; Wu, X.; Harris, J.; Hoffman, A. Graft copolymers of poly(ethylene glycol) (PEG) and chitosan. *Macromolecular Rapid Communication* 1997, 18, 547-550.

[40] Ohya, Y.; Cai, R.; Nishizawa, H.; Hara, K.; Ouchi, T. Preparation of PEG-grafted chitosan nanoparticles as peptide drug carriers. *STP Pharmaceutical Sciences* 2000, 10, 77-82.

[41] Park, K.R.; Nho, Y.C. Preparation and characterization of gelatin/chitosan hydrogel and PVP/gelatin/chitosan hydrogel by radiation crosslinkg. *Kongop Hwahak (Korean)* 2001, 12, 637-642.

[42] Shin, M.S.; Kim, S.J.; Park, S.J.; Lee, Y.H.; Kim, S.I. Synthesis and characteristics of the interpenetrating polymer network hydrogel composed of chitosan and polyallylamine. *Journal of Applied Polymer Science* 2002, 86, 498-503.

[43] Zhu, A.; Wang, S.; Cheng, D.; Chen, Q.; Lin, C.; Shen, J.; Lin, S. Attachment and growth of cultured fibroblast cells on chitosan/PHEA blended hydrogels. *Sheugwu Gongcheng Xuebao (Chinese)* 2002, 18, 109-111.

[44] Jameela, S. R., Misra, A.; Jayakrishnan, A. Cross-linked chitosan microspheres as carriers for prolonged delivery of macromolecular drugs. *Journal of Biomaterial Science. Polymer Edition* 1995, 6, 621-632.

[45] Kumbar, S.G.; Soppimath, K.S.; Aminabhavi, T.M. Synthesis and characterization of polyacrylamide-grafted chitosan hydrogel microspheres for the controlled release of indomethacin. *Journal of Applied Polymer Science* 2003, 87, 1525-1536.

[46] Mao, S. Chitosan copolymers for intranasal delivery of insulin: synthesis, characterization and biological properties. *Ph. D. Dissertation*, Philipps-Universität Marburg, Marburg, 2004.

[47] Jung, B.O.; Kim, C.H.; Choi, K.S.; Lee, Y.M.; Kim, J.J. Preparation of amphiphilic chitosan and their antimicrobial activities. *Journal of Applied Polymer Science* 1999, 72, 1713-1719.

[48] Lee, J.W.; Jung, M.C.; Park, H.D.; Park, K.D.; Ryu, G.H. Synthesis and characterization of thermosensitive chitosan copolymer as a novel biomaterial. *Journal of Biomaterial Science. Poylmer Edition* 2004, 15, 1065-1079.

[49] Lee, C.F.; Chia-Jen Wen, C.J.; Chia-Lung Lin, C.L.; Chiu, W.Y. Morphology and temperature responsiveness-Swelling relationship of poly(N-isopropylamide-chitosan) copolymers and their application to drug release. *Journal of Polymer Science Part A: Polymer Chemistry* 2004, 42, 3029-3037.

[50] Liu, L.; Wang, Y.; Shen, X.; Fang, Y. Preparation of chitosan-g-polycaprolactone copolymers through ring-opening polymerization of ε-caprolactone onto phthaloyl-protected chitosan. *Biopolymers.* 2005, 78, 163-170.

[51] Yao, K.D.; Peng, T.; Goosen, M.F.A.; Min, J.M.; He, Y.Y. pH-sensitivity of hydrogels based on complex forming chitosan: Polyether interpenetrating polymer network. *Journal of Applied Polymer Science* 1993, 48, 343-354.

[52] Yao, K.D.; Peng, T.; Feng, H.B.; He, Y.Y. Swelling kinetics and release characteristic of crosslinked chitosan: Polyether polymer network (semi-IPN) hydrogels. *Journal of Polymer Science Part A: Polymer Chemistry* 1994, 32, 1213-1223.

[53] Peng, T.; Yao, K.D.; Goosen, M.F.A. Structural changes of pH-sensitive chitosan/polyether hydrogels in different pH solution. *Journal of Polymer Science Part A: Polymer Chemistry* 1994, 32, 591-596.

[54] Yao, K.D.; Lin, J.; Zhao, R.Z.; Wang, W.H.; Wei, L. Dynamic water absorption characteristics of chitosan-based hydrogels. *Die Angewandte Makromolekulare Chemie* 1998, 255, 71-75.

[55] Yao, K.D.; Peng, T.; Xu, M.X.; Yuan, C.; Goosen, M.F.A.; Zhang, Q.; Ren, L. pH-dependent hydrolysis and drug release of chitosan/polyether interpenetrating polymer network hydrogel. *Polymer International* 1994, 34, 213-219.

[56] Singla, A.K.; Chawla, M. Chitosan: some pharmaceutical and biological aspects-an update. *Journal of pharmacy and pharmacology* 2001, 53, 1047-1067.

[57] Yao. K.D.; Peng. T.; Goosen, M.F.A.; Min, J.M..; He, Y. pH-sensitivity of hydrogels based on complex forming chitosan: Polyether interpenetrating polymer network. *Journal of Applied Polymer Science* 1993, 48, 343-354.

[58] Kumari, K.; Kundu, P. P. Semiinterpenetrating polymer networks of chitosan and L-alanine for monitoring the release of chlorpheniramine maleate. *Journal of Applied Polymer Science* 2007, 103, 3751-3757.

[59] Kumari, K.; Prasad, K.; Kundu, P. P. Optimization of chlorpheniramine maleate (CPM) delivery by response surface methodology-four component modeling using various response times and concentrations of chitosan-alanine, glutaraldehyde and CPM. *Express Polymer Letters* 2009, 3, 207-218.

[60] Kumari, K.; Raina, K.K.; Kundu, P. P. DSC studies on the curing kinetics of chitosan-alanine using glutaraldehyde as crosslinker. *Journal of Thermal Analysis and Calorimitry* 2009, 98, 469-476.

[61] Kumari, K.; Raina, K.K.; Kundu, P. P. Studies on the cure kinetics of chitosan-glutamic acid using glutaraldehyde as crosslinker through differential scanning calorimeter. *Journal of Applied Polymer Science* 2008, 108, 681-688.

[62] Kumari, K.; Kundu, P. P. Studies on in vitro release of CPM from semi-interpenetrating polymer network (IPN) composed of chitosan and glutamic acid. *Bulletin of Material Science* 2008, 31, 159-167.

[63] Kumari, K.; Kundu, P. P. Effect of drying processes and curing time of chitosan-lysine semi-IPN beads on chlorpheniramine maleate delivery. *Journal of Microencapsulation* 2009, 26, 54-62.

[64] Lee, Y.M.; Kim, S.S.; Kim, S.H. Synthesis and properties of poly(ethylene glycol) macromer/b-chitosan hydrogels. *Journal of Materials Science: Materials in Medicine* 1997, 8, 537-541.

[65] Kumar, M.N.V.R. A review of chitin and chitosan applications. *Reactive and Functional Polymers*, 2000, 46, 1-27.

[66] Gupta, K.C.; Kumar, M.N.V.R. Semi-interpenetrating polymer network beads of crosslinked chitosan–glycine for controlled release of chlorphenramine maleate. *Journal of Applied Polymer Science*, 2000, 76, 672-683.

[67] Gupta, K.C.; Kumar, M.N.V.R. Studies on semi-interpenetrating polymer network beads of chitosan-poly(ethylene glycol) for the controlled release of drugs. *Journal of Applied Polymer Science,* 2001, 80,639-649.

[68] Gupta, K.C.; Kumar, M.N.V.R. Drug release behavior of beads and microgranules of chitosan. *Biomaterials* 2000, 21, 1115-1119.

[69] Kim, S.J.; Park, S.J.; Kim, I.Y.; Shin, M.S.; Kim, S.I. Electric stimuli responses to poly(vinyl alcohol)/chitosan interpenetrating polymer network hydrogel in NaCl solutions. *Journal of Applied Polymer Science* 2002, 86, 2285-2289.

[70] Kim, S.I.; Kim, S.J.; Lee K.J.; Kim I.Y. Swelling kinetics of interpenetrating polymer hydrogels composed of poly(vinyl alcohol)/chitosan. *Journal of Macromolecular Science, Part A: Pure and Applied Chemistry* 2003, 40, 501-510.

[71] Kim, S.I.; Kim, S.J.; Park, S.J. S welling behavior of interpenetrating polymer network hydrogels composed of poly(vinyl alcohol) and chitosan. *Reactive and Functional Polymers* 2003, 55, 53-59.

[72] Saraydin, D.; Ekici, S. Synthesis, characterization and evaluation of IPN hydrogels for antibiotic release. *Drug Delivery* 2004, 11, 381-388.

[73] Patel, V.R.; Amiji, M.M. Preparation and characterization of freeze-dried chitosan-poly(ethylene oxide) hydrogels for site-specific antibiotic delivery in the stomach. *Pharmaceutical Research* 1996, 13, 588-593.

[74] Risbud, M.V.; Hardikar, A.A.; Bhat, S.V.; Bhonde, R.R. pH-sensitive freeze-dried chitosan–polyvinyl pyrrolidone hydrogels as controlled release system for antibiotic delivery. *Journal of Controlled Release* 2000, 68, 23-30.

[75] Chen, S.C.; Wu, Y.C.; Mi, F.L.; Yu-Hsin Lin, Y.H.; Yu, L.C.; Sung, H.W. A novel pH-sensitive hydrogel composed of N, O-carboxymethyl chitosan and alginate cross-linked by genipin for protein drug delivery. *Journal of controlled release* 2004, 96, 285-300.

[76] Khalid, M.N.; Agnely, F.; Yagoubi, N.; Grossiord, J.L.; Couarraze, G. Water state characterization, swelling behavior, thermal and mechanical properties of chitosan based networks. *European Journal of Pharmaceutical Sciences* 2002, 15, 425-432.

[77] Ren, L.; Yao, K.; Peng, T.; Xu, M.X.; , Chen Yuan, C.; Goosen, F. A.; Zhang, Q. pH-dependent hydrolysis and drug release of chitosan/polyether interpenetrating polymer network hydrogel. *Polymer International* 1994, 34, 213-219.

[78] Bhat, S.V.; Elangovan, T.K. Design and development of a site specific, stimuli sensitive, chitosan based novel drug delivery of dual therapy for inflammation of gut *Trends in Biomaterials and Artificial Organs*. 2002, 16, 38-42.

[79] Rao, P.K.; Shanmugasundaram, N.; Ravichandran, P.; Reddy, N.; Ramamurthy, N.; Pal, S. Collagen-chitosan polymeric scaffolds for the in vitro culture of human epidermoid carcinoma cells. *Biomaterials* 2001, 22, 1943-1951.

In: Focus on Chitosan Research
Editors: Arthur N. Ferguson and Amy G. O'Neill
ISBN 978-1-61324-454-8

Chapter 11

Biomedical Applications of Hydrogels Based in the Chemical Modification of Chitosan

E. A. Elizalde-Peña[1,2], G. Luna-Bárcenas*[1], E. Prokhorov[1], J. E. Gough[3], C. Velasquillo-Martínez[4], C. E. Schmidt[5] and I. Sanchez[2]

[1.] Centro de Investigación y de Estudios Avanzados del Instituto Politécnico Nacional Libramiento Norponiente No. 2000 Frac. Real de Juriquilla, Querétaro, , México
[2.] Department of Chemical Engineering, University of Texas at Austin, Austin, TX, USA
[3.] School of Materials: Materials Science Centre, The University of Manchester, Grosvenor Street, Manchester M1 7HS, United Kingdom
4. Instituto Nacional de Rehabilitación, Calz. México Xochimilco No. 289, Col. Arenal de Guadalupe, D, México
5. Department of Biomedical Engineering, University of Texas at ustin, Austin, TX, USA

Abstract

The interest for finding better materials with the objective to use them as implants, has led to search in the mixture of natural polymers for a source to satisfy this necessity. This chapter presents the information about the synthesis, characterization, and some applications of hydrogels based in the chemical reaction between the hybrid, natural-synthetic, Chitosan-g-Glycidyl Methacrylate (CTS-g-GMA), of cationic nature, with water-soluble anionic polymers, such as Xanthan gum (X) and Hyaluronic acid (HA). The formed polyelectrolyte complexes, due to electrostatic attraction between the polymers, have improved properties when compared to hybrid CTS-g-GMA, which provides a wide range of applications in the biomedical field. All materials have been characterized by different analytical techniques such as infrared spectroscopy (FTIR), X-ray diffraction (XRD), thermal analysis (DSC and TGA), and the results were compared to the precursor materials (chitosan, X, HA, and CTS-g-GMA). Due to the HA nature, the film obtained from this reaction has been assessed for use as a patch for wound healing; whereas the properties showed by the hydrogels obtained from the reaction with X, make them very promising for applications in the treatment of recovery of spinal cord injuries. Cell culture was performed in all materials; different cell types were seeded and the viability has been quantified by the DNA (proliferation) assay, over several time intervals. The analysis showed satisfactory results of the (CTS-g-GMA)-X when compared to pure chitosan. Peroxide and interleukin-1β (IL-1β) assays have been performed to analyze the inflammatory response caused by biomaterials. Results show a moderate inflammatory response of our hydrogels when compared to raw chitosan. The implant of the hydrogels [(CTS-g-GMA)-X] in Wistar rats was performed after injuring the spinal cord by a laminectomy. The somatosensory evoked potentials (SEP) obtained by electric stimulation onto peripheral nerves were registered in the corresponding central nervous system areas, showing a successful recovery after 30 days of the implant. The results are promising and strongly support the future use of these hydrogels as scaffolds for tissue engineering and recovering.

INTRODUCTION

Currently the natural polymeric materials has been received much attention to use them in vivo as implants. Biomaterials are defined as materials which are designed to restore, augment, or replace the natural functions of the living tissues or organs in the body. In simple words, a biomaterial is a material which becomes part of the body either temporally or permanently. Biomaterials should perform with an appropriate host response in a specific application without toxic, inflammatory, carcinogenic, and immunogenic responses [1, 2].

The first generation of biomaterials has been developed with the goal of combining physical and chemical properties to match those of the replaced tissue with a minimal toxic response in the host. Due the exigency of the new materials, the second generation was developed to increase bioactivity and the resorbable capacity. The third generation would effectively combine the properties of both generations to help the human body to heal in a short time [1].

Polysaccharides form a class of materials which have generally been underutilized in the biomaterials field. Recognition of the potential utility of this class of materials is however growing and the field of polysaccharide biomaterials is poised to experience rapid growth. Three factor have specifically contributed to this growing. First, the large and growing body of information points to the critical role of saccharide moieties in cell signaling schemes and in the area of immune recognition. Secondly, the recent development of powerful new synthetic techniques with the potential for automated synthesis of biologically active oligosaccharides. These techniques may allow us to finally decode an exploit the language of oligosaccharide signaling. The third factor is the explosion in tissue engineering research and the associated need for new materials with specific, controllable biological activity and biodegradability [3].

An alternative have been the hydrogels, these are used for biomaterials, such as soft contact lenses, artificial corneas, and artificial skins. Hydrogels are usually made of hydrophilic polymer molecules which are crosslinked by different chemical interactions. Hydrogels are elastic solids in the sense that there exists a remembered reference configuration to which the system returns even after being deformed for very long time. They can be of either chemical or physical nature; in the first case, the three-dimensional gel network is formed by the covalent crosslinking of the polymer chains, as consequence, these gels are not reversible. In the second case, the junctions between the chains are due to low energy interactions such as hydrogen bonding as well as Van der Waals or hydrophobic interactions. These gels are generally solvo- and thermo-reversible [2 - 5].

Some disadvantages of the hydrogels are their low hardness and low mechanical resistance after swelling in water. The mechanical properties of these materials can be substantially improved by adding a synthetic material with good mechanical properties. The implant of hydrogels over the biomaterial surfaces changes only their surface properties while the properties in the rest of the bulk remains unchanged. These implants can be made by physical absorption, couple insert, or polymerization [6 - 14].

Due to their size and surface properties, the hydrogels have been analyzed in biomedical studies, for therapeutic applications and clinic diagnosis. Hydrogels with high water content are difficult for cells to adhere. The development of hydrogels compatible with cell attachment and growth is a major goal of biomaterial research [4, 6].

Success in the application of biomaterials relies heavily on the biocompatibility of biomaterials. Biocompatibility is the appropriate biological performance, both local and systemic, of a given polymer in a specific application [2].

Numerous strategies currently used to engineer tissues depend on employing a material scaffold. These scaffolds serve as a synthetic extracellular matrix (ECM) to organize cells into a 3D architecture and to present stimuli, which direct the growth and formation of desire tissue. Depending on the tissue of interest and the specific application, the required scaffold material and its properties will be quite different [15, 16].

Natural and synthetic biomaterials serve as fundamental research and therapeutic tools to investigate and facilitate the repair of damaged or dysfunctional tissues, both in cell-based and acellular therapies [17].

The successful large-scale production of engineered tissues requires an adequate source of healthy expandable cells, the optimization of scaffolds, and the creation of bioreactors, which mimic the environment of the body and that are amenable to scale-up. Additional challenges include the preservation of the product so that it has a long self-life and the successful use of various approaches to prevent tissue rejection [16].

The potential of chitosan as a biomaterial stems from its cationic nature and high charge density in solution. The charge density allows chitosan to form insoluble ionic complexes or complex coacervates with a wide variety of water-soluble anionic polymers [3, 16].

Chitosan ([β-(1→4)-2- amine-2-desoxy-D-glucose]) is a natural polysaccharide that is formed by altering the N-deacetylation of its precursor, chitin. Chitin and chitosan can be represented by a unique structure, show in the figure 1. They can be considered as belonging to the family of glycosaminoglycans (GAG). GAGs are particularly interesting since they seem to be the fewest among the polysaccharides that express the property of bioactivity [5].[5]

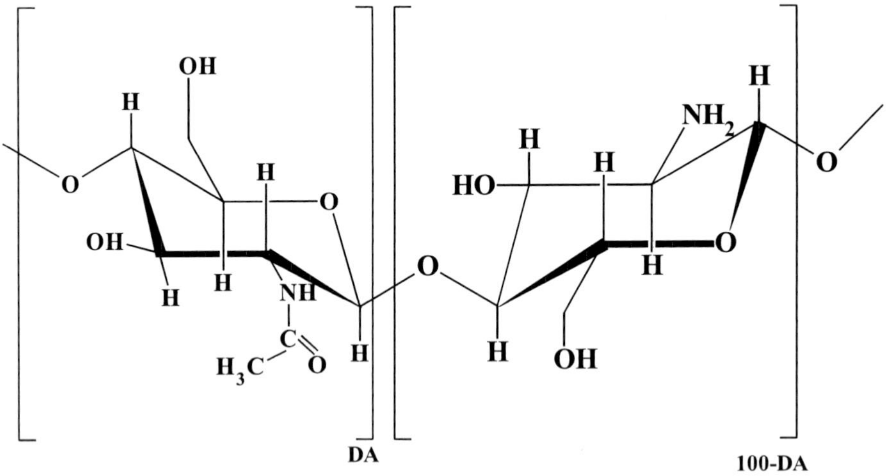

Figure 1. Chemical structure of chitin and chitosan. Note the similarity of both structures [5, 18].

The difference between chitin and chitosan is essentially related to the possibility to solubilize the polymer in dilute acidic media. Therefore the degree of acetylation (DA), which is related to the population balance of acetylated and deacetylated (100-DA) groups (left and right chemical structures in figure 1) is essential to define these two terms. When chitin is deacetyled in heterogeneous conditions, the solubility in aqueous acidic media is achieved for DA generally below 30%. Nevertheless, on reacetylating chitosan it is possible to observe a solubilization up to DA close to 60%. As a consequence, the frontier between chitin/chitosan can be located at a DA of 60%. An effect of the DA can be appreciated in the chitosan, which is a polymer semi-crystalline and the degree of crystallinity is a function of the degree of deacetylation [5 - 7].

Considering the chemical structure schematized in figure 1, the chitosan has a primary amino group and a hydroxyl group. Chitosan undergoes a host of chemical reactions under mild conditions and can be functionalized with great variety of atoms [5 - 7]. On other hand, the absence of chitosan in many living media must be regarded as an interesting opportunity for numerous possible applications, especially if we consider that it corresponds to a more or less charged polycation, a chemical structure not very common in nature [5].

Recent researches show that chitosan hydrogels have some properties that made them biocompatible materials, among those are: 1) they have low interfacial tension in the presence of the biological fluids of live tissue. Due to its large water content, the surface of the hydrogel is recognized as a diffuse super hydrophilic surface, 2) the hydrogels mimic some properties of the natural gels, cells and tissue, and 3) the soft and elastic nature of the hydrogels minimizes the mechanical tension in the surrounding tissue [6].

The chitosan hydrogels are generally prepared using cross-linking agents as glutaraldehyde or glioxal [19, 20]. Hydrogels based on chitosan have been prepared with no-covalent networks; however, the products obtained are rigid or fragile, depending on if they have a high or low molecular weight, respectively [21]. These hydrogels can be prepared using polyelectrolyte complexes from the chitosan and other natural polyanions like heparin, xanthan gum, dextran sulfate, carboximethyl dextran, alginic acid, carboximethyl cellulose. When two polyelectrolytes with opposite charges are mixed in aqueous solution, a polyelectrolyte complex is formed due to the electrostatic attraction between the polyelectrolytes, these materials are highly hydrophilic and form products with a high water absorption [22, 23].

Some hydrogels based on chitosan can be prepared using the xanthan gum, which has carboxyl groups like in the acidic polysaccharide and the chitosan that behaves as basic polysaccharide with amine groups. These hydrogels do not dissolve in alkaline solutions and are very sensitive to swelling in solutions with a pH in the range of 9 and 12 [1, 22].

These hydrogels in a solution with a pH around of 8 and in presence of some salt could have favorable characteristics to drug liberation in applications by oral route, because the hydrogels can react with the pH variation in the gastrointestinal tract [24 - 26].

Chitosan has been investigated for a variety of tissue engineering applications because it is structurally similar to naturally occurring glycosaminolglycans and is degradable by enzymes in humans, such as lysozyme, in which the kinetics of degradations are inversely related to the degree of crystallinity [27, 28]. In this topic, some works has been reported the use of lysozyme to determine the biodegradation of some chitosan derivates which the modification of the degree of acetylation provides a powerful means for controlling biodegradation and biocompatibility and can be optimized for tissue engineering applications; increasing the importance of this enzyme [9, 29].

Although considered as nontoxic, chitosan is often shown as being a strong elicitor of biological activity whether in plants or in animals. In both cases, the consequences of the contact between chitosan and living media have been extensively studied during the last 20 years. Some results in the case of plants could be use to understand some behaviors observed in animals [5].

The cytocompatibility of chitosan films at physiological pH, toward keratinocytes, fibroblast or chondrocytes has been recently studied in vitro [5]. Due to the chitosan properties, this contributes to wound healing through interaction with various cell types. Chitosan macromolecules can strengthen and accelerate cell proliferation and tissue organization of connective tissue comprising the supportive framework of an animal organ [30, 31].

Chitosan is also viable as a 3D matrix for cell encapsulation, which might be applied to implantable biomaterials. Because of its expanded structure, chitosan is suitable as a matrix for anchorage-dependent mammalian cell encapsulation [30]. Viable hybridoma (fused tumor and lymphocyte) cells were entrapped in chitosan-carboxymethylchitosan capsules. These cells exhibited healthy morphology, and displayed a tenfold increase in cell density and a threefold higher product concentration in comparison to a suspended culture [32].

Chitosan and Cell Culture

Fibroblasts

A fibroblast is a type of cell that synthesizes and maintains the extracellular matrix (ECM) of many animal tissues. Fibroblasts provide a structural framework (stroma) for many tissues, and play a critical role in wound healing. They are the most common cells of connective tissue in animals. Their main function is to maintain the structural integrity of connective tissue by continuously secreting precursors of the ECM. Fibroblasts secrete the precursors of all the components of ECM. The composition of ECM determines the physical properties of connective tissues [33].

Fibroblasts can migrate slowly over substratum as individual cells, in contrast to epithelial cells. While epithelial cells from the lining of body structures, it is fibroblasts and related connective tissues which sculpt the "bulk" of an organism [33].

One very important tissue engineering approach was developing skin substitutes with chitosan-based materials. The success of these applications lies in part due to chitosan's favorable interactions with fibroblasts and bioactive molecules intimately related to fibroblasts and wound healing [30].

Chitosan was successfully utilized for the regeneration of skin tissue; was inserted into a cut on the back of rats. A normal inflammatory reaction was observed after 2 days, followed by cell colonization after 7 days. A dermal equivalent with an average pore size of 100 pm provided an excellent environment for fibroblast growth and proliferation *in vitro*. This substrate was a mixture of bovine collagen types I and III, 85% weight per volume (w/v) chitosan extracted from shrimp shell, and GAGs of chondroitin-4 and chondroitin-6 sulfate, with a final composition of 72% (w/v) collagen, 20% (w/v) chitosan and 8% (w/v) GAGs. This study indicates that a minimum pore size is required as not to inhibit fibroblast migration, growth and metabolic activity, as well as the diffusion of nutrients in and out of the matrix [34].

Keratinocytes

The keratinocyte is the major cell type of the epidermis, making up about 90% of epidermal cells. The epidermis is divided into four or five layers (depending on the type of skin) based on keratinocyte morphology [35 - 36].

Keratinocytes originate in the basal layer from the division of keratinocyte stem cells. They are pushed up through the layers of the epidermis, undergoing gradual differentiation until they reach the stratum corneum where they form a layer of enucleated, flattened, highly keratinized cells called *squamous cells*. This layer forms an effective barrier to the entry of foreign matter and infectious agents into the body and minimizes moisture loss. Keratinocytes are shed and replaced continuously from the stratum corneum. The time of transit from basal layer to shedding is approximately one month although this can be accelerated in conditions of keratinocyte hyperproliferation such as psoriasis [35 - 36].

For the attachment in chitosan films of keratinocytes, the results revealed that whatever the DA is, all chitosan films are compatible for this types of cells. The adhesion tested showed that cell adhesion increases considerably on decreasing DA. The increase is important as long as the DA is low. In addition, for a given DA, fibroblast appeared to adhere twice as much as keratinocytes on these films, on the contrary, the proliferations is quite good of keratinocytes. This proliferation increases when the DA of chitosan decreases [37].

Nerve Cells

Neurons (also known as neurones and nerve cells) possess electrical excitability, the ability to respond to a stimulus and convert it into an action potential, which once begun, a nerve impulse travels rapidly and at a constant strength. Most notably, vertebrate electrical synapses are bidirectional. With their speed, simplicity, and reciprocity, electrical synapses are a unique feature of neuronal circuits in the mammalian brain. In vertebrate animals, neurons are the core components of the brain, spinal cord and peripheral nerves [38, 39].

The cell line NG108-15 was selected for this work due to presents many aspects similar to a motor neuron. The NG108-15 cell line is a neuroblastoma x glioma hybrid, which was derived by somatic cell hybridization [40 - 42].

Nerve cells are influenced by the presence of chitosan. Studies indicate that neurons cultured on a chitosan membrane can grow well and that chitosan conduit can greatly promote the repair of the peripheral nervous system. Studies considered the attachment, spreading and growth of gliosarcoma cells as a model of affinity of nerve cells to chitosan membranes. Chitosan coated with polylysine and a chitosan-polylysine mixture are even better materials than chitosan itself in nerve cell affinity, and are promising materials for nerve repair. Pre-coating materials with ECM molecules, especially laminin, can greatly improve their nerve cell affinity [30, 43].

Chitosan and Inflammatory Response

The inflammatory response is triggered whenever body tissues are injured by physical trauma (a blow), intense heat, irritant chemicals or infection by viruses, fungi, or bacteria. The inflammatory response enlists macrophages, mast cells, all type of white blood cells, and dozens of chemicals that kill pathogens and help repair tissue [33].

Macrophages are the chiefs of phagocytes, which derive from white blood cells called monocytes that leave the bloodstream, enter the tissues, and develop into macrophages. These cells belong to the body defense and they are known to be important mediators of inflammation and play an important role in immune regulation [33, 44].

Chitosan exhibits a positive effect on wound healing through its interaction with macrophages and leukocytes. Moreover, chitosan enhances the immune response, which is desirable for the application of drug carriers to tumor-bearing hosts, whose immunities are depressed [30].

Chitosan's positive effect on wound healing is also displayed through its interaction with macrophages. The activation of normal macrophages for the destruction of tumor cells occurs when they interact with activating agents, such as microorganisms, and with substances secreted by T cells in response to antigen stimulation. Upon activation, these macrophages can lyse tumor cells either by direct contact or through the release of diffusible cytotoxic molecules. Nitric oxide, Interleukin -1β (IL-1β), tumor necrosis α-factor and reactive oxygen intermediates are among the major cytotoxic molecules produced by activated macrophages for the lysis of tumor cells. Chitosan shows a biological aptitude for activating macrophages to destroy tumor cells and to produce IL-1β. IL-1β is a compound that regulates cell-mediated immune responses and other biological functions. These facts demonstrated that macrophages are activated as a consequence of direct interaction with polymer materials and produce IL-1β [44 - 46].

Spinal Cord and Biomaterials

The spinal cord is elastic, stretching but it is exquisitely sensitive to direct pressure. Any localized damage to the spinal cord or its roots leads to some functional loss, either paralysis (loss of motor function) or sensory loss. Severe damage to ventral root or ventral horn cells results in a flaccid paralysis of reach these muscles, which consequently cannot move either voluntary or involuntary [33]. The result of an incomplete or complete spinal cord lesion is either paraplegia (paralysis of lower body) or quadriplegia (paralysis of the body from the neck down), depending on whether the injury was sustained in the thoracic/lumbar region or neck region of the spinal column, respectively [47]. Anyone with traumatic spinal cord injury must be watched for symptoms of spinal shock, a transient period of functional loss that follows the injury. Spinal shock results in immediate depression of all reflex stop, blood pressure falls, and all muscles below the injury are paralyzed and insensitive. Neural function usually returns within a few hours injury. If function does not resume within 48 hours, paralysis is permanent in most cases [33].

Destruction of the spinal cord can be compared to a bomb exploding in a computer centre, and repairing the spinal cord is as complicated as trying to rebuild all of the computer connections. In the last years, there has been encouraging progress in animal models, with sufficient regeneration of the damaged spinal cord to enable some recovery of motor ability. When the spinal cord is injured, the first phase of injury involves mechanical tissue destruction. It is followed by a second phase of tissue loss, which is principally caused by a several local disturbance of the blood supply [47 - 49]. There have been attempts to minimize this secondary damage with neuroprotective agents, but, so far, only high-dose of a synthetic corticosteroid given within the first hours after injury is in use clinically [49, 50].

It is estimated that the annual incidence of spinal cord injury (SCI), not including those who die at the scene of the accident, is approximately 40 cases per million population in the U. S. or approximately 11,000 new cases each year. SCI primarily affects young adults, being this age of the general population of the United States has increased by approximately 8 years since the mid-1970's, the average age at injury has also steadily increased over time. Since 2000, the average age at injury is 38.0 years [51].

Injury to the spinal cord may involve the destruction of a substantial amount of tissue, including the white and gray matter, and blood vessels. This occurs following trauma, degenerative process or stroke, where the amount of tissue damage may increase with secondary pathophysiological changes, or following surgery, where sectioning neural tissue is unavoidable during elective oncological surgery which necessitates removal of a rim of vital tissue from around a tumor [51].

Whilst the functional deficit is related to the amount of tissue damage and to the interruption of the associated axonal pathways, it is also related to the inability of the adult mammalian central nervous system (CNS) to repair its own structures and to restore the tissue defect, as this occurs in the CNS of adult inframammalian vertebrates. Tissue repair does fact results in scarring, despite evidence that the CNS has significant repair potential following injury, and would be capable of restoring a cellular terrain for axonal growth [52, 53]..

For spinal cord, the recovery of them is complicated, because receives sensory information of almost tissues of the body. It transmits this information in the form of electrical impulses to the brain, along of millions of nerve fibers that are grouped together in bundles. A sharp blow in the spinal column can cause dislocation of individual vertebrae and severe damage to the spinal cord, including its complete severance [47].

Life expectancy is the average remaining years of life for an individual. Life expectancies for persons with SCI continue to increase, but are still somewhat below life expectancies for those with no spinal cord injury. Mortality rates are significantly higher during the first year after injury than during subsequent years, particularly for severely injured persons [51].

Based mainly on the advantages offered by chemical properties of chitosan, this work describes the synthesis, characterization and applications of a novel materials formed by a chemical reaction between the hybrid chitosan/glycidyl methacrylate (CTS-g-GMA) polymer with water-soluble anionic polymers, such as Hyaluronic acid (HA) and Xanthan gum (X). The resulting materials could be attractive due to the nature, and properties of both predecessors for biomedical applications such as implants.

Materials And Experimental Methods

The commercial reagents are from Aldrich Chemical used without additional purification. The chitosan had 87% deacetylation degree, determined by a titration method and an average molecular weight of 300 KDa, determined by viscosimetry in a 0.1M acetic acid and 0.2 M sodium chloride solution at 25 ºC. The Mark-Houwink constants, α and K were taken from Milas *et al* [54], for the xanthan (X) and from Kassai *et al.* for CTS [55]. The glycidyl methacrylate (GMA) liquid, had 97 % purity, and HA powder were used without additional purification.

Synthesis Methods

Chitosan-Glycidyl Methacrylate (CTS-g-GMA) Synthesis

This material was synthesized in two stoichiometric molar ratios CTS:GMA of 1:1 and 1:4, as described by Flores-Ramirez *et al*, the reaction proposed is presented in figure 2 [56].

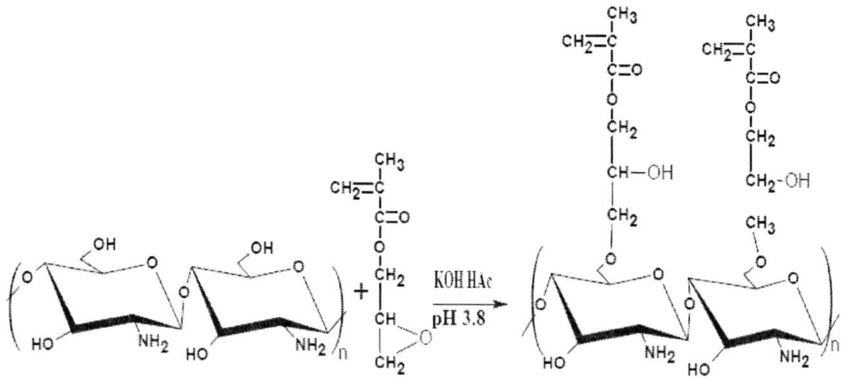

Figure 2. Proposed mechanism of the chemical reaction CTS-g-GMA as described elsewhere Flores-Ramirez *et al*, 2005 [56].

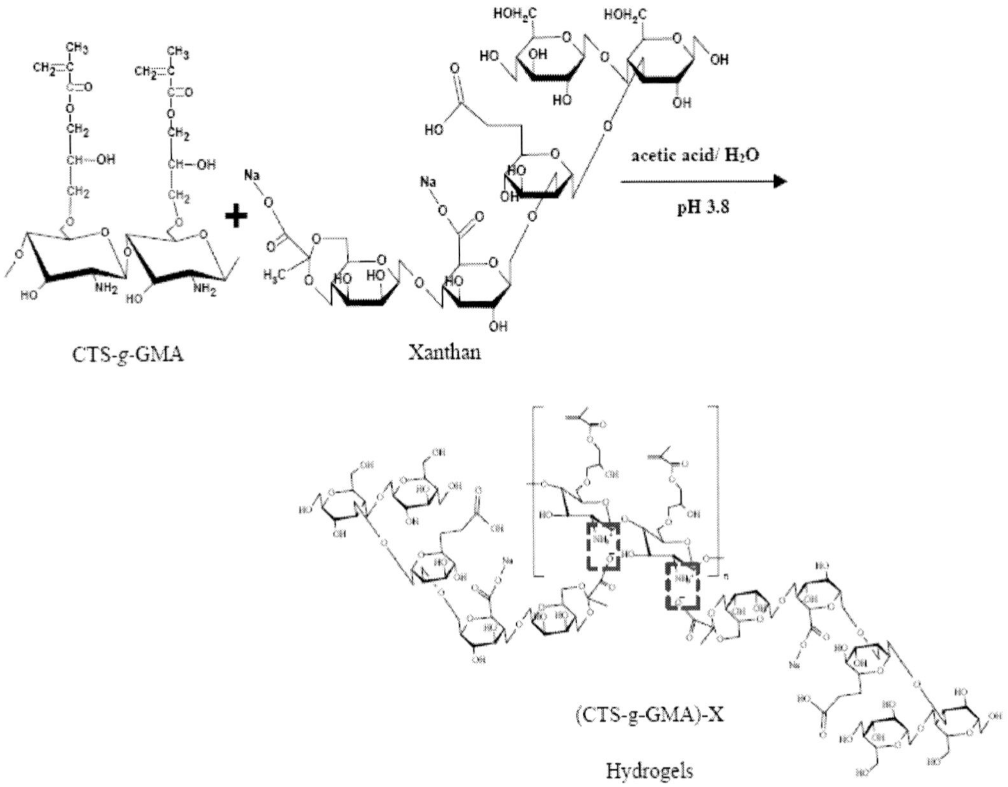

Figure 3. Mechanism proposer for the chemical reaction of the hybrid polymer CTS-g-GMA with Xanthan.

Chitosan-g-Glycidyl Methacrylate-Hyaluronic Acid Hydrogel Synthesis (CTS-g-GMA-HA)

Due to the electrolytic nature of both polymers in solution, anionic from HA and cationic from CTS-*g*-GMA, both solutions were performed separately.

A solution of HA was prepared using 0.25 g of HA which was dissolved in 30 mL of distilled water by magnetic agitation. Separately a solution of CTS-g-GMA was prepared using 0.25 g, of dry weight, which was dissolved in 40 mL of acetic acid solution 0.4 M by magnetic agitation. In a dish crystallizing HA solution was placed then the CTS-g-GMA solution was added slowly, to room temperature and magnetic stirred. After 1 hour, optimum reaction time, the agitation was stopped. The product was rinsed with a NaOH 0.2M solution until reach a pH around 6.8. The product was recovered and dried in an oven overnight to room temperature. The film obtained after dried was used for the physicochemical characterization and cell culture assays.

Chitosan-g-Glycidyl Methacrylate-Xanthan Hydrogel Synthesis [(CTS-g-GMA-)X]

The (CTS-*g*-GMA)-X hydrogels were synthesized as described in a previous work in 2009 [57]. Nevertheless this hydrogel has acid nature which involved the modification of characteristic morphology, and lessens the viability of the cells cultured onto them.

Table 1. General conditions of reaction for the materials Z, for both methods; A acidic media, and B neutral media

<table>
<tr><th rowspan="3">Material</th><th colspan="5">General conditions</th></tr>
<tr><th rowspan="2">Stoichiometric ratio (CTS:GMA)</th><th colspan="2">Solvent</th><th rowspan="2">Gravimetric Ratio (CTS-g-GMA):X</th><th rowspan="2">Reaction Time (hrs)</th></tr>
<tr><th>CTS-g-GMA</th><th>Xanthan</th></tr>
<tr><td>Z11A</td><td>1:1</td><td rowspan="2">Acetic Acid 0.4M</td><td rowspan="2">Distilled Water</td><td rowspan="2">1:1</td><td rowspan="2">1</td></tr>
<tr><td>Z14A</td><td>1:4</td></tr>
<tr><td>Z11B</td><td>1:1</td><td colspan="2" rowspan="2">Distilled Water</td><td rowspan="2">1:1</td><td rowspan="2">4</td></tr>
<tr><td>Z14B</td><td>1:4</td></tr>
</table>

Therefore the method was modified in two different ways to reach a pH similar to physiological pH: neutral and acid aqueous media. The (CTS-*g*-GMA)-X hydrogels were label as Z11*E*, and Z14*E* respectively. The number 1:1, and 1:4 corresponding to stoichiometric molar ratios of CTS:GMA. *E*=A for CTS-*g*-GMA materials dissolved in acid solution and *E*=B for a neutral dissolution, see table 1 for more details [56, 57].

In an Erlenmeyer flask, a known quantity, in dry weight, of CTS-*g*-GMA was placed, and then an acetic acid solution 0.4M (pH mixture 3.1) was added, for series A, or distilled water, for series B. When the CTS-*g*-GMA was dissolved, the xanthan solution was added slowly. The vial was closed and then a nitrogen flow was introduced, the temperature was 50 ± 1 °C, and constant magnetic agitation was used, a schematic drawing of the reaction is shown in figure 3.

After one hour (for series A) or four hours (for series B), the gas flow, heating and agitation were stopped. The product was cooled down in an ice bath for 15 minutes to complete the reaction.

For the polymers series A, NaOH 0.2 M solution was added until a pH around 6.8 was reached. Under these conditions, white pearls remain formed, immersed in a colourless solutions; the pearls were recovered by decantation. The pearls were rinsed several times in distilled water to eliminate the residual reactants. The product was dried in an oven at 45°C for one hour.

For the series B polymers, the material was recovered by decantation and washed several times in distilled water to eliminate the residual reactants. The product was dried in an oven for one hour at 45°C.

The films obtained after the drying process were used for the subsequent characterization and cell culture assays.

Characterization Techniques

FTIR Spectra Analysis

The FTIR spectra were recorded on a Perkin Elmer spectrometer (model Spectrum One), in the 4000 to 450 cm^{-1} range at a resolution of 4 cm^{-1} in the transmission mode using FTIR-grade KBr as supporting material.

X-Ray Diffraction Analysis (DRX)

X-ray diffraction measurements were performed in a Rigaku D/max-2100 diffractometer (Cu K_{α} radiation) at 30kV and 16 mA, with an angular resolution was of 0.02°.

Scanning Electron Microscope (SEM)

Micrographs were obtained in a JEOL and SIRION Scanning Electron Microscopes (JEOL).

Cell Culture

Pre-treatment for Hydrogels

All materials were fitted to provide the necessary conditions for cell culture. Discs of 10 mm diameter were placed in a multi-well plate. They were exposed 30 minutes to UV light, after that were washed twice with a sterile phosphate buffer solution (PBS), add 1 mL of corresponding medium, during 10 minutes, wash twice with PBS, repeat the procedure until pH remains neutral; once in this point, they were kept overnight in a sterile environment.

Chondrocytes

Bovine chondrocytes were used and their proliferation was assayed in T75 flasks with surface treatment for ensuring optimal cell attachment and growth, with DMEM:F12 (no glutamine, and supplemented with 5 % FBS and 1 % of antibiotic, 1mM HEPES and 250mg/L Ascorbic Acid) and maintained at 37 °C in a humidified incubator with 5 % CO_2.

Keratinocytes

Human epidermal keratinocytes (HEKa)were used and their proliferation was assayed in T75 flasks with surface treatment for ensuring optimal cell attachment and growth, with Dermalife basal medium (supplemented with DermaLife M LifeFactors Kit, Lifeline LL-0027) and maintained at 37 °C in a humidified incubator with 5 % CO_2.

Macrophages

The murine BALB/c monocyte macrophage cell line J774A.1 was obtained from the European Collection of Cell Cultures (No. 91051511). Cells were cultured in DMEM supplemented with 10 % FBS, and 1 % antibiotic (all from Life Technologies, Paisley, Scotland). Cells were cultured in Nunc tissue culture flasks and plates (Nunclon, Oxford, UK) and maintained at 37 °C in a humidified incubator with 5 % CO_2 [58].

Nerve Cells

Nerve cells line NG108-15 were used, their proliferation was assayed in T75 flasks with surface treatment for ensuring optimal cell attachment and growth, with DMEM high glucose (5% FBS, 1% HAT and 1% of antibiotic) media without antibiotic and maintained at 37°C in a humidified incubator with 5% CO_2.

DNA Assay for Counting Cells

The cells were seeded at density of 4 x 10^4 cells/mL in corresponding medium onto each disc of polymer, additionally glass coverslips were used as a positive control. The times for this assay were 1, 3, 5, 7, and 14 days, after which, media was removed and samples rinsed with PBS. Each sample was transferred to a separate multiwell plate. Distilled water was added to each new well and samples were freeze-thawed three times [59].

Aliquots of 100 μL of the samples, standards and blank (distilled water) were placed into a 96 well plate and 100 μL of Hoechst Stain was added to each well. The plate was shaken for 10 secondas and fluorescence measurements at 355 nm excitation and 460 nm emission were taken using a fluorescence plate reader [59].

Alamar Blue Assay for Cell Activity

The cells were seeded at density of 4 x 10^4 cells/mL in corresponding medium onto each disc of polymer, additionally glass coverslips were used as a positive control. The times for this assay were 1, 3, 5, 7, and 14 days, after which, media was replaced and after were added 200 μL of Alamar Blue stock solution; create once a blank samples. Incubate 2 hours at 37°C. After this take 200 μL of the media from the well and plate it into a 96 well plate; read using the fluorescent plate reader.

Collagen Assay

Chondrocytes were seeded at density of 4 x 10^4 cells/mL in corresponding medium onto each disc of polymer, additionally glass coverslips were used as a positive control. The times for this assay were 3, 7, and 14 days, after which, remove media and add 1 mL (100 mg/mL pepsin in 0.5M acetic acid, pH 3) for 16 h at 4°C; stir gently the samples in the well plate. Transfer 50 μL of test sample to an eppendorf, by duplicate, after add 50μl of distilled water to each eppendorf; for the blank add 100μl of distilled water to two eppendorfs. For the standard curve add 5 μL, 10 μL, 25 μL and 50 μL of collagen standard to eppendorfs. Make the volume up to 100μl with distilled water. To each of the tubes add 1 mL of Sircol Dye reagent. Close the lids and shake to mix dye. Place tubes in shaker for 30 minutes, after this transfer tubes to a micro-centrifuge and spin at >10000xg for 10 minutes. A pellet will form at the bottom of the eppendorf, remove the solution, and try not to touch the pellet. Use a cotton bud (or tissue paper) to get as much solution out as you can. Add 1ml of alkali reagent to each eppendorf, recap eppendorfs and shake to dissolve the pellet in the reagent, can support with a Vortex system, should dissolve within 10 minutes. Finally transfer 200µl of solution to a 96 well plate to measure absorbance at 540nm.

Inflammatory Response

Peroxide Assay

Dichlorofluorescein (Sigma, Dorset, UK) (DCF) stock solution diluted in methanol to a concentration of 2 mM was further diluted in Hank's balanced salt solution (HBSS) to a working ratio of 1:100.

Macrophages at a density of 2 x 10^5 cells/mL in DMEM were seeded onto each polymer disc, and a copper disc was used as a positive control for peroxide release. After one hour the media was removed and replaced with DCF. Fluorescence was measured at hourly intervals, by transferring 50 μL aliquots from all wells to a new 96 well plate. The readings were taken over six hours period. The fluorescence plate reader (Fluostar Optima) was adjusted at 485 nm excitation and 520 nm emission. After reading the samples were carefully returned to the original wells [58].

Interleukin 1β Assay (IL-1β)

The assay quantitatively determined the levels of IL-1β in cell culture supernatant. Cells were cultured for 2 and 48 hours on the material samples. Lipopolisaccharide (LPS) was used as a positive control for the IL-1β release at a range of concentrations. The assay was performed and optical density of each well was determined using a microplate reader at 450 nm (measurement) and 540 nm (reference) wavelengths [58].

Implants in Live Organisms

Male Wistar rats with an average weight of 300 g were used. A laminectomy was carried out on the rats at level vertebral body T8 to leave the spinal cord exposed and to injure it by means of a cut with bistouries. After this the polymer was placed on the injured zone (group one) and close the injury, a second group was used as control without polymer implant (group two).

Previously to the injury, left sciatic nerve dissection was made to stimulate it by means of hook electrodes with silver chloride wire connected to the stimulation system of a registry equipment (Cadwell 5200 A) for somatosensory evoked potentials (SEP).

The registry electrodes, were of silver in disc form of 4 mm of diameter, they were placed on the leather that covers the skull in the zone of both parietals, the earth electrode was fixed in the tail and the SEP were obtained with the following parameters: gain 20 microvolt by division, filter of discharges 500 Hertz, filter of losses 10 Hertz, scan speed 5 ms. The registry obtained was an average of 200 signals to stimuli of 0.5 volts of intensity and 0.2 ms of duration with 0.5 Hz of frequency. The same procedure was made after the injury on spinal cord.

The SEP measurements were obtained, by means of technique previously described, to 5, and 30 days after the injury. In each group, after obtaining the SEP the animals were sacrificed by means of an anesthesia overdose obtaining the spinal cord for its later histological analysis.

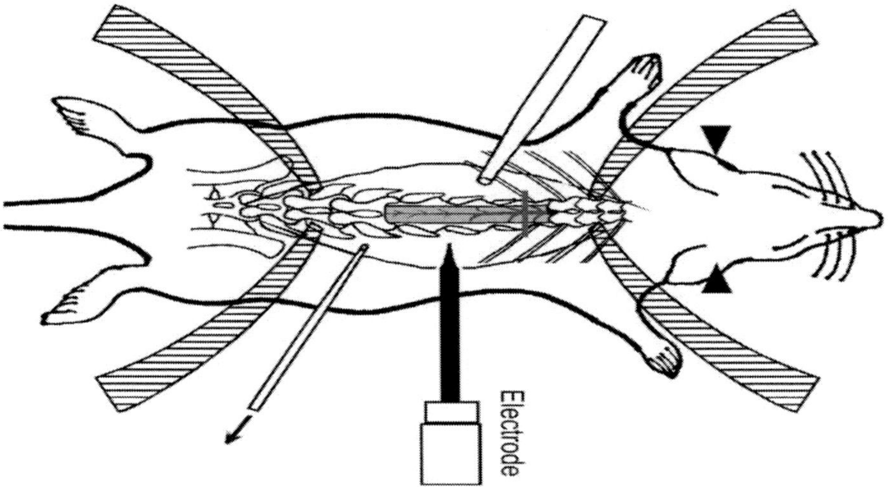

Figure 4. Representative scheme for the physical location of electrodes for SEP measurements, and injuried zone in the rat [60].

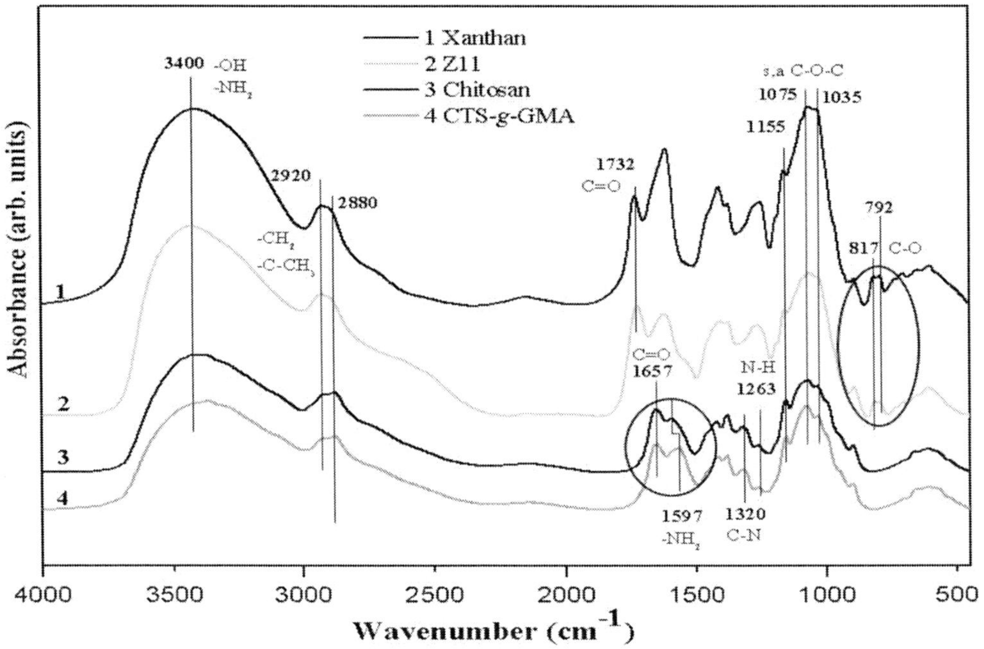

Figure 5. Infrared spectra of the chitosan, xanthan, CTS-g-GMA hybrid, and a representative sample of Z hydrogels [57].

Figure 4 shows a representative scheme for the measurement procedure to determine the somatosensory evoked potentials (SEP) to the rats and the physical location of the electrodes and injury. Special care was performed during the surgical procedure.

Results and Discussion

Characterization

Figure 5 shows the FTIR spectra from pure chitosan, xanthan, CTS-*g*-GMA and Z hydrogel. Due to nature of the molecules, polysaccharides, the spectra are very similar, mainly for chitosan and CTS-*g*-GMA, to each other except for the shift of the NH_2 peak (1597 cm^{-1} for chitosan and 1580 cm^{-1} for CTS-*g*-GMA). The decrease in the wavenumber could be associated to a protonation of the NH_2 group (larger effective mass) induced by the acid environment during the CTS-*g*-GMA synthesis. Since the intensity of that peak was kept approximately constant, we concluded that the functionalization of the chitosan by the GMA was not through the amine group. This suggests that the addition of the GMA was done through the meridional–OH group, which is the most reactive site in an acid environment.

The origin of most of the peaks shown in figure 5 has been reported elsewhere [61 - 68] except for the doublet at 817 and 792 cm^{-1}. Those could be a weak contribution due to the C-O link of the ring in the cyclic ether. The position suggests an ether molecule in the cis-position. In addition these bands could contribute to the swinging vibrations of the methylene groups from the cyclic ether.

The figure 6 shows the comparative spectra between both series, A and B, it is possible to observe that the spectra maintain the important bands compared with the material without neutralization [57]. Nevertheless, comparing the series B with A, it is evident that exist some differences between both series exist. The infrared spectra of Z series A, comparing with the series B, the important difference it is appreciated in the bands around 1695 and 1650 cm^{-1} assigned, respectively, to C=O and C=C stretching vibrations of GMA group attached to Chitosan.

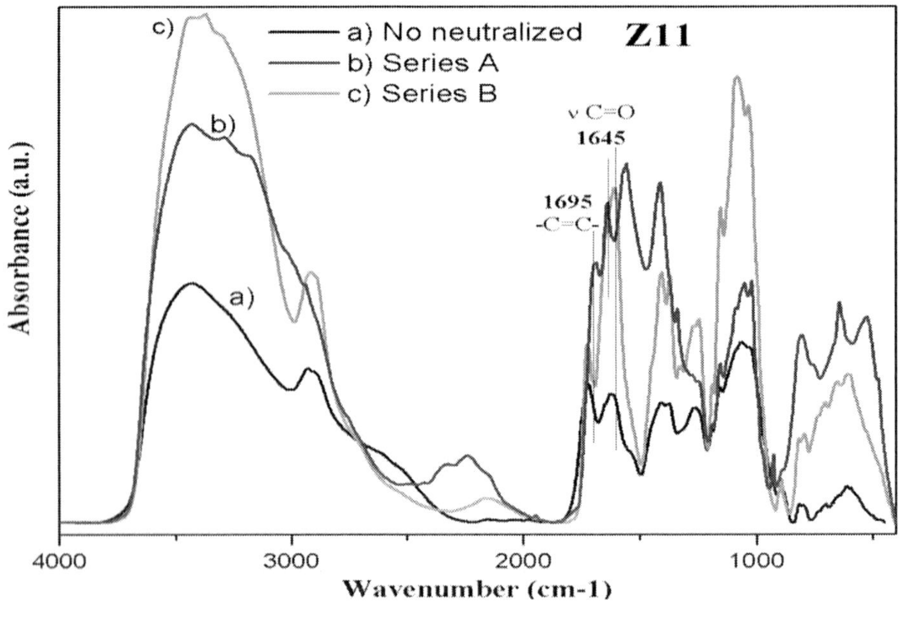

Figure 6. Comparative spectra for hydrogel Z11 both series, A and B, and compared with same material without neutralization, it is possible observe significant differences between them [57].

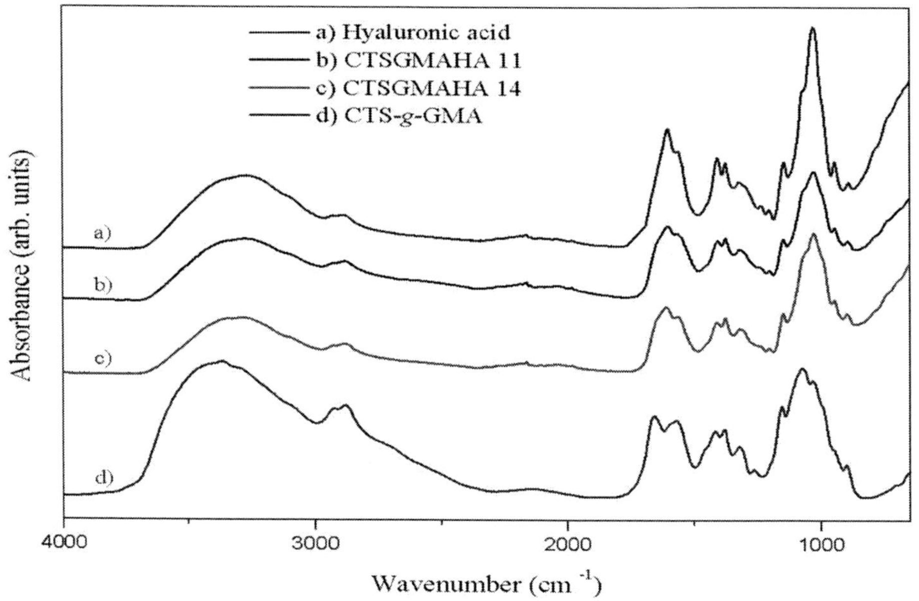

Figure 7. Slight shifts are showed in the infrared spectrum of hydrogels CTSGMAHA, compared with the precursors polymers.

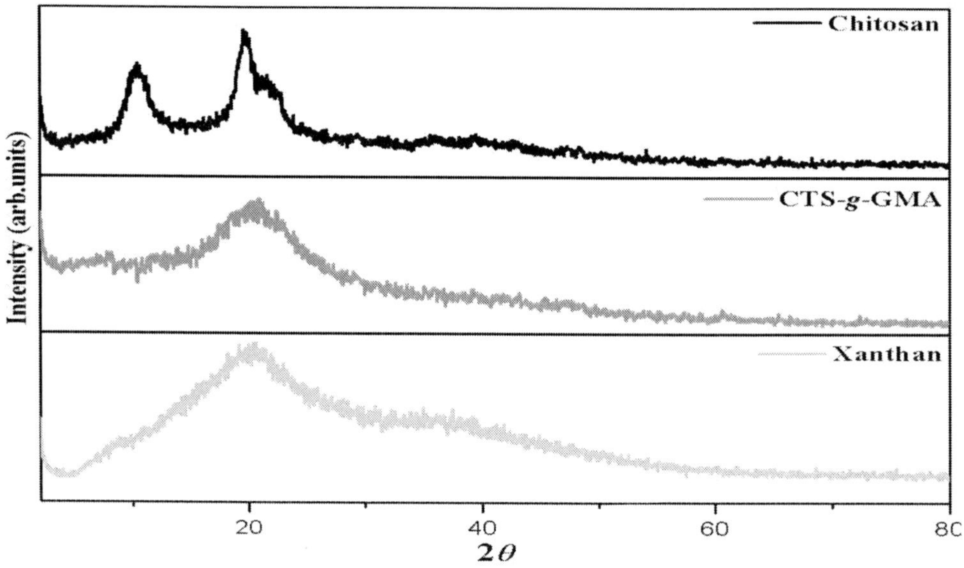

Figure 8. Ray X diffraction spectra for the Chitosan, Xanthan and CTS-*g*-GMA. Note that pure chitosan exhibit typical crystalline peaks at ca. 11 and 21 degrees.

This fact indicate that the chemical treatment used for the neutralization has degraded slightly the chemical structure of CTS-*g*-GMA moiety, like was reported in the previous work; where the affected groups in CTS-*g*-GMA are the terminal methyl-vinyl groups, which mainly are in contact with basic medium [56].

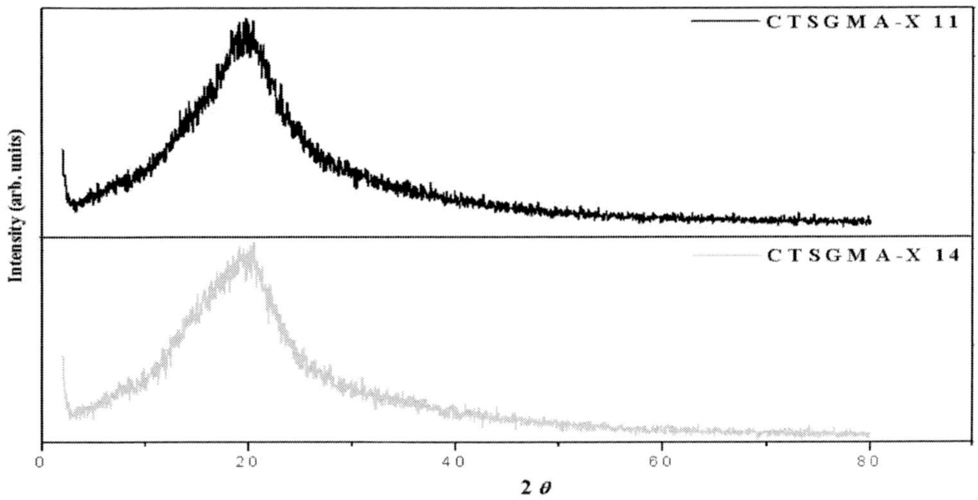

Figure 9. Ray X diffraction analysis for the (CTS-*g*-GMA)-X hydrogels. Note that both materials are amorphous.

The rest of the bands for the series A, as in the series B, show the same position as the un-neutralized biomaterials, described in previous work [56, 57].

For the CTSGMAHA hydrogels obtained, figure 7 shows the comparative spectrum of them against the precursors. It is possible to observe a slight enlargement in the width of the carboxylate band, around 1600 cm^{-1}; which could be attributed to decrease and displacing of the $-NH_2$ band and subsequent junction of bands.

Hydrogels CTSGMAHA 1:1 and 1:4 compared with the polymers precursors, the bands showed for these hydrogels have a similar behavior to HA, nevertheless the doublet defined at 1400-1370 cm^{-1} exhibit the trend of CTS-*g*-GMA for the hydrogel 1:1, while hydrogel 1:4 show increments in bands such as 1553 and 1407 cm^{-1}, assigned to carboxylate and hydroxyl groups [56, 69, 70].

The rest of bands shown a similar behavior for HA, this fact is attributed to the anionic nature of the HA in solution, in addition to their high molecular weight (1,000,000 Da); suggesting a molecular structure where the HA is wrapping the CTS-*g*-GMA through their crosslinking.

Figure 8 shows the X ray diffraction patterns for CTS, for CTS-*g*-GMA and xanthan. The CTS pattern shows broad but well defined diffraction lines at 12, 20 and 22.5 degrees in the 2 theta scale, these lines are on top of a diffuse background, this indicates that the structure of the material is formed by small crystalline regions embedded in amorphous regions. The patterns for CTS-*g*-GMA and xanthan show a broad band centered at about 20 degrees which has its second order at about 40 degrees; these two materials show a lower degree of crystallinity.

Figure 9 shows that the CTSGMA-X hydrogels have similar amorphous X-ray patterns, which probably reflect an important effect on the structural arrangement in the hydrogels due to X increase. In addition according to CTSGMA-X 14 more amorphous pattern is observed. This observation may explain the fact that the transestherification reaction provides more flexible bonds such as C-O-C.

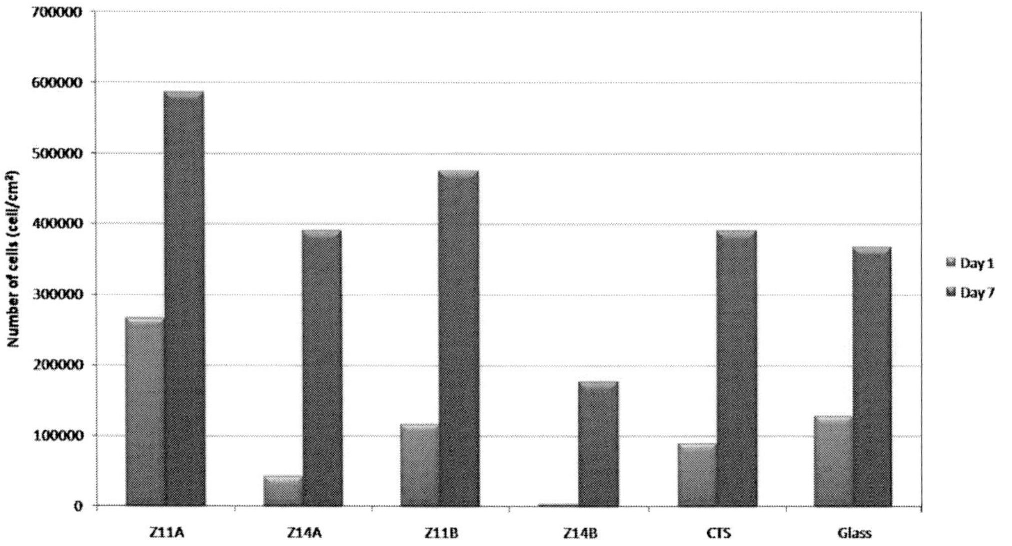

Figure 10. Cell growth of Human epidermal keratinocytes onto polymers, chitosan (precursor) and glass (positive control). From day 1 to day 7 there is significant cell growth onto our biomaterials.

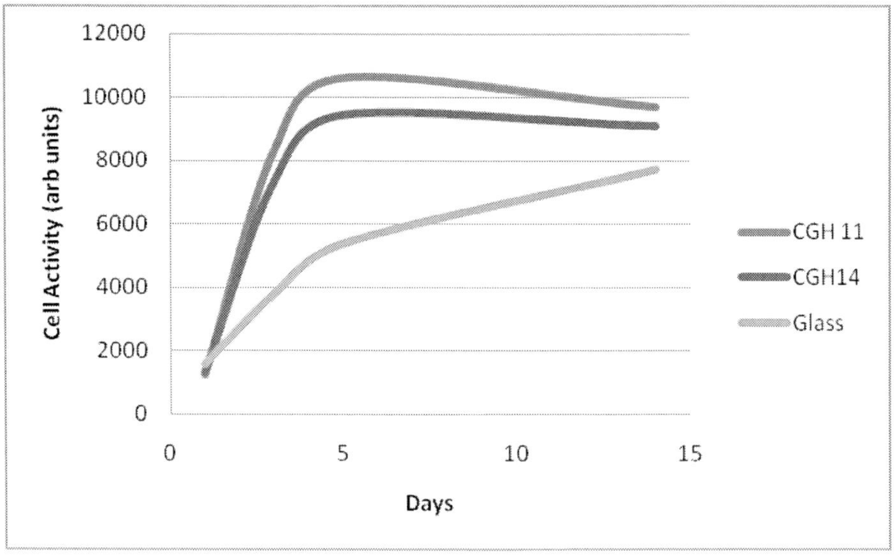

Figure 11. Keratinocytes cell activity obtained from the CGH materials and positive control (glass).

Cell Culture

The viability of HEKa has been quantified by the DNA (proliferation) assay. Figure 10 shows the growth onto biomaterials, the results indicate that most of samples, except Z14B, show a better behaviour to positive control (glass with a treated surface), increasing the cell number across the time.

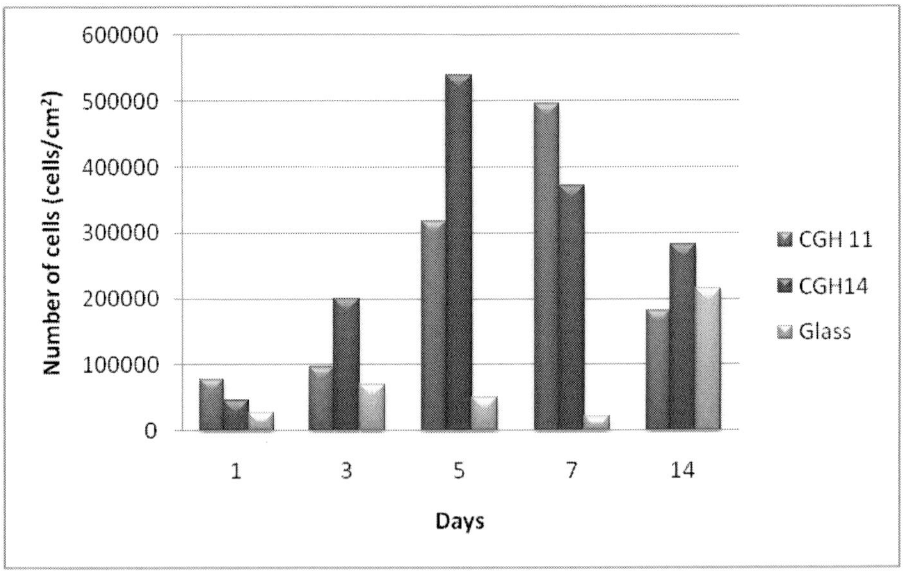

Figure 12. Cell growth of Human epidermal keratinocytes onto CGH polymers, and glass (positive control).

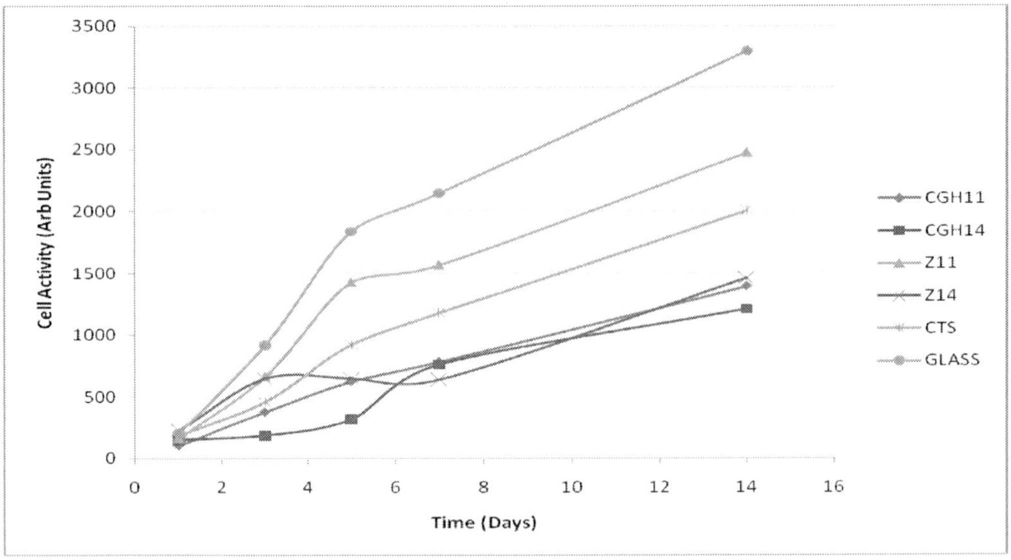

Figure 13. Bovine chondrocytes cell activity obtained from Alamar Blue assay, of all the polymers tested.

The polymers of series A show a better behaviour compared with the series B, being the best Z11A, which shows the greater growth after seven days. Statistical correlation analysis, T-test, shows that the results have no significant difference between most of polymers. However, there are significant differences in some of them between one and seven days.

Figures 11 and 12 shows the results for Alamar blue and DNA assay, respectively, we can observe the trend for materials (CTS-g-GMA)-HA (labeled as CGH 11 or 14 depending the stoichiometric ratio). For both assays the biopolymers CGH 11 and 14, the results trend is to increase the activity and growth of keratinocytes through time. For the CGH materials after 5 days the cell activity is maintained showed a slight decrease until 14 days; meanwhile the trend for positive control is to continue increasing.

The DNA results show that through time the keratinocytes increase their number, however after 5 and 7 days (for CGH14 and 11, respectively) the number of cells decrease, this fact could be due to effects of cell adhesion, the keratinocytes are unattached due to the surface degradation of the polymers, or that the surround is no friendly for the cells.

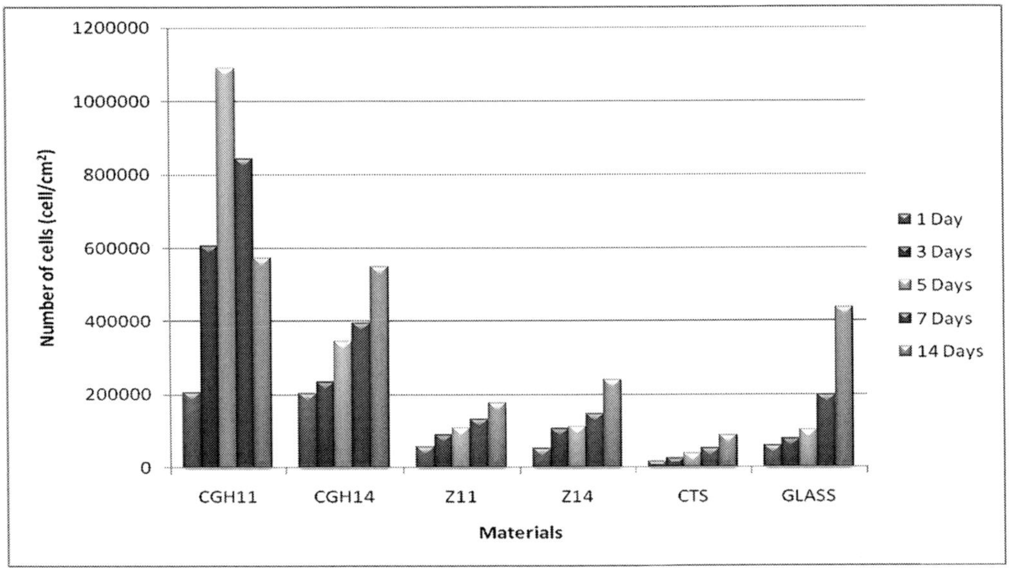

Figure 14. Cell growth of bovine chondrocytes onto CGH polymers, and glass (positive control). From all times (1 to 14 days) there is significant cell growth onto our biomaterials.

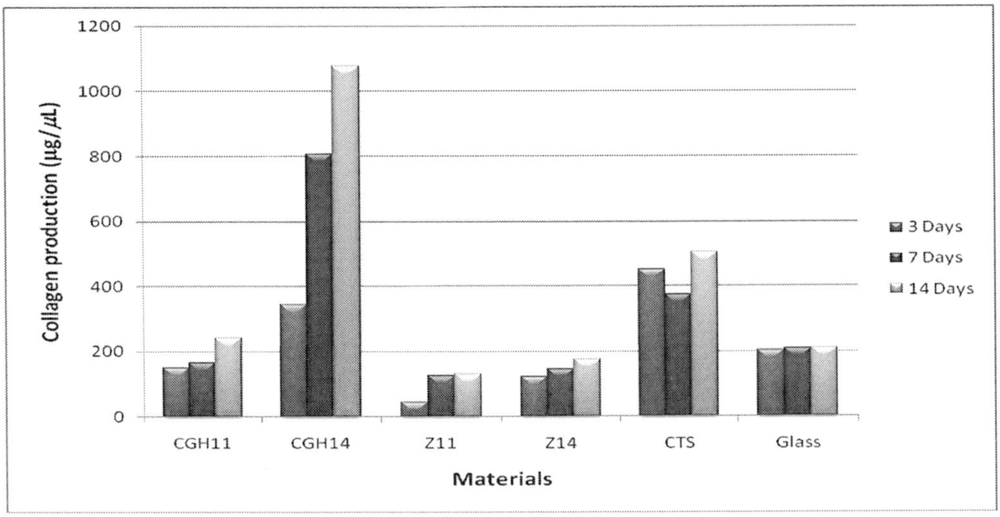

Figure 15. Collagen production from bovine chondrocytes onto CGH 11, CGH 14, and glass (positive control).

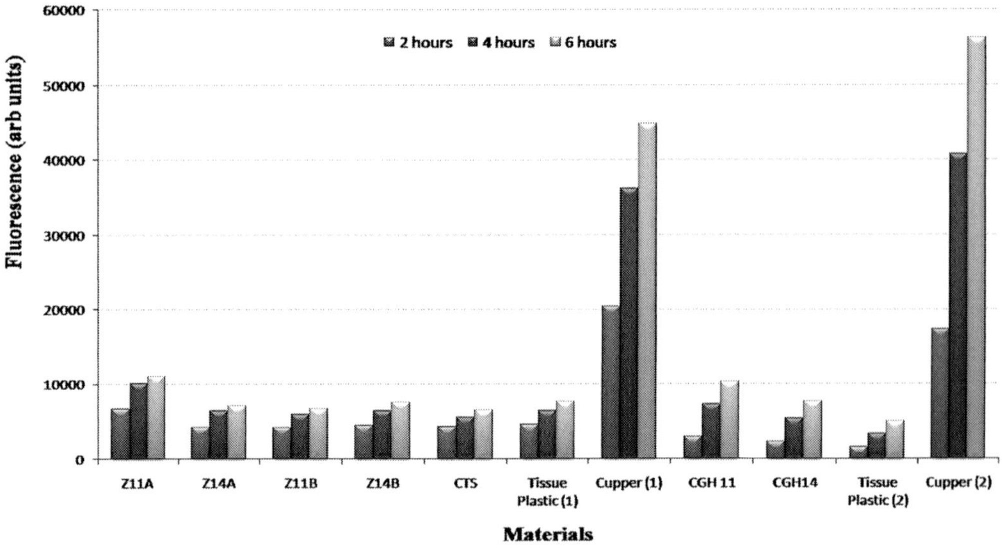

Figure 16. Representative histogram of peroxide assay, showing the materials with Xanthan and Hyaluronic acid, each batch with their respective negative and positive controls (tissue plastic and cupper, respectively).

Figure 13 shows the cell activity from chondrocytes obtained means Alamar blue assay, the results demonstrate an increase in the level through the time. However, this cannot be compared with the positive control (glass) which shows the double of activity compared with some polymers.

Figure 14 presents the chondrocytes growth quantified by DNA proliferation assay, which show a significant difference between the materials and the positive control. It is evident that in the CGH11 polymer after 5 days the cell number decreases this fact could be due probably, that the chondrocytes were stressed by lack of space, considering that sample area.

For this assay is evident that for polymers Z, the chondrocytes proliferation is similar if is compared with glass, or raw chitosan; however, the difference is significant when is compared with polymers CGH. This fact reveals that polymers CGH have characteristics that improve the chondrocytes adaptation, and their proliferation increase in a short time.

The results from collagen are showed in the figure 15, which is possible to observe that CGH14 promotes the production of collagen better than the CGH11 and glass, the maximum value for CGH, at 14 days, is clearly superior other polymers tested.

Figure 16 shows the results obtained from peroxide assay, to evaluate the inflammatory response, after 2, 4, and 6 hours. Peroxide assay results show a slight inflammatory response caused mainly for Z11A and CGH11, whereas the rest of polymers remain to lower levels similar to their respective negative control (tissue plastic).

In general, macrophages were not activated to release peroxide in response to the polymers, all values were very low compared with Cupper (positive control), after 6 hours of evaluation; this fact indicate that no one of polymers synthesized induce an aggressive inflammatory response by a host organism.

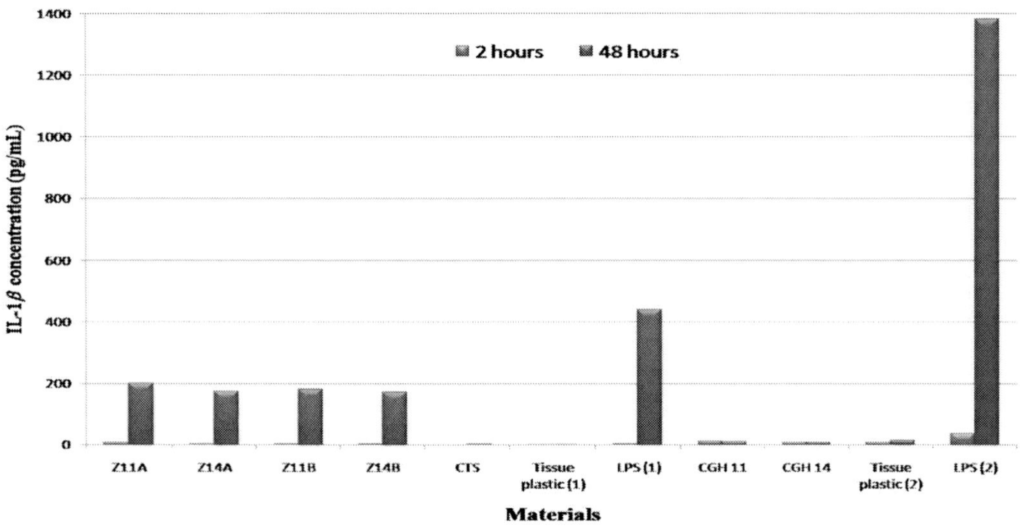

Figure 17. Interleukin-1β production by mouse macrophages J774A.1, after 2 and 48 hours, showing Z's and CGH's polymers, compared with their respective controls for each batch.

Table 2. Concentrations and values of IL-1β of different polymers based in chitosan compared with the materials

Material	Dose (μg/mL)	Incubation time (hrs)	IL-1β concentration (pg/mL)
Oligochitosan [71]	40	18	101.9 ± 46.6
CTS-X [44]	1000	24	70
S-DAC70 [72]	500	24	45
Chitosan-DNA[a] [73]	-	24	No detected
Z11A	1000	48	199.7 ± 6.1
Chitosan	1000	48	3.2 ± 0.5

[a] With 0.1, 1, 10 and 20 μg of DNA.

For this assay a calibration curve has been performed from recombinant mouse IL-1β standard. Figure 17 shows the comparative histogram for all polymers with their respective negative and positive controls. The histogram shows low concentrations of IL-1β after 2 hours of incubation; results showed similar values compared with controls, negative and positive (tissue plastic and lipopolysaccharide from E. Coli).

Nevertheless, after 48 hours an increase was observed in response to hydrogels Z, being the highest value for Z11A (199.7 pg/mL); similar values are shown by the materials obtained with X (hydrogels Z); while for polymers CGH the IL-1β concentration remain in low values, similar to the first value, which yield compare with their negative control.

All samples, inside each batch, show a similar behaviour when they were placed in contact with the macrophages, although all of them have high values compared with other authors. However, the weight of the samples was around 1 mg/mL, whereas other authors used 50 or 500 μg/mL of samples during 18 or 24 hours (see table 2) [71 - 73].

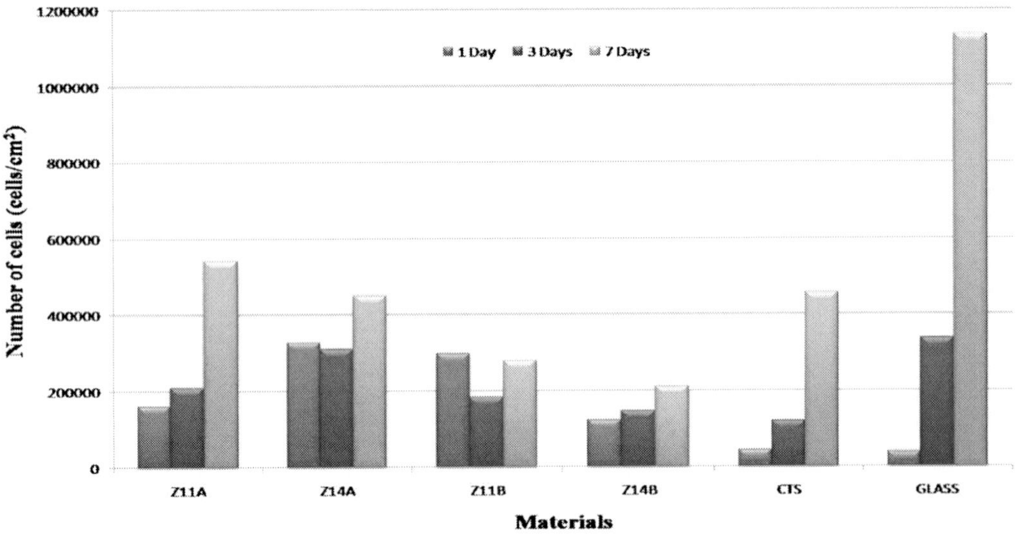

Figure 18. Cell growth of nerve cells NG108-15 onto biomaterials, chitosan (precursor) and glass (positive control). Note that in this assay cell growth steadily continues from day 1 to day 7.

Due to the properties exhibited by hydrogels Z, and the results obtained from immunocytochemistry, we decide that this materials are suitable for the growth of a speciality cell line such as nerve cells; thus cell line NG108-15 was tested. Figure 18 shows the proliferation behavior of this cell line onto our biomaterials; it is evident that most of the samples have a smaller growth compared with the positive control (glass with a treated surface).

Even if the growth is smaller than in the control specimens, considering the cell nature and special care for the cells, there is enough evidence for preservation and proliferation of them onto the biomaterials. Statistical correlation analysis, T-test, shows that the results have no significant difference between most of polymers. However, there are significant differences in some of them between one and seven days, especially in the positive control.

Figure 19 shows the results for live/dead assay after 5 days of the cell inoculation, the glass shows several nucleus (died cells) in red, immersed in a layer of live cells. On the biomaterials the trend continues to be to organise the cells in straight clusters, with a random orientation. However, in all the polymers, including chitosan film, it is evident the absence of dead cells in this clusters. This could be attributed to the lack of attachment of dead cells, and their possible discard in some of the washes, especially when considering that DNA assay shows a low number of cells onto biomaterials compared with the glass.

Having established that the hydrogels demonstrated were suitable for preservation of cells NG108-15, means DNA assay; was necessary promotes the cellular differentiation using enriched media. Neurofilament H stain reveals that both polymers of series Z allow the differentiation of a small number of cells, where as the polymers series B, and raw chitosan, do not. Figure 20 shows the representative cells differentiated to neurites in red, onto the glass (positive control), and hydrogels Z series A (Z11A and Z14A).

The conjunction of these results, for hydrogels Z, demonstrate that are adequate for the growth, preservation, proliferation and differentiation of nerve cells, and strongly support the future their use as scaffolds for cells and their application as implants in live organisms.

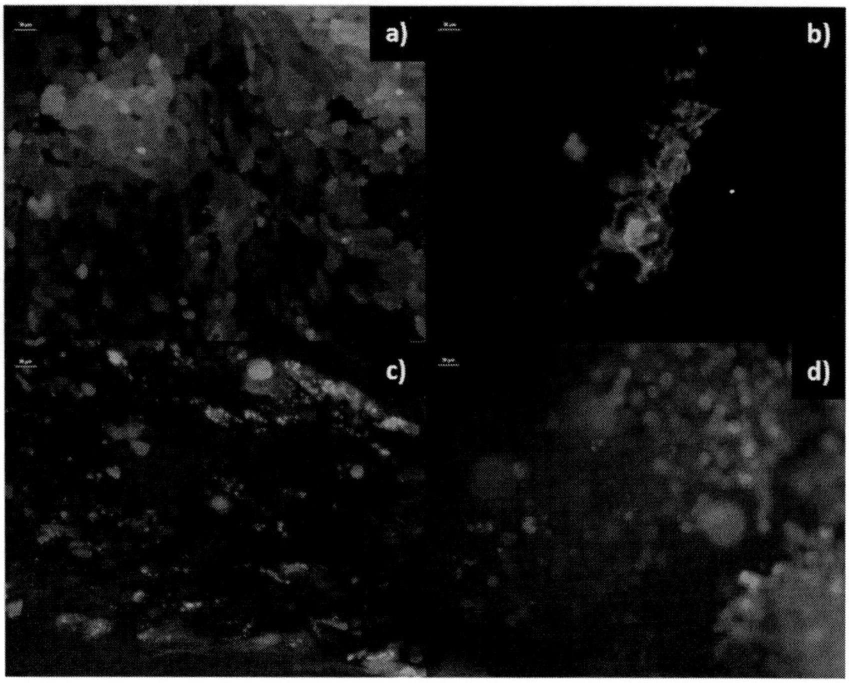

Figure 19. Representative results for Live/Dead assay after 5 days of cell incubation, a) glass (positive control), b) Chitosan, c) Z11B and d) Z14A. Original magnification 20x. Green fluorescent regions denote cell activity whereas red regions denote null cell activity.

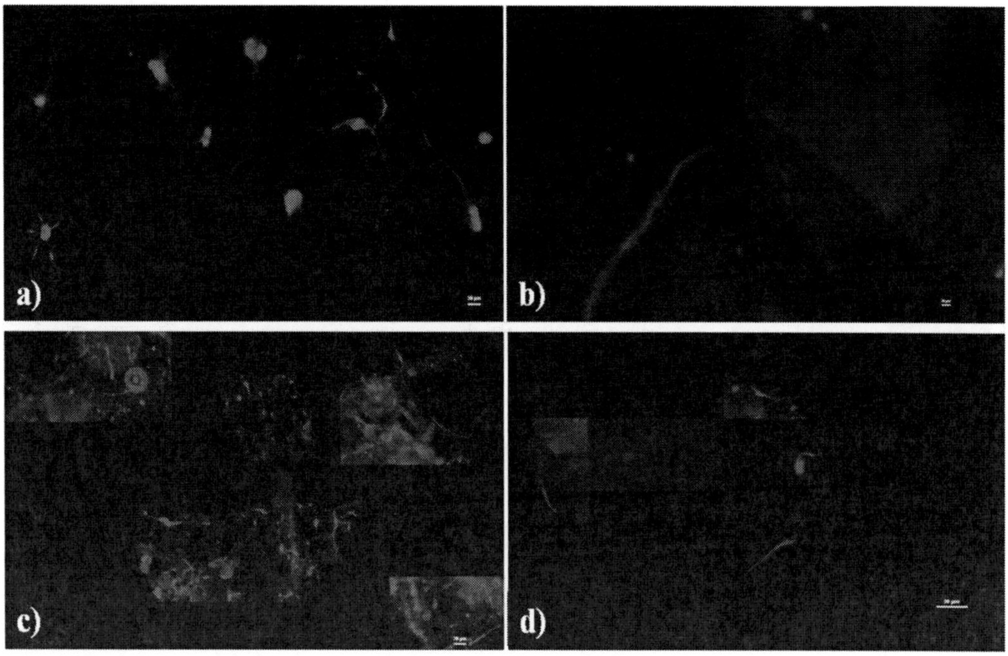

Figure 20. Images obtained from Neurofilament H stain, showing cells differentiated to neuritis in red and nucleus in blue. Original magnification a), b), and c) 10x; and d) 20x.

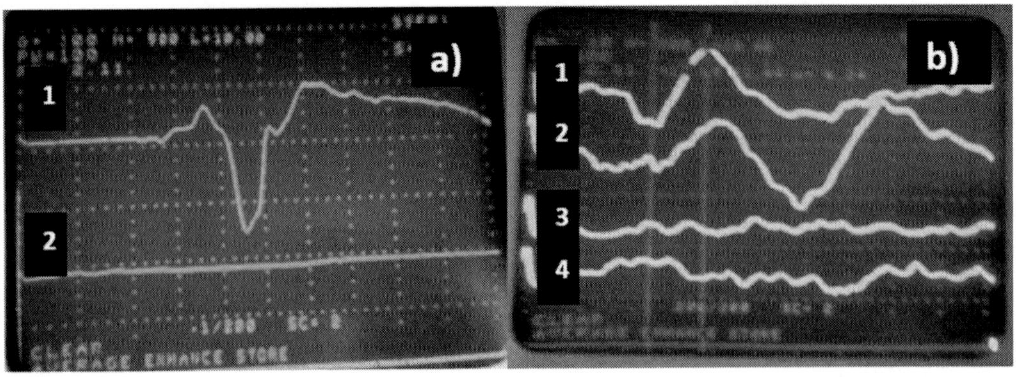

Figure 21. Somatosensory evoked potentials (SEP) obtained before the laminectomy. a) channel 1, and b) channels 1 and 2); and immediately after the injury a) channel 2, and b) channels 3 and 4). Note the electrical activity decay after the laminectomy.

Implants In Live Organisms

To implant the biopolymer onto spinal cord represents a significant advance, since if the polymer resist in a live organism without a visible degradation or severe damage in their structure, and without immunological rejection apparent; and considering the previous results, can support the use of our polymers as implants in live organism and they being able as a third generation biomaterial.

Figure 21 shows the SEP measurements in which are clearly evident the loss of components after the injury. It is clear for upper channels the existence of components with a correct definition in the measurement screen. For the channel 2 in figure 21 a) is possible to observe that the signal is almost a straight line after the injury which indicate that the animal loses the movement of its back legs and the sensibility of that part of their body due to the nonexistence of stimuli through spinal cord. A similar behavior can be observed for the figure 21 b).

After 5 days of the surgery, for the group one, it is possible to observe a recovery in the components of the SEP's, in addition these rats show a slight recovery compared with the group two. After 30 days of laminectomy, for the group two, it is not possible identify some SEP's components because continues being little definite, figure 22, and the animal shows no movements or signals of recovery.

Meanwhile, for the group one (figure 23), SEP's obtained 30 days after the surgery can be compared with the potentials observed in the upper channel of the figure 21 a) and b), which indicate a satisfactory recovery of the injury. In addition of the recovery of SEP's components the rats can move the back part of their body without signal of damage caused by the laminectomy. This effect in the recovery of the rats could be attributed to the presence of the polymer in the injured zone.

Comparing screens for SEP's, after 30 days, between the group one (figure 23) and two (figure 22), it is evident that group one shows a better recovery, almost complete, and identifying clearly the components of the SEP's; in addition the rats with implant show mobility in a similar fashion as before the surgery, whereas rats without implant still lack mobility.

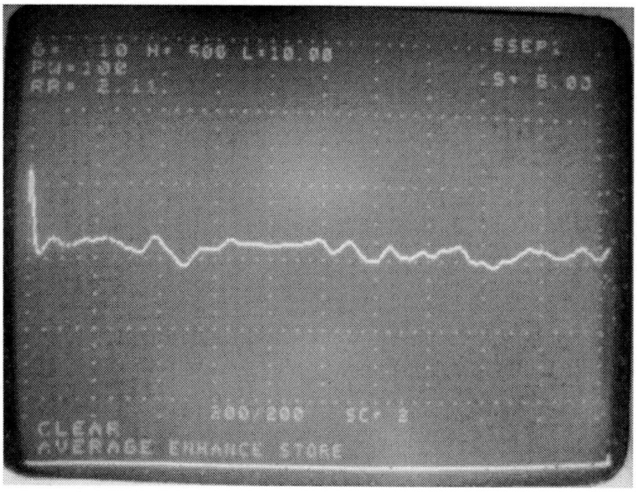

Figure 22. SEP obtained 30 days after the injury for the group two (no implant), note there is no electrical activity for rats without implant.

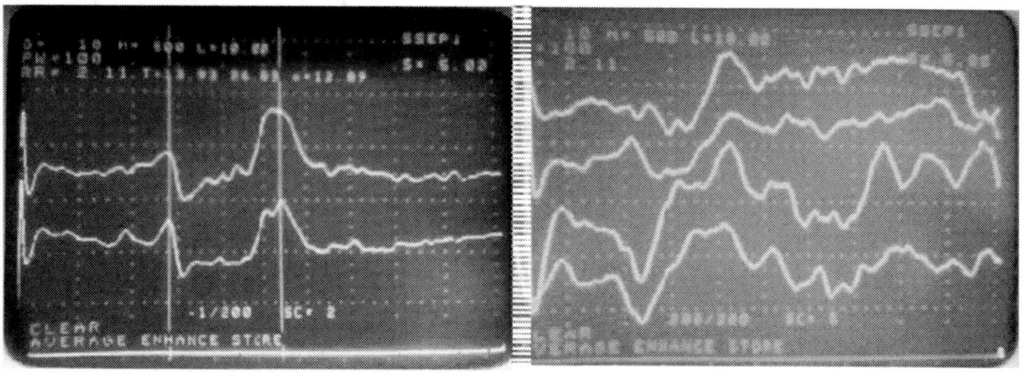

Figure 23. SEP obtained 30 days alter the injury for the group one (implant). There is significant electric activity such that the potentials resemble those initially taken (before surgery).

Conclusions

This study has shown that it is possible to carry out a chemical reaction between a hybrid biopolymer (CTS-*g*-GMA) with water-soluble anionic polymers, such as Xanthan gum and Hyaluronic acid, which could be corroborated by means of physicochemical characterization. FTIR analysis confirms the reaction between the species involved in each reaction, due to the presence of characteristic absorption bands.

After the cell culture, the synthesized materials exhibited convenient properties for cell viability, including cartilage, skin tissue and nerve cells. DNA and Alamar blue assays for HEKa and chondrocytes, shown high cell preservation, growth, and cell activity in most samples and can be compared satisfactory with the blank (glass with a treated surface).

In general, all polymers do not activate an aggressive inflammatory response. For peroxide assay all materials can be compared with negative control (tissue plastic); from IL-1β assay, the results show that the batch of hydrogels Z promotes an increase in the concentration of IL-1β, after 48 hours, whereas the batch of polymers CGH do not. These results indicate that no one of polymers synthesized induce an aggressive inflammatory response by a host organism.

All these results are critical if the materials are going to be successfully deployed in biomedical applications. Whether or not they could be biocompatible *in vivo* and useful for specific medical problems still remains to be explored.

All studies performed to nerve cells NG108-15 showed that hydrogels Z allow their growth, proliferation and differentiation, although these results are below those of glass (positive control).

In vivo laminectomy tests on Wistar rats showed a decrease in recovery time by 1/5 time, when our biomaterials were used as implants, to those untreated rats.

Results presented in this work are promising and strongly support the future use of these hydrogels as scaffolds for tissue engineering and recovering spinal cord and they can be considered as a third generation biomaterials.

REFERENCES

[1] Hench L. L., and Polak J. M., *Science.* 295, 1014-1017, 2002.

[2] Park K., Shalaby S. W. S., and Park H., *Biodegradable Hydrogels for Drug Delivery*, Technomic Publishing Co., Inc., Lancaster, PA, p. 1-72,1993.

[3] Francis Suh J. K., and Mathew H. W. T., *Biomaterials*, 21, 2589-2598, 2000.

[4] Koyano T., Minoura N., Nagura M., and Kobayashi K., *J. Biomed. Mater. Res*, 39, 486-490, 1998.

[5] Domard A., Domard M. Chitosan: Structure – Properties Relationship and Biomedical Applications. In: Dumitriu S. Ed., *Polymeric Biomaterials*, Marcel Dekker, New York, USA, p. 187–212 (2002).

[6] Muzzarelli R. A. A., *Chitin*, Pergamon, New York, USA (1977).

[7] Singh D. K., and Ray A. R., *JMS Rev. Macromol. Chem. Phys.*, C40 (1), 69-83, 2000.

[8] Domard, A., Roberts G. A. F., Varum K. M. (Eds), Advances in Chitin Science, Vol. II, Jaques André Press, Lyon, France (1998).

[9] Lee K. Y., Ha W. S., and Park W. H., *Biomaterials*, 16, 1211-1216, 1995.

[10] Hirano S., Inui H., Hutadilok N., Kosaki H., Uno Y., Toda T., *Polym. Mater. Sci. Eng*, 66, 348, 1992.

[11] Hirano S., Inui H., Iwata M., Yamanaka K., Tanaka H., Toda T., In: Progress in Clinical Biochemistry, Miyai K., Kanno T., Ischikawa E. (Eds.). Elsevier Applied Science, London, UK, p. 1009 (1992).

[12] Wan Y., Yu A., Wu H., Wang Z., and Wen D., *Journal of Materials Science: Materials In Medicine*, 16, 1017-1028, 2005.

[13] Chitin and Chitosan: Opportunities and Challenges, edited by Dutta P. K., SSM International Publication, Midnapore, India (2005).

[14] Chitin: Key to low-cost. Plentiful Biopolymers, emerging technologies, No. 54, Technical Insights, Inc, Englewood, USA.
[15] Drudy J. L., and Mooney D. J., *Biomaterials*, 24, 4337-4351, 2003.
[16] Griffith L. G., and Naughton G., *Science*, 295, 1009-1014, 2002.
[17] Salvay D. M., and Shea L. D., *Mol. BioSyst.*, 2, 36-48, 2006.
[18] Roberts, G.A.F.: Structure of chitin and chitosan. In: Chitin chemistry, edited by G.A.F. Roberts,. Mac Millan Press, Houndmills (1992).
[19] Nakatsuka S., and Andrady A. L., *J. Appl. Polym. Sci.*, 44, 17-28, 1992.
[20] De Angelis A. A., Capitani D., and Crescenzi V., *Macromolecules*, 31, 1595-1601, 1998.
[21] Kristl J., Smid-Korbar J., Struc E., Schara M., and Rupprecht H., *Int. J. Pharm.*, 99, 13-19, 1993.
[22] Martinez-Ruvalcaba A. Rhéologie des Solutions de Chitosane et des Hydrogles de Chitosane-Xantane. Thèse de doctorat de génie, Faculté de Génie, Université de Sherbrooke, Sherbrooke, Canada (2001).
[23] Dumitriu S., Magny P., Montane D., Vidal P. F., and Chornet E., *J. Bioactive and Compatible Polymers*, 9, 184-209, 1994.
[24] Chia-Hong Chu, Sakiyama T., and Yano T., *Biosci. Biotech. Biochem.*, 59 (4), 717-719, 1995.
[25] Ikeda S., Kumagai H., Sakiyama T., Chia-Hong Chu, and Nakamura K., *Biosci. Biotech. Biochem.*, 59 (8), 1422-1427, 1995.
[26] Qu X., Wirsén A., and Albertsson A. C., *Polymer*, 41, 4589-4598, 2000.
[27] Lee K. Y., Ha W. S., and Park W. H., *Biomaterials*, 16, 1211 (1995).
[28] Tomihata K., and Ikada Y., *Biomaterials*, 18, 567 (1997).
[29] Freiera T., Koha H. S., Kazaziana K. and Shoichet M. S., *Biomaterials*, 26, 5872-5878 (2005).
[30] Katalinich M. Characterization of Chitosan Films for Cell Culture Applications. Thesis for the Degree of Master in Science, The University of Maine, Maine, USA (2001).
[31] Muzzarelli R. A. A., Baldassarre V., Conti F., Ferrara P., Biagini G., Gazzanelli G., and Vasi V., *Biomaterials*, 9, 247-252, 1988.
[32] Yoshika T., Hirano R., Shioya T., and Kako M., *Biotechnology and Bioengineering*, 35, 66-72, 1990
[33] Marieb E. N. and Hoehn K. in: *Human Anatomy and Physiology*, p. 478-479, Pearson Benjamin Cummings, U. S. A. (2007).
[34] Cook J. R., Crute B. E., Patrone L. M., Gabriels J., Lane M. E., and Van Buskirk R. G., *In Vitro Cellular Developmental Biology*, 25, 914-922, 1989.
[35] Norris D.A , Shellman Y., and Bellus G. A., *Keratinocytes* in: Apoptosis and Inflammation, Winkler J. D. (Ed.), Birkhäuser Verlag, Basel, Switzerland pp. 121 (1994).
[36] Lebre M. C., Van der Aar A. M. G., Van Baarsen L., Van Capel T. M. M., Schuitemaker J. H. N., Kapsenberg M. L., and De Jong E. C., *Journal of Investigative Dermatology*, 127, 331–341, 2007.
[37] Malette W. G., Quigley H. J., and Adickes E. D.. In: Chitin in Nature and Technology, Muzzarelli R., Jeuniaux C., Gooday G. W. (Eds.). Plenum Press, New York, USA (1986).

[38] Tortora G. J. and Derrickson B., *Principles of Anatomy and Physiology*, pp. 406-407, John Wiley and Son, Inc, USA (2006).

[39] Connors B. W., and Long M. A., *Annu. Rev. Neurosci.*, 27, 393-418, 2004.

[40] Choi R. C. Y., Pun SDong., T. T. X., Wan D. C. C., and Tsim K. W. K., *Neuroscience Letters*, 236, 167-170, 1997.

[41] Pun S., Yang J. F., Ng Y. P., and Tsim K. W. K., *FEBS Letters*, 418, 275-281, 1997.

[42] Busis N. A., Daniels M. P., Bauer H. C., Pudimat P. A., Sonderegger P., Schaffner A. E., and Nirenberg M., *Brain Research.*, 324 201-210, 1984.

[43] Haipeng G., Yinghui Z., Jianchun L., Yandao G., Nanming Z., and Xiufang Z., *J Biomed Mater Res*, 52, 285-295, 2000.

[44] Chellat F., Tabrizian M., Dimitriu S., Chornet E., Magny P., Rivard C. H., L'Hocine Y., *J Biomed Mater Res*, 51, 107-116, 2000.

[45] Nishimura K., Nishimura S., Seo H., Nishi N., Tokura S., and Azuma I., *Journal of Biomedical materials Research*, 20,1359-1372, 1986.

[46] Peluso G., Petillo O., Ranieri M., Santin M., Ambrosio L., Daniela C., Avallone B., and Balsamo G., *Biomaterials*, 15 (15), 1215-1220, 1994.

[47] Shwab M. E., *Science*, 295, 1029-1031, 2002.

[48] Dumont R. J., Okonkwo D. O., Verma S., Hurlbert R. J., Boulos P. T., Ellegala D. B., and Dumont S. A., *Clin. Neuropharmacol.*, 24 (5), 254-264, 2001.

[49] Schwab M. E., and Bartholdi D., *Physiol. Rev.*, 76, 319-370, 1996.

[50] Bracken M. B., Shepard M. J., Holford T. R., Leo-Summers L., Aldrich E. F., Fazl M., Fehlings M. G., Herr D. L., Hitchon P. W., Marshall L. F., Nockels R. P., Pascale V., Perot P. L., Piepmeier J., Sonntag V. KH., Wagner F., Wilberger J. E., Winn H. R., and Young W., *J. Neurosurg.*, 89 (5), 699-706, 1998.

[51] Spinal Cord injury, form: National Spinal Cord injury Statistical Center, Birmingham, Alabama, In: www.spinalcord.uab.edu

[52] Woerly S., Pinet E., De Robertis L., Van Diep D., and Bousmina M., *Biomaterials.*, 22, 1095-1111, 2001.

[53] Larner J. A., Johnson A. R., and Keynes R. J., *Biol. Rev.*, 70 (4), 597-619, 1995.

[54] Milas M., Rinaudo M., and Tinland B., *Polym. Bull.*, 14, 157, 1985.

[55] Kassai M. R., Arul J., and Charlet G., *Journal of Polymer Science; Part B: Polymer Physics.*, 38, 2591, 2000.

[56] Flores-Ramírez N., Elizalde-Peña E. A., Vásquez-García S. R., González-Hernández J., Martinez-Ruvalcaba A., Sanchez I. C., Luna-Bárcenas G., and Gupta R. B., *J. Biomater. Sci. Polymer Edn*, 16 (4), 473-488 (2005).

[57] Elizalde-Peña E. A., Luna-Bárcenas G., Flores-Ramírez N., Vásquez-García S. R., Herrera-Gomez A., Nuño-Donlucas S., Gutiérrez-López A. A., González-Hernández J., Schmidt C., Gough J. E., and Sanchez I. C., Synthesis and Characterization of Chitosan-GMA-Xanthan Hydrogel, (2009) *submitted.*

[58] Gough J. E., Christian P., Unsworth J., Evans M. P., Scotchford C. A., and Jones I. A., *Journal of Biomedical Materials Research*, 69A (1), 17-25, 2004.

[59] Gough J. E., Christian P., Scotchford C. A., and Jones I. A., *Biomaterials*, 24 (27), 4905-4912, 2003.

[60] Sonohata M., Furue H., Katafuchi T., Yasaka T., Doi A., Kumamoto E., and Yoshimura M., *J. Physiol.*, 555.2, 515-526, 2003.

[61] Srivastava A., and Behari K., *J. Appl. Polym. Sci*, 104, 470 (2007).

[62] Su L., Ji W. K., Lan W. Z., and Dong X. Q., *Carbohydrate Polymers*, 53, 497, 2003.
[63] Hofmann R. and Posten C., *Eng. Life Sci.*, 2, 304 (2002).
[64] Lii C.-y., Liaw S. C., Lai V. M. -F., and Tomasik P., *Eur. Polym. J*, 38, 1377, 2002.
[65] Jampala S. N., Manolache S., Gunasekaran S., and Denes F. S., *J. Agric. Food Chem.*, 53, 3618, 2005.
[66] Soares R. M. D., Lima A. M. F., Oliveira R. V. B., Pires A. T. N., and Soldi V., *Polymer Degradation and Stability*, 90, 449, 2005.
[67] Basavaraju K. C., Demappa T., and Rai S. K., *Carbohydrate Polymers*, 69, 462, 2007.
[68] Mundargi R. C., Patil S. A., and Aminabhavi T. M., *Carbohydrate Polymers*, 69, 130, 2007.
[69] Lin-vien D., Colthup B. N., Ateley G. W., and Graseelli G. J. (EDS), in: The Handbook of Infrared and Raman Characteristics Frequencies of Organic Molecules, p. 227, Academic Press, U.S.A (1991).
[70] Pretsch E., Simon W., Seibl J., and Clerc T. (Eds), in: Tables of Spectral Data for Structure of Organic Compounds, p. I5-I265, Springer-Verlag, Germany (1989).
[71] Feng J., Zhao L., and Yu Q., *Biochemical and Biophysical Research Communications*, 31 (2), 414-420, 2004.
[72] Mori T., Irie Y., Nishimura S. I., Tokura S., Matsuura M., Okumura M., Kadosawa T., and Fujinaga T., *J. Biomed Mater Res. (Appl Biomater)*, 43 (4), 469-472, 1998.
[73] Chellat F., Grandjean-Laquerriere A., Le Naour R., Fernandes J., L'Hocine Yahia, Guenounou M., and Laurent-Maquind D., *Biomaterials*, 26 (9), 961-970, 2005.

In: Focus on Chitosan Research
Editors: Arthur N. Ferguson and Amy G. O'Neill
ISBN 978-1-61324-454-8

Chapter 12

A SPECTROSCOPIC STUDY OF THE INTERACTION OF METAL IONS WITH BIOMATERIALS

***Yassin Jeilani*[1], *Beatriz H. Cardelino*[2] and *Natarajan Ravi*[3,*]**

[1]. Environmental Science and Studies Program, Spelman College, Atlanta, Georgia
[2]. Chemistry Department, Spelman College, Atlanta, Georgia
[3]. Physics Department, Spelman College, Atlanta, Georgia

ABSTRACT

The fundamental electrical and magnetic interactions between iron ions and biomaterials were investigated using an experimental approach, ^{57}Fe Mössbauer spectroscopy, and ab initio computational methods. A conventional spin-Hamiltonian approach adopted for the data analysis of the Mössbauer data showed that the metal ion in the Fe-chitosan complex is in the high-spin ferric state and that it has an internal magnetic field of approximately 440 kG, at the nucleus. The magnitude of the internal field arises from the predominant Fermi-contact interaction of the high-spin ferric species with N/O ligands. Based on the analysis of the experimental data, a scheme for the Fe-chitosan complex has been proposed. Similar analyses of spectral data were performed for other biomaterials such as glucose, cellobiose, and glucosamine. Studies pertaining to metal ion interaction with monomers such as glucosamine, glucose, and chondritin sulfate indicate that the oxidation state of the metal ion does depend on the pH of the reaction medium. Stability of any oxidation state is attributed to the presence or absence of a glycosidic linkage between sugar units.

To further probe the geometry, energy, and details of bonding of these Fe complexes, density functional theory (DFT) computations were performed, using Becke's three-parameter functionals with Lee Yang-Parr's correlation correction (B3LYP), together with the largest suitable basis sets available in the Gaussian quantum mechanical program. Four hexa-coordinated iron compounds were studied: (a) Complex 1 - Fe(II)

* Corresponding author: email: nravi@spelman.edu.

with β-D-glucopyranose (glucose); (b) Complex 2 - Fe(II) with amino-2-deoxy-β-D-glucose chitosamine (glucosamine); (c) Complex 3 - Fe(II) with protonated glucosamine; and (d) Complex 4 - Fe(III) with protonated glucosamine. In all four cases, the iron atom was coordinated to oxygen atoms of two water molecules placed on an axial position. In addition, the iron atom was coordinated to two hydroxyl oxygen atoms of two glucose molecules in Complex 1, to two hydroxyl oxygen atoms of two protonated glucosamine molecules in Complexes 3 and 4, and with one hydroxyl oxygen and one amine nitrogen from two glucosamine molecules in Complex 2. In all four complexes, the two monosaccharides were rotated with respect to each other by 180°, both around the axis perpendicular to the molecular plane and around the molecular plane. Complex 1 was studied with low and high-spin electron configurations, whereas Complexes 2 and 3 were studied only with high-spin configurations. Predictions of Mössbauer chemical isomer shifts (δ_{Fe}) and electric quadrupole interaction (ΔE_Q) for these four complexes were obtained from standard curves based on experimental values.

INTRODUCTION

Biomolecules such as cellulose, glucose, chitosan, chitin, and chondritin sulfate form very stable complexes with a variety of metal ions, especially transition metals. These metal complexes have a wide range of applications, for example, in sewage system purification or in cancer research. Cellulose is a polymer of glucose with β(1→4) linkages ideal for making fibers since parallel polysaccharide chains can be readily linked by hydrogen bonds (Figure 1). Chitin, the second most abundant polysaccharide after cellulose, is the fibrous component of the exoskeleton of several arthropods: crabs, spiders, and insects [1,2]. This polymer, which resembles cellulose structurally and chemically, offers diverse applications due to the presence of an acetylamide group at one of the carbon atom positions. Chitosan, the N-deacetylated chitin, is similar to chitin but easier to work with due to its increased solubility in acetic acid [1]. The polar nature of the carbohydrates of these biopolymers, gives them the ability to chelate with metal ions. Such metal complexes structurally resemble the active sites of metalloenzymes due to the presence of O/N/S ligands [3]. The two polymers chitin and chitosan also act as substrates for the enzymes chitinase and chitosinase, respectively [4].

Chondritin sulfate, on the other hand, is an alternating copolymer of β(1→4)-D glucuronic acid and β(1→3)–N acetyl-D-galactosamine, which can be sulfated at C^4 and C^6 positions (labels for carbon positions shown in Figure 1). The presence of carboxyl (COOH) and sulfate (SO_4^{2-}) groups makes this polymer highly soluble in water. Chondritin forms networks with collagen in connective tissues and allows the transfer of globular proteins, stimulates the growth and differentiation of neurons, and has also been implicated in spinal cord injuries [2,5]. These polymers have been found to possess an excellent ability to chelate an array of transition metal ions, even though chitosan appears to be a better chelating agent [3]. This could be due to the presence of the amine (NH_2) group in chitosan.

The structural properties of transition metal complexes of chitin and chitosan have been studied to try to understand the mechanism of enzyme chelation [6,7]. Different models for the metal complexes of chitin and chitosan have been proposed. In the "pendant model" the single metal ion is bonded to an NH_2 group of chitosan, while in "the bridge or chelating model" the metal ions are proposed to be coordinated to several amino groups in the same or in different chitosan polymer chains [8]. Studies of Cu and Fe metal complexes have indicated that both NH_2 and hydroxyl (OH) groups may be bonded to the metal ions, and that more than one polymer chain may be involved in the formation of the complexes [9,10]. However, the way the metal binds to the carbohydrate matrix is still not clearly understood and hence requires further investigation.

Elucidation of the electronic and magnetic properties of a metal ion in metal complexes, metalloenzymes and synthetic analogues, have been pursued by probing the metal site with a variety of spectroscopic techniques [11]. Some of the magnetic resonance techniques, including Nuclear Magnetic Resonance (NMR), Electron Paramagnetic Resonance (EPR), Nuclear Quadrupole Resonance (NQR), and Mössbauer effect, are widely used [12]. For example, the Mössbauer effect has played a decisive role in understanding the structure-function relationship of several Fe containing redox enzymes such as hemoglobin, cytochromes, cytochrome *c* oxidase, ribonucleotide reductase, hemerythrin, nitrogenase, hydrogenase, and ferredoxin [13]. The experimental studies have often been intertwined with quantum mechanics or other simulations to model the experimental behavior [13].

In order to understand the nature of bonding, geometry, and magnetic characteristics of the Fe complexes with biopolymer materials, we have undertaken a Mössbauer study, since the technique is sensitive to hyperfine interactions of the order of 10^{-2} cm^{-1}. Hartree-Fock and density functional theory (DFT) methods were the theoretical approaches selected for predicting the geometries and energies of the metal complexes, as they have been implemented in the Gaussian 09W quantum mechanical program [14,15,16]. In this manuscript, we review ^{57}Fe Mössbauer findings and computational results of Fe complexes with several biomaterials.

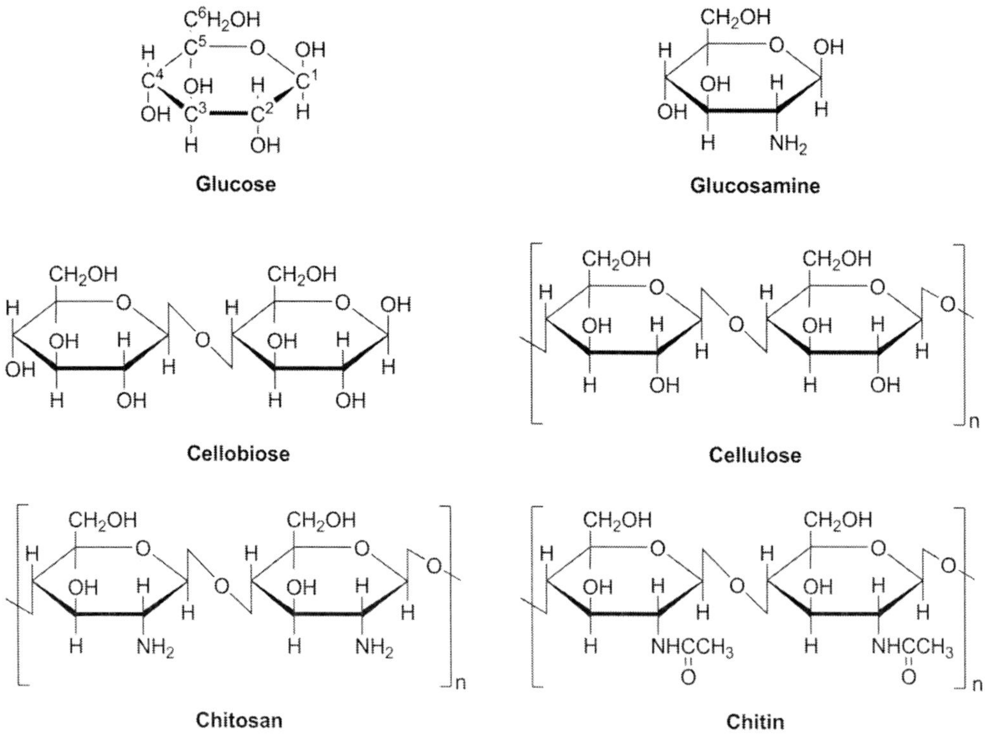

Figure 1. Chemical structures of glucose, glucosamine, cellobiose, cellulose, chitosan, and chitin.

Experimental Methods

Chitosan decamer (Wako, min. 80% deacetylation, molecular weight 1612 Da, and water content max. 10%) and polymer (Sigma chemicals 99.9% purity level, min. 85% deacetylation, molecular weight range: 310 - 375 kDa) were used without further purification. 0.2 g of chitosan decamer or polymer and anhydrous ferric chloride were mixed in the stoichiometric ratio in 200 ml of deionized water. Preparation of the Fe-complexes with the polymer, water-soluble polymer, and the monomeric forms of chitin and chitosan were undertaken in an acetic acid medium for the polymers and in aqueous medium for the monomer and chondritin sulfate. The chemicals glucosamine, N-acetyl glucosamine, gluconic acid, sulfonated N-acetyl glucosamine, chitin, chitosan, and chondritin sulfate were purchased from WAKO Bioproducts. In one set of experiments, a known amount of a given biopolymer was dissolved in acetic acid. The pH of the solution was adjusted to the desired level and an appropriate amount of the chloride salt of the metal ion was added to the solution. The mixture was stirred for approximately two hours, filtered, and was dried at 50^0C for approximately 12 hours. If no precipitate was formed at room temperature then the water was removed by warming the sample at approximately 50^0C until the solid was formed.

A Mössbauer sample was prepared in the solid state corresponding to a thickness of ~2 mg/cm^2 of ^{57}Fe in a shallow derlin sample cup with an inner cup sealed and frozen with liquid nitrogen. The Mössbauer spectrometer (Ranger model MS 1200, Ranger drive VA900) was used operating in a constant acceleration mode in a transmission geometry. For low temperature measurements, a Cryo Industries 8CN variable-temperature cryostat model was used with liquid He and the temperatures were measured by the calibrated silicon diode, sensitive to different temperature regions. The zero velocity of the Mössbauer spectra referred to the centroid of the room-temperature spectrum of a metallic Fe foil. The spectra were analyzed using the software WMOSS (WEB Research Co., Edina, MN).

Computational Methods

The calculations were performed using the Hartree-Fock method [14] and density functional theory (DFT) [15], as implemented in the Gaussian 09W quantum mechanical program [16]. B3LYP hybrid density functionals [17] were used for the DFT calculations. In the case of open-shell configurations, unrestricted Hartree-Fock and unrestricted B3LYP methods were employed. The following basis sets provided by Gaussian 09W were selected for the study: STO-3G; 6-311G [18]; 6-311G [19] with p-polarization functions for H, d polarization functions for C and O, and f polarization functions for Fe [20] [i.e., the 6-311G(f,d,p) basis sets]; and the configuration-consistent triple-zeta (cc-pVTz) [21] basis set on Fe. Hartree Fock calculations were performed using STO-3G basis sets, and DFT calculations using 6-311G, 6-311G(f,d,p) and cc-pVTz basis sets. The total number of basis functions and primitive-Gaussian functions for the four types of atoms of the study are shown in Table 1.

Table 1. Size of the basis sets. Labels: #BF = number of basis functions; #PF = number of primitive functions

Atom	STO-3G		6-311G		6-311G(f,d,p)		cc-pVTz	
	#BF	#PF	#BF	#PF	#BF	#PF	#BF	#PF
H	1	3	3	5	6	8		
C	5	15	13	26	18	32		
O	5	15	13	26	18	32		
Fe	18	57	39	71	46	81	68	351

Table 2. Typical Mössbauer parameters for Fe compounds in different oxidation and spin state

	Fe(II)		Fe(III)		Fe(IV)	
	low spin S=0	high-spin S=2	low-spin S= ½	high-spin S= 3/2	low-spin S= 1	high-spin S=2
Isomer Shift (δ_{Fe})	0.4-0.5	1.0-1.4	0.1-0.3	0.4-0.6	0-0.1	0.1-0.3
Quadrupole Splitting (ΔE_Q)	0.7-1.2	2.0-4.0	2.0-3.0	0.5-1.0	1.0	1.0

MÖSSBAUER SPECTROSCOPY

Mössbauer spectroscopy has played a very significant role in the structural elucidation of many Fe-containing metalloenzymes and the technique probes the metal sites directly to obtain information on the oxidation state, the geometry around the metal ion, and its magnetic properties. The Mössbauer parameter isomer-shift (δ_{Fe}), an oxidation state indicator, clearly identifies the number and the nature of ligands that surround the metal ion. For example, an Fe(II) and Fe(III) can be distinguished since they have characteristic values. In addition, Fe(III) in a tetrahedral sulfur environment can be distinguished from an octahedral O/N ligation as the isomer-shifts are distinctly different for these two situations. Another useful and important parameter is the quadrupole splitting (ΔE_Q) which gives a wealth of information regarding the extent of distortion of the coordination sphere. For example, an Fe(III) low-spin state is characterized by a large splitting value, while an Fe(III) high-spin state has a smaller splitting value. In addition, the magnetic properties arising from the presence of unpaired electrons on the transition metal ion offer insight into the nature of metal-metal interactions and illustrate spin-coupling mechanisms that may be present in the Fe-system under study. However, one should bear in mind that these parameters extracted from the Mössbauer spectrum depend on the experimental conditions such as temperature, and the strength of the applied magnetic field [16]. The range of isomer-shift and the quadrupole splitting values obtained, in general, for different Fe-species are listed in Table 2.

Interpretation and Analysis of the Mössbauer Data

It is important to know that the Mössbauer phenomenon rests on the fact that γ radiation can be emitted or absorbed without imparting recoil energy ($E_R = E_\gamma^2/2Mc^2 = 1.95 \times 10^{-3}$ eV, where E_γ = 14.4 keV and M is the mass of the ^{57}Fe nucleus). In a solid, most of the recoil energy is converted into lattice vibration energy. Mössbauer has shown, however, that there is a certain probability, described by the recoil-free fraction f, that γ emission and absorption take place in solids without recoil. In order to observe the resonance effect, the ^{57}Fe nucleus must be placed in a solid or frozen in a solution matrix. ^{57}Fe is a stable isotope with 2.2% natural abundance. In ^{57}Fe Mössbauer spectroscopy, transitions between the nuclear ground state of ^{57}Fe (nuclear spin I_g = ½; nuclear g-factor g_g = 0.181) and a nuclear excited state at 14.4 keV (I_e = 3/2, g_e = -0.106, nuclear quadrupole moment Q) are observed. In the principal axis system of the electric-field-gradient (EFG) tensor, the off-diagonal elements vanish and, since the EFG tensor is traceless, only two independent components need to be specified. By convention, these parameters are V_{zz} and the symmetry parameter η, defined by $\eta = (|V_{xx}| - |V_{yy}|)/|V_{zz}|$. By choosing the coordinate system such that $|V_{zz}| \geq |V_{yy}| \geq |V_{xx}|$, the asymmetry parameter can be restricted to $0 \leq \eta \leq 1$. The conventional Hamiltonian of the quadrupole interaction for the ^{57}Fe nuclear excited state is

$$H_Q = \frac{eQV_{zz}}{12}[I_z^2 - I_e(I_e+1) + \eta(I_x^2 - I_y^2)] \qquad (1)$$

where e is the electron charge, and I_z, I_x, I_y are spin operators. The interaction due to H_Q splits the nuclear excited state into two degenerate doublets. For $\eta = 0$ these doublets are labeled by the magnetic quantum numbers $\pm^3/_2$ and $\pm^1/_2$; for $V_{zz}>0$ the $^3/_2$ levels have the higher energy. The two doublets are separated in energy by the quadrupole splitting

$$\Delta E_Q = \frac{eQV_{zz}}{2}\sqrt{1+\frac{\eta^2}{3}} \qquad (2)$$

The nuclear ground and excited states of the ^{57}Fe nucleus has magnetic moments which can interact with a magnetic field H. The interaction is described by

$$H = -\mu \bullet H = -g_n \beta_n H \bullet I \qquad (3)$$

where g_n is the nuclear g-factor and β_n the nuclear magneton, and I the nuclear spin. In the absence of quadrupolar interactions, the Hamiltonian splits the nuclear states into equally spaced levels of energy $E(m) = -g_n\beta_n H(m)$. The quadrupolar and magnetic interactions are normally combined. A conventional approach followed in spectroscopic investigations of metalloenzymes is to adopt a spin-Hamiltonian formalism for obtaining the fine and hyperfine parameters [42,43]. This approach appears to offer information pertinent to the electronic and magnetic properties of the metal sites involved in the redox process. This information is vital for understanding the redox states and the characteristics of the metal sites. The general Hamiltonian for such a situation is given by

$$H_e = g_e\beta_e\vec{H}\bullet\vec{S} + D[S_z^2 - S(S+1) + \frac{E}{2D}(S_+^2 + S_-^2)] - g_n\beta_n\vec{H}\bullet\vec{I} + \vec{I}\bullet\tilde{A}\bullet\vec{S} + H_Q \quad (4)$$

where H_Q is the term due to quadrupole interaction; g_e the magnetic moment of the electron; β_e the electron Bohr magneton; $\vec{H}$ the applied magnetic field; $\vec{S}$ the spin of the system; S_z the component of the spin along the z-direction of the reference frame; S_+ and S_- the step-up and step-down operators related to the x and y components of the spin by $S_x = \frac{1}{2}(S_+ + S_-)$ and $S_y = \frac{1}{2i}(S_+ - S_-)$; D and E the axial and rhombic zero-field splitting parameters respectively; g_n and β_n the nuclear g-factor and nuclear Bohr magneton respectively; $\vec{I}$ the nuclear spin; and $\tilde{A}$ the hyperfine coupling tensor [22]. This Hamiltonian is widely used for most spin-resonance techniques, such as Electron Paramagnetic Resonance, Magnetic Circular Dichroism, and Mössbauer Spectroscopy [23].

Results and Discussion

The following complexes have been studied by Mössbauer spectroscopy: Fe with glucose, Fe with cellobiose, Fe with glucosamine, Fe with chitosan, and Fe with chitin. The following complexes have been investigated using computational methods: Fe(II) with glucose, Fe(II) with glucosamine, Fe(II) with protonated glucosamine, and Fe(III) with protonated glucosamine.

I. Experimental Results

(A) Fe-glucose and Fe-cellobiose

Cellobiose is the dimer of glucose and the Mössbauer spectra of Fe-glucose and Fe-cellobiose, measured at room temperature, show a simple quadrupole doublet with an isomer shift δ_{Fe} and a quadrupole splitting ΔE_Q of 1.30 mm/s and 2.90 mm/s, respectively. Although the Mössbauer parameters are virtually identical for both complexes, the line width of the Fe-cellobiose complex appears to be broader than that of the Fe-glucose complex. However, both systems exhibit Mössbauer parameters typical of high-spin Fe(II). It is not uncommon to have a large electric field gradient at the Fe nucleus, especially for an Fe(II) high-spin state because of its d^6 electron configuration (see table 2 for the range of values). The presence of such a combination of large δ_{Fe} and ΔE_Q values unambiguously identifies the metal ion to be in a high-spin ferrous state [^{5}D state of Fe(II)]. A least-squares fit of the spectral data results on a line-width Γ (full-width at half maximum – FWHM) of nearly 0.40 mm/s, a value somewhat larger than the natural line-width (0.25 mm/s), suggesting a plausible second species with a slight different ligation but with the same metal oxidation and spin state [24].

(B) Fe-glucosamine

The Mössbauer spectra of the Fe complexes with water soluble chitosan and with D-glucosamine monomer were measured at 4.2 K and are shown in parts A and B of Figure 2, respectively. A visual examination of the spectrum of the Fe complex with water-soluble chitosan clearly indicates that the spectrum is composed of two components, a magnetic part and a quadrupole doublet, whereas the spectrum of the Fe complex with monomer glucosamine shows a pure quadrupole doublet. Comparison of the line positions of the magnetic spectrum of the Fe-water-soluble chitosan complex with earlier reported data for the Fe(III)-water-insoluble chitosan complex reveals that the magnetic component of the spectra are virtually identical and it accounts for about 35 ± 5% of the total intensity of the spectrum. After removal of this component, the remaining central doublet yields a set of Mössbauer parameters δ_{Fe} = 1.39 mm/s and ΔE_Q = 2.89 mm/s. The spectral data of the Fe-glucosamine-monomer complex, on the other hand, shows a pure quadrupole doublet with an apparent isomer shift δ_{Fe} and quadrupole splitting ΔE_Q of 1.37 and 3.01 mm/s, respectively. The parameters obtained from the least-squares fit of the experimental data of the Fe-glucosamine-monomer complex unequivocally indicate the presence of high-spin Fe(II). On the basis of an analysis of the low-temperature spectral data, it can be concluded that a mixture of Fe(II) and Fe(III) ions is formed when mixing the Fe(II) salt with water-soluble chitosan, while the D-glucosamine complex showed only the presence of Fe(II). It should be pointed out that, in both samples, the Mössbauer parameters are identical within the experimental uncertainties [25,26,27].

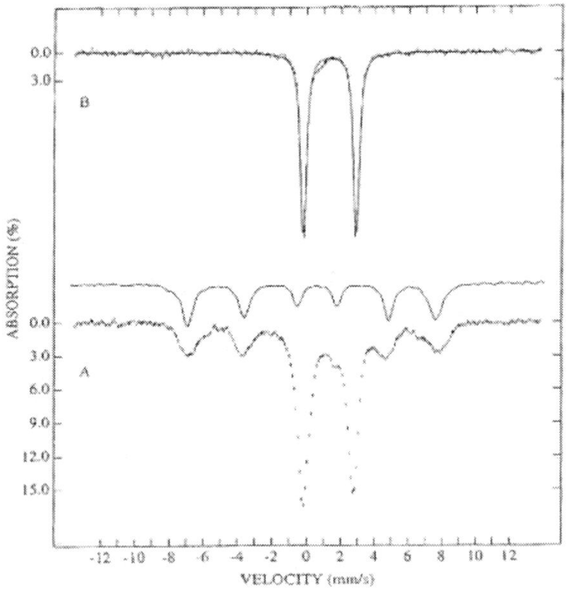

Figure 2. 4.2 K Mössbauer spectra of crystalline powder of (A) Fe-water-soluble chitosan sample. The solid line plotted above spectrum A is the experimental data of the Fe-chitosan polymer complex scaled to 30% of absorption. (B) Fe-glucosamine-monomer sample.

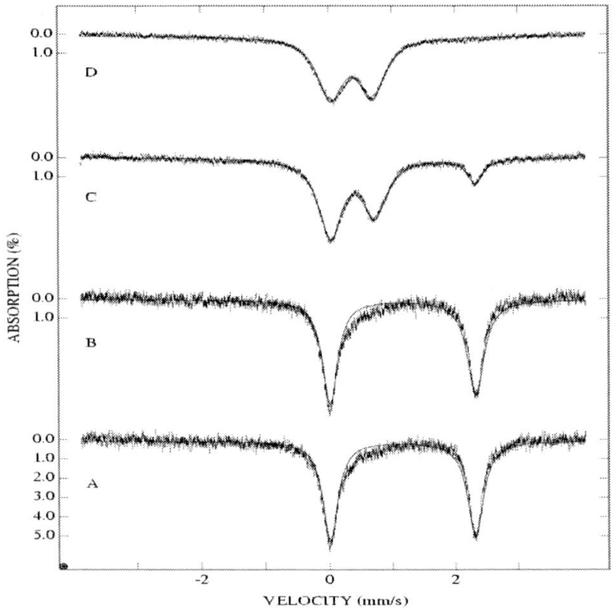

Figure 3. Mössbauer spectra of crystalline powder of Fe-glucosamine samples prepared at different pH values, measured at 293 K. pH values: (A) 1.6; (B) 4.8; (C) 9.2; (D) 11.5.

To further probe the structural details, Fe complexes of D-glucosamine were prepared at various pH values. Figure 3 shows the Mössbauer spectra of these complexes, measured at 293 K, and at various pH values. These spectra clearly show a quadrupole doublet. Furthermore, the Mössbauer parameters correspond to Fe(II), Fe(III), or Fe(II)/Fe(III), depending on the pH of the reaction mixture. At low pH (1.5-4.8) a clear Fe(II) quadrupole doublet is seen, while at high pH (9.2), a mixture of Fe(II) and Fe(III) species is observed. A further increase of the pH to 11.5 leads to a sole Fe(III) species. The isomer shift, quadrupole splitting, line width, and the corresponding absorption areas determined by a least-squares fit of the spectral data, indicate the presence of both Fe(II) and Fe (III) states, depending upon the pH of the reaction mixture. It is interesting to note that a reaction between D-glucosamine and Fe(III) chloride (in an acidic pH range) does not yield an Fe(III) species but shows an Fe(II) species as observed in the Mössbauer spectrum with δ_{Fe} = 1.35 mm/s and ΔE_Q = 3.01 mm/s. This observation is in conformity with the fact that glucosamine acts as a reducing agent. Therefore, it can be concluded that, regardless of the oxidation state of the precursor Fe material, one often ends up with an Fe(II) state if the reaction is maintained in an acidic pH range where glucosamine acts as a reducing agent.

This important observation should be compared with the fact that a reaction between an iron salt and water-insoluble chitosan always yields an Fe(III)-containing Fe-chitosan complex. This suggests that the coordination properties and the redox potential of the iron sites in complexes with different polymerization degree are the more likely factors that directly influence the oxidation state of the metal ion.

A working structural model for the Fe-chitosan complex has been reported [25] based on the results obtained from the investigations. While the data suggested that there is at least five ligands around Fe in the +3 state, a four coordinated Fe(II) state could not be completely ruled out. Involvement of the amide nitrogen in the bonding with the metal ion was also proposed. A plausible explanation for the stabilization of the respective oxidation states was offered using simple carbohydrate chemistry, where the presence of β(1-4) glycosidic linkage could be invoked as a possible mechanism.

(C) Fe-chitosan and Fe-chitin

The Mössbauer spectra of Fe complexes with chitosan (decamer and polymer), measured at 4.2 K, are shown in Figure 4. As the two spectra are virtually identical and superimposable, the results presented here are applicable to both systems. The spectrum is reminiscent of a classic six line hyperfine signature, with line positions at -6.76, -3.51, -0.51, +1.74, +4.72, and +7.47 mm/s. The Mössbauer spectra measured at 4.2 K, with a magnetic field of 50 mT applied parallel and perpendicular (data not shown) to the incoming γ-rays, are identical, indicating the uniaxial nature of this magnetic system. Any uniaxial system can be characterized by a six-line Mössbauer spectrum with the intensity ratio of 3:2:1:1:2:3 for the six lines respectively [28].

The least-squares fit of the experimental data, the line width and the shape of each peak clearly indicate the presence of a second species amounting to approximately 10% of the total absorption intensity of the spectrum, which we are currently attributing to an adventitiously bound Fe or the metal ion bonded to the acetylated part of the chitosan (~ max. 20%). However, this spectrum has a magnetic splitting almost identical to the metal ion bonded to the deacetylated part of the chitosan (~ 80%) species and is characteristic of high-spin Fe(III) species. From the line positions obtained by a least-squares fit of the experimental data of the predominant species, and by averaging the line positions, an isomer-shift δ_{Fe} of 0.49 mm/s and an internal magnetic field at the nucleus of 441 kG (H_{int}) were deduced. The quadrupole splitting ΔE_Q can only be indirectly found from this spectrum. The formal relationship between ε and ΔE_Q is

$$\varepsilon = \frac{e^2 qQ}{4} \frac{(3\cos^2\theta - 1)}{2} \tag{5}$$

where ε is the energy difference between the lines 1, 2 (Δ_{12}) and 5, 6 (Δ_{56}), and θ is the angle between the principal component of the EFG tensor and the direction of the internal magnetic field [28]. By comparing to the quadrupole splitting obtained from the high-temperature spectral data presented below, it can be inferred that the principal component of the EFG tensor and the direction of the magnetic field are neither perpendicular nor parallel to each other.

These parameters, particularly the isomer-shift δ_{Fe}, are typical of the Fe(III) high-spin state with five or six O/N ligands [29]. An Fe(III) high-spin consists of 5 unpaired electrons (d^5), giving rise to a contribution of 110 kG/unpaired electron. The internal magnetic field in this case is solely due to Fermi-contact interaction, since the orbital angular momentum L is equal to 0. It should be pointed out that the measured internal magnetic field at the Fe nucleus is significantly smaller than any high-spin Fe(III) ion [30]. The difference can only be attributed to the large extent of covalency between the metal and the surrounding ligands. The uniaxial property of the system can be used to parameterize the hyperfine interactions by using a spin-Hamiltonian approach. Using this Hamiltonian, an attempt to simulate the 4.2 K,

50 mT, spectrum yields the parameters $\Delta E_Q = -0.70$ mm/s , $\delta_{Fe} = 0.54$ mm/s, $\eta = -5.0$, hyperfine coupling tensor elements $A_{x,y,z} = -179$ kG, $\Gamma = 0.38$ mm/s, $g_e = 2.0$, $D = -3.0$ cm^{-1}, $E/D = 0.23$, confirming the Fe(III) high-spin state of the Fe-chitosan complex.

An analysis of the Mössbauer spectra of the Fe-chitosan complex, measured at different temperatures to extract the isomer shift δ_{Fe} at different temperatures, unambiguously establishes the Fe(III) high-spin state of the complex and supports the validity of the procedure adopted for analyzing the 4.2 K hyperfine spectrum. Our current data show a transition from a magnetic to a paramagnetic phase around 100 K and does not indicate any other structural phase transitions in the temperature range 100 to 300 K. The trend in the isomer shift δ_{Fe} *vs* temperature is in conformity with expected behavior of a Debye solid. In general, the temperature dependence of the isomer shift δ_{Fe} is due to the second-order Doppler shift and theoretically it is found to be linear if the forces coupling the atoms are assumed to be harmonic. The slope of such a linear plot is given by

$$\frac{d\delta_{Fe}(T)}{dT} = -\frac{3\,k\,E_\gamma}{2\,M\,c^2} \qquad (6)$$

where c is the velocity of light, M is the atomic mass of the Mössbauer nucleus, k is the Boltzmann constant, and E_γ is the γ-ray energy, which in this case is 14.4 keV [18]. The slope given by the above equation can be calculated to be -7.31×10^{-4} mm/s.K for ^{57}Fe atoms. A plot of isomer shift δ_{Fe} *vs* temperature for the Fe-chitosan system shows a linear relationship throughout the temperature range studied, with a slope of 6.0×10^{-3} mm/s.K, in reasonable agreement with the predicted value. This suggests that the assumptions under which Equation (6) is derived are valid, i.e., that the forces coupling the atoms in the metal complex are harmonic to a good approximation, and suggests the Debye nature of the solid.

From all the experimental observations and analysis of the data, we proposed a working scheme for the Fe-binding site of chitosan [25], as depicted below in Figure 5.

A similar metal-chitosan complex study has been reported with Cu [31]. It is worthwhile comparing the reported structure with the present results. The proposed structure for the Cu-chitosan complex indicates that Cu is likely to bond to three oxygen atoms and one nitrogen ligand, in a square-planar or a tetrahedral geometry [31]. In this proposed structure, two bonded oxygen atoms and the nitrogen atom are believed to emerge from the monosaccharide group and two such groups are involved in forming the metal coordination sphere. Monterio and Airoldi [31] also argue against another plausible structure in which one bonded oxygen atom and one nitrogen atom arise from the monosaccharide group, while the other two oxygen atoms stem from the water molecules of hydration [31]. From the amino groups contained in chitosan, the amount of water, and the Fe present in the complex, it has been concluded that for each Fe(III) ion there are 2 moles of amino groups and four moles of oxygen. On the basis of this stoichiometry, the structure $[Fe(H_2O)_{4-x}(Glu)_2Cl_x]Cl_{3-x}xH_2O$ has been proposed [4], where Glu represents glucosamine.

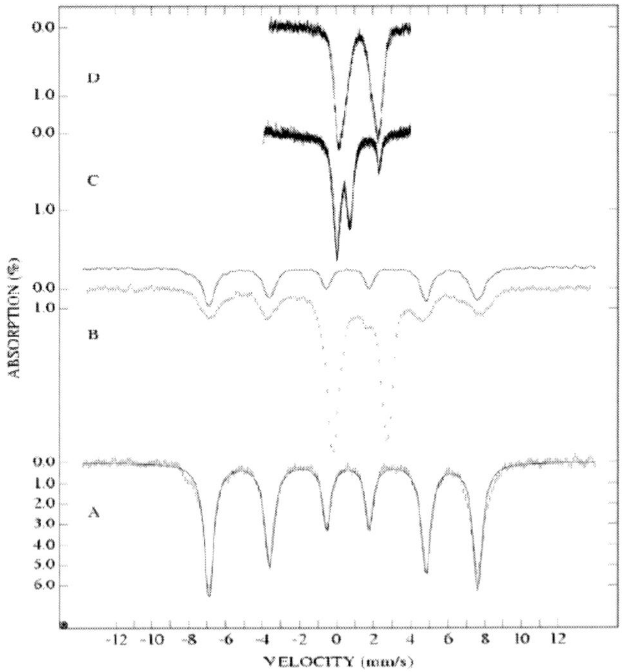

Figure 4. Mössbauer spectra of crystalline powder of (A) Fe-chitosan sample measured at 4.2 K. (B) Fe-water-soluble-chitosan sample measured at 4.2 K. The solid line plotted above spectrum B is the experimental data of the Fe-chitosan-polymer complex, scaled to 30% of absorption. (C) Fe-glucosamine (monomer) at pH 9.2, measured at 293 K, and (D) Fe-chondritin-sulfate complex at pH 4.0 at 293 K.

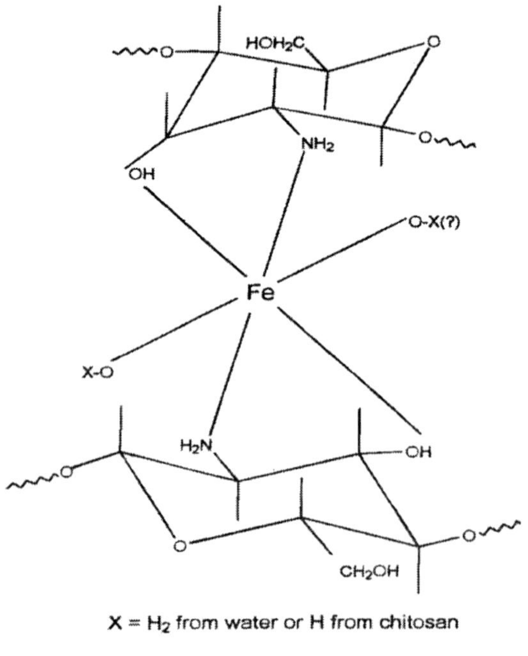

Figure 5. A proposed schematic representation of the Fe-chitosan complex.

Our current study clearly indicates that the Fe-chitosan complex has either a penta- or a hexa-coordinated Fe(III), and it is tempting to speculate that there is at least one molecule of water arising from hydration, and that the remaining N/O ligands are part of two saccharide units of chitosan. This suggests that Fe(III) is indeed coordinated to an amino group (NH_2) of chitosan. The same conclusion can be arrived by comparing the absorptions in the 1500-1700 cm^{-1} region, which are assigned to the bending mode of $-NH_2$ in chitosan [4a]. The shift of ~200 cm^{-1} for -OH group absorption upon coordination of the Fe(III) to chitosan indicates that Fe(III) is also complexing with -OH groups. Thus, the infrared data indeed lends support and corroborates the proposed working scheme.

II. Computational Results

Low and high-spin complexes (singlet and quintet multiplicities, respectively) of Fe(II) were studied, with β-D-glucopyranose (glucose) (Complex 1), 2-amino-2-deoxy-β-D-glucose chitosamine (glucosamine) (Complex 2), and protonated glucosamine (Complex 3). A high-spin (hextet multiplicity) was calculated for the Fe(III) complex with protonated glucosamine (Complex 4). Based on Pauling's theory that metal complexes approximate the electronic configuration of noble gases, Fe(II) and Fe(III), with six and five d electrons, respectively, would tend to form hexa-coordinated complexes and thus achieve 6 + 12 electrons (18 electrons) or 5+12 (17 electrons), respectively. The nature of the ligands and the oxidation state of the metal contribute to the crystal-field splitting (Δ_{oct}) of the d orbitals of the transition metal. When Δ_{oct} is large, the complexes have low spin and when it is small, high spin is preferred. Low spin Fe(II) has no unpaired electrons and Fe(III) has one unpaired electron; high spin Fe(II) has four unpaired electrons and Fe(III) has five unpaired electrons [32].

The molecular structures were optimized by minimizing the electronic energy. Computations of the normal vibrational frequencies were carried out to corroborate that stable structures were achieved. Since the model chemistries selected were Hartree-Fock and DFT methods, the force constants for the vibrational analysis were computed analytically. Natural bond orbital analyses (NBO) [33] were performed and the results were compared to Mulliken populations [34]. The nuclear magnetic resonance (NMR) shielding tensors were also computed with the gauge-invariant atomic orbital (GIAO) method [35]. The GaussianView [36] computer program was used as the graphical user interface.

Three types of nuclear interactions are important in Mössbauer spectroscopy: (1) the chemical isomer chemical shift (δ_{Fe}); (2) the electric quadrupole splitting (ΔE_Q); and (3) the hyperfine (or magnetic) splitting [37,38,39]. The theory behind the spectroscopic technique and the relevant parameters have been previously described in the Mösbauer spectroscopy section. In this study, predictions only of the first two parameters were attempted, based on standard curves obtained from experimental values. In previous studies [40], good correlations were obtained between experimental Mössbauer chemical shifts (δ_{Fe}) and various ways of calculating effective charge on Fe. In particular, Sadoc et al [40 h], correlated δ_{Fe} values with the difference between formal ionic charge and charge calculated using localized atomic orbitals; in that study, charge transfer contribution from the ligands was also considered. In the present study, experimental δ_{Fe} values for five high-spin hexa-coordinated complexes of Fe(II) and Fe(III) were correlated to natural orbital contributions from: (a) the s electrons on Fe, (b) the p and d electrons on Fe, and (c) the sum of the valence p electrons on the ligands multiplied by the formal charge on Fe. Multiplying the ligand p electrons by the formal charge on Fe weighed differently the attraction that the Fe atoms in the compounds have on the ligand p electrons. In a way, performing a multiple linear regression on separate terms allowed for weighing differently the various types of electron density contributions. No relativistic correction was performed on the calculations of electron density; this was the case, both for the molecules used to generate the standard curve, as well as for the molecules for which values of δ_{Fe} were predicted.

The standard curve for predicting electric quadrupole splitting (ΔE_Q) was obtained from experimental values as a function of the EFG, as expressed in Equation 2. The experimental values chosen were all for high-spin, hexa-coordinated, neutral molecules, extrapolated to 0K.

Input structures. The initial structures contained two β-D-glucopyranose (glucose) monomers (Complex 1), two 2-amino-2-deoxy-β-D-glucose chitosamine (glucosamine) monomers (Complex 2), or two protonated glucosamine monomers (Complex 3 and 4). Glucose and glucosamine are shown in Figure 6. The glucose figure displays the hydroxyl group at C^1 cis to the CH_2OH group attached to C^6 for the β anomer. The O^6-C^6-C^5-O^5 and O^6-C^6-C^5-C^4 torsion angles shown are 60° and 180°, respectively, making them gauche-trans rotamers. The glucosamine figure shows the amine group attached to C^2. For the complex of protonated glucosamine monomers, an additional H atom was attached to N^2; thus, the complex consisted of protonated glucosamine monomer with charge of +1.

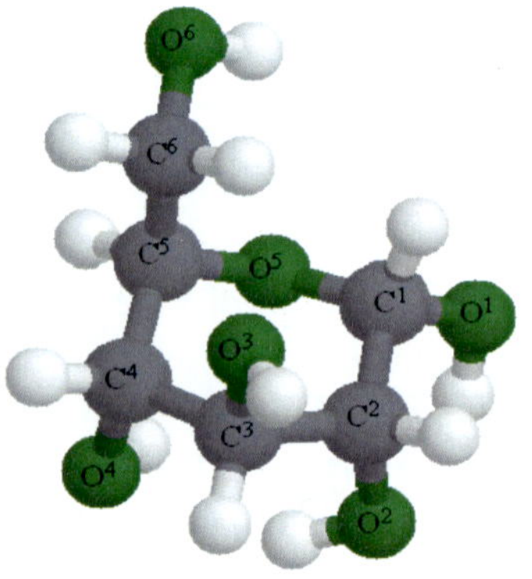

a.

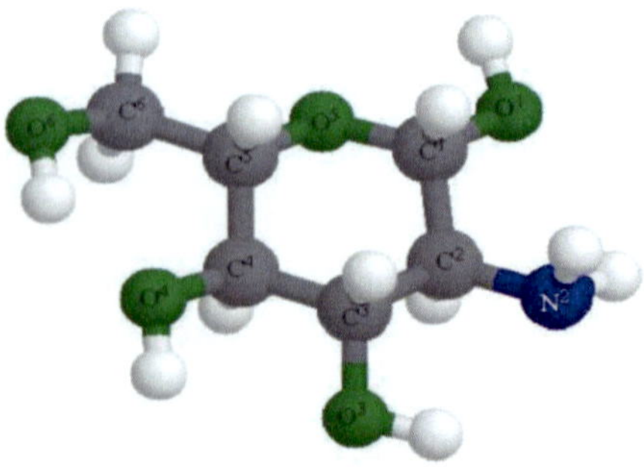

b.

Figure 6. Structures of (a) β-D-glucopyranose (glucose) monomers and (b) β-D-glucose chitosamine (glucosamine) monomers. Green represents the O atoms, gray the C atoms, white the H atoms, and blue the N atom.

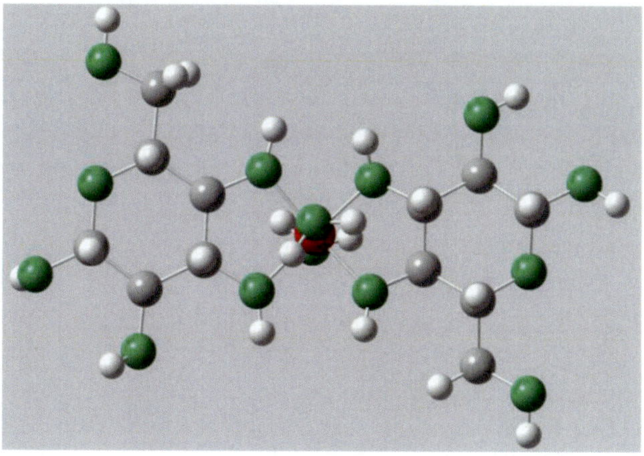

a.

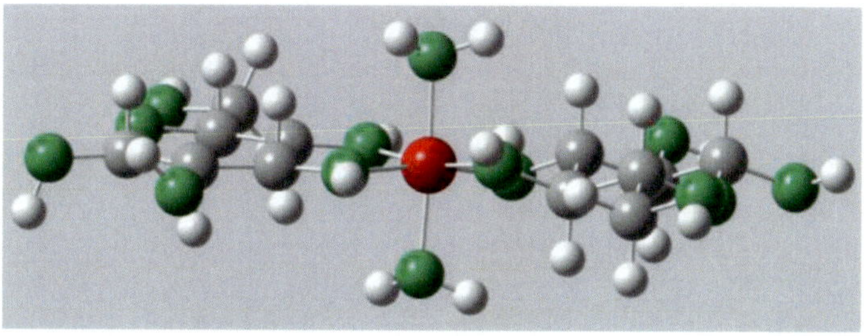

b.

Figure 7. Initial structure of the Fe(II) complex, hexa-coordinated to O^2 and O^3 of two glucose molecules and two H_2O molecules (Complex 1). Green represents the O atoms, gray the C atoms, white the H atoms, and red the Fe atom. (a) top view; (b) side view.

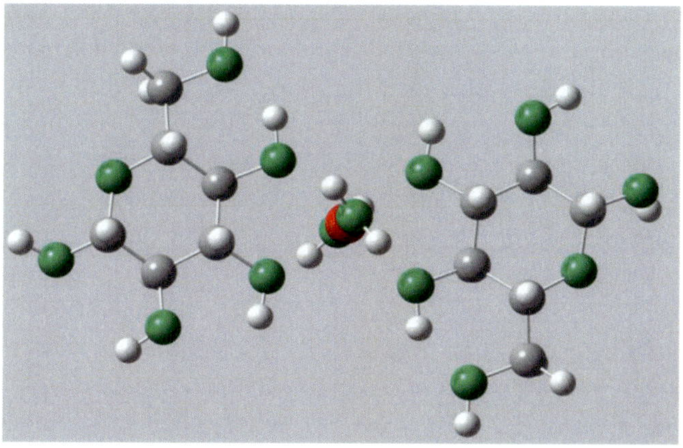

Figure 8. Top view of the high-spin Fe(II) complex with two β-D-glucopyranose (glucose) molecules and two H_2O molecules (Complex 1). Green represents the O atoms, gray the C atoms, white the H atoms, and red the Fe atom.

In the complexes, the two monosaccharide molecules were rotated by 180° with respect to each other, both with respect to an axis perpendicular to the main molecular plane and with respect to the molecular plane. In the Fe(II) complexes with glucose and glucosamine, the octahedral Fe atoms were hexacoordinated to the O^3 atoms of the two monosaccharides, to the O atoms of two water molecules placed in the axial position, and to the O^4 atoms of glucose or the N_4 atoms of glucosamine (Complex 1 and 2, respectively). In the Fe (II) and Fe(III) complexes with protonated glucosamine (Complex 3 and 4, respectively), the Fe atoms were coordinated to O^2 and O^3, in addition to the O atoms of two water molecules placed in the axial position. Figure 7 displays the Fe(II) complex with two glucose molecules and two water molecules (Complex 1).

Final structures. As an example, Figure 8 displays a top view of the final structure of the high spin Fe(II) complex with two glucose molecules and two water molecules (Complex 1), as calculated using B3LYP with 6-311G basis sets. The two water molecules have their H atoms staggered with respect to each other. The two O-Fe distances for the water molecules are both 2.02 Å. The four hydroxyl groups that complex with the Fe have their H atoms away from the Fe atom. The two O^3-Fe distances are 2.16 Å, and the two O^4-Fe distances are 2.14 Å.

Binding energies. Table 3 summarizes the 0K energies for the Fe(II) complex with glucose (Complex 1), Fe(II) complex with glucosamine (Complex 2), Fe(II) complex with protonated glucosamine (Complex 3), and Fe(III) complex with protonated glucosamine (Complex 4). In all cases, the iron atom coordinated with two monosaccharide monomers and two molecules of water. The calculations were done at the Hartree-Fock level with STO-3G basis sets, and at the B3LYP level with 6-311G basis sets or with 6-311G(f,d,p) basis sets. High- and low-spin calculations (quintet and singlet multiplicities, respectively) were performed on Complex 1; only high-spin calculations were performed on Complexes 2 and 3 (quintet multiplicity), and on Complex 4 (hextet multiplicity). The binding energies were obtained from the 0K electronic energy of the following reaction:

$$\text{Fe(II)} + 2 \text{ sugar molecules} + 2 \text{ water molecules} \rightarrow \text{Complex} \quad (7)$$

As can be seen from Table 3: (a) The complexation reactions are exergonic, with binding energies decreasing with increasing level of calculation and with increasing size of basis set. (b) The energy of the quintet is lower than the energy of the singlet, both for the free Fe(II) atoms as well as for the glucose complex (Complex 1), at all levels of calculation. For the DFT calculations, the energy of the quintet is lower by about 130 kJ mol^{-1} for the complex and by 380 kJ mol^{-1} for the free Fe(II) atom, for both 6-311G and 6-311G(f,d,p) basis sets. (c) The DFT calculations on the glucose complex (Complex 1), shows binding energies for the singlet of 2.0 and 1.8 MJ mol^{-1} for the 6-311G and 6-311G(f,d,p) basis sets, respectively. The binding energies for the quintet are 1.7 and 1.6 MJ mol^{-1} for the two different basis sets, respectively. The binding energies for the two DFT calculations are within 7% for the singlet and 8% for the quintet. (d) In the case of the glucosamine complex (Complex 2), the binding energies of the quintet at the DFT level are 1.7 and 1.6 MJ mol^{-1} for the 6-311G and 6-311G(f,d,p) basis sets, respectively, i.e., within 7% of each other. (e) At the DFT/6-311G(f,d,p) level, the high-spin (quintet) complex of Fe(II) with protonated glucosamine (Complex 3) has the lowest binding energy of -344 kJ mol^{-1}, whereas the binding energy for the high-spin (hextet) Fe(III) with protonated glucosamine (Complex 4) is 1.5 MJ mol^{-1}.

Table 3. 0K electronic energies for hexa-coordinated iron complexes with two water molecules and two monosaccharides: Fe(II) with glucose (Complex 1), Fe(II) with glucosamine (Complex 2), Fe(II) with protonated glucosamine (Complex 3), and Fe(III) with protonated glucosamine (Complex 4). The 0K electronic energies of the free components are also listed. All energies are in atomic units, except when specified otherwise. BE = binding energy

	HF/STO-3G		B3LYP/6-311G		B3LYP/6-311G(f,d,p)	
	Low spin	High spin	Low spin	High spin	Low spin	High spin
Fe(II)	-1247.6908	-1248.0715	-1262.5983	-1262.7430	-1262.5986	-1262.7437
Fe(III)			-1261.3568	-1261.5901	-1261.3576	-1261.5901
Glucose	-674.4781		-687.1418		-687.3704	
Glucosamine	-654.9658					
Protonated glucosamine	-667.8786					
H_2O	-74.9608		-76.4159		-76.4474	
Complex 1	-2747.5653	-2747.9494	-2790.4590	-2790.5111	-2790.9289	-2790.9771
Complex 2		-2708.8816		-2750.7861		-2751.2490
Complex 3						-2751.5268
Complex 4						-2750.8074
BE 1	-0.9965	-0.9899	-0.7453	-0.6527	-0.6946	-0.5978
BE 2		-0.9468		-0.6478		-0.6018
BE 3						-0.1311
BE 4						-0.5652
BE 1 (kJ mol^{-1})	-2,591	-2,599	-1,957	-1,714	-1,824	-1,569
BE 2 (kJ mol^{-1})		-2,486		-1,701		-1,580
BE 3 (kJ mol^{-1})						-344
BE 4 (kJ mol^{-1})						-1,484

Crystal Field Splitting

The electron configurations for free Fe(II) is $1s^2\ 2s^2\ 2p^6\ 3s^2\ 3p^6\ 3d^6$ and for free Fe(III) is $1s^2\ 2s^2\ 2p^6\ 3s^2\ 3p^6\ 3d^5$. Natural bond orbital analyses were performed to calculate orbital populations and to estimate the octahedral crystal field splitting (Δ_{oct}). Table 4 displays the electronic occupation obtained for the d orbitals on the iron atom in the four complexes studied. As can be seen, the total d orbital population was close to 6, for all complexes, even for the Fe(III) complex (Complex 4). Complex 4 had a total charge of +5 due to the two $-NH_3^+$ groups of the protonated glucosamines. Those $-NH_3^+$ groups accounted for +1.4 of the charge. In addition, the sugar rings in Complex 4 carried +1.2 higher charge than the Fe(II) complexes, and the C_6 group in Complex 4 carried +0.6 higher charge than the Fe(II) complexes.

Table 4. Calculated octahedral crystal field splitting (Δ_{oct}) for the hexa-coordinated iron complexes with two water molecules and two monosaccharides: Fe(II) complexe with glucose (Complex 1), Fe(II) with glucosamine (Complex 2), and Fe(III) with protonated glucosamine (Complex 3). pop = electron population; ε = atomic orbital energy

	Complex 1 - singlet		Complex 1 - quintet				Complex 2 - quintet			
			α spin		β spin		α spin		β spin	
type	pop	ε (au)	pop	ε (au)	pop	ε (au)	pop	ε (au)	pop	ε (au)
d_{xy}	0.238	-0.293	0.995	-0.541	0.059	-0.221	0.994	-0.536	0.177	-0.244
d_{xz}	1.989	-0.455	0.998	-0.579	0.020	-0.267	0.997	-0.579	0.016	-0.268
d_{yz}	1.988	-0.452	0.998	-0.576	0.016	-0.262	0.997	-0.575	0.010	-0.261
d_{x2y2}	1.974	-0.455	0.992	-0.517	0.974	-0.425	0.991	-0.517	0.871	-0.399
d_{z2}	0.224	-0.289	0.994	-0.574	0.074	-0.270	0.994	-0.576	0.073	-0.272
Total pop.	6.413		4.976		1.143		4.973		1.148	
Δ_{oct} (au)		0.163		0.047				0.051		
Δ_{oct} (cm^{-1})		35,820		10,333				11,092		
Δ_{oct} (nm)		279		968				902		
			Complex 3 - quintet				Complex 4 - quintet			
			α spin		β spin		α spin		β spin	
		type	pop	ε (au)	pop	ε (au)	pop	ε (au)	pop	ε (au)
		d_{xy}	0.993	-0.778	0.056	-0.459	0.996	-0.896	0.047	-0.576
		d_{xz}	0.998	-0.813	0.024	-0.503	0.998	-0.928	0.023	-0.618
		d_{yz}	0.998	-0.806	0.013	-0.491	0.998	-0.921	0.013	-0.605
		d_{x2y2}	0.992	-0.751	0.954	-0.656	0.992	-0.868	0.947	-0.771
		d_{z2}	0.995	-0.806	0.099	-0.508	0.993	-0.920	0.107	-0.624
		Total pop.	4.976		1.146		4.977		1.136	
Δ_{oct} (au)				0.044				0.041		
Δ_{oct} (cm^{-1})				9,667				9,009		
Δ_{oct} (nm)				1,034				1,110		

The hydroxyl groups of the rings and the water molecules had very similar charges in all four complexes. Thus, the additional +1 charge of Fe(III) with respect to Fe(II) of Complex 4 became distributed among the atoms in the rings and the C^6 groups.

The Δ_{oct} values were calculated as the difference between average t_{2g} orbital energies and average e_g orbital energies on Fe. In the case of unrestricted open-shell calculations, only the α orbital energies were used to calculate Δ_{oct}.

Properties Related to Mössbauer Spectroscopy

Mulliken atomic charges and natural atomic orbital charges provide useful information about the electron environment of the Fe atom, and are listed in Table 5. Table 5 displays the effective atomic charges for the iron and the ligand atoms of the four complexes studied, based on DFT calculations, using the 6-311G basis sets.

The water oxygens (O^w) are more negative than the sugar oxygens (O^s) or the sugar nitrogens (N^s), creating an anisotropy around the Fe atom. The Fe-O^w distances in the glucose complex (Complex 1) are both 2.10 Å; in the glucosamine complex (Complex 2) are 2.14 and 2.18 Å; in the protonated glucosamine complex with Fe(II) (Complex 3) are 2.11 and 2.12 Å, and in the protonated glucosamine complex with Fe(III) (Complex 4) both are 2.13 Å.

Table 5. Atomic charges for iron and its ligand atoms on the high-spin Fe(II) complex with two glucose molecules and two H_2O molecules (Complex 1), the Fe(II) complex with two glucosamine and two H_2O molecules (Complex 2), the Fe(II) complex with two protonated glucosamines and two H_2O molecules (Complex 3), and the Fe(III) complex with two protonated glucosamines and two H_2O molecules (Complex 4). Nat.=Natural atomic orbital charge; Mul.=Mulliken atomic charge

Complex 1	Atomic charges		Complex 2	Atomic charges	
	Nat.	Mul.		Nat.	Mul.
Sugar 1: O^3	-0.85	-0.80	Sugar 1: O^3	-0.84	-0.80
Sugar 2: O^3	-0.85	-0.79	Sugar 2: O^3	-0.84	-0.80
Sugar 1: O^2	-0.82	-0.78	Sugar 1: N^2	-0.92	-0.89
Sugar 2: O^2	-0.82	-0.78	Sugar 2: N^2	-0.92	-0.89
Fe(II)	1.66	1.61	Fe(II)	1.63	1.59
Water 1: O	-0.97	-0.84	Water 1: O	-0.97	-0.84
Water 2: O	-0.97	-0.84	Water 2: O	-0.97	-0.84
Complex 3	Atomic charges		Complex 4	Atomic charges	
	Nat.	Mul.		Nat.	Mul.
Sugar 1: O^3	-0.79	-0.52	Sugar 1: O^3	-0.79	-0.53
Sugar 2: O^3	-0.79	-0.52	Sugar 2: O^3	-0.79	-0.53
Sugar 1: O^4	-0.86	-0.61	Sugar 1: O^4	-0.83	-0.58
Sugar 2: O^4	-0.86	-0.61	Sugar 2: O^4	-0.83	-0.58
Fe(III)	1.64	1.52	Fe(III)	1.66	1.52
Water 1: O	-0.98	-0.60	Water 1: O	-0.99	-0.61
Water 2: O	-0.98	-0.59	Water 2: O	-0.99	-0.61

The two equivalent Fe-O^s distances in the glucose complex are 2.14 and 2.16 Å (Complex 1); in the glucosamine complex (Complex 2) the Fe-O^s distances are 2.13 Å and the Fe-N^s distances are 2.24 Å; in the protonated glucosamine with Fe(II) (Complex 3) are 2.10 and 2.36 Å, and in the protonated glucosamine with Fe(III) (Complex 4), the Fe-O^s distances are 2.13 and 2.16 Å.

Table 6 contains data for eight high- and low-spin hexa-coordinated Fe(II) and Fe(III) complexes and their chemical isomer shift [40h] that were used to obtain a standard curve for predicting δ_{Fe}. A multiple-regression line was obtained for the experimental isomer shift (δ_{Fe}) versus (a) Fe s electron contribution, (b) Fe p and d electron contribution, (c) ligand p-valence electron contribution multiplied by the change in the formal charge on Fe, calculated from the natural charge. Figure 9 displays the experimental and fitted δ_{Fe} values.

Table 6. Experimental isomer shift values (δ_{Fe}), change in the formal charge (formal charge minus natural charge) (CC), and calculated electron contributions for hexa-coordinated Fe complexes. Results of the multiple regression on δ_{Fe} values versus (a) Fe s electron contribution, (b) Fe p and d electron contribution, and (c) ligand (L) p electron contributions multiplied by the change in the formal charge on Fe

	δ_{Fe}	Fe	Fe natural atomic orbitals					L.-p-val.
Species	mm s^{-1}	CC	s-core	s-val.	p-core	p-val.	d-val.	
$[Fe(H_2O)_6]^{+2}$	1.39	-0.84	6.000	0.219	11.999	0.397	6.190	31.205
$[Fe(H_2O)_6]^{+3}$	0.50	-1.51	6.000	0.273	11.999	0.481	5.713	31.100
$[FeF_6]^{-4}$	1.34	-1.67	6.000	1.015	12.000	0.545	5.673	34.685
$[FeF_6]^{-3}$	0.48	-1.67	6.000	0.285	12.000	0.526	5.676	34.657
$[Fe(CN)_6]^{4-}$	-0.02	-3.57	5.990	0.449	11.992	1.362	7.686	14.685
$[Fe(CN)_6]^{3-}$	-0.13	-4.17	5.990	0.455	11.992	1.412	7.242	15.265
$[Fe(NH_3)_6]^{+2}$	1.23	-1.11	6.000	0.281	11.999	0.552	6.830	26.790
$[Fe(CO)_6]^{2+}$	0.15	-3.85	5.991	0.502	11.993	1.501	7.821	12.091

	Independent Term	Contribution		
		s	p-d	L x CC
Regression coefficients	-1.0208	1.3102	-0.2599	0.0364
Coefficient of determination	0.952			
Uncertainty in δ_{Fe} estimate	0.180 mm s^{-1}			

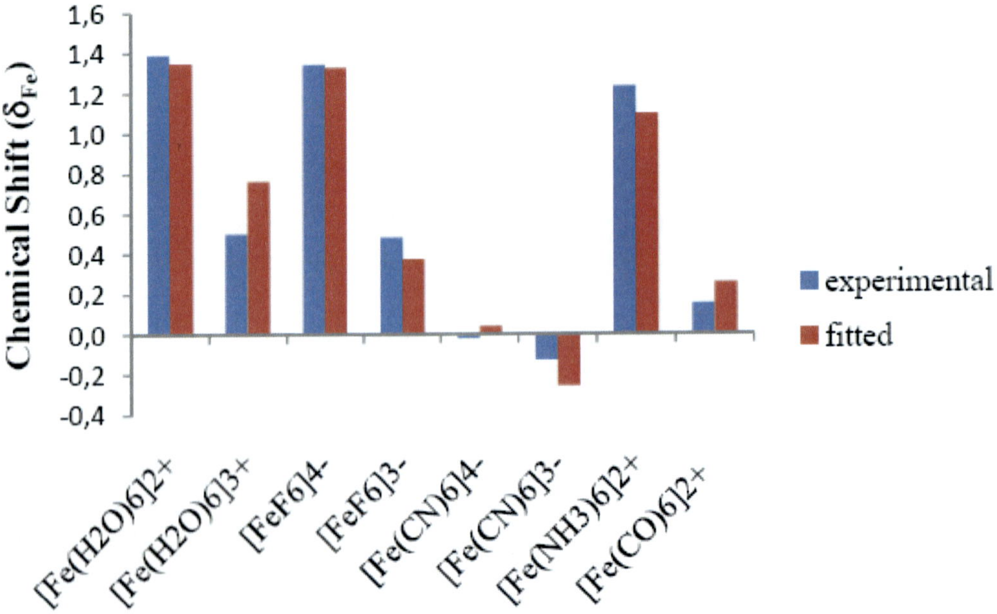

Figure 9. Comparison between experimental [40h] and calculated chemical isomer shifts (δ_{Fe}) for eight high- and low-spin hexa-coordinated Fe(II) and Fe(III) complexes.

Table 7. Data for the prediction of δ_{Fe} for four high-spin complexes using a standard curve. Complex 1 – Fe(II) with two glucose and two water molecules; Complex 2 – Fe(II) with two glucosamine and two water molecules; Complex 3 – Fe(II) with two protonated glucosamine and two water molecules; Complex 4 – Fe(III) with two protonated glucosamine and two water molecules

	Complex 1	Complex 2	Complex 3	Complex 4
Change in formal charge of Fe (based on natural orbitals)	-0.341	-0.366	-0.357	-0.344*
Multiplicity	quintet	quintet	quintet	hextet
Atomic orbital type	Natural orbital occupancy			
Fe-s-core	5.999	5.999	5.999	5.999
Fe-s-valence	0.212	0.235	0.224	0.221
Fe-p-core	11.998	11.998	11.998	11.998
Fe-p-valence	0.007	0.007	0.006	0.006
Fe-d-valence	6.118	6.121	6.122	6.113
Ligands p-valence	31.049	29.697	31.044	31.016
Predicted δ_{Fe} (mm s^{-1})	2.02 ± 0.18	2.04 ± 0.18	2.02 ± 0.18	2.03 ± 0.18

* Please see discussion in the text.

Table 7 summarizes the data used to predicted δ_{Fe} values for the four complexes studied: Complex 1 – Fe(II) with two glucose and two water molecules; Complex 2 – Fe(II) with two glucosamine and two water molecules; Complex 3 – Fe(II) with two protonated glucosamine and two water molecules; Complex 4 – Fe(III) with two protonated glucosamine and two water molecules. The change in the formal charge of iron was taken as the difference between the calculated natural orbital charge and the formal charge. But, the formal charge for all four complexes was taken to be +2. In the case of Complex 4, where the Fe had an expected formal charge of +3, it was still taken as +2, because the third positive charge of the iron became distributed within the sugar rings and the C_6 groups of the protonated glucosamines. Please see the discussion under the section Crystal field splitting.

The following hexa-coordinated, high spin, neutral iron complexes were selected to obtain a standard curve for predicting electric quadrupole splitting ΔE_Q: $FeCl_2(H_2O)_4$, $FeBr_2(H_2O)_4$, and $Fe(HCOO)_2(H_2O)_2$. These species were selected because their counter-ions are not part of the octahedral arrangement. The experimental values [41] for $FeCl_2(H_2O)_4$, and $Fe(HCOO)_2(H_2O)_2$ were extrapolated to 0K using the curves shown in Figure 10. $FeBr_2(H_2O)_4$ was also extrapolated to 0K using the $FeCl_2(H_2O)_4$ parameters.

DFT calculations with a cc-pVTZ basis set on Fe and 6-311g(f,d,p) basis sets on all atoms were performed to obtain values of EFG for $FeCl_2(H_2O)_4$, $FeBr_2(H_2O)_4$, and $Fe(HCOO)_2(H_2O)_2$, as well as for the three molecules studied, in the high spin conformation. Table 9 summarizes the data and Figure 11 displays the experimental values, the fitting curve, and the predicted values of ΔE_Q for the molecules studied.

The large ΔE_Q values predicted are a consequence of considerable deformations of the octahedra. For example, Figure 12 displays the final structure obtained for Complex 1, Fe(II) with glucose and water ligands, and Complex 2, Fe(II) with glucosamine and water ligands. Table 9 lists some angles and dihedral angles of the final structures of the complexes. The atoms have been labeled with two numbers; the one on the left corresponds to the standard atom numbering used in Figure 6, and the one on the right distinguishes between the two monosaccharides. For Complex 1, the distortion of the octahedron occurs through the water molecules, as can be seen from the listed angles in Table 9. For Complex 2, the distortion of the octahedron occurs through the two sugar ligands, whereas the dihedral angles of the water molecule maintain symmetry.

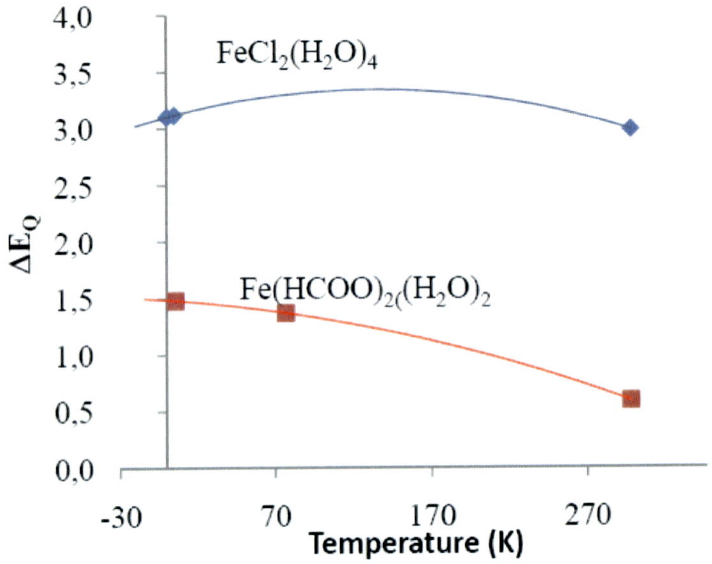

Figure 10. Extrapolation to 0K of experimental values [41] of ΔE_Q for $FeCl_2(H_2O)_4$, and $Fe(HCOO)_2(H_2O)_2$.

Table 8. Experimental values [41] of ΔE_Q extrapolated to 0K for $FeCl_2(H_2O)_4$, $FeBr_2(H_2O)_4$, and $Fe(HCOO)_2(H_2O)_2$; predicted values of ΔE_Q (0K) for the following high-spin complexes: Complex 1 – Fe(II) with two glucose and two water molecules; Complex 2 – Fe(II) with two glucosamine and two water molecules; Complex 3 – Fe(II) with two protonated glucosamine and two water molecules; Complex 4 – Fe(III) with two protonated glucosamine and two water molecules

	ΔE_Q (0K) (mm s^{-1})	V_{zz} (au)	η	V_{zz} (1 + $\eta^2/3)^{1/2}$ (au)
$FeCl_2(H_2O)_4$	3.10	-2.06	0.07	-2.06
$FeBr_2(H_2O)_4$	2.72	-1.95	0.13	-1.96
$Fe(HCOO)_2(H_2O)_2$	1.49	-2.53	0.14	-2.54
	Predicted ΔE_Q (0K) (mm s^{-1})	V_{zz} (au)	η	V_{zz} (1 + $\eta^2/3)^{1/2}$ (au)
Complex 1	2.17 ± 0.46	-2.285	0.118	-2.290
Complex 2	1.01± 0.46	-2.743	0.153	-2.754
Complex 3	7.37± 0.46	-0.197	0.616	-0.209
Complex 4	3.36± 0.46	-1.744	0.487	-1.812

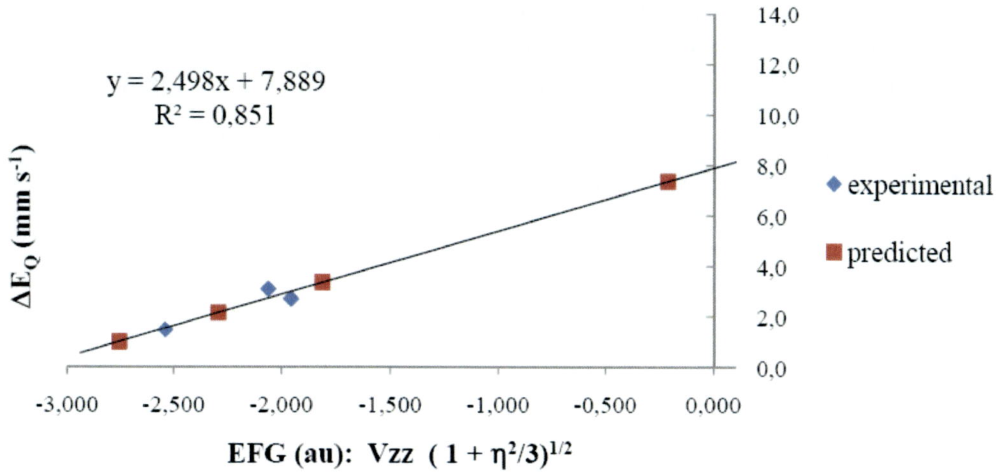

Figure 11. Experimental values [41] of ΔE_Q extrapolated to 0K for $FeCl_2(H_2O)_4$, $FeBr_2(H_2O)_4$, and $Fe(HCOO)_2(H_2O)_2$; predicted values of ΔE_Q (0K) for the following high-spin complexes: Complex 1 – Fe(II) with two glucose and two water molecules; Complex 2 – Fe(II) with two glucosamine and two water molecules; Complex 3 – Fe(II) with two protonated glucosamine and two water molecules; Complex 4 – Fe(III) with two protonated glucosamine and two water molecules.

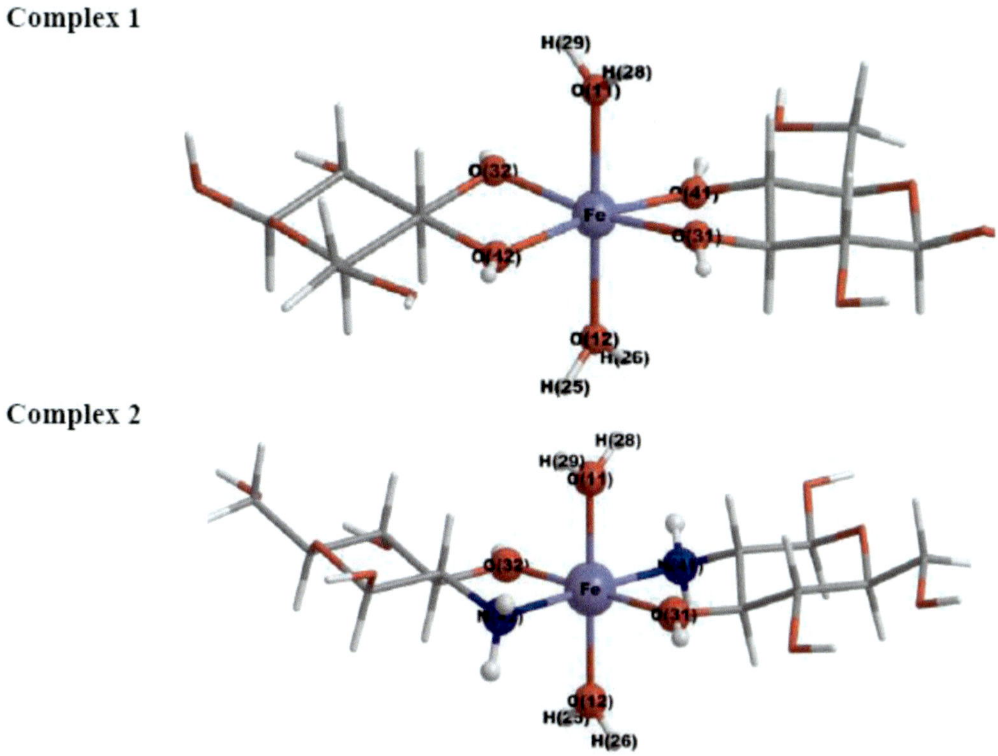

Figure 12. Octahedra distortion in Complex 1 [Fe(II) with glucose and water ligands] and Complex 2 [Fe(II) with glucosamine and water ligands]. Atoms attached to Fe have been labeled with two numbers, the one on the left corresponding to the standard atom numbering used in Figure 6, and the one on the right distinguishes the monosaccharide or water molecule.

Table 9. Angles and dihedral angles of the final structures of Complex 1 [Fe(II) with two water molecules and two glucose molecules] and Complex 2 [Fe(II) with two water molecules and two glucosamine molecules]. Labels are shown in Figure 12

Complex 1			Complex 2		
	Angle	Dihedral angle		Angle	Dihedral angle
O11-Fe-O42	95		O11-Fe-N42	91	
O11-Fe-O41	95		O11-Fe-N41	91	
O12-Fe-O41	85		O12-Fe-N41	89	
O12-Fe-O42	85		O12-Fe-N42	89	
O11-Fe-O31	85		O11-Fe-O31	90	
O11-Fe-O32	85		O11-Fe-O32	90	
O12-Fe-O31	95		O12-Fe-O31	90	
O12-Fe-O32	95		O12-Fe-O32	89	
O11-Fe-O12	180		O11-Fe-O12	90	
H29-O11-Fe-O12		56	H29-O11-Fe-O12		90
H28-O11-Fe-O12		-123	H28-O11-Fe-O12		90
H25-O12-Fe-O11		-157	H25-O12-Fe-O11		90
H26-O12-Fe-O11		24	H26-O12-Fe-O11		90

The difference between these two complexes is either an NH_2 or a OH group attached to the C^4 of the monosaccharides. The NH_2 group has an additional hydrogen atom that may hinder motion of the water molecules. It is possible that the presence of the extra hydrogen of the NH_2 group forces the two water molecules into very symmetric positions in Complex 2, but in Complex 1 the two water molecules may move with more freedom.

CONCLUSIONS

The following are the principal findings of this investigation.

1. The Mössbauer data of the Fe-incorporated chitosan indicate that the material is magnetic with an unusually small internal field of 441 kG at 4.2 K for a high-spin Fe (III) system, and magnetic spectrum disappears and a pure quadrupole doublet is seen around 100 K.
2. An analysis of the temperature dependence of the isomer-shift suggests that the material behaves like a Debye solid, and the vibration between the atoms is harmonic to a reasonable approximation.
3. Mössbauer data analysis of Fe complexes of chitin, and chitosan, and glucosamine clearly indicates that the metal ion in the 3+ state is stabilized in the case of polymer, while the 2+ state is partially stabilized in the case of water-soluble chitosan, and that complete stabilization is achieved with D-glucosamine which is the monomer of chitosan. The pertinent information about the oxidation state of the metal ion was obtained by using the isomer-shift deduced from the Mössbauer spectra.

4. Binding energies were computed as the 0K energy difference between the complex and the components (iron ion plus two water molecules, plus two monosaccharides). All binding energies for the four complexes studied by theoretical means were found to be exergonic. The best calculations indicate a binding energy of 1.8 MJ mol^{-1} for low-spin Fe(II)-glucose, 1.6 MJ mol^{-1} for high-spin Fe(II)-glucose, but the high-spin complex was found to be 126 kJ mol^{-1} more stable than the low-spin complex. The best calculations for high-spin Fe(II)-glucosamine complex resulted in a binding energy of 1.6 MJ mol^{-1} for Fe(II)-protonated-glucosamine 340 kJ mol^{-1}, and 1.5 MJ mol^{-1} for Fe(III)-protonated-glucosamine.
5. All high-spin complexes were found to have octahedral crystal field splitting (Δ_{oct}) of about 10,000 cm^{-1}. The low-spin Fe(II)-glucose complex had $\Delta_{oct} = 36{,}000$ cm^{-1}. Iron ion in all complexes had natural atomic charges of about +1.6. The extra charge in Fe(III) became distributed within the sugar rings and the C^6 of the protonated-glucosamines. All high-spin complexes had d^6 configurations.
6. A standard curve based on the δ_{Fe} experimental values of eight Fe complexes was generated, based on a multiple least-squares regression on the following three parameters: (a) s-type natural atomic orbital (NAO) population on Fe; (b) p and d-type NAO population on Fe; (c) p-type NAO populations on the ligands, multiplied by the change in formal charge on the Fe. Using the standard curve for δ_{Fe}, all high-spin complexes studied were predicted to have δ_{Fe} values of 2.0 ± 0.2 mm s^{-1}. The predicted values of δ_{Fe} were within 0.7 mm s^{-1} from the experimental values.
7. Three high-spin hexa-coordinated experimental values of ΔE_Q were used to obtain a standard curve by linear least-squares regression on $V_{zz} (1 + \eta^2/3)^{1/2}$ values. The experimental values were extrapolated to 0K. The predicted values of ΔE_Q for the high-spin complexes were: (2.2 ± 0.5) mm s^{-1} for the Fe(II)-glucose complex; (1.0 ± 0.5) mm s^{-1} for the Fe(II)-glucosamine complex; (7.4 ± 0.5) mm s^{-1} for the Fe(II)-protonated-glucosamine complex; and (3.4 ± 0.5) mm s^{-1} for the Fe(III)-protonated-glucosamine complex. The values suggest that the predicted structure for the Fe(II)-glucosamine complex is much more distorted than the real complex. For the Fe(II)-glucose complex, the distortion occurs through the water molecules, whereas for the Fe(II)-glucosamine complex, much smaller distortions were seen only on the ligand position of sugar portion. The predicted value of ΔE_Q for the Fe(II)-glucose complex was 7% smaller than the experimental value; on the other hand, the predicted ΔE_Q value was 50% smaller for Fe(II)-glucosamine, and was double for Fe(II)-protonated-glucosamine. The discrepancies between the experimental and predicted values will be improved in the future by using standard curves generated from similar chemical systems.

ACKNOWLEDGMENT

Natarajan Ravi acknowledges the Department of Energy (DOE) Grant No. DE-FG52-09NA29518.

References

[1] Brine, C. J. In *Chitin, Chitosan and Related Enzymes*; Zikakis, J. P. Ed.; Academic Press: NY 1984.

[2] S.E. Tully, R. Mabon, G.I. Gama, S. Tsai, X. Liu, and Hseieh-Eilson, *J. Am. Chem. Soc.* 126 (2004) 7736.

[3] (a) Ohtakara, A. *Met. Enz.* 1988, 161, 505; Y. Arakai, Y.; Ito, E. *Met. Enz.* 1988, 161, 510; (b) Sanford, P. A. In *Chitin and Chitosan*; G. Skjak-Braek, G.; Anthonsen, T.; Sanford., P., Eds.; Elsevier: Amsterdam 1989.

[4] Nieto, J. M.; Peniche-Covas, C.; Del Bosque, J. *Carb. Pol.* 1992, 18, 221-224; Muzzarelli, R. A. A. In *Chitin*; Pergamon Press; NY 1973; (b) Muzzarelli, R. A. A. In *Natural chelating polymer*; Pergamon Press: Oxford, U.K. 1973; (d) Roberts, G. A. F. In *Chitin Chemistry*, MacMillan, London, 1992.

[5] E.J. Bradbury, L.S.F. Moon, R.J. Popat, V.R. King, G.S. Bennett, P.N. Patel, J.W. Fawcett, S.B. McMohan, *Nature* 416 (2002) 636.

[6] Chiessi, E.; Paradossi, G.; Venanzi, M.; Pispisa, B. *Int. J. Biol. Macro.* 1993, 15, 150.

[7] Gamblin, B. E.; Stevens, J. G.; Wilson, K. L. *Hyp. Int.* 1998, 112, 117.

[8] Mazeau, K.; Winter, W. T.; Chanzy, H. *Macromolecules* 1994, 27, 7606.

[9] Ogawa, K.; Oka, K.; Miyanishi, T.; Hirano, S. In *chitin related enzymes*; Zikakis, J.P. Ed.; Academic Press: Orlando, FL 1984.

[10] (a) Ogawa, K. *Nippon Nogeikagaku Kaishi* 1988, 62, 12225; (b) Schlick, S. *Macromolecules* 1986, 19, 192.

[11] Bernstein, T.; Koetzle, T. F.; Williams, G. J. B.; Meyer, E.; Brice, M. D.; Rodgers, J. R.; Kennard, O.; Shimoaouchi, T.; Tasumi, M. The Protein Data Bank: A Computer-based Archival File for Macromolecular Structures. *J. Mol. Biol.* 1977, 112, 535.

[12] Chemical Reviews, *Bioinorganic Enzymology*, Holm, R. I.; Solomon, E. Guest Eds.; ACS publication, 1996, 96.

[13] (a) Trautwein, A. X.; Bill, E.; Bominaar, E.; Winkler, H. *Struc. and Bonding* 1991, 71, 1; (b) E. MÜnck, E.; Surerus, K. K.; Hendrich, M. P. *Met. in Enzymology* 1993, 227, 463; (c) Huynh, B. H. *Met. in Enzymology* 1994, 243, 523; (d) J.B. Lynch, C. Juarez-Garcia, E. Münck, L. Que, *J. Biol. Chem.* 264 (1989) 8091; (e) P. Nordlund, H. Eklund, *Curr. Opin. Struc. Biol.* 5 (1955) 758.

[14] Cramer, Christopher J. *Essentials of Computational Chemistry*. Chichester: John Wiley and Sons, Ltd. 2002; pp. 153–189. ISBN 0-471-48552-7.

[15] Parr, R. G.; Yang, W. *Density-functional theory of atoms and molecules*. Oxford Univ. Press, Oxford, 1989. ISBN 0-19-504279-4.

[16] Gaussian 09, Revision A.1, Frisch, M. J., Trucks, G. W., Schlegel, H. B., Scuseria, G. E., Robb, M. A., Cheeseman, J. R., Scalmani, G., Barone, V., Mennucci, B., Petersson, G. A., Nakatsuji, H., Caricato, M., Li, X., Hratchian, H. P., Izmaylov, A. F., Bloino, J., Zheng, G., Sonnenberg, J. L., Hada, M., Ehara, M., Toyota, K., Fukuda, R., Hasegawa, J., Ishida, M., Nakajima, T., Honda, Y., Kitao, O., Nakai, H., Vreven, T., Montgomery, Jr., J. A., Peralta, J. E., Ogliaro, F., Bearpark, M., Heyd, J. J., Brothers, E., Kudin, K. N., Staroverov, V. N., Kobayashi, R., Normand, J., Raghavachari, K., Rendell, A.,

Burant, J. C. Iyengar, S. S. Tomasi, J. Cossi, M. Rega, Millam, N. J., Klene, M. Knox, J. E., Cross, J. B., Bakken, V., Adamo, C., Jaramillo, J., Gomperts, R. E. Stratmann, O. Yazyev, A. J. Austin, R. Cammi, C. Pomelli, J. W. Ochterski, R. Martin, R. L., Morokuma, K., Zakrzewski, V. G., Voth, G. A., Salvador, P., Dannenberg, J. J., Dapprich, S., Daniels, A. D., Farkas, O., Foresman, J. B., Ortiz, J. V., Cioslowski, J., and Fox, D. J., Gaussian, Inc., Wallingford CT, 2009.

[17] (a)Becke, D. "Density-Functional Exchange-Energy Approximation with Correct Asymptotic Behavior". *Physical Review A38* 1988, 3098-3100. (b) Lee, C.; Yang, W.; Parr, R. G. "Development of the Colle-Salvetti Correlation-Energy Formula into a Functional of the Electron Density". *Physical Review B* 1988, 37, 785-789.

[18] Hehre, W. J.; Ditchfield, R.; Pople, J. A.; "Self-consistent molecular orbital methods. 12. Further extensions of Gaussian-type basis sets for use in molecular-orbital studies of organic molecules"; *J. Chem. Phys.* 56 1972, 2257-2261.

[19] Ditchfield, R.; Hehre, W. J.; Pople, J. A.; "Self-consistent molecular orbital methods. 9. Extended Gaussian-type basis for molecular-orbital studies of organic molecules"; *J. Chem. Phys.* 1971, 54, 724-728

[20] Hay, P. J.; "Gaussian basis sets for molecular calculations-representation of 3D orbitals in transition-metal atoms"; *J. Chem. Phys.* 1977, 66, 4377-4384.

[21] Kendall, R. A.; Dunning Jr., T. H.; and Harrison, R. J. "Electron affinities of the first-row atoms revisited. Systematic basis sets and wave functions," *J. Chem. Phys.*, 96 (1992) 6796-806.

[22] Greenwood, N. N.; Gibb, T. C. In *Mössbauer Spectroscopy*, Chapman and Hall Ltd., London 1971.

[23] Abraham, A; Bleany, B. In *Electron Paramagnetic Resonance of Transition ions*, Dover Publications 1970.

[24] Yassin Jeilani, Beatriz H. Cardelino, and Natarajan Ravi (unpublished results)

[25] S. C. Bhatia, and N. Ravi, *Biomacromolecules* 1 (2000) 413.

[26] S.C. Bhatia, and N. Ravi, *Biomacromolecules* 4 (2003) 723.

[27] S.C. Bhatia, B. Cardelino, and N. Ravi, *Hyp. Int.* 165 (2005) 339 .

[28] Kundig, W. *Nucl. Instr. Met.* 1967, 48, 219.

[29] Moura, I.; Tavares, P.; Moura, J. J. G.; Ravi, N.; Huynh, B. H.; Liu, M-Y.; Le Gall, J. *J. Biol. Chem.* 1992, 267, 4489.

[30] Tavares, P.; Ravi, N.; Moura, J. J. G.; LeGall, J.; Huang, Y-H.; Crouse, B. R.; Johnson, M. K.; Huynh, B. H.; Moura, I. *J. Biol. Chem.* 1994, 269, 10504.

[31] Monteiro, O. C., Jr.; Airoldi, C. *J. Coll. Int. Sci.* 1999, 212, 212.

[32] Wiberg, E.; Wiberg, N.; Holleman, A. F. *Inorganic chemistry*. Academic Press, San Diego, CA, 2001, p. 1180. ISBN 0-12-352651-5.

[33] (a) NBO Version 3.1; Glendening, E. D.; Reed, A. E.; Carpenter, J. E.; Weinhold, F.; QCPE Bull 1990, 10, 58. (b) Foster, J. P.; Weinhold, F.; "Natural hybrid orbitals"; J.Amer.Chem.Soc. 1980, 102, 7211-7218. (c) Reed, A. E.; Curtiss, L. A.; Weinhold, F.; "Intermolecular interactions from a natural bond orbital, donor-acceptor viewpoint"; *Chem.Rev.* 1988, 88, 899-926.

[34] Mulliken, R. S.; "Electronic Population Analysis on LCAO—MO Molecular Wave Functions. I"; *J. Chem. Phys.* 1955, 23, 1833-1840. (b) Mulliken, R. S.; "Criteria for the

Construction of Good Self-Consistent-Field Molecular Orbital Wave Functions, and the Significance of LCAO-MO Population Analysis"; *J. Chem. Phys*. 1962, 36, 3428-3440.

[35] Ditchfield, R. "Self-consistent perturbation theory of diamagnetism. I A gauge-invariant LCAO method for N.M.R. chemical shifts." *Mol. Phys*. 1974, *27*, 789-807.

[36] Gaussian View version 4.1.2. Gaussian Inc.

[37] Gütlich, P. "The Principle of the Mössbauer Effect and Basic Concepts of Mössbauer Spectrometry". http://pecbip2.univ-lemans.fr/~moss/webibame/. Last accessed 03/30/2010.

[38] Debrunner, P. G. "Mössbauer spectroscopy of iron porphyrins". In *Iron Porphyrins*. Lever, A. B. P.; Gray, H. B.; editors. VCH Publishers: New York, 1989; Vol. 3, pp.137-234.

[39] Zhang, Y.; Mao, J.; Godbout, N.; Oldfield, E. "Mössbauer Quadrupole Splittings and Electronic Structure in Heme Proteins and Model Systems: A Density Functional Theory Investigation". *J. Am. Chem. Soc.* 2002, *124*, 13921-13930.

[40] (a) Blomquist, J.; Roos,B.O.; Sundbom, M. "Interpretation of the ^{57}Fe Isomer Shift by Means of Atomic Hartree–Fock Calculations on a Number of Ionic States". *J. Chem. Phys*. 1971, *55,* 141-145. (b) Duff, K. J."Calibration of the isomer shift for ^{57}Fe"; *Phys. Rev. B* 1973, *9*, 66-72. (c) Nieuwpoort, W.C.; Post, D.; van Duijnen, P.Th. "Calibration constant for ^{57}Fe Mössbauer isomer shifts derived from ab initio self-consistent-field calculations on octahedral FeF_6 and $Fe(CN)_6$ clusters". *Phys. Rev. B* 1978, *17*, 91-98. (d) Neese, F. "Prediction and interpretation of the ^{57}Fe isomer shift in Mossbauer spectra by density functional theory". *Inorg. Chim. Acta* 2002, *337* 181-192. (e) Zhang, Y.; Mao, J.; Oldfield, E. "^{57}Fe Mössbauer Isomer Shifts of Heme Protein Model Systems: Electronic Structure Calculations". *J. Am. Chem. Soc*. 2002, *124*, 7829-7839. (f) Liu, T.; Lowell, T.; Han, W.-G.; Noodleman, L. "DFT" *Inorg. Chem*. 2003, *42*, 5244-5251. (g) Sinnecker, S.; Slep, L. D.; Bill, E.; Neese, F. "Performance of Nonrelativistic and Quasi-Relativistic Hybrid DFT for the Prediction of Electric and Magnetic Hyperfine Parameters in ^{57}Fe Mössbauer Spectra". *Inorg. Chem*. 2004, *44*, 2245-2254. (h) Sadoc, A.; Broer, R.; de Graaf, C. "CASSCF study of the relation between the Fe charge and the Mössbauer isomer shift"; *Chemical Physics Letters* 2008, *454*, 196–200.

[41]Hoy, G. R.; Barros, F. de S. "Mössbauer studies of inequivalent ferrous ion sites in ferrous formate". *Phys. Rev*. 1965, *139*, A929-A934.

In: Focus on Chitosan Research
Editors: Arthur N. Ferguson and Amy G. O'Neill
ISBN 978-1-61324-454-8

Chapter 13

CHITOSAN: PROPERTIES AND ITS PHARMACEUTICAL AND BIOMEDICAL ASPECTS

Ali Demir Sezer
Marmara University, Faculty of Pharmacy,
Haydarpaşa, Istanbul, Turkey

ABSTRACT

Chitosan, a natural based nontoxic cationic polysaccharide polymer obtained by alkaline deacetylation of chitin, presents excellent properties such as biodegradability, antibacterial and wound-healing activity and immunological properties. These properties make chitosan a good candidate for the development of conventional and novel drug delivery systems. Chitosan has been used as a polymer for a controlled delivery system, gene delivery, scaffold, haemostatic action in wound healing, cell culture and cosmetic applications. Chitosan has become the focus of major interest in recent years because it has applications in not only the drug industry but also the agriculture, textile and paper industries. On the other hand, use of this biopolymer as a pharmaceutical excipient by different dose and for a number of applications is not new, but it still appears to be present in marketed drugs. The development of new delivery systems for controlled release of drugs is one of the most interesting areas of research in the pharmaceutical sciences. Nano and microparticles can be used for controlled release of different biological substances and drugs such as plasmids, hormones, peptide and proteins, antibiotics and vaccines. In this field, chitosan particular systems and chitosan matrixes, prepared by using different methods, can be used for encapsulating the drugs. Moreover, there has been interest in the chemical modification and PEGylated chitosan in order to improve its solubility and applications. Representatives of these novel chitosan modified polymers are carboxymethyl chitosan, thiolated chitosan, succinate and phthalate of chitosan salts and trimethyl-chitosan. The main chemical modifications of chitosan that have been proposed in the literature are reviewed in this chapter. Furthermore, recent studies suggested that chitosan and its derivatives are promising candidates as a supporting material for tissue engineering applications owing to their porous structure, gel forming properties, ease of chemical modification, high affinity to in vivo macromolecules. In general, this review provides an overview of using chitosan in pharmaceutical and biomedical applications.

1. The Structure and Manufacture of Chitosan

Chitosan began to be used in the pharmaceutical industry in the early 1970s, with chitin. The history of chitosan dates back to the discovery of this polymer by Rouget in 1859. Rouget obtained chitosan from chitin, using a treatment with concentrated KOH solution in an acidic environment, and called this structure modified chitin. Modified chitin is stained a violet color in acid and diluted iodine solutions; however, normal chitin turns to brown. In 1894, Hoppe-Seyler studied chitin again and called this structure chitosan. Bartnicki and Garcia managed to isolate chitosan from the cell wall of *Mucor rouxii* in 1968 [1-3].

Chitosan is obtained from decalcification of many arthropods like chitin. Chitosan is the polymeric molecules of long linear chained β-(1-4)-linked glucans. Despite the hydrophobic property of chitin, which is not soluble in water and many organic solvents, chitosan obtained by deacetylation has a hydrophilic property and its solubility depends on the degree of deacetylation [4, 5].

Figure 1 indicates the chemical Formula of chitosan with the structure of [(1-4)-2-amino-2-deoksi-β-D-glucan] [6, 7].

As indicated in Figure 2, the first stage of obtaining chitosan involves obtaining chitin by the extraction of crushed arthropod shells. This is followed by deacetylation. The parts rich in chitin are first treated with alkali and then with acid, and inorganic substances are removed from these parts. Chitosan is obtained as a result of the deacetylation of the dried residue with concentrated sodium hydroxide [2].

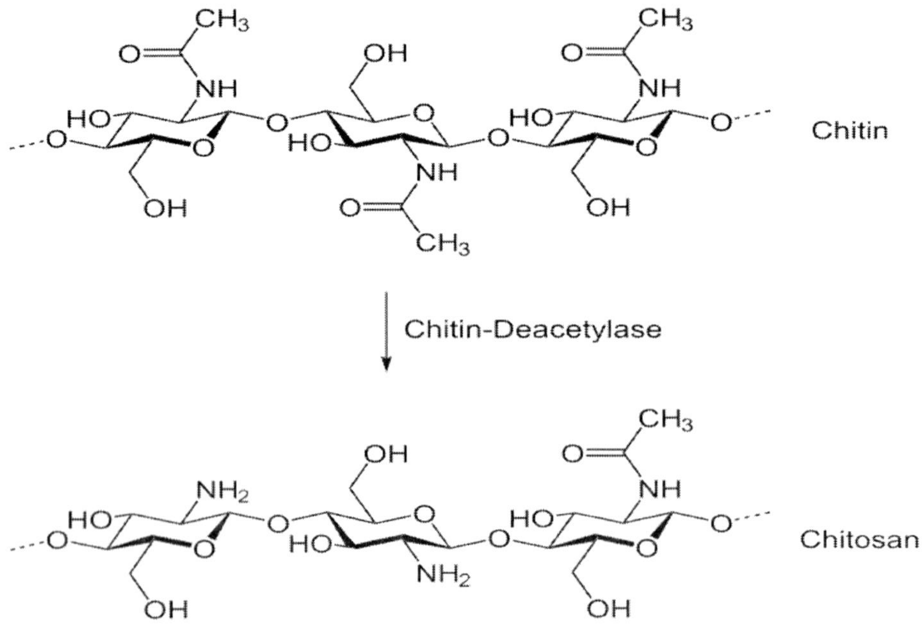

Figure 1. Chemical structure of chitosan.

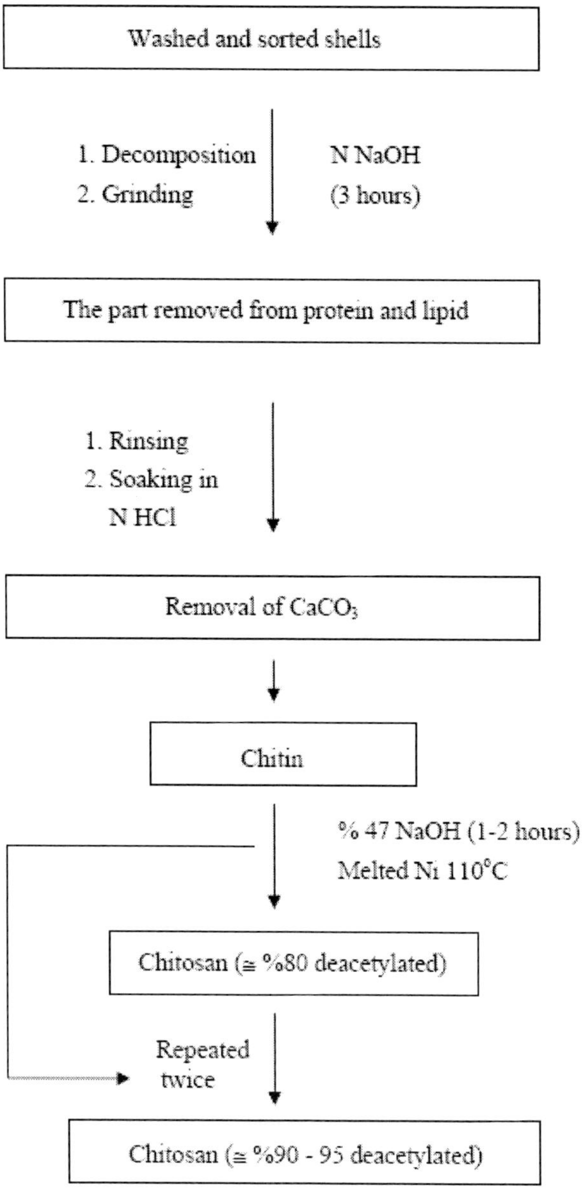

Figure 2. Schema of chitosan production from crustaceans.

2. Physicochemical Properties of Chitosan

Chitosan is a white or cream-colored biopolymer of different sizes, which can be dissolved in weak acid solutions [7,8].

In chitosan, the mole fraction of *glucosamine* is defined as deacetylation degree. This value generally varies between 70–90%. Chitosan contains 10% or less moisture, and the amount lost during drying is 2% or lower. In order to make chitosan soluble in water, the salts should be prepared. For this, chitosan is first dissolved in an appropriate acid, such as acetic acid. It is then filtered and purified, sprayed and dried and chitosan salts, which can be dissolved in water, are obtained. Although chitosan can be dissolved in organic acids such as acetic acid, lactic, propionic and butyric acid, it is partially dissolved in mineral acids such as percloric, hydrochloric and nitric acid and is not dissolved in sulfuric and phosphoric acid [3,9].

Average molecular weight of chitosan varies between 1.10^5 - 3.10^5, depending on production conditions. In parallel to this, it has a large viscosity range in a diluted acid environment [7].

As chitosan has a weak alkali property, to convert glucosamine units into a positive charged, water-soluble state, acid of a particular concentration should be added [10].

The polycationic property of chitosan varies according to the degree of deacetylation. The hydrophobic property of chitosan increases with the inclusion of N- acetyl groups in the molecule. Chitosan can 50% tolerate organic solvents such as ethanol, acetone and methanol depending on the N-acetyl groups [2,11,12].

Unlike other polysaccharides, chitosan is a cationic polymer. It does not dissolve in water, due to free amino groups in neutral or alkali environments. However, in acidic environments, it dissolves in water due to protonation of amino groups. Solubility depends on the distribution of free amino and N- acetyl groups. In acidic pH, chitosan is a polysaccharide with a high charge density, in which each glucosamine unit has a positive charged ion. pH and ionic power significantly affect the viscosity of polyelectro liquids. For example, acidic solutions of chitosan react with anionic solutions such as heparin, sodium alginate, carboxyl cellulose and polyacrylic acid to form polyelectrolyte complexes [7,13,14].

The problem of sterilization of chitosan in pharmaceutical and biomedical applications has been analyzed using UV, ionization radiation and chemical substances. The use of an appropriate sterilization method is of great importance to avoid degradation of the structure of the biomaterial. A review of the literature indicates that the use of high temperature or β and ϒ irridation to sterilize chitosan can damage the physical properties of the polymers [15-19].

3. The Pharmacologic Properties of Chitosan

In the human body, chitosan is biodegraded by the enzyme lysozyme. The product of this degradation is glucosamine, which is a natural monosaccharide. The molecular weight of this biodegradable polysaccharide varies between 50 – 2000 KD [2].

Chitosan and its biodegradation products are non-toxic. It was found that the oral LD_{50} value in rats was 16 g/kg; and the interperitoneal LD_{50} value in mice was more than 350 mg/kg. These values are the same as lethal doses of salt and sugar, and correspond to 1/10 of lethal doses of acrylamide and methacrylate. Similar to these values, the reliability of chitosan, which is used as a basic material in food additives and cosmetics, was tested on mice and rats. It was reported that, in acute toxicity test, the LD_{50} dosage was more than 10 g/kg in mice subcutaneously; more than 5.2 g/kg in mice and more than 3.0 g/kg in rats intraperitoneally [20]. On the other hand, studies on dogs reported different results. After subcutaneous application of varying amounts of chitosan (10-200 mg/kg) in dogs, doses of more than 50 mg/kg and 150 mg/kg resulted in anorexia and mortality, respectively. Hematologic findings indicated that there was a significant increase in leucocytosis, serum

LDH_2 and LDH_3 iso-enzymes. In autopsy findings, acute hemorrhagic pneumonia was observed in all killed dogs and thus it was reported that chitosan can cause fatal pneumonia in dogs [21].

It is known from previous studies that chitosan has anti-carcinogenic, anti-inflammatory and analgesic effects [22, 23]. In addition, positive outcomes were obtained in ulcer, hyperbilirubinemia and cholesterol treatments. In the literature, it is reported that chitosan is selectively accumulated in L-1210 leukemia cells and prevented cell proliferation. Chitosan has a homeostatic effect and is used in treatment of wounds and burns [24-26].

The effects of chitosan on tumor growth and metastasis are under research. With this aim, sarkom-180, Dunn osteosarcoma, B16F1 melanoma, P815 mast cell tumor and KLN205 pulmonary squamous tumor cells were subcutaneously injected to the back area of mice. It was reported that the tumors shrank after the intratumoral injection of 10 mg/kg or 100 mg/kg chitosan, the environment rich in macrophages was particularly stimulated by adding chitosan; and that cell proliferation of tumor cells was inhibited by *in vitro* chitosan addition. However, the treatment on mice with, KLN205 pulmonary squamous cell tumor increased lung metastasis ratio; and the treatment in the mice with P815 mast cell tumor facilitated tumor growth and liver metastasis. The number of capillary veins was observed to increase in both tumor tissues. In conclusion, it was reported that angiogenic inductive and activation property of chitosan in macrophages increased the antitumor effect of the polymer [27].

It was found that, due to the mucoadhesive properties of cationic biopolymers like chitin and chitosan in weak alkali and neural environments, they prevented formation of fungi and increased the effect of antifungal and antimicotic formulations. In addition, low molecular weight chitosan types activate ciliary movement [28-30].

4. Compatibility and Stability of Chitosan

Chitosan is compatible with many cationic and non-ionic substances. It forms complexes with some metals and jellifies by cross bonding with valuable anions [31-34].

Due to its high molecular weight and branched structure, chitosan is a good viscosity-increasing polymer. It was found that, in high molecular weight types, viscosity decrease and structure degradation is rapid; and that chitosan solutions decreased inversely proportionally with mixing speed, and thus it showed pseudo plastic flow [35-37]. Chitosan, which is bonded to proteins by hydrogen bonds, is used as a thickener and emulsifier regulator in pharmaceutical formulations, due to its hydrocolloid properties [14, 38].

5. The Use of Chitosan in Drug Delivery Systems

General uses of chitosan and its derivates are given in Table 1; Their various pharmaceutical uses are shown in Table 2. For pharmaceutical purposes, chitosan is used in the preparation of solid dispersions for increasing the solubility of hard-to-dissolve drugs; in tablet formulations, as direct compression material and inactive ingredients (filling material, adsorbent, bond and lubricant); in preparation of sustained release granules; film covering; controlled drug release preparations (nano-microsphere, bead, mikrosponge, implant, film etc.); in covering of granules and liposome as a hydro gel and ointment base [1, 2, 5, 10, 14, 25, 38-45].

Table 1. General Uses of Chitosan

• For biomedical purposes
* As degradable suture, operation and eye bandage
* As artificial skin and renal membranes
* Production of hard and soft lenses
* As a covering material in wounds and burns
* Specific protein/enzyme absorption
* Cell culture
* Preparation of dentures in dentistry and orthopedics
* As a polymer in pharmaceutical applications
* Lowering cholesterol
• In immobilization of living cells and enzymes
• In break-up of protein and amino acids or in specific protein/ enzyme adsorption
• As an adsorbent substance in removal of toxic substances
• In cosmetics industry (skin and hair care products)
• In wastewater treatment
• As a bonding agent for paints
• Ultra-filtration, gas and liquid release agent
• Chromatographic support material
• In paper and textile industry
• In agriculture and animal feed (seed covering, soil improvement)
• In fiber industry

Table 2. Pharmaceutical Uses of Chitosan

• *In Conventional formulations*
* As direct compression material in tablets
* As polymer in controlled release matrix tablets
* As a bond in wet granulation
* Gel
* Film
* As a polymer in emulsion formulations
* Moisturizer
* As coating compound in classic dosage
* Preparation of nano and microparticles
• *New Applications*
* Transmucosal drug applications
* Preparation of bioadhesive systems
* Peptide formulations
* Vaccine technology
* Gene treatment

Various studies were carried out to enhance the bioavailability of hard-to-dissolve drugs such as phenytonin, griseofulvine and prednisolone. Powdered mixtures were prepared and it was found that, after the use of chitin and chitosan in these formulations, depending on the decrease in crystal size of active ingredients, dissolution rates increased [46-48].

For the development of implantable sustained release formulations, Machida et al. used chitosan and its water-soluble derivative *hydroxypropyl* chitosan as a polymer. They reported that, when compared to chitosan, *in vitro* and *in vivo* degradation of *hydroxypropyl* chitosan was slower in enzymatic reactions. This was because large groups like the *hydroxypropyl* group, which substitute the main structure, prevent contact of enzymes and polymers [49].

Chitosan is preferred as an inactive ingredient when manufacturing tablets using direct compression techniques, since it increases powder flowability and facilitates compression. It is also used in wet granulation formulations, due to its good bond properties [46, 48, 50].

As a result of the cross-linking of polymers of various properties with chitosan, hydro gels were prepared and the properties of many high and low molecular weight drugs, particularly antibiotics, were analyzed.

In a study carried out with chitosan hydrogels, hydrogels containing pilocarpine were prepared using chitosan, β-glycerol phosphate and hydroxyethyle cellulose; the effects of gel composition, temperature and rheological properties of the gels on jellification rate and elastic resistance were analyzed [51].

Cerchiara et al. [52] prepared hydrogels of chitosan, which cross-link with lauric, miristic, palmitic and stearic acids by a freezing and drying method, and used hydrophilic propranolol hydrochloride as the model drug.

The formulation of antibiotics with chitosan hydrogels is also frequently analyzed. In a previous study, cross-linked hydro gels of amoxicillin and metronidazole with chitosan-polyethylene oxide were prepared by freezing and drying method and using swelling property of polymer network structure specific to pH, release amount of antibiotics in gastric fluid was analyzed. The study reported that 65% of amoxicillin and 59% of metronidazole were released. As a result, it was reported that chitosan-PEO hydro gels prepared by the freeze and dry method allow the local application of antibiotics in low pH environments like gastric fluid [53].

Another use of chitosan in drug delivery systems is preparation of membrane and films. Films are prepared using chitosan, preferably as polyanionic or neural polymer, or as a cross-link and thus membranes with high mechanical resistance are obtained.

Shu et al. [54] prepared chitosan- citrate films and found that swelling of the films was affected by pH and ionic strength. In addition, they reported that, as sodium chloride in the environment weakens the electrostatic interaction between citrate and chitosan, drug release from the films accelerated and swelling increased.

In a study that analyzed the effect of drying techniques on the membranes which were prepared with chitosan with a deacetylation degree lower than 80%, it was found that, compared to films prepared in an incubator and exposed to IR, the films dried in room temperature had higher water vapor and oxygen permeability [55].

Lopez and Bodmeier [56] prepared cross-linked chitosan and alginate films using chitosan glutamate and sodium alginate. When compared to dry films, wet alginate films had low resistance but higher elasticity; the swelling of alginate films in an acidic environment was not affected by $CaCl_2$ concentration; swelling and permeability properties of chitosan films depended on the pH level and cross-link.

Chitosan is a biopolymer, which is used in preparation of nano-microparticles, due to its poly cationic property, biocompatibility and inertness, permeability with many substances and easy biodegradation. Chitosan particles are prepared using methods such as conservation (simple and complex), solvent evaporation, suspension polymerization, ionotropic jellification and spray drying.

Bodmeier et al. [10] dripped a chitosan, solution containing active ingredient into anionic triolyphosphate (TPP) solution. A three-dimensional network structure formed as a result of ionotropic jellification and the active ingredient was retained in this structure. In another study, Bodmeier and Chen [57] prepared chitosan beads using water soluble and non-water-soluble active ingredients. They reported that, as the water solubility of the substances increased, release rate also increased; microparticles prepared using only chitosan are more appropriate for non-water-soluble drugs.

Chitosan nano and microparticles were also prepared, utilizing the bioadhesive property of chitosan, to enhance intestinal absorption of peptide and proteins [58, 59].

Many drugs of various groups were encapsulated with chitosan nano-microparticles and their treatment effectiveness was analyzed. The use of chitosan in anticancer drugs improved drug targeting and controlled release of the drug in the target area [60, 62].

Since chitosan is a biopolymer compatible with biologic environment, it is inert, biodegradable and its degradation products are non-toxic, considering that it can increase the effectiveness of vaccines and provide ease of use in biotechnology, it has been used in vaccine technology [63].

Illum et al. [64] made use of these bioadhesive properties and its extension of intercellular contact points. They prepared chitosan complexes containing influenza, whooping cough and diphtheria vaccines. Nasal application on test animals resulted in strong immune response and increased vaccination effectiveness. In addition, chitosan was used as a cationic polymer in non-viral gene applications. It was found that its effectiveness in gene expression depended on the dimension and stability of the complex, and the DNA protection capacity of the structure against DNaz enzyme and intracellular traffic [65, 66].

6. Biomedical Uses of Chitosan

The fact that chitosan has many applications in various pharmaceutical forms, such as film, hydro gel, fiber and nano particle; low toxicity and a biocompatibility with blood and tissue and many other properties, draws attention to commercial and biomedical applications. Many potential applications of this biopolymer are being analyzed in other fields [15, 16, 24, 67-78].

The use of chitin and derivatives as wound healing materials began with a study by Prudden et al. [43]. Prudden et al. found that shark cartilage accelerated healing of wounds, and suggested that glucosamine, which is one of the components of this structure, had an accelerating effect on wound healing [11]. Therefore, the researchers defined N-acetyl-D-glucosamine (GlcNAc) derived from glucosamine as a structure that has a role in wound healing. On the other hand, chitosan is a polysaccharide containing β-(1→4)-linked GlcNAc and D-glucosamine (GlcN) units. It was observed that topical healing findings of the wound treated with chitosan accelerated granulation [24]. In large and open wounds, the process of formation of new tissue is defined as granulation, which should form before wound epithelialization. According to this process, chitosan is a suitable polymer for wound treatment. Wound healing mechanisms of chitosan were investigated *in vitro* and *in vivo*. The literature suggests that chitosan is an effective material in treatment of burns [34]. Furthermore, commercial preparations of chitosan as suture [1] and wound covering material [27] are available.

Chitosan, which is mainly composed of is chitin, is known to have a haemostatic property. Although the mechanism of chitosan in wound and burn treatment has not been fully determined, it is considered that chitosan accelerates fibroblast formation in wound healing and increases early phase reactions related to healing. Research on various animal models reported that chitosan accelerated homeostasis, decreased fibrous tissue formation, facilitated osteogenesis and enhances tissue regeneration [11, 79].

Chitosan facilitated clotting when combined with heparinize or defibrinated blood and erythrocytes, and is therefore a good haemostatic agent that can be used in open wounds and burns in surgical operations [67, 70, 80, 81]. In addition, various animal models showed that chitosan was effective and reliable when used as a topical haemostatic or a wound healer biopolymer [16, 68, 69, 82]. This biopolymer, which has a high molecular weight, is thought to provide homeostasis by adhering erythrocyte cell membranes [16, 78, 83-85].

To analyze the effect of chitosan on wound healing, experimental open skin wounds were formed on dorsal region of dogs. Chitosan was applied to the wounds for 15 days and the results were evaluated in immunohistochemical terms. On the third day, the wounds treated with chitosan exhibited severe polymorphonuclear leukocyte (PMN) infiltration compared to the control. Migration of macrophages and foreign giant cells was accelerated by chitosan treatment at the third day. These findings indicate that, in early phases of healing, chitosan facilitated the migration of inflammatory cells, which can provide a considerable part of the production and secretion of growth factors. Granulation further accelerated at days 9 and 15 of chitosan treatment. The amount of collagen observed in experimental animals was higher than the controls. Immunohistochemical findings of collagen I, III and IV indicated that Type III collagen increased in the chitosan group [24]. Another group of researchers found that, in addition to accelerating the migration of PMN to the wound area, chitosan had a significant role in secretion of inflammatory mediators, such as tumor necrosis factor (TNF-α), interleukins (IL-1, IL-8 and IL-12) and macrophage inflammatory protein (MIP-1α and MIP-1β) [86, 87].

The literature contains data that indicate chitosan accelerates osteopontin production with PMN. Osteopontin (OPN), which is a ligand of $\alpha_V\beta_3$ integrin, is a glycosylated phosphoprotein. It facilitates proliferation or diffusion of various types of cell. In a previous study, it was observed that, in the presence of chitosan, the role of OPN in granulosus inflammation increased like other inflammatory cytokines [88].

As described above, Ueno et al. [80] conducted immunohistopathological analysis of the presence of OPN in the same granulation tissues using OPN antibody. The findings for granulation tissues at day 6 after injury indicated that, in immunohistochemical terms, when compared to tissues not treated with chitosan, OPN was more dominant in granulation tissues treated with chitosan. In conclusion, it was found that, when compared to tissues that were not treated with chitosan, wound tissues treated with chitosan had more OPN-positive PMN.

The effects of chitosan on macrophage properties were also analyzed. Chitosan is composed of GlcN and GlcNAc. It was found that macrophages express the receptors for mannose and GlcNAc- glycoprotein, which mediate glycoprotein intake. It is thought that bonding of GlcNAc to specific receptors is a precondition of increasing macrophage activation [89-92]. Macrophages have a significant role in wound healing. Activation of macrophages enables the release of various biological mediators and phagocytosis of foreign bodies. To determine the activation mechanism of chitosan on macrophages, peritoneal macrophages were prepared and the effects of polysaccharides containing chitin, chitosan and derivatives were analyzed. The results indicated that the expression of activation determinants, such as main histocompatibility complex (MHC) class I, class II, Fc receptors and mannose receptor, was induced after chitosan treatment. These findings support the suggestion that chitosan induces activation in peritoneal macrophages [93].

The literature contains findings suggesting that chitosan accelerates production of biological mediators [93-95]. Monocytes in circulation and differentiated active macrophages migrate to injury sites after stimulation. These macrophages synthesize the growth factors effecting the healing of the wound [93]. Glucan is a macrophage activating agent and accelerates cytokine production from macrophages. Since chitosan is a polysaccharide with glucan structure, it was reported to have a role in production of cytokines, which facilitate tissue repair [94, 95].

Nishimura et al. [96] reported that chitosan stimulated IL-1 production with macrophages. In addition to IL-6 and TNF α, the proliferation of fibroblast and collagen synthesis are both positively and negatively affected by IL-1 [93]. The macrophages stimulated with chitosan can produce other growth factors having a positive effect on ECM production. Therefore, transforming growth factor (TGF-β1) and platelet derivative growth factor (PDGF) production were measured from human monocyte derivative macrophages cultured with chitosan. Furthermore, mRNA expressions of TGF-β1 and PDGF stimulated with chitosan were detected *in vitro*. While it was observed that chitosan facilitates TGF-β1 and PDGF production, these results indicate that ECM production can be increased by growth factors [85].

The effects of chitin and chitosan on fibroblast activation and cytokine production in wound treatment have been widely analyzed [1, 85, 97]. In a previous study, chitin and derivatives were applied to rat dermal fibroblast cell culture. While it was reported that chitin and derivatives did not stimulate IL-6 production, production of IL-1α, IL-1β and TNF-α increased. The data confirms that chitin increased the production of cytokines, which have a role in angiogenesis and neutrophil migration. However, it was reported that chitin and derivatives did not have an accelerating effect on the proliferation of cultured fibroblasts [97].

In another study, which analyzed the effects of chitosan on L929 fibroblast cells and ECM production, L929 cells were cultured with chitosan and ECM production was evaluated *in vitro*. Type I and III collagen and fibronectin were secreted by L929, both with and without chitosan; however, it was reported that there was no significant difference between control and chitosan groups in terms of ECM amount. These results indicate that chitosan does not accelerate ECM production with fibroblasts, but can increase ECM production with growth factors such as TGF-α1 and PDGF [85].

Many researchers analyzed the effects of chitosan on wound healing in various animal models, and its accelerating effect on wound healing was analyzed in detail. In their collected work, Minami et al., [11] analyzed fields of application of different types of chitosans:

Chitipack S, which is prepared for traffic accident injuries; Chitipack P, developed as a covering material for large diameter fissures and cuts; and Chitipack C, developed as a covering material for burns and traumas. The common characteristic of these covering materials is that they accelerate homeostasis and proliferation, and accelerate healing of wounds or burns. Another collected work discussed the effectiveness of chitin/polyester, chitin/ cotton and chitosan/cotton wound covering materials on wound treatment in various types of animals. They reported accelerated tissue regeneration without scar tissue formation [1]. Muzzareli et al. [98] reported that, as a result of the clinical applications of chitin and chitosan, granulation developed rapidly in human chronic leg ulcers, epidermis formation increased and healing at an early phase was achieved.

Due to their accelerating effects in wound healing, such as optimal oxygen permeability, water absorption, moisturizing property, slow enzymatic degradation, haemostatic, antibacterial and re-epithelialization effects, film and hydro gels prepared with chitosan eliminate the need for repeated dressings and thus provide ease of use. These advantages led to the use of chitosan based film and hydro gels in wound and burn treatment [1]. With this purpose, Özmeriç et al. [99] prepared a film containing chitosan, which is known to have haemostatic and antibacterial effects, and taurine, an amino acid that regulates inflammation process. They analyzed the effectiveness of the film in dogs with induced teeth bone defects. Neutrophil and macrophage counts indicated that chitosan films prepared with taurine statistically accelerated healing at an earlier phase when compared to chitosan films without taurine content, and thus the healing effect of chitosan was increased.

A previous study compared Omiderm®, a commercial wound dressing product, with chitosan films prepared with two different solvents, acetic acid and lactic acid on the rats with an incision scar. The films were evaluated in terms of adhesion to the wound area, ease of application, elasticity and re-epithelialization properties in histological terms. The analysis indicated that the wounds treated with lactic acid and acetic acid achieved a healing similar to Omiderm®. In particular, the film prepared with acetic acid was found to be more elastic and easily applicable to the wound surface [100].

Güvercin et al. used a membrane form of chitosan to determine the effect on soft tissue healing. They studied wound healing with secondary epithelialization in Wistar-Alibino rats with an experimental incision scar on the back. The findings indicated that, compared to the controls, the groups treated with chitosan membranes achieved a better healing outcome in a shorter time [101].

Mi et al. [102] prepared chitosan films containing silver sulfadiazine, which were developed for use in burn treatment. They analyzed the antibacterial and treatment effectiveness of the films on bacterial culture containing *P. aeruginosa* and *S. aureus* and in 3T3 fibroblast cell culture. Chitosan film containing silver sulfadiazine had a long-term antibacterial effect; silver decreased potential toxic effects and they indicated that chitosan film containing silver sulfadiazine could be used as an effective covering material in treatment of infected burns.

Wang et al. [103] prepared chitosan-alginate films and tested the effect of the films, firstly *in vitro*, in rat and human fibroblast cell culture, and later *in vivo* , on rats with an incision scar. Macroscopic and histopathologic evaluations indicated that film applied wounds healed faster. After 14 days, it was observed that the wounds completely closed; at 7 and 14 days, fibroblast and thus re-epithelialization increased.

Macro porous sponge-like chitosan films were prepared using a phase precipitation method. It was reported that the oxygen permeability and swelling property of the film made it suitable for use as a wound covering. Micro porous films were applied to dermal wounds formed on the back of Wistar rats. It was observed that antimicrobial effect on the wound surface increased and no microorganism reproduction was observed. It was observed that, with increased chitosan, fibroblasts proliferation in the wound surface increased, with a corresponding increase in re-epithelialization [104].

By preparing sponge-like poly vinyl alcohol (PVA) chitosan /fibroin (PCF) films, deep dermal wounds were formed in Spraque-Dawley rats in such a way to include the dermis. To determine the treatment value in the formed wounds different combinations of films were used. Macroscopic and histopathologic evaluations of the experimental and control groups indicated the following wound healing process: PCF < chitosan / fibroin ≤ fibroin < PVA / chitosan < chitosan films [105].

Due to the above-mentioned properties, chitosan is used in tissue engineering and wound treatment. Another application is bone formation and the treatment of bone defects. Lee et al. [106] prepared chitosan sponges cross-linked with tricalcium phosphate, added the prepared sponges to rat osteoblastic cell culture and evaluated cell proliferation, alkali phosphate activity (ALPase) and calcium storage levels at 1, 7, 14, 28 and 56.days in histopathologic terms. The results indicated that cross-linked sponges stimulated the proliferation of osteoblastic cells by increasing ALPase activity, and that the cells stored calcium more effectively. The same researchers analyzed the effect on bone regeneration of cross-linked chitosan - tricalcium phosphate sponges containing platelet derivative growth factor-BB (PDGF-BB). The research marked the sponges containing PDGF-BB with I^{125} and implanted these into rats with an 8 mm bone defect at weeks 2 and 4 of treatment. Histopathologic analysis indicated that the sponges containing PDGF-BB increased bone regeneration, and a newly shaped healthy bone formed in the defect area [107].

The effect of chitosan and derivatives was analyzed on diabetic wounds formed on diabetic rats. Diabetic wounds are known to require a longer healing period than normal wounds. By cross-linking chitosan and PVP glutaraldehyde, semi-interpenetrated polymer network structures (semi-IPN) were formed. After *in vitro* controls, hydro gels with optimal characteristics were applied to rats with diabetic wounds. It was found that, in chitosan-PVP treated wounds, there was an increase in fibroblast proliferation and collaged accumulation. It was reported that, due to its wound-healing properties, chitosan-PVP hydro gel could be used as a covering material in diabetic wound healing [108]. The effect of FGF-chitosan films on wound models formed on diabetic rats was analyzed. Chitosan films containing basic fibroblast growth factor (bFGF) were prepared and the wound healing effectiveness of chitosan films containing bFGF was analyzed in diabetic rats. The results indicated that at the end of 21-day treatment, the wounds treated with chitosan film and bFGF containing chitosan film had similar histopathologic findings after 5-day treatment; however at 20-day treatment b FG-chitosan films were observed to be more effective, when considering only granulation tissue formation [109]. In a similar study, chitosan hydro gels prepared with UV-irridation and containing fibroblast growth factor-2 (FGF-2) were applied to wounds formed on diabetic and healthy rats. The data indicated that, with chitosan hydro gels, the wound surface closed faster in both diabetic and healthy rats, when compared to the controls. Data also showed that the addition of FGF-2 to chitosan hydro gels increased wound healing in terms of granulation tissue, capillary and epithelialization formation when compared to chitosan hydro gels [110].

The effectiveness of chitosan hydro gels prepared with UV-irradiation was also analyzed. Hydro gels were applied to wounds formed in the back of test rats and the results were compared with the control group. It was reported that wound narrowing and wound closing was observed in the wounds on which chitosan hydro gel was applied. It was found that, between the 2nd and 4th Days of treatment, granulation tissue began to form. In addition, it was reported that photo cross-linked chitosan hydro gels increased fibroblast formation in the wound area [111].

The accelerating effect of chitosan on wound healing is generally explained by the prevention of fibrous tissue formation and facilitating regeneration of organized tissue. Chitosan was applied to skin and subcutaneous incisions formed in the back of rats and healing was observed without fibroplasias or scar tissue formation. It was reported that, as healing with chitosan accelerates re-epithelialization, the tissue adapts to its normal structure in a short time [68]. Minoura et al. [112] analyzed the properties of hydro gels prepared using polyvinyl alcohol (PVA) and chitosan. They found that, in dermal incisions in rats, the hydro gels increased distention resistance of the wound at early phases of healing. Chung et al. [113] analyzed the effects of fungal chitin and chitosan obtained from cell culture on opossum L929 fibroblast cells, and reported that fungal chitin and chitosan accelerated healing, increased fibroblast migration and wound distention.

Hagiwara et al. [84] analyzed the proliferation of progenitor endothelial cells in chitosan treated Gore-tex vascular grafts. They observed that a clot layer formed, which increased proliferation; however, this layer did not form in control group grafts. While the surface of Gore-tex grafts treated with chitosan was fully covered with a clot layer, no clot layer was observed in the control group. Control group grafts were observed to be biocompatible however; these grafts were covered with fibrous ligament tissue. In conclusion, it is thought that chitosan prevents fibroplasias and helps the regeneration of normal tissues via the progenitor cells in the edge of the wound.

Muzzarelli et al. [98] carried out a series of studies on regeneration, bone defects and wound healing in dogs. Wounds were formed on radius bones, starting from the cortex and extending to the bone marrow. While the wounds treated with serum healed after callus formation with osteoblast-osteoclast activity, chitosan treated wounds healed by cortical bone formation without callus formation. The same researchers reported that chitosan derivatives stimulate re-growth of bone and chitosan can be an alternative for reconstructive purposes. The researchers reported that, in terms of bioactivity, chitosan andmethylpyrrolidone chitosan can be used as a potential osteoconductive agent in dental surgery and bone defects [114]. This chitosan derivative was hydrophilically modified. It is thought that this derivative bonded to fibroblast growth factors that stimulate angiogenesis with N-acetyleglucosamine units, which biologically resemble glucosamineglucans, and osteoblast-like cell proliferation. In another study [115], the same researchers analyzed the effect of chitosan in human oral injuries. The researchers applied chitosan to periodontal wounds and observed that fibroplasias decreased in the wound, and cell proliferation and tissue organization increased in periodontal soft tissue. Chitosan ascorbate gel was used to increase regeneration of periodontal tissue in humans, and it was reported that normal cell proliferation and organization increased in the damaged cell. Similarly, chitosan was used to treat defects formed in jawbone due to exodontia and apical resection. Radiography and biopsy techniques indicated that chitosan accelerated normal bone formation [116].

Based on the acceleration of re-epithelialization and wound healing of chitosan, it was thought that chitosan could be effective in post-operative adhesion formation, repair of surgical incisions and in anastomosis. When carboxymethyl chitosan and derivatives were applied to post-operative rats in the form of solution or hydro gels, it was observed that, in the small intestine and uterine horn lacerations, it increased the number, size and power of peritoneal adhesion, and was more effective than hyaluronic acid when used for this purpose [117]. On the other hand, it was reported in the literature that, when chitosan was applied to abdominal skin incisions, abdominal-aortic anastomosis and large intestine anastomosis, it had no adverse effect on the healing of the repaired incision wounds [118].

Recently, dermal substitution and wound healing have been among the research areas of medicine in which there have been many recent advances, but neither the commercially available products nor the materials currently described in experimental studies are able to fully substitute for natural living skin. On the other hand, healing of dermal wounds with macromolecular agents such as natural polymers is preferred to skin substitutes for many advantages such as biocompatibility, non-irritant and non-toxic properties, easy and safeapplication on dermis. With this aim, Sezer et al. prepared different pharmacutical dosage forms containing chitosan and fucoidan such as film [119], hydrogel [120] and microsphere [121] and evaluated their healing effect on dermal burns on a rabbit. In conclusion, the in vitro and in vivo studies investigating the efficacy of these three forms in the treatment of experimental dermal burns demonstrated that the use of chitosan in the treatment of burns was due to its ability to re-epithelize and encourage fibroblast migration. Although fucoidan solution alone was not as effective as chitosan, the burn healing effect of fucoidan-chitosan films, hydrogels and microspheres were observed to be better than chitosan dosage forms.

REFERENCES

[1] Shigemasa Y., Minami S.: Applications of chitin and chitosan for biomaterials. *Biotechnol. Genet. Eng. Rev.*, 13: 383-420, 1995.

[2] Singla A.K., Chawla M.: Chitosan: some pharmaceutical and biological aspect-an update. *J. Pharm. Pharmacol.,* 53: 1047-1067, 2000.

[3] Chitosan. Ed: Budavari S.: The Merck Index, s. 342, Merck and Co., Whitehouse Station, Newjersy, 1996.

[4] Muzzarelli C., Muzzarelli R.A.A.: Natural and artificial chitosan-inorganic composites. *J. Inorg. Biochem.*, 92: 89-94, 2002.

[5] Krajewska B.: Application of chitin and chitosan based materials for enzyme immobilizations: a review. *Enzyme Microb. Tech.*, 35: 126-139, 2004.

[6] Chitosan. Ed: Rowe R.C., Sheskey P.J., Weller P.J.: Handbook of Pharmaceutical Excipients. p. 132-135, Pharmaceutical Press, London, 2003.

[7] Domard A., Domard M.: Chitosan: structure-properties relationship and biomedical applications. Ed: Severian D., *Polymeric Biomaterials*. p. 187-212, Marcel Dekker, New York, 2002.

[8] Natural condensation polymer feedstocks. Ed: Carraher C.E., Swift G.G., *Functional Condensation Polymers*. p. 162-165, Kluwer Academic, Hingham, 2002.

[9] Synowiecki J., Al-Khateeb N.A.: Production, properties, and some new applications of chitin and its derivatives. *Crit. Rev. Food Sci.*, 43: 145-171, 2003.

[10] Bodmeier R., Oh K.H., Pramar Y.: Preparation and evaluation of drug-containing chitosan beads. *Drug Dev. Ind. Pharm.*, 15: 1475-1494, 1989.

[11] Minami S., Okamoto Y., Hamada K., Fukumoto Y., Shigemasa Y.: Veterinary practice with chitin and chitosan. Ed: Jolles P., Muzzarelli R.A.A.: Chitin and Chitinases. p. 265-277, Birkhauser Verlag, Basel, 1999.

[12] Sashiwa H., Aiba S.: Chemically modified chitin and chitosan as biomaterials. *Prog. Polym. Sci.,* 29: 887-908, 2004.

[13] Ravindra R., Krovvidi K.R., Khan A.A.: Solubility parameter of chitin and chitosan. *Carbohyd. Polym.*, 36: 121-127, 1998.

[14] Dodane V., Vilivalam V.D.: *Pharmaceutical applications of chitosan.* PSTT, 1: 246-253, 1998.

[15] Berthod F., Saintigny G., Chretien F., Hayek D., Collombel C., Damour O.: Optimization of thickness, pore size and mechanical properties of a biomaterial designed for deep burn coverage. *Clin. Mater.*, 15: 259-265, 1994.

[16] Rao S.B., Sharma C.P.: Use of chitosan as a biomaterials: studies on its safety and hemostatic potential. *J. Biomed. Mater. Res.*, 34: 21-28, 1997.

[17] Jumaa M., Müller B. W.: Physicochemical properties of chitosan-lipid emulsions and their stability during the autoclaving process. *Int. J. Pharm.,* 183: 175-184, 1999.

[18] Zahraoui C., Sharrock P.: Influence of sterilization on injectable bone biomaterials. *Bone*, 25: 63-65, 1999.

[19] Larena A., Caceres D.A., Vicario C., Fuentes A.: Release of a chitosan-hydroxyapatite composite loaded with ibuprofen and acetyl-salicylic acid submitted to different sterilization treatments. *Appl. Surf. Sci.*, 238: 518-522, 2004.

[20] Erden N., Çelebi N.: Applied of chitin and chitosan on pharmaceutical technology. *FABAD J. Pharm. Sci.*, 15: 277-287, 1990.

[21] Usami Y., Minami S., Okamoto Y., Matsuhashi A., Shigemasa Y.: Influence of chain length of N-acetyl-D-glucosamine and D-glucosamine residues on direct and complement-mediated chemotactic activities for canine polymorphonuclear cells. *Carbohyd. Polym.*, 32: 115-122, 1997.

[22] Okamoto Y., Kawakami K., Miyatake K., Morimoto M., Shigemasa Y., Minami S.: Analgesic effects of chitin and chitosan. *Corbohyd. Polym.*, 49: 249-252, 2002.

[23] Miyatake K., Okamoto Y., Shigemasa Y., Tokura S., Minami S.: Anti-inflammatory effect of chemically modified chitin. *Carbohyd. Polym.*, 53: 417-423, 2003.

[24] Ueno H., Yamada H., Tanaka I., Kaba N., Matsuura M., Okumura M., Kadosawa T., Fujinaga T.: Accelerating effects of chitosan for healing at early phase of experimental open wound in dogs. *Biomaterials*, 20: 1407-1414, 1999.

[25] Khor E., Lim L.Y.: Implantable applications of chitin and chitosan. *Biomaterials,* 24: 2339-2349, 2003.

[26] Şenel S., McClure S.J.: Potential applications of chitosan in veterinary medicine. *Adv. Drug. Deliver. Rev.*, 56: 1467-1480, 2004.

[27] Ueno H., Mori T., Fujinaga T.: Topical formulation and wound healing applications of chitosan. *Adv. Drug Deliver. Rev.*, 52: 105-115, 2001.

[28] Lehr C.M., Bouwstra J.A., Schacht E.H., Junginger H.E.: In vitro evaluation of mucoadhesive properties of chitosan and some other natural polymers. *Int. J. Pharm.,* 78: 43-48, 1992.
[29] Knapczyk J., Macura A.B., Pawlik B.: Simple tests demonstrating the antimycotic effect of chitosan. *Int. J. Pharm.*, 80: 33-38, 1992.
[30] Aspden T.J., Adler J., Davis S.S., Skaugrud Ø., Illum L.: Chitosan as a nasal delivery system: evaluation of the effect of chitosan on mucociliary clearance rate in the frog palate model. *Int. J. Pharm.*, 122: 69-78, 1995.
[31] Shiraishi S., Imai T., Otagiri M.: Controlled release of indomethacin by chitosan-polyelectrolyte complex: optimization and in vivo/invitro evaluation. *J. Control. Release,* 25: 217-225, 1993.
[32] Braier N.C., Jishi R.A.: Density functional studies of Cu^{2+} and Ni^{2+} binding to chitosan. *J. Mol. Struc.*, 499: 51-55, 2000.
[33] Varma A.J., Deshpande S.V., Kennedy J.F.: Metal complexation by chitosan and its derivatives: a review. *Carbohyd. Polym.,* 55: 77-93, 2004.
[34] Guibal E.: Interactions of metal ions with chitosan-based sorbents: a review. *Sep. Purif. Technol.*, 38: 43-74, 2004.
[35] Zhang H., Neau S.: In vitro degradation of chitosan by a commercial enzyme preparation: effect of molecular weight and degree of deacetylation. *Biomaterials,* 22: 1653-1658, 2001.
[36] Mao S., Shuai X., Unger F., Simon M., Bi D., Kissel T.: The depolymerization of chitosan: effects on physicochemical and biological properties. *Int. J. Pharm.*, 281: 45-54, 2004.
[37] Montembault A., Viton C., Domard A.: Physico-chemical studies of the gelation of chitosan in a hydroalcoholic medium. *Biomaterials*, 26: 933-943, 2005.
[38] Janes K.A., Calvo P., Alonso M.J.: Polysaccharide colloidal particles as delivery systems for macromolecules. *Adv. Drug. Deliver. Rev.*, 47: 83-97, 2001.
[39] Akbuğa J.: A biopolymer: chitosan. *International Journal of Pharmaceutical Advances*, 1, 3-18, 1995.
[40] Akbuğa J., Özbaş-Turan S., Erdoğan, N.: Plasmid-DNA loaded chitosan microspheres for *in vitro* IL-2 expression. *European Journal of Pharmaceutics and Biopharmacutics*, 58: 501-507, 2004.
[41] Sezer A.D., Akbuğa J.: Fucosphere-new microsphere carriers for peptide and protein delivery: preparation and in vitro characterization. *Journal of Microencapsulation*, 23, 513-522, 2006.
[42] Kaş H. S.: Chitosan: properties, preparation and application to microparticulate systems. *J. Microencapsul.*, 14: 689-711, 1997.
[43] Kumar M.N.V.R.: A review of chitin and chitosan applications. *React. Funct. Polym.*, 46: 1-27, 2000.
[44] Hejazi R., Amiji M.: Chitosan-based gastrointestinal delivery systems. *J. Control. Release*, 89: 151-165, 2003.
[45] Sinha V.R., Singla A.K., Wadhawan S., Kaushik R., Kumria R., Bansal K., Dhawan S.: Chitosan microspheres as a potential carrier for drugs. *Int. J. Pharm.*, 274: 1-33, 2004.
[46] Sawayanagi Y., Nambu N., Nagai T.: Permeation of drugs through chitosan membrans. *Chem. Pharm. Bull.*, 30: 3297-3301, 1982.

[47] Sawayanagi Y., Nambu N., Nagai T.: Dissolution properties and bioavailability of phenytoin from ground mixtures with chitin or chitosan. *Chem. Pharm. Bull.*, 31: 2064-2068, 1983.
[48] Sawayanagi Y., Nambu N., Nagai T.: Enhancement of dissolution properties of prednisolone from ground mixtures with chitin or chitosan. *Chem. Pharm. Bull.*, 31: 2507-2509, 1983.
[49] Machida Y, Nagai T., Abe M., Sannan T.: Use of chitosan and hydroxypropylchitosan in drug formulations to effect sustained release. *Drug Des. Deliv.*, 1: 119-130, 1986.
[50] Knapczyk J.: Excipient ability of chitosan for direct tableting. *Int. J. Pharm.*, 89: 1-7, 1993.
[51] Li J., Xu Z.: Physical characterization of a chitosan-based hydrogel delivery system. *J. Pharm. Sci.*, 91: 1669-1677, 2002.
[52] Cerchiara T., Luppi B., Bigucci F., Orienti I., Zecchi V.: Physically cross-linked chitosan hydrogels as topical vehicles for hydrophilic drugs. *J. Pharm. Pharmacol.*, 54: 1453-1459, 2002.
[53] Patel V.R., Amiji M.M.: Preparation and characterization of freeze-dried chitosan-poly (ethylene oxide) hydrogels for site-specific antibiotic delivery in the stomach. *Pharm. Res.,* 13: 588-593, 1996.
[54] Shu X.Z., Zhu K.J., Song W.: Novel pH-sensitive citrate cross-linked chitosan film for drug controlled release. *Int. J. Pharm.*, 212: 19-28, 2001.
[55] Srinivasa P.C., Ramesh M.N., Kumar K.R., Tharanathan R.N.: Properties of chitosan films prepared under different drying conditions. *J. Food. Eng.*, 63: 79-85, 2004.
[56] Lopez C.R., Bodmeier R.: Mechanical, water uptake and permeability of crosslinked chitosan glutamate and alginate films. *J. Control. Release*, 44: 215-225, 1997.
[57] Bodmeier R., Chen H., Paeratakul O.: A novel approach to the oral delivery of micro or nanoparticles. *Pharm. Res.*, 6: 413-417, 1989.
[58] Pan Y., Li Y., Zhao H., Zhen J., Xu H., Wei G., Hao J., Cui F.: Bioadhesive polysaccharide in protein delivery system: chitosan nanoparticles improve the intestinal absorption of insulin in vivo. *Int. J. Pharm.*, 249: 139-147, 2002.
[59] Xu Y., Du Y.: Effect of molecular structure of chitosan on protein delivery properties of chitosan nanoparticles. *Int. J. Pharm.,* 250: 215-226, 2003.
[60] Wang Y.M., Sato H., Adachi I., Horikoshi I.: Optimization of the formulation design of chitosan microspheres containing cisplatin. *J. Pharm. Sci.*, 85: 1204-1210, 1996.
[61] Chandy T., Das G.S., Rao G.H.R.: 5-fluorouracil-loaded chitosan coated polylactic acid microspheres as biodegradable drug carriers for cerebral tumours. *J. Microencapsul.*, 17: 625-638, 2000.
[62] Mitra S., Gaur U., Ghosh P.C., Maitra A.N.: Tumour targeted delivery of encapsulated dextran-doxorubicin conjugate using chitosan nanoparticles as carrier. *J. Control. Release,* 74: 317-323, 2001.
[63] Calvo P., Lopez C. R., Jato J.L.V., Alonso M.J.: Chitosan and chitosan/ethylene oxide-propylene oxide block copolymer nanoparticles as novel carriers for proteins and vaccines. *Pharm. Res.*, 14: 1431-1436, 1997.
[64] Illum L., Jabbal-Gill I., Hinchcliffe M., Fisher A.N., Davis S.S.: Chitosan as a novel nasal delivery system for vaccines. *Adv. Drug. Deliver. Rev.*, 51: 81-96, 2001.
[65] Borchard G.: Chitosan for gene delivery. *Adv. Drug. Deliv. Rev.*, 52: 145-150, 2001.

[66] Liu W.G., Yao K.D.: Chitosan and its derivatives-apromising non-viral vector for gene transfection. *J. Control. Release*, 83: 1-11, 2002.

[67] Tan W., Krishnaraj R., Desai T.A.: Evaluation of nanostructured composite collagen-chitosan matrices for tissue engineering. *Tissue Eng*., 7: 203-210, 2001.

[68] Cho Y.W., Cho Y.N., Chung S.H., Yoo G., Ko S.W.: Water-soluble chitin as a wound healing accelerator. *Biomaterials*, 20: 2139-2145, 1999.

[69] Howling G.I., Dettmar P.W., Goddard P.A., Hampson F.C., Dornish M., Wood E.J.: The effect of chitin and chitosan on the proliferation of human skin fibroblast and keratinocytes in vitro. *Biomaterials,* 22: 2959-2966, 2001.

[70] Ishihara M., Nakanishi K., Ono K., Sato M., Kikuchi M., Saito Y., Yura H., Matsui T., Hattori H., Uenoyama M., Kurita A.: Photocrosslinkable chitosan as a dressing for wound occlusion and accelerator in healing process. *Biomaterials,* 23: 833-840, 2002.

[71] Klokkevold P.R., Lew D.S., Ellis D.G., Bertolami C.N.: Effect of chitosan on lingual hemostasis in rabbits. *J. Oral Maxillofac. Surg*., 49: 858-863, 1991.

[72] Klokkevold P.R., Subar P., Fukayama H., Bertolami C.N.: Effect of Chitosan on lingual hemostasis in rabbits with platelet dysfunction induced by epoprostenol. *J. Oral Maxillofac. Surg*., 50: 41-45, 1992.

[73] Okamoto Y., Shibazaki K., Minami S., Matsuhashi A., Tanioka S., Shigemasa Y.: Evaluation of chitin and chitosan on open wound healing in dogs. *J. Vet. Med. Sci*., 57: 851-854, 1995.

[74] Minami S., Masuda M., Suzuki H., Okamoto Y., Matsuhashi A., Kato K., Shigemasa Y.: Subcutaneous injected chitosan induces systemic activation in dogs. *Carbohyd. Poly*., 33: 285-294, 1997.

[75] Klokkevold P.R., Fukayama H., Sung E.C., Bertolami C.N.: The effect of chitosan (poly-N-acetylglucosamine) on lingual hemostasis in heparinized rabbits. *J. Oral Maxillofac. Surg*., 57: 49-52, 1999.

[76] Suh J.K.F., Matthew H.W.T.: Application of chitosan-based polysaccharide biomaterials in cartilage tissue engineering: a review. *Biomaterials*, 21: 2589-2598, 2000.

[77] Kim I., Park J. W., Kwon I.C., Baik B.S., Cho B.C.: Role of BMP βig-h3, and chitosan in early bony consolidation in distraction osteogenesis in a dog model. *Plast. Reconstr. Surg*., 109: 1966-1977, 2002.

[78] Mi F., Wu Y., Shyu S., Schoung J., Huang Y., Tsai Y., Hao J.: Control of wound infections using a bilayer chitosan wound dressing with sustaniable antibiotic delivery. *J. Biomed. Mater. Res*., 59: 438-449, 2002.

[79] Paul W., Sharma C.P.: Chitosan and alginate wound dressings: a short review. *Trends Biomater. Artif. Organs,* 18: 18-23, 2004.

[80] Ueno H., Murakami M., Okumura M., Kadosawa T., Uede T., Fujinaga T.: Chitosan accelerates the production of osteopontin from polymorphonuclear leukocytes. *Biomaterials*, 22: 1667-1673, 2001.

[81] Kweon D., Song S., Park Y.: Preparation of water-soluble chitosan/heparin complex and its application as wound healing accelerator. *Biomaterials*, 24: 1595-1601, 2003.

[82] Tamai Y., Miyatake K., Okamoto Y., Takamori Y., Sakamoto H., Minami S.: Enhanced healing of cartilaginous injuries by glucosamine hydrochloride. *Carbohyd. Poly*., 48: 369-378, 2002.

[83] Suzuki Y., Okamoto Y., Morimoto M., Sashiwa H., Saimoto H., Tanioka S.I., Shigemasa Y., Minami S.: Influence of physico-chemical properties of chitin and chitosan on complement activation. *Carbohyd. Polym*. 42: 307-310, 2000.
[84] Hagiwara K., Kuribayashi Y., Iwai H., Azuma I., Tokura S., Ikuta K., Ishihara C.: A sulfated chitin inhibits hemagglutination by Theileria sergenti merozoites. *Carbohyd. Polym.*, 39: 245-248, 1999.
[85] Ueno H., Nakamura F., Murakami M., Okumura M., Kadosawa T., Fujinaga T.: Evaluation effects of chitosan for the extracellular matrix production by fibroblasts and the growth factors production by macrophages. *Biomaterials,* 22: 2125-2130, 2001.
[86] Lloyd A.R., Oppenhiem J.J.: Poly's lament: The neglected role of the polymorphonuclear neutrophil in the afferent limb of the immune response. *Immunol. Today,* 13: 169-172, 1992.
[87] Cassatella M.A.: The production of cytokines by polymorphonuclear neutrophils. Immunol. *Today,* 16: 21-26, 1995.
[88] Senger D.R., Perruzzi C.A., Papadopoulos-Sergiou A., Water L.V.: Adhesive properties of osteopontin: regulation by a naturraly occurring thrombin-cleavage in close proximity to the GRGDS cell-binding domain. *Mol. Biol. Cell*, 5: 565-574, 1994.
[89] Peluso G., Petillo O., Ranieri M., Santin M., Ambrosio L., Calabro D., Avallone B., Balsamo G.: Chitosan-mediated stimulation of macrophage function. *Biomaterials,* 15: 1215-1220, 1994.
[90] Warr G.A.: A macrophage receptor for (mannose/glucosamine)-glycoproteins of potential importance in phagocytic activity. *Biochem. Bioph. Res.* Co., 93: 737-745, 1980.
[91] Porporatto C., Bianco I.D., Riera C.M., Correa S.G.: Chitosan induces different L-arginine metabolic pathways in resting and inflammatory macrophages. *Biochem. Bioph. Res.* Co., 304: 266-272, 2003.
[92] Feng J., Zhao L., Yu Q.: Receptor-mediated stimulatory effect of oligochitosan in macrophages. *Biochem. Bioph. Res.* Co., 317: 414-420, 2004.
[93] DiPietro L.A.: Wound healing: the role of the macrophage and other immune cells. *Shock*, 4: 233-240, 1995.
[94] Gonzalez J.A., Digby J.D., Rice P.J., Breuel K.F., DePonti W.K., Kalpfleisch J.H., Browder W., Williams D.L.: At low serum glucan concentrations there is an inverse correlation between serum glucan and serum cytokine levels in ICU patients with infections. *Int. Immunopharmacol.*, 4: 1107-1115, 2004.
[95] Berner M.D., Sura M.E., Alves B.N., Hunter K.W.: IFN-γ primes macrophages for enhanced TNF-α expression in response to stimulatory and non-stimulatory amounts of microparticulate β-glucan. *Immunol. Lett.*, 98: 115-122, 2005.
[96] Nishimura K., Ishihara C., Ukei S., Tokura S., Azuma I.: Stimulation of cytokine production in mice using deacetylated chitin. *Vaccine*, 4: 151-156, 1986.
[97] Mori T., Okumura M., Matsuura M., Ueno K., Tokura S., Okamoto Y., Minami S., Fujinaga T.: Effects of chitin and its derivatives on the proliferation and cytokine production of fibroblasts in vitro. *Biomaterials*, 18: 947-951, 1997.
[98] Muzzarelli R.A.A, Mattioli-Belmonte M., Tietz C., Biagini R., Ferioli G., Brunelli M.A., Fini M., Giardino R., Ilari P., Biagini G.: Stimulatory effect on bone formation exerted by a modified chitosan. *Biomaterials*, 15: 1075-1081, 1994.

[99] Özmeriç N., Özcan G., Haytaç C.M., Alaaddinoğlu E.E., Sargon M.F., Şenel S.: Chitosan film enriched with an antioxidant agent, taurine in fenestration defects. *J. Biomed. Mater. Res.*, 51: 500-503, 2000.

[100] Khan T.A., Peh K.K.: A preliminary investigation of chitosan film as dressing for punch biopsy wounds in rats. *J. Pharm. Pharmaceut. Sci.*, 6: 20-26, 2003.

[101] Güvercin M.: Experimental investigation of chitosan effects on soft tissue and evaluation of cell culture. Marmara University Health Sciences Institute, Ph.D. Thesis, İstanbul, Turkey 2004 (Supervisor: Prof. E. Göker and J. Akbuğa).

[102] Mi F., Wu Y., Shyu S., Chao A., Lai J., Su C.: Asymmetric chitosan membranes prepared by dry/wet phase separation: a new type of wound dressing for controlled antibacterial release. *J. Membrane Sci.*, 212: 237-254, 2003.

[103] Wang L., Khor E., Wee A., Lim L. Y.: Chitosan-alginate PEC membrane as a wound dressing: assessment of incisional wound healing. *J. Biomed. Mater. Res.* (Appl. Biomater.), 63: 610-618, 2002.

[104] Mi F., Shyu S., Wu Y., Lee S., Shyong J., Huang R.: Fabrication and characterization of a sponge-like asymmetric chitosan membrane as a wound dressing. *Biomaterials,* 22: 165-173, 2001.

[105] Yeo J.H., Lee K.G., Kim H.C., Oh Y.L., Kim A., Kim S.Y.: The effects of PVA/chitosan/fibroin (PCF)-blended spongy sheets on wound healing in rats. *Biol. Pharm. Bull.,* 23: 1220-1223, 2000.

[106] Lee Y., Park Y., Lee S., Ku Y., Han S., Choi S., Klokkevold P.R., Chung C.: Tissue engineered bone formation using chitosan/tricalcium phosphate sponges. *J. Periodontol.*, 71: 410-417, 2000.

[107] Lee Y., Park Y., Lee S., Ku Y., Han S., Klokkevold P.R., Chung C.: The bone regenerative effect of platelet-derived growth factor-bb delivered with a chitosan/tricalcium phosphate sponge carrier. *J. Periodontol.*, 71: 418-424, 2000.

[108] Karslı S.: Investigation of the usage of chitosan hydrogels in diabetic wound animals. Marmara University Health Sciences Institute, M.Sc. Thesis, İstanbul, 2006 (Supervisor: Prof. J. Akbuğa).

[109] Mizuno K., Yamamura K., Yano K., Osada T., Saeki S., Takimoto N., Sakurai T., Nimura Y.: Effect of chitosan film containing basic fibroblast growth factor on wound healing in genetically diabetic mice. *J. Biomed. Mater. Res.*, 64A: 177-181, 2003.

[110] Obara K., Ishihara M., Ishizuka T., Fujita M., Ozeki Y., Maehara T., Saito Y., Yura H., Matsui T., Hattori H., Kikuchi M., Kurita A.: Photocrosslinkable chitosan hydrogel containing fibroblast growth factor-2 stimultes wound healing in healing-impaired db/db mice. *Biomaterials,* 24: 3437-3444, 2003.

[111] Ishihara M., Ono K., Sato M., Nakanishi K., Saito Y., Yura H., Matsui T., Hattori H., Fujita M., Kikuchi M., Kurita A.: Acceleration of wound concentration and healing with a photocrosslinkable chitosan hydrogel. *Wound Rep. Reg.*, 9: 513-521, 2001.

[112] Minoura N., Koyano T., Koshizaki N., Umehara H., Nagura M., Kobayashi K.: Preparation, properties, and cell attachment/growth behavior of PVA/chitosan-blended hydrogels. *Mat. Sci. Eng.* C., 6: 275-280, 1998.

[113] Chung L.Y., Schmidt R.J., Hamlyn P.F., Sagar B.F., Andrews A.M., Turner T.D.: Biocompatibility of potential wound management products: hydrogen peroxide generation by fungal chitin/chitosan and their effects on the proliferation of murine L929 fibroblast in culture. *J. Biomed. Mater. Res.*, 39: 300-307, 1998.

[114] Muzzarelli R., Baldassarre V., Conti F., Ferrara P., Biagini G., Gazzanelli G., Vasi V.: Biological activity of chitosan: ultrastructural study. *Biomaterials*, 9: 247-252, 1988.
[115] Muzzarelli R., Biagini G., Pugnaloni A., Filippini O., Baldassarre V., Castaldini C., Rizzoli C.: Reconstruction of parodontal tissue with chitosan. *Biomaterials*, 10: 598-603, 1989.
[116] Muzzarelli R.A.A., Biagini G., Bellardini M., Simonelli L., Castaldini C., Fratto.: Osteoconduction exerted by methylpyrrolidinone chitosan used in dental surgery. *Biomaterials,* 14: 39-43, 1993.
[117] Kennedy R., Costain D.J., McAlister V.C., Lee T.D.G.: Prevention of experimental postoperative peritoneal adhesions by N,O-carboxymethyl chitosan. *Surgery,* 120: 866-870, 1996.
[118] Costain D.J., Kennedy R., Ciona C., McAlister V.C., Lee T.D.G.: Prevention of postsurgical adhesions with N,O-carboxymethyl chitosan: examination of the most efficacious preparation and the effect of N,O-carboxymethyl chitosan on postsurgical healing. *Surgery,* 121: 314-319, 1997.
[119] Sezer A.D., Hatipoğlu F., Cevher E., Oğurtan Z., Baş A.L., Akbuğa J.: Chitosan films containing fucoidan as a wound dressing for dermal burn healing: preparation and in vitro/in vivo evaluation. *AAPS PharmSciTech*, 8:2, Article 39, E1-E8, 2007.
[120] Sezer A.D., Cevher E., Hatipoğlu F., Oğurtan Z., Baş A.L., Akbuğa J.: Preparation of fucoidan-chitosan hydrogel and its application as burn healing accelerator on rabbit. *Biological and Pharmaceutical Bulletin*, 31: 2326-2333, 2008.
[121] Sezer A.D., Cevher E., Hatipoğlu F., Oğurtan Z., Baş A.L., Akbuğa J.: The use of fucosphere in the treatment of dermal burns in rabbits. *European Journal of Pharmaceutics and Biopharmaceutics*, 69: 189-198, 2008.

In: Focus on Chitosan Research
Editors: Arthur N. Ferguson and Amy G. O'Neill
ISBN 978-1-61324-454-8

Chapter 14

COMBINED USE OF BIOPOLYMER CHITOSAN AND ENZYME TYROSINASE FOR REMOVAL OF BISPHENOL A AND ITS DERIVATIVES FROM AQUEOUS SOLUTIONS

***Kazunori Yamada*[1], *Mizuho Suzuki*[1], *Ayumi Kashiwada*[1], and *Kiyomi Matsuda*[2]**
[1.] Department of Applied Molecular Chemistry,
College of Industrial Technology,
Nihon University, Narashino, Chiba, Japan.
[2.] Department of Sustainable Engineering,
College of Industrial Technology,
Nihon University, Narashino, Chiba, Japan.

ABSTRACT

In this chapter, the availability of chitosan was systematically investigated for removal of bisphenol A (BPA, 2,2-bis(hydroxyphenyl)propane) through the tyrosinase-catalyzed quinone oxidation. First, the process parameters, such as the hydrogen peroxide (H_2O_2)-to-BPA ratio, pH value, temperature, and tyrosinase dose, were discussed in detail for the enzymatic quinone oxidation of BPA. Tyrosinase-catalyzed quinone oxidation of BPA was found to be effectively enhanced by adding H_2O_2, and the optimum conditions for BPA at 0.3 mM were determined to be pH 7.0 and 40°C in the presence of H_2O_2 at 0.3 mM ($[H_2O_2]/[BPA]=1.0$). Removal of BPA from aqueous solutions was accomplished by adsorption of enzymatically generated quinone derivatives on chitosan beads. The use of chitosan in the form of porous beads was found to be more effective than that in the form of powders and solutions because heterogeneous removal of BPA with chitosan beads was much faster than homogeneous removal of BPA with chitosan solutions, and quinone conversion for the heterogeneous system with chitosan beads was a little higher than that for the heterogeneous system with chitosan powders. The removal efficiency was enhanced by increasing the amount of chitosan beads dispersed in the BPA solutions and BPA was completely removed by quinone adsorption in the presence of chitosan beads more than 0.10 cm^3/cm^3. In addition, a variety of bisphenol derivatives were completely or effectively removed by the procedure constructed in this study, although the enzyme dose and/or the amount of chitosan beads were further increased as necessary for some of the bisphenol derivatives used.

1. Introduction

Water Pollution and Endocrine-Disrupting Chemicals

Recently, pollution of surface water such as river, lake, and swamp, has been a major concern that needs to be solved. Various substances are discharged into the environment through industrial effluents. Among them are substances referred to as environmental hormones, or to be accurate an endocrine disrupting chemicals such as polychlorinated biphenyl, nonylphenol, octylphenol, and bisphenol A (2,2-bis(hydroxyphenyl)propane, BPA). When these substances enter the body through the mouth, they serve a similar function to female hormone [1,2]. These environmental pollutants can cause adverse effects on human [3-6] and wildlife such as fish, amphibians, and birds even in small amounts.

BPA and Bisphenol Derivatives

BPA as one of the endocrine disrupting chemicals is a ubiquitous substance used mainly in the production of epoxy resins and polycarbonate plastics or as an antioxidant [7-11]. The annual production of BPA stood at 520,000 ton in Japan as of in 2005, and have continued to increase. Estrogenic activity of BPA was published by Dodds et al., in 1930s [12]. Thereafter, the estrogenic effects of humans and animals have been reported in many articles [13-16]. BPA is frequently detected in effluent samples of wastewater treatment plants and also found in sediments and fish. Unfortunately, one of the most likely sources is the leachate from hazardous waste landfill. When waste plastics produced from BPA are buried in a landfill, a hydrolytic or leaching process may occur to release BPA to the leachate [17-19].

Recently the bisphenol derivatives such as bisphenol B (BPB) [20], bisphenol C (BPC) [21-25], bisphenol E (BPE) [26-28], bisphenol F (BPF) [27-34], bisphenol O (BPO) [35], bisphenol S (BPS) [36-40], bisphenol T (BPT) [41], and bisphenol Z (BPZ) [42,43] shown in Figure 1 have come into wide use for synthesis of specialized epoxy and polycarbonate resins with more high-performance compared with ordinary ones or other modifications. Although these bisphenol derivatives have two phenol groups in common, the chemical structure between two phenol groups is different. Some of these bisphenol derivatives are reported to exhibit estrogenic activity in human breast cancer cell line MFC-7 or inhibitory effects on the androgenic activity of 5α-dihydrotestrosterone in mouse fibroblast cell line NIH3T3. Their results suggested that the para position of phenolic OH group was required for hormonal activities, and substituents at meta positions and bridging alkyl moiety influenced the activity [44-46]. However, further details have yet to be revealed.

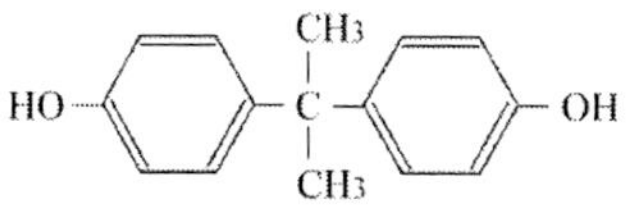

bisphenol A (BPA)
2,2-bis(4-hydroxyphenyl)propane

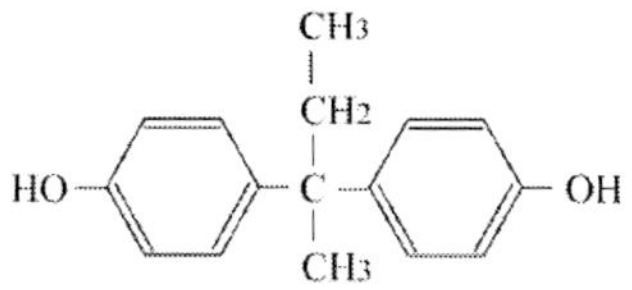

bisphenol B (BPB)
2,2-bis(4-hydroxyphenyl)butane

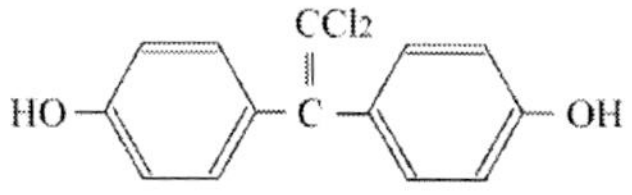

bisphenol C (BPC)
1,1-dichloro-
2,2-bis(4-hydroxyphenyl)ethylene

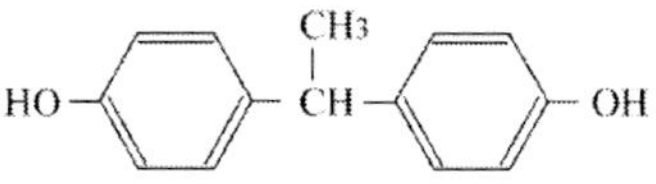

bisphenol E (BPE)
4,4'-ethylidenebisphenol

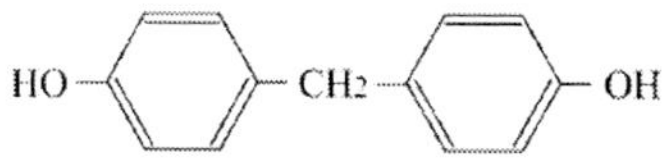

bisphenol F (BPF)
4,4'-dihydroxydiphenylmethane

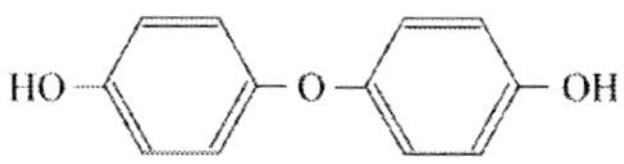

bisphenol O (BPO)
4,4'-dihydroxydiphenyl ether

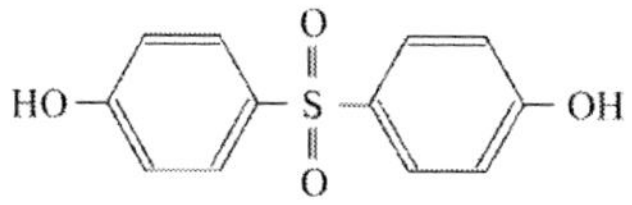

bisphenol S (BPS)
4,4'-dihydroxydiphenyl sulfone

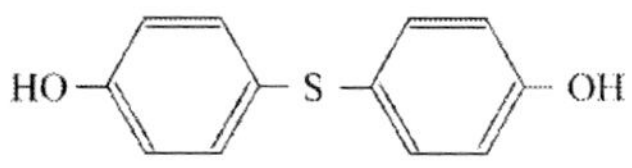

bisphenol T (BPT)
bis(4-hydroxyphenyl)sulfide

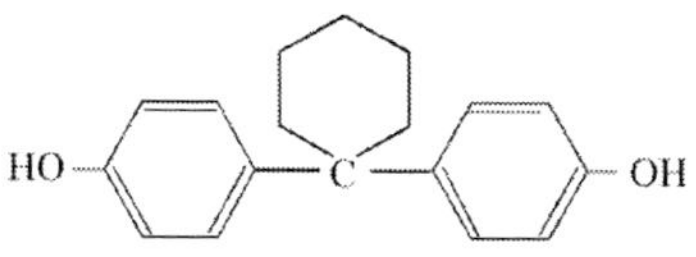

bisphenol Z (BPZ)
1,1-bis(4-hydroxyphenyl)cyclohexane

Figure 1. Chemical structures of BPA and bisphenol derivatives used in this study.

Enzymatic Treatment of Phenolic Compounds

Chemical procedures such as adsorptive behavior [47-51], membrane separation [52-54], electrochemical polymerization [55], photolysis and photooxidation [56-59], or biological procedures by use of microalgae and bacteria [60-63], have been constructed to detoxify or degrade BPA. Alternatively, much attention has been paid to the potentials of enzymes to specifically catalyze the transformation of phenol compounds [64-67]. The use of enzymes has many advantages over the above conventional procedures which are effective but suffer from high cost, incomplete purification, formation of hazardous byproducts and applicability to only a limit concentration range.

The enzymes can catalyze the conversion of phenolic compounds include peroxidase, laccase, and tyrosinase. Peroxidases from horseradish [68-75], soybean [76-79], turnip [80], and microbe [81,82], have been more frequently used to catalyze radicalization of BPA [71,73,75,80-82] as well as alkylphenols [68,72,76,80] and chlorophenols [68-71,74,76-80] in the presence of hydrogen peroxide (H_2O_2). The radicals enzymatically generated spontaneously react to form water-insoluble polymers [75]. Alternatively, many studies on tyrosinase-catalyzed quinone oxidation of phenolic compounds have been wildly reported in many articles. Tyrosinase has two catalytic functions, that is, *o*-hydroxylation of monophenols to *o*-diphenols in the presence of molecular oxygen (cresolase activity) and a two-electron oxidation of *o*-diphenols to *o*-quinones (catecholase activity) [83,84].

Commercially available mushroom tyrosinase has been more frequently for treatment of recalcitrant phenolic compounds. *n*-Alkylphenols and chlorophenols undergo tyrosinase-catalyzed quinone oxidation without any other additives [85-94]. In addition, Jiménez et al. reported that 4-*tert*-butylphenol (4TBP), which underwent no quinone oxidation by mushroom tyrosinase without H_2O_2, was enzymatically converted to 4-*tert*-butyl-*o*-benzoquinone in the presence of H_2O_2 [95]. However, according to the report by Yoshida et al., BPA undergo no quinone oxidation by commercially available mushroom tyrosinase in the absence of H_2O_2 [92]. Therefore, little was reported on tyrosinase-catalyzed treatment of BPA.

Use of Chitosan for Removal of Phenolic Compounds

Since quinone derivatives enzymatically generated from phenolic compounds have a high reactivity, they can undergo different chemical reactions. The use of chitosan as an amino-group-containing biopolymer can be cited as a procedure to remove alkylphenols and chlorophenols through the reaction of the quinone derivatives [85,86,88,93,94].

When chitosan solutions were added to aqueous solutions of alkylphenols or chlorophenols containing tyrosinase, quinone derivatives enzymatically generated reacted with chitosan's amino groups and consequently water-insoluble aggregates were generated [96]. On the other hand, the techniques to remove phenolic compounds from aqueous solutions through the reaction of quinone derivatives generated by tyrosinase with chitosan were systematically carried out by the study group of G. F. Payne et al. in 1990s [85-89].

Thereafter, we reported that when porous chitosan beads were dispersed in solutions of alkylphenols or chlorophenols and then the enzymatic reaction was initiated by adding tyrosinase, enzymatically generated quinone derivatives were effectively chemisorbed on the chitosan beads [93,94]. It should be noted that adsorption of quinone derivatives on chitosan beads is enhanced by increasing the amount of added chitosan beads.

Chitosan is produced in large amounts by deacetylation of chitin that is contained in the shells of the crustaceans such as crabs and prawns, the tissue of the mollusks such as cuttlefishes and shellfish, and the outer shells of the insects. However, the use of chitosan is limited mainly to wastewater treatment to capture heavy metal ions. Most of commonly used chitin is obtained from the shells of the crabs and prawns. Therefore, if chitosan beads are used as an adsorbent for effective removal of BPA, an alternative usage of chitosan will come out.

In this chapter, removal of bisphenol A is described through tyrosinase-catalyzed quinone oxidation and subsequent nonenzymatic quinone adsorption on chitosan beads. First, the effects of the process parameters such as the H_2O_2 concentration, pH value, temperature, and enzyme dose, were investigated for quinone conversion of BPA by tyrosinase. In particular, the effect of the addition of chitosan was discussed in the form of beads, solutions, and powder for removal of BPA. In addition, removal of bisphenol derivatives, which have two phenol groups in common, but different chemical structures between phenol groups, was also estimated under the optimum conditions determined for treatment of BPA.

2. Tyrosinase-Catalyzed Quinone Oxidation of BPA

2.1. Effect of H_2O_2 Concentration on Quinone Oxidation

H_2O_2 was added to a BPA solution in a pH 7.0 buffer at 40°C so as to reach the final concentration to 0.3 mM ($[H_2O_2]/[BPA]=1.0$), and then the enzymatic reaction was initiated by adding tyrosinase solution to the mixture solutions. The solution were continuously stirred and the UV-visible spectra were recorded at predetermined time intervals.

Figure 2 shows the time course of the absorbance at 385 nm and conversion % value for tyrosinase-catalyzed quinone oxidation of BPA in the presence or absence of H_2O_2 (0.3 mM) at pH 7.0 and 40°C. An absorbance at 385 nm gradually increased over the reaction time for in the absence of H_2O_2. The conversion % value was limited to only 29 %, even when the reaction time was prolonged to 3 hr. An increase in the absorbance at 385 nm refers to the enzymatic generation of quinone derivatives from BPA. This peak is absent for the original BPA solution containing only tyrosinase. Even if H_2O_2 was added to a BPA solution without tyrosinase, no quinone derivatives were generated and the reaction solution remained transparent (The symbols of △, were overlapped with those of ▲), indicating that BPA isn't directly oxidized by H_2O_2 alone.

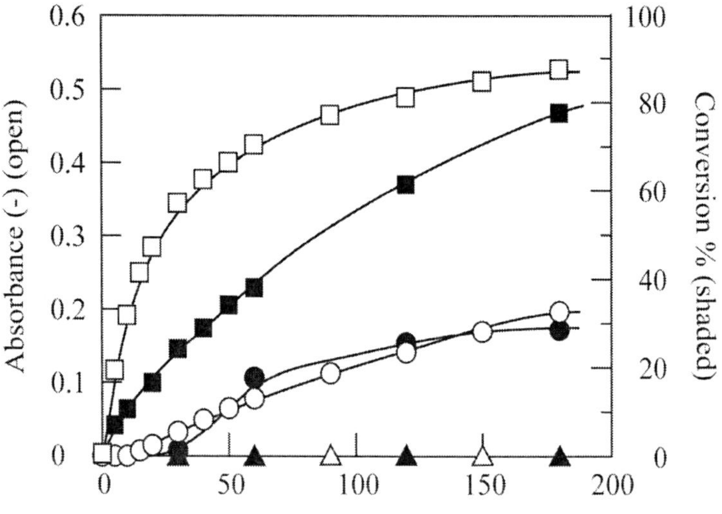

Figure 2. The time course of the absorbance at 385 nm (open) and conversion % value (shaded) for BPA solutions (0.3 mM) containing tyrosinase (200 U/cm^3) (open circle, shaded circle), tyrosinase and H_2O_2 (0.3 mM) (open square, shaded square), and H_2O_2 (0.3 mM) (open triangle, shaded triangle) at pH 7.0 and 40°C.

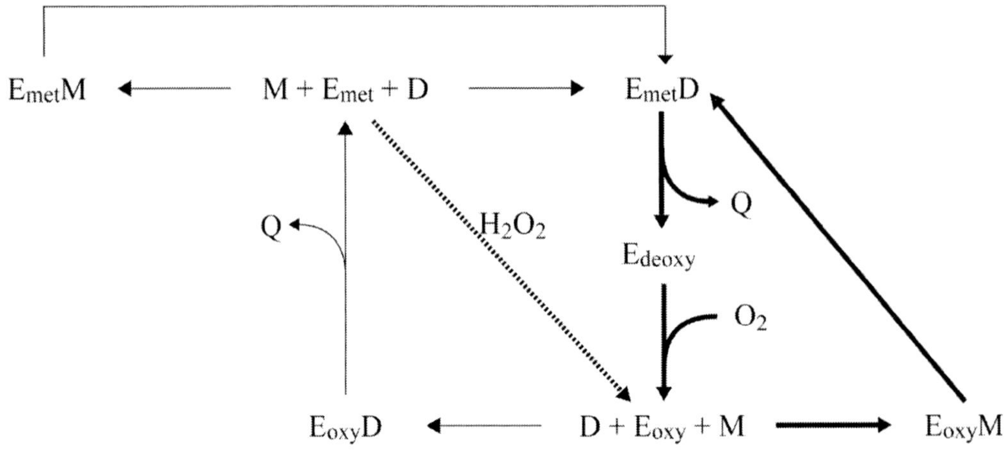

Scheme 1. Catalytic cycle for oxidation of monomphenol (M) and diphenol (D) substances to quinone derivatives (Q) by tyrosinase (E) and transformation of tyrosinase in the presence of H_2O_2.

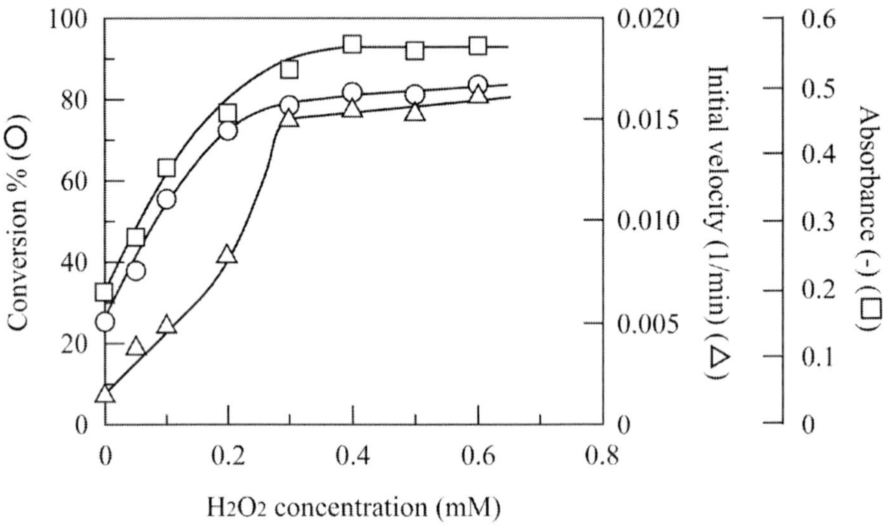

Figure 3. The effect of H_2O_2 on tyrosinase-catalyzed (200 U/cm^3) quinone oxidation of BPA (0.3 mM) at pH 7.0 and 40°C. The enzymatic reaction time was 3 hr.

Generally, the peak assigned to quinone derivatives appears at 350-400 nm and the position of the peak depends on the chemical structure of the original phenolic compounds [85-89]. The peak position at 385 nm obtained in this study was in agreement with those assigned to quinone derivatives enzymatically generated from alkylphenols [93]. The emergence of the peak at 385 nm and color development of the solution support tyrosinase-catalyzed quinone generation from BPA. Here, a bisquinone derivative can be possibly generated in addition to a monoquinone derivative with an unreacted phenolic –OH group as shown in Scheme 1, since BPA has two phenolic groups. According to the HPLC analytic investigation by Yoshida et al, a small amount of a bisquinone derivative was detected in addition to a monoquinone derivative for quinone oxidation of BPA with tyrosinase purchased from Worthington Biochemical Co. [92].

When H_2O_2 was added to a BPA solution containing tyrosinase at 200 U/cm^3, both the absorbance at 385 nm and the conversion % value sharply increased in the first 1 hr and the solutions became yellow, then orange-red, and finally dark brown [85]. Payne et al., described that the cresolase activity of mushroom tyrosinase is rather slower than the catecholase activity in the absence of H_2O_2 for tyrosinase-catalyzed quinone oxidation of different phenolic compounds [85-89]. In addition, 4TBP and 4-*tert*-pentylphenol (4TPP) undergo quinone oxidation in the presence of H_2O_2 but not in the absence of H_2O_2 [92,93]. Here, it is well known that tyrosinase has three forms, met-, oxy-, and deoxy-forms, depending on the presence or absence of H_2O_2 as shown in Scheme 1. The portion of oxy-form tyrosinase is so small in the absence of H_2O_2 that the no quinone oxidation of 4TBP and 4TPP occurred [95]. Ikehata and Nicell reported that when molecular oxygen binds to deoxy-form tyrosinase, it is brought to oxy-form tyrosinase [91,97]. Some of met-form tyrosinase is also converted into oxy-form tyrosinase in the presence of H_2O_2. The enzymatic quinone oxidation of BPA is considered to be accelerated by the addition of H_2O_2. [95].

Alternatively, it was reported that the cresolase activity of mushroom tyrosinase was also enhanced by adding 3-hydroxyanthranilic acid (HAA) for tyrosinase-catalyzed quinone oxidation of *N*-acetyl-*L*-tyrosine and 4TBP [98]. However, since HAA is referred to as a carcinogen, the use of HAA is not favorable in the viewpoint of environmental conservation. Subsequently, the effect of the H_2O_2 concentration on tyrosinase-catalyzed quinone oxidation of BPA was investigated.

Figure 3 shows the changes in the absorbance and conversion % value at the reaction time of 3 hr and the initial velocity for BPA solutions containing H_2O_2 of different concentrations at pH 7.0 and 40°C. Here, the initial velocity was determined from the slope of the absorbance against the reaction time in the initial reaction stage. An increase in the H_2O_2 concentration led to an increase in the absorbance in the range of H_2O_2 concentrations below 0.3 mM and the conversion % value increased with an increase in the H_2O_2 concentration. However, the increment of the absorbance and the conversion % value tended to level off at further increased H_2O_2 concentrations. Here, there was a distinct difference in the H_2O_2-dependent tyrosinase-catalyzed quinone oxidation between BPA and branched *p*-alkylphenols such as 4TBP and 4TPP. 4TBP and 4TPP underwent no tyrosinase-catalyzed quinone oxidation in the absence of H_2O_2 [94,95]. On the other hand, BPA was gradually converted into quinone derivatives by tyrosinase even in the absence of H_2O_2 as shown in Figure 2. The cycle of tyrosinase-catalyzed quinone oxidation of phenol, *p*-*n*-alkylphenols, and chlorophenols has been deeply investigated by many researchers [66,83,91,97,99]. According to the enzymatic reaction mechanism shown in Scheme 1, tyrosinase-catalyzed quinone oxidation of BPA can be also explained like other phenolic compounds above-mentioned.

Mushroom tyrosinase has the catecholase activity as well as the peculiar characteristics of cresolase activity. Therefore, phenol, *p*-*n*-alkylphenols, and chlorophenols were successfully quinone-oxidized through formation of the corresponding diphenols even in the absence of H_2O_2. BPA gradually underwent quinone oxidation by mushroom tyrosinase in the absence of H_2O_2 as shown in Figure 2 unlike 4TBP and 4TPP. However, tyrosinase-catalyzed quinone oxidation of BPA was considerably slower than that of *p*-*n*-alkylphenols and chlorophenols. The binding of branched alkylphenols such as 4TBP and 4TPP to met-form tyrosinase is considered to scavenge a portion of tyrosinase from the catalytic turnover as a dead-end complex in the cresolase activity [50,53,57,67]. On the other hand, oxy-form tyrosinase generated from met-form tyrosinase in the presence of H_2O_2 can convert BPA into the corresponding quinone derivatives. The optimum H_2O_2 concentration was determined to be 0.3 mM, since the concentration of remaining H_2O_2 should be as low as possible in the viewpoint of environmental conservation.

2.2. Effects of pH and Temperature

Here, the effects of the process parameters such as the pH value and temperature on enzymatic quinone oxidation of BPA were investigated because the optimum pH and temperature of the tyrosinase activity are slightly different from one article to another and depend on the origin of enzymes and the experimental conditions such as the kind and concentration of phenol compounds, kind and salt concentration of buffers, and the presence or absence of H_2O_2 [91,93,100].

First, tyrosinase-catalyzed quinone oxidation of BPA (0.3 mM) was estimated in buffers of different pH values at 40°C in the presence of H_2O_2. Figure 4 shows the effect of the pH value on tyrosinase-catalyzed quinone oxidation of BPA. Tyrosinase was deactivated in a short reaction time at pH 4.0 and the quantity of approximately 75 % of BPA was left unreacted in the solution. The initial velocity was sharply increased at pH 5.0 and 6.0, but the absorbances 3 hr after the enzymatic reaction was initiated were relatively low. This indicates that tyrosinase was gradually deactivated in the solution. On the other hand, although the initial velocity at pH 7.0 was lower than those at pH 5.0 and 6.0, the absorbance increased over the reaction time and the absorbance at pH 7.0 was higher than those at pH 5.0 and 6.0. However, when the pH value was further increased, the activity of tyrosinase gradually decreased. Since it was found from the above results that the activity of tyrosinase was successfully kept going and a high conversion efficiency was obtained at pH 7.0 compared with at other pH values, the optimum pH value was determined to be 7.0.

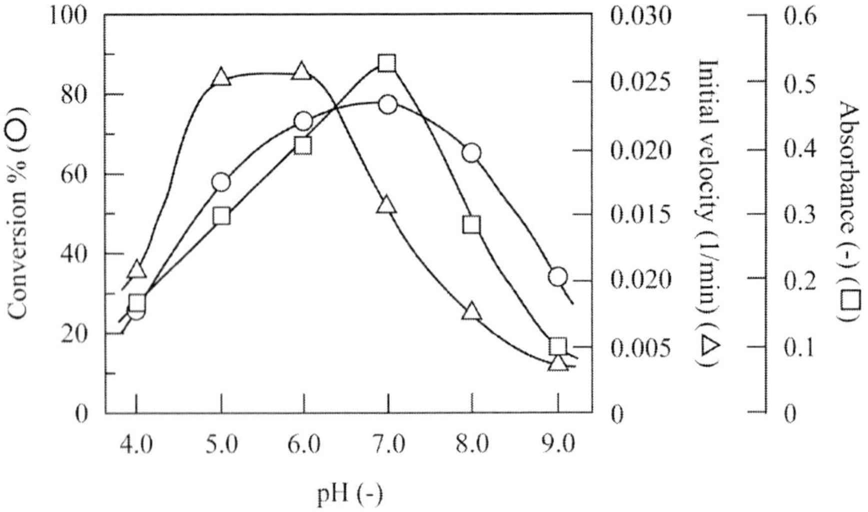

Figure 4. The effect of pH value on tyrosinase-catalyzed (200 U/cm^3) quinone oxidation of BPA (0.3 mM) in the presence of H_2O_2 of 0.3 mM at 40°C. The enzymatic reaction time was 3 hr.

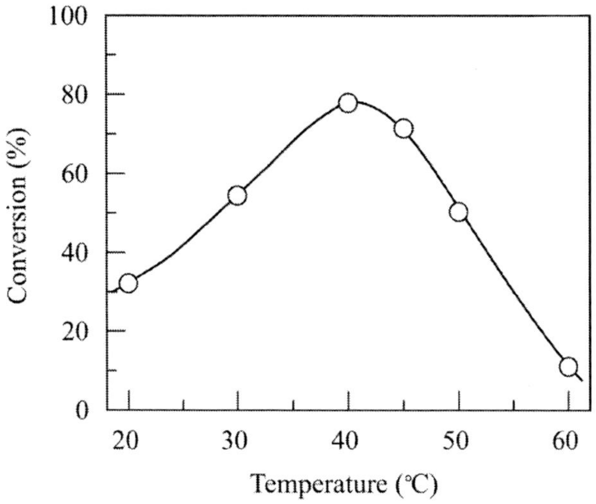

Figure 5. The effect of the temperature on tyrosinase-catalyzed (200 U/cm^3) quinone oxidation of BPA (0.3 mM) in the presence of H2O2 of 0.3 mM at pH 7.0. The enzymatic reaction time was 3 hr.

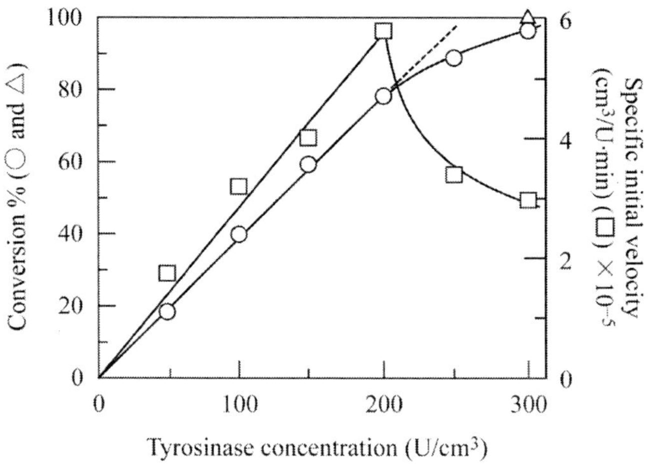

Figure 6. The effect of the tyrosinase concentration on tyrosinase-catalyzed (200 U/cm^3) quinone oxidation of BPA (0.3 mM) in the presence of H2O2 of 0.3 mM at pH 7.0 and 40°C. The enzymatic reaction time (hr)- circle: 3, triangle: 5.

Subsequently, the effect of the temperature on the tyrosinase-catalyzed quinone oxidation of BPA was investigated at pH 7.0. As shown in Figure 5, quinone oxidation of BPA increased by increasing the reaction temperature and the activity of tyrosinase passed through the maximum value at 40°C. The activity of tyrosinase sharply decreased at temperatures higher than 40°C probably due to thermal denaturation as observed for other enzymes [101,102]. Therefore, the optimum temperature was determined to be 40°C.

It was found from the above results that tyrosinase-catalyzed quinone oxidation of BPA was accelerated by the addition of H2O2 and the optimum conditions for quinone oxidation of BPA at 0.3 mM were determined to be 0.3 mM for H2O2 ([H2O2]/[BPA]=1.0) at pH 7.0 and 40°C.

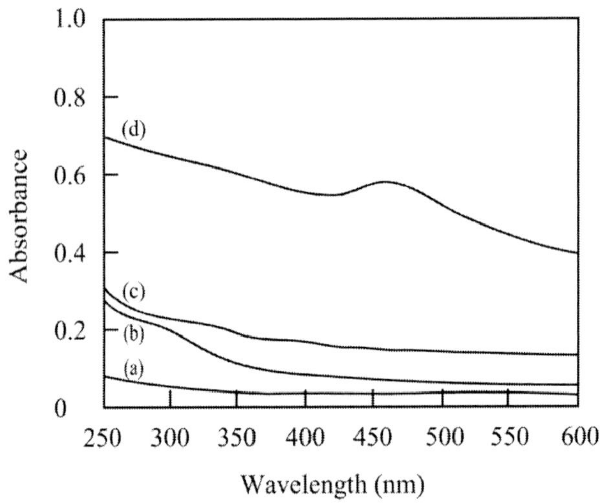

Incubation conditions- (a) untreated chitosan film, (b) cellulose film in BPA with H_2O_2 and tyrosinase, (c) chitosan film in BPA with H_2O_2, (d) chitosan film in BPA with H_2O_2 and tyrosinase.

Figure 7. UV-visible spectra of chitosan films incubated in a BPA solution containing H_2O_2 and tyrosinase and in various control solutions.

2.3. Effect of Enzyme Dose on Quinone Oxidation

On the basis of the optimum conditions determined in the above section, the enzymatic reaction was initiated for BPA at 0.3 mM with tyrosinase of different concentrations in the presence of H_2O_2 at 0.3 mM. The specific initial velocity was calculated from the initial velocity obtained from the slope of the absorbance against the reaction time and the tyrosinase concentration.

Figure 6 shows the effect of the tyrosinase concentration on quinone oxidation of BPA in the presence of H_2O_2 at 0.3 mM at pH 7.0 and 40ºC. The tyrosinase-catalyzed quinone oxidation of BPA was proportional to the tyrosinase concentration in the tyrosinase concentration range below 200 U/cm^3 and the quantity of 78 % of BPA was converted into quinone derivatives at 200 U/cm^3. In addition, the specific initial velocity reached the maximum value at 200 U/cm^3 and sharply decreased at further increased tyrosinase concentrations. We empirically demonstrated that quinone oxidation of BPA was accelerated by increasing the tyrosinase concentration up to 200 U/cm^3. An increase in the tyrosinase concentration at a constant H_2O_2 concentration led to an increase in the specific initial velocity. This behavior is considered to be attributed to an increase in the oxy-form tyrosinase concentration. However, it is our upcoming challenge to estimate the enzymatic quinone oxidation quantitatively. In addition to the above results, from the fact that the conversion % value was negatively deviated downward from the linear relationship at higher than 200 U/cm^3, the optimum tyrosinase concentration was determined to be 200 U/cm^3. The importance to determine the optimum conditions as forementioned is derived from our consideration that the activities of many enzymes can decrease due to the presence of enzymatically generated chemical species or under unfavorable conditions such as high temperatures and lower and higher pH values. The enzymatic activity can be effectively developed by determining the above-mentioned process parameters, leading to a reduction in the time required for the enzymatic reaction.

3. Reaction of Quinone with Chitosan's Amino Group

The reactivity of quinone derivatives enzymatically generated from BPA with amino groups was estimated with chitosan films. A chitosan (chitosan 300, Wako Pure Chemicals) film (average thickness : 24 μm) was incubated in a BPA solution containing tyrosinase and H_2O_2 at pH 7.0 and 40°C for 2 hr. After incubation, the chitosan film was throughout washed with water, and then dried in an oven at 50°C. The UV-visible spectrum of the incubated chitosan film was recorded on a spectrophotometer. The incubated chitosan film was placed perpendicular to the light path such that the light passed directly through the film in the spectrophotometer [93,94].

Figure 7 shows the UV-visible spectra of chitosan films incubated in BPA solutions containing both tyrosinase and H_2O_2 and containing either tyrosinase or H_2O_2. The absorbance at 460 nm emerged for the chitosan film incubated in a BPA solution containing both tyrosinase and H_2O_2. This peak didn't appear for the chitosan films incubated in BPA solutions containing either tyrosinase or H_2O_2. In addition, this peak was not also observed for a control cellulose film incubated with a BPA solution containing both tyrosinase and H_2O_2 [93]. This difference reveals that an amino group is involved in quinone binding to chitosan. According to some articles, this reaction is considered to occur through either Schiff base or Michael-type addition [85,93,103]. The position of the peak at 460 nm is assigned to the binding of quinone derivatives generated from BPA to amino groups. This value was in good agreement with the position of the peak assigned to amino groups reacted with quinone derivatives generated from *p*-cresol under a similar condition [93]. It was found from these results that quinone derivatives enzymatically generated from BPA successfully reacted with chitosan's amino groups under a relatively mild condition. Therefore, we estimated quinone adsorption on chitosan beads in the next section.

4. Removal of BPA

4.1. Usage of Chitosan Beads

Since it was found in the previous section that quinone derivatives enzymatically generated from BPA reacted with chitosan's amino group, removal of BPA was investigated by combined use of quinone oxidation of BPA by tyrosinase and subsequent quinone adsorption on chitosan beads.

After H_2O_2 was added to BPA solutions in the presence of different amounts of chitosan beads and then the temperatures of the solutions were adjusted to 40°C, the enzymatic reaction was initiated by adding tyrosinase. Figure 8(a) shows the time course of the absorbance of the BPA solutions containing tyrosinase (200 U/cm^3) and H_2O_2 (0.3 mM, $[H_2O_2]/[BPA]=1.0$) in the presence of different amounts of chitosan beads at pH 7.0 and 40°C. The increase in the absorbance at 385 nm was considerably depressed by addition of chitosan beads because quinone derivatives enzymatically generated were chemically adsorbed on chitosan beads. However, the quantity of 2.3 % of BPA was left unreacted, even when the reaction time was prolonged to 5 hr at the amount of chitosan beads of 0.025 cm^3/cm^3. In the presence of chitosan beads at 0.050 cm^3/cm^3, a small amount of quinone derivatives were left in the solution and the solution remained a little colored after the reaction for 3 hr. The conversion % value increased up to 95.8 % at 3 hr with almost complete quinone conversion at 5 hr. At 0.10 cm^3/cm^3, BPA was completely removed for 5 hr through the quinone oxidation and subsequent quinone adsorption on chitosan beads. In addition, when the amount of chitosan beads was increased up to 0.15 cm^3/cm^3, the time required to remove BPA completely was shortened to 3 hr. This mean that an increase in the amount of chitosan beads dispersed in the solution led to a reduction in the time required to remove BPA [92-94,103]. Alternately, the BPA concentration was estimated for a BPA solution without tyrosinase but in the presence of chitosan beads (0.10 cm^3/cm^3) at pH 7.0 and 40°C as a control experiment. The BPA concentration gradually decreased probably due to physical adsorption on chitosan beads. Then, even if the reaction time was prolonged to 5 hr, the adsorption % was limited to be only 15 %.

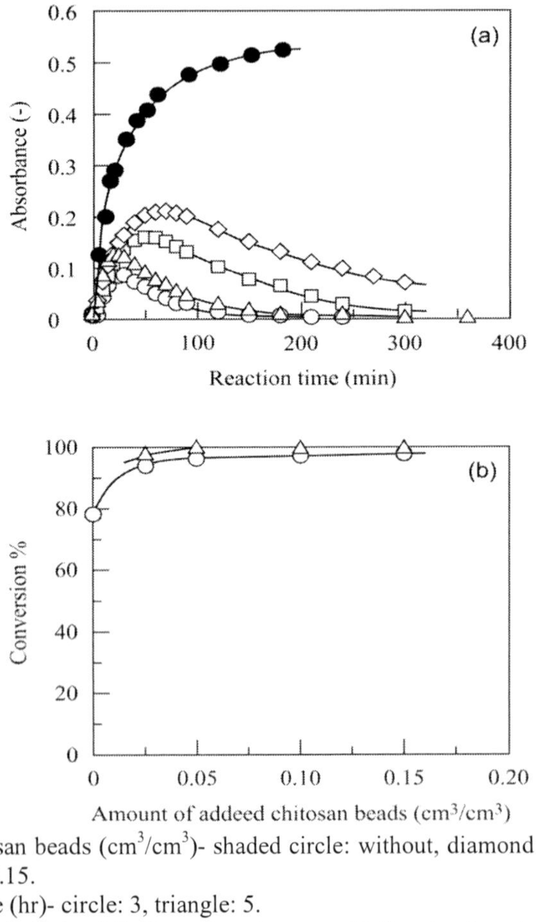

(a) Amount of added chitosan beads (cm^3/cm^3)- shaded circle: without, diamond: 0.025, square: 0.050, triangle: 0.10, circle: 0.15.

(b) Enzymatic reaction time (hr)- circle: 3, triangle: 5.

Figure 8. Removal of BPA (0.3 mM) through quinone oxidation by tyrosinase (200 U/cm^3) and subsequent quinone adsorption on chitosan beads in the presence of H2O2 of 0.3 mM at pH 7.0 and 40°C.

4.2. Kinetic Analysis of Quinone Oxidation and Adsorption

The concentrations of components related with the quinone oxidation and subsequent quinone adsorption were calculated as a function of the reaction time from the conversion % value and the absorbances in the presence of chitosan beads at 0.10 cm^3/cm^3. First, the concentration of BPA enzymatically converted into quinone derivatives, $[BPA]_{converted}$, corresponding to the concentration of quinone derivatives enzymatically generated, $[Q]_{generated}$, was calculated from the peak area at time *t* obtained by the HPLC measurements. BPA was completely converted into quinone derivatives for the enzymatic reaction of 5 hr at 300 U/cm^3 as shown in Figure 6, when the solution had the absorbance of 0.827. Therefore, the apparent concentration of remaining quinone derivatives, $[Q]^{app}_{remained}$, at time *t* was estimated from equation 2.

$$[Q]^{app}_{remained} = 0.30 \cdot \frac{Abs_t}{0.827} \tag{2}$$

where Abs_t is the absorbance at time *t* in the presence of chitosan beads. Since the difference between the net concentration of quinone derivatives enzymatically generated and apparent concentration of remaining quinone derivatives, $[Q]_{generated} - [Q]^{app}_{remained}$, results from quinone adsorption on chitosan beads, the adsorption % value can be calculated by

$$\text{Adsorption\%} = \frac{[Q]_{generated} - [Q]^{app}_{remained}}{[Q]_{generated}} \cdot 100 \tag{3}$$

Finally, the removal % value was calculated form both conversion % and adsorption % values using equation 4.

$$\text{Removal \%} = \frac{\text{Conversion \%} \cdot \text{Adsorption\%}}{100} \tag{4}$$

Figure 9 shows the time course of the concentration of quinone derivatives enzymatically generated, the apparent concentration of remaining quinone derivatives, and the values of conversion %, adsorption %, and removal %. The BPA concentration decreased over the reaction time and BPA was completely converted into quinone derivatives at 5 hr in the presence of chitosan beads at 0.10 cm^3/cm^3. In the range of this reaction time, the removal % value also continuously increased. Here, the initial velocity calculated from a slope in the concentration of remaining BPA against the enzymatic reaction time in the initial stage in the presence of chitosan beads was 4.32×10^{-3} mM/min. This value was 2.45 times higher than the initial velocity of 1.77×10^{-3} mM/min in the absence of chitosan beads determined from Figure 2. This indicates that the decrease in the quinone concentration in the reaction solution through quinone adsorption on chitosan beads lead to the increase in the enzymatic quinone oxidation. It was noted that although the apparent concentration of remaining quinone derivatives vanished at 4 hr, removal of BPA was still uncompleted because there remained a small amount of unreacted BPA in the solution. Then, the absorbance disappeared and the complete conversion was reached at 5 hr. This shows that BPA was completely removed from the solution.

From the above results, we can note that the combined use of tyrosinase-catalyzed quinone oxidation in the presence of H_2O_2 and quinone adsorption on chitosan beads was a very effective means to remove BPA from solutions. As shown in Figure 8(b), tyrosinase-catalyzed quinone oxidation of BPA was enhanced by increasing the amount of added chitosan beads. In other words, a decrease in the quinone concentration in solutions through chemical adsorption on chitosan beads accelerated the enzymatic quinone oxidation of BPA. This indicates that quinone adsorption on chitosan beads would suppress the unfavorable quinone-related inactivation of tyrosinase such as contact between quinone derivatives and tyrosinase molecules or their active sites.

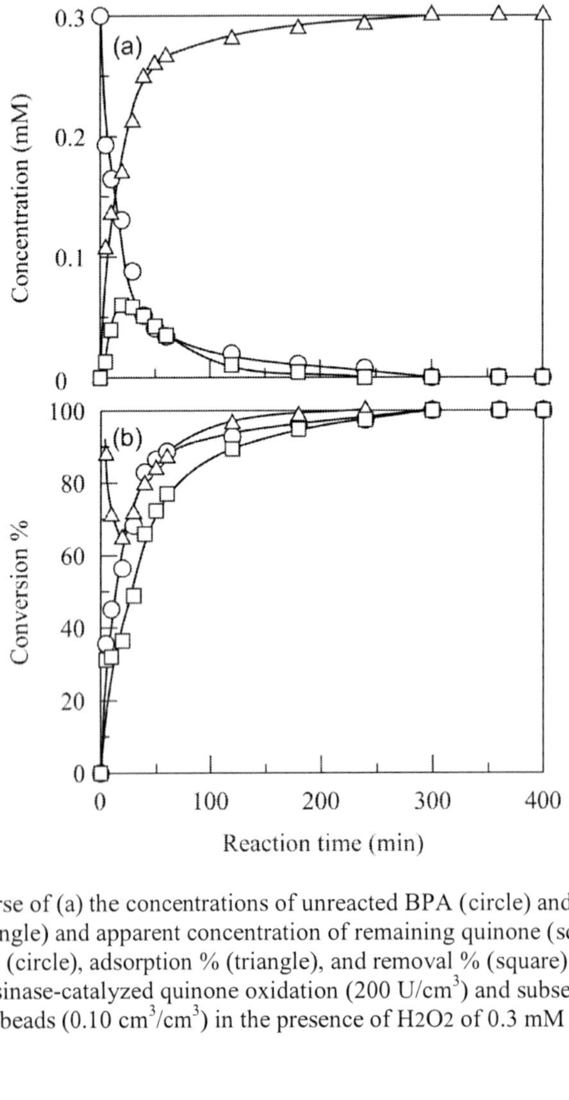

Figure 9. The time course of (a) the concentrations of unreacted BPA (circle) and enzymatically generated quinone (triangle) and apparent concentration of remaining quinone (square) and (b) the values of conversion % (circle), adsorption % (triangle), and removal % (square) for removal of BPA (0.3 mM) through tyrosinase-catalyzed quinone oxidation (200 U/cm^3) and subsequent quinone adsorption on chitosan beads (0.10 cm^3/cm^3) in the presence of H_2O_2 of 0.3 mM at pH 7.0 and 40°C.

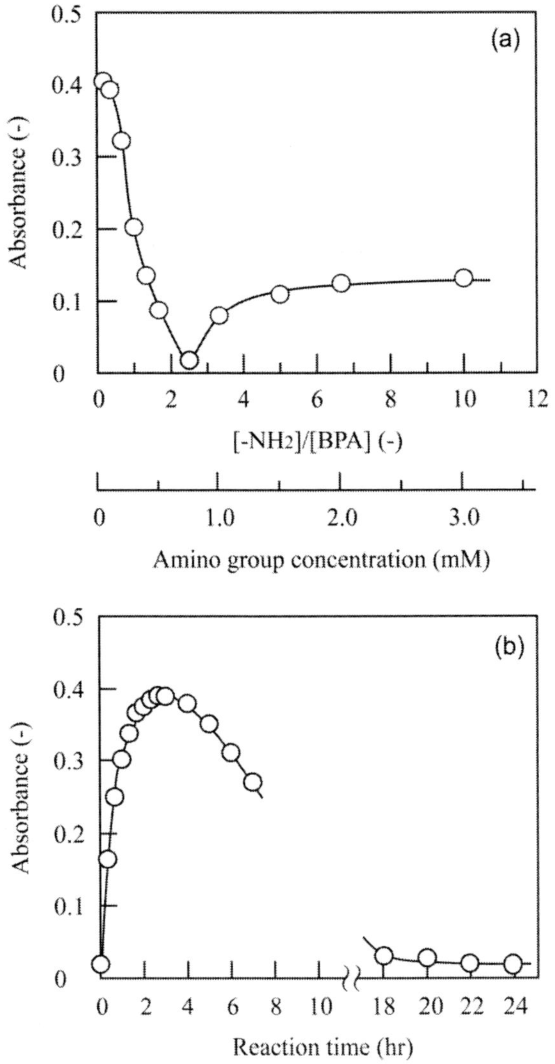

Figure 10. Removal of BPA (0.3 mM) through tyrosinase-catalyzed quinone oxidation (200 U/cm^3) and subsequent homogeneous quinone reaction with tyrosinase in the presence of H_2O_2 of 0.3 mM at pH 6.0 and 40°C.

4.3. Usage of Chitosan Powders

As another control experiment, chitosan powders were used for the adsorbent of enzymatically generated quinone derivatives in place of chitosan beads. Here, 0.30 g of chitosan powders (particle size: 75-100 μm), which corresponded to the quantity of 0.40 cm^3 (0.10 cm^3/cm^3) of chitosan beads equilibrated in a pH 7.0 buffer, were dispersed in a BPA solution containing H_2O_2 and tyrosinase and then the solution was constantly stirred at pH 7.0 and 40°C.

As shown in Figure 10, the increase in the absorbance at 385 nm was depressed by adding chitosan powders in the BPA solution in the same manner as chitosan beads. The absorbance gradually decreased over the reaction time through the maximum value. Hereat, the conversion % value was 95.1•% at the reaction time of 5 hr. A small amount of BPA was left untreated in the reaction solution. This value was a little higher than the conversion value of 78 % for tyrosinase-catalyzed quinone conversion of BPA in the absence of chitosan beads and chitosan powder, but lower than the the conversion value of 100 % obtained for chitosan beads at 0.10 cm^3/cm^3.

On the assumption that the chitosan powders used are in a perfect sphere with the diameter of 100 μm and the density is 1.0 g/cm^3, the gross surface area of 0.30 g of chitosan powder is calculated to be 180 cm^2, which corresponds to the specific surface area of 60 cm^2/g. This value is much lower than that of the chitosan beads used in this study. Therefore, we can safely say that a high porosity of the chitosan beads used, or a large specific surface area, is one of the important factors to enhance the rate of quinone conversion of BPA, which leads to an increase in the quinone adsorption on chitosan beads.

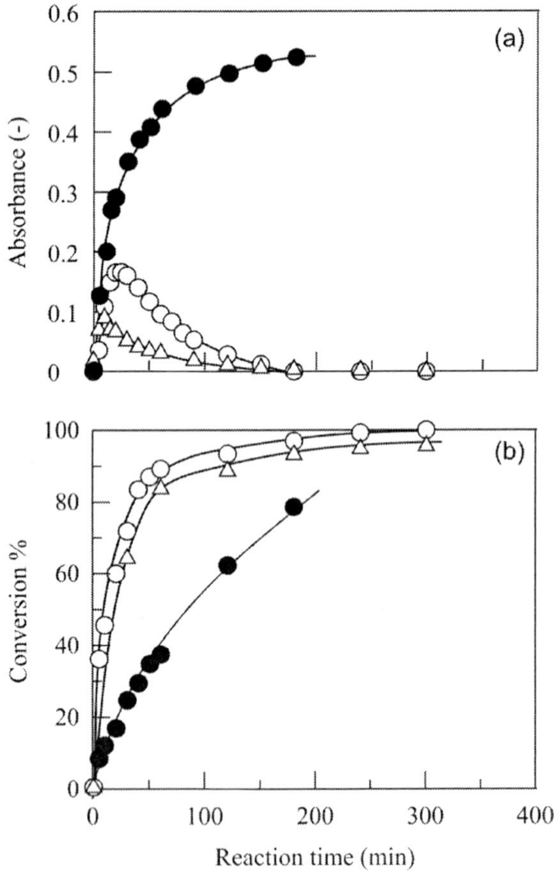

Form of chitosan- shaded circle: without, triangle: chitosan powders (0.0075 g/cm^3, 0.30 g), open circle: chitosan beads (0.10 cm^3/cm^3).

Figure 11. Removal of BPA (0.3 mM) through quinone oxidation by tyrosinase (200 U/cm^3) and subsequent quinone adsorption on chitosan powders or chitosan beads in the presence of H2O2 of 0.3 mM at pH 7.0 and 40°C.

4.4. Usage of Chitosan Solutions

As another control experiment, removal of BPA with chitosan solutions was also estimated. Chitosan solutions were added to a BPA solution containing tyrosinase (200 U/cm^3) and H2O2 (0.3mM) at pH 6.0 and 40°C so as to reach final amino group concentration from 0.2 to 10 mM, and then the solutions were moderately stirred for 24 hr. Here, the pH value was adjusted to 6.0 because chitosan was insoluble in a pH 7.0 buffer.

When tyrosinase was added to a BPA solution containing H2O2 without chitosan, the color was developed by enzymatic quinone generation. In this case, even when the stirring was continued for 24 hr, the BPA solution was kept brown and no precipitates were generated. Contrary to this, when both tyrosinase and chitosan were added to a BPA solution containing H2O2, water-insoluble aggregates were generated and the solutions were gradually decolorized irrespective of the amino group concentration. The aggregates generated were filtered out with a 5C filter paper, and then the absorbance of the filtrates was measured at 385nm. As shown in Figure 11(a), the absorbance at 385nm sharply decreased with an increase in the chitosan's amino group concentration. The decrease in the absorbance indicates that the concentration of water-soluble quinone derivatives in solutions was decreased. The reaction solution was fairly decolorized by stirring for 24 hr at the amino group concentration of 0.75 mM, or [-NH2]/[BPA]=2.5. However, when the amino group concentration further increased, the absorbance gradually increased and the generation of aggregates decreased because chitosan chains with unreacted free amino groups were soluble in the solutions [96]. It is considered that the presence of H2O2 has an insignificant influence on chitosan, for example chain cleavage, because H2O2 was consumed by the transformation of tyrosinase into oxy form from met form as shown in Scheme 1. In fact, a considerable change wasn't observed for a BPA solution containing H2O2 and chitosan.

Figure 11(b) showed the time course of the absorbance at [-NH2]/[BPA]=2.5 where the minimum absorbance was obtained. The absorbance was gradually decreased after passing through the maximum value, which indicates that quinone derivatives reacted with amino groups of chitosan dissolving in the solution. Most of BPA was removed from solution by filtering out water-insoluble aggregates generated, when the solution was stirred for 24 hr. Unfortunately, the absorbance of 0.02 at 24 hr showed that a small amount of quinone derivatives remained in the solution. For the homogeneous reaction with chitosan solutions, it was found that the generation of water-insoluble precipitates was considerably slow beyond our expectation. The absorbance of 0.386 was obtained and no water-insoluble aggregates were generated after stirring for 3 hr. On the other hand, BPA was completely removed from solutions at 3-5 hr by the procedure developed in this study as shown in Figure 7, although the time required to remove BPA completely depended on the amount of added chitosan beads. This comparison insists that the heterogeneous procedure with chitosan beads constructed in this study is much effective in removing BPA from aqueous solutions.

Table 1. Removal of bisphenol derivatives at pH 7.0 and 40°C as the optimum conditions determined for BPA

Bisphenol derivatives	Inital conc. (mM)	Tyrosinase conc. (U/cm^3)	Chitosan beads (cm^3/cm^3)	Conversion (%)	Conversion time (hr)	Adsorption (%)	Adsorption time (hr)	Removal (%)
BPA	0.3	200	0.10	100	5	100	3	100
BPB	0.3	200	0.10	72.4	5	85.0	5	61.5
		200	0.20	78.4	5	91.3	5	71.6
		300	0.20	87.9	5	98.8	3.5	86.8
		400	0.20	97.9	5	75.1	5.5	73.5
		400	0.20	98.9	5	99.3	46	98.2
BPC	0.05	200	0.10	57.9	5	—	5	57.9
		200	0.20	60.8	5	—	5	60.8
		300	0.20	63.4	5	—	5	63.4
BPE	0.3	50	0.10	100	5	98.9	5	98.9
		50	0.20	100	5	100	5.5	100
BPF	0.3	30	0.10	100	2	100	2	100
BPO	0.3	150	0.10	100	1	100	2	100
BPS	0.3	200	0.10	3.4	5	6.0	5	0.20
BPT	0.3	150	0.10	100	4	100	0.3	100
BPZ	0.02	200	0.10	64.3	5	—	5	64.3
		200	0.20	66.3	5	—	5	66.3
		300	0.20	86.0	5	—	5	86.0

5. Removal of Bisphenol Derivatives

5.1. Effect of Enzyme Dose and Amount of Chitosan Beads

The experiments of tyrosinase-catalyzed quinone oxidation and quinone adsorption on chitosan beads were applied to removal of bisphenol derivatives. The concentration of remaining bisphenol derivatives were determined by the HPLC measurements. The volume composition of aqueous acetonitrile solutions as the mobile phase and the wavelength of the UV-spectrophotometer depended on the type of the bisphenol derivatives used. In addition, Since BPC and BPZ had lower solubility in a pH 7.0 buffer, the tyrosinase-catalyzed treatment of BPC and BPZ was carried out at the initial concentrations of 0.05 and 0.02 mM, respectively.

The experiments of removal of bisphenol derivatives shown in Figure 1 was carried out at pH 7.0 and 40°C as the optimum conditions determined for quinone oxidation of BPA. The results of removal of bisphenol derivatives were summarized in Table 1. Of seven kinds of bisphenol derivatives used in this study, BPS underwent little tyrosinase-catalyzed quinone oxidation. BPE, BPF, and BPT were effectively converted into the corresponding quinone derivatives at lower tyrosinase concentrations than BPA. Since quinone adsorption of quinone derivatives generated from BPE and BPO on chitosan beads was a little low, we tried to remove both bisphenol derivatives by increasing the amount of chitosan beads to 0.20 cm^3/cm^3. On the other hand, either tyrosinase-catalyzed quinone oxidation or quinone adsorption on chitosan beads for BPB, BPC, and BPZ was lower compared with BPA. Therefore, the tyrosinase concentration and amount of chitosan beads were increased so as to enhance both quinone oxidation and quinone adsorption. Consequently, in the presence of chitosan beads at 0.20 cm^3/cm^3, BPB was almost completely removed at the tyrosinase concentration of 400 U/cm^3 and the removal % value increased up to 98.9% by prolonging the reaction time to 20 hr. Compared with these bisphenol derivatives, the removal % values for BPC and BPZ was limited to 63.1 and 86.0 %, respectively, even if the tyrosinase concentration was increased to 300 U/cm^3 in the presence of chitosan beads at 0.20 cm^3/cm^3. It was found from the above results that the bisphenol derivatives except for BPC, BPS, and, BPZ were completely or effectively removed by increasing either the tyrosinase concentration or the amount of chitosan beads at pH 7.0 and 40°C.

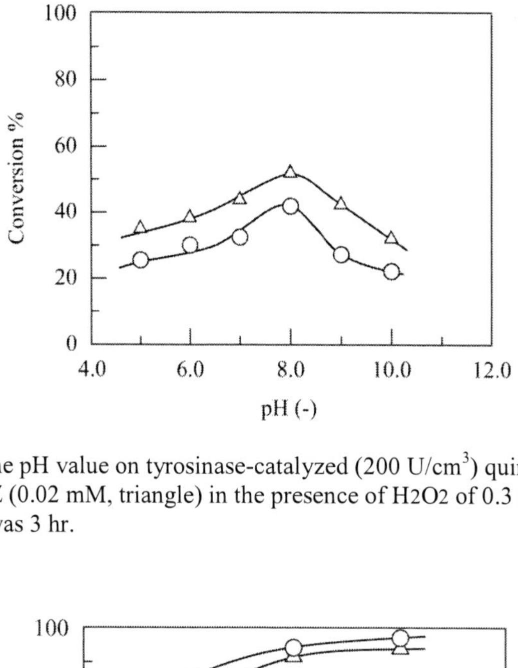

Figure 12. The effect of the pH value on tyrosinase-catalyzed (200 U/cm^3) quinone oxidation of BPC (0.05 mM, open) and BPZ (0.02 mM, triangle) in the presence of H2O2 of 0.3 mM at 40°C. The enzymatic reaction time was 3 hr.

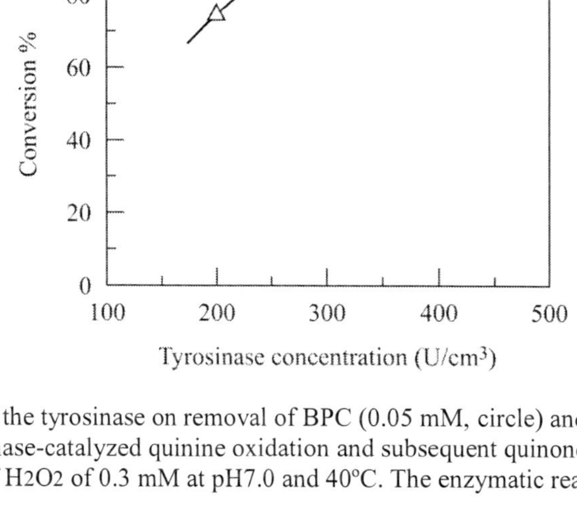

Figure 13. The effect of the tyrosinase on removal of BPC (0.05 mM, circle) and BPZ (0.02 mM, triangle) through tyrosinase-catalyzed quinine oxidation and subsequent quinone adsorption on chitosan beads in the presence of H2O2 of 0.3 mM at pH7.0 and 40°C. The enzymatic reaction time was 5 hr.

5.2. Effect of Enzyme Dose and Amount of Chitosan Beads

The removal % values of BPC and BPZ were lower than those of other bisphenol derivatives at pH 7.0. In addition, their adsorption % values were lower than the conversion % value. Therefore, the effect of the pH value on quinone oxidation for these two bisphenol derivatives was estimated at $200U/cm^3$. As shown in Figure 12, the conversion % had the maximum value at pH 8.0. Then, the removal value increased with an increase in the tyrosinase concentration as shown in Figure 13, and the removal values of BPC and BPZ went up to 97.0 and 94.2 %, respectively.

For bisphenol derivatives which had lower removal efficiency or little underwent tyrosinase-catalyzed quinone oxidation, one of the solutions is the usage of tyrosinases derived from different origins, for example *Aspergillus oryzae* tyrosinase because the substrate selectivity of enzymes varies by their origin to a greater or lesser.

6. Conclusions

In this chapter, the availability of chitosan was investigated for removal of bisphenol A through the tyrosinase-catalyzed quinone oxidation and subsequent nonenzymatic quinone adsorption on chitosan beads. First, the tyrosinase-catalyzed quinone oxidation of BPA was systematically estimated as a function of the H_2O_2-to-BPA ratio, pH value, temperature, and tyrosinase dose. Tyrosinase-catalyzed quinone oxidation was accelerated by the addition of H_2O_2 and the optimum conditions for BPA at 0.3 mM were determined to be the presence of H_2O_2 of 0.3 mM at pH 7.0 and 40°C.

When chitosan beads were dispersed in BPA solutions containing tyrosinase and H_2O_2, quinone derivatives enzymatically generated were effectively chemisorbed on chitosan beads. The removal efficiency increased by increasing the amount of chitosan beads added to BPA solutions, and BPA was completely removed in the presence of chitosan beads at 0.10 cm^3/cm^3. The time required to completely remove BPA was shortened by increasing the amount of chitosan beads. It was found from the a control removal experiment with chitosan solutions that the removal time for the heterogeneous systems with chitosan beads was considerably shortened compared with the homogeneous systems with chitosan solutions. In addition, for the homogeneous systems with chitosan solutions, quinone derivatives enzymatically generated from BPA were a little left in the reaction solutions even at the optimum amino group concentration of 0.75 mM, that is, $[-NH_2]/[BPA]=2.5$. At least the reaction times of 20 to 24 hr were required to effectively remove BPA from solutions because the formation of precipitates through the reaction of quinone derivatives with chitosan's amino groups was rather slow. Contrary to this, BPA was completely removed only for 3 to 5 hr through the tyrosinase-catalyzed quinone oxidation and subsequent nonenzymatic quinone adsorption on chitosan beads. In addition, it was found from the a control removal experiment with chitosan powders that the porosity of the chitosan beads used, or the large specific surface area, is one of the important factors to enhance the removal efficiency.

Bisphenol derivatives, which are suspected as endocrine disrupting chemicals, were completely or effectively removed by this procedure, although either the tyrosinase concentration or the amount of added chitosan beads is required to be increased for some bisphenol derivatives that had low removal efficiency. In addition, we can safely say from the above results that an alternative usage of chitosan came out. Our study on removal of BPA provides the results comparable to or better than studies reported previously because alternative separation techniques for removing colored quinone derivatives enzymatically generated from BPA would be operationally complex and would require capital investments. Although we focused on the technical feasibility in this study, the practical application of this procedure may require the immobilization of tyrosinase.

REFERENCES

[1] Vos, J. G., Dybring, E., Greim, H. A., Ladefoged, O., Lambré, C., Tarazona, J. V., Brandt, I., and Vethaak, A. D. (2000). Health effects of endocrine-disrupting chemicals on wildlife, with special reference to the European situation, *Crit. Rev. Toxicol.*, *30*, 71-133.

[2] Rahman, M. F., Yanful, E. K., and Jasim, S. Y. (2009). Occurrences of endocrine disrupting compounds and pharmaceuticals in the aquatic environment and their removal from drinking water: challenges in the context of the developing world, *Desalination*, *248*, 578-585.

[3] Díaz-Ferrero, J., Rodríguez-Larena, M. C., Comellas, L., and Jiménez, B. (1997). Bioanalytical methods applied to endocrine disrupting polychlorinated biphenyls, polychlorinated dibenzo-*p*-dioxins and polychlorinated dibenzofurans. A review. *Trends Anal. Chem.*, *16*, 563-573.

[4] Szelci, J., Soto, A. M., Geck, P., Desronvil, M., Prechtl, N. V., Weill, B. C., and Sonnenshein, C. (2000). Identification of human estrogen-inducible transcripts that potentially mediate the apoptotic response in breast cancer, *J. Steroid Biochem. Molec. Biol.*, *72*, 89-102.

[5] Rice, C., Birnbaum, L. S., Cogliano, J., Mahaffey, K., Needham, L., Rogan, W. J., and vom Saal, F. S. (2003). Exposure assessment for endocrine disruptors: Some considerations in the design studies, *Environ. Health Perspect.*, *111*, 1683-1690.

[6] Kang, J. H., Kondo, F., and Katayama, Y. (2006). Human exposure to bisphenol A, *Toxicol.*, *226*, 79-89.

[7] Suzuki, Y., Nakagawa, Y., Takano, I., Yaguchi, K., and Yasuda, K. (2004). Environmental Fate of bisphenol A and its biological metabolites in river water and their xeno-estrogenic activity, *Environ. Sci. Technol.*, *38*, 2389-2396.

[8] Guillette Jr., L. J. and Gunderson, M. P. (2001). Alterations in development of reproductive and endocrine systems of wildlife populations exposed to endocrine-disrupting contaminants, *Reproduction*, *122*, 857-864.

[9] Gunderson, M. P., LeBlanc, G. A., and Guillette Jr., L. J. (2001). Alterations in sexually dimorphic biotransformation of testosterone in juvenile American alligators (*Alligator mississippiensis*) from contaminated lakes, *Environ. Health Perspect.*, *109*, 1257-1264.

[10] Chikae, M., Ikeda, R., Hasan, Q., Morita, Y., and Tamiya, E. (2003). Effect of alkylphenols on adult male medaka: plasma vitellogenin goes up to the level of estrous female, *Environ. Toxicol. Pharmacol.*, *15*, 33-36.

[11] Clotfelter, E. D., Bell, A. M., and Levering, K. R. (2004). The role of animal behaviour in the study of endocrine-disrupting chemicals, *Anim. Behav.*, *68*, 665-676.

[12] Dodds, E. C. and Lawson, W. (1936). Synthetic estrogenic agents without the phenanthrene nucleus, *Nature* (London), *137*, 996-997.

[13] Schafer, T. E., Lapp, C. A., Hanes, C. M., Lewis, J. B., Wataha, J. C., and Shuster, G. S. (1999). Estrogenicity of bisphenol A and bisphenol A dimethacrylate *in vitro*, *J. Biomed. Mater. Res.*, *45*, 192-197.

[14] Kubo, T., Maezawa, N., Osada, M., Katsumura, S., Funae, Y., and Imaoka, S. (2004). Bisphenol A, an environmental endocrine-disrupting chemical, inhibits hypoxic response via degradation of hypoxia-inducible factor 1□ (HIF-1□): structural requirement of bisphenol A for degradation of HIF-1□, *Biochem. Biophy. Res. Commun.*, *318*, 1006-1011.

[15] Singleton, D. W., Feng, Y., Chen, Y., Busch, S. J., Lee, A. V., Puga, A., and Khan, S. A. (2004). Bisphenol-A and estradiol exert novel gene regulation in human MCF-7 derived breast cancer cells, *Mol. Cell. Endocrinol.*, *221*, 47-55.

[16] Heneweer, M., Muusse, M., Dingemans, M., de Jong, P. C., van den Berg, M., and Sanderson, J. T. (2005). Co-culture of primary human mammary fibroblasts and MCF-7 cells as an *in vitro* breast cancer model, *Toxicol. Sci.*, *83*, 257-263.

[17] Yasuhara, A., Shiraishi, H., Nishikawa, M., Yamamoto, T., Uehiro, T., Nakasugi, O., Okumura, T., Kenmotsu, K., Fukui, H., Nagase, M., Ono, Y., Kawagoshi, Y., Baba, K., and Noma, Y. (1997). Determination of organic components in leachates from hazardous waste disposal sites in Japan by gas chromatography-mass spectrometry, *J. Cromatogr. A*, *774*, 321-332.

[18] Yamamoto, T., Yasuhara, A., Shiraishi, H., and Nakasugi, O. (2001). Bisphenol A in hazardous waste handfill leachates, *Chemosphere*, *42*, 415-418.

[19] Kurata, Y., Ono, Y., and Ono, Y. (2008). Occurrence of phenols in leachates from municipal solid waste handfill sites in Japan, *J. Mater. Cycles Waste Manag.*, *10*, 144-152.

[20] Kitano, H., Hirabayashi, T., Ide, M., and Kyogoku, M. (2003). Complexation of bisphenol A with calyx[6]arene-polymer conjugates, *Macromol. Chem. Phys.*, *204*, 1419-1427.

[21] Gadhia, S. T., Joshi, J. K., and Parsania, P. H. (2005). Thermal analysis of cured halogenated bisphenol-C-epoxy resins, *Polym. Degrad. Stab.*, *88*, 217-223.

[22] Joshi, J. K., Gadhia, S. T., and Parsania, P. H. (2005). Thermal properties of cured bisphenol-C-epoxy resins, *J. Polym. Mater.*, *22*, 133-143.

[23] Brzozowski, Z. K., Staszczak, S. K., Hadam, L. K., Rupinski, S., Bogdal, D., and Gorczyk, J. (2006). Synthesis of solid epoxy resins base on BPC II and bisphenol A using conventional heating equipment and microwave, *J. Appl. Polym. Sci.*, *100*, 3850-3854.

[24] Mavarti, S. I., Mehta, N. M., and Parsania, P. H. (2006). Synthesis and physicochemical study of bisphenol-C-formaldehyde-toluene diisocyanate polyurethane-jute and jute-rice Husk/Wheat husk composites, *J. Appl. Polym. Sci.*, *101*, 2363-2370.

[25] Lyon, R. E., Speitel, L., Filipczak, R., Walters, R. N., Crowley, S., Stoliarov, S. I., Castelli, L., and Ramirez, M. (2007). Fire smart DDE polymers, *High Perform. Polym.*, *19*, 323-355.

[26] Sheng, X., Akinc, M., and Kessler, M. R. (2008). Cure kinetics of thermosetting bisphenol E cyanate ester, *J. Therm. Anal. Cal.*, *93*, 77-85.
[27] Takeda, K. and Kobayashi, T. (2006). Hybrid molecularly imprinted membranes for targeted bisphenol derivatives, *J. Membr. Sci.*, *275*, 61-69.
[28] Takeda, K. and Kobayashi, T. (2005). Bisphenol A imprinted polymer adsorbents with selective recognition and binding characteristics, *Sci. Technol. Adv. Mater.*, *6*, 165-171.
[29] Poustkova, I., Dobias, J., Steiner, I., Poustka, J., and Voldrich, M. (2004). Stability of bisphenol A diglycidyl ether and bisphenol F diglycidyl ether in water-based food simulants, *Eur. Food Res. Technol.*, *219*, 534-539.
[30] Gao, J. G., Xia, L. Y., and Liu, Y. F. (2004). Structure of boron-containing bisphenol-F formaldehyde resin and kinetics of its thermal degradation, *Polym. Degrad. Stab.*, *83*, 71-77.
[31] Dang, W. R., Kubouchi, M., Sembokusa, H., and Tsuda, K. (2005). Chemical recycling of glass fiber reinforced epoxy resin cured with amine using nitrid acid, *Polymer*, *46*, 1905-1912.
[32] Cho, H. S., Liang, K. W., Chatterjee, S., and Pittman Jr., C. U. (2005). Synthesis, morphology, and viscoelastic properties of polyhedral oligomeric silsesquiozane nanocomposites with epoxy and cyanate ester matrices, *J. Inorg. Organomer. Polym. Mater.*, *15*, 541-553.
[33] Malshe, V. C. and Waghoo, G. (2006). Weathering characteristics of epoxy-ester paints, *Progr. Org. Coat.*, *56*, 131-134.
[34] Barone, L., Carciotto, S., Cicala, G., and Recca, A. (2006). Thermomechanical properties of epoxy/poly(epsilon-caprolactone) blends, *Polym. Eng. Sci.*, *46*, 1576-1582.
[35] Sun, S. J., Liao, Y. C., and Chang, T. C. (2000). Studies on the synthesis and properties of thermotropic liquid crystalline polycarbonates. VII. Liquid crystalline polycarbonates and poly(ester-carbonate)s derived from various mesogenic qroups, *J. Polym. Sci. Polym. Chem.*, *38*, 1852-1860.
[36] Liaw, D. J. and Chang, P. (1997). Synthesis and characterization of polycarbonates based on bisphenol S by melt transesterification, *J. Polym. Sci. Polym. Chem.*, *35*, 2453-2460.
[37] Choi, J. H., Kwak, S. Y., Kim, S. Y., Kim, J. S., and Kim, J. J. (1998). Effect of molecular structure of polyarylates on the compatibility in polyarylate poly(vinyl chloride) blends, *J. Appl. Polym. Sci.*, *70*, 2173-2180.
[38] Gao, J. G. and Li, Y. F. (2000). Curing kinetics and thermal property characterization of a bisphenol-S epoxy resin and DDS system, *Polym. Int.*, *49*, 1590-1595.
[39] Shen, S. G., Li, Y. F., Gao, J. G., and Sun, H. W. (2001). Curing kinetics and mechanism of bisphenol S epoxy resin with 4,4'-diaminodiphenyl ether or phathlic anhydride, *Int. J. Chem. Kinet.*, *33*, 558-563.
[40] Lu, H. and Zheng, S. X. (2005). Miscibility and intermolecular specific interactions in thermosetting blends of bisphenol S epoxy resin with poly(ethylene oxide), *J. Polym. Sci. Polym. Phys.*, *43*, 359-367.
[41] Berti, C., Celli, A., Marianucci, E., and Vannini, M. (2007). Preparation and characterization of novel random copoly(arylene ether-thioether ketone)s containing 2,2-bis(4-phenylene)propane units, *Eur. Polym. J.*, *43*, 2453-2461.

[42] Chen, J. R. Li, Y. S., and Chang, F. C., (2001). Synthesis and physical properties of linear and branch phenoxies, *J. Polym. Res. Taiwan*, *8*, 225-233.

[43] Matoliukstyte, A., Grazulevicius, J., Reina, J. A., Jankauskas, V., and Montrimas, E. (2006). Synthesis and properties of glass-forming condensed aromatic amines with reactive functional groups, *Mater. Chem. Phys*, *98*, 324-329.

[44] Kitamura, S., Suzuki, T., Sanoh, S., Kohta, R., Jinno, N., Sugihara, K., Yoshihara, S., Fujimoto, N., Watanabe, H., and Ohta, S. (2005). Comparative study of the endocrine-disrupting activity of bisphenol A and 19 related compounds, *Toxicol. Sci.*, *84*, 249-259.

[45] Kubo, T., Maezawa, N., Osada, M., Katsumura, S., Funae, Y., and Imaoka, S. (2004). Bisphenol A, an environmental endocrine-disrupting chemical, inhibits hypoxic response via degradation of hypoxia-inducible factor 1□ (HIF-1□): structural requirement of bisphenol A for degradation of HIF-1□, *Biochem. Biophy. Res. Commun.*, *318*, 1006-1011.

[46] Schafer, T. E., Lapp, C. A., Hanes, C. M., Lewis, J. B., Wataha, J. C., and Schuster, G. S. (1999). Estrogenicity of bisphenol A and bisphenol A dimethlacrylate *in vitro*, *J. Biomed. Mat. Res.*, *45*, 192-197.

[47] Nakanishi, A., Tamai, M., Kawasaki, N., Nakamura, T., and Tanada, S. (2002). Adsorption characteristics of bisphenol A onto carbonaceous materials produced from wood chips as organic waste, *J. Colloid Interface Sci.*, *252*, 393-396.

[48] Kitaoka, M. and Hayashi, K. (2002). Adsorption of bisphenol A by cross-linked □-cyclodextrin polymer, *J. Incl. Phenom. Macrocycl. Chem.*, *44*, 429-431.

[49] Tsai, W. T., Hsu, H. C., Su, T. Y., Lin, K. Y., and Lin, C. M. (2006). Adsorption characteristics of bisphenol-A in aqueous solutions onto hydrophobic zeolite, *J. colloid Interface Sci.*, *209*, 513-519.

[50] Li, Y., Li, N., Chen, D., Wang, X., and Xu, Z. (2009). Bisphenol A adsorption onto metals oxides and organic materials in the natural surface coatings samples (NSCSs) and surficial sediments (SSs): Inhibition for the importance of Mn oxides, *Water Air Soil Polut.*, *196*, 41-49.

[51] Lin, G., Ma, J., Li, X., and Qin, Q. (2009). Adsorption of bisphenol A from aqueous solution onto activated carbons with different modification treatments, *J. Hazard. Mater.*, *164*, 1275-1280.

[52] Wintgens, T., Gallenkemper,M., and Melin, T. (2004). Removal of endocrine disrupting compounds with membrane processes in wastewater treatment and reuse, *Water Sci. Technol.*, *50*, 1-8.

[53] Zhang, Y., Causserand, C., Aimar, P., and Cravedi, J. P. (2006). Removal of bisphenol A by a nanofiltration membrane in view of drinking water production, *Water Res.*, *40*. 3793-3799.

[54] Shengji, X., Xing, L., Yao, J., Bingzhi, D., and Juanjuan, Y. (2008). Application of membrane techniques to produce drinking water in China, *Desalination*, *222*, 497-501.

[55] Kuramitz, H., Matsushita, M., and Tanaka, S. (2004). Electrochemical removal of bisphenol A based on the anodic polymerization using a column type carbon fiber electrode, *Water Res.*, *38*, 2331-2338.

[56] Rosenfeldt, E. J. and Linden, K. G. (2004). Degradation of endocrine disrupting chemicals bisphenol A, ethinyl estradiol, and estradiol during UV photolysis and advanced oxidation processes, *Environ. Sci. Technol.*, *38*, 5476-5483.

[57] Chen, P. J., Linden, K. G., Hinton, D. E., Kasgiwada, S., Rosenfeldt, E. J., and Kullman, S. W. (2006). Biological assessment of bisphenol A degradation in water following direct photolysis and UV advanced oxidation, *Chemosphere*, *65*, 1094-1102.

[58] Rosu, D., Cascaval, C. N., and Rosu, L. (2006). Effect of UV radiation on photolysis of epoxy maleate of bisphenol A, *J. Photochem. Photobiol. A*: *Chem*., *177*, 218-224.

[59] Wang, G., Qi, P., Xue, X., Wu, F., and Deng, N. (2007). Photodegradation of bisphenol Z by UV irradiation in the presence of □-cyclodextrin, *Chemosphere*, *67*, 762-769

[60] Kang, J. H. and Kondo, F. (2002). Bisphenol A degradation by bacteria isolated from river water, *Arch. Environ. Contam. Toxicol*., *43*, 265-269.

[61] Hirooka, T., Akiyama, Y., Tsuji, N., Nakamura, T., Nagase, H., Hirata, K., and Miyamoto, K. (2003). Removal of hazardous phenols by microalgae under photoautotrophic conditions, *J. Biosci. Bioeng*., *95*, 200-203.

[62] Ike, M., Chen, M. Y., Danzl, E., Sei, K., and Fujita, M. (2006). Biodegradation of a variety of bisphenols under aerobic and anaerobic conditions, *Water Sci. Technol*., *53*, 153-159.

[63] Yamanaka, H., Moriyoshi, K., Ohmoto, T., Ohe, T., and Sakai, K. (2008). Efficient microbial degradation of bisphenol A in the presence of activated carbon, *J. Biosci. Bioeng*., *105*, 157-160.

[64] Gianfreda, L., Xu, F., and Bollag, J. M. (1999). Laccases: A useful group of oxidoreductive enzymes, *Bioremediat. J*., *3*, 1-25.

[65] Durán, N. and Esposito, E. (2000). Potential applications of oxidative enzymes and phenoloxidase-like compounds in wastewater and soil treatment: a review, *Appl. Catal., B*: *Environ*., *28*, 83-99.

[66] Durán, N., Rosa, M. A., D'Annibale, A., and Gianfreda, L. (2002). Applications of laccases and tyrosinases (phenoloxidases) immobilized on different supports: a review, *Enzyme Microb. Technol*., *31*, 907-931.

[67] Ikehata, K., Buchanan, I. D., and Smith, D. W. (2004). Recent developments in the production of extracellular fungal peroxidases and laccases for waste treatment, *J. Environ. Eng. Sci*., *3*, 1-19.

[68] Nicell, J. A., Bewtra, J. K., Taylor, K. E., Biswas, N., and St. Pierre, C. (1992). Enzyme polymerization and precipitation of aromatic compounds from wastewater, *Water Sci. Technol*., *25*, 157-164.

[69] Nicell, J. A., Bewtra, J. K., Biswas, N., and Taylor, E. (1993). Reactor development for peroxidase catalyzed polymerization and precipitation of phenols from wastewater, *Water Sci. Technol*., *27*, 1629-1639.

[70] Nicell, J. A. (1994). Kinetics of horseradish peroxidase-catalyzed polymerization and precipitation of aqueous 4-chlorophenol, *J. Chem. Tecnol. Biotechnol*., *60*, 203-215.

[71] Xu, Y. P., Huang, G. L., and Yu, Y. T. (1995). Kinetics of phenolic polymerization catalyzed by peroxidase in organic media, *Biotechnol. Bioeng*., *47*, 117-119.

[72] Tonami, H., Uyama, H., Kobayashi, S., and Kubota, M. (1999). Peroxidase-catalyzed oxidative polymerization of *m*-substituted phenol derivatives, *Macromol. Chem. Phys*., *200*, 2365-2371.

[73] Huang, Q. and Weber Jr., W. J. (2005). Transformation and removal of bisphenol A from aqueous phase via peroxidase-mediated oxidative coupling reactions: Efficary, products, and pathways, *Environ. Sci. Technol*., *39*, 6029-6036.

[74] Kim, E. Y., Chae, H. J., and Chu, K. H. (2007). Enzymatic oxidation of aqueous pentachlorophenol, *J. Environ. Sci.*, *19*, 1032-1036.

[75] Yamada, K., Ikeda, N., Takano, Y., Kashiwada, A., Matsuda, K., and Hirata, M. Determination of optimum process parameters for peroxidase-catalyzed treatment of bisphenol A and application of removal of bisphenol derivatives, *Environ. Technol.*, *in press*.

[76] Wright, H. and Nicell, J. A. (1999). Characterization of soybean peroxidase for the treatment of aqueous phenols, *Bioresour. Technol.*, *70*, 69-79.

[77] Flock, C., Bassi, A., and Gijzen, M. (1999). Removal of aqueous phenol and 2-chlorophenol with purified soybean peroxidase and raw soybean hulls, *J. Chem. Technol. Biotechnol.*, *74*, 303-309.

[78] Bódalo, A., Gómez, J. L., Gómez, E., Hidalgo, A. M., Gómez, M., and Yelo,M. (2006). Removal of 4-chlorophenol by soybean peroxidase and hydrogen peroxide in a discontinuous tank reactor, *Desalination*, *195*, 51-59.

[79] Bódalo, A., Gómez, J. L., Gómez, E., Hidalgo, A. M., Gómez, M., and Yelo,M. (2007). Elimination of 4-chlorophenol by soybean peroxidase and hydrogen peroxidase: Kinetic model and intrinsic parameters, *Biochem. Eng. J.*, *34*, 242-247.

[80] Duarte-Vázquez, M. A., Ortega-Tovar, M. A., García-Almendarez, B. E., and Regalado, C. (2002). Removal of aqueous phenolic compounds from a model system by oxidative polymerization with turnip (*Brassica napus* L var purple top white globe) peroxidase, *J. Chem. Technol. Biotechnol.*, *78*, 42-47.

[81] Sakurai, A., Toyoda, S., and Sakakibara, M. (2001). Removal of bisphenol A by polymerization and precipitation method using *Coprinus cinereus* peroxidase, *Biochem. Lett.*, *23*, 995-998.

[82] Sakurai, A., Toyoda, S., Masuda, M., and Sakakibara, M. (2004). Removal of bisphenol A by peroxidase-catalyzed reaction using culture broth of *Coprinus cinereus*, *J. Chem. Eng. Jpn.*, *37*, 137-142.

[83] Espín, J. E., Varón, R., Fenoll, L. G., Gilabert, M. A., García-Ruíz, P. A., Tudela, J., and García-Cánovas, F. (2000). Kinetic characterization of the substrate specificity and mechanism of mushroom tyrosinase, *Eur. J. Biochem.*, *267*, 1270-1279.

[84] Nappi, A. J. and Vass, E. (2001). The effect of Nitric oxide on the oxidations of L-dopa and dopamine mediated by tyrosinase and peroxidase, *J. Biolog. Chem.*, *276*, 11214-11222.

[85] Sun, W. Q., Payne, G. F., Moas, M. S. G. L., Chu, J. H., and Wallance, K. K. (1992). Tyrosinase reaction/chitosan adsorption for removing phenols from wastewater, *Biotechnol. Prog.*, *8*, 179-186.

[86] G. F. Payne, Sun, W. Q., and Sohrabi, A. (1992). Trosinase reaction/chitosan adsorption for selectively removing phenols from aqueous mixtures, *Biotechnol. Bioeng.*, *40*, 1011-1018.

[87] Payne, G. F. and Sun, W. Q. (1994). Tyrosinase reaction and subsequent chitosan adsorption for selective removal of a contaminant from a fermentation recycle stream, *Appl. Environ. Microbiol.*, *60*, 397-401.

[88] Sun, W. Q. and Payne, G. F. (1996). Tyrosinase-containing chitosan gels: a combined catalyst and sorbent for selective phenol removal, *Biotechnol. Bioeng.*, *51*, 79-86.

[89] Payne, G. F., Chaubal, M. V., and Barbari, T. A. (1996). Enzyme-catalyzed polymer modification: reaction of phenolic compounds with chitosan films, *Polymer*, *37*, 4643-4648.
[90] Švitel, J. and Miertus, S. (1998). Development of tyrosinase-based biosensor and its application for monitoring of bioremediation of phenol and phenolic compounds, *Environ. Sci. Technol.*, *32*, 828-832.
[91] Ikehata, K. and Nicell, J. A. (2000). Characterization of tyrosinase for the treatment of aqueous phenols, *Bioresource Technol.*, *74*, 191-199.
[92] Yoshida, M., Ono, H., Mori, Y., Chuda, Y., and Onishi, K. (2001). Oxidation of bisphenol A and related compounds, *Biosci. Biotechnol. Biochem.*, *65*, 1444-1446.
[93] Yamada, K., Akiba, Y., Shibuya, T., Kashiwada, A., Matsuda, K., and Hirata, M. (2005). Water purification through bioconversion of phenol compounds by tyrosinase and chemical adsorption by chitosan beads, *Biotechnol. Prog.*, *21*, 823-829.
[94] Yamada, K., Inoue, T., Akiba, Y., Kashiwada, A., Matsuda, K., and Hirata, M. (2006). Removal of *p*-alkylphenols from aqueous solutions by combined use of mushroom tyrosinase and chitosan beads, *Biosci. Biotechnol. Biochem.*, *70*, 2467-2475.
[95] Jiménez, M. and Carcía-Carmona, F. (1996). Hydrogen peroxide-dependent 4-*t*-butylphenol hydroxylation by tyrosinase: a new catalytic activity. *Biochim. Biophy. Acta*, *1297*, 33-39.
[96] Wada, S., Ichikawa, H., and Tatsumi, K. (1995). Removal of phenols and aromatic amines from wastewater by a combination treatment with tyrosinase and a coagulant. *Biotechnol. Bioeng.*, *45*, 304-309.
[97] Ikehata, K and Nicell, J. A. (2000). Color and toxicity removal following tyrosinase-catalyzed oxidation of phenols, *Biotechnol. Prog.*, *16*, 533-540.
[98] Rescigno, A., Sanjust, E., Soddu, G., Rinaldi, A. C., Sollai, F., Curreli, N., and Rinaldi, A. (1998). Effect of 3-hydroxyanthanilic acid on mushroom tyrosinase activity, *Biochim. Biophys. Acta*, *1384*, 268-276.
[99] Espín, J. C., Varón, R., Tudela, J., and García-Cánovas, F. (1997). Kinetic study of the oxidation of 4-hydroxyanisole catalyzed by tyrosinase, *Biochem. Molecul. Biol. Int.*, *41*, 1265-1276.
[100] Xuan, Y.J., Endo, Y., and Fujimoto, K. (2002). Oxidative degradation of bisphenol A by crude enzyme prepared from potato, *J. Agric. Food Chem.*, *50*, 6575-6578.
[101] Yamada, K., Nakasone, T., Nagano, R., and Hirata, M. (2003). Retention and reusability of trypsin activity by covalent immobilization onto grafted polyethylene plates, *J. Appl. Polym. Sci.*, *89*, 3574-3581.
[102] Yamada, K., Iizawa, Y., Yamada, J., and Hirata, M. (2006). Retention and reusability of trypsin activity by covalent immobilization onto grafted polyethylene plates, *J. Appl. Polym. Sci.*, *102*, 4886-4896.
[103] Kumar, G., Bristow, J. F., Smith, P. J., and Payne, G. F. (2000). Enzymatic gelation of the natural polymer chitosan, *Polymer*, *41*, 2157-2168.

In: Focus on Chitosan Research
Editors: Arthur N. Ferguson and Amy G. O'Neill
ISBN 978-1-61324-454-8

Chapter 15

CHITOSAN USAGE IN SYNTHESIS OF HYDROGEL-BIOMATERIALS WITH CONTROLLED PROPERTIES FOR BIOMEDICAL/BIOTECHNOLOGICAL APPLICATIONS

Eugenia Dumitra Teodor[1,*], Simona Carmen Lițescu[1], Cristian Petcu[2,†] and Gabriel Lucian Radu[3,~]
[1] National Institute for Biological Sciences, Centre of Bioanalysis, Bucharest
[2] National Institute for Chemistry and Petrochemistry, Bucharest
[3] Politehnica University of Bucharest, Bucharest

ABSTRACT

Over the past 40 years a greater attention has been focused on development of controlled and sustained drug delivery systems. The goals in designing these systems are to reduce the frequency of dosing or to increase effectiveness of the drug by localization at the site of the action, decreasing the dose required or providing uniform drug delivery. Polymers have been the keys to the great majority to design biomaterials for drug delivery systems. Chitosan, a natural biopolymer, has been extensively used for its potential in the development of biomaterials due to the polymeric cationic character and gel and film forming properties.

Hydrogels preformed by chemical cross-linking or physical interactions form three-dimensional, hydrophilic, polymeric networks capable of imbibing large amounts of water or biological fluid. The association of magnetic nanoparticles, which could be

* Bucharest 6, 296 Spl. Independentei, tel/fax: +40212200900 e-mail: eu_teodor@yahoo.com.
† Bucharest 6, 202 Spl. Independentei, tel: +40213163093.
~ Bucharest 6, 313 Spl. Independentei, tel: +40214023802.

controlled by an exterior magnetic field and have dimensions to facilitate their penetration in cells/tissues, with hydrogel type biopolymeric shells confer them compatibility and the capacity to retain and deliver bioactive substances.

In this chapter we present a synthesis of our works, from the last 5 years, in this domain of chitosan usage for synthesis of biomaterials suitable in obtaining of drug delivery systems. In our works we used mainly chitosan with slight addition of hyaluronic acid for obtaining biomaterials with controlled properties. Chitosan and hyaluronic acid are two natural biopolymers with similar structure and special biological characteristics, easily to process in porous scaffolds, films/pellicles and beads. They are biocompatible, biodegradable, promote cell migration and cell adhesion and electrostatic interact, having a great potential in the development of drug delivery systems and in tissue engineering.

In a first stage, we obtained biopolymeric pellicles by casting method using mixtures of chitosan and hyaluronic acid solutions. Chitosan/hyaluronic acid pellicles crosslinked with sodium citrate could be used for enzymes immobilization on the surface or by entrapping in the polymer network. The pellicles with immobilized enzymes are stable during several months and could be used multiple times without a notable decrease of the immobilized enzymatic activity. The ultrastructure of simple chitosan pellicles, chitosan/hyaluronic acid pellicles and enzymes immobilized on chitosan/hyaluronic acid pellicles were examined by confocal scanning laser microscopy.

More recently, in other works, we developed a new system based on magnetic nanoparticles covered in a layer-by-layer technique with a biocompatible hydrogel from chitosan and hyaluronic acid capable of vectoring support for biologic active agents (L-asparaginase, protease inhibitor, e.g.). Characterization of size and morphology of the obtained hydrogel-magnetic nanoparticles with entrapped L-asparaginase/protease inhibitor was made using dynamic light scattering method, transmission electron microscopy and confocal microscopy. The structure of magnetic nanoparticles coated with hydrogel was characterized by Fourier transformed infrared spectroscopy. The biocompatibility of nanoparticles was evaluated and also the interactions with microorganisms.

The characteristics of developed pellicles and nanocomposite materials made them suitable to be used for medical or biotechnological purposes.

INTRODUCTION

Chitosan (Ch), a compound with excellent biodegradable and biocompatible characteristics, is a naturally occurring polysaccharide. Chitosan has been extensively studied for its potential in the development of drug delivery systems due to the polymeric cationic character and its gel and film forming properties [1,2]. Chitosan films were usually prepared by chemical crosslinking with bifunctional agents such as glutaraldehyde, glioxal, polyvinylalcohol [3-5] etc. The chemical crosslinking agents induce possible toxicity and other undesirable effects (neurotoxic, mutagenic). To overcome these disadvantages, reversible physical crosslinking by electrostatic interaction was used as an alternative method in the preparation of chitosan film [5]. For example, Yao et al. reported the preparation of pectin/chitosan films by dissolving this polyelectrolyte complex in formic acid and then evaporating the solvent, Chu et al. prepared xanthan/chitosan complex film by the evaporation method using concentrated sodium chloride [6]. The use of low molecular weight ions to prepare an ionic crosslinking polymeric matrix proved to be a simple and mild procedure, involving just the soaking of the polymer films in ion solution [7,8].

Shu and Zhu reported for the first time that there is some electrostatic interaction between sodium citrate and chitosan, and they used sodium citrate to prepare citrate-crosslinked chitosan beads, microspheres and films [9-11].

Hyaluronic acid (hyaluronan, HA), an abundant non-sulfated glycosaminoglycan component of synovial fluid and extracellular matrices, is another attractive building block for new biocompatible and biodegradable biomaterials, which have applications in drug delivery, tissue engineering, and viscosupplementation [12-16]. Hyaluronic acid and chitosan are two natural biopolymers with similar structure (two polysaccharides formed from repetitive disaccharide units based on glucuronic acid and N-acetyl glucosamine and, respectively from randomly distributed N-acetyl glucosamine and glucosamine) and special biological characteristics, therefore they are very suitable for biomaterials obtaining. They are biocompatibles, biodegradable, and highly hydrophilic, promote cell migration, cell adhesion and they electrostatic interact. The hyaluronic acid macromolecule is highly anionic and forms polyelectrolyte complexes with chitosan, which is cationic [17-19]. They have a great potential in the development of drug delivery systems and in tissue engineering [10,14,15].

There are several possible definitions of a hydrogel, but simply we defined hydrogels as macromolecular networks swollen in water or biological fluids [20]. Based on this definition hydrogels are often divided into three classes depending of the nature of their network, namely entangled network, covalently cross-linked network and network formed by secondary interactions. This classification is not entirely suitable and there are no strict borders between these classes. Therefore, beside classifications, hydrogels are biomaterials which through their high water content and soft rubber consistency can give them a certain resemblance to living tissue [21,22].

Hydrogels have three main applications, namely: 1) as drug delivery systems allowing release of bioactive materials by diffusion, (2) as scaffolds in cell culture and implants, and (3) as biocompatible and biodegradable wound dressing materials.

There is a deal of interest for generation of hydrogels with various applications, but the drug controlled and sustained delivery aspect is probably the most studied. The range of drugs may be very wide and, the same, the polymers which may be used in hydrogel generation.

A first aim in this chapter is to describe our results to obtain new non-toxic biocompatible hydrogels, based on chitosan and hyaluronic acid and stabilized by physics or ionic crosslinking agents, with good swelling capacity and degradability for generation of sustained drug delivery systems or/and wound dressing materials.

We obtained citrate crosslinked chitosan/acid hyaluronic pellicle-type hydrogels for immobilization of invertase on the hydrogel surface or by entrapping (e.g. the protease inhibitor). The pellicle-type hydrogels with immobilized enzymes are stable during several months and could be used multiple times without a notable decrease of the immobilized enzymatic activity [23,24]. Confocal scanning laser microscopy (CSLM) was used to analyze the surface of various pellicle-type hydrogels obtained.

Another aspect described in this chapter is the association of magnetic nanoparticles with hydrogel type biopolymeric shells, which confers to composite material bio-compatibility and the capacity to retain and deliver bioactive substances. This is supported by the fact that hydrogel-magnetic nanoparticles could be controlled by a magnetic field in order to facilitate their penetration in target tissues. It is possible to obtain in this way “smart” nanostructured biomaterials, which could support different biotechnological and biomedical applications. Moreover, it should be underlined that by this way it is attained the main goal of nanoparticles employment in therapeutics, namely to improve drug solubility and bioavailability [25], due to the fact that there is a need to develop suitable drug delivery systems that distribute the bioactive molecule only to the site of action, without affecting healthy organs and tissues [26].

Ferromagnetic nanoparticles applicability in therapeutics known an increasing interest due to their biocompatibility, targeting action, and subcellular size [27,28]. There are used either as contrast [29] and labeling agents [30] or as drug delivery systems [31-33], in magnetic bioseparation, magnetic resonance- imaging contrast enhancement, and hyperthermia treatment of cancer due to their properties of superparamagnetism, high saturation magnetization, highmagnetic susceptibility, and low toxicity [34,35]. Magnetic nanoparticles have been used as support material for binding of enzymes including yeast alcohol dehydrogenase [36] lipase [37] glucose oxidase [38,39] cholesterol oxidase [40] directly via carbodiimide activation or by functionalization of magnetic nanoparticles. The immobilization is commonly accomplished through a surface coating with polymers, the use of coupling agents or crosslinking reagents, and encapsulation.

Interest in nanosized drug delivery systems based on magnetic nanoparticles has increased in the past few years. Recent studies describe an approach in the configuration of a novel hydrogel nanocomposite with superparamagnetic property based on magnetic nanoparticles suspension mixed with different polymers and cyclic oligosaccharide [41]. Novel magnetic hybrid hydrogels were fabricated by *in situ* embedding of magnetic iron oxide nanoparticles into the porous hydrogel networks, this magnetic hydrogel material was found to hold a potential application in magnetically assisted bioseparation [42].

Based on our works, we described the synthesis of magnetic nanoparticles (MP), with enhanced biocompatibility obtained by hydrogel biomaterial covering of nanoparticles, which preserve the nanometric dimensions, and moreover are nontoxic and noninteractive with pathogen bacteria. Magnetic nanoparticles were obtained by co-precipitation method [43] and were covered with hydrogel-type biopolymer shell according to layer-by-layer technique using chitosan and hyaluronic acid [44].

The developed nanocomposite materials could be used for immobilization of various enzymes. We used L-asparaginase as model drug, a chemotherapy agent for the treatment of acute lymphoblastic leukaemia (ALL). Free L-asparaginase, in some cases it presents toxicity, the main side effect being an allergic or hypersensitivity reaction. Enzyme immobilization into a polymeric shell could improve their delivery and eliminate the allergic reaction [45]. Previous other works present the feasibility of intravenous PEG-asparaginase (a form a Escherichia coli L-asparaginase covalently linked to polyethylene glycol) administration [46], covalent immobilization of L-asparaginase on the microparticles of the natural silk sericin protein [47], or immobilization of L-asparaginase on various supports to develop a asparagines biosensor for leukemia [48]. In our experiments, L-asparaginase (E.C.3.5.1.1.) synthesized and purified from *E. coli* genetic modified strain (BIOTEHGEN-Centre of Microbial Biotechnologies Bucharest, collection) [49] was immobilized by entrapment in hydrogel coated magnetic nanoparticles. The residual activity of the immobilized L-asparaginase was determined spectrometrically at 450 nm with Nessler reagent method [50].

Characterization of size and size distribution of the hydrogel-magnetic nanoparticles with entrapped L-asparaginase was made using dynamic light scattering (DLS) method and transmission electron microscopy (TEM). The morphology of the particles was investigated with transmission electron microscopy and confocal microscopy (CSLM). The structure of magnetic nanoparticles coated with hydrogel was characterized by Fourier transform infrared spectroscopy (FT-IR) [51].

Our goal was to obtain nanosize biocompatible materials, with L-asparaginase entrapped, capable to penetrate cells/tissues and deliver L-asparaginase [52]. In addition, the obtained nanostructures present different behavior to adherence capacity of bacteria and yeasts and could be used in bioseparation [52]. Finally, the obtained results are promising for using the synthesized hydrogel-magnetic nanoparticles for immobilization and delivery for a wide class of therapeutic agents.

1. Pellicle-Type Hydrogel Biomaterials Based on Chitosan and Hyaluronic Acid

Chitosan (Ch, MW 600 000) was purchased from Fluka and used without further purification. Medium-molecular-weight (200 000 Da) hyaluronic acid (HA) was extracted from the bovine vitreous body using our previous developed method [53]. Invertase (3800 U/mg) was obtained from Serva, gelatin was purchased from Merck and sodium citrate from Sigma. All chemicals were of high analytical grade purity.

Methods

Hydrogel Pellicles Synthesis

Different pellicle-type hydrogels (Table 1) were obtained from 2% (w/v) chitosan solution in 1% (w/v) acetic acid, 1% (w/v) HA solution in 0.2 M NaCl and 2% (w/v) gelatin solution in bi-distilled water by mixing in different ratios: a) Ch/HA (9:1, 3:1 and 2:1); b) Ch/HA/Gelatin (13:2:5 and 4:1:5). 20 mL of biopolymeric mixture were casted into 50 mm Petri dishes and were dried for 24 hours at 37^0C. The dried pellicles were cut into 1x1 cm pieces.

In order to improve the qualities of pellicles it were used physic cross-linking agents (UV and thermal radiations) and/or ionic substances (sodium citrate, calcium chloride). Citrate crosslinked hydrogels were prepared by soaking the Ch/HA/gelatin hydrogel pieces in sodium citrate solution (50 mL solution 2.5 and 5% (w/v) sodium citrate in bidistilled water), for 2-4 hours at room temperature. The obtained hydrogels were washed extensively with distilled water and dried for 24 hours at 37^0C.

Swelling Test

The obtained pellicle-type hydrogels (~ 150 mg dry weight pieces) were suspended in tubes containing 50 mL of PBS (pH 7.2 at 20^0C). At an appropriate time interval, the pellicles were taken out, and the excess buffer was removed carefully with filter paper and then weighted immediately. The swelling capacity was calculated according to the following equation [54]:

$$\%S = \frac{m_w - m_i}{m_i} x100, \tag{1}$$

where $\%S$ is swelling ratio, m_w is the weight of samples after swelling test performing (buffer immersion) and m_i is initial weight.

In Vitro Degradation Study

Hydrogels were put into Falcon tubes containing 50 mL PBS (pH 7.2 at 37^0C). At predetermining time points, hydrogels were collected and excess buffer was removed from the surface by using filter paper. Samples were weighted by using an analytical balance with ± 0.1 mg accuracy. After the equilibration time of swelling in PBS, the degradation ratios were calculated according to the following equations [55]:

$$\%D = \frac{m_0 - m_t}{m_0} x100, \tag{2}$$

where $\%D$ represent degradation ratio, m_0 is the original weight after equilibration time of swelling in PBS and m_t is the weight at time t.

Immobilization of Invertase

Invertase (EC.3.2.1.26) was immobilized by soaking the citrate crosslinked Ch/HA/gelatin pellicles in tubes with invertase solution (5 mL solution 50 µg/mL of invertase) for 24 hours. The residual enzyme activity was tested after different periods of times (1 h, 2 h, 3 h, 4 h, and 24 h). The catalytic active hydrogel samples were washed with distilled water, dried, weighted and then tested for immobilized invertase activity.

Invertase Assay

Invertase catalyzes the hydrolysis of sucrose to glucose and fructose, which can be determined by UV-Vis spectrometry using the 3,5-dinitrosalicylic acid (DNS) method [56]. To test the activity of the immobilized enzyme we used reaction mixtures containing 0.5 mL 0.25 M sucrose in 0.02 M acetate buffer pH 4.5 and small sections (ca 0.5x0.5 cm) of immobilized invertase pellicles immersed in 0.5 mL acetate buffer. In control reactions, we replaced the immobilized enzyme with 0.5 mL enzyme solution in water. Free and immobilized enzyme reaction mixes were treated with 1 mL DNS reagent after 15 minutes of incubation at 37^0C. The mixtures were then boiled 5 minutes, cooled and spectrometrically measured at 540 nm with a Jasco UV-Vis spectrometer. The enzymatic activity was expressed as nmols inverted sugar/minute/at 37^0C.

The activity of the immobilized invertase was determined after several consecutive cycles of enzymatic reactions to determine the operational activity (repeated use capability). In addition, we tested the pH (between 3.4 and 6.2) and temperature influence (between 25^0C-60^0C) on the activity of both the free and the immobilized enzyme.

Surface Observation

The Confocal Scanning Laser Microscopy (CSLM) could be used to analyze biological or biotechnological samples without preliminary processing and provides reliable results. CSLM can provide a better understanding of pellicle structure, and it has been used extensively in cell biology studies [57]. CSLM is completely non-destructive, enabling analysis of live biological samples even if the sample is thick. By moving the sample in the axial-direction, stacks of optical sections can be generated and it can be used to localize labeled structures in three dimensions [58].

The image acquisition was performed using a Confocal Spectral Laser Scanning Microscope (LEICA TCS SP), equipped with an Arion laser tuned on 488 nm wavelength. By using objectives of 10X, 0.3 NA, and 20X, 0.4 NA the resolution was ranging between 300-500 nm, approximately. Epi-reflection was used for the hydrogels examination.

Results and Discussions

Pellicle-Type Hydrogels Generation

The polyblend hydrogels based on chitosan, hyaluronic acid, gelatin, were prepared using a casting method. We tested different ratio of biopolymers and different cross-linking agents to obtain hydrogels with good consistency, high mechanical strength, good swelling capacity and biodegradability [23]. The data from Table 1 show different types of prepared hydrogels, biopolymeric ratios, cross-linking agent types and some observations about characteristics of obtained pellicle-type hydrogels. The pellicles made by mixing chitosan/hyaluronic acid and chitosan/hyaluronic acid/gelatin are homogeneous and easy to detach from dish surface. About cross-linking agents, the best results were obtained by pellicles immersion in sodium citrate solution or by alternative exposure to UV and thermal radiations.

Table 1. The influence of the composition of different Ch/HA/gelatin hydrogels types to the invertase immobilization

Support (pellicle)	Hydrogel characteristics	U invertase immobilized/ mg support
Ch; 2.5 % sodium citrate	Transparent, elastic, no adhesive	20.87
Ch/HA (9:1); 5% sodium citrate	Homogenous, elastic, no adhesive	33.06
Ch/HA (9:1); 2.5 % sodium citrate	Homogenous, elastic, no adhesive	45.40
Ch/HA (3:1); 5% sodium citrate	Homogenous, elastic, no adhesive	89.98
Ch/HA (2:1); UV radiations + exposure at 370C	Homogenous, elastic, no adhesive	91.03
Ch/HA (3:1); 2.5% sodium citrate	Homogenous, elastic, no adhesive	111.20
Ch/HA/gelatin (13:2:5); 5% sodium citrate	Slight inelastic and adhesive	53.00
Ch/HA/gelatin (4:1:5); 2.5% sodium citrate	Slight inelastic and adhesive	64.89

Swelling and Degradation Tests

The swelling capacity of obtained hydrogels was tested only for samples with good consistency, elasticity and mechanical strength and is shown in the figures 1. Hydrogels based on chitosan and HA and cross-linked with sodium citrate solution had similar swelling behavior, swelling capacity increasing with hyaluronic acid content (Figure 1).

The *in vitro* degradation behavior at pH 7.2 of different hydrogels, such as simple chitosan, chitosan and gelatin and chitosan and hyaluronic acid (different ratios) was investigated (Figure 2). There were significant differences between degradation ratios and time of degradation for examined hydrogels. The increasing of hyaluronic acid amount determined an increasing of degradation time and degradation ratio. Gelatin adding increased the degradation ratio but hydrogels become brittle in short time.

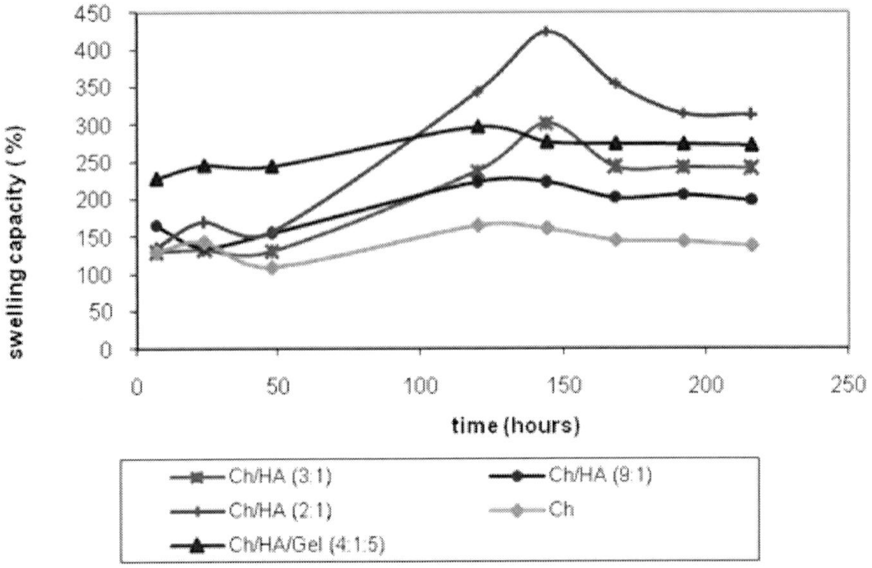

Figure 1. Swelling capacity of hydrogels based on chitosan and hyaluronic acid (cross-linking agent sodium citrate), in PBS pH 7.2, at 20^0C.

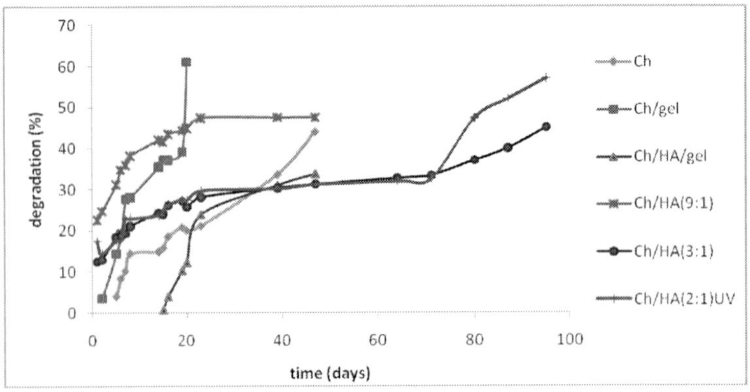

Figure 2. The degradation percentage of some hydrogels obtained, in PBS pH 7.2 at 37^0C.

The biodegradbility of hydrogels at pH 7.2 was better to those that contain gelatin and with increased ratio of hyaluronic acid. The equilibration time of swelling in PBS of hydrogels was also different, 5 hours generally, and more days for hydrogels with higher amount of chitosan. The sample made only from 2% chitosan cross-linked with sodium citrate did not suffer any degradation at pH 7.2 during three months. Sodium citrate was more efficient like cross-linking agent comparison with calcium chloride or KIO_4. The sample obtained by UV radiations exposure alternating with thermal radiations exposure of chitosan and HA (2:1) had good consistency, stability, swelling capacity and biodegradability.

Immobilization of Invertase

Invertase was immobilized by simply soaking the pellicle-type hydrogels in the invertase solution. In parallel tubes, the soaking solution was removed after different periods of time, and the invertase activity was determined. After 2 hour of soaking, the enzymatic activity began to decrease, and several samples presented a total loss of activity after 24 hour of soaking. After this period of time the pellicle samples were washed, dried and tested for enzymatic activity. The specific activity of immobilized enzyme was different function of hydrogel composition (Table I). A specific activity between 89.98 and 111.2 U/mg support for immobilized invertase was obtained with samples with higher concentration of hyaluronic acid. The specific activity for hydrogels with higher content of chitosan was between 33.06 and 45.4 U/mg support and only 20.87 U/mg for simple chitosan hydrogel.

Both the free and the immobilized enzyme displayed approximately the same optimal pH and temperature (4.5 and 45^0C respectively) [24]. In addition, the immobilized invertase was more active in comparison with the free enzyme at pH 4.5 and at any tested temperature. The optimum concentration of the crosslinking agent, sodium citrate, was found to be 2.5% (w/v) and the optimal crosslinking time was 2 hours.

After five consecutive reaction cycles the activity of the immobilized enzyme decreased only with 15%. The invertase activity of sample with highly activity decreased with 10% after one year of storing the pellicle-type hydrogel at 4^0C.

Surface Aquisition

The experiments by CSLM were developed in order to compare the surface structure of simple citrate/chitosan hydrogel, citrate/Ch/HA hydrogel and citrate/Ch/HA with immobile-ized invertase. The micrographies presented in Figures 3-4 illustrate some structural aspects of surface of the investigated pellicle-type hydrogels.

The results obtained by confocal microscopy confirmed the macroscopic observations regarding the quality of chitosan/HA hydrogels. The addition of hyaluronic acid modifies the structural features of chitosan, namely its plasticity. The micrographies obtained with confocal microscopy show the smooth surface of chitosan hydrogels (Figure 3a) different from the rough surface of Ch/HA hydrogel (Figure 3b) with interpenetrating macromolecules. On the surface of Ch/HA hydrogels were observed the crystals of the sodium citrate used as crosslinking agent (Figure 4).

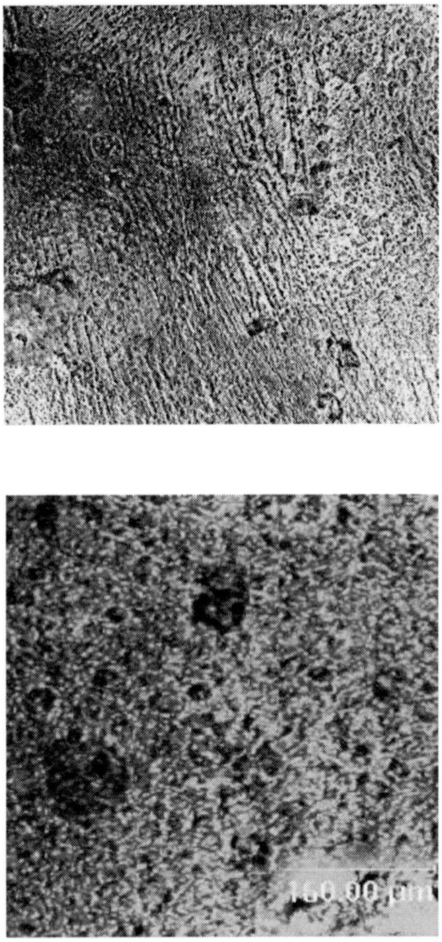

a.

b.

Figure 3. Microscopy analysis by reflection of simple citrate chitosan (a) and citrate chitosan/HA (3:1) (b) hydrogels.

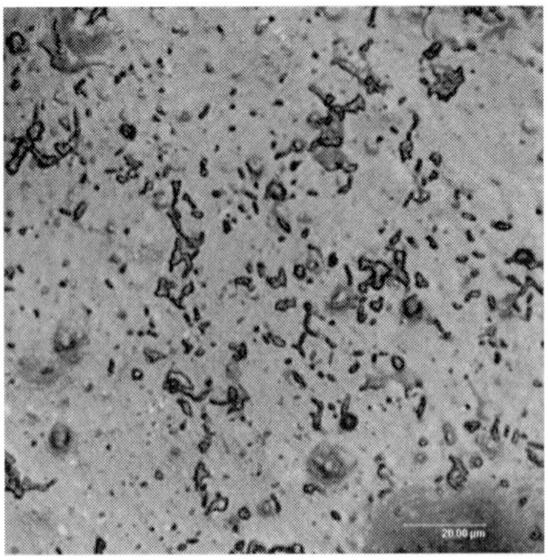

Figure 4. Microscopy analysis by reflection of chitosan/HA (3:1) hydrogel treated with sodium citrate and with immobilized invertase.

The examination of Ch/HA hydrogels with immobilized invertase by confocal microscopy revealed the presence of enzyme molecules attached to the film surface (Figure 4). The enzyme presence was observed by CSLM even one year after the immobilization process was performed. Furthermore, the invertase activity determined after one year was 90% from the initial activity. This is a strong indicator of the hydrogel quality in preserving the stability of the enzyme.

Our macroscopic and microscopic observations showed that HA has the capacity to improve the quality of chitosan hydrogels (consistency, elasticity). In addition, our data indicate that invertase could strongly binds to the surface of citrate/Ch/HA pellicle-type hydrogels. The samples with higher HA content (3:1, 2:1) displayed a highly capacity of binding invertase (Table 1) in comparison with simple chitosan samples or with Ch/HA/gelatin hydrogels. Similar data were obtained when laccase or protease inhibitors (data not shown) were tested. The addition of gelatin to chitosan/HA hydrogels does not improve their capacity to immobilize invertase or the characteristics of hydrogels. The specific activity of 111.2 U/mg pellicle for immobilized invertase on Ch/HA (3:1) pellicle was much higher at that time (2005) than previously reported in the literature for invertase adsorbed on inorganic materials [59], and is still a very good value in comparison with more recent studies [60].

2. Hydrogel-Magnetic Nanoparticles Based on Chitosan and Hyaluronic Acid

Methods

Synthesis of Hydrogel-Magnetic Nanoparticles

Water-dispersible magnetic nanoparticles (MP) were obtained according to previous studies, using an adjusted Massart method [43]. Briefly, the magnetic nanoparticles were precipitate from aqueous mixture of Fe^{2+} and Fe^{3+} salts (1:2 molar ratio), and treated with NH_4OH at 75 ^{0}C. Subsequent to precipitation the magnetic nanoparticles were encapsulated in bio-polymer shells.

Two different biopolymeric materials were tested namely chitosan (Ch) from crab shells (Sigma) and hyaluronic acid (HA) (extract from bovine vitreous, own extraction method), [53]. Two covering procedure were performed, namely (i) layer-by-layer coating - using 2% chitosan solution and 1% hyaluronic acid solution - and (ii) hybrid polymer coating. The hybrid polymer consisted from Chit-HA hydrogels obtained by physical mixing of 2% chitosan solution and 1% HA solution. Different ratios of chitosan solution and hyaluronic acid solution were employed in covering magnetic nanoparticles. Finally, the obtained hydrogel-magnetic nanoparticles were treated with 2.5% (w/v) sodium citrate solution for 2 hours, and then washed with bidistilled water.

Characterization of Hydrogel-Magnetic Nanoparticles

Size distribution and characterization of bare and encapsulated magnetic nanoparticles was acquired using dynamic light scattering (DLS) technique, at room temperature using

Malvern instrument (Nicomp 270, laser source, λ 632.8 nm) operating in the range 1 nm-1 μm.

The morphology of the particles was investigated with Transmission Electron Microscopy (TEM) and Confocal Microscopy, using a Philips EM 208, and a Confocal Spectral Laser Scanning Microscope (LEICA TCS SP).

The structure of hydrogel coated magnetic nanoparticles was characterized by Fourier Transform Infrared Spectroscopy (FT-IR) technique using a Bruker Tensor 27 device.

Biocompatibility Studies

The swelling tests and degradation tests were performed, for both bare and covered nanoparticles after the same procedure used to pellicle-type hydrogels (see above).

The obtained hydrogel-magnetic nanoparticles (~ 10 mg dry weight) were suspended in tubes containing 10 mL of PBS (pH 7.2 at 20^0C). At a defined time period (3h, 6h, 12h, 24h etc) the excess buffer was carefully removed, using a magnetic separator and the hydrogel-magnetic nanoparticles were weighted immediately. The swelling capacity was calculated according to the equation 1 (see above).

For degradation studies, hydrogel-magnetic nanoparticles were suspended into Falcon tubes containing 10 mL PBS (pH 7.2 at 37^0C). At predetermining time points, hydrogels-magnetic nanoparticles were collected with a magnetic separator and excess buffer was removed from the tubes. Samples were weighted by using an analytical balance with ± 0.1 mg accuracy. After the equilibration time of swelling in PBS, the degradation ratio was calculated according to the equation 2 (see above).

Cytotoxicity of nanostructures was determined by MTT Cell Proliferation Assay [61], a quantitative, convenient method to evaluate a cell population's response to external factors. The key component is (3-[4,5-dimethylthiazol-2-yl]-2,5-diphenyl tetrazolium bromide) or MTT. Mitochondrial dehydrogenases of viable cells cleave the tetrazolium ring, yielding purple formazan crystals which are insoluble in aqueous solutions. The resulting purple solution is spectrophotometrical measured. An increase or decrease in cell number results in a concomitant change in the amount of formazan produced, indicating the degree of cytotoxicity caused by the test material.

MTT test was done on Vero cells (kidney epithelial cells from African green monkey). These were seeded into 24-well plates at a density of $5x10^4$ cells /well and were cultured 24 hours in Dulbecco's modified Eagles medium/10% FBS (DMEM). After 24 hours, the medium was replaced with different samples of obtained hydrogel-magnetic nanoparticles (conc. 2-12 ng/cell).

After 2 days exposure of cells to nanoparticles the cells were washed with phosphate buffer and 500 μL MTT solution (0.5 mg/mL) was added in each well. The cells were incubated for 3 hours at 37^0C and the formazan crystals formed in living cells were solubilized in isopropanol. The absorbance was measured at 570 nm with a Jasco UV-Vis spectrometer. The viability of the treated cultures was expressed as a percentage of the control, untreated cells.

Immobilization of L-asparaginase

L-Asparaginase was immobilized by entrapment in hydrogel layer of obtained nanostructures. L-asparaginase was obtained according previous studies [47], by biosynthesis from *Escherichia coli* using a recombinant strain of *E.coli* with improved capacity of producing isoenzyme EC 2 with anti-tumor activity (kindly supplied from BIOTEHGEN collection). After biosynthesis, the enzyme was purified by adsorption chromatography (bentonite), ammonium sulphate fractionation and ionic exchange chromatography (DEAE Sephadex A50). The resulted purified enzyme (150 U/mg) was precipitated with ethanol and dried in vacuum [47].

Asparaginase Assay

The residual activity of the immobilized L-asparaginase was determined spectrometrically at 450 nm with Nessler reagent method (essentially that of Mashburn and Wriston, 1963) where the rate of hydrolysis of asparagine is determined by measuring released ammonia [50]. One unit releases one micromole of ammonia per minute at 37°C and pH 5 under the specified conditions.

Results and Discussions

Synthesis of Hydrogel-Magnetic Nanoparticles

Magnetic nanoparticles (MP) were obtained by co-precipitation of iron oxides. MP from solutions of iron II and III, were covered with successive layers of different chitosan and hyaluronic acid ratios, both in a layer-by layer (l-b-l) technique and in a hybrid polymer covering technique, resulting different variants of hydrogel-magnetic nanoparticles.

Table 2. Composition and characteristics of some synthesized hydrogel-magnetic nanoparticles

Sample no.	Sample composition	Diameter media (nm)	Zeta Potential (mV)	Obs.
6	Dispersion of magnetic nanoparticles (MP)	66.9-169.3	(-34) -(-50.6)	Magnetic, black
13	20 mL MP+ 10 mL 2% Ch + 10 mL 1% HA +10 mL 2% Ch	147 - 816 187 - 444 181 - 263	3.74-43.7 (-43.7)-7.13 12.7-47.6	Magnetic, stable
20	20 mL MP + 1.5 mL 2% Ch + 5 mL 1% HA + 1.5 mL 2% Ch	222 - 816 241 - 444 240 - 264	(-0.23)-41.5 (-3.26)-(-14.7) 12.7-46.1	Magnetic, stable
22	20 mL MP + 3 mL [mixed 1% HA and 2% Ch; 1:3, (v:v)]	2289-6237	(-12.5)-(-15)	
30	20 mL MP + 3 mL 2% Ch + 3 mL 1% HA + 3 mL 2% Ch	1225-2662 916-1530 254-264	(-14.6)-3.74 (- 9.29)-(-19) 28.9-31.6	Magnetic, stable
38	20 mL MP + 1 mL 2% Ch	741 - 816	(-24.8)-41.5	

Thirty-eight variants were tested, some of them, the most important ones, being presented in table 2. Sample 22 exemplified in the table was obtained by covering magnetic nanoparticles with a pre-formed mixture of Ch and HA. Samples 13, 20 and 30 were obtained using three successive layers of Ch/HA/Ch (l-b-l technique) and sample 38 was obtained from MP covered with single layer of chitosan. In order to decide if the obtained nanoparticles are suitable for biological applications, characterization in terms of size, size distribution and morphology were performed.

Characterization of Hydrogel-Magnetic Nanoparticles

The hydrogel-magnetic nanoparticles were characterized by DLS and zetametry to determine the size, size distribution and zeta potential (Table 2). As could be noticed from the values of the particles sizes (in swelled stage), the layer-by-layer covering technique with 3 successive layers of Ch/HA/Ch seemed to provide the most suitable nanoparticles dimensions (180-264 nm), ensuring a degree of covering of the MP and a compact structure. The zeta potential values alternating from negative to positive values proved that the MP covering is efficient and, moreover, the final nanostructures obtained are stable, taking into account that the values are higher than 30mV.

Nanoparticles covered with one, or two layers of polymer, and those covered with mixed polymers finally presented too large dimensions detected by DLS measurements (Table 2).

The synthesized covered magnetic nanoparticles were subjected to confocal laser microscopy analysis to provide some images of the covered MP, the results proving that a high degree of MP spherical conformation is obtained by l-b-l method with 3 layers of Ch/HA/Ch (Figure 5). The samples obtained by hybrid polymer covering present clusters and a low degree of dispersion and have micrometric dimensions in swelled stage (Figure 6 and Table 1).

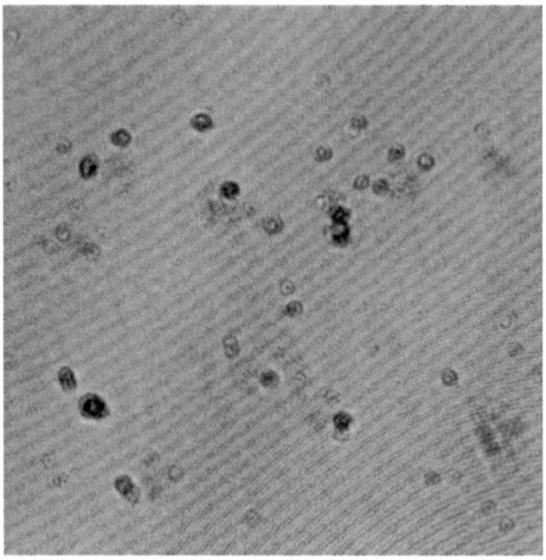

Teodor et al., Nanoscale Res Lett, 2009, 4:546; with kind permission of Springer Science and Business Media.

Figure 5. Confocal microscope micrograph of hydrogel-magnetic nanoparticles obtained by layer-by-layer technique (sample 13), (250x250 μm).

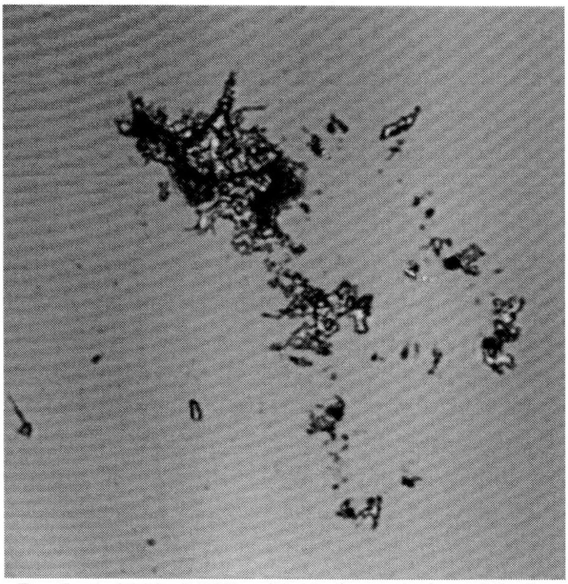

Teodor et al., Nanoscale Res Lett, 2009, 4:547; with kind permission of Springer Science and Business Media.

Figure 6. Confocal microscope micrograph of hydrogel-magnetic nanoparticles obtained by hybrid polymer covering technique (sample 22), (250x250 μm).

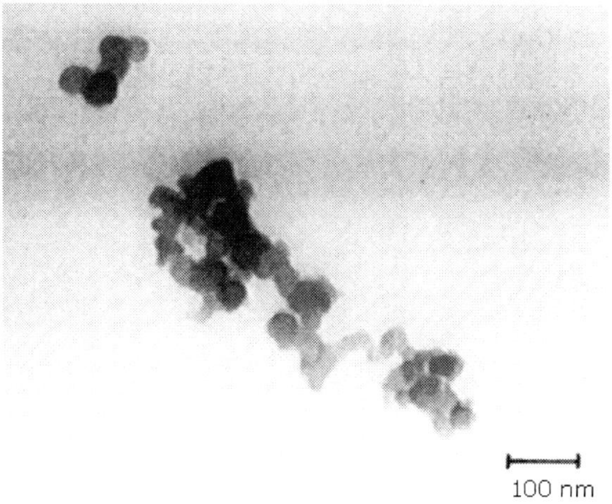

Teodor et al., Nanoscale Res Lett, 2009, 4:547; with kind permission of Springer Science and Business Media.

Figure 7. Transmission electron microscope micrograph of hydrogel-magnetic nanoparticles (sample 13).

Further studies were done only with samples suitable for applications in bio-medical area (especially for delivery systems).

The morphology of nanostructures synthesized by l-b-l technique was studied using TEM, the obtained images (Figure 7), demonstrating a homogenous distribution and a spherical shape of obtained nanostructures, conclusion that agrees to that rose from confocal analysis (Figure 5). In addition, the micrographs obtained from TEM (Figure 7) show that the dimensions of nanoparticles, bare or encapsulated, *are between 20 nm and 50 nm in dried stage.* It is known that nanosized delivery systems should have dimensions ranged between 1 and 100 nm [62].

The structure characterization of the obtained covered MP was performed by FTIR, from the obtained spectra being obvious that the suitable covering of magnetic nanoparticle surface was performed, the structural pattern of chitosan and hyaluronic acid being observed on the NP surface. As could be noticed from the Figure 8, the presence of hydroxyl -OH and $-NH_2$- groups from chitosan and HA on the covered nanoparticles is obvious, the specific bands at the wave numbers 3200 cm^{-1}, respectively 3680 cm^{-1}, with their confirmation in the region 1790-1520 cm^{-1}, being easily noticed in the nanostructures spectra. The slightly shift on the wave numbers values registered between chitosan, respectively HA itself and covered MP it is ascribable to the link of shells to the magnetic nanoparticles.

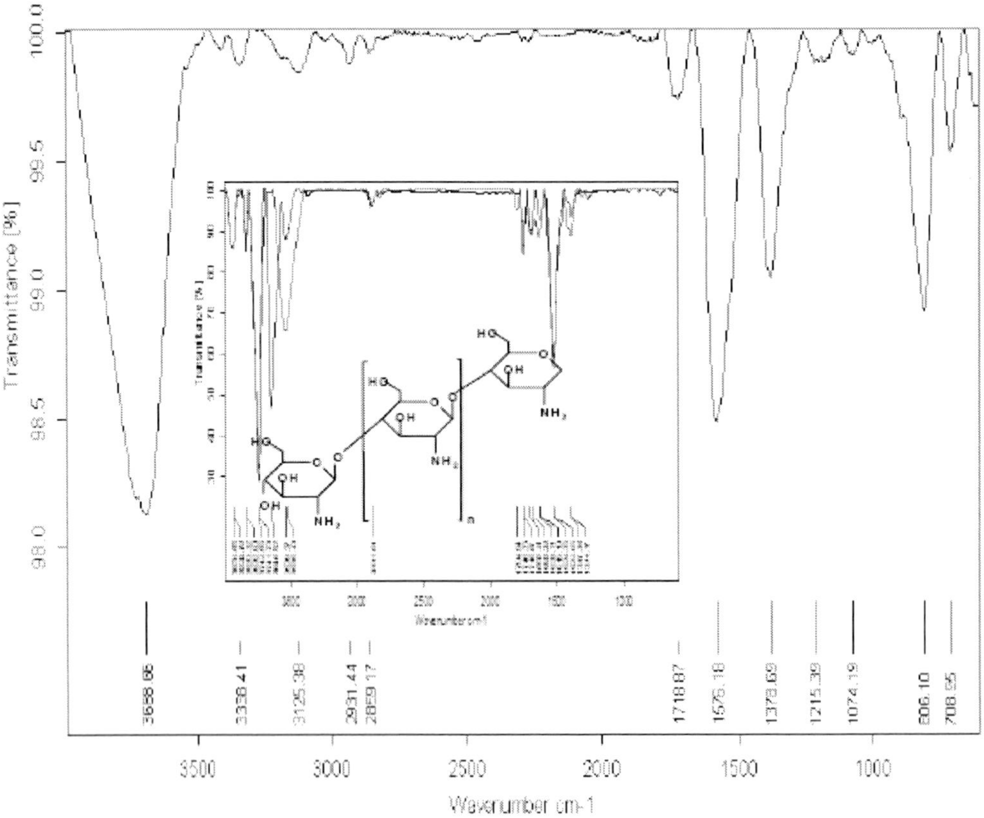

Teodor et al., Nanoscale Res Lett, 2009, 4:548; with kind permission of Springer Science and Business Media.

Figure 8. FTIR spectra of layer-by-layer Ch-HA covered magnetic nanoparticles; inset overlaid FTIR spectra of chitosan and hyaluronic acid.)

Biocompatibility Tests

With the argument of appropriate size, good covering and suitable end groups able to bind an active principle, several tests of biocompatibility were performed on the obtained nanostructures, the first step being that of swelling behavior.

Experiments shown that the obtained nanostructures had similar swelling behavior each other, the swelling capacity increasing with the hyaluronic acid/chitosan ratio enhancing. The obtained swelling capacity for five types of nanoparticles is presented in figure 9, the higher value of the swelling capacity being obtained after 48 hours.

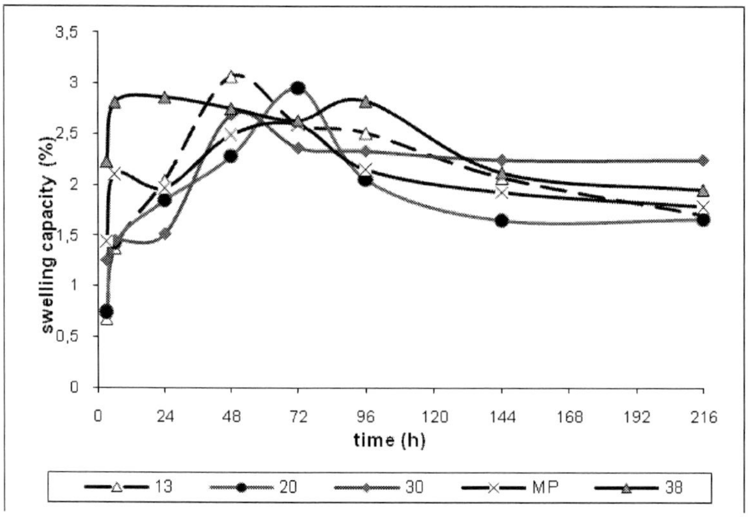

Teodor et al., Nanoscale Res Lett, 2009, 4:548; with kind permission of Springer Science and Business Media.

Figure 9. Swelling behavior of covered magnetic nanoparticles.

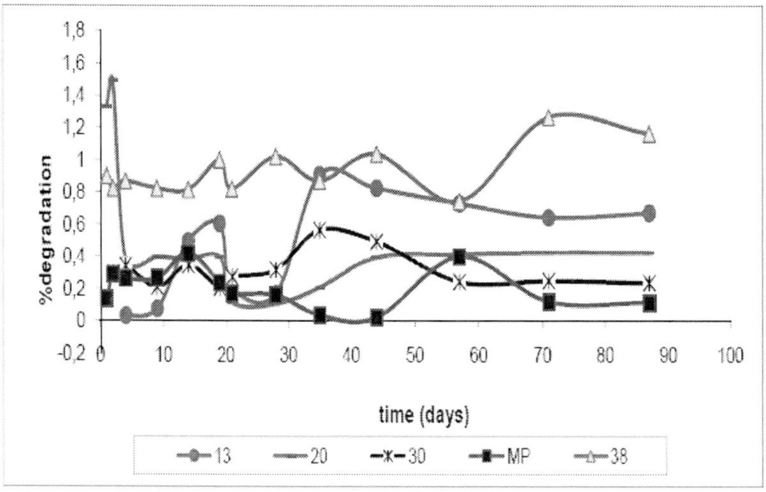

Teodor et al., Nanoscale Res Lett, 2009, 4:548; with kind permission of Springer Science and Business Media.

Figure 10. Degradation of Ch-HA hydrogels – magnetic nanoparticles.

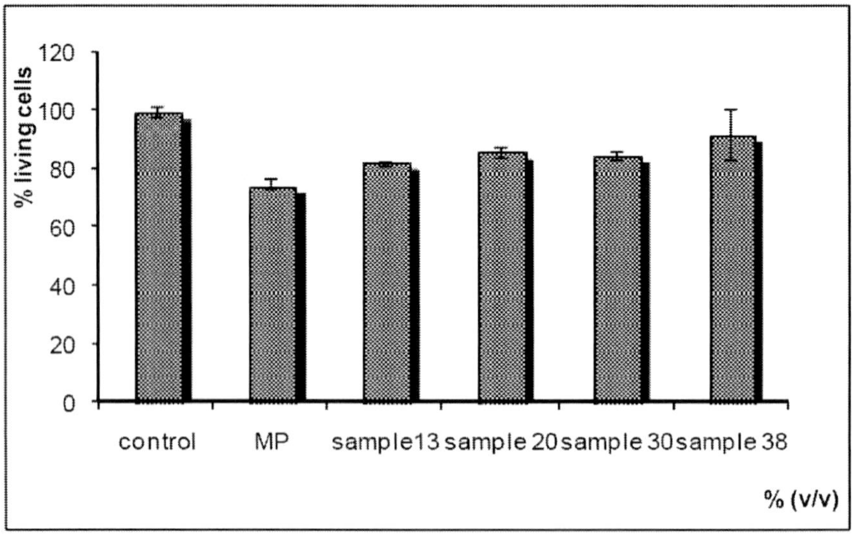

Teodor et al., Nanoscale Res Lett, 2009, 4:548; with kind permission of Springer Science and Business Media.

Figure 11. Biocompatibility /cytotoxicity of obtained nanoparticles cultured 48 hours with Vero cells (nanoparticles conc. was 12 ng/cell). The absorbance at 570 obtained for control was considered 100%. The results for the treated cells were expressed as percentage from the control, untreated culture (mean value ±SD).

The *in vitro* degradation studies for the encapsulated MP were performed in phosphate buffer solution for 80 days. The synthesized hydrogel-magnetic nanoparticles presented a great stability in neutral media, with respect to our previous studies about pellicle-type hydrogels degradation (see above Figure 2). If for the pellicle-type hydrogels the degradation percent was between 30 and 50 in 40 days (Figure 2), the nanostructured MP stability increased, the maximum degradation percent attained in 80 days being < 2% (Figure 10).

The second experiment of biocompatibility concerned the cytotoxicity tests of nanostructures that were determined using MTT cell proliferation assay. The tests for *in vitro* biocompatibility were performed in triplicates on Vero cells cultivated with different concentrations of magnetic nanostructures (between 2 and 12 ng/cell) As could be observed from Figure 11, the obtained nanostructures are highly biocompatible, no cytotoxicity was detected in cells cultured, after 48 hours, with highest concentrations of hydrogel-magnetic nanoparticles (12 ng/cell), the cell phenotype being normal, and the cell viability upon 75% (Figure 11). At higher concentrations (20 ng/cell), the viability decreases, but it maintains over 65% (data not shown).

Immobilization of L-Asparaginase

Hydrogel-magnetic nanoparticles were used for immobilization of biological active substances (e.g. L-asparaginase). According to our studies, the three layers of successive chitosan-hyaluronic acid-chitosan ensure the final nanometric dimensions and stability of obtained nanostructures (see table 2). L-asparaginase was immobilized by entrapment in hyaluronic acid (middle) layer or in external chitosan layer (Table 3). The hydrogel-magnetic nanoparticles with entrapped L-asparaginase were characterized by DLS and zetametry to determine the size, size distribution and zeta potential (Table 3).

Table 3. Dimensions (swelled stage) and zeta potentials of hydrogel magnetic nanoparticles obtained by alternation of chitosan and hyaluronic acid layer on magnetic nanoparticles (layer-by-layer technique); L-asparaginase is entrapped in middle or external layer

Sample No.	Sample Type (composition)	D media (nm)	Zeta Potential (mV)
0	Magnetic nanoparticles (MP) suspension	25.21	- 45.4
1	20 mL MP		
	+1.5 mL Ch	349.5	- 47.9
	+ 5 mL HA (and 5 mg asparaginase BIOTEHGEN)	320	- 49.7
	+ 1.5 mL Ch	274	- 31.7
2	20 mL MP		
	+ 3 mL Ch	418.7	- 32.4
	+ 3 mL HA (and 5 mg asparaginase BIOTEHGEN)	539.2	- 38
	+ 3 mL Ch	261.8	+ 2.44
4	20 mL MP		
	+ 3 mL Ch	768.2	- 23
	+ 3 mL HA	835.5	- 4.09
	+ 3 mL Ch (and 5 mg asparaginase BIOTEHGEN)	324.9	+ 8.15
5	20 mL MP		
	+1.5 mL Ch	304.4	- 32.5
	+ 3 mL HA (and 5 mg asparaginase BIOTEHGEN)	676.9	- 18.2
	+ 1.5 mL Ch	289.1	+ 3.49

Teodor et al., J. Mater Sci: Mater Med, 2009, 20: 1308; with kind permission of Springer Science and Business Media.

The studies of microscopy confirmed that obtained nanostructures are homogenous in shape and dimensions, the diameters of nanoparticles been around 20-30 nm in dried stage (Figures 12,13).

FT-IR spectrometry analyses proved the encapsulation of magnetic nanoparticles with chitosan and hyaluronic acid. As could be observed from Figure 14 (14b, overlaid spectra) the bare nanoparticles themselves are characterized mainly of four absorption bands: first band, at 3125 cm-1, is of medium intensity, being ascribable to free HO- groups; the second band, at about 1623 cm^{-1} is a weak, wide band, being the result of vibration frequency of NH_3^+ remained from the synthesis processes, this ascription being confirmed by the third band, that occurs at 1401 cm^{-1}, that is specific for ammonium salts. The magnetite formation is supported by the presence of a strong, wide absorption band at 580 cm^{-1}, confirmed by the weak band occurring at 400-450 cm^{-1}. When the layer-by- layer covering is performed using chitosan, hyaluronic acid and entrapped L-asparaginase, in all the obtained samples were noticed the specific bands of bio-polymers, at the wave numbers 3200 cm^{-1}, respectively 3680 cm^{-1} being confirmed on 1790-1520 cm^{-1} region. Chitosan, respectively HA, as any macromolecular carbohydrate polymer, rises at 3508 cm^{-1} a wide band, due to the stretching vibrations of H-O bonds, which are present even on the spectra of covered nanoparticles: These prove the presence of the chitosan and HA structural pattern in the obtained hydrogel-magnetic nanoparticles, supporting the efficiency of the covering. In addition, the changes occurred in the shape, position and intensity of 1700 cm^{-1} – 500 cm^{-1} and 3520 – 3100 cm^{-1} bands (Figure 14 a,b) allowed us to presume that H-bonds or even covalent bonds are formed between magnetic nanoparticles and hydrogel layers. It was noticed that the amide band is present, both in chitosan and in covered nanoparticles spectra, at 1650 cm^{-1} in the chitosan case, and slightly shifted toward smaller wavenumbers in the case of biocompatible nanoparticles, thus confirming once more the covalent link between the biocompatible polymer and the magnetic nanoparticle.

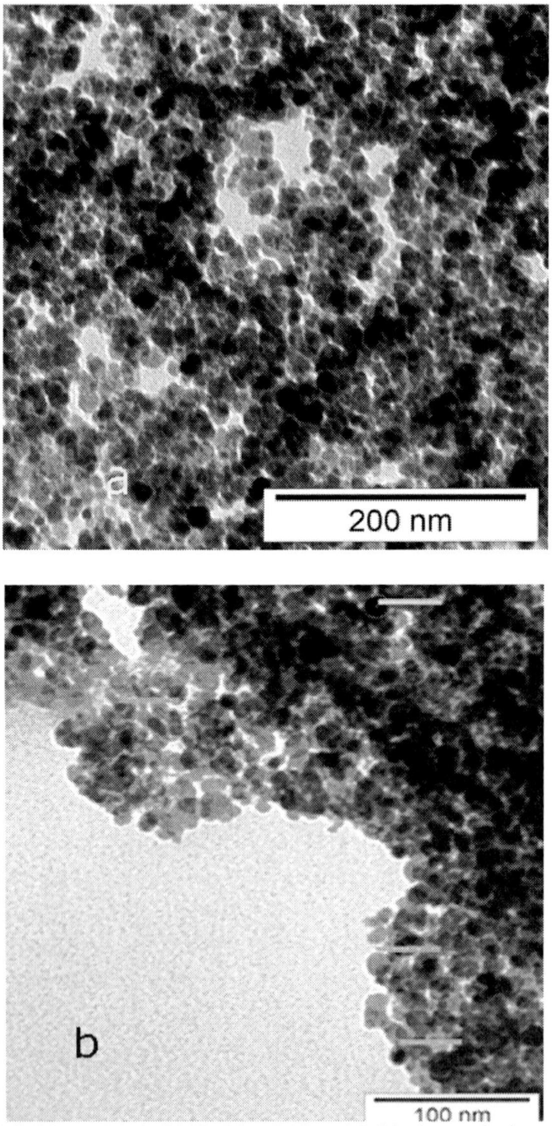

Teodor et al., J. Mater Sci: Mater Med, 2009, 20: 1310; with kind permission of Springer Science and Business Media.

Figure 12. (a) TEM images of uncovered (bare) MP (d=15.97 nm); (b) MP covered with hydrogel with L-asparaginase (sample 5, d=17.51 nm).

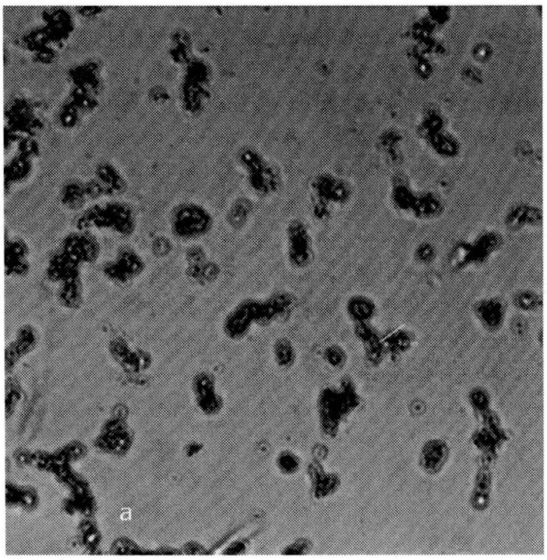

a.

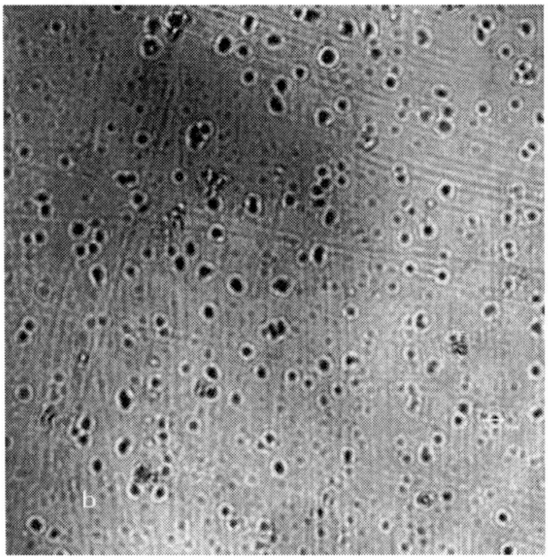

b.

Teodor et al., J. Mater Sci: Mater Med, 2009, 20: 1310; with kind permission of Springer Science and Business Media.

Figure 13 (a) Confocal microscope images (250x250 μm) of uncovered (bare) MP (dx=90 nm); (b) MP covered with hydrogel with L-asparaginase (sample 1) (dx=169 nm).

The magnetic susceptibility measurements of the synthesised samples (Figure 15) showed a decrease of magnetic susceptibility with the increase of magnetic field frequency, specific for superparamagnetic particles (less than 20 nm).

The residual activity of L-asparaginase entrapped in hydrogel-magnetic nanoparticles is presented in Table 4. The analysis was made in duplicates in the moment of synthesis, and at 3 and 6 months after synthesis. The residual activity remains about the same in samples 4 and 5, or is slightly diminished in sample 1 and 2.

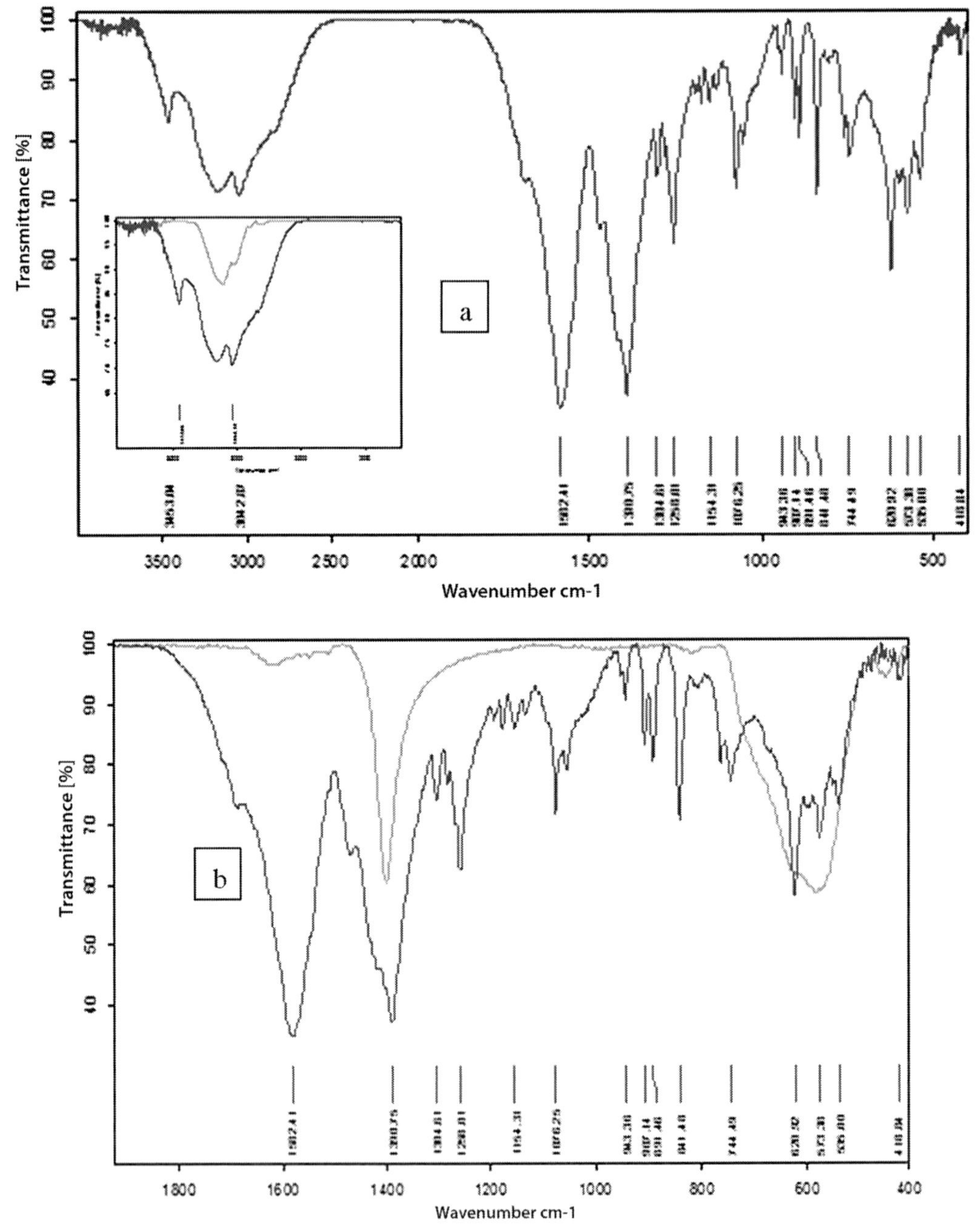

Teodor et al., J. Mater Sci: Mater Med, 2009, 20: 1311; with kind permission of Springer Science and Business Media.

Figure 14. a) FTIR spectra of sample 2; inset –overlay region 2500 –3700 cm^{-1} for sample 2 and bare magnetic nanoparticles; (b) overlay region 1900 –400 cm^{-1} for sample 2 and bare magnetic nanoparticles.

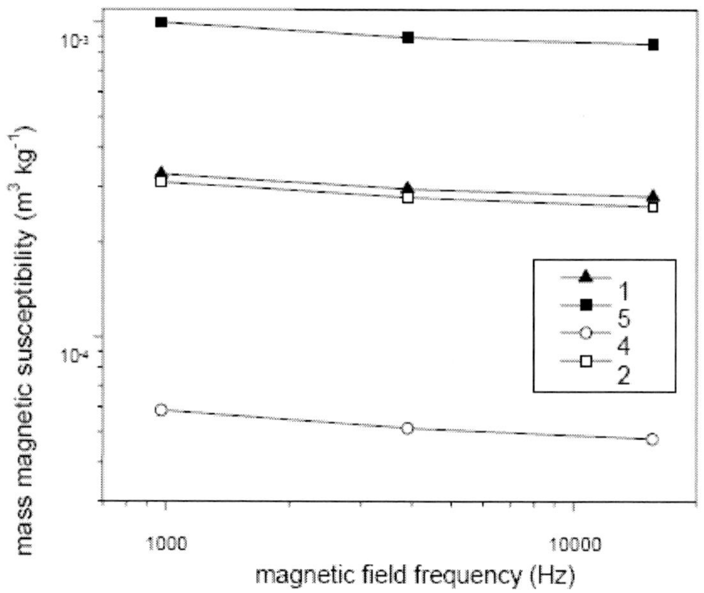

Teodor et al., J. Mater Sci: Mater Med, 2009, 20: 1312; with kind permission of Springer Science and Business Media.

Figure 15. Magnetic susceptibility of synthesized hydrogel magnetic nanoparticles (samples 1, 2, 4 and 5) function of the magnetic field frequency.

Table 4. L-asparaginase residual activity of obtained nanostructures

Sample	Abs., 450 nm	μmoli NH_3 /sample	U/mL	Total U immobilized	Immobilizatio n yield (%)
1	0.0965-0,0762	0.45-0.34	1.5-1.13	225	43.02
2	0.1121-0,0780	0.53-0.35	1.77-1.16	265.5	50.73
4	0.2198-0.188	1.13-0.95	3.76-3.17	475	90.76
5	0.1307-0.1147	0.64-0.55	2.13-1.83	319.5	61.05

Teodor et al., J. Mater Sci: Mater Med, 2009, 20: 1312; with kind permission of Springer Science and Business Media.

The tests for *in vitro* biocompatibility were performed in triplicates on normal Vero cells cultivated with different concentrations of magnetic nanostructures (between 2 and 12 ng/cell). All the tested samples present no toxicity even at higher concentrations of nanoparticles and the phenotype of the Vero cells remains normal (data not shown, the results are similar with those presented above for MP without enzyme).

The obtained nanostructures did not exhibit any antimicrobial or antifungal activity, but in exchange they influence the expression of bacterial adhesions, as demonstrated by the results of the adherence showing differences in the behavior of microorganisms in the presence of magnetic nanostructures [51]. These results are interesting and required further studies.

Magnetic nanoparticles continue to receive attention from scientists. Recent studies describe an approach to synthesize superparamagnetic iron oxide nanoparticles in the presence of polymerized lactic acid or PVA, which can be used for developing target specific MRI contrast agents, ferrogels or drug delivery systems [63,31] or PVP hydrogel magnetic nanospheres which exhibited passive drug release that could be exploited to enhance therapeutic efficacy [32], or in the formation of a novel hydrogel nanocomposite with superparamagnetic property [40]. Novel magnetic hybrid hydrogels were fabricated by the in situ embedding of magnetic iron oxide nanoparticles into the porous hydrogel networks, this magnetic hydrogel material was found to hold a potential application in magnetically assisted bioseparation [41].

Nanomedicines are defined as delivery systems in the nanometer size range (preferably 1 to 100 nm) containing encapsulated, dispersed, adsorbed, or conjugated drugs and imaging agents [62]

We obtained biocompatible hydrogel-magnetic nanoparticles with L-asparaginase entrapped, with sizes below 300 nm in swelled stage, and below 30 nm in dried stage, capable to penetrate the cells and tissues (especially tumor tissues). The majority of solid tumors exhibit a vascular pore cutoff size between 380 and 780 nm [64], although tumor vasculature organization may differ depending on the tumor type, its growth rate and microenvironment [64,65].

The residual activity of the immobilized L-asparaginase remains the same after 6 months, stored at 4^0C. The immobilization yield of the L-asparaginase in different samples was between 43-90%. The obtained nanostructures are biocompatible and could be used for delivery of L-asparaginase in tumor tissues. A recent work presents a similar biomaterial obtained by successfully immobilization of L-asparaginase in carboxymethyl konjac glucomannan-chitosan nanocapsules prepared under mild conditions by electrostatic complexation [66]. The magnetic nanostructures described in this chapter are superparamagnetic particles and could be controlled by a magnetic field.

In addition, from adherence studies of microorganisms in the presence of obtained hydrogel-magnetic nanostructures, we can also use them in magnetic bioseparation, because the different capacity of bacteria or yeast to adhere, or not, to inert substrate treated with hydrogel-nanoparticles. The bare magnetic nanoparticles have a similar behaviour with hydrogel-magnetic nanoparticles, so simple magnetic nanoparticles could be used for magnetic separation of some strains from different media.

CONCLUSION

Biomaterials based on chitosan and slight addition of hyaluronic acid (pellicle-type hydrogels and hydrogel-nanoparticles) have characteristics which make them promising for biomedical or biotechnological purposes.

The pellicle-type hydrogels obtained with any ratio of hyaluronic acid added to chitosan had better consistency, elasticity and higher mechanical strength than those based on single chitosan or chitosan with carrageenan or gelatin. Hydrogels obtained by increasing ratios of hyaluronic acid had better elasticity, mechanical strength, swelling capacity and degradability in neutral media. In addition, these hydrogels are suitable for enzyme immobilization on surface or by entrapping in hydrogel network.

The second types of biomaterials, the hydrogel-magnetic nanoparticles with a magnetic core (Fe_3O_4) encapsulated in layer-by layer chitosan-hyaluronic acid hydrogel were synthesized and characterized, proving to be suitable to cellular wall penetration due to their dimensions (between 180-264 nm in swelled stage, and between *20-50 nm in dried stage*), spherical shape, homogenous distribution and swelling capacity.

Performed biocompatibility tests proved that the hydrogel-magnetic nanoparticles resulted from our experiments are biocompatible and relative inert to microorganisms, so they are suitable to be used for loading and delivery of active compounds. The experiments with L-asparaginase entrapped in hydrogel layer were promising, the resulted nanostructures being stable and capable to load any others enzymes or drugs.

REFERENCES

[1] Yao, K. D., Peng, T., Yin, Y. J., Xu, M. X. (1995) Microcapsules/microspheres related to chitosan. *Macromol. Chem. Phys*, C35, 155–180.

[2] Illum, L. (1998) Chitosan and its use as a *pharmaceutical* excipient. *Pharm Res*, 15, 1326–1331.

[3] Nakatsuka, S., Andrady, A. L. (1992) Permeability of vitamin B-12 in chitosan membranes. I: Effect of crosslinking and blending with poly(vinyl alcohol) on permeability. *J. Appl. Polym. Sci,* 4, 17–28.

[4] Thacharodi, D., Rao, K. P. (1993) Propranolol hydrochloride release behaviour of crosslinked chitosan membranes. *J. Chem. Tech. Biotechnol*, 58, 177–181.

[5] Yao, K. D., Liu, J., Cheng, G. X., Lu, X. D., Tu, H. L, Da Silva J. A. L. (1996) Swelling behaviour of pectin/chitosan complex films. *J. Appl. Polym. Sci*, 60, 279–283.

[6] Chu, C-H., Sakiyama, T., Yano, T. (1995) pH-sensitive swelling of a polyelectrolyte complex gel prepared from xhantan and chitosan. *Biosci. Biotech. Biochem*, 59, 717–719.

[7] Al-Musa, S., Fara, D. A., Badwan, A. A. (1999) Evaluation of parameters involved in preparation and release of drug loaded in crosslinked matrices of alginate. *J. Control Release,* 57, 223–232.

[8] Holte, O., Onsoyen, E., Myrvold, R., Karlsen, J. (2003) Sustained release of water-soluble drug from directly compressed alginate tablets. *Eur. J. Pharm. Sci*, 20, 403-407.

[9] Shu, X. Z., Zhu, K. (2000) A novel approach to prepare tripolyphosphate/chitosan complex beads for controlled release drug delivery. *Int. J. Pharm*, 201, 51–58.

[10] Shu, X. Z., Zhu, K. J. (2001) Chitosan/gelatin microspheres prepared by modified emulsification and ionotropic gelation. *J. Microencapsulation*, 18, 237-245.

[11] Shu, Z. S., Zhu, K. J., Song, W. (2001) Novel pH-sensitive citrate cross-linked chitosan film for drug controlled release. *Int. J. Pharmaceutics*, 212, 19-28.

[12] Prestwich, G. D., Marecak, D. M., Marecek, J. F., Vercruysse, K. P., Ziebell, M. R. (1997) Controlled chemical modification of hyaluronic acid: synthesis, applications and biodegradation of hydrazide derivatives. *J. Control Release*, 53, 99-102.

[13] Vercruysse, K. P., Prestwich, G. D. (1998) Hyaluronate derivatives in drug delivery. *Crit. Rev. Therapeut Carrier Syst*, 15, 513-555.
[14] Luo, Y., Kirker, K., And Prestwich, G. (2000) Cross-linked hyaluronic acid hydrogel films: new biomaterials for drug delivery. *J. Controlled Release,* 69, 169-184.
[15] Abatangelo, G., Weigel, P., New Frontiers in Medical Sciences: Redefining Hyaluronan; Elsevier: Amsterdam, (2000)
[16] Benedetti, L., Cortivo, R., Berti, T., Berti, A., Pea, F., Mazzo, M., Moras, M., Abatangelo, G. (1993) Biocompatibility and biodegradation of different hyaluronan derivatives (HYAFF) implanted in rats. *Biomaterials,* 14, 1154-1160.
[17] Masteikova, R., Chalupova, Z., Sklubalova, Z. (2003) Stimuli-sensitive hydrogels in controlled and sustained drug delivery. *Medicina*, 39, 19-24.
[18] Mao, J. S., Iiu, H. F., Yin, Y. J., Yao, K. D. (2003) The properties of chitosan–gelatin membranes and scaffolds modified with hyaluronic acid by different methods. *Biomaterials*, 24, 1621-1629.
[19] Baran, E. T. , Mano, J. F., Reis, R. L. (2004) Starch-chitosan hydrogels prepared by reductive alkylation crosslinking. *J. Mat. Sci, Mat. in Medicine*, 18, 759-765.
[20] Peppas, N. A. Hydrogels in Medicine and Pharmacy: Fundamentals. vol. I, N. A. Peppas (Ed.), CRC Press, Boca Raton, Florida (1986)
[21] Ross-Murphy, S. B. (1994) Rheological characterization of polymer gels and networks. *Polym. Gels Netw*, 2, 229-237
[22] Berger, J., Reist, M., Mayer, J. M., Felt, O., Peppas, N. A., Gurny, R. (2004) *Structure and interactions* in covalently and ionically crosslinked chitosan hydrogels for biomedical applications. *Eur. J. Pharm. Biopharm*, 57, 19-34.
[23] Teodor, E., Rugina, A., Radu, G. L. (2005) Hydrogel-biomaterials based on polysaccharides for biomedical applications. *Rev de Chimie*, 56, 1211-1214.
[24] Teodor, E., Radu, G. L., Dan, C., Stanciu, G. A. (2006) Chitosan-hyaluronic acid films for enzyme immobilization studied by Confocal Scanning Laser Microscopy. *Rev de Chimie*, 57, 211-215.
[25] Pankhurst, Q. A., Connolly, J., Jones S. K. Dobson, J. (2003) Applications of magnetic nanoparticles in biomedicine. *Appl. Phys*, 36R, 167-181.
[26] Ulijn, R.V. (2006) Enzyme-Responsive Materials: A New Class of Smart Biomaterials. *J. Mater. Chem,* 16, 2217-2225.
[27] Kumar, M. N. V. R. (2000) Nano and microparticles as controlled drug delivery devices. *J. Pharm. Sci*, 3, 234-258.
[28] Cui, D., Gao, H. (2003) Advance and Prospect of Bionanomaterials. *Biotechnol Progress,* 19, 683-693.
[29] Bulte, J. W. M. (2005) Magnetic nanoparticles as markers for cellular MR imaging. *J. Magn. Magn. Mater*, 289, 423-427.
[30] Salata, O. V. (2004) Application of nanoparticles in biology and medicine. *J. Nanobiotech,* 2, 1-6.
[31] Saiyed, Z. M., Telang, S. D., Ramchad, C. N. (2003) Application of magnetic techniques in the field of drug discovery and biomedicine. *BioMagn. Res. Technol*, 1, 1-8.
[32] Liu, T. Y., Hu, S. H., Liu, K. H., Liu, D. M., Chen, S. Y. (2008) Study on controlled drug permeation of magnetic-sensitive ferrogels: effect of Fe3O4 and PVA. *J. Control Release*, 20, 228-236.

[33] Guowei, D., Adriane, K., Chen, X., Jie, C., Yinfeng, L. (2007) PVP magnetic nanospheres: biocompatibility, in vitro and in vivo bleomycin release. *Int. J. Pharm,* 2, 78-85.

[34] Dresco, P. A., Zaitsev, V. S., Gambino, R. J., Chu, B. (1999) Preparation and properties of magnetite and polymer magnetite nanoparticles. *Langmuir*, 15, 1945-1951.

[35] Tartaj, P., Morales, M. P., Veintemillas-Verdaguer, S., Gonzalez-Carreno, T., Serna, C. J. (2003) The preparation of magnetic nanoparticles for applications in biomedicine. *J. Phys.* D. Appl Phys, 36, R182-R197.

[36] Liao, M. H., Chen, D. H. (2001) Immobilization of yeast alcohol dehydrogenase on magnetic nanoparticles for improving its stability. *Biotechnol. Lett*, 23, 1723-1727.

[37] Huang, S. H., Liao, M. H., Chen, D. H. (2003) Direct Binding and Characterization of Lipase onto Magnetic Nanoparticles. *Biotechnol. Prog*, 19, 1095-1100.

[38] Kulla, K. C., Gooda, M. D., Thakur, M. S., Karanth, N. G. (2004) *Enhancement of stability* of immobilized glucose oxidase by modification of free tiols. *Biosens Bioelectron,* 19, 621-625.

[39] Przybyt, M. (2003) Behaviour of glucose oxidase during formation and ageing of silica gel studied by fluorescence spectroscopy. *Mater Sci*, 21, 398-415.

[40] Kouassi, G. K., Irudayaraj, J., McCarty, G. (2005) Examination of Cholesterol oxidase attachment to magnetic Nanoparticles. *J. Nanobiotech*, 3, 1-6.

[41] Ma, D., Zhang, L. M. (2008) Fabrication and modulation of magnetically supramolecular hydrogels. *J. Phys .Chem*, B 22, 6315-6322.

[42] Liang, Y. Y., Zhang, L. M., Jiang, W., Li, W. (2007) Embedding magnetic nanoparticles into polysaccharide-based hydrogels for magnetically assisted bioseparation. *Chemphyschem*, 12, 2367-2372

[43] Massart, R. (1981) Preparation of aqueous magnetic liquids in alkaline and acidic media, IEEE Trans Magn, MAG-17, 1247.

[44] Teodor, E., Litescu, S. C., Petcu, C., Mihalache, M., Somoghi, R. (2009) Nanostructured Biomaterials with Controlled Properties Synthesis and Characterization. *Nanoscale Res. Lett*, 4, 544-549.

[45] Verma, N., Kumar, K., Kaur, G., Anand, S. (2007) L-asparaginase: a promising chemotherapeutic agent. *Critical Reviews in Biotechnology*, 27, 45-62.

[46] Fu, C. H., Sakamoto, K. M. (2007) PEG-asparaginase. *Expert Opin. Pharmacoter*, 8, 1977-1984.

[47] Zhang, Y. Q., Tao, M. L., Shen, W. D., Zhou, Y. Z., Ding, Y., Ma, Y., Zhou, W. L. (2004) Immobilization of L-asparaginase on the microparticles of the natural silk sericin protein and its characters. *Biomaterials*, 25, 3751-3759.

[48] Verma, N., Kumar, K., Kaur, G., Anand, S. (2007) *E. coli* K-12 asparaginase-based asparagine biosensor for leukemia. *Artif Cells Blood Substit. Immobil. Biotechnol*, 35, 449-456.

[49] Cornea, C. P., Lupescu, I., Vatafu, I., Caraiani, T., Savoiu, V. G., Campeanu, G., Grebenisan, I., Negulescu, G. P., Constantinescu, D. (2002) *Rom Biotechnol Letters*, 7, 717-722.

[50] Mashburn, L., Wriston, J. (1963) Tumor inhibitory effect of l-asparaginase. *Biochem. Biophys. Res. Commun,* 12, 50-56.

[51] Teodor, E., Litescu, S. C., Lazar, V., Somoghi, R. (2009) Hydrogel-magnetic nanoparticles with immobilized L-Asparaginase for biomedical applications. *J. Mater Sci: Mater Med,* 20, 1307-1314.

[52] Teodor, E., Lupescu, I., Truica, G. (2008) The release of L-asparaginase from hydrogel-magnetic nanoparticles. *Rom Biotechnol Letters*, 13, 3907-3913.

[53] Teodor, E., Cutaş, F., Moldovan, L., Tcacenco, L., Caloianu, M. (2003) Characterisation of hyaluronic acid extracted from swine vitreous humor by liquid chromatography. *Rom Biol. Sci*, I, 35-46.

[54] Mao, J. S., Iiu, H. F., Yin Y. J., Yao, K. D. (2003) The properties of chitosan-gelatin membranes and scaffolds modified with hyaluronic acid by different methods. *Biomaterials,* 24, 1621-1629.

[55] Xu, Y., Du, Y. (2003) Effect of molecular structure of chitosan on protein delivery properties of chitosan nanoparticles. *Int. J. Pharmaceutics*, 250, 215-226.

[56] Summer, J., Howell, S. (1935) A method for the determination of invertase activity. *J. Biol. Chem*, 108, 51-54.

[57] Wright, S. J., Centonze, V. E., Stricker, S. A., De Vries, P. J., Schatten, S. W., Paddock, G. (1993) Cell biological applications of confocal microscopy. *Methods Cell Biol*, 38, 1-45.

[58] Neut, D., Hendriks, J. G. E., Van Horn, J. R., Van Der Mei, H. C., Busscher, H. J. (2005) Pseudomonas aeruginosa biofilm formation and slime excretion on antibiotic-loaded bone cement. *Acta Orthopaedica,* 76**,** 109–114.

[59] Prodanovic, R. M., Simic, M. B., Vujcic, Z. M. (2003) Immobilization of periodate oxidized invertase by adsorption on sepiolite**.** *J. Serb. Chem. Soc*, 68, 819-824.

[60] Milovanovic, A., Bozic, N., Vujcic, Z. (2007) Cell wall invertase immobilization within calcium alginate beads. *Food Chemistry*, 104, 81-86.

[61] Mosmann, T. (1988) Rapid colorimetric assay for cellular growth and survival: application to proliferation and cytotoxic assays. *J. Immunol. Methods*, 65, 55-63.

[62] Koo, O. M., Rubinstein, I., Onyuksel, H. (2005) Role of nanotechnology in targeted drug delivery and imaging: a concise review. *Nanomedicine: Nanotechnology, Biology and Medicine,* 1, 193-212.

[63] Liu, S., Wei, X., Chu, M., Peng, J., Xu, Y. (2006) Synthesis and characterization of iron oxide/polymer composite nanoparticles with pendent functional groups. *Colloids Surf.* B Biointerfaces, 51, 101-106.

[64] Hobbs, S. K., Monsky, W. L., Yuan, F., Roberts, W. G., Griffith, L., Torchilin, V. P. (1998) Regulation of transport pathways in tumor vessels: role of tumor type and microenvironment. *Proc. Natl. Acad. Sci. USA*, 95, 4607-12.

[65] Jain, R. K. (1998) Delivery of molecular and cellular medicine to solid tumors. *J. Control Release,* 53, 49-67.

[66] Wang, R., Xia, B., Li, B. J., Peng, S. L., Ding, L. S., Zhang, S. (2008) Semi-permeable nanocapsules of konjac glucomannan-chitosan for enzyme immobilization. *Int. J. Pharm*, 364, 102-107.

INDEX

A

ab initio computational methods, xii, 349
abatement, 12
accelerator, 81, 113, 120, 190, 394, 397
access, 79, 128, 129, 182
accessibility, 21, 123, 211
accounting, 27
acetic acid, 5, 6, 61, 62, 65, 67, 93, 95, 96, 118, 124, 130, 131, 134, 148, 187, 204, 229, 235, 238, 301, 303, 304, 325, 326, 327, 329, 350, 352, 380, 387, 431
acetone, 90, 380
acetonitrile, 416
acetylamide group, 350
acetylation, viii, 3, 63, 79, 107, 115, 116, 130, 131, 136, 137, 186, 218, 225, 230, 320, 321
ACF, 104
acidic, 31, 51, 57, 93, 107, 165, 175, 201, 203, 206, 216, 234, 244, 259, 267, 273, 294, 296, 304, 305, 306, 307, 308, 309, 320, 321, 327, 357, 378, 380, 383, 453
acidity, 3
acrylate, 159
acrylic acid, 66, 257
acrylonitrile, 279
action potential, 120, 322
activated carbon, 422, 423
activation energy, ix, 118, 141, 157, 160, 161, 162, 166, 167, 302, 303
active compound, 200, 451
active site, 50, 350, 412
actuation, 274
actuators, 266, 274
acupuncture, 96, 97, 99
acylation, 50, 209, 245
adaptation, 216, 338
additives, 402
adenocarcinoma, 184
adhesion, xiv, 57, 58, 60, 61, 66, 67, 68, 69, 70, 71, 82, 85, 86, 88, 91, 100, 108, 112, 116, 120, 136, 187, 188, 190, 204, 208, 214, 242, 322, 337, 387, 390, 428, 429
adhesions, 397, 449
adhesive properties, 100, 211
adhesives, 65
adipose, 104, 113
adsorption, vii, viii, xiii, 1, 2, 3, 4, 5, 6, 7, 8, 11, 12, 17, 18, 19, 20, 21, 22, 23, 24, 25, 26, 27, 28, 29, 30, 31, 32, 33, 34, 35, 36, 37, 38, 39, 40, 42, 43, 44, 45, 58, 68, 69, 78, 82, 85, 86, 87, 88, 118, 131, 175, 190, 203, 294, 382, 399, 402, 409, 410, 411, 412, 414, 416, 417, 418, 422, 424, 425, 439, 454
adsorption isotherms, 5, 7, 8, 19, 20, 21, 22, 23, 24, 26, 27, 28, 29, 30, 32, 33, 34, 45
advancement, 281
advancements, 281
adverse conditions, 200
adverse effects, 185, 400
AFM, 126, 127, 239
agar, 123, 126
age, 90, 212, 324
agglutination, 65, 66
aggregation, 61, 66, 67, 69, 81, 82, 117, 120, 125, 135, 187, 190
agriculture, x, xiii, 171, 172, 377, 382
AIBN, 258
alanine, 301, 302, 304, 314
albumin, 52, 86, 102, 122, 124
algae, 172
alkaline media, 90, 305
alkylation, 50, 89, 279, 452
allergic reaction, 430
allergy, 213
ametropia, 214
amine, xiii, 39, 51, 53, 58, 117, 123, 153, 177, 201, 229, 232, 237, 241, 245, 249, 258, 319, 321, 332, 350, 362, 421

amine group, 51, 53, 153, 177, 229, 232, 237, 241, 245, 249, 258, 321, 332, 362
amines, 50, 51, 58, 159, 176, 206, 233, 262, 422, 425
amino, vii, viii, xii, 2, 3, 4, 15, 23, 25, 32, 35, 39, 40, 41, 42, 43, 45, 50, 67, 84, 88, 101, 123, 125, 126, 131, 148, 156, 172, 174, 180, 201, 207, 224, 256, 258, 261, 262, 264, 290, 296, 298, 299, 302, 308, 309, 320, 350, 359, 361, 362, 378, 380, 382, 387,뱀402, 409, 415, 418
amino acid, 125, 382, 387
amino acids, 125, 382
amino groups, 3, 4, 32, 35, 39, 40, 41, 42, 43, 45, 67, 101, 123, 126, 131, 175, 207, 224, 258, 261, 264, 296, 298, 299, 302, 308, 309, 350, 359, 380, 402, 409, 415, 418
amino-groups, vii, 2, 23, 40
ammonia, 62, 65, 103, 116, 439
ammonium, 52, 65, 173, 221, 257, 265, 266, 267, 274, 298, 299, 439, 445
ammonium salts, 445
amorphous phases, 10, 145
amorphous polymers, 145, 146
amorphous-crystalline polymers, 24
amphibians, 400
amplitude, 14, 148, 149, 150
angiogenesis, 117, 139, 183, 195, 196, 386, 389
angiotensin converting enzyme, 188, 194
angiotensin II, 188
aniline, 267, 268, 274
anisotropy, 366
annealing, 149, 152, 153, 159, 164, 166, 247, 248
annihilation, 300
anomalous diffusion, 306
anorexia, 380
ANOVA, 302
antibiotic, 205, 306, 308, 309, 315, 328, 329, 393, 394, 454
antibody, 220, 265, 266, 280, 385
anti-cancer, 184, 209, 310
anticancer activity, 124, 136
anticancer drug, 52, 56, 82, 124, 384
anticoagulant, 66, 67, 81, 187, 192
anticoagulation, 101
antidepressant, 68
antigen, 128, 182, 204, 211, 212, 221, 323
antioxidant, x, 121, 122, 123, 133, 139, 171, 174, 179, 180, 197, 206, 217, 396, 400
antitumor, 60, 72, 102, 136, 172, 174, 182, 183, 185, 189, 192, 195, 197, 198, 202, 381
antitumor agent, 182, 189
apoptosis, 119, 122, 182, 183, 192
aptitude, 323
aqueous solutions, xiii, 8, 67, 132, 208, 234, 235, 237, 258, 299, 305, 399, 402, 415, 422, 425, 438
arginine, 181, 203
argon, 4, 19, 149
aromatic compounds, 423
aromatic rings, 68, 78
arrest, 122
Arrhenius dependence, 157
Arrhenius equation, 118
arteries, 63, 100
arteriosclerosis, 188
arteriovenous shunt, 101
arthropods, 350, 378
articular cartilage, 75
ascites, 183, 184
asparagines, 430
aspartate, 207
aspartic acid, 74
assessment, 60, 92, 104, 120, 137, 190, 396, 419, 423
asymmetry, 42, 354
atherosclerotic plaque, 69
atmosphere, 149
atomic force, 120, 174, 191, 281
atomic orbitals, 361, 368
atoms, xii, 4, 143, 320, 350, 352, 359, 362, 363, 364, 366, 367, 369, 372, 374, 375
ATP, 213
attachment, 58, 86, 88, 89, 90, 91, 100, 101, 105, 112, 134, 224, 242, 319, 322, 323, 328, 329, 340, 396, 453
attribution, 147
autolysis, 125
automated synthesis, 318
autopsy, 381
avoidance, 267

B

bacteria, 85, 89, 126, 127, 136, 137, 139, 172, 174, 175, 176, 177, 178, 189, 190, 191, 202, 203, 206, 215, 323, 401, 423, 430, 431, 450
bacterial strains, 135
bacteriophage, 133, 139
bacteriostatic, 224
bacterium, 174, 175, 176, 177
barriers, x, 199, 200, 202, 211
basal layer, 322
base, x, 50, 65, 223, 266, 270, 273, 280, 281, 382, 409, 420
behaviors, 39, 57, 87, 143, 312, 321
Beijing, 83
bending, 12, 13, 233, 235, 305, 306, 361
benefits, 57, 116, 120, 174, 206, 214, 310

beverages, 214
bias, 150, 151
binding energies, 364, 373
binding energy, 364, 365, 373
bioactive materials, 110, 429
bioassay, 60
bioavailability, x, 173, 199, 200, 206, 210, 211, 216, 310, 383, 393, 430
biochemistry, 124
biocompatibility, viii, x, xv, 49, 55, 58, 59, 83, 84, 85, 87, 88, 89, 91, 92, 96, 101, 106, 107, 108, 111, 128, 137, 184, 199, 202, 204, 216, 224, 250, 256, 257, 275, 293, 298, 309, 319, 321, 384, 390, 428, 430, 443, 444, 449, 451, 453
Biocompatibility, 319, 396, 438, 443, 444, 452
biocompatibility test, 451
biocompatible materials, 234, 241, 250, 320, 431
bioconversion, 425
biodegradability, viii, xiii, 49, 58, 59, 81, 83, 84, 85, 91, 95, 105, 184, 201, 202, 207, 216, 242, 256, 257, 319, 377, 433, 435
biodegradation, x, 87, 116, 171, 321, 380, 384, 451, 452
biological activities, x, 133, 137, 172, 174, 188, 189, 192, 202, 203, 204, 224
biological activity, 85, 104, 205, 319, 321
biological fluids, 320, 429
biological media, 386
biological processes, 241
biological samples, 433
biological systems, 116, 191
biomaterials, viii, xii, xiv, 49, 50, 57, 63, 64, 68, 71, 72, 76, 81, 87, 89, 92, 105, 106, 108, 111, 139, 190, 195, 204, 217, 242, 266, 274, 295, 311, 318, 319, 321, 334, 335, 337, 340, 344, 349, 351, 390, 391, 394, 427, 428, 429, 451, 452
biomedical applications, x, xiii, 67, 77, 81, 85, 128, 137, 173, 188, 202, 207, 223, 227, 234, 241, 257, 325, 344, 378, 380, 384, 390, 429, 452, 454
biomolecules, viii, 49, 51, 52, 72, 200, 265, 266
biopolymer, xiii, xiv, 21, 63, 142, 143, 153, 162, 165, 166, 250, 256, 274, 280, 342, 343, 351, 352, 377, 379, 384, 385, 392, 402, 427, 430
biopolymers, xiv, 63, 137, 147, 151, 210, 211, 257, 336, 350, 381, 428, 429, 433
biopsy, 104, 389, 396
bioremediation, 425
bioseparation, 430, 431, 450, 453
biosynthesis, 125, 438
biotechnology, x, 143, 153, 171, 172, 384
birds, 400
Birmingham, Alabama, 346
bisphenol, xiii, 399, 400, 401, 402, 416, 418, 419, 420, 421, 422, 423, 424, 425
bleeding, 61, 62, 63, 64, 65, 67, 72, 77, 78, 79, 214, 215
bleeding time, 61, 65
blend films, 85, 217, 234, 235, 238, 240, 274
blends, 74, 113, 164, 204, 217, 227, 228, 234, 242, 243, 244, 245, 247, 248, 249, 273, 274, 275, 290, 295, 301, 421
blood, viii, x, 49, 55, 59, 61, 62, 63, 64, 65, 66, 67, 68, 69, 70, 71, 76, 77, 79, 80, 81, 82, 90, 100, 111, 116, 136, 171, 174, 186, 187, 188, 191, 193, 194, 195, 204, 211, 212, 214, 215, 277, 292, 293, 311, 324, 384, 385
blood circulation, x, 171
blood clot, 61, 62, 65, 67, 68, 69, 188
blood pressure, 188, 324
blood supply, 324
blood transfusion, 215
blood transfusions, 215
blood vessels, 100, 324
bloodstream, 323
body fluid, 257, 294
body weight, 120, 183
Boltzmann constant, 359
bonding, xii, 4, 13, 40, 173, 207, 216, 269, 319, 349, 351, 357, 381, 382, 386
bonds, 4, 10, 12, 13, 39, 40, 41, 123, 132, 159, 233, 262, 291, 309, 334, 445
bone, viii, ix, x, 49, 58, 60, 76, 83, 84, 88, 89, 91, 92, 93, 105, 107, 108, 109, 110, 115, 138, 191, 199, 204, 275, 387, 388, 389, 391, 395, 396, 454
bone biology, 108
bone cells, 275
bone form, 58, 76, 93, 110, 388, 389, 395, 396
bone marrow, 58, 76, 191, 389
bones, 85, 89, 91, 92, 103, 389
bounds, viii, 2, 43, 45
bowel, 210
bradykinin, 188
brain, 65, 78, 93, 323, 324
branching, 229, 230
breakdown, 72, 101
breast cancer, 194, 210, 281, 400, 419, 420
breast carcinoma, 184
brittleness, 239
burn, 103, 104, 113, 385, 387, 390, 391, 397
by-products, 293

C

caffeine, 299
calcium, 50, 58, 65, 69, 70, 89, 93, 104, 108, 119, 136, 190, 265, 297, 388, 431, 435, 454

calcium channel blocker, 297
caliber, 111
calibration, 339
calorimetric measurements, 145
calyx, 420
cancer, 81, 123, 124, 183, 184, 209, 210, 220, 281, 350, 430
candidates, xiii, 51, 210, 215, 224, 270, 309, 377
capillary, 78, 104, 203, 307, 381, 388
carbohydrate, 109, 194, 224, 351, 358, 445
carbohydrates, 125, 350
carbon, 7, 21, 22, 27, 88, 180, 201, 274, 279, 350, 422
carbon nanotubes, 274
carbonyl groups, 132
carboxyl, 67, 69, 85, 87, 192, 245, 260, 262, 269, 321, 350, 380
carboxylic acid, 174, 261, 262, 308
carboxylic acids, 262
carboxylic groups, 53, 57, 245
carcinogen, 405
carcinoma, 51, 124, 183, 184, 195, 197, 309, 315
cardiac catheterization, 62
cardiovascular disease, 123
cartilage, viii, 49, 59, 60, 71, 83, 84, 91, 107, 109, 110, 138, 234, 241, 343, 384, 394
cartilaginous, 394
casting, xiv, 61, 204, 235, 238, 274, 428, 433
castor oil, 311
catalysis, 225, 277
catalyst, 52, 299, 424
cation, viii, 2, 4, 8, 34, 35, 38, 39, 41, 43, 45, 117, 268, 269
cauterization, 214, 215
cDNA, 281
cell biology, 433
cell culture, xi, xiii, 84, 85, 95, 101, 136, 202, 289, 326, 327, 328, 330, 343, 377, 386, 387, 388, 389, 396, 429
cell cycle, 104, 122, 124
cell death, 123, 125, 126, 177
cell division, 91, 125
cell line, 103, 105, 123, 184, 185, 192, 195, 323, 328, 340, 400
cell lines, 103, 105, 123, 184, 185
cell membranes, 126, 385
cell metabolism, 117
cell signaling, 318
cell surface, 65, 125, 174, 176, 187
cellulose, 3, 50, 62, 65, 66, 81, 90, 145, 146, 157, 172, 201, 217, 224, 225, 226, 227, 321, 350, 351, 380, 383, 408, 409
central nervous system, xii, 93, 318, 324
ceramic, 60, 71, 76
ceramic materials, 60
cerebral cortex, 84, 86
chain scission, 130, 131, 132, 237
challenges, viii, 49, 105, 265, 319, 419
charge density, 57, 319, 380
cheese, 190
chelating model, 350
chemical bonds, 66
chemical characteristics, 189
chemical degradation, 160
chemical inertness, 256
chemical interaction, 319
chemical properties, 24, 133, 143, 186, 196, 216, 226, 256, 318, 325, 395
chemical reactions, 50, 226, 266, 320, 402
chemical structures, 320, 403
chemicals, 3, 5, 100, 209, 323, 352, 400, 418, 422, 431
chemotherapeutic agent, 453
chemotherapy, 196, 209, 220, 430
China, 83, 111, 422
chitinase, 119, 138, 350
chitosan-based hydrogels, 313
chitosinase, 350
chloroform, 249
cholera, 175
cholesterol, 122, 174, 188, 213, 381, 382, 430
chondritin sulfate, xii, 349, 350, 352
chondrocyte, 91, 110
chondroitin sulfate, 57, 74, 91, 106, 197
chromatograms, 130
chromatography, 131, 210, 420, 439
cimetidine, 300, 309
circulation, 51, 52, 122, 128, 211, 386
City, 62
classes, 93, 429
classification, 20, 429
cleavage, 309, 395, 415
clinical application, 87, 102, 211, 387
clinical trials, x, 200, 212, 215, 222
closure, 65, 104, 117
cluster model, 21
clustering, 126, 159
clusters, 21, 25, 127, 340, 376, 440
CMC, 91
CNS, 324
CO2, 328, 329
coatings, 204, 205, 217, 422
coenzyme, 125
collagen, viii, 50, 57, 58, 60, 62, 66, 69, 74, 75, 76, 77, 78, 83, 85, 86, 91, 93, 99, 100, 101, 104, 106,

111, 112, 113, 182, 186, 194, 195, 204, 309, 322, 329, 338, 350, 385, 386, 394
collagen sponges, 78
collateral, x, 199, 216
collateral damage, x, 199, 216
colon, 57, 74, 183, 192, 205, 206, 207, 210, 219
colon cancer, 192
colonization, 215, 222, 322
commercial, xi, 62, 71, 84, 104, 118, 124, 133, 134, 137, 172, 206, 241, 242, 255, 277, 290, 325, 384, 385, 387, 392
communication, 127, 279
comparative method, vii, 2, 7, 21, 27, 45
compatibility, xiv, 67, 68, 77, 80, 81, 101, 102, 116, 191, 195, 204, 207, 224, 225, 234, 242, 304, 421, 428, 429
compensation, 150
complement, 181, 186, 187, 189, 193, 196, 391, 395
complementary DNA, 281
compliance, 173, 211, 293
composites, x, 66, 74, 81, 82, 89, 93, 105, 106, 108, 110, 128, 138, 149, 223, 224, 234, 249, 266, 273, 274, 275, 390, 420
composition, xi, 58, 60, 92, 130, 172, 188, 207, 220, 223, 234, 237, 241, 248, 250, 258, 261, 274, 277, 295, 298, 322, 383, 416, 433, 435, 439, 445
compounds, xii, 39, 41, 43, 121, 128, 159, 180, 183, 185, 186, 189, 200, 204, 211, 212, 216, 219, 224, 225, 226, 234, 241, 267, 290, 349, 353, 362, 401, 402, 406, 419, 422, 423, 425
compression, 10, 63, 64, 65, 78, 226, 381, 382, 383
computer, 96, 97, 98, 324, 361
computer-aided design, 96, 98
condensation, 24, 67, 390
conditioning, 149
conductivity, 107, 147, 153, 154, 155, 156, 157, 158, 160, 161, 257, 266, 273, 274, 277, 280
configuration, 43, 52, 187, 319, 352, 355, 361, 430
conformity, 357, 359
Congress, iv, 190, 252
connective tissue, 60, 85, 91, 321, 322, 350
consent, 215
conservation, 384, 405, 406
constituents, 89, 177, 302
consumption, 227, 249
control group, 388, 389
controversial, 142, 146, 166, 183
convention, 354
COOH, 66, 266, 350
cooling, 63, 95, 149, 247, 248, 257
coordination, 4, 39, 41, 43, 353, 357, 359, 361
copolymer, 2, 50, 73, 103, 106, 172, 209, 235, 237, 239, 240, 244, 258, 261, 264, 265, 267, 272, 281, 296, 298, 299, 306, 313, 350, 393
copolymerization, 50, 234, 256, 257, 258, 265, 266, 267, 268, 282, 299
copolymers, xi, 55, 58, 223, 224, 225, 234, 237, 240, 241, 242, 257, 258, 261, 263, 266, 289, 295, 296, 298, 299, 310, 311, 312, 313
copper, viii, 2, 3, 4, 5, 8, 31, 32, 33, 34, 35, 36, 37, 38, 39, 40, 41, 42, 43, 44, 45, 185, 190, 198, 330
coronary artery disease, 213
correlation, xii, 4, 122, 300, 336, 340, 349, 395
correlation analysis, 336, 340
correlations, 361
cortex, 389
cortical bone, 389
corticosteroids, 205, 213
cosmetic, xiii, 377
cosmetics, 380, 382
cost, xi, 101, 200, 204, 205, 215, 223, 256, 293, 345, 401
cotton, 66, 81, 186, 329, 387
covalency, 358
covalent bond, 57, 291, 446
covering, 103, 104, 381, 382, 385, 387, 388, 430, 437, 439, 440, 441, 442, 443, 445
crabs, 5, 84, 290, 350, 402
cristallinity, 6, 143
crystal structure, 11
crystalline, 3, 8, 10, 12, 14, 117, 132, 143, 145, 232, 234, 240, 242, 248, 249, 256, 320, 333, 334, 356, 357, 360, 421
crystalline solids, 143
crystallinity, 3, 8, 10, 11, 12, 32, 87, 129, 132, 228, 232, 244, 248, 296, 304
crystallites, 11, 232, 249
crystallization, 10, 190, 247
crystals, 130, 435, 438
CSF, 180, 189, 218
cultivation, 59, 224, 241
culture, xii, 60, 79, 84, 86, 88, 91, 92, 93, 102, 104, 109, 110, 116, 120, 123, 124, 125, 202, 208, 263, 310, 315, 318, 321, 328, 382, 387, 396, 420, 424, 444
cure, 66, 302, 309, 314
curing process, 302
cycles, 65, 165, 237, 432, 435
cytochrome, 102, 351
cytochromes, 351
cytocompatibility, 85, 86, 104, 107, 321
cytokines, 181, 182, 183, 185, 186, 194, 204, 385, 386, 395
cytometry, 104, 119, 122

cytoplasm, 183
cytoskeleton, 203
cytotoxicity, 95, 102, 105, 121, 128, 190, 263, 298, 438, 444

D

damping, 164
data analysis, xii, 349
deacetylation, xi, xiii, 2, 5, 50, 51, 62, 65, 66, 67, 69, 71, 73, 77, 87, 107, 117, 126, 129, 130, 133, 137, 139, 148, 149, 156, 165, 172, 174, 186, 187, 189, 193, 197, 200, 224, 225, 226, 227, 228, 229, 232, 233, 234, 236, 241, 249, 256, 289, 290, 310, 319, 320, 325,뫰352, 377, 378, 379, 380, 383, 392, 402
decomposition, 101, 148
deconvolution, 244
defect site, 92
defects, 59, 91, 92, 99, 104, 108, 117, 387, 388, 389, 396
deformability, 181, 192, 225, 237
deformation, xi, 10, 39, 40, 41, 61, 66, 187, 223, 225, 226, 249
degenerate, 354
degradation, 50, 53, 58, 62, 72, 87, 90, 102, 107, 108, 116, 119, 128, 129, 130, 131, 132, 133, 134, 135, 136, 139, 155, 160, 161, 166, 167, 197, 200, 209, 211, 212, 242, 293, 294, 305, 337, 342, 380, 381, 383, 384, 387, 392, 420, 422, 423, 425, 432, 434, 435, 438, 444
degradation mechanism, 131
degradation rate, 50, 87, 131, 209
degree of crystallinity, viii, 50, 115, 227, 228, 232, 240, 320, 321, 334
dehydration, 5, 10, 186, 302
denaturation, 93, 407
dendritic cell, 204
density functional theory, xii, 349, 351, 352, 376
dentures, 382
Department of Energy, 373
depolarization, 125
depolymerization, 62, 116, 128, 129, 130, 131, 132, 134, 138, 139, 392
deposition, 58, 89, 93, 187, 205
deposits, 125, 204
depression, 99, 124, 324
dermis, 103, 113, 186, 388, 390
desorption, 20, 27
destruction, 323, 324
detachment, 291
detection, 102, 117, 121, 122, 161, 266, 277, 279, 280, 281
detoxification, 68, 78, 101
deviation, 29, 37, 150
DFT, xii, 349, 351, 352, 361, 364, 366, 369, 376
diabetes, 208, 213, 217
diabetic patients, 204
dialysis, 79
diaphragm, 101
dibenzo-p-dioxins, 419
dichotomy, 182
dielectric constant, 146, 153, 276
dielectrics, 168
diet, 197, 213
dietary fat, 189
differential scanning, 131, 142, 149, 302, 309, 314
differential scanning calorimeter, 302, 314
differential scanning calorimetry, 131, 142, 149, 309
diffraction, 4, 8, 10, 13, 44, 333, 334
diffuse reflectance, 40, 41
diffusion, 10, 17, 18, 19, 38, 57, 90, 98, 203, 233, 249, 274, 291, 294, 296, 297, 298, 306, 309, 322, 385, 429
diffusion process, 38
diffusion time, 18, 19, 294
digestion, 101, 188, 192, 205
dimethacrylate, 420
dipole moments, 166
discharges, 330
discomfort, 62, 293
diseases, 92, 122, 135, 208, 210, 216, 263, 281
dislocation, 325
dispersion, xi, 145, 223, 244, 249, 440
displacement, 18
dissociation, 53
distilled water, 6, 148, 270, 326, 327, 329, 431, 432
distortions, 373
distribution, x, 15, 17, 18, 51, 55, 71, 95, 96, 97, 98, 124, 125, 128, 146, 171, 174, 182, 208, 228, 229, 237, 380, 430, 437, 440, 442, 444, 451
DMA analysis, 142, 143, 162, 166
DMF, 118
DNA, xii, 50, 51, 53, 73, 75, 76, 103, 123, 124, 126, 127, 135, 177, 179, 180, 183, 195, 203, 212, 216, 221, 263, 265, 266, 281, 287, 318, 329, 335, 336, 337, 338, 339, 340, 343, 384, 392
dogs, 65, 99, 111, 120, 185, 193, 197, 380, 385, 387, 389, 391, 394
DOI, 192, 197
donors, 70, 101, 159
dopamine, 424
doping, 267, 272, 273, 274
dosage, x, 199, 203, 205, 206, 211, 216, 219, 311, 380, 382, 390
dosing, xiv, 173, 213, 292, 427
double helix, 10

down-regulation, 119, 125
drawing, 94, 189, 327
dressing material, 224, 429
dressings, 62, 63, 64, 68, 78, 79, 80, 113, 137, 215, 241, 387, 394
drinking water, 419, 422
drug carriers, x, 199, 204, 261, 295, 311, 313, 323, 393
drug delivery, viii, ix, x, xi, xiii, xiv, 49, 50, 51, 55, 57, 60, 72, 73, 74, 102, 107, 115, 120, 171, 190, 195, 199, 200, 201, 202, 205, 206, 207, 209, 210, 211, 212, 216, 218, 219, 220, 234, 241, 242, 256, 258, 262, 289, 290, 291, 292, 293, 294, 296, 299, 310, 311, 312, 315, 377, 383, 427, 428, 429, 430, 450, 451, 452, 454
drug discovery, 281, 452
drug release, xi, 52, 53, 57, 75, 200, 203, 206, 208, 209, 219, 224, 261, 263, 289, 291, 296, 297, 298, 300, 301, 303, 304, 307, 309, 310, 312, 313, 314, 315, 381, 383, 450
drugs, x, xiii, 50, 52, 53, 55, 56, 67, 73, 183, 199, 200, 202, 203, 205, 207, 209, 210, 211, 216, 217, 218, 262, 263, 292, 295, 296, 300, 304, 305, 310, 313, 314, 377, 381, 383, 384, 392, 393, 429, 450, 451
drying, vii, 1, 2, 5, 6, 10, 11, 12, 13, 15, 19, 21, 22, 23, 26, 30, 32, 35, 39, 44, 45, 62, 63, 65, 95, 101, 102, 104, 126, 134, 144, 162, 204, 209, 210, 304, 307, 308, 314, 327, 379, 383, 384, 393
DSC, xii, 142, 144, 147, 149, 160, 161, 166, 235, 241, 245, 246, 247, 248, 263, 300, 301, 302, 309, 310, 314, 318
dyes, 308
dynamic mechanical analysis, 161, 165, 166

E

E.coli, 439
ECM, 58, 86, 88, 91, 101, 102, 117, 319, 321, 323, 386
elastomers, 290
elderly population, 213
electric charge, 291
electric field, 306, 355
electrical conductivity, 153, 265, 273, 274, 276
electrical properties, 153, 266, 274
electrochemical impedance, 281
electrode surface, 267
electrodes, 8, 134, 149, 265, 266, 267, 277, 279, 281, 330, 331
electrolysis, ix, 116, 134
electrolyte, 280, 306
electromagnetic, 276
electron, ix, xiii, 8, 68, 78, 90, 102, 109, 116, 123, 124, 125, 127, 130, 177, 183, 204, 226, 257, 265, 272, 275, 277, 301, 304, 350, 354, 355, 358, 361, 362, 365, 366, 367, 368, 402, 430, 441
electron microscopy, ix, 8, 102, 116, 124, 127, 430
Electron Paramagnetic Resonance, 351, 355, 375
electron state, 226
electrons, 272, 277, 353, 358, 361
electrospinning, xi, 102, 105, 249
ELISA, 88, 104
elongation, 85, 98, 142, 178, 239, 248
elucidation, 353
e-mail, 427
emboli, 214
embolization, 214
emission, 329, 330, 354
employment, 430
emulsions, 391
encapsulation, 52, 60, 234, 296, 321, 430, 445
encoding, 53, 76, 221
endocrine, 400, 418, 419, 420, 422
endocrine system, 419
endocrine-disrupting chemicals, 419, 420
endoskeleton, 91
endothelial cells, 100, 112, 117, 122, 138, 275, 389
endothelium, 100, 112
endothermic, 145, 245
endotherms, 245
energy, ix, xi, xii, 4, 29, 30, 33, 34, 35, 38, 90, 125, 130, 142, 161, 167, 225, 255, 256, 282, 303, 319, 349, 354, 358, 359, 361, 364, 366, 373
engineering, viii, xi, xii, xiii, xiv, 49, 50, 51, 57, 58, 60, 72, 75, 76, 83, 84, 87, 88, 89, 90, 91, 92, 93, 96, 100, 101, 102, 103, 105, 107, 108, 109, 110, 111, 112, 113, 116, 136, 202, 204, 207, 224, 234, 242, 257, 277, 289, 318, 321, 322, 344, 377, 388, 394, 428, 429
enlargement, 334
entrapment, 75, 296, 430, 438, 444
entropy, 5
environment, ix, x, xi, 51, 93, 101, 103, 104, 107, 115, 127, 146, 162, 175, 179, 180, 199, 200, 202, 203, 206, 207, 209, 242, 289, 293, 294, 308, 310, 319, 322, 328, 332, 353, 366, 378, 380, 381, 383, 384, 400, 419
environmental conditions, 176
environmental protection, x, 171, 172
enzymatic activity, xiv, 408, 428, 429, 432, 435
enzyme, xiv, 50, 58, 102, 133, 188, 194, 265, 266, 277, 306, 321, 350, 380, 382, 384, 390, 392, 400, 403, 425, 432, 435, 437, 439, 449, 451, 452, 454
enzyme chelation, 350
enzyme immobilization, 390, 451, 452, 454

enzymes, ix, xiv, 50, 87, 116, 125, 126, 134, 179, 200, 206, 216, 224, 258, 277, 321, 350, 351, 374, 381, 382, 383, 401, 406, 407, 408, 418, 423, 428, 430, 451
epidermis, 176, 322, 387
epithelia, 129, 207
epithelial cells, 104, 203, 322, 438
epithelium, 55, 196, 200, 209, 211, 220, 221
epoxy resins, 400, 420
EPR, 195, 351
EPS, 127
equilibrium, 5, 7, 8, 10, 34, 36, 37, 39, 57, 150, 258, 298, 305, 306
equipment, 330, 420
erythrocytes, 65, 66, 180, 187, 191, 385
ester, 247, 248, 421
ester bonds, 247, 248
estrogen, 210, 419
ethanol, 68, 124, 134, 234, 259, 380, 439
etherification, 50, 123, 256
ethyl acetate, 234
ethylcellulose, 195
ethylene, 3, 5, 6, 7, 55, 66, 81, 173, 208, 298, 304, 305, 306, 308, 312, 314, 315, 393, 421
ethylene glycol, 3, 5, 6, 7, 55, 66, 81, 173, 298, 304, 305, 314
ethylene oxide, 208, 306, 308, 312, 315, 393, 421
EU, 113
evacuation, 6, 19, 23
evaporation, 24, 152, 153, 209, 384, 428
evidence, 4, 125, 127, 142, 147, 148, 160, 165, 166, 302, 324, 340
evoked potential, xii, 318, 330, 331, 342
evolution, 229
excision, 65
excitability, 322
excitation, 329, 330
excretion, 188, 454
exertion, ix, 115
exoskeleton, 84, 172, 256, 350
exothermic peaks, 247
experimental condition, 131, 263, 353, 406
experimental design, 121
exploitation, 135
exposure, 143, 219, 419, 433, 435, 438
external environment, 294
extracellular matrix, 58, 76, 88, 90, 91, 103, 112, 117, 319, 321, 395
extraction, 65, 91, 126, 129, 137, 143, 200, 234, 378, 437
extrusion, x, 95, 204, 223, 227, 228, 229, 233, 234, 237, 245, 248, 249, 250
exudate, 186

F

fabrication, 91, 95, 102, 112, 142, 224, 242, 256, 275, 277, 308
Fabrication, 81, 396, 453
facial nerve, 99
fatty acids, 188, 213
Fe metal complexes, 350
Fe-chitosan complex, xii, 349, 357, 359, 360, 361
fermentation, 201, 424
Fermi-contact interaction, xii, 349, 358
ferredoxin, 351
ferric species, xii, 349
ferric state, xii, 349
ferrous ion, 121, 376
fiber, 10, 11, 50, 60, 62, 63, 72, 86, 91, 96, 104, 106, 111, 382, 384, 421, 422
fibers, 66, 75, 77, 81, 91, 96, 117, 120, 203, 234, 265, 274, 350
fibrin, 62, 63, 67, 70, 78
fibrinogen, 62, 67, 69, 187
fibrinolytic, 67, 187
fibroblast growth factor, 58, 388, 389, 396
fibroblast proliferation, 388
fibroblasts, ix, 85, 101, 103, 107, 112, 115, 117, 120, 121, 123, 181, 186, 189, 197, 322, 386, 388, 395, 420
fibrosis, 93, 214
fibrous tissue, 92, 95, 385, 389
fibula, 110
filament, 62, 66, 81
film thickness, 155
filtration, 134, 192, 228, 235, 382
financial, 282
financial support, 282
first generation, 318
fish, 190, 197, 400
fixation, 10
flavonoids, 217
flexibility, 52, 105, 107, 205
flocculation, 125, 203
Flory-Huggins equation, 24, 26, 27
fluctuations, 146, 166
fluid, xiv, 64, 129, 204, 306, 308, 383, 427
fluorescence, 102, 122, 126, 329, 330, 453
food, x, 46, 128, 143, 153, 171, 172, 191, 204, 256, 281, 308, 380, 421
food additives, 380
food industry, 143, 153
force, 58, 126, 136, 308, 361
force constants, 361
formaldehyde, 68, 259, 420, 421
formula, 201, 265, 290

fragments, 125, 130, 131, 134
France, 344
free radicals, ix, 115, 130, 131, 180, 193, 226, 257
free volume, 20, 24, 159
freedom, 17, 372
freezing, 10, 208, 237, 308, 383
freshwater, 130, 172
FTIR, xii, 5, 8, 12, 88, 117, 132, 149, 227, 229, 232, 233, 235, 236, 273, 274, 281, 318, 328, 332, 343, 442, 448
FTIR spectroscopy, 229
functional food, 121, 192
functionalization, 332, 430
fungi, 84, 124, 172, 178, 179, 201, 202, 256, 290, 298, 323, 381
fungus, 179
fungus growth, 179

G

gamma radiation, 84
gamma rays, 84
gastrointestinal tract, 213, 321
Gaussian 09W quantum mechanical program, 351, 352
gel, xiii, xiv, 3, 6, 64, 68, 75, 81, 82, 101, 104, 110, 128, 131, 183, 186, 190, 206, 207, 208, 220, 262, 265, 269, 270, 298, 299, 305, 308, 309, 319, 377, 382, 383, 384, 388, 389, 427, 428, 451, 453
gel permeation chromatography, 131
gelation, 51, 207, 209, 220, 257, 263, 392, 425, 451
gene expression, 51, 86, 88, 384
gene regulation, 420
gene therapy, x, 103, 116, 138, 171, 196, 209, 220
genes, x, 124, 125, 199, 211, 216
genetic background, 182
genetic mutations, 281
geometry, xii, 4, 18, 98, 349, 351, 352, 353, 359
Georgia, 349
Germany, 226, 228, 249, 347
germination, 178
glass transition, ix, 85, 141, 142, 143, 144, 145, 146, 147, 148, 153, 155, 157, 158, 159, 160, 161, 164, 165, 166, 242, 245
glass transition temperature, ix, 85, 141, 142, 143, 144, 147, 148, 155, 159, 160, 164, 165, 166
glasses, 143
glassy polymers, 24
glaucoma, 263
glioma, 323
glucose, xii, 55, 118, 122, 172, 177, 201, 212, 277, 279, 319, 329, 349, 350, 351, 355, 361, 362, 363, 364, 365, 366, 367, 369, 370, 371, 372, 373, 430, 432, 453
glucose oxidase, 279, 430, 453
glucoside, 132
glue, 62, 67
glutamate, 119, 136, 140, 203, 207, 383, 393
glutamic acid, 301, 302, 303, 304, 314
glutamine, 328
glutathione, ix, 115, 122
glycerol, 164, 239, 242, 383
glycine, 305, 314
glycogen, 101
glycol, 55, 74, 85, 195, 298, 312, 430
glycoproteins, 291, 395
glycosaminoglycans, 58, 100, 256, 319
glycoside, 73
GPC, 134, 229, 235, 240, 241
grades, 71
grafted copolymers, 173, 298
grain size, 228
granules, 64, 68, 80, 86, 205, 381
gravimetric analysis, 310
gray matter, 324
growth factor, 58, 60, 75, 76, 120, 181, 183, 186, 197, 385, 386, 388, 395, 396
growth rate, 125, 450
guidance, 76, 96, 97, 98, 99, 116
guidelines, 213

H

H. pylori, 308, 309
haemostatic agent, 385
hair, 382
half-life, 134
Hamiltonian, xii, 349, 354, 355, 358
hard tissues, 89
hardness, 60, 319
Hartree-Fock, 351, 352, 361, 364
hazardous waste, 400, 420
healing, xiii, 81, 82, 100, 104, 113, 117, 119, 120, 181, 185, 186, 189, 197, 204, 207, 214, 377, 384, 385, 386, 387, 388, 389, 390, 391, 394, 395, 396, 397
health, 90, 116, 121, 174, 212
health care, 212
heat capacity, 142, 144, 261
heat shock protein, 104
heat transfer, 95
heating rate, 150, 302
heavy metals, 3, 5
height, 86, 93, 302
helical conformation, 10
Helicobacter pylori, 309
hemerythrin, 351
hemocompatibility, 68, 80

hemodialysis, 64, 122
hemoglobin, 120, 215, 351
hemophilia, 214
hemorrhage, viii, 49, 51, 61, 62, 63, 64, 65, 66, 67, 68, 69, 72, 78, 79, 80, 82, 311
hemorrhage control, viii, 49, 51, 62, 63, 64, 65, 67, 68, 72, 79
hemostasis, 61, 62, 63, 64, 65, 68, 69, 77, 78, 79, 80, 82, 186, 187, 193, 195, 197, 215, 394
hepatic injury, 64, 79, 311
hepatitis, 211, 213, 221
hepatocellular carcinoma, 102, 124, 184
hepatocytes, 101, 112, 124, 139, 179, 197
hepatoma, 102, 184
herbal medicine, 308
heterogeneity, 19
heterogeneous systems, 418
high blood pressure, 174
high density lipoprotein, 122
histamine, 309
histochemistry, 105
histogram, 338, 339
histology, 105
history, 110, 378
HIV, 213
HLA, 221
homeostasis, 385, 387
homogeneity, 302
homopolymerization, 258
homopolymers, 227, 237, 241
hormone, 119, 400
hormones, xiii, 377, 400
host, 105, 181, 318, 320, 338, 344
hot spots, 225
human, ix, x, 51, 58, 60, 61, 64, 66, 67, 68, 69, 70, 76, 79, 81, 89, 92, 100, 104, 107, 110, 112, 115, 120, 122, 123, 124, 127, 128, 133, 134, 135, 136, 138, 181, 183, 184, 191, 194, 195, 197, 199, 207, 208, 210, 212, 216, 224, 234, 241, 257, 281, 309, 315, 318, 380, 386, 387, 389, 394, 400, 419, 420
human body, 128, 234, 241, 257, 318, 380
human health, x, 199, 216, 281
humidity, 119, 149, 219, 235
Hungary, 301
Hunter, 395
hybrid, x, xi, 66, 76, 102, 105, 109, 112, 223, 225, 227, 241, 242, 250, 255, 261, 282, 299, 317, 323, 325, 326, 331, 343, 352, 375, 430, 437, 439, 440, 441, 450
hybridization, 281, 323
hybridoma, 321
hydrocarbons, 19
hydrogen, ix, xiii, 2, 4, 10, 12, 13, 15, 25, 39, 40, 44, 116, 122, 123, 124, 130, 133, 138, 139, 159, 161, 165, 173, 197, 207, 216, 227, 234, 268, 269, 291, 308, 319, 350, 372, 381, 396, 399, 402, 424
hydrogen atoms, 15
hydrogen bonds, 2, 10, 12, 13, 25, 39, 44, 159, 161, 227, 234, 291, 308, 350, 381
hydrogen peroxide, ix, xiii, 116, 122, 123, 124, 130, 133, 138, 139, 197, 396, 399, 402, 424
hydrogenase, 351
hydrolysis, 50, 85, 124, 131, 132, 133, 134, 140, 185, 233, 241, 242, 274, 300, 309, 314, 315, 432, 439
hydroperoxides, 122
hydrophilicity, 55, 87, 88, 175, 227, 250, 264
hydrophobicity, 85, 88, 123, 270, 305
hydrosphere, 190
hydroxide, 52, 228, 266
hydroxyapatite, viii, 58, 83, 108, 109, 110, 391
hydroxyl, xii, 3, 25, 42, 50, 67, 84, 108, 121, 123, 130, 180, 256, 257, 280, 299, 305, 306, 308, 320, 334, 350, 351, 362, 364, 366, 442
hydroxyl groups, 25, 42, 50, 84, 180, 256, 257, 305, 334, 364, 366
hyperbilirubinemia, 381
hypercholesterolemia, 202, 213
hyperfine interaction, 351, 358
hypersensitivity, 430
hypertension, 188, 194
hyperthermia, 430
hypothesis, 188
hypoxia, 420, 422
hypoxia-inducible factor, 420, 422
hysteresis, 263

I

ibuprofen, 391
ideal, 51, 57, 90, 146, 213, 214, 291, 304, 350
identification, 142
IFN, 183, 185, 395
IL-8, 120, 186, 385
image, 96, 433
images, 8, 9, 54, 56, 98, 99, 126, 130, 239, 263, 274, 275, 440, 442, 446, 447
imbalances, 177
immersion, 66, 305, 306, 432, 433
immobilization, xiv, 3, 88, 265, 277, 280, 281, 382, 419, 425, 428, 429, 430, 431, 433, 437, 444, 450, 454
immobilized enzymes, xiv, 428, 429
immune regulation, 323
immune response, 90, 92, 129, 182, 210, 211, 212, 220, 323, 384, 395

immune system, 101, 129, 185
immunity, 116
immunization, 210, 212, 220
immunofluorescence, 96
immunogenicity, 85, 212, 221
immunomodulatory, 204
immunoreactivity, 96
immunostimulant, 133
immunostimulatory, 189
implants, xi, 51, 57, 202, 203, 317, 318, 319, 325, 340, 342, 344, 429
improvements, viii, 49, 133
impulses, 93, 324
impurities, 130
in transition, 375
in vitro, ix, 51, 55, 57, 59, 63, 65, 67, 68, 72, 73, 80, 87, 92, 93, 100, 102, 104, 107, 110, 111, 112, 116, 117, 120, 121, 122, 123, 124, 127, 129, 134, 139, 140, 183, 187, 192, 197, 200, 203, 208, 209, 216, 218, 219, 224, 258, 263, 299, 306, 310, 314, 315, 321,뱀322, 381, 383, 385, 386, 387, 388, 390, 392, 394, 395, 397, 420, 422, 434, 444, 449, 453
in vivo, xiii, 50, 51, 55, 72, 73, 75, 80, 87, 92, 93, 96, 100, 101, 104, 106, 107, 111, 117, 122, 124, 134, 180, 184, 187, 200, 209, 210, 216, 218, 263, 312, 318, 344, 377, 383, 385, 387, 390, 392, 393, 397, 453
incidence, 324
incubator, 328, 329, 383
India, 289, 344
induction, 10, 76, 122, 182, 183, 194, 198
industries, xiii, 174, 204, 377
industry, xiii, 1, 3, 172, 201, 216, 290, 310, 377, 378, 382
infancy, 310
infection, 65, 174, 180, 181, 207, 308, 323
inferior vena cava, 64, 80
inflammation, ix, 93, 115, 205, 210, 315, 323, 385, 387
inflammatory bowel disease, 210
inflammatory cells, 105, 120, 181, 189, 385
inflammatory disease, 210
inflammatory mediators, 186, 189, 385
influenza, 212, 221
influenza a, 212
informed consent, 214, 215
infrared spectroscopy, xii, xv, 130, 207, 281, 306, 318, 428, 431
infrastructure, 282
ingredients, 381, 383, 384
inhibition, 55, 89, 102, 119, 122, 127, 132, 133, 140, 177, 178, 180, 185, 187, 188, 192, 209, 298
inhibitor, xiv, 122, 428, 429
injuries, viii, xii, 49, 72, 82, 103, 105, 120, 318, 350, 387, 389, 394
injury,63, 64, 69, 77, 79, 80, 82, 90, 93, 101, 104, 106, 110, 113, 120, 136, 137, 138, 182, 191, 324, 325, 330, 331, 342, 343, 346, 385, 386
innate immunity, 119, 190
inoculation, 340
insects, 2, 172, 200, 350, 402
insulation, 103
insulin, 55, 74, 203, 208, 212, 218, 220, 221, 298, 313, 393
integration, 117, 177
integrin, 88, 385
integrins, 88, 108
integrity, 68, 125, 126, 322
intensive care unit, 215
interface, 88, 150, 275, 291, 361
interference, 125, 276
internal field, xii, 349, 372
internalization, 52, 124, 180, 181, 189
intervention, 78
intestine, 122, 174, 216
intraocular, 211
intravenously, 68, 292
intrinsic viscosity, 130, 203
inversion, 55, 298
investments, 419
iodine, 378
ion adsorption, 35
ionization, 307, 380
ions, ix, xii, 3, 8, 36, 37, 38, 39, 40, 41, 118, 125, 135, 141, 147, 161, 166, 177, 206, 256, 265, 272, 349, 350, 351, 356, 369, 375, 428
IR spectra, 13, 39
IR spectroscopy, vii, 1, 38, 42, 44, 227, 235
iron, xii, 349, 357, 364, 365, 366, 367, 369, 373, 376, 430, 439, 450, 454
irradiation, viii, 63, 83, 84, 90, 109, 129, 130, 131, 132, 133, 137, 140, 208, 220, 237, 261, 270
IR-spectra, 232, 233
IR-spectroscopy, 19
islands, 89
isolation, 59, 172, 196, 201
isomerization, 236
isoniazid, 305
isotherms, vii, 1, 2, 5, 20, 22, 24, 25, 26, 29, 30, 31, 33, 34, 35
isotope, 354
issues, 138
Italy, 115, 135, 251

J

Japan, 8, 67, 168, 255, 282, 399, 400, 420
joints, 91

K

K^+, 126
KBr, 149, 328
keratinocyte, 117, 322
keratinocytes, 107, 321, 322, 328, 335, 336, 337, 394
ketones, 237
Keynes, 346
kidney, 51, 213, 438
kill, 185, 323
kinetic curves, 36, 38
kinetic model, vii, 2
kinetic parameters, 39
kinetics, vii, ix, 2, 5, 36, 37, 38, 39, 113, 116, 117, 124, 129, 130, 208, 209, 226, 265, 296, 298, 299, 302, 306, 309, 313, 314, 321, 421

L

labeling, 430
laboratory tests, 219
lactate dehydrogenase, 177
lactic acid, 50, 52, 66, 73, 75, 76, 100, 112, 203, 209, 242, 387, 450
lactose, 67, 207
lakes, 419
laminectomy, xii, 318, 330, 342, 344
large intestine, 205, 390
L-arginine, 191, 194, 395
larva, 61, 65, 77
lead, 4, 35, 71, 90, 130, 132, 135, 172, 176, 187, 205, 242, 274, 309, 310, 412
leakage, ix, 64, 116, 124, 177, 183
lesions, 61, 65
leukemia, 123, 136, 184, 381, 430, 453
liberation, 128, 321
life sciences, 90, 103
lifetime, 279, 294, 300
ligament, 127, 389
ligand, 359, 362, 366, 367, 368, 373, 385
light, xiv, 63, 90, 128, 131, 133, 139, 203, 244, 265, 359, 409, 428, 430, 437
light scattering, xiv, 128, 131, 203, 244, 428, 430, 437
linear dependence, 157
linear function, 37
linear polymers, 295
linoleic acid, 210
lipid peroxidation, 122, 180
lipids, 116, 125, 188, 211
liposomes, 211, 295
liquid chromatography, 131, 454
liquid phase, 33, 232
liquids, 157, 158, 380, 453
Listeria monocytogenes, 176
liver, viii, 49, 62, 63, 64, 80, 83, 84, 90, 101, 109, 112, 116, 180, 213, 381
liver cells, 103
liver disease, 101
local anesthetic, 215
localization, xiv, 210, 427
low temperatures, 147
lubricants, 213
lymphatic system, 210, 220
lymphocytes, 182, 193
lymphoid, 194
lymphoid tissue, 194
lysine, 86, 106, 107, 301, 304, 314
lysis, 125, 323
lysozyme, 88, 116, 224, 321, 380

M

machinery, 282
macromolecular chains, 187
macromolecular networks, 429
macromolecules, viii, x, xiii, 2, 5, 15, 18, 19, 25, 35, 43, 44, 45, 51, 55, 57, 130, 131, 143, 174, 199, 204, 216, 268, 298, 321, 378, 392, 435
macrophage inflammatory protein, 186, 385
macrophages, 119, 120, 121, 137, 180, 181, 182, 183, 185, 187, 189, 191, 194, 197, 198, 323, 338, 339, 381, 385, 386, 395
magnetic characteristics, 351
magnetic field, xi, xii, xiv, 10, 15, 255, 349, 353, 354, 355, 358, 428, 429, 447, 449, 450
magnetic moment, 354, 355
magnetic properties, 351, 353, 354
magnetic resonance, 351, 430
magnetization, 430
magnitude, xii, 70, 127, 266, 349
Maillard reaction, 118, 136
majority, xiv, 11, 72, 281, 427, 450
malignancy, 213
maltose, 118
mammalian brain, 323
mammalian cells, 275
mammalian tissues, 185
mammals, 103, 188
management, 78, 137, 214, 396
mandible, 110
manipulation, 207, 216, 310
mannitol, 126

manufacturing, 143, 205, 383
marine diatom, 172
Mark-Houwink equation, 229
mass, 15, 16, 25, 26, 36, 101, 116, 134, 149, 172, 228, 242, 244, 249, 301, 332, 354, 359, 420
mass loss, 116
mass spectrometry, 134, 420
mast cells, 182, 323
material surface, 187
materials science, 143, 153
matrix, xi, 3, 52, 58, 73, 74, 75, 76, 88, 93, 105, 110, 124, 153, 159, 183, 197, 206, 219, 237, 242, 248, 258, 266, 277, 279, 280, 289, 293, 294, 296, 304, 307, 309, 321, 322, 351, 354, 382, 428
matrix metalloproteinase, 73, 124
matrixes, xiii, 185, 280, 377
matter, iv, 119, 142, 322
mean arterial pressure, 64
measurement, ix, 8, 63, 104, 116, 119, 124, 131, 145, 149, 263, 298, 330, 331, 342
measurements, 7, 8, 16, 18, 19, 125, 126, 130, 131, 142, 144, 148, 149, 150, 153, 157, 159, 160, 161, 162, 164, 165, 166, 167, 229, 232, 274, 275, 280, 309, 328, 329, 330, 331, 342, 352, 411, 416, 440, 447
mechanical properties, xi, 3, 58, 72, 81, 84, 85, 87, 88, 90, 91, 92, 97, 102, 107, 108, 110, 219, 223, 224, 234, 237, 247, 248, 250, 274, 304, 309, 315, 319, 391
mechanical stress, 143
mechanochemistry, 249
media, 85, 135, 165, 225, 296, 308, 320, 321, 327, 329, 330, 340, 423, 439, 444, 445, 450, 451, 453
medical, x, xi, xv, 57, 62, 63, 71, 90, 100, 106, 194, 199, 200, 202, 214, 216, 224, 250, 255, 311, 344, 428, 441
medical assistance, 216
medical history, 214
medication, 209
medicine, 73, 75, 90, 102, 120, 172, 221, 252, 390, 391, 452, 454
medium composition, 294
melanoma, 381
melt, xi, 85, 223, 225, 227, 242, 247, 248, 421
melting, 145, 225, 226, 241, 242, 245, 246, 249
melting temperature, 225
melts, 242
membranes, 84, 85, 100, 107, 117, 126, 139, 216, 256, 275, 279, 291, 308, 309, 323, 382, 383, 387, 396, 421, 451, 452, 454
mercury, 7
mesenchymal stem cells, 59, 75, 93, 110
metabolic pathways, 181, 194, 395
metabolism, 117, 125, 211, 242
metabolites, 256, 419
metal complexes, 8, 39, 197, 350, 351, 361
metal ion, xii, 3, 4, 5, 36, 38, 116, 177, 180, 203, 349, 350, 351, 352, 353, 355, 357, 358, 372, 392, 402
metal ions, 3, 4, 5, 38, 116, 177, 203, 350, 392, 402
metal nanoparticles, 3
metalloenzymes, 350, 351, 353, 354
metals, vii, 2, 4, 35, 92, 143, 177, 185, 225, 266, 282, 381, 422
metastasis, 120, 124, 134, 139, 185, 194, 195, 381
methacrylic acid, 264
methanol, 62, 116, 134, 301, 303, 330, 380
methodology, 132, 167, 173, 301, 314
methyl methacrylate, 52
methylcellulose, 206, 217
MHC, 386
Miami, 62
mice, 67, 90, 93, 102, 109, 117, 120, 124, 129, 139, 184, 185, 186, 190, 192, 193, 194, 196, 197, 210, 212, 220, 221, 380, 381, 395, 396
microcrystalline, 65, 80, 196
microelectronics, 266
microgels, 257, 263
micrometer, 149
microorganism, 127, 175, 388
microorganisms, xv, 125, 128, 174, 177, 191, 277, 323, 428, 449, 450, 451
microscope, 8, 98, 127, 130, 208, 301, 304, 440, 441, 447
microscopy, xiv, 120, 123, 126, 129, 136, 174, 191, 227, 281, 428, 429, 430, 435, 437, 440, 445, 454
microspheres, 55, 56, 60, 68, 73, 76, 77, 85, 92, 107, 110, 209, 211, 217, 279, 296, 297, 298, 310, 312, 313, 390, 392, 393, 429, 451
microstructure, 104, 116, 259
microstructures, 112, 259
microwave heating, 132
migration, xiv, 57, 60, 89, 98, 105, 117, 139, 142, 186, 189, 322, 385, 386, 389, 390, 428, 429
military, 64
mineralization, 58, 86, 89, 92
MIP, 186, 385
mixing, 121, 225, 226, 249, 263, 302, 356, 381, 431, 433, 437
MMA, 264
MMP, 117, 124
MMP-9, 117, 124
model system, 191, 200, 424
models, vii, 1, 2, 4, 5, 24, 25, 36, 37, 44, 45, 61, 63, 64, 68, 78, 137, 151, 207, 209, 302, 324, 350, 385, 386, 388

modifications, viii, xi, xiii, 3, 24, 49, 50, 55, 58, 72, 77, 81, 83, 84, 105, 109, 135, 138, 139, 173, 188, 255, 312, 377, 400
modified polymers, xiii, 108, 377
modules, 163
modulus, 85, 87, 88, 102, 142, 147, 248
moisture, ix, xi, 118, 141, 147, 148, 153, 155, 156, 157, 158, 159, 160, 161, 162, 163, 164, 165, 166, 167, 228, 234, 248, 255, 322, 379
moisture content, ix, 141, 147, 148, 153, 155, 156, 157, 158, 159, 160, 161, 162, 163, 164, 165, 166, 167, 228, 234
molar ratios, 325, 327
mold, 93, 94, 95, 96, 97
molds, 111, 174, 178, 189
mole, 129, 379
molecular dynamics, 142, 146, 147, 166
molecular mass, 5, 117
molecular mobility, 14, 16
molecular oxygen, 402, 405
molecular structure, 173, 174, 334, 361, 393, 421, 454
molecular weight, viii, x, xi, 51, 57, 62, 63, 65, 67, 71, 77, 85, 87, 88, 108, 115, 116, 120, 122, 129, 130, 131, 132, 133, 134, 135, 139, 140, 143, 148, 171, 172, 187, 192, 193, 194, 196, 197, 198, 201, 218, 227, 229, 247, 248, 289, 290, 298, 307, 308, 320, 325, 334, 352, 380, 381, 383, 385, 392, 428
molecular weight distribution, 88, 108, 134
mollusks, 402
momentum, 358
monolayer, 24, 104, 205
monomers, xii, 227, 257, 264, 270, 298, 306, 349, 362, 364
monosaccharide, 230, 359, 363, 364, 371, 380
Moon, 76, 168, 198, 374
morbidity, 110
morphogenesis, 178
morphology, xiv, 56, 67, 82, 93, 96, 98, 100, 101, 104, 108, 117, 120, 126, 136, 209, 220, 224, 242, 248, 249, 257, 265, 266, 274, 299, 303, 304, 310, 321, 322, 326, 421, 428, 430, 438, 440, 442
mortality, 65, 380
Moscow, 1, 139, 223, 252, 253
Mössbauer data, xii, 349, 372
Mössbauer effect, 351
MRI, 450
mRNA, 117, 120, 177, 183, 266, 386
mucin, 291
mucosa, 129, 182, 206, 209, 214, 220
mucus, 210, 291
multidimensional, x, 171
multilayered structure, 100
multiple regression, 368
municipal solid waste, 420
muscles, 103, 256, 296, 324
mutant, 210, 220
MWD, 88
myelin, 120
myocardial infarction, 188

N

N/O ligands, xii, 349, 361
NaCl, 61, 188, 194, 305, 314, 431
NADH, 122
nanocomposites, 225, 421
nanofibers, 91, 102, 105, 117, 139, 224
nanoindentation, 127, 191
nanometer, 450
nanoparticles, x, xiv, xv, 52, 60, 67, 73, 74, 76, 77, 102, 112, 121, 128, 138, 139, 183, 184, 195, 199, 200, 203, 211, 212, 218, 221, 224, 275, 295, 312, 313, 393, 427, 428, 429, 430, 431, 437, 438, 439, 440, 441, 442, 443, 444, 445, 447, 448, 449, 450, 451, 452, 453, 뱀454
nanoscale structures, 249
nanostructures, 263, 275, 431, 438, 440, 442, 443, 444, 445, 449, 450, 451
Nanostructures, 116
nanotechnology, 221, 454
National Institutes of Health, 222
natural killer cell, 182, 185
natural polymers, viii, x, xi, 19, 83, 91, 223, 227, 261, 263, 295, 296, 317, 390, 392
natural resources, 71
N-deacetylated chitin, 350
necrosis, 124, 182, 183, 185, 323
negative effects, 70
nephrectomy, 64, 80
nerve, viii, 49, 59, 65, 75, 76, 83, 86, 87, 93, 94, 95, 96, 97, 99, 105, 106, 107, 108, 111, 116, 120, 136, 138, 186, 202, 274, 322, 323, 324, 330, 340, 343, 344
nerve fibers, 95, 120, 324
nervous system, 93, 135
neuritis, 341
neuroblastoma, 323
neurodegenerative disorders, 136
neuronal cells, 96
neuronal circuits, 323
neurons, 58, 86, 93, 96, 119, 140, 323, 350
neuroprotective agents, 324
neurosurgery, 77
neurotoxicity, 119, 135, 136, 140

neutral, xi, 52, 67, 72, 90, 93, 95, 96, 148, 174, 176, 189, 207, 224, 228, 235, 241, 250, 256, 305, 327, 328, 362, 369, 380, 444, 451
neutrophils, 181, 182, 395
New Zealand, 92, 104
NH2, 3, 12, 13, 35, 39, 40, 43, 44, 50, 66, 88, 235, 244, 256, 262, 266, 268, 332, 334, 350, 361, 372, 415, 418, 442
nickel, viii, 2, 5, 8, 30, 31, 33, 34, 35, 36, 45
NIR, 42
nitric oxide, ix, 115, 122, 123, 180, 181, 185, 189, 198
nitric oxide synthase, 122, 181, 185
nitrite, ix, 62, 116, 128, 129
nitrogen, vii, xiii, 1, 2, 4, 5, 7, 19, 20, 95, 117, 128, 149, 177, 266, 327, 350, 352, 357, 359
nitrogenase, 351
NMR, vii, 1, 5, 10, 14, 15, 16, 116, 117, 131, 132, 227, 229, 230, 231, 232, 243, 244, 273, 312, 351, 361
novel materials, 310, 325
NQR, 351
nuclear magnetic resonance, 361
Nuclear Magnetic Resonance, 351
Nuclear Quadrupole Resonance, 351
nucleating agent, 247
nuclei, 15, 92, 177
nucleic acid, 125, 177, 266, 277
nucleophiles, 262
nucleotides, 125
nucleus, xii, 340, 341, 349, 354, 355, 358, 359, 420
nuclides, 4
null, 341
nutrient, 98
nutrients, 90, 93, 177, 322
nutrition, x, 92, 171

O

O/N/S ligands, 350
obesity, 116
obstacles, x, 223, 227, 250
occlusion, 8, 81, 394
ocular diseases, 208
OH, 3, 4, 12, 13, 40, 41, 43, 50, 121, 130, 179, 241, 266, 268, 332, 351, 361, 372, 400, 404, 442
oil, 138, 296, 297
oleic acid, 204, 218
oligomers, x, 118, 119, 121, 123, 124, 127, 128, 133, 140, 171, 177, 180, 182, 184, 187, 188, 189, 192, 194
oligosaccharide, 123, 124, 133, 136, 140, 174, 186, 198, 318, 430
operations, 55, 79, 205, 385
opportunities, 204, 216
optical density, 126, 330
optical microscopy, 120
optical properties, 276
optimization, 51, 89, 134, 216, 319, 392
oral cavity, 128, 138, 208, 216
organ, 90, 101, 103, 213, 251, 321
organelles, 277
organic compounds, 201, 225, 226
organic polymers, 143
organic solvents, x, 90, 118, 139, 201, 223, 244, 250, 378, 380
organism, ix, 105, 115, 174, 175, 322, 338, 342, 344
organize, 319
organs, viii, 49, 50, 91, 100, 103, 211, 318, 430
ornithine, 86
oxidation, xii, xiii, 3, 122, 179, 237, 256, 257, 271, 279, 281, 349, 353, 355, 357, 358, 361, 372, 399, 402, 403, 404, 405, 406, 407, 408, 409, 410, 411, 412, 413, 414, 416, 417, 418, 422, 423, 424, 425
oxidative damage, 122, 195
oxidative stress, 121, 122, 135, 197
oxide nanoparticles, 430, 450
oxygen, xii, 4, 21, 118, 125, 204, 350, 359, 383, 387, 388
oyster, 172, 190

P

PAA, 66
paclitaxel, 55, 73
PACs, 21, 24, 27
paints, 382, 421
palate, 392
parallel, 10, 71, 98, 350, 358, 380, 435
paralysis, 324
partial thromboplastin time, 188
participants, 214
partition, 200
pathogens, ix, 103, 115, 128, 185, 211, 323
pathology, 206
pathophysiological, 324
pathways, 124, 125, 182, 324, 423, 454
PCR, 91, 117, 120
pellicle, 429, 431, 433, 435, 437, 438, 444, 450
pendant model, 350
pepsin, 329
peptic ulcer, 309
peptide, viii, xiii, 58, 83, 88, 100, 207, 210, 212, 216, 220, 310, 313, 377, 384, 392
peptides, x, 75, 89, 99, 106, 108, 111, 199, 203, 208, 210, 212, 216, 224
percolation, 275
periodontal, 75, 127, 389

peripheral nervous system, 323
permeability, 55, 126, 175, 177, 192, 202, 203, 216, 383, 384, 387, 388, 393, 451
permeation, x, 199, 216, 290, 452
permission, 292, 293, 295, 297, 300, 301, 303, 305, 307, 439, 440, 441, 442, 443, 444, 445, 446, 447, 448, 449
permit, 90, 128, 247
permittivity, 146, 154
peroxidation, 179
peroxide, 131, 132, 133, 330, 338, 344, 425
phagocytosis, 120, 186, 386
pharmaceutical, ix, x, xiii, 3, 78, 85, 115, 118, 128, 137, 138, 190, 199, 201, 202, 203, 204, 216, 218, 224, 258, 277, 279, 293, 299, 308, 312, 314, 377, 378, 380, 381, 382, 384, 390, 391, 451
pharmaceuticals, 419
pharmacology, 314
phase inversion, 6
phase transitions, 359
PHB, viii, 83, 85, 106, 249
phenol, 400, 401, 403, 405, 406, 423, 424, 425
phenolic compounds, 401, 402, 404, 405, 424, 425
phenotype, 91, 444, 449
phenotypes, ix, 115
phenytoin, 393
Philadelphia, 168
phosphate, 50, 60, 68, 69, 76, 93, 95, 108, 110, 119, 126, 190, 265, 267, 298, 310, 328, 383, 388, 396, 438, 444
photographs, 126, 183, 310
photolysis, 121, 401, 422, 423
photooxidation, 401
photopolymerization, 261
physical and mechanical properties, 296
physical interaction, xiv, 427
physical properties, 58, 76, 85, 88, 106, 142, 147, 153, 207, 212, 217, 234, 266, 322, 380, 422
physical structure, 62, 71
physicochemical characteristics, 74, 87, 106, 129, 263
physicochemical properties, 87, 107, 200, 242, 266, 277, 295
physics, 24, 142, 429
pilot study, 78
placebo, 213
plants, 321, 400
plaque, 133
plasma membrane, 125
plasma proteins, 69, 180, 187
plasmid, 53, 76, 209
plastic deformation, 225, 234
plasticity, 85, 227, 237, 238, 435
plasticization, 159, 160, 164
plasticizer, 153, 239
plastics, 227, 290, 400
platelet aggregation, 69, 71, 187
platelets, 66, 68, 69, 78, 186, 187
platform, 173, 204, 212, 221
PMMA, 50
pneumonia, 120, 181, 185, 193, 381
polar, vii, 1, 4, 20, 145, 153, 350
polar groups, vii, 1, 4
polarity, 175
polarization, 150, 153, 154, 155, 352
pollutants, 90, 400
pollution, 172, 400
poly(3-hydroxybutyrate), 106
poly(vinyl chloride), 421
polyacrylamide, 296, 297, 313
polyarylate, 421
polyarylates, 421
polycarbonate, 400
polycarbonate plastics, 400
polycarbonates, 421
polychlorinated biphenyl, 400, 419
polychlorinated dibenzofurans, 419
polycondensation, 226
polydispersity, viii, 115, 121, 130, 131, 135
polyelectrolyte complex, xii, 68, 74, 75, 81, 106, 107, 174, 205, 207, 218, 257, 318, 321, 380, 392, 428, 429, 451
polyesters, 224, 241, 242, 244, 250, 311
polymer blends, 225
polymer chain, 4, 5, 10, 12, 13, 24, 35, 44, 45, 88, 129, 159, 187, 236, 237, 256, 266, 291, 294, 296, 319, 350
polymer chains, 4, 5, 10, 12, 13, 24, 35, 44, 45, 88, 129, 159, 187, 236, 256, 291, 294, 319, 350
polymer films, 428
polymer materials, 256, 323
polymer matrix, 144, 164, 294
polymer molecule, 12, 129, 319
polymer network structures, 388
polymer networks, 308, 311, 314
polymer solutions, 24, 25, 208, 257
polymer structure, 12, 227
polymer systems, 159, 290
polymeric chains, 160
polymeric materials, xi, 85, 145, 223, 244, 257, 318
polymeric matrices, 219
polymeric membranes, 224
polymerization, 62, 71, 133, 134, 135, 174, 226, 267, 274, 275, 280, 298, 313, 319, 357, 384, 401, 422, 423, 424
polymerization process, 280

polymorphism, 227
polypropylene, 66, 80
polysaccharide, viii, x, xiii, 4, 49, 57, 67, 72, 73, 84, 109, 111, 153, 172, 173, 187, 194, 195, 200, 201, 205, 223, 225, 226, 227, 242, 244, 250, 256, 290, 318, 319, 321, 350, 377, 380, 384, 386, 393, 394, 428, 453
polysaccharide chains, 350
Polysaccharides, 73, 135, 146, 219, 318
polystyrene, 311
polyurethane, 311, 420
polyvinyl alcohol, 24, 389
polyvinylalcohol, 428
pools, 125
population, 127, 320, 324, 365, 366, 373, 438
porosity, 4, 18, 19, 46, 89, 95, 306, 307, 308, 414, 418
porous media, 17
porphyrins, 376
Portugal, 171
positron, 300
potassium, 129, 149, 177, 228
potato, 425
precipitation, vii, 1, 3, 6, 21, 31, 52, 124, 204, 209, 244, 261, 388, 423, 424, 430, 437, 439
precursor cells, 59
preservation, 119, 140, 190, 197, 319, 340, 343
preservative, 195
President, 250
prevention, 124, 130, 134, 309, 389
priming, 182
principles, 90, 103
probability, 354
probe, xii, 281, 349, 357
probiotic, 172
process duration, 227, 249
prodrugs, 312
product market, 214
progenitor cells, 389
progesterone, 99, 111
programming, 119
pro-inflammatory, 119
project, 250, 282
proliferation, ix, xii, 57, 58, 60, 75, 76, 85, 86, 87, 88, 89, 91, 93, 96, 100, 104, 107, 111, 115, 117, 123, 124, 127, 136, 182, 185, 195, 224, 318, 321, 322, 328, 329, 335, 338, 340, 344, 381, 385, 386, 387, 388, 389, 394, 395, 396, 444, 454
propagation, 93
propane, xiii, 399, 400, 421
propranolol, 383
propylene, 208, 312, 393
prostheses, 99, 111
prosthesis, 99, 111
protease inhibitors, 437
protection, ix, x, 50, 53, 103, 115, 116, 135, 174, 179, 180, 199, 200, 265, 299, 384
protein oxidation, 122, 180
protein structure, 136
protein synthesis, 101, 125, 177
proteins, xiii, 50, 52, 57, 76, 88, 93, 110, 120, 122, 128, 130, 177, 180, 203, 205, 207, 208, 212, 216, 258, 263, 350, 377, 381, 384, 393
proteoglycans, 58
proteolytic enzyme, 212, 216
prothrombin, 188
proton pump inhibitors, 309
protons, 14, 15, 16, 17, 157, 230
Pseudomonas aeruginosa, 175, 454
psoriasis, 322
PTT, 188
purification, 3, 148, 210, 220, 228, 229, 230, 258, 325, 350, 352, 401, 425, 431
purity, 117, 129, 325, 352, 431
PVA, 105, 207, 208, 220, 233, 234, 235, 236, 237, 238, 239, 240, 241, 248, 305, 306, 388, 389, 396, 450, 452
PVA films, 238
PVAc, 233, 234, 235, 236, 237, 241
PVP, 308, 313, 388, 450, 453
pyrolysis, 131

Q

quality of life, 90
quantification, 279
quantum mechanics, 351
quartz, 7, 66
quaternary ammonium, 55
quinone, xiii, 279, 399, 402, 403, 404, 405, 406, 407, 408, 409, 410, 411, 412, 413, 414, 415, 416, 417, 418, 419
quinones, 402

R

radiation, 8, 84, 132, 193, 228, 261, 313, 328, 354, 380
Radiation, 219, 285
radical mechanism, 237
radical polymerization, 257, 258, 259, 262, 297, 306
radicalization, 402
radicals, 121, 122, 129, 132, 179, 180, 193, 268, 269, 402
radius, 131, 203, 389
radius of gyration, 131, 203
raw materials, 63, 201

reactants, 327
reaction mechanism, 405
reaction medium, xii, 226, 249, 349
reaction rate, 264
reaction temperature, 133, 258, 407
reaction time, 132, 133, 134, 258, 326, 403, 404, 405, 406, 407, 408, 410, 411, 414, 417, 418
reactions, xi, 67, 101, 119, 129, 132, 136, 183, 223, 224, 225, 226, 227, 236, 248, 256, 273, 299, 364, 383, 385, 423, 432
reactive groups, 256
reactive oxygen, 122, 185, 217, 323
reactivity, 71, 224, 402, 409
reading, 330
reagents, 5, 86, 227, 249, 325, 430
receptors, ix, 65, 103, 115, 119, 181, 277, 386
reciprocity, 322
recognition, 58, 89, 108, 123, 256, 318, 421
reconstruction, 44, 104, 110
recovery, xii, 65, 105, 132, 196, 279, 318, 324, 342, 344
recurrence, 309
recycling, 421
red blood cells, 64, 69, 71, 79, 101, 120, 187, 214
reference frame, 355
refractive index, 241, 276
regenerate, 91
regeneration, viii, 49, 51, 57, 58, 59, 60, 75, 87, 91, 92, 93, 95, 96, 98, 99, 100, 102, 104, 105, 106, 107, 108, 109, 110, 111, 112, 113, 120, 136, 185, 186, 189, 202, 224, 249, 275, 322, 324, 385, 387, 388, 389
regression, 24, 29, 37, 362, 367, 373
regression line, 367
regression method, 24
rejection, 319, 342
relaxation, ix, 10, 14, 15, 17, 18, 19, 40, 57, 141, 142, 143, 144, 145, 146, 147, 148, 150, 153, 155, 156, 157, 158, 159, 160, 161, 163, 164, 165, 166, 167, 224, 225, 242, 247, 248, 249, 296, 297, 306, 309
relaxation process, ix, 141, 142, 143, 145, 146, 147, 155, 156, 157, 159, 160, 161, 162, 164, 166, 167, 248, 309
relaxation processes, ix, 141, 142, 145, 146, 155, 160, 162, 166
relaxation properties, 142, 224, 242, 247, 249
relaxation rate, 145, 296
relaxation times, 10, 14, 17, 18, 19, 146, 161, 225
relevance, 200
reliability, 143, 380
repair, ix, 57, 59, 60, 64, 65, 80, 87, 91, 93, 104, 113, 115, 138, 177, 185, 319, 323, 324, 386, 390
repression, 183
reproduction, 388
repulsion, 264, 269, 308
requirements, ix, 51, 72, 115
researchers, viii, 49, 51, 52, 53, 55, 60, 174, 211, 274, 310, 384, 385, 386, 388, 389, 405
resection, 67, 389
residuals, 302
residues, 50, 65, 134, 172, 177, 180, 201, 307, 391
resins, 400
resistance, 3, 85, 128, 149, 150, 152, 153, 154, 161, 205, 207, 280, 319, 383, 389
resolution, 89, 126, 149, 232, 328, 433
resources, 241
respiratory problems, 206
response, ix, xii, 62, 85, 99, 102, 112, 116, 117, 124, 125, 126, 129, 132, 145, 146, 150, 151, 152, 153, 180, 181, 182, 190, 196, 202, 203, 210, 212, 256, 266, 274, 277, 279, 280, 298, 300, 301, 306, 314, 318, 323, 338, 339, 344, 395, 419, 420, 422, 438
response time, 274, 277, 314
responsiveness, 313
restrictions, 233, 249
restructuring, x, 199
retardation, 145
retina, 221
reusability, 425
ribonucleotide reductase, 351
rights, iv
rings, 13, 35, 45, 159, 266, 365, 366, 369, 373
risk, 92, 93, 188, 213
risk factors, 188
RNA, 117, 125, 126
rods, 127, 204
room temperature, 6, 8, 55, 95, 96, 147, 153, 201, 208, 228, 257, 258, 274, 326, 352, 355, 383, 431, 437
root, 324
roots, 324
rubber, ix, 67, 141, 142, 144, 145, 147, 148, 161, 165, 207, 429
rubbers, 143
rubbery state, 145
Russia, 1, 5, 139, 223, 225, 227, 252, 253

S

safety, 77, 190, 200, 204, 212, 213, 309, 391
salt concentration, 406
salts, xiii, 5, 132, 172, 188, 203, 207, 226, 228, 377, 379, 437
saturation, 24, 177, 430
savings, 310
scaling, ix, 115

scanning electron microscopy, 303
scar tissue, 103, 387, 389
scatter, 302
scattering, 130, 227, 228, 248
scavengers, 180, 189
school, 250
science, x, 106, 139, 146, 171, 199, 202
seafood, 200, 213, 290
second generation, 318
secrete, 100, 186, 189, 322
secretion, 101, 102, 104, 119, 124, 137, 180, 191, 385
sedimentation, 203
sediments, 400, 422
seed, 382
seeding, 90
selectivity, 225, 266, 281, 298, 418
self-assembly, 102, 277
SEM micrographs, 66, 117, 239, 303
semicircle, 151, 154, 155
semiconductor, 266
semiconductors, 266
semi-crystalline polymers, 142
sensation, 103
sensing, 265, 266, 277
sensitivity, 7, 215, 258, 259, 261, 262, 266, 270, 277, 280, 281, 299, 307, 309, 313, 314
sensors, 256, 265, 266, 276, 277, 286
septic shock, 182
sequencing, 281, 287
serum, 96, 122, 135, 181, 186, 210, 213, 263, 308, 380, 389, 395
serum albumin, 122, 135, 308
services, iv
sewage, 225, 350
shape, 19, 20, 56, 126, 145, 203, 304, 358, 442, 445, 446, 451
shear, x, 129, 193, 223, 225, 226, 228, 245, 249, 250
shear deformation, x, 223, 225, 226, 245, 249, 250
sheep, 68, 82
shelf life, 118
shellfish, 84, 172, 192, 290, 402
shock, 324
shortage, 101
showing, xii, 60, 96, 182, 188, 229, 278, 302, 318, 338, 339, 341, 449
shrimp, 84, 129, 136, 155, 196, 197, 201, 256, 322
side chain, 51, 53, 55, 58, 132, 145, 257, 262, 267, 299, 310
side effects, 72, 184, 192, 209, 210, 309
signal transduction, 88, 119
signals, 10, 14, 15, 69, 230, 244, 256, 330, 342
signs, 182, 186
silica, 68, 82, 453
silicon, 69, 352
silk, 66, 102, 104, 112, 113, 201, 430, 453
silkworm, 217
silver, 5, 8, 32, 35, 36, 45, 188, 330, 387
simulations, 351
skeleton, 58
skin, viii, 49, 59, 83, 84, 85, 103, 107, 113, 138, 185, 186, 190, 195, 202, 204, 205, 218, 234, 322, 343, 382, 385, 389, 390, 394
small intestine, 390
smart materials, xi, 255, 256, 257, 282
smooth muscle, 100, 104
smooth muscle cells, 100
SO42-, 350
sodium, ix, xiv, 62, 66, 74, 75, 95, 96, 116, 128, 132, 134, 204, 206, 209, 211, 218, 219, 229, 234, 298, 325, 378, 380, 383, 428, 429, 431, 433, 434, 435, 436, 437
sodium hydroxide, 95, 96, 204, 234, 378
soft matter, 224
software, 232, 352
solid phase, 55
solid state, x, 74, 145, 223, 225, 226, 249, 250, 352
solid tumors, 450, 454
solidification, 91
solubility, x, xi, xiii, 51, 52, 55, 58, 67, 72, 118, 135, 171, 173, 174, 175, 208, 224, 234, 241, 243, 244, 250, 256, 258, 291, 292, 294, 296, 298, 299, 305, 320, 350, 377, 378, 381, 384, 416, 430
solvents, xi, 90, 118, 201, 223, 226, 227, 234, 242, 249, 250, 387
somatic cell, 323
sorption, 24, 31, 32, 38, 159, 224, 227, 239
sorption kinetics, 38
sorption process, 32
species, 36, 39, 41, 42, 43, 122, 126, 129, 155, 156, 180, 206, 217, 267, 272, 279, 306, 343, 353, 355, 357, 358, 369, 408
specific surface, 4, 19, 37, 63, 126, 264, 414, 418
spectrophotometry, 122, 263
spectroscopic techniques, 351
spectroscopy, ix, 5, 8, 39, 42, 43, 47, 116, 127, 131, 141, 142, 143, 146, 149, 159, 166, 280, 281, 300, 353, 355, 361, 376, 453
spiders, 350
spin, xii, xiii, 10, 14, 15, 17, 18, 19, 329, 349, 350, 353, 354, 355, 356, 358, 359, 361, 362, 363, 364, 365, 366, 367, 368, 369, 370, 371, 372, 373
spinal cord, xii, 93, 106, 318, 323, 324, 325, 330, 342, 344, 350
spinal cord injury, 106, 324, 325
spine, 110

spleen, 62
sponge, 60, 63, 65, 72, 76, 77, 103, 388, 396
spore, 177, 178
squamous cell, 322, 381
stability, 69, 102, 107, 118, 121, 128, 133, 200, 208, 265, 294, 384, 391, 435, 437, 444, 453
stabilization, 52, 358, 372
stable complexes, 4, 350
stable radicals, 122
starch, 123, 132, 136, 146, 164, 225, 227
state, x, xii, 10, 13, 18, 24, 25, 26, 81, 85, 89, 142, 144, 160, 164, 165, 180, 187, 223, 224, 225, 226, 227, 228, 232, 233, 234, 237, 241, 242, 248, 249, 250, 251, 256, 258, 263, 266, 277, 300, 307, 308, 315, 349, 353, 354, 355, 357, 358, 359, 361, 372, 380
states, 182, 273, 308, 354, 357, 358
steel, 93, 94, 95, 101, 134, 226
stem cells, 96, 104, 111, 113, 322
stent, 100, 112
sterile, 186, 328
stimulus, xi, 55, 145, 173, 255, 274, 306, 322
stoichiometry, 359
stomach, 67, 184, 205, 216, 306, 315, 393
storage, 101, 135, 163, 204, 248, 388
stress, xi, 101, 122, 125, 134, 136, 138, 147, 176, 228, 238, 255, 291
stretching, 12, 13, 39, 40, 41, 232, 233, 324, 332, 445
stroke, 188, 324
stroma, 321
stromal cells, 58, 76
strong interaction, 274
structural changes, vii, 2, 4, 299, 309
structural characteristics, 182, 189, 204, 256, 275
structural defects, 226
structural transformations, 224, 228, 249
subcutaneous tissue, 92
substance use, 21, 400
substitutes, viii, 49, 90, 92, 103, 110, 322, 390
substitution, 67, 88, 103, 123, 124, 185, 188, 256, 298, 390
substitution reaction, 88
substrate, 66, 92, 101, 102, 235, 278, 291, 310, 322, 418, 424, 450
substrates, 66, 87, 102, 133, 350
sucrose, 432
suicide, 68
sulfate, 5, 8, 55, 65, 73, 100, 123, 187, 192, 212, 219, 321, 322, 350, 352, 360
sulfur, 353
Sun, 47, 73, 74, 77, 104, 109, 113, 123, 137, 139, 253, 285, 286, 421, 424
supplier, 148
suppression, 177, 185, 209
surface area, 19, 64, 68, 98, 304, 414
surface chemistry, 86, 107
surface modification, 58, 108
surface properties, 55, 85, 86, 319
surface structure, 120, 435
surface tension, 218
surface treatment, 106, 328, 329
surfactant, 195, 237
surfactants, 294
surrogates, 133
survival, 61, 63, 64, 65, 79, 101, 124, 126, 311, 454
survival rate, 61
susceptibility, 125, 130, 430, 447, 449
suspensions, 228, 234
suture, 66, 187, 382, 385
swelling, 4, 5, 17, 19, 20, 21, 24, 51, 57, 85, 87, 88, 95, 207, 220, 233, 234, 240, 249, 256, 257, 258, 259, 261, 262, 264, 265, 269, 270, 274, 294, 296, 299, 301, 304, 305, 306, 307, 308, 309, 310, 315, 319, 321, 383, 388, 429, 431, 432, 433, 434, 435, 438, 443,뱀451
swelling kinetics, 300, 306, 309
swelling process, 262, 306, 309
swelling processes, 309
Switzerland, 345
symmetry, 11, 42, 354, 370
symptoms, 65, 213, 214, 324
syndrome, 213
synergistic effect, 133, 192
synovial fluid, 429
synthesis, x, xi, xiv, 4, 89, 101, 124, 142, 155, 156, 172, 180, 185, 188, 190, 197, 219, 223, 224, 225, 227, 228, 232, 233, 234, 235, 248, 249, 251, 264, 267, 268, 273, 275, 290, 297, 306, 313, 317, 325, 332, 386, 400, 421, 428, 430, 445, 447, 451
synthetic analogues, 351
synthetic polymers, x, 57, 100, 129, 206, 223, 224, 225, 227, 234, 250

T

T cell, 185, 323
Taiwan, 422
tamoxifen, 210, 220
target, x, 52, 55, 57, 127, 134, 149, 174, 175, 176, 189, 199, 200, 216, 295, 310, 384, 429, 450
target organs, 52, 57
techniques, xii, 64, 84, 102, 117, 131, 142, 145, 148, 204, 225, 227, 235, 274, 295, 296, 300, 304, 309, 318, 351, 355, 383, 389, 402, 419, 422, 452
technologies, 310

technology, x, 74, 132, 199, 201, 202, 205, 209, 218, 220, 226, 261, 281, 282, 382, 384, 391
teeth, 89, 92, 387
TEG, 63, 67, 68, 69, 70, 71
TEM, 117, 127, 183, 430, 442, 446
ce, ix, 141, 146, 147, 155, 157, 160, 162, 167, 359, 372
tendon, 75, 99, 106
tensile strength, 85, 87, 185, 186, 238, 247, 248
tension, 320
testing, 72, 102, 214
testosterone, 419
tetanus, 211
Tetanus, 128
textiles, 191
texture, xi, 255
TGA, xii, 149, 153, 156, 161, 162, 165, 166, 167, 274, 310, 318
TGF, 59, 60, 92, 110, 181, 186, 386
theoretical approaches, 351
therapeutic agents, 57, 200, 263, 431
therapeutics, 200, 211, 430
therapy, 92, 103, 135, 204, 206, 209, 213, 215, 220, 309, 315
thermal activation, 146
thermal analysis, xii, 131, 142, 145, 147, 208, 318
thermal degradation, ix, 141, 148, 155, 160, 161, 165, 166, 167, 261, 421
thermal expansion, 142, 145
thermal relaxation, ix, 141, 147, 156
thermal stability, 118, 225, 262, 274
thermal treatment, 124
thermodynamical stability, 245
thermogravimetric analysis, 149, 309
thermoplastic polyurethane, 66, 80
thin films, 143, 235, 276
thrombin, 62, 67, 187, 188, 395
thrombosis, 69
thrombus, 67, 187
thymus, 195
tibialis posterior, 120
TIMP, 117
TIMP-1, 117
tin, 268, 281, 299
tin oxide, 268
tissue homeostasis, 119
titanium, 113
TMC, 221, 298
TNF, 105, 119, 180, 182, 185, 186, 189, 385, 386, 395
TNF-alpha, 105, 119
TNF-α, 385, 386, 395
toluene, 299, 420
topology, 111
torsion, 362
total cholesterol, 213
toxic effect, 65, 117, 121, 387
toxic substances, 382
toxicity, viii, x, 49, 55, 58, 83, 84, 85, 91, 121, 124, 183, 185, 191, 192, 199, 204, 216, 221, 234, 290, 293, 380, 384, 425, 428, 430, 449
Toyota, 374
trafficking, 119
transcription, 127
transcripts, 419
transducer, 277, 278
transection, 68
transesterification, 421
transfection, 51, 53, 73, 103, 191, 221, 263, 394
transformation, 3, 43, 206, 226, 309, 401, 404, 415
transformations, x, 223, 225, 226
transforming growth factor, 59, 75, 76, 92, 120, 181, 186, 386
transfusion, 215
transition metal, 3, 5, 11, 350, 353, 361
transition metal ions, 5, 350
transition temperature, 142, 148, 159, 160, 164, 165, 257, 258, 261, 263, 264
translocation, 203
transmission, xiv, 8, 69, 205, 228, 328, 352, 428, 430
transmission electron microscopy, xiv, 69, 428, 430
Transmission Electron Microscopy, 438
Transmission Electron Microscopy (TEM), 438
transplantation, 100, 101, 224
transport, 100, 119, 125, 129, 177, 200, 203, 292, 296, 454
trauma, viii, 49, 64, 69, 80, 82, 92, 180, 207, 323, 324
trial, 78, 212, 213, 214, 228, 242
triglycerides, 213
trypsin, 425
tuberculosis, 128, 212, 221
tumor, ix, 73, 92, 102, 115, 116, 117, 120, 124, 134, 139, 181, 182, 183, 184, 185, 186, 191, 195, 196, 198, 202, 204, 209, 321, 323, 324, 381, 385, 439, 450, 454
tumor cells, ix, 73, 115, 124, 181, 182, 183, 185, 323, 381
tumor growth, 120, 124, 139, 185, 195, 209, 381
tumor necrosis factor, 185, 186, 191, 198, 385
tumors, 102, 184, 381
tumours, 393
Turkey, 377, 396
turnover, 405
tyrosine, 405

U

UK, 64, 138, 328, 330, 344
ulcer, 309, 381
ulcerative colitis, 210
ultrasound, 129, 130, 131, 132, 136, 139
ultrastructure, xiv, 428
ultraviolet irradiation, 67
uniform, xi, xiv, 55, 63, 92, 125, 223, 226, 227, 237, 248, 250, 294, 427
United, 317, 324
United Kingdom, 317
United States, 324
universal gas constant, 34
urea, 101, 124, 262, 279
USA, 7, 67, 68, 138, 139, 168, 285, 286, 287, 290, 317, 344, 345, 346, 454
uterus, 55
UV, 5, 8, 39, 42, 43, 67, 125, 131, 132, 207, 261, 263, 274, 301, 304, 305, 328, 380, 388, 403, 408, 409, 416, 422, 423, 431, 432, 433, 435, 438
UV irradiation, 67, 207, 304, 305, 423
UV light, 328
UV radiation, 261, 423, 433, 435
UV spectrum, 132
UV-irradiation, 67, 389

V

vaccinations, 212
vaccine, 129, 139, 204, 210, 211, 212, 216, 220, 221, 384
vacuole, 125
vacuum, 7, 20, 21, 134, 147, 149, 150, 153, 154, 159, 439
valence, 361, 367, 369
valve, 100, 111
vapor, vii, 1, 7, 20, 23, 24, 26, 44, 143
variables, 70, 130, 131, 180, 209, 299, 301
variations, 129, 153, 246
vascular diseases, 100
vascular surgery, 111
vascularization, 117
vasculature, 450
vasoconstriction, 69
vasopressin, 203
vector, 15, 53, 55, 73, 203, 263, 394
vegetables, 191
vehicles, 202, 207, 216, 393
vein, 62, 68, 100, 122, 138
velocity, 352, 359, 405, 406, 408, 411
versatility, 203, 204, 216
vertebrae, 325
vertebrates, 91, 101, 324
vesicle, 224
vessels, 60, 64, 65, 100, 454
vibration, 207, 354, 372, 445
vinyl monomers, 257, 261
viral gene, 211, 384
viruses, 133, 202, 323
viscoelastic properties, 421
viscosity, x, 118, 130, 135, 142, 146, 171, 172, 174, 196, 203, 229, 232, 234, 242, 304, 380, 381
vitamin C, 122
Volunteers, 213, 222

W

Washington, 46, 168, 193, 312
waste, 172, 192, 290, 400, 422, 423
waste treatment, 423
wastewater, 256, 382, 400, 402, 422, 423, 424, 425
water absorption, 299, 313, 321, 387
water clusters, vii, 1
water evaporation, 152, 156
water sorption, 24, 159, 240
water vapor, vii, 1, 2, 4, 5, 6, 7, 20, 21, 22, 23, 24, 25, 27, 28, 29, 44, 383
water-soluble polymers, 270
wave number, 442, 445
wavelengths, 330
WAXS, 227, 228, 229, 232, 233, 235, 239, 240, 247, 248
weak interaction, 53
wealth, 353
weight gain, 68
weight loss, 88, 153, 162
weight ratio, 262, 264, 265, 299, 304
welding, 65, 80
wells, 84, 330
western blot, 203
Western blot, 120
wettability, 86, 107
white blood cells, 66, 91, 323
whooping cough, 384
workers, 181, 182, 185, 186
worldwide, 90, 204
wound healing, ix, x, xii, xiii, 67, 69, 104, 115, 116, 117, 139, 172, 185, 186, 189, 190, 193, 197, 207, 318, 321, 322, 323, 377, 384, 385, 386, 387, 388, 389, 390, 391, 394, 396
wound infection, 394

X

xanthan gum, 321
xenografts, 92

X-ray diffraction, vii, xii, 1, 5, 8, 10, 11, 12, 15, 19, 88, 117, 208, 237, 318, 328
X-ray diffraction (XRD), xii, 318
X-ray diffraction data, 15

Y

yeast, 430, 450, 453
yield, 104, 118, 129, 132, 133, 134, 158, 225, 226, 228, 232, 237, 242, 249, 293, 339, 357, 449, 450
young adults, 324

Z

zinc, 121
zinc oxide, 121